ACKNOWLEDGMENTS

Authors

Sharon L. Senk
Associate Professor of Mathematics, Michigan State University, East Lansing, MI

Denisse R. Thompson
Assistant Professor of Mathematics Education, University of South Florida, Tampa, FL

Steven S. Viktora
Chairman, Mathematics Department, New Trier High School, Winnetka, IL

Zalman Usiskin
Professor of Education, The University of Chicago

Nils P. Ahbel
Mathematics Teacher, Kent School, Kent, CT (Second Edition only)

Suzanne Levin
UCSMP (Second Edition only)

Marcia L. Weinhold
Mathematics Teacher, Kalamazoo Area Mathematics and Science Center, Kalamazoo, MI (Second Edition only)

Rheta N. Rubenstein
Mathematics Department Head, Renaissance H.S., Detroit, MI (First Edition only)

Judith Halvorson Jaskowiak
Mathematics Teacher, John F. Kennedy H.S., Bloomington, MN (First Edition only)

James Flanders
UCSMP (First Edition only)

Natalie Jakucyn
UCSMP (First Edition only)

Gerald Pillsbury
UCSMP (First Edition only)

UCSMP Production and Evaluation

Series Editors: Zalman Usiskin, Sharon L. Senk
Director of Second Edition Studies:
  Gurcharn Kaeley
Director of First Edition Studies: Sandra Mathison
  (State University of New York, Albany), Assistant
  to the Director: Catherine Sarther
Technical Coordinator: Susan Chang
Second Edition Teacher's Edition Editor:
  David Witonsky

We wish also to acknowledge the generous support of the **Amoco Foundation** and the **Carnegie Corporation of New York** for the development, testing, and distribution of the First Edition of these materials, and the continuing support of the **Amoco Foundation** for the Second Edition.

We wish to thank the many editors, production personnel, and design personnel at ScottForesman for their magnificent assistance.

Design Development

Curtis Design

Multicultural Reviewers for ScottForesman

Winifred Deavens
St. Louis Public Schools, St. Louis, MO

Seree Weroha
Kansas City Public Schools, Kansas City, KS

Efraín Meléndez
Los Angeles Unified School District, CA

Linda Skinner
Edmond Public Schools, Edmond, OK

UCSMP
SCOTTFORESMAN

The University of Chicago School Mathematics Project

Advanced Algebra

Second Edition

About the Cover Ellipses and other conic sections are important topics in this book. *UCSMP Advanced Algebra* emphasizes facility with algebraic expressions and forms, powers and roots, and functions based on these concepts. Logarithmic, trigonometric, polynomial, and other special functions are studied both for their abstract properties and as tools for modeling real-world situations.

Authors

Sharon L. Senk   Denisse R. Thompson   Steven S. Viktora
Zalman Usiskin   Nils P. Ahbel   Suzanne Levin
Marcia L. Weinhold   Rheta N. Rubenstein   Judith Halvorson Jaskowiak
James Flanders   Natalie Jakucyn   Gerald Pillsbury

ScottForesman

A Division of HarperCollins*Publishers*

Editorial Offices: Glenview, Illinois
Regional Offices: Sunnyvale, California • Tucker, Georgia
Glenview, Illinois • Oakland, New Jersey • Dallas, Texas

It is impossible for UCSMP to thank all the people who have helped create and test these books. We wish particularly to thank Carol Siegel, who coordinated the use of the test materials in schools; Lianghuo Fan, Alev Yalman, Scott Anderson, Jason Goeppinger, Eric Landahl, Jim Guszcza, and Colleen Webb of our editorial staff; Eric Chen, Timothy Day, Thao Do, Tony Ham, Zenobia K. Mehta, Jeong Moon, Adil Moiduddin, Antoun Nabhan, Young Nam, and Wei Tang of our technical staff; and Kwai Ming Wa, Sungeum Choi, Merilee Maeir, Catherine Moushon, Raymound Moushon, Noah Berlatsky, and Claudia Ceccarelli of our evaluation staff.

A first draft of *Advanced Algebra* was written and piloted during the 1985–1986 school year, revised and tested during the 1986–1987 school year, and again revised and tested during the 1987–1988 school year. We appreciate the assistance of the following teachers who taught these preliminary versions, participated in the research, and contributed ideas to help improve the text:

Rita Belluomini
Rich South High School
Richton Park, Illinois

Timothy Craine
Renaissance High School
Detroit, Michigan

Mary Crisanti
Lake Park West High School
Roselle, Illinois

Joe DeBlois
West Genessee High School
Camillus, New York

Cynthia Harris
Taft High School
Chicago Public Schools

Marilyn Hourston
Whitney Young High School
Chicago Public Schools

Marvin Koffman
Kenwood Academy
Chicago Public Schools

Sharon Llewellyn
Renaissance High School
Detroit, Michigan

Kenneth Lucas
Glenbrook South High School
Glenview, IL

Donald Thompson
Hernando High School
Brooksville, Florida

Jill Weitz
Brentwood School
Los Angeles, California

Since the ScottForesman publication of the First Edition of *Advanced Algebra* in 1990, thousands of teachers and schools have used the materials and have made additional suggestions for improvements. The materials were again revised, and the following teachers and schools participated in field studies in 1993–1994:

Leah Regulinski
Boulder High School
Boulder, Colorado

Sally Richardson
Mt. Zion High School
Mt. Zion, Illinois

Bryan Friddle
Pontotoc High School
Pontotoc, Mississippi

Maria Saucedo
Hanks High School
El Paso, Texas

Ray Thompson
Thornton Fractional South
High School, Lansing, IL

Lynne G. Rees
Lassiter High School
Marietta, Georgia

Debra L. Schmeltzer
Argo Community High School
Summit, Illinois

Karen Kuchenbrod Umbaugh
Sentinel High School
Missoula, Montana

Marcia M. Booth
Framingham High School
Framingham, Massachusetts

Al Schectman
Steinmetz Academic Center
Chicago Public Schools

Julie Knittle
Shawnee Mission NW High
School, Shawnee, Kansas

Carla Randall
Lake Oswego High School
Lake Oswego, Oregon

Robert Young
Springfield High School
Springfield, Pennsylvania

We wish also to acknowledge the contribution of the text *Advanced Algebra with Transformations and Applications,* by Zalman Usiskin (Laidlaw, 1975), to some of the conceptualizations and problems used in this book.

THE UNIVERSITY OF CHICAGO SCHOOL MATHEMATICS PROJECT

The University of Chicago School Mathematics Project (UCSMP) is a long-term project designed to improve school mathematics in grades K–12. UCSMP began in 1983 with a 6-year grant from the Amoco Foundation. Additional funding has come from the National Science Foundation, the Ford Motor Company, the Carnegie Corporation of New York, the General Electric Foundation, GTE, Citicorp/Citibank, and the Exxon Education Foundation.

UCSMP is centered in the Departments of Education and Mathematics of the University of Chicago. The project has translated dozens of mathematics textbooks from other countries, held three international conferences, developed curricular materials for elementary and secondary schools, formulated models for teacher training and retraining, conducted a large number of large and small conferences, engaged in evaluations of many of its activities, and through its royalties has supported a wide variety of research projects in mathematics education at the University. UCSMP currently has the following components and directors:

Resources	Izaac Wirszup, Professor Emeritus of Mathematics
Elementary Materials	Max Bell, Professor of Education
Elementary Teacher Development	Sheila Sconiers, Research Associate in Education
Secondary	Sharon L. Senk, Associate Professor of Mathematics, Michigan State University Zalman Usiskin, Professor of Education
Evaluation Consultant	Larry Hedges, Professor of Education

From 1983 to 1987, the director of UCSMP was Paul Sally, Professor of Mathematics. Since 1987, the director has been Zalman Usiskin.

Advanced Algebra

The text *Advanced Algebra* has been developed by the Secondary Component of the project, and constitutes the core of the fourth year in a six-year mathematics curriculum devised by that component. The names of the six texts around which these years are built are:

Transition Mathematics
Algebra
Geometry
Advanced Algebra
Functions, Statistics, and Trigonometry
Precalculus and Discrete Mathematics

The content and questions of this book integrate geometry, discrete mathematics, and statistics together with algebra. Pure and applied mathematics are also integrated throughout. It is for these reasons that the book is deemed to be part of an integrated series. However, algebra is the trunk from which the various branches of mathematics studied in this book emanate, and prior exposure to a year of algebra and a year of geometry or their equivalent is assumed. It is for this reason that we call this book simply *Advanced Algebra*.

The First Edition of *Advanced Algebra* introduced many features that have been retained in this edition. There is **wider scope,** including significant amounts of geometry and statistics. These topics are not isolated as separate units of study or enrichment. They are employed to motivate, justify, extend, and otherwise enhance important concepts of algebra. The geometry is particularly important because many students have in the past finished geometry and never seen that important content again. A **real-world orientation** has guided both the selection of content and the approaches allowed the student in working out exercises and problems, because being able to do mathematics is of little use to an individual unless he or she can apply that content. We require **reading mathematics,** because students must read to understand mathematics in later courses and must learn to read technical matter in the world at large. The use of **up-to-date technology** is integrated throughout, with *automatic graphers* assumed.

Four dimensions of understanding are emphasized: skill in carrying out various algorithms; developing and using mathematics properties and relationships; applying mathematics in realistic situations; and representing or picturing mathematical concepts. We call this the SPUR approach: **S**kills, **P**roperties, **U**ses, **R**epresentations.

The **book organization** is designed to maximize the acquisition of both skills and concepts. Ideas introduced in a lesson are reinforced through Review questions in the immediately succeeding lessons. This daily review feature allows students several nights to learn and practice important concepts and skills. Then, at the end of each chapter, a carefully focused Progress Self-Test and a Chapter Review, each keyed to objectives in all the dimensions of understanding, are used to solidify performance of skills and concepts from the chapter so that they may be applied later with confidence. Finally, to increase retention, important ideas are reviewed in later chapters.

Since the ScottForesman publication of the First Edition of *Advanced Algebra* in 1990, the entire UCSMP secondary series has been completed and published. Thousands of teachers and schools have used the first edition and some have made

suggestions for improvements. There have been advances in technology and in thinking about how students learn. As such, many revisions have been made to improve the materials. We have moved many lessons and reorganized others. We have added many new applications, updated others, and introduced lessons on linear, quadratic, and exponential modeling. We deleted the chapter dealing with higher dimensions because so few teachers reached this material and we felt stronger attention needed to be given to the earlier chapters.

Those familiar with the First Edition will note a rather significant reorganization of the treatment of functions. The material in Chapter 7 of the First Edition has been dispersed. The study of functions begins in Chapter 1 in the Second Edition, and function language and notation appear throughout the course. We were encouraged to do this by the widespread availability of graphing technology which makes it much easier to approach functions. Consequently we introduce **automatic graphers,** as we did in the First Edition, but in this edition they are *required* because they are used throughout as a pattern-finding, concept-developing, and problem-solving tool.

There are also a number of features new to this edition, including the following: **Activities** have been incorporated into many lessons to help students develop concepts before or as they read. There are **projects** at the end of each chapter because in the real world much of the mathematics done requires a longer period of time than is customarily available to students in daily assignments. There are many more questions requiring **writing,** because writing helps students clarify their own thinking, and writing is an important aspect of communicating mathematical ideas to others.

Comments about these materials are welcomed. Please address comments to:
UCSMP, The University of Chicago,
5835 S. Kimbark, Chicago, IL 60637.

CONTENTS

CHAPTER 1   4

FUNCTIONS

CHAPTER 2   70

VARIATION AND GRAPHS

LINEAR FUNCTIONS

MATRICES

GETTING STARTED

Welcome to UCSMP Advanced Algebra.
We hope you enjoy this book—it was written for you.

What Is Advanced Algebra?

Advanced Algebra might be best described as "what every high school graduate should know about mathematics that has not been learned in previous courses." It contains the mathematics that educated people around the world use in conversation, and that most colleges want or expect you to have studied. But it is not a hodge-podge of topics. Familiar ideas such as properties of numbers, graphs, expressions, equations, and inequalities appear throughout the book. In addition, you will study many new topics, including matrices, logarithms, trigonometry, and conic sections.

Throughout the course we will use the concept of function to help organize ideas.

The name *Advanced Algebra* may sound as if this class will be like algebra—but more difficult. That is half true. The content of this book is related to what you learned in first-year algebra. However, this book is not necessarily more difficult. Some questions are harder, but you know a lot more than you did then. In particular, you know much more geometry, and you have had a year of practice with algebra.

Studying Mathematics

An important goal of this book is to help you learn mathematics on your own, so that you will be able to deal with the mathematics you see in newspapers, magazines, on television, on any job, and in school. The authors, who are all experienced teachers, offer the following advice:

1 You can watch basketball hundreds of times on television. Still, to learn how to play basketball, you must have a ball in your hand and actually dribble, shoot, and pass it. Mathematics is no different. You cannot learn much algebra just by watching other people do it. You must participate. You must think through and work problems. Some teachers have this slogan:

Mathematics is not a spectator sport.

2 You are expected to read each lesson. Sometimes you may do this as a class or in a small group; other times you will do the reading on your own. No matter how you do the reading, it is vital for you to understand what you have read. Here are some ways to improve your reading comprehension:

Read slowly and thoughtfully, paying attention to each word, graph, and symbol.

Look up the meaning of any word you do not understand.

Work examples yourself as you follow the steps in the text.

Draw graphs by hand or on your automatic grapher when following a complicated example.

Reread sections that are unclear to you.

Discuss difficult ideas with a fellow student or your teacher.

3 Writing can help you understand mathematics, too. You will often be asked to justify your solution to a problem, to write a formal argument, or to make up your own example to illustrate an idea. Writing good explanations takes practice. You should use solutions to the examples in each lesson to guide your writing.

Equipment Needed for this Course

To be a successful student in any high school mathematics course, you need notebook paper, graph paper, sharpened pencils, good erasers, and a ruler. It is best if the ruler is made of transparent plastic, and marked in both centimeters and inches.

For this book, you need to have technology (a graphics calculator or a computer with suitable software) throughout the year while reading the lessons, doing homework, participating in class, and taking tests and quizzes. With technology, you will be able to solve realistic problems without having to do tedious calculations, and you will be able to learn more mathematics than would otherwise be possible. You will also find such technology useful in other courses. If you do not have technology available for your use at all times, you will not be able to do some of the questions, and others will be very time consuming.

Your technology must be able to:

1. deal with arithmetic operations ($+$, $-$, $\times$, $\div$), the numbers pi (π) and e, square roots ($\sqrt{\ }$), reciprocals $\left(x^{-1} \text{ or } \frac{1}{x}\right)$, powers ($x^y$ or $x^\wedge y$), trigonometric functions (*sin, cos,* and *tan*), and logarithm functions (*log* and *ln*);

2. graph functions of one variable involving the above expressions;

3. find lines of best fit and other regression equations;

4. accommodate the writing, editing, storing, and running of programs.

Your school may give you technology to use or advise you on which technology to purchase. If you buy a graphics calculator, we recommend that you get one that is also capable of generating tables, operating on lists, and performing matrix operations.

Getting Off to a Good Start

It is always helpful to spend some time getting acquainted with your textbook. The questions that follow are designed to help you become familiar with *UCSMP Advanced Algebra*.

We hope you join the thousands of students who have enjoyed this book. We wish you much success.

QUESTIONS

Covering the Reading

1. Name four topics that Advanced Algebra includes.

2. What tools other than paper and pencil are needed for this course?

3. How can the statement "Mathematics is not a spectator sport" be applied to the study of Advanced Algebra?

4. Identify two strategies that you might use to improve your reading comprehension.

5. What kinds of writing will you be asked to do in Advanced Algebra?

Knowing Your Textbook

In 6–13, answer the questions by looking at the Table of Contents, the lessons and chapters of the textbook, or material at the end of the book.

6. Refer to the Table of Contents. What lesson would you read to learn about step functions?

7. What are the four categories of questions in each lesson?

8. Suppose you just finished the questions in Lesson 2-8.
   a. On what page can you find answers to check your work?
   b. Which answers are given?

9. At the end of Question 25 in Lesson 1-4, you see *(Lesson 1-3)*. What does this mean?

10. Refer to the Chapter Review at the end of a chapter. What does SPUR mean?

11. This book has some Appendices. How many are there and what do they cover?

12. Use the Index in the back of your book to find *decibel*.
    a. In what lesson is this discussed?
    b. What is the approximate relative intensity in decibels of a very loud rock concert?

13. Each chapter is introduced with an application of a major idea from the chapter. Read page 5. The relation between time and value given in the table and graph is an example of a __?__.

Functions?

CHAPTER 1

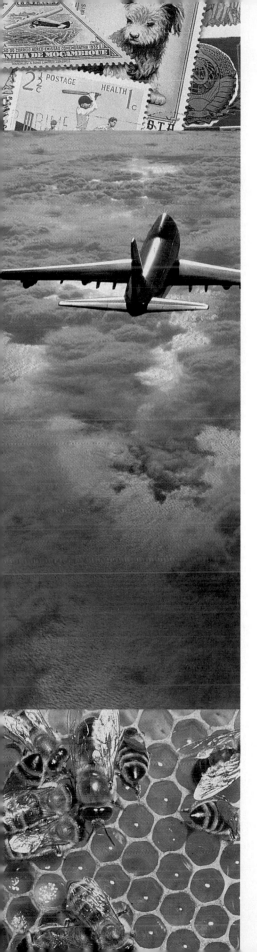

FUNCTIONS

Many people are avid collectors. People all over the world collect things ranging from antique jewelry to comic books to stamps. Some people are collectors because they use the items or find them interesting. Other people collect things as investments, hoping they will increase in value over time. For instance, a *Superman* comic which sold in 1970 for 25¢ now sells for a price between $12.00 and $30.00, depending on the condition and content of the comic book.

Of course, not all items continue to increase in value over time. The collector's edition of *The Death of Superman* originally sold for $2.95 on November 17, 1992, the day it was released. The next day, after all issues had been sold, the comic book was worth $20.00. By December 18, the price had skyrocketed to $100.00. But then the price began to drop. On January 10, 1993, the collectors edition was worth only $75.00 and by June 15, 1993, the comic book's value had decreased to $30.00, and had been stable for some time. By June 15, 1994, the price for a copy of *The Death of Superman* in mint condition was $10.00.

The six pairs of dates and values given above can be represented in a table or in a graph.

Days after release	Value in $
0	2.95
1	20.00
31	100.00
54	75.00
210	30.00
575	10.00

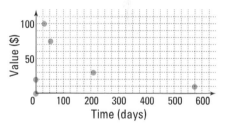

It is difficult to predict the future value of the comic book from these points. However, when a graph or table has a regular pattern, you can often estimate values in the future with some confidence.

The relation between time and value is an example of a *function*. In this course you will study many kinds of functions, their properties, and their uses. In this chapter you will learn how to describe functions with words, tables, graphs, and symbols. Because an understanding of functions depends on your ability to use algebraic expressions and sentences correctly, you will also review these important ideas in this chapter.

1-1

The Language of Algebra

"That's some of my earlier work."

The language of algebra uses numbers and variables. A **variable** is a symbol that can be replaced by any member of a set of numbers or other objects. When numbers and variables are combined using the operations of arithmetic, the result is called an **algebraic expression,** or simply an **expression.**

The expression πr^2 uses the variable r and the numbers π and 2. An algebraic expression with two variables is $a + b$.

An **algebraic sentence** consists of expressions related with a verb. The most common verbs in algebra are = (is equal to), < (is less than), > (is greater than), ≤ (is less than or equal to), ≥ (is greater than or equal to), ≠ (is not equal to), and ≈ (is approximately equal to). Some algebraic sentences are $A = \pi r^2$, $a + b = b + a$, and $3x + 9 < 22$.

Algebra is the study of expressions, sentences, and other relations involving variables. Because many expressions and sentences are based on patterns in arithmetic, algebra sometimes is called generalized arithmetic.

Writing Expressions and Sentences

From your earlier study of algebra, you should know how to write expressions and sentences describing real situations, and how to evaluate expressions or sentences.

Example 1

Express the cost of *y* cans of orange juice at *x* cents per can.

Solution 1

Use a special case. 5 cans at 60¢ per can would cost
5 · 60¢ = 300¢ = $3.00. That suggests multiplication.
So y cans at x cents per can will cost xy cents.

Solution 2

Recognize a general model for multiplication. The unit "cents per can," which can be written as $\frac{cents}{can}$, signals a *rate*. Multiply *y* cans by the rate factor $x \frac{cents}{can}$, to obtain the total cost. The unit of *xy* is the "product" of the units.

$$x \frac{cents}{can} \cdot y \text{ cans} = xy \text{ cents}$$

You may have noticed that part of the solution in Example 1 is written **using this typestyle.** This style is used to indicate what you might write on your homework paper as the solution to the problem.

Example 2

Kim collects stamps. She now has 9000 stamps. If Kim buys 40 stamps each month, how many stamps will she have after *m* months?

Solution

Make a table. Notice in this table that the arithmetic is not carried out. This makes the pattern easier to see.

Months from now	Number of stamps
1	9000 + 1 · 40
2	9000 + 2 · 40
3	9000 + 3 · 40
4	9000 + 4 · 40

The number in the left column, which gives the number of months, is always in a particular place in the expressions at the right. That shows a pattern.

m	9000 + m · 40

Kim will have 9000 + 40m stamps after m months.

Check

To check the expression 9000 + 40*m*, pick a value for *m* not in the table. We pick *m* = 5, indicating 5 months from now. Then substitute this value into the expression 9000 + 40*m* = 9000 + 40 · 5 = 9200. Now extend the table to *m* = 5, by adding 40 as you go from one row to the next. As you can see, it checks.

Months from now	Number of stamps
1	9040
2	9080
3	9120
4	9160
5	9200

You could also describe the situation in Example 2 with a sentence. Suppose S = the number of stamps Kim has after m months. Then you could write $S = 9000 + 40m$.

As Examples 1 and 2 show, there is often more than one way to translate situations into algebra. You should try to learn a variety of ways. The expression you get a second way can be used to check the expression you got the first way.

Evaluating Expressions and Formulas

Substituting for the variables in an expression and calculating a result is called **evaluating the expression.** In order to evaluate expressions, you must use the rules for grammar and punctuation of the language of algebra. The following rules for *order of operations* are used to evaluate expressions worldwide.

Rules for Order of Operations
1. Perform operations within parentheses ( ), brackets [ ], or other grouping symbols, like square root symbols or fraction bars, from the inner set of symbols to the outer set. Inside grouping symbols, use the order given in Rules 2, 3, and 4.
2. Take powers.
3. Multiply and divide in order from left to right.
4. Add and subtract in order from left to right.

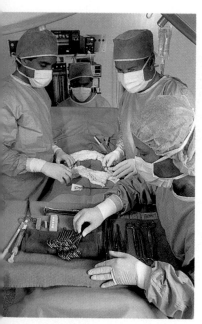

Doctor's orders. *The order in which operations are carried out is important in medicine— as well as in most everyday procedures that are done.*

An **equation** is a sentence stating that two expressions are equal. A **formula** is an equation stating that a single variable is equal to an expression with one or more different variables on the other side. Below are some examples.

$$A = \pi r^2 \qquad \text{both an equation and a formula}$$
$$S = 2w\ell + 2wh + 2\ell h \qquad \text{both an equation and a formula}$$
$$a + b = b + a \qquad \text{an equation that is not a formula}$$

Formulas are useful because they express important ideas with very few symbols and because they can be applied easily to many situations. Here is an example of evaluating an expression in a science formula with more than one variable in the expression. It also illustrates how we work with units in formulas.

Example 3

The formula $d = \frac{1}{2}gt^2$ tells how to find d, the distance an object has fallen during time t, when it is dropped in free fall near the Earth's surface. The variable g represents the acceleration due to gravity. Near the Earth's surface, $g = 9.8 \frac{m}{sec^2}$. About how far will a rock fall in 5 seconds if it is dropped near the Earth's surface?

▶

▶ **Solution**

Substitute $g = 9.8 \frac{m}{sec^2}$ and $t = 5$ sec into the formula.

$$d = \frac{1}{2}gt^2$$
$$= \frac{1}{2}\left(9.8 \frac{m}{sec^2}\right)(5 \text{ sec})^2$$
$$= \frac{1}{2} \cdot 9.8 \frac{m}{sec^2} \cdot 25 \text{ sec}^2$$
$$= 4.9 \cdot 25 \text{ m}$$
$$d = 122.5 \text{ m}$$

In 5 seconds, a rock dropped near the Earth's surface will fall about 122.5 meters.

Check

The time units cancel out and you are left with meters. This is an appropriate measure for distance, so the unit checks. Does the distance seem reasonable to you?

Super hero. *Superman first appeared in the comics in 1938. Although he "died" in the November 17, 1992 issue, he was "brought back" in the April 15, 1993 issue.*

QUESTIONS

Covering the Reading

These questions check your understanding of the reading. If you cannot answer a question, you should go back and reread the lesson to help you find an answer.

In 1 and 2, refer to the information about the value of *The Death of Superman*.

1. How much did the comic book increase in value between November 17 and December 18, 1992?

2. **a.** How many variables are described in the table and graph?
   b. Name them.

3. Name all the variables in $2\pi r$.

4. Tell whether $2\pi r$ is an example of each of the following.
   a. an equation   **b.** an expression
   c. a formula   **d.** a sentence

In 5–7, translate into words.
   5. $\leq$ 6. $\neq$ 7. $\approx$

8. Give an example of an algebraic expression not found in the reading.

9. Give an example of an algebraic sentence.

10. **a.** What is the cost of 10 cans of tomato juice costing c cents per can?
    b. What is the cost of m cans of tomato juice costing c cents each?

11. A person now owns 25 stamps and is buying 3 stamps a week. How many stamps will there be after w weeks?

12. To evaluate $3 \cdot 5^{(4-2)}$, you must do these steps. Put them in order.
(1) Multiply.   (2) Take the power.   (3) Do the subtraction.

In 13–15, a sentence is given. **a.** Is the sentence an equation? **b.** Is the sentence a formula?

13. $A = \pi r^2$ **14.** $6s^2 > 0$ **15.** $2(L + W) = 2L + 2W$

In 16 and 17, refer to Example 3. How far will a rock fall in four seconds if it is dropped from the given spot?

16. near the Earth's surface

17. near the surface of the moon, where $g = 1.6 \frac{m}{sec^2}$

Applying the Mathematics

These questions extend the concepts of the lesson. You should take your time, study the examples and explanations, and try a variety of methods. Check your answers to odd-numbered questions with the ones in the back of the book.

18. Suppose a collector now owns K kachina dolls and is buying 6 dolls per year. How many dolls will the collector have after y years?

In 19–23, tell which expression, (a) $x + y$, (b) $x - y$, (c) $y - x$, (d) xy, (e) $\frac{x}{y}$, or (f) $\frac{y}{x}$, will lead to a correct answer.

19. You give a friend y dollars. You had x dollars. What do you have left?

20. Mrs. Bell is y years old. A friend is x years older. How old is the friend?

21. You drove x miles in y hours. What was your rate?

22. You buy x granola bars at y cents per bar. What is the total cost?

23. A picture of a building is x times actual size. The height of the building is y. What is the height of the building in the picture?

Children of the Hopi tribe learn to identify kachinas—powerful spirits of the earth, sky, and water—by playing with kachina dolls.

In 24–27, make up one example of a situation different from those in this lesson that can lead to the expression.

24. $x + y$ **25.** $x - y$ **26.** xy **27.** $\frac{x}{y}$

In 28–30, evaluate each expression when $x = 15$, $y = -3$, and $z = 2$.

28. $\frac{x}{y} - z^3$ **29.** $\frac{x}{(y - z)^3}$ **30.** $\frac{x}{y - z^3}$

31. Young's formula, $C = \left(\frac{g}{g + 12}\right)A$, has been used to determine how much medicine C to give to a child under age 13 when the adult dosage A is known. Here g is the child's age measured in years. Suppose an adult dosage for a medicine is 600 milligrams. What is the dosage for a 3-year-old according to this formula? (Caution! Do not apply this formula yourself. Medicines should only be taken under the supervision of a physician or pharmacist.)

32. Dennis plans to use 1150 ft of fence to enclose the rectangular pasture shown below. One side borders a river where there is already a thick hedge. It needs no fencing.

a. Let x be the width of the pasture as labeled. Write an expression for the length in terms of x.
b. Write an expression for the area of the pasture in terms of x.
c. Suppose the pasture must enclose at least 60,000 square feet. Write a sentence relating the area expression in part **b** to the area the pasture must enclose.

Review

Every lesson contains review questions to practice on ideas you have studied earlier. If the idea is from an earlier course, the question is designated as being from a previous course.

33. *Multiple choice.* Which sentence correctly relates the angle measures in the figure at the left? *(Previous course)*
  (a) $x + y + z = 180$ (b) $x + y = 180 + z$
  (c) $y + z = x + y$ (d) none of these

34. What is a polygon with six sides called? *(Previous course)*

35. a. Solve $3x = 5x + 18$ for x.
   b. Check your work. *(Previous course)*

Exploration

These questions extend the concepts of the lesson. Often they have many possible answers. Sometimes they require you to use dictionaries, encyclopedias, or other sources.

36. Many integers can be written with three 9s. For instance,

$$4480 = 9! \div (9 \times 9) \text{ and}$$
$$720 = 9^{\sqrt{9}} - 9.$$

What integers from 1 to 25 can be written with three 9s? Allow yourself the operations of $+, -, \times, \div, \sqrt{\ }, !$, decimal points, powers, 99, and parentheses.

Net pay. *The person shown at the lower right is a tennis instructor at a summer camp in California. An instructor's total pay is often a function of the number of hours worked.*

The Definition of Function

Bjorn has a summer job from which he earns $5.00 per hour. His weekly pay in dollars is given by the formula

$$P = 5H,$$

where H is the number of hours worked.

Below is a table of possible values for H and P.

H	1	2	3	6.5	10	20
P	5.00	10.00	15.00	32.50	50.00	100.00

In the equation $P = 5H$, P is called the **dependent variable** because its value (Bjorn's pay) *depends* on the number of hours worked. The variable H is called the **independent variable.**

> **Definition**
> A **function** is a correspondence or pairing between two variables such that each value of the first (independent) variable corresponds to exactly one value of the second (dependent) variable.

In the formula $P = 5H$, each value that is substituted for H results in just one value of P. Hence, the relationship described by this equation is a function.

When the relationship between two variables is a function, we say that the dependent variable **is a function of** the independent variable. In the formula above we say that P is a function of H.

Example 1

The table below shows a relation between the year Y and the percent P of public high schools in the United States with desktop computers available for student use.

Y	1981	1982	1983	1984	1985	1986	1987	1988	1989	1990	1991
P	42.7	57.8	86.1	94.6	97.4	98.7	99.0	99.1	99.1	98.8	99.4

a. Is P a function of Y?
b. Is Y a function of P?
Justify your answers.

Solution

a. Think: Is any value of Y paired with more than one value of P? No. Each value of Y is paired with only one value of P; so P is a function of Y.
b. Y is not a function of P, because P = 99.1 is paired with two different years, Y = 1988 and Y = 1989.

In Example 1, because P is a function of Y, Y is the independent variable and P is the dependent variable in the function.

As indicated above, functions can be described with formulas or with tables. In Lesson 1-4, you will see how functions can also be described with graphs.

Domain and Range of a Function

The **domain of a function** is the set of values which are allowable substitutions for the independent variable. The **range of a function** is the set of values of the dependent variable that can result from the substitutions for the independent variable. The substitutions for the independent variable are often called **input,** and the resulting values of the dependent variable are often called **output.**

Example 2

Refer to Example 1, where P is a function of Y. Give the domain and range of the function.

Solution

The domain is the set of all values of the independent variable Y. So The domain is {1981, 1982, 1983, 1984, 1985, 1986, 1987, 1988, 1989, 1990, 1991}. The range is {42.7, 57.8, 86.1, 94.6, 97.4, 98.7, 99.0, 99.1, 98.8, 99.4}.

Notice in the range of Example 2, the element 99.1 is listed once. Recall that when listing elements of a set, each element needs to be listed only once, and you may list the elements in any order.

Some sets of numbers are used frequently as domains.

The set of **natural numbers** or **counting numbers** is {1, 2, 3, 4, 5, . . . }.

The set of **whole numbers** is {0, 1, 2, 3, 4, 5, . . . }.

The set of **integers** is {. . . , -3, -2, -1, 0, 1, 2, 3, . . .}.

The set of **rational numbers** is the set of numbers that can be represented as ratios of the form $\frac{a}{b}$, where a and b are integers and $b \neq 0$.

Samples: $0, 1, -7, \frac{2}{3}, 1\frac{9}{11}, -\frac{34}{10}, 0.0004, 9.6\overline{18}, \sqrt{16}$

The set of **real numbers** is the set of numbers that can be represented by decimals.

Samples: $0, 1, -7, 35$ million, $2.34, \pi, \sqrt{5}$

Example 3

The area A of a circle with radius r is given by $A = \pi r^2$.
a. Is A a function of r? Why or why not?
b. Identify the domain and range of the function.

Solution

a. Each value of r substituted into the equation gives just one value of A. So, A is a function of r.
b. The domain of this function is the set of possible values of r. The range is the set of possible values of A. Radii and areas of circles must be positive numbers. So Both the domain and range of this function are the set of positive real numbers

Notice that if we think of r and A as measures of length and area as in Example 3, we could also say that r is a function of A. For every area A there is a unique radius r. However, if we think of the formula $A = \pi r^2$ simply as an equation, and allow r to be negative as well, then r is not a function of A. For instance, the value $A = 25\pi$ corresponds to two different values of r: $r = 5$ and $r = -5$.

$$25\pi = \pi(5)^2 \text{ and } 25\pi = \pi(-5)^2$$

Functions on a Calculator

Many of the rules programmed into a calculator are designed to give you a single answer (the output) when you enter a value (the input) in the domain of the function. Here are some calculator keys, the related functions, their domains, and the values which produce error messages.

Key	Function	Domain	Inputs Giving Errors
x^2 , the squaring key	$y = x^2$	set of all real numbers	none
$\sqrt{\ }$, the square root key	$y = \sqrt{x}$	the set of nonnegative real numbers	$x < 0$
1/x , or x^{-1} , the reciprocal key	$y = \frac{1}{x}$	the set of nonzero real numbers	$x = 0$

On some calculators, to evaluate expressions using squaring, square roots, or reciprocals, you press the function key followed by the value. Graphics calculators often work this way. So to evaluate $\sqrt{50}$ you might press the following:

$$\boxed{\sqrt{}}\ 50\ \boxed{\text{EXE}}\ .$$

On other scientific calculators, to evaluate expressions using function keys you press the number followed by the function key. On such calculators to evaluate $\sqrt{50}$ you might enter

$$50\ \boxed{\sqrt{}}\ .$$

Activity 1

Use the function keys on your calculator to evaluate these expressions. Record the key sequence you use, as well as your answer.

a. $\sqrt{65}$ **b.** $(-1.4142)^2$ **c.** $\frac{1}{2000}$

Recall that calculators often display very large and very small numbers using scientific notation. For instance, some calculators display the result of $(1,500,000)^2 = 2,250,000,000,000$ as $\boxed{\text{2.25E 12}}$.

The E in such a display does not mean error. Rather, the display means $2.25 \cdot 10^{12}$.

Activity 2

Evaluate the expressions using a calculator. For each, copy the display and rewrite it in scientific notation.

a. $(260,000)^2$ **b.** $\frac{.0005}{(2500)^2}$ **c.** $\frac{\sqrt{5}}{\sqrt{80,000,000}}$

QUESTIONS

Covering the Reading

1. Define *function*.

2. Consider the table below.

x	0	1	2	3	4	5
y	5	7	4	4	-6	7

   a. Is y a function of x? Explain your answer.
   b. Is x a function of y? Explain your answer.

3. The volume of a cube with each edge of length e is given by $V = e^3$. Is V a function of e? Justify your answer.

In 4 and 5, define each term.

4. domain of a function   5. range of a function

6. *True or false.* Every function has both a domain and range.

In 7 and 8, use the table below, and consider T = the average length of a school term (in days) as a function of a year Y.

Y	1910	1920	1930	1940	1950	1960	1970	1980	1990
T	157.5	161.9	172.7	175.0	177.9	178.0	178.9	178.5	179.8

7. Which is the independent variable?

8. What is the range of the function?

9. List the set of positive integers less than 7.

In 10–12, identify each number as an integer, a rational number, or a real number. A number may belong to more than one set.

10. 1437

11. $\frac{15}{19}$

12. $-\sqrt{11}$

13. Name a real number that is not an integer.

14. Name a rational number that is
    a. a whole number.
    b. not a whole number.

In 15–17, consider the 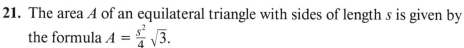 keys on your calculator. Which, if any, produce an error message when the given value is input?

15. -25

16. 0

17. 6.5

18. What answers did you get for Activity 1?

19. What answers did you get for Activity 2?

Applying the Mathematics

20. Consider the function $y = x^2$.
    a. What is its domain?
    b. What is its range? (Hint: Think about the kind of number that results when you square either a positive or negative number.)

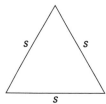

21. The area A of an equilateral triangle with sides of length s is given by the formula $A = \frac{s^2}{4}\sqrt{3}$.
    a. State the domain for s.
    b. Find the area of an equilateral triangle with sides of 10 cm. Round your answer to the nearest tenth.

22. The formula $T = 180(n - 2)$ describes T as a function of n. Here T is the total number of degrees in the interior angles of a polygon with n sides. (Formulas from geometry are summarized in Appendix A.)
    a. What is the domain of the function?
    b. Find the total number of degrees in the interior angles of a hexagon.
    c. Draw a hexagon. Check that your answer to part **b** is correct either by using a protractor, or by giving a logical argument based on the Triangle Sum Theorem.

MEDIAN AGE AT FIRST MARRIAGE		
Year	Males	Females
1900	25.9	21.9
1910	25.1	21.6
1920	24.6	21.2
1930	24.3	21.3
1940	24.3	21.5
1950	22.8	20.3
1960	22.8	20.3
1970	23.2	20.8
1980	24.7	22.0
1989	26.2	23.8
1990	26.1	23.9
1991	26.3	24.1
1992	26.5	24.4

23. Refer to the table at the left. Let Y = the year, M = median age at first marriage for males, and F = median age at first marriage for females.
   a. Is M a function of Y? Explain your answer.
   b. Is F a function of M? Why or why not?

24. The word "function" was most likely a part of your vocabulary before you read the definition given in this section. Describe how the word "function" is used in normal conversation. Compare your use to the one given here. What commonalities exist? What differences?

In 25 and 26, evaluate each function for the given value of the independent variable.

25. $p = 300(2)^n$, $n = 5$ **26.** $w = 4t^2 + 7t - 2$, $t = 3$

Review

A lesson reference following a review question indicates a place where the idea of the question is discussed.

In 27–30, tell whether the answer is $a + b$, $a - b$, $b - a$, ab, $\frac{a}{b}$, or $\frac{b}{a}$. *(Lesson 1-1)*

27. A football player gains a yards on one play and gains b yards on the next. What is the total gain?

28. A football player gains a yards on one play and loses b yards on the next. What is the total gain?

29. How many different outfits can be made from b skirts and a shirts?

30. Suppose you spend b dollars to buy a grams of perfume. What is the cost per gram?

31. *Multiple choice.* Which property is illustrated by the statement $(4 \cdot x) \cdot y = 4 \cdot (x \cdot y)$? *(Previous course)*
   (a) Commutative Property of Multiplication
   (b) Associative Property of Multiplication
   (c) Distributive Property
   (d) Inverse Property

Exploration

32. Look through a newspaper, magazine or one of your other texts for a table of pairs of numbers or objects. Does the table describe a function? Why or why not?

*Grouping
Symbols
and
Calculators*

IN·CLASS
A C T I V I T Y

Work in small groups. Each person should have a calculator. Do each question individually; then share your solution with others in your group.

In 1 and 2, recall that when using a calculator to evaluate expressions with square roots, you may need parentheses where none are written.

1 **a.** Evaluate $\sqrt{9 + 4^2}$ in your head.

b. Evaluate $\sqrt{9 + 4^2}$ on your calculator. Write the key strokes which give you the correct answer.

2 The formula $d = \sqrt{12{,}800h + h^2}$ gives the distance (in km) to the horizon from a point h km above the surface of the Earth. When $h = 4300$, $d \approx 8600$. Find the distance to the horizon from a point that is 1300 km above the surface of the Earth.

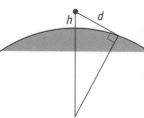

In 3 and 4, recall that the fraction bar also acts as a grouping symbol.

3 **a.** Evaluate $y = x + \dfrac{x^2}{20}$ when $x = 75$.

b. Evaluate $y = \dfrac{x + x^2}{20}$ when $x = 75$.

(The answers to parts **a** and **b** are not equal.)

4 **a.** Evaluate $y = \dfrac{1}{x^2 - 3}$ when $x = 0.7$.

b. Evaluate $y = \dfrac{1}{x^2} - 3$ when $x = 0.7$.

(Again the answers to parts **a** and **b** are not equal.)

5 Graphics calculators sometimes give different answers than other calculators to expressions involving constants that are juxtaposed (written next to each other). For instance, to find the volume of a beach ball with a radius of 5 inches, you could evaluate $V = \frac{4}{3}\pi r^3$ when $r = 5$. Input each expression below into your calculator as written. (y^x or $\wedge$ is the powering key.)

a. $4 \div 3 \times \pi \times 5 \; y^x \; 3$

b. $4 \div 3\pi \times 5 \; y^x \; 3$

c. $4 \times \pi \times 5 \; y^x \; 3 \div 3$

d. $(\; 4 \div 3 \;) \times \pi \times 5 \; y^x \; 3$

e. $(\; 4 \div 3\pi \;) \times 5 \; y^x \; 3$

Which give(s) the correct results on your calculator? Explain why the other answers are wrong or why you get an error message.

Drive right? *Shown are several of London's famous cabs. Because cars in Great Britain travel on the left side of the road, they are manufactured with the controls on the right side.*

The table of values below was distributed several years ago by the Highway Code of Great Britain. It gives information about three distances (thinking, braking, and stopping distances), each of which is a function of a car's speed.

Speed of car	mph	10	20	30	40	50	60
Thinking distance	ft	10	20	30	40	50	60
Braking distance	ft	5	20	45	80	125	180
Stopping distance	ft	15	40	75	120	175	240

The *thinking distance* is the distance a car travels after a driver decides to stop but before he or she applies the brakes. The *braking distance* is the distance the car needs in order to come to a complete stop after the driver applies the brakes. The *stopping distance* is the sum of the thinking and braking distances.

The data in the table above can be described with variables.

Suppose the car's speed (in mph) $= x$.

Then the thinking distance (in feet) $= x$,

and the braking distance (in feet) $= \dfrac{x^2}{20}$.

So the stopping distance (in feet) $= x + \dfrac{x^2}{20}$.

We could describe each of these functions as a formula with y and x. However, if we let $y = x$, $y = \frac{x^2}{20}$, and $y = x + \frac{x^2}{20}$, all the x's and y's would be too confusing. So instead, we use a notation called $f(x)$ **notation.**

$f(x)$ Notation

The symbol $f(x)$ is read "f of x." This notation is attributed to Leonard Euler (pronounced "oiler"), a Swiss mathematician who lived from 1707 to 1783. Euler wrote one of the most influential algebra books of all time. The parentheses in Euler's notation do not stand for multiplication. Instead, they enclose the independent variable.

We let $T(x)$ be the thinking distance at speed x.

$$T(x) = x, \qquad \text{read "} T \text{ of } x \text{ equals } x \text{."}$$

We let $B(x)$ stand for the braking distance at speed x.

$$B(x) = \frac{x^2}{20}, \qquad \text{read "} B \text{ of } x \text{ equals } \frac{x^2}{20} \text{."}$$

And we let $S(x)$ denote the stopping distance at speed x.

$$S(x) = x + \frac{x^2}{20}, \quad \text{read "} S \text{ of } x \text{ equals } x + \frac{x^2}{20} \text{."}$$

The letters T, B, and S name functions. The variable x is called the **argument** of the function. The numbers $T(x)$, $B(x)$, and $S(x)$ stand for the **values** of these functions, that is, the values of the dependent variable. The argument is the input to a function. The value of the function is its output.

Example 1

Evaluate $B(45)$.

Solution

Substitute 45 for x in the equation $B(x) = \frac{x^2}{20}$.

$$B(45) = \frac{45^2}{20} = 101.25$$

Check

Refer to the table on page 19. The braking distance for a car going 45 mph should be between that for cars going 40 mph and cars going 50 mph; that is, between 80 ft and 125 ft. So 101.25 feet is a reasonable value of the function at $x = 45$.

Euler's function notation is very efficient. Writing $B(45) = 101.25$ is much shorter than writing **the value of the function B at x = 45 is 101.25,** or **the braking distance of a car traveling 45 mph is about 101.25 feet.**

Mapping Notation

In addition to formulas and Euler's notation, functions may be described by **arrow** (or **mapping**) **notation.** Like Euler's notation, mapping notation states both the name of the function and the independent variable. A description of the functions T, B, and S in mapping notation is shown below.

$$T: x \rightarrow x, \qquad \text{read "}T \text{ maps } x \text{ onto } x\text{."}$$

$$B: x \rightarrow \frac{x^2}{20}, \qquad \text{read "}B \text{ maps } x \text{ onto } \frac{x^2}{20}\text{."}$$

$$S: x \rightarrow x + \frac{x^2}{20}, \quad \text{read "}S \text{ maps } x \text{ onto } x + \frac{x^2}{20}\text{."}$$

Example 2

Use the function S defined above to complete the statement:

$$S: 57 \rightarrow (\underline{\quad ? \quad}).$$

Solution

Substitute 57 for x in the mapping notation for stopping distance.

$$S: x \rightarrow x + \frac{x^2}{20}$$
$$S: 57 \rightarrow 57 + \frac{57^2}{20}$$
$$S: 57 \rightarrow 219.45$$

This is read "S maps 57 onto 219.45."

Check

According to the table on page 19, the stopping distance should be between 175 ft and 240 ft, so 219.45 feet is a reasonable value of the function.

You should be able to read and express formulas for functions using either Euler's notation or mapping notation.

Example 3

The sum of the angle measures in a polygon with n sides is given by the formula $S = 180(n - 2)$. Rewrite the formula
a. using Euler's notation.
b. using mapping notation.

Solution

The independent variable is n. Pick a name for the function, say f. Replace the dependent variable with $f(n)$.
a. $f(n) = 180(n - 2)$
b. $f: n \rightarrow 180(n - 2)$
Notice that $S = f(n)$.

Covering the Reading

In 1–3, how is each read?

1. $f(x)$

2. $B: x \rightarrow \dfrac{x^2}{20}$

3. $S(x) = x + \dfrac{x^2}{20}$

In 4–6, refer to the functions in this lesson.

4. $B: 50 \rightarrow (\underline{})$

5. $S(55) = (\underline{})$

6. $T(10) = (\underline{})$

7. The value of a function is a value of which variable, dependent or independent?

In 8 and 9, refer to Example 3.

8. Evaluate $f(12)$, and explain what you have just calculated.

9. State whether n, the number of sides, is the dependent or independent variable.

In 10 and 11, suppose that $g: x \rightarrow 5x - 2$.

10. Describe this function using Euler's notation.

11. $g: 3 \rightarrow (\underline{})$

Applying the Mathematics

12. Refer to the table at the beginning of the lesson. The usual "rule of thumb" is to maintain one car length (about 16 feet) between your car and the car in front of you for every 10 mph. How realistic is this estimate at a speed of 60 mph? Write a sentence or two explaining your conclusion.

13. Suppose $g(x) = 2^x$ and $h(x) = x^2$. Which is greater, $g(15)$ or $h(15)$? Justify your answer.

14. Suppose $g: x \rightarrow \sqrt{x}$. Fill in the blanks.
 a. $g: 9 \rightarrow \underline{}$
 b. $g: \underline{} \rightarrow 9$

In 15 and 16, Margo painted her bedroom. It took 10 hours. Each hour she wrote down the percent of the job she thought she had finished.

Hours	1	2	3	4	5	6	7	8	9	10
Percent finished	5	20	35	50	50	65	70	80	95	100

15. If $P(h)$ is the percent finished after h hours, what is $P(5)$?

16. If Margo began at 8 A.M., in which hour did she not work at all?

In 17–19, use the table below, in which x is the year and $p(x)$ is the average price of one gallon of unleaded gasoline (in cents).

x	1976	1978	1980	1982	1984	1986	1988	1990	1992
$p(x)$	61.4	67.0	124.5	129.6	121.2	92.7	94.6	116.4	112.7

17. **a.** Evaluate $p(1980) - p(1978)$.
    b. Explain in words what you have just calculated.

18. **a.** For what values of x is $p(x) > 100$?
    b. What does the answer to part **a** mean about the price of gasoline?

19. Solve $p(x) = 121.2$.

20. If $f(x) = \frac{9 + x}{x^2}$, evaluate each of the following.
    a. $f(4)$ **b.** $f(-4)$ **c.** $f(n)$ **d.** $f(4n)$

Number lines. *In 1979, a gas shortage prompted many states to adopt odd-even rationing (based on license-plate numbers) to determine what days gas was available to each driver.*

Review

21. Is y a function of x? Why or why not? *(Lesson 1-2)*

x	0	1	-1	2	-2
y	0	1	1	4	4

22. Given the function f defined by $f(x) = \frac{1}{x}$, what is the domain of f? *(Lesson 1-2)*

23. Ozair is saving money for college. He now has $278 in his account. He plans to save $25 each week. How much will he have
    a. after 6 weeks?   **b.** after n weeks? *(Lesson 1-1)*

In 24 and 25, an equation is given. **a.** Solve the equation. **b.** Check your answer. *(Previous course)*

24. $-3x + 19 = 74$ 25. $n + 10 = 3n + 9$

26. **a.** What is a horizontal line?
    b. What is a vertical line?
    c. Give one example of a horizontal line and one example of a vertical line found in your classroom. *(Previous course)*

Exploration

27. Euler was one of the greatest mathematicians of all time. Find out some of his contributions to mathematics.

1-4

Graphs of Functions

Because a function is a set of ordered pairs, when both coordinates are real numbers, we can associate each ordered pair with a point in the plane. That is, we can make a graph of the function.

Recall that each point in the coordinate plane has (x, y) coordinates. If $y = f(x)$, then you can also say that each point on the graph has coordinates of the form $(x, f(x))$. Below is the part of the table used in Lesson 1-3 for a stopping-distance function.

Speed of car (mph) = x	10	20	30	40	50	60
Stopping distance (ft) = $S(x)$	15	40	75	120	175	240

The graph on the left below is a graph of the stopping-distance function S for values of x from 0 to 60. All the points on it are of the form $(x, S(x))$. To find the value of $S(40)$ from the graph, start at 40 on the x-axis. Read up to the curve of the function S and then across to find the value on the y-axis. So $S(40) = 120$. Notice that this agrees with the values in the table above.

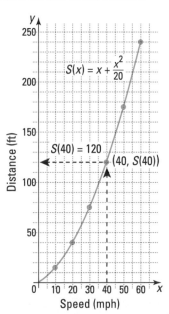

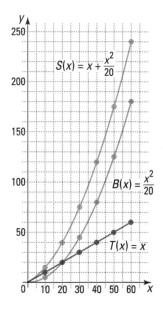

In the graph on the right above, we have plotted points to represent the three functions T, B, and S studied in the previous lesson. Notice how Euler's notation helps to distinguish the y-coordinates of the graphs. Euler's $f(x)$ notation is very handy when more than one function is being studied.

Finding Domain and Range from a Graph

Recall that the domain of a function is the set of allowable values for the independent variable. For the functions T, B, and S, the situation and the table determine the domain. The allowable speeds are from 0 to 60 mph. So the domain of each function is $\{x: 0 \le x \le 60\}$. This is read "the set of all x from 0 to 60."

The range of a function is the set of values of the dependent variable that result from all possible substitutions for the independent variable. The range of the function B can be found by examining the graph at the bottom of page 24. When x has values from 0 to 60, the values of $B(x)$ range from 0 to 180. So the range of B is $\{y: 0 \le y \le 180\}$.

If a graph is plotted in the (x, y) coordinate plane, then x is considered the independent variable, and y is the dependent variable. If other coordinates are used, the independent variable is the first coordinate and is plotted along the horizontal axis. The dependent variable is the second coordinate and is plotted along the vertical axis.

Example 1

Refer to the graph below. The oven temperature T varies with the length of time t the oven has been on. The oven, whose initial temperature was 80°, was set for 350°. The actual temperature was measured and then graphed over a 30-minute interval.

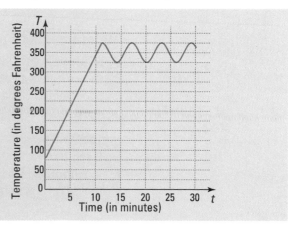

Cooking up a storm.
Pictured is a commercial oven. Widespread use of gas ranges began in the early 1900s. Electric ranges became popular after 1930.

a. Estimate the oven temperature when the oven had been on for 20 minutes.
b. Estimate how much time it took the oven to first reach 350°.
c. Identify the independent and dependent variables.
d. State the domain and range of the function.

Solution

a. Use the graph and read the value of T when t is 20. The temperature is about 325°F.
b. Read up to 350° on the vertical axis. Read across until you first reach the graph. Now read down from this point. It took about 10 minutes for the oven to first reach 350°. ▶

c. The independent variable is graphed on the horizontal axis and the dependent variable is graphed on the vertical axis. Hence, Time is the independent variable, and oven temperature is the dependent variable.

d. The domain is the set of possible values of t, the time (in minutes) when the temperature was monitored. The temperature was monitored during 30 minutes. So The domain is $\{t: 0 \leq t \leq 30\}$. The range is the set of all values of T, the temperature in degrees Fahrenheit. On the graph notice that the values of T vary from 80° to 375°.

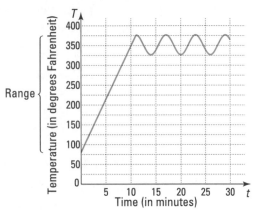

Thus, The range of the function is $\{T: 80 \leq T \leq 375\}$.

Relations and Functions

A **relation** is any set of ordered pairs. Any correspondence or pairing between two variables can be written as a set of ordered pairs. Thus every function is a relation. Using the language of relations, we can reword the definition of function given in Lesson 1-2.

Definition

A **function** is a relation in which no two different ordered pairs have the same first coordinate.

As shown in Example 2 not all relations are functions. Not all graphs represent functions.

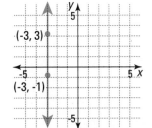

Example 2

Consider the vertical line graphed at the left. Is the line the graph of a function? Justify your answer.

Solution

Two different points on the graph have the same first coordinate, for instance (-3, -1) and (-3, 3). By definition, in a function there cannot be two ordered pairs with the same first coordinate unless the second coordinates are also the same. Therefore, this line is not the graph of a function.

When different points have the same first coordinate, they lie on the same vertical line. This simple idea shows that you can tell whether a relation is a function from its graph.

> **Theorem (Vertical-Line Test for Functions)**
> No vertical line intersects the graph of a function in more than one point.

A function *can* have two ordered pairs with the same second coordinate. For instance, the oven-temperature function graphed in Example 1 contains several points with $T = 325°$. This means that a horizontal line can intersect the graph of a function more than once.

QUESTIONS

Covering the Reading

In 1–3, refer to the graphs of functions *T, B,* and *S* in this lesson.

1. Give their common domain.

2. Give the range of *S*.

3. a. Explain how to estimate $B(45)$ from the graph.
  b. Evaluate $B(45)$ using an equation.

In 4–6, refer to Example 1. Call this function *f*.

4. What was the temperature of the oven when it was turned off?

5. Estimate $f(5)$ and explain what it means.

6. What is the maximum value of *f*?

In 7 and 8, the graph of a function is given. State **a.** its domain, and **b.** its range.

7.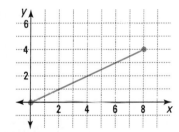

8.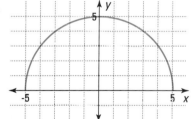

In 9–11, a relation is graphed. Is the relation a function? How can you tell?

9.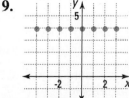

10.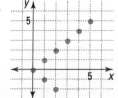

11.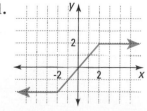

Applying the Mathematics

In 12–16, refer to the graph below. Let $A(t)$ be Alice's weight at age t and $B(t)$ be Bill's weight at age t. The domain of both these functions is $\{t: 0 \le t \le 20\}$.

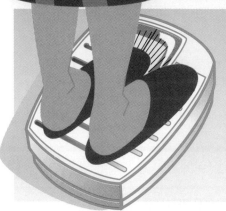

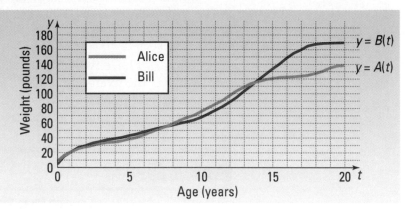

12. What is the range of A?

13. Estimate $B(5)$ from the graph.

14. Approximate $B(10) - A(10)$ from the graph, and explain what your answer means.

15. For what three values of t is $A(t) = B(t)$?

16. During what age intervals did Bill weigh more than Alice?

In 17 and 18, two pairs of numbers satisfying the equation are given.
a. Find three more pairs of numbers that satisfy the sentence.
b. Plot these points.
c. Is y a function of x?

17. $y = x^2$

x	y
−1	1
−2	4

18. $x = y^2$

x	y
1	−1
4	−2

In 19 and 20, refer to the table below, which gives the values (in dollars) of the Corvette Coupe Hatchback V-8 and the Mercedes Benz Sedan 300E in 1986, their year of manufacture, and every year thereafter. (Data are from the NADA Official Used Car Guide.)

Year	1986	1987	1988	1989	1990	1991	1992	1993
Corvette	27,027	22,775	20,600	18,750	16,950	14,025	13,050	12,200
Mercedes	34,700	31,675	29,650	26,350	24,100	19,400	16,775	15,450

Let $x =$ the year, $C(x) =$ the value of the Corvette in that year, and $M(x) =$ the value of the Mercedes in that year.

19. Graph the points $(x, C(x))$, and connect them with a smooth curve. Graph the points $(x, M(x))$ on the same set of axes and connect them using a smooth curve. Label each graph.

20. Which car do you think has held its value better? Use the table or your graph to justify your answer.

In 21–24, refer to the graphs below. The two graphs give information about the number of farms and the average size (in acres) of these farms across the United States from 1983 to 1993.

Farms: Larger, but Fewer

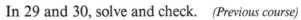

21. Estimate the average size of a farm in 1990.

22. If the average-size function is called A, then $A: 1985 \rightarrow \underline{\quad?\quad}$.

23. If the number-of-farms function is called N, find $N(1991)$.

24. a. Calculate $A(1993) \cdot N(1993) - A(1983) \cdot N(1983)$.
   b. Explain in words what the computation in part **a** represents.

Review

25. Let $f(x) = 1 + 3x^3$. Evaluate each expression.   *(Lesson 1-3)*
   a. $f(2)$ **b.** $f(-2)$ **c.** $f\left(\frac{n}{2}\right)$

26. The function *CTOF* converts degrees Celsius to degrees Fahrenheit. If $CTOF: x \rightarrow 1.8x + 32$, then $CTOF(100) = \underline{\quad?\quad}$.   *(Lesson 1-3)*

27. Let n = the number of sides of a polygon and $f(n)$ = the number of diagonals. The figures at the left show that $f(4) = 2$ and $f(5) = 5$. *(Lessons 1-2, 1-3)*
   a. Find $f(3)$ and $f(6)$.   **b.** What is the domain of f?

28. Carla and Darla own a total of 150 books. If Carla owns b books, how many does Darla own?   *(Lesson 1-1)*

In 29 and 30, solve and check.   *(Previous course)*

29. $\frac{1}{2}n - 72 = 50$ **30.** $x = \frac{2}{3}x + 8$

Exploration

31. Find a coordinate graph in a newspaper or magazine. Does it represent a function? Why or why not? If it is a function, identify its domain and range.

A bird in the hand. *This parrot, in São Paulo, Brazil, will choose the ticket for a player in a lottery game. See Example 2.*

Suppose you know that y is a function of x. Two key questions often arise.

1. Given the value of x, what is the value of y?

2. Given the value of y, what is the value of x?

If the function is described with an equation, then answering the first question usually involves evaluating an expression. Answering the second question usually involves solving an equation. Example 1 illustrates these two situations.

Example 1

Lionel just bought a used car with 33,000 miles on it. He expects to drive it about 800 miles/month. If $m =$ the number of months Lionel has had the car, then $f(m) = 33{,}000 + 800m$ is the expected total number of miles the car will have been driven after m months.
a. How many miles will be on the car after Lionel has had it for a year?
b. After how many months will the car be due for its 50,000-mile check-up?

Solution

a. One year equals 12 months. You are asked to find $f(m)$ when $m = 12$. Evaluate $f(12)$.

$$f(m) = 33{,}000 + 800m$$
$$f(12) = 33{,}000 + 800(12)$$
$$= 33{,}000 + 9600$$
$$= 42{,}600$$

The car will have about 42,600 miles on it after one year.

b. You must find m when $f(m) = 50{,}000$. Solve the equation $33{,}000 + 800m = 50{,}000$.

$$800m = 17{,}000 \quad \text{Subtract 33,000 from each side.}$$
$$m = 21.25 \quad \text{Divide each side by 800.}$$

After about 21 months the car will be due for its 50,000-mile check-up.

Using the Distributive Property

Recall that when you subtract (or add the opposite of) a number from each side of an equation, you are applying the Addition Property of Equality to find an *equivalent* but simpler equation. This is one of the properties of real numbers used in algebra that are listed in Appendix B. You should skim this Appendix now, and refer to it whenever you need to check the meaning or name of a particular property. In the next Example, another property, the Distributive Property, is applied to solve an equation containing parentheses.

Distributive Property
For all real numbers a, b, and c, $c(a + b) = ca + cb$.

Example 2

Suppose you win a lottery and suddenly have $50,000 to invest. You decide to put part of the money in a CD (certificate of deposit) which pays 6% annual interest, and put the rest in a savings account which pays 4%. If d dollars are invested at 6%, then

$$E(d) = .06d + .04(50,000 - d)$$

gives the interest earned in one year. How much should you put in each place to earn $2400 per year?

Solution 1

You are given that $E(d) = 2400$, and asked to find d. Solve an equation.

$.06d + .04(50,000 - d) = 2400$	
$.06d + 2000 - .04d = 2400$	Distribute the .04.
$2000 + .02d = 2400$	Add like terms.
$.02d = 400$	Subtract 2000 from each side.
$d = 20,000$	Divide each side by .02.
$50,000 - d = 30,000$	

You should put $20,000 in the CD at 6%, and $30,000 in the savings account at 4%.

Solution 2

You can avoid working with decimals by multiplying each side by 100. This clears the decimals.

$100[.06d + .04(50,000 - d)] = 100 \cdot 2400$	Multiplication Property of Equality
$6d + 4(50,000 - d) = 240,000$	Distribute the 100.
$6d + 200,000 - 4d = 240,000$	Distribute the 4.
$2d + 200,000 = 240,000$	Add like terms.
$2d = 40,000$	Subtract 200,000 from each side.
$d = 20,000$	Multiply each side by $\frac{1}{2}$.
$50,000 - d = 30,000$	

You should put $20,000 in the CD at 6% and $30,000 in the savings account at 4%.

▶

▶ **Check**

First, check the equation.
Does $.06 \cdot 20{,}000 + .04(50{,}000 - 20{,}000) = 2400$? Yes.
Now, check the situation.
If $20,000 is invested at 6%, the interest is $1200. If $30,000 is
invested at 4%, the interest is $1200. The total interest earned is $2400.

In Example 2, it might seem silly to invest some funds at a lower rate.
But money from a CD cannot be withdrawn early without paying a
penalty. Also, a higher interest rate often means a higher risk. So it is
wise not to put all your money in one place.

Clearing Fractions in Equations

If you want to eliminate or "clear" an equation of fractions, multiply
each side of the equation by a common multiple of the denominators. If
there is more than one term on either side of the equation, you will then
need to apply the Distributive Property. Example 3 illustrates this
procedure.

Example 3

Stuart Dent works part-time to earn spending money and to save for
college. Each month he plans to save $\frac{1}{4}$ of his earnings, to spend $\frac{1}{5}$ of his
earnings on clothes, and to spend $\frac{1}{6}$ of his earnings on transportation. The
remainder will be used for entertainment. Last month Stuart had $46 to
spend on entertainment. How much did he earn last month?

Solution

Write a sentence to describe the situation. **Let E equal Stuart's
monthly earnings.** The total of Stuart's savings and his expenses
equals this monthly earnings. So,

savings   expenses   earnings

$$\frac{1}{4}E + \overbrace{\frac{1}{5}E + \frac{1}{6}E + 46} = E$$

Solve the equation. To "clear" the fractions, multiply each side of the
equation by 60, the least common denominator of $\frac{1}{4}$, $\frac{1}{5}$, and $\frac{1}{6}$.

$$60\left(\frac{1}{4}E + \frac{1}{5}E + \frac{1}{6}E + 46\right) = 60E$$

$15E + 12E + 10E + 2760 = 60E$ Distribute the 60.

$\qquad\qquad 37E + 2760 = 60E$ Add like terms.

$\qquad\qquad\qquad\qquad 2760 = 23E$ Subtract $37E$ from each side.

$\qquad\qquad\qquad\qquad\quad 120 = E$ Multiply by $\frac{1}{23}$.

Stuart earned $120 last month.

Check

Does $\frac{1}{4}(120) + \frac{1}{5}(120) + \frac{1}{6}(120) + 46 = 120$?
$30 + $24 + $20 + $46 = 120. It checks.

R and R. *During a recent
year, people under the age
of 25 spent an average of
$233 on entertainment
fees and admissions.
Some of that money was
spent on concerts given by
entertainers such as
Gloria Estefan.*

Opposite of a Sum Theorem

The Opposite of a Sum Theorem is derived from the Distributive Property.

> **Opposite of a Sum Theorem**
> For all real numbers a and b, $-(a + b) = -a + -b$.

In Example 4, this theorem is applied to solve an equation.

Example 4

Suppose $f(m) = 6m - (5 - 9m)$. For what value of m is $f(m) = 13$?

Solution

Solve $6m - (5 - 9m) = 13$.

$$
\begin{array}{ll}
6m - 5 + 9m = 13 & \text{Opposite of a Sum Theorem} \\
15m - 5 = 13 & \text{Add like terms.} \\
15m = 18 & \text{Add 5 to each side.} \\
m = \frac{18}{15} = \frac{6}{5} & \text{Multiply each side by } \frac{1}{15}.
\end{array}
$$

Check

Substitute $\frac{6}{5}$ for m and follow the order of operations.

Does $6\left(\frac{6}{5}\right) - \left(5 - 9 \cdot \frac{6}{5}\right) = 13$?

Does $\frac{36}{5} - \left(5 - \frac{54}{5}\right) = 13$?

Does $\frac{36}{5} - 5 + \frac{54}{5} = 13$?

Does $\frac{90}{5} - 5 = 13$? Yes.

QUESTIONS

Covering the Reading

In 1 and 2, Lucinda bought a used car with 14,000 miles on it. She expects to drive it about 600 miles/month. So $f(m) = 14,000 + 600m$ gives the number of miles she can expect to see on the odometer at the end of m months.

1. How many miles should Lucinda expect to have on her car after 2 years?

2. After how many months will the car be due for a 40,000-mile check-up?

In 3 and 4, refer to Example 2.

3. a. Evaluate $E(5000)$.
   b. Explain the meaning of your answer to part **a** in terms of investments.

4. a. Solve $E(d) = 3000$.
   b. Explain the meaning of your answer to part **a** in terms of investments.

In 5 and 6, an equation is given.
a. Solve the equation and check your solution.
b. Make up a question that could be answered by solving the equation.

5. $.07x + .05(100,000 - x) = 6500$ 6. $\frac{1}{2}x + \frac{1}{5}x + 63 = x$

In 7 and 8, **a.** identify the least common multiple of the denominators.
b. Solve the equation.

7. $\frac{m}{3} + \frac{m}{7} = 1$ 8. $\frac{1}{6}x + \frac{2}{3}x = 30$

9. **a.** According to the Opposite of a Sum Theorem, $-(-2x + 9) = $ _____.
   b. Solve the equation $12x - (-2x + 9) = 26$.

In 10 and 11, **a.** solve each equation, and **b.** check your work.

10. $4x - (x - 1) = 7$ 11. $3n - (9 + 5n) = 18$

Applying the Mathematics

In 12–14, a bowler's *handicap* is a bonus given to some bowlers in a league. The handicap H is a function of A, the bowler's average score, and is sometimes determined by the formula $H = .8(200 - A)$, when $0 < A < 200$.

12. What is the handicap for a bowler whose average is 120?

13. If a bowler has a handicap of 30, what is the bowler's average?

14. What is the domain of the function?

In 15–19, solve.

15. $3y + 60 = 5y + 42$ 16. $2z + 2 = 2 - 2z$ 17. $\frac{12}{y} = 5$

18. $\frac{2}{3}x + 80 = x + 25$ 19. $0.05x + 0.1(2x) + 0.25(100 - 3x) = 20$

20. A farmer grows three crops: wheat, corn, and alfalfa. He farms all the land he owns. On his farm, $\frac{1}{3}$ of the land is planted with wheat, $\frac{2}{5}$ is planted with corn, and 60 acres are planted with alfalfa.
   a. How many acres of crops are on this farm?
   b. How many acres of wheat are there?
   c. How many acres of corn are there?

21. Suppose $f(n) = \frac{1}{n + 2}$.
   a. What is $f(18)$?   **b.** If $f(n) = \frac{1}{99}$, find n.

22. Tell whether the sentence is equivalent to $20(x + 5) = 55$.
   a. $4(x + 5) = 11$ **b.** $x + 5 = \frac{11}{4}$ **c.** $4(x + 1) = 11$

Time to spare. *Bowling can be traced to the Middle Ages in Germany, England, and the Netherlands. Native Americans were also known to have various forms of bowling. The American Bowling Congress was organized in 1895.*

In 23–26, refer to the graph below. In the graph, x = the year, $I(x)$ = the value in billions of dollars of imports into the United States, and $E(x)$ = the value of exports from the United States, also in billions of dollars. *(Lessons 1-3, 1-4)*

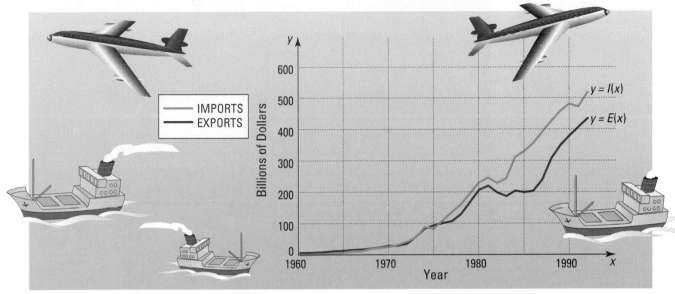

23. Estimate $I(1975)$.

24. In what unit is the dependent variable of the function E measured?

25. In what year(s) was $E(x) > \$100$ billion?

26. A negative *balance of trade* exists when a country's imports are greater than its exports. Write a few sentences about the balance of trade in the U.S. from 1960 to 1992.

27. A cylindrical column of a building has a lateral area of 320 ft². If its radius is 5 ft, what is its height? (Refer to the Appendix of Geometry Formulas if necessary.) *(Previous course)*

Exploration

28. The Greek mathematician Diophantus, who lived in the second century A.D., was the first person to replace unknowns by single letters. There is a famous problem by which you can calculate how long he lived. The *Greek Authority* states: "Diophantus passed one sixth of his life in childhood, one twelfth in youth, and one seventh more as a bachelor. Five years after his marriage was born a son who died four years before his father, at half his father's final age." How long did Diophantus live?

Rewriting Formulas

Formula driven. *Pictured is a* Seahawk I *jet of* Alaska Airlines. *Most commercial airplanes have on-board computers that continually calculate the estimated flight time based on distance and rate (including wind speed).*

Consider the formula

$$d = rt,$$

where d is the distance an object travels, r is the rate at which it travels, and t is the time the object travels. If you want to calculate the distance traveled on an airplane going at a rate of $650 \frac{\text{miles}}{\text{hour}}$ for 2.5 hours, you can use this formula. It gives d *in terms of r and t,* the variables for which you have values.

$$d = (650 \text{ miles/hour}) \cdot (2.5 \text{ hours}) = 1625 \text{ miles}$$

But suppose you want to know how much time you would need to travel 380 miles if you drive a car at an average rate of $60 \frac{\text{miles}}{\text{hour}}$. It might be helpful to have a formula that gives t *in terms of d and r,* that is, a formula that is solved for t. Using properties in Appendix B, you can write:

$$d = rt$$
$$\frac{d}{r} = \frac{rt}{r} \qquad \text{Divide both sides by } r.$$
$$\frac{d}{r} = t \qquad \text{Simplify.}$$

To find t, substitute the values for d and r.

$$t = \frac{d}{r} = \frac{380 \text{ mi}}{60\frac{\text{miles}}{\text{hour}}} \approx 6.3 \text{ hours}$$

You would need about 6.3 hours to travel 380 miles.

The formulas $d = rt$ and $t = \frac{d}{r}$ are equivalent as long as $r \neq 0$. The first is solved for d; the second is solved for t. Notice that when a formula is **solved for a variable,** that variable has a coefficient and an exponent equal to one. That variable is said to be written **in terms of** the others.

The most useful version of a formula depends upon the situation.

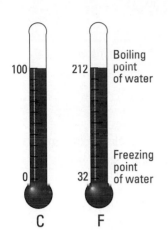

C

F

100 | 212 — Boiling point of water

0 | 32 — Freezing point of water

Example 1

Pierre lives in New Orleans, where he measures temperature using the Fahrenheit scale. When he visited his cousin Rae in Montreal, Canada, he found that temperature was reported in degrees Celsius. Because Celsius temperature readings didn't mean much to him, Pierre converted temperatures in Celsius C to Fahrenheit F using this formula:

$$F = 32 + 1.8C.$$

Rae visited Pierre the following summer. Rewrite the formula so she can use it to convert degrees Fahrenheit to Celsius.

Solution

$F = 32 + 1.8C$	Given
$F - 32 = 1.8C$	Subtract 32 from both sides.
$\dfrac{F - 32}{1.8} = C$	Divide both sides by 1.8.

So $C = \dfrac{F - 32}{1.8}$ is an equivalent formula that is suitable for Rae.

Check

Evaluate the formula for a pair of temperatures you know are equivalent. The boiling point of water is 212°F or 100°C.

Does $100 = \dfrac{212 - 32}{1.8}$? Yes.

Notice that when using the formula $F = 32 + 1.8C$, Pierre thinks of F as a function of C. When using the formula $C = \dfrac{F - 32}{1.8}$, Rae thinks of C as a function of F.

Example 2

Scuba divers sometimes use the formula $t = \dfrac{33v}{x + 33}$ to determine the time t (in minutes) they can dive with a given volume v of air compressed into tanks (in cubic feet) to a depth of x feet below sea level. Rewrite the formula for v in terms of x and t.

Solution

$t = \dfrac{33v}{x + 33}$	
$t(x + 33) = 33v$	Multiply each side by $x + 33$.
$\dfrac{t(x + 33)}{33} = v$	Divide each side by 33.

The formula is solved for v. If you wish to simplify it to contain fewer constants, you can rewrite it as follows.

$\dfrac{tx + 33t}{33} = v$	Use the Distributive Property.
$\dfrac{tx}{33} + \dfrac{33t}{33} = v$	Addition of Fractions Theorem
$\dfrac{tx}{33} + t = v$	Simplify the fraction.

Sometimes you may want to rewrite a formula without solving for a particular variable. Consider the next example.

Example 3

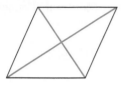

When a quadrilateral has perpendicular diagonals with lengths d_1 and d_2, then its area A satisfies $A = \frac{1}{2}d_1 d_2$. What is always true about the product of the lengths of the diagonals?

Solution

The product of the diagonal lengths is $d_1 d_2$. Multiplying both sides of $A = \frac{1}{2}d_1 d_2$ by 2 solves the equation for this expression. So $2A = d_1 d_2$.

Thus, In a quadrilateral with perpendicular diagonals, the product of the lengths of its diagonals is twice its area.

Three kinds of quadrilaterals with perpendicular diagonals are kites, rhombuses, and squares, so the theorem in Example 3 has many applications.

QUESTIONS

Covering the Reading

In 1–3, refer to the formula $d = rt$ at the beginning of the lesson.

1. How far can a race car traveling 190 mph go in 1.4 hr?

2. The formula $t = \frac{d}{r}$ is solved for ___?___ .

3. Find an equivalent formula that is solved for r.

In 4 and 5, complete the sentence "___ is written in terms of ___" for the given formula.

4. $V = s^3$

5. $A = \frac{1}{2} bh$

6. Refer to Example 1. Find the Celsius temperature equivalent to 98.6°F.

7. Refer to Example 2. Solve the formula for x.

8. Refer to Example 3. If the area of a rhombus is 24 cm^2, find the product of its diagonals.

Applying the Mathematics

9. *Multiple choice.* Which formula is easiest to use if you are given L and h and want to find N?
   (a) $N = 7Lh$ (b) $L = \frac{N}{7h}$ (c) $h = \frac{N}{7L}$

10. *Multiple choice.* Which equation is easiest to use if you are given L and want to find T?
(a) $L = 100 + .0004T$ (b) $T = 2500(L - 100)$
(c) $.0004T = L - 100$

11. The formula $C = \pi D$ gives the circumference of a circle in terms of its diameter D.
a. Solve this formula for π.
b. Use your result in part **a** to write a sentence that gives the definition of π.

12. The formula $C = 2\pi r$ gives the circumference of a circle in terms of its radius r. Find the *ratio* of C to r.

In 13–15, the *pitch* P of a gabled roof is a measure of the steepness of the slant of the roof. **Pitch** is the ratio of the vertical rise R to half the span S of the roof. That is, $P = \frac{R}{.5S}$.

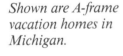

13. The picture below shows the framing of a plankhouse built by Native Americans on the Pacific coast of North America. Measure the rise and span of the building and estimate its pitch.

14. a. Solve the pitch formula for R.
b. A family is planning to build an A-frame beach house like one of those shown below. The builder proposes a span of 10 meters and a pitch of 0.9. What is the rise of the roof?

Shown are A-frame vacation homes in Michigan.

15. a. Solve the pitch formula for S.
b. If a builder wants a roof to have a pitch of $\frac{4}{12}$ and a rise of 10 feet, what must be the span of the building?

16. In the late 1660s, Isaac Newton discovered the Law of Universal Gravitation described by the formula

$$F = \frac{Gm_1 m_2}{d^2}$$

where F is the force between two bodies with masses m_1 and m_2, G is the gravitational constant, and d is the distance between the bodies.
  a. Solve for m_1.
  b. Solve the formula for the product of the masses.

17. Solve the formula $t = a + (n - 1)d$ for n.

Review

18. *Skill sequence.* Solve each equation. *(Lesson 1-5)*
  a. $\frac{1}{3}n = 60$ **b.** $\frac{1}{3}n + 15 = 60$ **c.** $\frac{1}{3}n + \frac{1}{2}n + 15 = 60$

19. Solve $5 - 2(x - 7) - (3 - x) = 4$. *(Lesson 1-5)*

20. Draw a graph of a function that has domain equal to $\{x: 2 \le x \le 8\}$ and range equal to $\{y: 0 \le y \le 4\}$. *(Lesson 1-4)*

In 21 and 22, use the formula

$$T = 2\pi \sqrt{\frac{L}{g}}$$

which gives the time T (in seconds) for one complete swing of a pendulum that is L meters long near the surface of the Earth. (g is the acceleration due to gravity, 9.8 m/sec^2.) *(Lessons 1-1, 1-2)*

21. Find the time for a pendulum 50 cm long to complete one swing.

22. a. *True or false.* T is a function of L.
  b. If the statement in part **a** is true, identify the domain of the function.

Swing time. *Because changes in temperature expand or contract a pendulum's rod, some grandfather clocks come with mechanisms to keep the total length of the pendulum constant.*

23. *Multiple choice.* Which of the following does not equal the complex fraction $\dfrac{\frac{a}{b}}{\frac{c}{d}}$? *(Previous course)*

  (a) $\frac{ad}{bc}$ (b) $\frac{a}{b} \div \frac{c}{d}$ (c) $\frac{ac}{bd}$ (d) $\frac{a}{b} \cdot \frac{d}{c}$

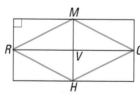

24. Let d_1 and d_2 be the lengths of the diagonals of rhombus *RHOM*. Use the diagram at the left and write a few sentences to explain why the area A of the rhombus equals $\frac{1}{2}d_1 d_2$. *(Previous course)*

Exploration

25. Look in a science book to find a formula with three variables not mentioned in this lesson. Explain what the variables represent. Solve the formula for each variable.

Patterns and Sequences

IN-CLASS

ACTIVITY

Work on this activity in small groups. Each group will need paper and a calculator.

Informally, a *sequence* is an ordered list. Here is a sequence of rectangular figures made from dots •.

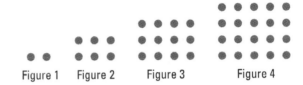

Figure 1   Figure 2   Figure 3   Figure 4

The number of dots in each of the figures produces a sequence of numbers:

$$2, 6, 12, 20, \ldots$$

1 **a.** Draw the 5th figure.
b. How many dots are needed to make the 5th figure?

2 **a.** Complete a table of values with the following entries.

n	number of dots in Figure n
1	2
2	6
3	12
4	20
5	
6	

b. Is the set of ordered pairs of the form (n, number of dots in nth figure) a function? Why or why not?
c. If so, what is the domain of this function?

3 How many dots are needed to make the 10th figure? Explain how you got your answer.

4 How many dots are needed to make the 55th figure? (Hint: Look for patterns.) Explain how you got your answer.

5 How many dots are needed to make the nth figure? Justify your answer.

Sequences on cue. *Most pocket billiard games start with 15 balls racked in a triangular array. The array is a representation of the fifth triangular number. In some games, the numbered balls must be pocketed in sequence.*

What Is an Explicit Formula for a Sequence?

Here is another sequence of dots.

The number of dots in each of the figures above produces a sequence of numbers:

$$1, 3, 6, 10, 15.$$

These numbers are often called the *triangular numbers* because of the way they can be represented by the triangular figures of dots.

Each number in the sequence is called a **term.** Thus,

> the 1st term (of the sequence) is 1;
> the 2nd term is 3;
> the 3rd term is 6;
> the 4th term is 10;
> the 5th term is 15.

Activity

a. Draw a picture to represent the 6th triangular number.
b. What is the 6th term of the sequence of triangular numbers?

Here is one way to find a formula for *generating* all terms of the sequence of triangular numbers. Notice that if you take a duplicate of each triangular figure and place it as shown below, a rectangular array is formed.

These are the same rectangular figures you used in the In-class Activity on page 41. Thus, each triangular number can be represented by half a rectangular array. For instance, the number of dots representing the 4th triangular number is half the number of dots in a 4 by 5 rectangular array.

$$\frac{1}{2} \cdot 4 \cdot 5 = 10$$

This idea can be used to develop a general formula for any triangular number.

Number of Term	Term (Number of Dots)
1	$\frac{1}{2} \cdot 1 \cdot 2 = 1$
2	$\frac{1}{2} \cdot 2 \cdot 3 = 3$
3	$\frac{1}{2} \cdot 3 \cdot 4 = 6$
4	$\frac{1}{2} \cdot 4 \cdot 5 = 10$
5	$\frac{1}{2} \cdot 5 \cdot 6 = 15$
6	$\frac{1}{2} \cdot 6 \cdot 7 = 21$
$\vdots$	$\vdots$
n	$\frac{1}{2} \cdot n \cdot (n + 1)$

Thus, the number of dots in the *n*th triangular figure is

$$\frac{1}{2} \cdot n \cdot (n + 1) = \frac{n(n + 1)}{2}.$$

If we let $t(n)$ = the *n*th triangular number, then

$$t(n) = \frac{n(n + 1)}{2}.$$

The domain of this function is the set of natural numbers $\{1, 2, 3, \ldots\}$; the range is the set of terms $\{1, 3, 6, 10, 15, 21, \ldots\}$.

The sentence $t(n) = \frac{n(n + 1)}{2}$ is called an **explicit formula for the *n*th term** of the sequence 1, 3, 6, 10, 15, 21, . . . because we use it to calculate the *n*th term explicitly, or directly, by substituting a value for *n*. In the next lesson, you will learn about another type of formula for generating terms of a sequence. Explicit formulas are important because they can be used to calculate any term in the sequence by substituting a particular value for *n*.

Example 1

What is the 20th triangular number?

Solution

To find the 20th term in the sequence, evaluate $t(20)$.

$$t(20) = \frac{20(20 + 1)}{2} = \frac{20 \cdot 21}{2}$$

The 20th triangular number is 210.

Notation for Sequences

In general, a **sequence** is a function whose domain is the set of natural numbers or the natural numbers from 1 to n. A special notation is often used with sequences. Instead of writing $t(10) = 55$ to indicate that the 10th term is 55, we can also write

$$t_{10} = 55.$$

This is read "t sub 10 equals 55." The number 10 is called a *subscript* or *index*. It is called a **subscript** because it is written below and to the right of a variable. t_1 is read "t sub 1" and t_n is read "t sub n." Both t_1 and t_n are called **subscripted variables.** The subscript of t_1 is 1; the subscript of t_n is n.

The subscript is often called an **index** because it *indicates* the position of the term in the sequence. For example, if t_n represents the nth term of the triangular-number sequence, then

$$t_1 = 1, \quad t_2 = 3, \quad t_3 = 6, \quad t_4 = 10, \quad t_5 = 15, \text{ and } t_n = \frac{n(n + 1)}{2}.$$

Using Explicit Formulas to Generate Terms of a Sequence

Example 2

Consider the formula $t_n = 3n$, for integers $n \geq 1$.
a. What are the first four terms of the sequence it defines?
b. Evaluate t_{20}, and explain what it represents.

Solution

a. Substitute $n = 1, 2, 3$, and 4 one at a time into the formula for t_n.
$t_1 = 3 \cdot 1 = 3$
$t_2 = 3 \cdot 2 = 6$
$t_3 = 3 \cdot 3 = 9$
$t_4 = 3 \cdot 4 = 12$
b. Substitute $n = 20$ into the formula.
$t_{20} = 3 \cdot 20 = 60$
This means that the 20th term of the sequence
3, 6, 9, 12, . . . is 60.

Sequences arise naturally in many situations in science, business, finance, and other areas. The next example looks at a sequence in biology.

Example 3

A microbe reproduces by splitting to make 2 cells. Each of these cells then splits in half to make a total of 4 cells. Each of these splits to make a total of 8, and so on. Each splitting is called a *generation.* If a colony begins with 500 microbes, the following equation gives the number of microbes in the *n*th generation (assuming no microbes die).

$$P_n = 500(2)^{n-1}$$

a. Write the first term of the sequence of populations of microbes.
b. Write the fifth term of the sequence.

Solution

Notice that the variable is in the exponent.
a. For the population in the first generation, substitute 1 for *n* in the formula.
$$P_1 = 500(2)^{1-1} = 500(2)^0 = 500(1) = 500$$
This checks with the given information. There are 500 microbes in the first generation.
b. For the population in the fifth generation, use $n = 5$ in the formula.
$$P_5 = 500(2)^{5-1} = 500(2)^4 = 500(16) = 8000$$
There are 8000 microbes in the fifth generation.

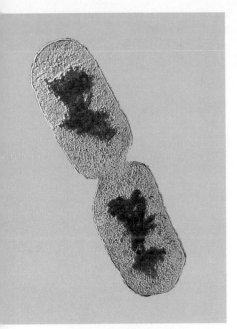

Time to split. *Shown is the bacterium* Escherichia coli *splitting into two cells called daughter cells.* E. coli *bacteria have played a major role in genetic engineering.*

QUESTIONS

Covering the Reading

1. An ordered list of items is called a ___?___.

2. What is each item in a sequence called?

3. Explain why the numbers 1, 3, 6, 10, and 15 are called triangular numbers.

4. **a.** What answers did you get for the Activity in the lesson?
   b. Show how to use the formula $t_n = \dfrac{n(n + 1)}{2}$ to check these answers.

5. Find the 25th triangular number.

6. **a.** Draw the next term in the sequence.

   ●   ●●   ●●●   ●●●●
        ●●   ●●●   ●●●●
            ●●●   ●●●●
                   ●●●●

   b. Give a formula for S_n, the number of dots in the *n*th term.

In 7–9, consider the sequence *c* whose first five terms are 1, 8, 27, 64, 125.

7. What number is the 4th term?

8. How is the sentence "$c_3 = 27$" read?

9. $c_5 = $ ___?___

In 10–12, consider the sentence $a_4 = 15$.

10. Which number is the subscript?

11. The number 15 is the __?__ term of a sequence.

12. Rewrite the sentence using Euler's notation.

13. Use sequence notation to rewrite the sentence "The 6th term of sequence t is 23."

In 14 and 15, an explicit formula for a sequence is given for integers $n \geq 1$. Write the first four terms of the sequence.

14. $a_n = 5n - 3$

15. $S_n = \dfrac{n(3n - 1)}{2}$

16. Refer to Example 3.
   a. Copy the table below and use the explicit formula to fill in the missing numbers.

Generation	Number of Microbes
1	500
2	
3	
4	
5	
6	

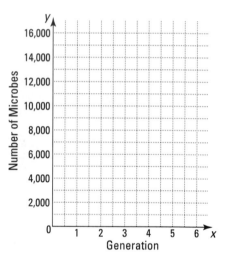

   b. Graph the values from the table.
   c. Should the points on the graph be connected? Why or why not? Relate your answer to the domain of the function.

Applying the Mathematics

In 17–19, Yoshi started with a company at an annual salary of $18,000. He expects an increase of 5% at the end of each year. Then the formula $s_n = 18,000(1.05)^{n-1}$ gives Yoshi's salary in his nth year.

17. Calculate the first three terms of the sequence.

18. At this growth rate, what would Yoshi's salary be in his 30th year with this company?

19. Given the function with equation $s_n = 18,000(1.05)^{n-1}$, identify:
   a. the independent variable; **b.** the dependent variable; and
   c. the domain of the function.

20. Let t be the sequence whose first seven terms are 93, 86, 79, 72, 65, 58, and 51. What do you think t_8 should equal? Why?

In 21 and 22, *multiple choice.* Which is a formula for the nth term of the sequence?

21. 2, 4, 8, 16, 32, . . .
    (a) $t_n = 2n$ (b) $t_n = n^2$ (c) $t_n = 2^n$

22. 2, 9, 28, 65, 126, . . .
    (a) $t_n = 7n - 5$ (b) $t_n = 7n^2 - 2$ (c) $t_n = n^3 + 1$

Review

23. A formula for the sum S of the angle measures in an n-gon is $S = 180(n - 2)$. Solve this formula for n. *(Lesson 1-6)*

24. When the Mustafas bought their new home, they were not told the capacity of the heating oil tank. When the tank was $\frac{1}{10}$ full, they had 280 gallons of oil delivered. Then the tank was $\frac{8}{10}$ full.
  a. Let $x =$ the capacity of the oil tank. Write an equation that can be used to find the capacity of the tank.
  b. Solve and check the equation in part **a.** *(Lesson 1-5)*

25. Let $P(x)$ be the perimeter of the triangle at the left. *(Lessons 1-1, 1-5)*
  a. Write a formula for $P(x)$ in terms of x.
  b. Evaluate $P(40)$.
  c. Solve $P(x) = 40$.

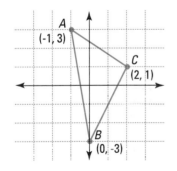

26. A carnival charges a $3.00 admission fee and $1.50 for each ride ticket. If S is the total spent at the carnival for admission and rides, then S is a function of the number N of tickets bought, and can be written as $S = 1.5N + 3$. *(Lessons 1-2, 1-4)*
  a. Specify the domain of this function.
  b. Write the four smallest numbers in the range of this function.
  c. Graph this function.

27. What percent of 80 is 56? *(Previous course)*

28. Draw $\triangle ABC$ as shown at the left on graph paper. Draw its image under a size change of magnitude 3. *(Previous course)*

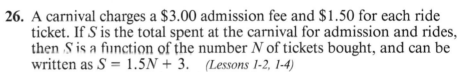

Exploration

29. The array at the left is part of an infinite pattern called *Pascal's triangle*. The first and last terms of each row are 1. Each other term, from the third row on, is the sum of the two numbers diagonally above it.
  a. Write the next four rows in the array.
  b. Where can you find the number sequence that comes from the dot pattern at the start of this lesson?
  c. Describe at least one other interesting sequence in this array.

```
            1
          1   1
        1   2   1
      1   3   3   1
    1   4   6   4   1
  1   5  10  10   5   1
  -   -   -   -   -   -   -
-   -   -   -   -   -   -   -
```

LESSON

1-8

Recursive Formulas for Sequences

Guess my age. *The age of this frog in amber can be found by carbon dating. After each 5700-year period, half of the ^{14}C carbon atoms decay. The sequence a_n of fractions of ^{14}C left after n 5700-year periods is $\frac{1}{2}, \frac{1}{4}, \frac{1}{8}, \ldots$*

As you saw in the last lesson, if you know an explicit formula for a sequence, then you can write the terms of a sequence rather quickly. For instance, the formula $a_n = -3 + 5n$ generates the following terms.

$$a_1 = -3 + 5 \cdot 1 = 2$$
$$a_2 = -3 + 5 \cdot 2 = 7$$
$$a_3 = -3 + 5 \cdot 3 = 12$$
$$a_4 = -3 + 5 \cdot 4 = 17$$

We say that the formula $a_n = -3 + 5n$ *generates the sequence*

$$2, 7, 12, 17, 22, \ldots.$$

Generating Sequences Using a Calculator

You can also use a calculator to generate this sequence. Notice that the sequence begins with 2 and that each term after that is obtained by adding 5 to the previous term. Try the following key strokes on your calculator and check that your display matches the one below.

Key strokes: 2 `+` 5 `ENTER` `+` 5 `ENTER` `+` 5 `ENTER` `+` 5 `ENTER`
Display: `2`   `7`   `12`   `17`   `22`

If your calculator does not have an `ENTER` key, you should be able to use the same key strokes if you replace `ENTER` by `EXE` or `=`.

Activity 1

Use your calculator to generate the sequence
$$40, 20, 10, 5, 2.5, \ldots.$$
What key strokes did you use?

What Is a Recursive Formula?

Notice that no explicit formula was used to generate the sequence in Activity 1. Instead, each term of the sequence was derived from the preceding one. For instance, to find the 6th term of the sequence

$$40, 20, 10, 5, 2.5, \ldots,$$

you could divide the preceding term (the 5th term) by 2 or multiply the preceding term by 0.5. Each of these ideas involves thinking *recursively*.

A **recursive formula** or **recursive definition** for a sequence is a set of statements that
a. indicates the first term (or first few terms), and
b. tells how the nth term is related to one or more of the previous terms.

So the sequence $40, 20, 10, 5, 2.5, \ldots$ could be described recursively as

$$\begin{cases} \text{first term} = 40 \\ \text{new term} = \text{previous term divided by 2} \\ \qquad\qquad \text{for all terms after the first} \end{cases}$$

The brace { indicates that both lines are needed for the recursive definition.

Example 1

Consider the sequence defined as follows.

$$\begin{cases} t_1 = 1 \\ t_n = (\text{previous term})^2 + 3, \text{ for integers } n \geq 2 \end{cases}$$

Write the first four terms of the sequence.

Solution

The first term is given.
$$t_1 = 1$$
According to the second line of the definition,
$$t_2 = (\text{previous term})^2 + 3.$$
But the previous term is 1. So
$$t_2 = 1^2 + 3 = 4.$$
To find t_3, use the second line of the definition again.
$$t_3 = (\text{previous term})^2 + 3$$
Now the previous term, t_2, is equal to 4. So
$$t_3 = 4^2 + 3 = 19.$$
Finally, $t_4 = (\text{previous term})^2 + 3.$ So
$$t_4 = 19^2 + 3 = 364.$$
The first four terms are 1, 4, 19, 364.

Recursive Formulas on Calculators

On many graphics calculators, the ANS key is used to refer to the result of the previous calculation. So the ANS key can be used in the recursive definition to refer to the previous term. Using this key, you could write the sequence 40, 20, 10, 5, 2.5, . . . in Activity 1 as follows.

$$\begin{cases} a_1 = 40 \\ a_n = \dfrac{\text{ANS}}{2}, \text{ for integers } n \geq 2 \end{cases}$$

Activity 2

a. Clear the screen on a graphics calculator. Press 40 and then ENTER or EXE. This stores the first term or *initializes* the value of a_1.

b. The second line tells you how to find a_2 and all subsequent terms. After initializing a_1 on the calculator display, press ANS ÷ 2, followed by ENTER or EXE. What is displayed?

c. Press ENTER again. The calculator will use the second term as the value for ANS and divide this value by 2 to generate the third term.

d. Keep pressing the ENTER or EXE key. How many terms of the sequence

$$\begin{cases} a_1 = 40 \\ a_n = \dfrac{\text{ANS}}{2}, \text{ for integers } n \geq 2 \end{cases}$$

are greater than 0.1?

As shown in Example 2, the ANS key may be combined with several other operations.

Example 2

Consider the sequence defined by the recursive formula below.

$$\begin{cases} t_1 = 6 \\ t_n = 3 \cdot \text{ANS} - 5, \text{ for integers } n \geq 2 \end{cases}$$

Find the first five terms of this sequence.

Solution 1

Initialize the first term on a graphics calculator by pressing 6 and then ENTER or EXE. Now enter the formula to find the other terms:

3 × ANS − 5.

As you press ENTER or EXE, you will generate the terms

$$6, 13, 34, 97, 286.$$

▶

Solution 2

Evaluate the formula by hand.

$t_1 = 6$ is given.

[ANS] refers to the previous term t_1 of the sequence.

So, $t_2 = 3 \cdot t_1 - 5$

$= 3 \cdot 6 - 5$

$= 13$.

To evaluate t_3, think of $t_2 = 13$ as the previous term.

Thus, $t_3 = 3 \cdot t_2 - 5$

$= 3 \cdot 13 - 5$

$= 34$.

Now use $t_3 = 34$ as the value of [ANS].

$t_4 = 3 \cdot t_3 - 5$

$= 3 \cdot 34 - 5$

$= 97$

Finally, $t_5 = 3 \cdot t_4 - 5$

$= 3 \cdot 97 - 5$

$= 286$.

Notice that the first term of the sequence in Example 3 is 6, the term given to you in the definition. Although 13 is the first term given by the rule for t_n, it is the second term of the sequence. Remember, t_n represents the nth term.

Activity 3

Find the first five terms of the sequence defined by

$$\begin{cases} t_1 = 1 \\ t_n = 2 \cdot \boxed{\text{ANS}} - 4, \text{ for integers } n \geq 2. \end{cases}$$

Writing Recursive Formulas

If you can describe a sequence in words, then you can use that description to write a formula for the sequence.

Example 3

When Jennifer started her new job, she enrolled in a payroll deduction plan to save for a car. The first month she had $25 deducted. Thereafter, she decided to deduct $50 each month. The sequence

25, 75, 125, 175, 225, . . .

gives the amount deducted after n months.
a. Use words to write a recursive definition of the sequence.
b. Write a recursive formula for this sequence using the [ANS] key.

Solution

a. Identify the first term, and how each term after the first is related to previous terms. The first term is 25. Each term after the first is found by adding 50 to the previous term.

b. Translate the words in part **a** into symbols.
$$\begin{cases} t_1 = 25 \\ t_n = \boxed{\text{ANS}} + 50, \text{ for integers } n \geq 2 \end{cases}$$

QUESTIONS

Covering the Reading

In 1 and 2, write the first four terms of the sequence generated by the calculator instructions.

1. Begin with 24; repeatedly press $\boxed{\times}$ 1.5 $\boxed{\text{ENTER}}$ (or $\boxed{\text{EXE}}$).

2. Begin with 4; repeatedly press $\boxed{-}$ 9 $\boxed{\text{ENTER}}$ (or $\boxed{\text{EXE}}$).

In 3 and 4, refer to Activity 1 and the sequence 40, 20, 10, 5, ...

3. Write calculator instructions as in Questions 1 and 2 using multiplication to generate the first four terms.

4. The 10th term of this sequence is .078125. What is the 11th term?

In 5–7, write the first five terms of the sequence defined by the recursive formula.

5. The first term is 7; each term after the first is 10 more than the previous term.

6. $\begin{cases} t_1 = -20 \\ t_n = \frac{1}{4} \cdot \text{previous term, for integers } n \geq 2 \end{cases}$

7. $\begin{cases} t_1 = 25 \\ t_n = 5 \cdot \boxed{\text{ANS}} + 6, \text{ for integers } n \geq 2 \end{cases}$

8. *Multiple choice.*   Which rule states that the nth term is seven less than the previous term?
   (a) $t_n = 7 \cdot \boxed{\text{ANS}}$ (b) $t_n = 7 - \boxed{\text{ANS}}$
   (c) $t_n = \boxed{\text{ANS}} + 7$ (d) $t_n = \boxed{\text{ANS}} - 7$

9. Refer to Activity 2. How many terms of the sequence
$$\begin{cases} a_1 = 40 \\ a_n = \dfrac{\boxed{\text{ANS}}}{2}, \text{ for integers } n \geq 2 \end{cases}$$
are greater than 0.1?

10. What is the answer to Activity 3?

11. Consider the sequence that begins 100, 94, 88, 82, 76, ...

    a. What is the first term?

    b. From the second term on, each term is __?__ the previous term.

    c. Write a recursive formula for the sequence.

12. Consider the sequence of odd numbers beginning with 1, 3, 5, 7, 9, ... Write a recursive definition for the sequence in words.

Applying the Mathematics

In 13 and 14, suppose that in a movie theater, the first row has 15 seats. Each succeeding row has 2 more seats than the row in front of it.

13. **a.** Write a recursive formula for a sequence that gives the number of seats S_n in row n.

    b. Find the number of seats in the tenth row.

14. **a.** Is the set of ordered pairs of the form (n, S_n) a function? Why or why not?

    b. If this set is a function, what is its domain?

15. **a.** Write the first four terms of the sequence defined by $x_n = 3(4)^{n-1}$.

    b. Write the first four terms of the sequence defined by

$$\begin{cases} y_1 = 3 \\ y_n = 4 \cdot \boxed{\text{ANS}}, \text{ for integers } n \geq 2. \end{cases}$$

    c. *True or false.* The sequences defined in parts **a** and **b** have the same terms.

16. **a.** Write the first six terms of the sequence defined recursively as

$$\begin{cases} t_1 = 1 \\ t_n = \boxed{\text{ANS}} + n, \text{ for integers } n \geq 2. \end{cases}$$

    b. Write the first six terms of the sequence defined explicitly as

$$t_n = \frac{n(n + 1)}{2}.$$

    c. What do you conclude from your answers to part **a** and part **b?**

17. Explain in your own words why a recursive formula must have two parts.

18. The table at the left gives the postage rate for letters up to 11 ounces in 1995.

    a. Describe in words the sequence that gives the postage for an integer weight of n ounces.

    b. Write a recursive formula for the sequence that gives the postage rate for a letter weighing n ounces.

Weight Not Over (oz)	Rate ($/letter)
1	.32
2	.55
3	.78
4	1.01
5	1.24
6	1.47
7	1.70
8	1.93
9	2.16
10	2.39
11	2.62

19. In the first year of its existence, a company had a gross income of $100,000. Each year, the company hopes to increase its gross income by 50%. The sequence $c_n = 100,000 \, (1.5)^{n-1}$ gives the yearly gross income hoped for in year n. *(Lesson 1-7)*
    a. Find c_3.
    b. What gross income does the company hope to have in the 10th year of its existence?

20. The force F required to accelerate a mass of m kilograms a meters per second squared is given by $F = ma$. Here, force is measured in newtons. *(Lessons 1-5, 1-6)*
    a. Solve this formula for a.
    b. If a force of 225 newtons is applied to an object with mass 25 kilograms, what is the acceleration of the object?

21. The formula $P = 21,782 + 450x$ gives an estimate of the population of Peru (in thousands) from 1990 to the year 2025, where x is the number of years after 1990.
    a. Estimate the population of Peru in 2010.
    b. In what year will the estimated population exceed 29,000,000? *(Lesson 1-2, Previous course)*

22. a. Make up a set of five ordered pairs that represents a function.
    b. Make up a set of five ordered pairs that does not represent a function. *(Lesson 1-2)*

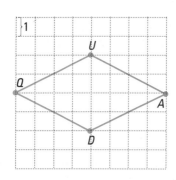

23. Refer to the quadrilateral at the left. *(Previous course)*
    a. Find its area.
    b. Find its perimeter.
    c. Copy the figure, and on a coordinate grid, draw a quadrilateral that has the same area, but a larger perimeter.

24. Suppose n is an integer.
    a. What is the next larger integer?
    b. What is the next smaller integer? *(Previous course)*

Shown are students from Santa Ana High School in Cuzco, Peru at the school's 100th anniversary. Cuzco was the capital of the Inca Empire.

Exploration

25. Consider the sequence defined recursively by

$$\begin{cases} t_1 = 1000 \\ t_n = .5 \left(\boxed{\text{ANS}} + \dfrac{a}{\boxed{\text{ANS}}} \right), \text{ for integers } n \geq 2. \end{cases}$$

   a. Suppose $a = 25$. Evaluate the terms of this sequence until all the terms appear to be the same value or close to the same value. This value is called the *limit* of the sequence.
   b. Repeat part **a** with $a = 49$.
   c. Repeat part **a** with $a = 2$.
   d. Use your results from parts **a** through **c** to make a conjecture about what this sequence does.

54

LESSON
1-9

Notation for Recursive Formulas

The sequence 17, 12, 7, 2, -3, . . . can be defined recursively with words. In words you can write:

> The first term is 17.
> Each term after the first is found by subtracting 5 from the previous term.

The equations below generate this sequence on a graphics calculator.

$$\begin{cases} t_1 = 17 \\ t_n = \boxed{\text{ANS}} - 5, \text{ for integers } n \geq 2 \end{cases}$$

You know that the symbol t_n denotes the nth term. The term that precedes the nth term is the $(n-1)$st term. (Think: one less than n is $n-1$.) Thus t_{n-1} denotes the $(n-1)$st term. So the symbol t_{n-1} can be used in place of $\boxed{\text{ANS}}$ to denote the previous term. Thus, the sequence above can also be defined as follows.

$$\begin{cases} t_1 = 17 \\ t_n = t_{n-1} - 5, \text{ for integers } n \geq 2 \end{cases}$$

The next two examples show how to use t_{n-1} in descriptions of sequences.

Example 1

Consider the sequence defined as follows.

$$\begin{cases} t_1 = 60 \\ t_n = 2 \cdot t_{n-1}, \text{ for integers } n \geq 2 \end{cases}$$

a. Describe the sequence in words.
b. Find the first four terms of the sequence.

Solution

a. You need to identify the first term, and the rule for generating all following terms. The statement $t_n = 2 \cdot t_{n-1}$ means that to get t_n, the previous term t_{n-1} must be multiplied by 2. The first term is 60. Each term after the first is found by multiplying the previous term by 2.

b. The first term is $t_1 = 60$. To find the next three terms, substitute $n = 2, 3,$ and 4 in the rule for t_n.
The second term is t_2. $t_2 = 2 \cdot t_{2-1} = 2t_1 = 2 \cdot 60 = 120$
The third term is t_3. $t_3 = 2 \cdot t_{3-1} = 2t_2 = 2 \cdot 120 = 240$
The fourth term is t_4. $t_4 = 2 \cdot t_{4-1} = 2t_3 = 2 \cdot 240 = 480$
The first four terms of the sequence are 60, 120, 240, 480.

Example 2

Consider the sequence defined recursively as

$$\begin{cases} T_1 = 1 \\ T_n = T_{n-1} + n, \text{ for integers } n \geq 2. \end{cases}$$

Find T_2, T_3, and T_4.

Solution

The first term is $T_1 = 1$. To find T_2, substitute 2 for n in the rule for T_n.

When n = 2, $T_2 = T_{2-1} + 2$
$= T_1 + 2$
$= 1 + 2$
$= 3.$

To find T_3, substitute 3 for n in the rule for T_n.

When n = 3, $T_3 = T_{3-1} + 3$
$= T_2 + 3$
$= 3 + 3$
$= 6.$

To find T_4, substitute 4 for n in the rule for T_n.

When n = 4, $T_4 = T_{4-1} + 4$
$= T_3 + 4$
$= 6 + 4$
$= 10.$

1

3

6

10

$t_n = \frac{n(n+1)}{2}$

Recursive and Explicit Formulas

The recursive formula in Example 2 above generates the terms

$$1, 3, 6, 10, 15, 21, \ldots.$$

These are the triangular numbers, which you studied in Lesson 1-7. There, we found that an explicit formula for this sequence is

$$t_n = \frac{n(n+1)}{2}, \text{ for integers } n \geq 1.$$

Thus, the triangular numbers can be generated both explicitly and recursively.

Recursive and explicit formulas are useful at different times. An explicit formula lets you calculate a specific term without having to calculate all the previous terms. For instance, you can find the 100th triangular number by evaluating

$$t_{100} = \frac{100(100 + 1)}{2} = 5050.$$

You do not need to know any other term of the sequence.

A recursive formula is useful when an explicit formula is not known or when an explicit formula is difficult to determine. The sequence in Example 3 illustrates such a situation. It also shows you how the nth term of a sequence may be described using more than just the previous term.

Example 3

A male bee develops from an unfertilized egg. That is, a male bee has a mother, but no father. A female bee develops from a fertilized egg. That is, a female bee has both a mother and a father.

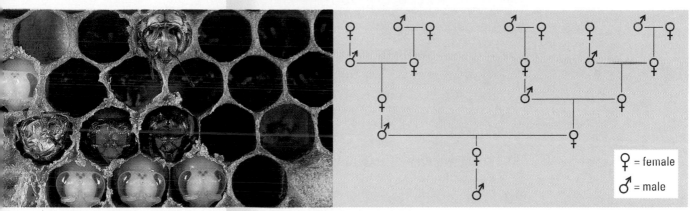

Honey, I'm home.
Pictured in this honeycomb are honeybee pupae at various stages of development.

The figure above shows the ancestors of a male bee. Counting symbols in each row of the figure from the bottom up, notice that the number of bees in each of the first six generations is 1, 1, 2, 3, 5, 8.

The first two terms are equal to 1. Beginning with the third term, each term is found by adding the previous two terms.

a. Find the 7th term of the sequence.

b. Using a_n to represent the nth term, write a recursive formula for the sequence.

Solution

a. The 7th term is the sum of the two previous terms. So the 7th term is 5 + 8, or 13.

b. If a_n is the nth term, then a_{n-1} is the $(n - 1)$st term, that is, the term before the nth term. The notation a_{n-2} denotes the $(n - 2)$nd term, that is, two terms before the nth term. So a recursive formula is

$$\begin{cases} a_1 = 1 \\ a_2 = 1 \\ a_n = a_{n-1} + a_{n-2}, \text{ for } n \geq 3. \end{cases}$$

Check

Use the formula to generate the first few terms of the sequence.

$a_1 = 1$
$a_2 = 1$
$a_3 = a_{3-1} + a_{3-2} = a_2 + a_1 = 1 + 1 = 2$
$a_4 = a_{4-1} + a_{4-2} = a_3 + a_2 = 2 + 1 = 3$
$a_5 = a_{5-1} + a_{5-2} = a_4 + a_3 = 3 + 2 = 5$
$a_6 = a_{6-1} + a_{6-2} = a_5 + a_4 = 5 + 3 = 8$

The first six terms of the sequence are 1, 1, 2, 3, 5, 8. It checks.

The sequence in Example 3 is called the Fibonacci (pronounced "Fee-boh-NOTCH-ee") sequence. It is named after Leonardo of Pisa who wrote under the name Fibonacci in the 12th century.

QUESTIONS

1. Suppose t_n denotes the nth term of a sequence. What does t_{n-1} denote?

2. If a_n is the nth term of a sequence, what is the previous term?

3. *Multiple choice.* Which rule states that the nth term is five times the previous term?

    (a) $t_n = t_{n-1} - 5$ (b) $t_n = \dfrac{t_{n-1}}{5}$

    (c) $t_n = t_{n-1} + 5$ (d) $t_n = 5 \cdot t_{n-1}$

4. Rewrite a recursive definition for the sequence below using the t_{n-1} notation.

$$\begin{cases} t_1 = 15 \\ t_n = 5 \cdot \boxed{\text{ANS}} + 7, \text{ for integers } n \geq 2 \end{cases}$$

5. The first term of a sequence is 5. Each term after the first is found by multiplying the previous term by -3.
    a. Write the first four terms of the sequence.
    b. Let T_n be the nth term of the sequence. Write a recursive definition for this sequence.

6. Refer to Example 1. Find the fifth and sixth terms of the sequence.

7. Consider the sequence defined recursively by

$$\begin{cases} s_1 = -3 \\ s_n = s_{n-1} - 4, \text{ for integers } n \geq 2. \end{cases}$$

    a. Describe the sequence in words.
    b. Write the first five terms of the sequence.

8. a. Refer to Example 2. Use the recursive formula to find T_5.
    b. What explicit formula could you evaluate to find the same term?

9. Write the first ten terms of the Fibonacci sequence.

10. Describe a situation that generates the Fibonacci sequence.

"This must be Fibonacci's."

In 11 and 12, write the first five terms of the sequence.

11. $\begin{cases} t_1 = 3 \\ t_n = 7 \cdot t_{n-1} + 4, \text{ for integers } n \geq 2 \end{cases}$

12. $\begin{cases} s_1 = 5 \\ s_n = s_{n-1} + s_{n-1}, \text{ for integers } n \geq 2 \end{cases}$

Applying the Mathematics

13. Consider the sequence 3, 14, 25, 36, 47, ...
   a. Describe the sequence in words.
   b. Write a recursive definition for this sequence.

14. Consider the sequence defined recursively as
$$\begin{cases} a_1 = 1 \\ a_2 = 3 \\ a_n = a_{n-1} + 2a_{n-2}, \text{ for integers } n \geq 3. \end{cases}$$
   a. Describe the sequence in words.
   b. Write the first five terms of the sequence.

15. Consider the sequence defined as follows.
$$\begin{cases} t_1 = 100 \\ t_n = t_{n-1} + n, \text{ for integers } n \geq 2. \end{cases}$$
   a. Write the first five terms of the sequence.
   b. How is this sequence related to the sequence of triangular numbers?

16. A dairy farmer wants to build a herd of dairy cows. The first year the farmer has 100 cows. Each year the farmer wants to increase the herd by 10%.
   a. Write a recursive definition that describes the sequence giving the number of cows c_n in year n.
   b. In how many years will the farmer have more than 400 cows?

17. The Lucas sequence begins with 1 and 3. After that, each term is the sum of the two preceding terms. The first six Lucas numbers are 1, 3, 4, 7, 11, 18.
   a. Write the next four Lucas numbers
   b. Let L_n be the nth Lucas number. Write a recursive formula for the Lucas numbers.

Review

18. A sequence to be evaluated by a graphics calculator is defined as
$$\begin{cases} t_1 = 2 \\ t_n = 5 \cdot \boxed{\text{ANS}}, \text{ for integers } n \geq 2. \end{cases}$$
   a. Write the first five terms of the sequence.
   b. What is the first term that is greater than 10^5? *(Lesson 1-8)*

19. Antonio opens a savings account. He puts in $10 per week. So, the formula $a_n = 10n$ gives the amount in his account after n weeks.
  a. Evaluate a_4, and explain what it means.
  b. Which term of the sequence is 500? *(Lessons 1-5, 1-7)*

20. A formula for the area of a triangle is $A = \frac{1}{2} bh$. *(Lesson 1-6)*
  a. Solve for b.
  b. Suppose the area of a triangle is 147 square centimeters. Determine the length of the base if the height measures seven centimeters.

21. The formula $P = \frac{nRT}{V}$ is used in chemistry. Solve this formula
  a. for V.   **b.** for T. *(Lesson 1-6)*

22. Solve and check. $4c - (5c - 1) = 7$ *(Lesson 1-5)*

23. Refer to the graph at the left. State **a.** the domain, and **b.** the range of this relation. *(Lesson 1-4)*

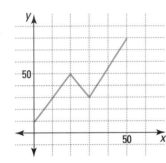

In 24 and 25, let $m: x \to 2|x| + 3$. *(Lessons 1-2, 1-3)*

24. a. $m: 3 \to$ ___?___
  b. $m: -2 \to$ ___?___
  c. $m: 0 \to$ ___?___

25. State the domain of m.

26. Recall that the distance between two points with coordinates (x_1, y_1) and (x_2, y_2) is $\sqrt{(x_2 - x_1)^2 + (y_2 - y_1)^2}$, and the midpoint of the line segment joining them has coordinates $\left(\frac{x_1 + x_2}{2}, \frac{y_1 + y_2}{2}\right)$.
  a. Find the distance between $(-4, -7)$ and $(8, -2)$.
  b. Find the coordinates of the midpoint of the line segment with endpoints $(-4, -7)$ and $(8, -2)$. *(Previous course)*

Exploration

27. The formula

$$t_n = \frac{\left(\frac{1 + \sqrt{5}}{2}\right)^n - \left(\frac{1 - \sqrt{5}}{2}\right)^n}{\sqrt{5}}$$

is an explicit formula for the Fibonacci sequence.
  a. Use a calculator to give the value of t_n when $n = 1, 2, 3, 4,$ and 5.
  b. Does this formula generate the same terms as the recursive definition given in Example 3?
  c. Which formula for the sequence do you prefer? Why do you prefer it?

PROJECTS
CHAPTER ONE
1

A project presents an opportunity for you to extend your knowledge of a topic related to the material of this chapter. You should allow more time for a project than you do for typical homework questions.

1 Values over Time

Recall that the chapter opener describes the values of *The Death of Superman* comic over time. Find another object whose value fluctuates up and down, for instance, a classic car or a share of stock. Track the value of that object over a reasonable domain. Record the values in a table and a graph. Write a paragraph explaining what might cause the value to fluctuate.

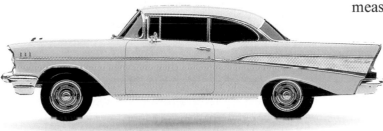

1957 Bel-Air

2 Functions and Graphs in the Newspaper

Use a newspaper that is published daily. Read it for several days.

a. Find 6 to 8 graphs you think are interesting. Look for a variety of subjects and shapes of graphs.

b. Tell whether each graph represents a function. Identify its domain and range. Make up at least one question that can be answered by the graph.

3 Temperature Formulas

In Lesson 1-6, a formula is given relating temperatures measured in the Celsius and Fahrenheit scales. In *War and Peace*, Leo Tolstoy describes a calm frost measured in "degrees Réaumur."

a. Find out when and where the Réaumur scale was used.

b. Find formulas relating degrees Réaumur to each of degrees Fahrenheit and degrees Celsius.

c. Explain how to use the formulas to compare the freezing and boiling points of water on each of these three scales.

1956 Thunderbird

4 Sums of Squares of Fibonacci Numbers

The Fibonacci sequence, 1, 1, 2, 3, 5, 8, 13, 21, . . . , defined in Lesson 1-9, has many interesting properties. One involves the sums of the squares of the Fibonacci numbers.

a. Copy and finish the table below:

$$
\begin{aligned}
F_1^2 &= 1^2 = 1 = 1 \times 1 \\
F_1^2 + F_2^2 &= 1 + 1^2 = 2 = 1 \times 2 \\
F_1^2 + F_2^2 + F_3^2 &= 2 + 2^2 = 6 = 2 \times 3 \\
F_1^2 + F_2^2 + F_3^2 + F_4^2 &= 6 + 3^2 = 15 = 3 \times \underline{\quad} \\
F_1^2 + F_2^2 + F_3^2 + F_4^2 + F_5^2 &= 15 + 5^2 = \underline{\quad} = \underline{\quad} \times \underline{\quad} \\
F_1^2 + F_2^2 + F_3^2 + F_4^2 + F_5^2 + F_6^2 &= \underline{\quad} + \underline{\quad} = \underline{\quad} = \underline{\quad} \times \underline{\quad}
\end{aligned}
$$

b. Look at the pattern in the two columns containing factors of the sum. What sequence do you see in the columns?

c. Make a conjecture in words about the sum of the squares of the first n Fibonacci numbers:

$$F_1^2 + F_2^2 + F_3^2 + \ldots + (F_{n-1})^2 + (F_n)^2 = \underline{\quad} \times \underline{\quad}.$$

d. Write the conjecture in symbols, so each product on the right side is in terms of Fibonacci numbers.

e. Examine the sum of the squares of any two consecutive Fibonacci numbers $(F_{n-1})^2 + (F_n)^2$. Make a table, look for a pattern, and make a conjecture about this sum.

5 Applications of Fibonacci Numbers

The Fibonacci numbers 1, 1, 2, 3, 5, 8, 13, 21, . . . occur naturally in many contexts. One, the family tree of a male bee, is described in Lesson 1-9. Look in reference books to find some other places where Fibonacci numbers occur, and prepare a report on what you find.

6 Pascal's Triangle Revisited

In the Exploration of Lesson 1-7, diagonals of Pascal's triangle were investigated. Here we look at the pattern that even and odd numbers form in the triangle.

a. Copy the first six rows of Pascal's triangle, identifying each even number as shown below.

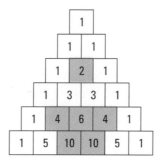

b. Continue writing rows of the triangle until you have completed a minimum of 15 rows.

c. Shade in all squares with even numbers. What patterns do you see?

d. One pattern of even numbers is associated with a Polish mathematician named Sierpinski. Find out when this mathematician lived, and what some of his mathematical work involved.

SUMMARY

The language of algebra uses numbers and variables. These are combined in expressions. Two expressions connected with a verb make up a sentence.

Formulas are equations stating that a single variable is equal to an expression with one or more variables on the other side. Formulas are evaluated or rewritten using the rules for order of operations.

A function is a correspondence between two variables in which each value of the independent variable corresponds to exactly one value of the dependent variable. If a function is defined as a set of ordered pairs, each first coordinate is paired with exactly one second coordinate. The domain of a function is the set of possible values for the independent variable, while the range is the set of values obtained for the dependent variable.

Functions are often named by a single letter. The mapping notation $f: x \rightarrow x + 20$, Euler's notation $f(x) = x + 20$, and the formula $y = x + 20$ all describe the same function. When a function is graphed, no vertical line will intersect the graph in more than one point.

A sequence is a function whose domain is the set of natural numbers. Sequences may be defined explicitly or recursively. With an explicit formula, the nth term can be calculated directly from n. With a recursive formula, the nth term is calculated from one or more previous terms. Recursive formulas can be evaluated easily on a graphics calculator by using the ANS key.

VOCABULARY

Below are the most important terms and phrases for this chapter.
You should be able to give a general description and a specific example of each and a precise definition for those marked with an asterisk (*).

Lesson 1-1
*variable
algebraic expression
expression
algebraic sentence
evaluating an expression
order of operations
*formula, equation

Lesson 1-2
*function
*independent variable
*dependent variable
is a function of
*domain of a function
*range of a function
input, output
*natural numbers
*counting numbers
*whole numbers
*integers
rational numbers
real numbers

Lesson 1-3
*$f(x)$ notation
argument of a function
values of a function
arrow, or mapping, notation

Lesson 1-4
relation
Vertical-Line Test

Lesson 1-5
Distributive Property
"clearing" fractions
Opposite of a Sum Theorem

Lesson 1-6
solved for a variable
in terms of
pitch

Lesson 1-7
*sequence
term of a sequence
triangular numbers
*explicit formula
subscript
subscripted variable
index
generate the terms of a
    sequence

Lesson 1-8
*recursive formula
recursive definition
ANS key

Lesson 1-9
Fibonacci sequence

PROGRESS SELF-TEST

Take this test as you would take a test in class. Then check your work with the solutions in the Selected Answers section in the back of the book.

In 1–4, use the two sequences defined here.

Sequence A
$$t_n = 5 + 7n$$

Sequence B
$$\begin{cases} S_1 = 5 \\ S_n = S_{n-1} + 7, \\ \quad \text{for integers } n \geq 2 \end{cases}$$

1. Write the first four terms of sequence A.

2. **a.** Write the first four terms of sequence B.
   b. Give a recursive definition of sequence B in words.

3. Find t_8.

4. Calculate S_8.

5. If $f(x) = 9x^2 - 11x$, find $f(3)$.

6. Suppose $T: n \to \dfrac{n(n + 1)}{2}$. Then $T(12) = \underline{\ ?\ }$.

7. *Multiple choice.* Which equation below is solved for d?
   (a) $2A = d_1 d_2$ (b) $d = rt$
   (c) $d^2 = (x_1 - x_2)^2 + (y_1 - y_2)^2$

8. Suppose you travel 12 miles in t hours. Write a formula that gives your speed s in miles per hour.

In 9–11, solve.

9. $4(3x - 8) = 11$

10. $1.7y = 0.9 + 0.5y$

11. $.12p + .08(15{,}000 - p) = 1480$

In 12–14, use the formula $V = \frac{1}{3}\pi r^2 h$ for the volume of a cone.

12. Find the volume of the cone at the right to the nearest cubic centimeter.

6 cm
4 cm

13. What is a reasonable domain for r?

14. Rewrite this equation for h in terms of V and r.

15. A function f contains only the points (1, 2), (3, 4), and (5, 6). Find its domain and range.

16. Refer to the graphs below. Which are graphs of functions?

(a)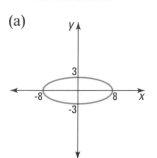

(b)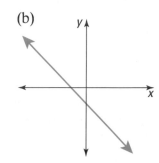

(c)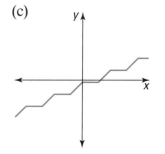

(d)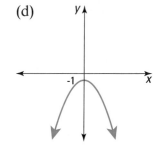

In 17 and 18, determine whether the relation is a function. Justify your answer.

17. $\{(95, -5), (4, 4), (5, 5)\}$

18.

x	0	1	1	2	2
y	0	1	-1	1.4	-1.4

(Consider x to be the independent variable.)

19. Cherlyn's grandparents gave her $20 on her first birthday, $30 on her second, $40 on her third, and so on, increasing by $10 each year. Let a_n = the amount received on her nth birthday. Write a recursive formula to describe this situation.

20. Consider the equation $\frac{1}{2}z + \frac{1}{5}z + 90 = z$.
    a. Solve for z.
    b. Make up a question about a real situation that could be answered by solving this equation.

PROGRESS SELF-TEST

21. The function $P(L) = 2\pi \sqrt{\dfrac{L}{g}}$, with $g =$ 9.8 m/sec^2 gives the period of a pendulum as a function of its length L. Find the period of a pendulum with length 1 meter to the nearest second.

22. A reading program at an elementary school encourages children to read books. When a class reads 250 books, the entire class gets a pizza party. Suppose a class reads an average of 45 books a week. Then the equation $45w = B$ gives the number B of books read after w weeks. If this reading trend continues, after how many weeks will the class deserve a pizza party?

In 23–25, refer to the graph below which shows the enrollment in public and private high schools (grades 9–12) and colleges in the U.S. from 1970 to 1990 and projected through the year 2000. Let $H(x) =$ the high school enrollment and $C(x) =$ the college enrollment during the years 1970 to 2000.

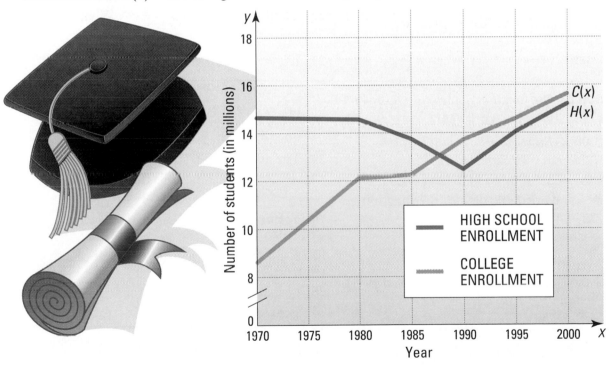

23. Estimate $C(1975)$ and describe what this value represents.

24. Estimate the solution to $H(x) = C(x)$, and explain what the solution means.

25. What is the domain of the function H?

CHAPTER REVIEW

Questions on SPUR Objectives

SPUR stands for **S**kills, **P**roperties, **U**ses, and **R**epresentations. The Chapter Review questions are grouped according to the SPUR Objectives for this chapter.

SKILLS DEAL WITH THE PROCEDURES USED TO GET ANSWERS.

Objective A: *Evaluate expressions and formulas, including correct units in answers.* (Lessons 1-1, 1-2)

1. In the formula $d = \frac{n(n-3)}{2}$, find d when $n = 17$.

2. If $d = \frac{1}{2} gt^2$, find d when $g = 32$ ft/sec^2 and $t = 2.5$ sec.

3. Evaluate $20,000(.9)^n$ to the nearest hundredth when $n = 10$.

4. Evaluate $2^{n-1} + (n-1)^2$ when $n = 4$.

Objective B: *Use function notation.* (Lesson 1-3)

5. If $f(x) = 2x - 3$, what is $f(4)$?

6. Suppose $t: n \to 5 - 4n^2$. Then $t: -2 \to$ ___?___.

7. Let $h: a \to a^5$. Then $h: -3 \to$ ___?___.

8. If $M(x) = 3x^2 + 4$, then $M(2k) = $ ___?___.

9. If the following table defines the function g, find $g(5)$ and $g(6)$.

x	1	2	3	4	5	6
$g(x)$	3	5	7	9	11	13

10. This table defines the function p. Find $p(-1)$.

a	-2	-1	0	1	2
$p(a)$	7	16	4	-6	-1

Objective C: *Solve and check linear equations.* (Lesson 1-5)

In 11–18, solve and check.

11. $\frac{3}{2}x = 9$

12. $\frac{3}{10}(t - 20) = \frac{6}{5}$

13. $\frac{6}{U} = 8$

14. $4 = 6a - (2 - 2a)$

15. $3 - (m + 2) = 4m$

16. $.05(4500 - x) + .08x = 1200$

17. $\frac{x}{2} + \frac{x}{3} + 10 = x$

18. $3y - 2(y + 5) = 10y$

Objective D: *Rewrite formulas.* (Lesson 1-6)

19. Solve for n in the formula $t = 4 - 5n$.

20. The measure θ of an exterior angle of a regular polygon is given by $\theta = \frac{360}{n}$ where n is the number of sides. Solve for n in terms of θ.

21. Recall the formula $d = \frac{1}{2} gt^2$. What is the ratio of d to t^2?

22. If $x = 3y$, then $\frac{x}{y} = $ ___?___.

23. Which equation(s) is (are) solved for t?
    (a) $t = \frac{D}{R}$ (b) $10t - 5t^2 = h$
    (c) $A = \frac{1}{2} ht$ (d) $180(n - 2) = t$

24. The area A of a trapezoid is given by $A = \frac{1}{2} h(b_1 + b_2)$. Solve this equation for h.

Objective E: *Evaluate sequences.* (Lessons 1-7, 1-8, 1-9)

In 25–28, write the first five terms of the sequence.

25. $t_n = 20 + 3n$, for integers $n \geq 1$

26. $s_n = n^2 - 1$

27. $\begin{cases} t_1 = 10 \\ t_n = t_{n-1} - 7, \text{ for integers } n \geq 2 \end{cases}$

28. $\begin{cases} a_1 = 2 \\ a_n = -2 + 3a_{n-1}, \text{ for integers } n \geq 2 \end{cases}$

Objective F: *Write a recursive definition for a sequence.* *(Lessons 1-8, 1-9)*

In 29 and 30, a sequence is defined recursively to be evaluated on a graphics calculator.
a. Write the first five terms of the sequence.
b. Rewrite the definition using sequence notation.

29. $\begin{cases} t_1 = 16 \\ t_n = (\boxed{\text{ANS}}) - 5, \text{ for integers } n \geq 2 \end{cases}$

30. $\begin{cases} a_1 = -1 \\ a_n = (\boxed{\text{ANS}})^2 - 1, \text{ for integers } n \geq 2 \end{cases}$

In 31 and 32, a sequence is given. **a.** Describe the sequence in words. **b.** Write a recursive definition for the sequence. (Hint: Only one operation is needed.)

31. 6, 12, 18, 24, . . .

32. -2, -2.75, -3.5, -4.25, . . .

PROPERTIES DEAL WITH THE PRINCIPLES BEHIND THE MATHEMATICS.

Objective G: *Determine whether a relation defined by a table, a list of ordered pairs, or a simple equation is a function.* *(Lesson 1-2)*

33. Does the table below describe y as a function of x? Justify your answer.

x	-1	-1	-4	0	-16
y	1	-1	2	0	-4

34. Does the set below describe a function? Why or why not? {(1, 2), (2, 3), (3, 4), (4, 1)}

In 35–37, an equation is given. **a.** Make a table of four pairs of numbers that satisfy the equation. **b.** Does the equation describe a function?

35. $x = y^2$ 36. $xy = 15$ 37. $x = -5$

Objective H: *Determine the domain and range of a function defined by a table, a list of ordered pairs, or a simple equation.* *(Lesson 1-2)*

38. What real number is not in the domain of f, where $f(x) = \frac{1}{x^2}$?

In 39 and 40, give **a.** the domain, and **b.** the range of the function described.

39. $\{(2, -2), (3, -3), (-9, 9), \left(-\frac{1}{2}, \frac{1}{2}\right)\}$

40.

a	-2	-1	0	1	2
$m(a)$	1	-1	1	-1	1

41. Consider $y = x^4$.
    a. What is the domain of the function?
    b. Are there any values of x for which x^4 is negative?
    c. What is the range of $y = x^4$?

42. Let $f(x) = -\sqrt{x}$.
    a. What is the domain of f?
    b. Are there any values of x for which $-\sqrt{x}$ is positive? Explain your answer.

USES DEAL WITH APPLICATIONS OF MATHEMATICS IN REAL SITUATIONS.

Objective I: *Use addition, subtraction, multiplication, and division to write expressions which model real-world situations.* *(Lesson 1-1)*

43. The dimensions of a building are 100 times as large as the dimensions of its model. If a floor on the model is x cm long, how long is the floor on the building?

44. There are s students per bus and b buses. How many students are there in all?

45. Carol takes M minutes to walk B blocks. What is her walking speed?

46. Jamal can do six problems in 5 minutes. How long will it take him to do P problems?

47. Luis has 20 sticks of gum. He gives one to each of c friends. How many sticks of gum does he have left?

48. Juana can mow the lawn around the family business in c hours with the riding mower. Her sister Rosa can mow the lawn in d hours with the walking mower.
  a. What fraction of the lawn can Juana mow in one hour?
  b. What fraction of the lawn can Rosa mow in one hour?
  c. If they work together, what fraction of the lawn can they mow in one hour?

Objective J: *Use functions to solve real-world problems.* *(Lessons 1-2, 1-3, 1-4, 1-7, 1-8, 1-9)*

49. Sandra's annual salary is $26,000. She gets an increase of 6% at the end of each year. The sentence $a_n = 26{,}000(1.06)^{n-1}$ gives Sandra's salary at the end of n years.
  a. Write Sandra's salary for the first five years.
  b. At this growth rate, what would Sandra's salary be after 20 years with this company?

50. The famous scientist Galileo found a relationship between the distance $d(t)$ a dropped object falls (in feet) in time t (in seconds). (Of course he used different units.) The function is defined by the rule $d(t) = 16t^2$. What is the approximate distance fallen when $t = 1.5$ sec?

51. The function S defined by $S(x) = x + \frac{x^2}{20}$ relates the stopping distance in feet to the car's speed x in miles per hour. How many feet does it take for a car traveling at 70 mph to stop?

52. Todd works out at a gym 3 times a week. Every Saturday his trainer measures the circumference of his biceps to show Todd his progress. At the beginning, the measurement was 12″, but he put on $\frac{1''}{4}$ every month.
  a. Write an equation that gives the circumference of his biceps after m months.
  b. What would the measurement be at the end of 2 years if this trend continued?
  c. Do you think this equation would continue to describe what's happening after 5 years of working out? Explain your answer.

In 53–55, let $P(x)$ and $B(x)$ be the populations of Philadelphia and Baltimore, respectively, in year x.

	Philadelphia	Baltimore
1900	1,290,000	509,000
1950	2,070,000	950,000
1980	1,688,000	787,000
1990	1,586,000	736,000

53. What does $B(1950)$ represent?

54. a. Calculate $P(1980) - B(1980)$.
  b. What does part **a** represent?

55. a. Calculate $B(1990) - B(1980)$.
  b. Write in words what the calculation in part **a** represents.

56. Devin opened a savings account with $50. Each month he adds $15 to his account.
  a. Write the amounts in this account for the first six months.
  b. Write a recursive formula that generates the sequence giving his savings account balance at the end of each month.

Objective K: *Use linear equations to solve real-world problems.* *(Lessons 1-5, 1-6)*

57. Suppose a baby blue whale weighs 4000 lb at birth and gains 200 lb a day while nursing. A formula that gives its weight W after d days of nursing is $W = 4000 + 200d$.
  a. Write an equation that can be used to find the number of days a young blue whale has been nursing if it weighs 14,000 lb. (Baby blue whales nurse for 5 to 7 months.)
  b. Solve this equation.

58. At Central High School all students are in grade 10, 11, or 12. This year $\frac{2}{5}$ of the students are in grade 10, $\frac{1}{3}$ are in grade 11, and 320 are in grade 12. How many students are at Central High this year?

59. When Melissa came home after a week's vacation, the temperature in her apartment had gone up to 85°. Half an hour after she turned on the air conditioner, the temperature was 82°. Suppose the equation $T = 85 - .1m$ gives the temperature, in degrees, after the air conditioner has been on m minutes. How long will it take the temperature to fall to 68°?

60. David has a savings account with which to buy a car. After the down payment his account balance is $2400. His monthly car payment is $215 per month. Thus after x payments his account balance will be $2400 - 215x$. After how many months will David run out of money?

REPRESENTATIONS DEAL WITH PICTURES, GRAPHS, OR OBJECTS THAT ILLUSTRATE CONCEPTS.

Objective L: *Determine the domain, range, and values of a function from its graph.*
(Lesson 1-4)

In 61 and 62, find the domain and range of the function whose graph is shown.

61.

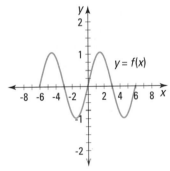

62.

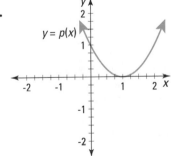

63. From the graph in Question 61, find the approximate value of $f(3)$.

64. For what x value is $p(x) = 0$ in Question 62 above?

Objective M: *Apply the Vertical-Line Test for a function.* *(Lesson 1-4)*

In 65–68, determine whether the given graph represents the graph of a function. Justify your answer.

65.

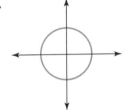

66.

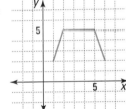

67.

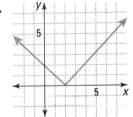

68.

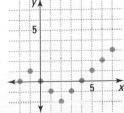

CHAPTER

2

VARIATION AND GRAPHS

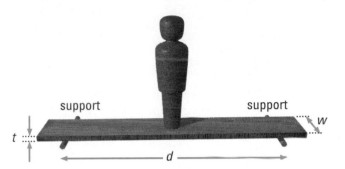

support   support

t

w

d

A construction worker walking on a board of a scaffold knows that too much weight will break the board. The largest weight that can be safely supported by a board depends on its width w, its thickness t, and the distance d between the board's supports.

Wider or thicker boards make the scaffold stronger. Increasing the distance between the supports weakens the scaffold. Thus, the strength varies as the dimensions of the board change or as the distance between supports changes. But how does it vary? In this chapter, you will study variation, that is, how one quantity changes as others are changed.

The chapter begins with a simple relationship called *direct variation.*

You can save. *More than half the beverage cans used in the U.S. are recycled. One recycled aluminum can saves enough energy to keep a 100-watt light bulb burning for about 3.5 hours.*

In many places, you can get refunds for returning aluminum cans. For example, since 1987 in New York, you can get a 5¢ refund for each can returned. Thus, if r is the total refund in cents and c is the number of cans you return, then

$$r = 5c.$$

Doubling the number of cans returned doubles your refund. Tripling the number of cans triples your refund. We say that r **varies directly as** c.

In Michigan, you can get 10¢ per can returned, so $r = 10c$. Again, r varies directly as c.

Recall the formula $A = \pi r^2$ for the area of a circle. In this formula, as the radius r increases, the area A also increases. In this case, A varies directly as r^2. Often this wording is used: The area A varies directly as the *square* of r.

$A = \pi r^2$

Direct-Variation Functions

The formulas $r = 5c$, $r = 10c$, and $A = \pi r^2$ are all of the form $y = kx^n$, where k is a nonzero constant, called the **constant of variation,** and n is a positive number. These formulas all describe *direct-variation functions*.

> **Definition**
> A **direct-variation function** is a function with a formula of the form $y = kx^n$, with $k \neq 0$ and $n > 0$.

When y varies directly as x^n we also say that y is **directly proportional to** x^n. For instance, the formula $A = \pi r^2$ can be read "the area of a circle is directly proportional to the square of its radius." Here $n = 2$ and $k = \pi$, so π is the constant of variation. In the formulas $r = 5c$ and $r = 10c$, $n = 1$. The constants of variation are 5 for New York and 10 for Michigan.

Full-grown giraffes range from 4.3 to 5.9 meters in height. Taller giraffes weigh more than shorter ones, following the ideas in Example 1.

Example 1

The weight w of an adult animal of a given species is known to vary directly with the cube of its height h.
a. Write an equation relating w and h.
b. Identify the dependent and independent variables.

Solution

a. An equation for the direct variation is $w = kh^3$.
b. Because w is given in terms of h, The dependent variable is w and the independent variable is h.

Solving Direct-Variation Problems

Direct-variation functions arise in many real-world situations. For instance, after applying brakes, the braking distance d a car travels before coming to a stop is directly proportional to the square of its speed s.

$$d = ks^2$$

The value of k depends upon the type of car, the condition of its brakes, and the condition of the road. In Example 2, we illustrate how to find the value of k from known information and how to use that value to make a prediction.

Example 2

A certain car needs 25 ft to come to a stop after the brakes are applied at 20 mph. Braking distance d (in ft) is directly proportional to the square of the speed s (in mph). What distance is needed to stop this car after the brakes are applied at 60 mph?

Solution

1. Find an equation relating d and s. From the given information, $d = ks^2$.
2. Determine the constant of variation. You are given that $d = 25$ ft when $s = 20$ mph. To find k, substitute these values into $d = ks^2$.

$$25 \text{ ft} = k \cdot (20 \text{ mph})^2$$
$$25 \text{ ft} = (400 \text{ mph}^2) \cdot k$$

So
$$k = \frac{1}{16} \frac{\text{ft}}{(\text{mph})^2}.$$

3. Substituting $k = \frac{1}{16}$ into $d = ks^2$ gives

$$d = \frac{1}{16} s^2$$

as a formula relating speed and braking distance for this situation.
4. Evaluate the formula when $s = 60$ mph.

$$d = \left(\frac{1}{16} \frac{\text{ft}}{(\text{mph})^2}\right) \cdot (60 \text{ mph})^2$$
$$d = \frac{1}{16} \frac{\text{ft}}{(\text{mph})^2} \cdot 3600 \text{ mph}^2$$
$$d = 225 \text{ ft}$$

The car will need 225 ft to come to a stop after the brakes are applied at 60 mph.

Notice that to use variation functions to predict values, you carry out four steps.

1. Write an equation that describes the variation.
2. Find the constant of variation.
3. Rewrite the variation function using the constant of variation.
4. Evaluate the function for the desired value of the independent variable.

This four-step algorithm is used again in Example 3.

Example 3

Suppose z varies directly as the fourth power of w. If $z = 80$ when $w = 2$, find z when $w = 3$.

Solution

1. Write an equation that describes the variation.
$$z = kw^4$$
2. Find the constant of variation. You are given that $z = 80$ when $w = 2$. Substitute these values into the variation formula to find k.
$$80 = k \cdot 2^4$$
$$80 = k \cdot 16$$
$$5 = k$$
3. Rewrite the variation formula using the constant of variation.
$$z = 5w^4$$
4. Evaluate the formula for the desired value of the independent variable.
$$z = 5 \cdot 3^4$$
$$= 5 \cdot 81$$
$$z = 405$$

When $w = 3$, $z = 405$.

QUESTIONS

Covering the Reading

1. Name three variables that affect the weight a board on a scaffold can support.

2. Give an example of a direct-variation function from geometry.

3. In the function $y = 3x^5$, __?__ varies directly as __?__ and __?__ is the constant of variation.

In 4 and 5, suppose $y = -4.3x$.

4. Find y when $x = 3.1$

5. Is this an example of a direct-variation function? How can you tell?

In 6 and 7, assume that y is directly proportional to the square of x.

6. *Multiple choice.* Which equation represents this situation?
   (a) $y = 2x$
   (b) $y = kx^2$
   (c) $x = ky^2$
   (d) $y = 2x^k$

7. Which is the dependent variable?

In 8 and 9, refer to Example 2.

8. Find the distance the car travels before coming to a stop if its brakes are applied at 40 mph.

9. Suppose that some other car needs 30 ft to stop if its brakes are applied at 20 mph.
   a. Find k.
   b. Write the variation function relating s and d.
   c. How far would the car travel before coming to a stop if its brakes were applied at 60 mph?

10. Suppose W varies directly as the fifth power of z, and $W = 96$ when $z = 2$.
    a. Find the constant of variation.
    b. Find W when $z = 10$.

11. Suppose y varies directly as the cube of x, and $y = 1000$ when $x = 5$. Find y when $x = 9$.

12. Suppose $f(x) = \frac{x^2}{9}$. Is this a direct-variation function? Write a sentence or two justifying your conclusion.

Applying the Mathematics

13. The power P generated by a windmill is directly proportional to the cube of the wind speed w.
    a. Write an equation relating P and w. Identify the dependent and the independent variables.
    b. If a 10 mph wind generates 150 watts of power, how many watts will a 6 mph wind generate?

14. When lightning strikes in the distance, you do not see the flash and hear the thunder at the same time. You first see the lightning; then you hear the thunder.
    a. Write an equation to express this situation: "The distance d from the observer to the flash varies directly as the time t between the observer's seeing the lightning and hearing the thunder."
    b. Suppose that lightning strikes a known point 4 miles away, and that you hear the thunder 20 seconds later. Then, how far away has lightning struck if 30 seconds pass between the time you see the flash and hear its thunder?

Wind power. *Shown is a windmill "farm" in Kern County, California. The more than 16,000 windmills in the U.S. in 1990 generated about 2.5 million kilowatt-hours of electricity—enough to meet the needs of a city the size of San Francisco for a year.*

15. Refer to the formula $A = \pi r^2$ for the area of a circle with radius r.
   a. Complete the table below, leaving all answers in terms of π.

r	1	2	3	4	5	6	7	8	9	10
A	π									

   b. The area when $r = 4$ is how many times as large as the area when $r = 2$?
   c. The area when $r = 6$ is how many times as large as the area when $r = 3$?
   d. The area when $r = 10$ is how many times as large as the area when $r = 5$?
   e. Make a conjecture. When the radius doubles, the area __?__.
   f. Follow a similar procedure to complete the following conjecture. When the radius triples, the area __?__.

Review

In 16–18, find the absolute value.   *(Previous course)*

16. $|7.4|$ **17.** $|-3.9|$ **18.** $|6 - 10|$

In 19–21, write as a power of 3.   *(Previous course)*

19. $3^2 \cdot 3^4$ **20.** $(3 \cdot 3)^5$ **21.** $(3^3)^5$

22. Rewrite 395,000,000,000 in scientific notation.   *(Previous course)*

In 23–25, refer to the graph below. It shows a firm's profits for the years 1990 to 1995.   *(Previous course)*

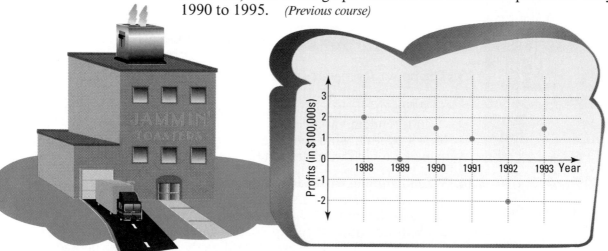

23. About how much profit did the company make in 1990?

24. In what year(s) did the company lose money?

25. During which intervals did the profits decline?

In 26 and 27, copy the triangle below. Graph the image of the triangle under the given transformation.   *(Previous course)*

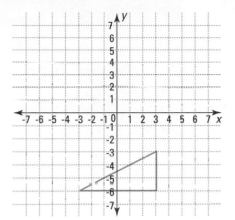

Music to your ears. *A carillon, like this one atop Lupton Hall at Oglethorpe University in Atlanta, is a set of 23 or more bells, usually having diameters that range from 3.5 in. to 10.25 ft. Carillons are popular on college campuses because the sounds produced can be heard over a wide area.*

26. reflection over the *x*-axis

27. size change of magnitude $\frac{1}{3}$

Exploration

28. The speed of sound in air is about 1088 feet per second.
  a. Convert the speed of sound to miles per second.
  b. Use your answer from part **a** to find the time it takes sound to travel four miles. Compare this answer to the values in Question 14.
  c. What environmental conditions affect the speed of sound?

29. Find a manual for learning how to drive. Often these manuals contain stopping and braking distances. Do the braking distances given in the manual vary directly as the square of the speed?

Division of labor. *The amount of time each student must work tends to vary inversely as the number of students.*

Metro Car Sales hires students to wash cars in the company's display lots. The manager knows from experience that one student can wash all the cars in 36 working hours. When more students work, each student needs to work fewer hours. If s equals the number of students who work and t equals the time (in hours) each student needs to work, then the product st is the total number of hours worked, and

$$st = 36, \text{ or } t = \frac{36}{s}.$$

Some combinations of s and t that might be used to finish the job are given below.

s	4	8	9	12	16
t	9	$4\frac{1}{2}$	4	3	$2\frac{1}{4}$

Definition of Inverse Variation

The formula $t = \frac{36}{s}$ which determines the values in the table has the form $y = \frac{k}{x^n}$ where $n = 1$. This is an instance of *inverse variation*. We say t **varies inversely as** s. In this example, the constant of variation k is 36.

> **Definition**
> An **inverse-variation function** is a function with a formula of the form $y = \frac{k}{x^n}$, with $k \neq 0$ and $n > 0$.

When y varies inversely as x^n, we also say that y **is inversely proportional to** x^n. As with direct variation, inverse variation occurs in many kinds of situations.

Example 1

The number n of oranges you can pack in a box is approximately inversely proportional to the cube of the average diameter d of the oranges. Write an equation to express this relation.

Solution

The cube of the diameter is d^3. So, $n \approx \dfrac{k}{d^3}$.

Solving Inverse-Variation Problems

Many scientific principles involve inverse-variation functions. For instance, the *Law of the Lever* states that to balance a given person seated on a seesaw, the distance d the other person is from the pivot (or fulcrum) is inversely proportional to that person's weight w. That is, $d = \dfrac{k}{w}$.

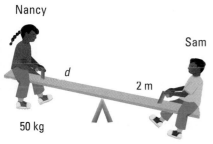

Nancy

Sam

d

2 m

50 kg

55 kg

Example 2

Nancy and Sam are trying to balance on a seesaw. Sam, who weighs 55 kilograms, is sitting 2 meters from the fulcrum. Nancy weighs 50 kilograms. How far away from the fulcrum must she sit to balance Sam?

Solution

We use the same four steps as for direct-variation problems.
Let d = a person's distance in meters from the fulcrum.
Let w = a person's weight in kilograms.
1. Write an equation relating d and w. From the Law of the Lever,
$$d = \frac{k}{w}.$$

2. To find k, substitute Sam's weight and distance into $d = \dfrac{k}{w}$.
$$2 = \frac{k}{55}$$
$$110 = k$$

3. The work in step 2 tells you that the variation formula for this situation is
$$d = \frac{110}{w}.$$

4. Evaluate this formula when $w = 50$ kg.
$$d = \frac{110}{50}$$
$$d = 2.2$$
Nancy must sit 2.2 meters away from the fulcrum.

Check

Does 2 meters · 55 kilograms = 2.2 meters · 50 kilograms? Yes, the numbers and the units agree.

Caution: If a different person is seated on the same seesaw, the value of k in the Law of the Lever may change.

The weight W of a body above the surface of the Earth varies inversely as the square of its distance d from the center of the Earth. That is, $W = \frac{k}{d^2}$. This instance of an *inverse-square variation* explains why astronauts are almost weightless in space.

Example 3

If an astronaut weighs 165 pounds on the surface of the Earth, what will the astronaut weigh 18,000 miles above the Earth's surface? (The radius of the Earth is approximately 4000 miles.)

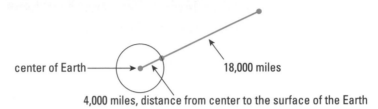

center of Earth — 18,000 miles

4,000 miles, distance from center to the surface of the Earth

Solution

1. Write an equation that describes the variation. Let W = the weight of the astronaut in pounds and d = the distance from the center of the Earth to the astronaut. Because W varies inversely as the square of the distance,

$$W = \frac{k}{d^2}.$$

An astronaut 18,000 miles above the Earth's surface is 22,000 miles from the center of the Earth. We need to find W when d = 22,000.

2. Find k, the constant of variation, by substituting the values W = 165 lb and d = 4000 mi. This indicates that on the surface of the Earth, the astronaut weighs 165 lb.

$$165 = \frac{k}{(4000)^2}$$
$$165 \cdot 16,000,000 = k$$
$$2.64 \cdot 10^9 = k$$

3. Rewrite the variation function using the constant of variation.

$$W = \frac{2.64 \cdot 10^9}{d^2}$$

4. Substitute d = 22,000 into the inverse-square function.

$$W = \frac{2.64 \cdot 10^9}{(22,000)^2}$$
$$\approx 5.5 \text{ lb}$$

At 18,000 miles above the Earth's surface, the astronaut weighs only about 5.5 pounds.

But can he keep the weight off? *On June 3, 1965, Astronaut Edward H. White II became the first American to walk in space. White was the pilot of the* Gemini IV *space mission.*

Covering the Reading

In 1 and 2, refer to the Metro Car Sales problem on page 78.

1. The time to finish the job varies inversely as the ___?___.

2. Only 12 students are found to work. How long will it take them to complete the job?

3. The equation $y = \frac{k}{x^3}$ means y varies inversely as ___?___.

4. Suppose x varies inversely as the 4th power of t. Write an equation to describe the variation.

5. *Multiple choice.* Assume k is a nonzero constant. Which equation does not represent an inverse variation?
   (a) $y = kx$ (b) $y = \frac{k}{x}$ (c) $xy = k$ (d) $y = \frac{k}{x^2}$

6. Refer to Example 2. If Sam sits 2.5 yards from the fulcrum, how far away from the fulcrum must Nancy sit to balance him?

7. Find the distance needed to balance on the seesaw shown below.

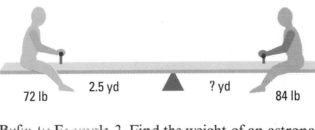

72 lb   2.5 yd   ? yd   84 lb

8. Refer to Example 3. Find the weight of an astronaut in a space lab 300 miles above the Earth if the astronaut weighs 150 lb on Earth.

Applying the Mathematics

9. Translate this statement into a variation equation. The time t an appliance can be run on 1 kilowatt-hour of electricity is inversely proportional to the wattage rating w of the appliance.

10. If y varies inversely as x^3, and $y = 10$ when $x = 4$, find the value of y when $x = 2$.

Lights, camera, . . . *Most professional photographers use hand-held light meters to measure light intensity, thereby enabling them to determine the correct exposure.*

11. The intensity I of light varies inversely as the square of the observer's distance D from the light source.
   a. Translate this statement into a variation equation.
   b. Suppose that the light intensity is 30 lumens when the observer is 6.7 meters from the light. Find the constant of variation.
   c. Find the light intensity when the distance between the observer and the light is 20 meters.

12. Refer to Example 1. Why do you think a correct variation equation is $n \approx \frac{k}{d^3}$ and not some other equation such as $n \approx kd^3$ or $n \approx \frac{k}{d^2}$?

In 13 and 14, complete the sentence with the word "directly" or "inversely."

13. The volume of a sphere varies __?__ as the cube of its radius.

14. The number of tiles needed to tile a floor varies __?__ as the square of the length of a side of the tile.

15. Consider again the Metro Car Sales problem on page 78.
  a. Complete the table below.

s	1	2	3	4	5	6	7	8	9	10	11	12
t				9				$4\frac{1}{2}$	4			3

  b. Compare the values of t when $s = 2$ and when $s = 4$, when $s = 4$ and when $s = 8$, and when $s = 6$ and when $s = 12$. Make a conjecture. When the number of people working doubles, the time __?__.
  c. Follow a similar procedure to complete the following conjecture. When the number of people working triples, the time __?__.

7'6"er. Shawn Bradley, shown with the ball, was the number two pick in the 1993 NBA draft. He was signed by the Philadelphia 76ers.

Review

16. At some restaurants, the price of a pizza varies directly with the square of its diameter. If you pay $5.95 for a cheese pizza with a 10-inch diameter, how much should a 14-inch-diameter cheese pizza cost? *(Lesson 2-1)*

17. If two people have the same shape, then their weight varies directly as the cube of their height. In 1994, the basketball player Shawn Bradley was 7'6" tall, and he weighed about 250 pounds. How much would you expect a person 5'10" tall with Shawn's shape to weigh? *(Lesson 2-1)*

18. If $f(p) = 5p^2$, find $f(3x)$. *(Lesson 1-3)*

19. Line ℓ is parallel to line m in the figure at the left. The expressions represent angle measures. Find y. *(Previous course)*

In 20–22, simplify. *(Previous course)*

20. $x^{10} \cdot x^3$ **21.** $\frac{x^{12}}{x^4}$ **22.** $(2x)^3$

ℓ $(2x - 2)°$

m $(4x - 104)°$

$y°$

23. a. Plot the point $P = (2, -5)$.
  b. Write an equation for the horizontal line through P.
  c. Write an equation for the vertical line through P. *(Previous course)*

Exploration

24. Consult a science or economics text to find some other examples of inverse variation.

**Introducing
Lesson 2-3**

*Functions
of
Variation*

IN·CLASS
ACTIVITY

In Lesson 2-1, Question 15, you explored the effects of doubling or tripling the radius r on the area A of a circle. In Lesson 2-2, Question 15, you explored the effects of doubling or tripling the values of s on the value of t in the equation $t = \frac{36}{s}$.

In this Activity, you will explore how changes in the independent variable of two other variation functions result in changes in the dependent variable. Work on this Activity in small groups.

1 Consider the direct-variation formula $y = x^3$.
a. Each person in your group should choose a different value for the independent variable. Double, triple, and quadruple the original value and record the values in a chart like the one shown here. Then find the corresponding values of y for each x and record them in a table.

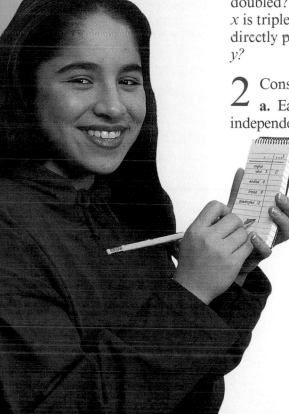

b. Draw Conclusions. Discuss your results with other members of your group and answer these questions. What happens to y when x is doubled? What happens to y when x is tripled? What happens to y when x is multiplied by four? If y is directly proportional to x^3, when x is multiplied by c, what happens to y?

2 Consider the inverse-variation formula $y = \frac{1}{x^3}$.
a. Each person in your group should choose a different value for the independent variable. Double, triple, and quadruple that original value and make a chart as in part **1**. Find the corresponding values of y for each x and record them in a table.
b. Draw Conclusions. Discuss your results with other members of your group and answer these questions. What happens to y when x is doubled? What happens to y when x is tripled? What happens to y when x is multiplied by four? If y is inversely proportional to x^3, when x is multiplied by c, what happens to y?

The
*Fundamental
Theorem of
Variation*

Modern cat. *White Bengal tigers may attain a shoulder height of 1 meter and weigh as much as 225 kg. Tigers such as this one may have descended from the extinct Saber-toothed cat. See Example 2.*

Recall that the area A of a circle varies directly with the square of its radius r. $A = \pi r^2$. Some values of r and A are given in the following table.

r	1	2	3	4	5	6	8
A	π	4π	9π	16π	25π	36π	64π

What happens to the area of a circle when the radius is tripled? One way to answer this question is to compare values from the table when the radius is tripled. For example,

> if $r = 1$, then $A = \pi$, and
> if $r = 3$, then $A = 9\pi$.

Notice that when the radius is tripled, the area is multiplied by nine. This pattern also holds if you compare the ordered pairs $(2, 4\pi)$ and $(6, 36\pi)$. Based on the two instances above, it seems reasonable to conjecture that if you triple a circle's radius, the area is multiplied by nine. In geometry, we proved this using properties of similarity. The following example shows how to prove this using properties of algebra.

Example 1

Given the direct-variation formula $A = \pi r^2$, prove that if r is tripled, A is multiplied by nine.

Solution

Let A_1 be the original area (before tripling the radius). Let A_2 be the area after tripling the radius. To find A_2, r must be tripled. So replace r by $3r$. Here is a proof given in two-column form.

1. $A_1 = \pi r^2$ — given
2. $A_2 = \pi(3r)^2$ — substitution
3. $ = \pi \cdot 9r^2$ — Power of a Product Property
4. $ = 9\pi r^2$ — Associative and Commutative Properties of Multiplication
5. $ = 9A_1$ — substitution (step 1 into step 4)

What Is the Fundamental Theorem of Variation?

In the In-class Activity on page 83, you studied the variation equations $y = x^3$ and $y = \frac{1}{x^3}$. You should have found that in $y = x^3$, when x is tripled y is multiplied by 27; and in $y = \frac{1}{x^3}$, when x is tripled, y is divided by 27.

Example 1 and the problems in the Activity are instances of the Fundamental Theorem of Variation. We state the theorem and give the proof of part **a.** You are asked to prove part **b** in the Questions.

The Fundamental Theorem of Variation

a. If y varies directly as x^n (That is, $y = kx^n$.), and x is multiplied by c, then y is multiplied by c^n.

b. If y varies inversely as x^n (That is, $y = \frac{k}{x^n}$.), and x is multiplied by a nonzero constant c, then y is divided by c^n.

Proof of Part a:
Let y_1 = original value before multiplying x by c.
Let y_2 = value when x is multiplied by c.
To find y_2, x must be multiplied by c.

$y_1 = kx^n$ — definition of direct variation
$y_2 = k(cx)^n$ — definition of y_2
$y_2 = k(c^n x^n)$ — Power of a Product Property
$y_2 = c^n(kx^n)$ — Associative and Commutative Properties of Multiplication
$y_2 = c^n y_1$ — substitution

Applications of the Fundamental Theorem of Variation

The Fundamental Theorem of Variation can be applied in many situations.

Example 2

If we assume that animals of a given species are similar, then we can apply a theorem from geometry which says that the weight w of an animal of the species varies directly with the cube of its height h. If an ancient cat was 1.8 times as tall as a modern one, how many times as great was the weight of the ancient cat than the weight of the modern cat?

Solution 1

Set the problem up as in Example 1.
An equation for the variation function is $w = kh^3$.
Let w_1 = the weight of the modern cat, and
w_2 = the weight of the cat with height $1.8h$.

$$w_1 = kh^3$$
$$\begin{aligned} w_2 &= k(1.8h)^3 \\ &= k(1.8)^3 h^3 \\ &= (1.8)^3(kh^3) \\ &= (1.8)^3 w_1 \\ &= 5.832 w_1 \end{aligned}$$

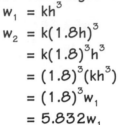

So $w_2 \approx 5.8 w_1$. An ancient cat weighed about 5.8 times as much as a modern cat.

Solution 2

Because weight varies directly as the cube of the height, An equation for the variation function is $w = kh^3$. Now apply the Fundamental Theorem of Variation. This is a direct-variation function with $n = 3$. So, When h is multiplied by 1.8, w is multiplied by $(1.8)^3 = 5.832$. Thus, an ancient cat weighed about 5.8 times as much as a modern cat.

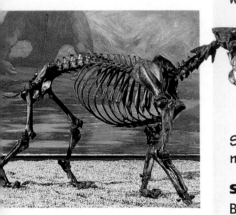

Ancient cat. *Pictured is a skeleton of the extinct Saber-toothed cat. Saber-toothed cats are named for the pair of elongated, blade-like canine teeth—up to 20 cm long—they had in their upper jaw.*

Example 3

Recall that $d = \frac{k}{w}$ is the Law of the Lever. Nathan weighs twice as much as his daughter Stephie. Compare their distances from the fulcrum when they are balanced on a seesaw. Justify your answer.

Solution

Apply the Fundamental Theorem of Variation. Because Nathan weighs twice as much as Stephie, we must find the effect of replacing w with $2w$ in the formula $d = \frac{k}{w}$. This is an inverse-variation function with $n = 1$. So when w is multiplied by 2, d is divided by 2. Thus Nathan's distance from the fulcrum is half of Stephie's distance.

QUESTIONS

Covering the Reading

In 1 and 2, consider the formula $A = \pi r^2$ for the area of a circle as a function of its radius.

1. The pairs $(3, 9\pi)$ and $(6, 36\pi)$ illustrate the pattern that if the circle's radius is doubled, the area is multiplied by __?__.

2. **a.** Find two pairs of numbers that illustrate this result: if the radius is multiplied by five, then the circle's area is multiplied by 25.
   b. Follow the solution to Example 1 and prove the result in part **a.**

3. If $y = kx^n$ and x is multiplied by c, then y is __?__.

4. If $y = \frac{k}{x^n}$ and x is multiplied by c $(c \neq 0)$, then y is __?__.

5. Refer to Example 2. Suppose an ancient cat was 0.9 times as tall as a modern one. Compare the weight of the ancient cat to the weight of a modern cat.

6. Refer to Example 3. Suppose Nathan weighs three times as much as his niece Oprah.
   a. Compare Nathan's and Oprah's distances from the fulcrum when they are balanced.
   b. Justify your answer.

Applying the Mathematics

In 7 and 8, suppose $y = 5x^4$.

7. Describe the change in y when x is tripled. Explain your reasoning.

8. What happens to y when x is divided by three? Find a set of ordered pairs to illustrate your answer.

In 9–11, state the effect that halving the x-values $\left(\text{multiplying them by } \frac{1}{2}\right)$ has on the y-values.

9. $y = 10x$ 10. $y = 10x^2$ 11. $y = \frac{10}{x}$

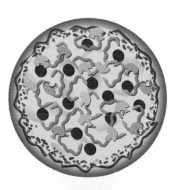

12. For a pizza party, Becki planned to order five 7-inch-diameter pizzas. At the last minute, extra guests were invited and she needed to increase the order.
    a. Find the number of square inches of pizza in the original order.
    b. Suppose Becki doubles the order, that is, she orders ten pizzas instead of five. How many square inches of pizza will she have?
    c. Suppose Becki doubles the size of the pizza to be ordered. That is, she orders five 14-inch-diameter pizzas. How many square inches of pizza will she have?
    d. Explain the difference in parts **b** and **c** in relation to the Fundamental Theorem of Variation.

In 13 and 14, refer to the logos at the left. The radius of the larger logo is twice the radius of the smaller one.

13. What is the ratio of the larger circumference to the smaller?

14. What is the ratio of the larger area to the smaller?

15. How is the volume of a sphere affected if its radius is doubled? Explain how you got your answer.

16. Complete the proof of part **b** of the Fundamental Theorem of Variation.

Review

17. Use the equation $W = \frac{k}{d^2}$, where W = weight of a body and d = its distance from the center of the Earth. Suppose a chimpanzee weighs 50 lb on the surface of the Earth. How much will it weigh when orbiting in the space shuttle *Discovery* 200 miles above the Earth's surface? (Remember, the radius of the Earth is about 4000 miles)   *(Lesson 2-2)*

18. Suppose r varies directly as the third power of s. If $r = 24$ when $s = 8$, find r when $s = 5$.   *(Lesson 2-1)*

19. *Multiple choice.*   Most of the power of a boat's motor goes into generating the wake (the track left in the water). The engine power P used to generate the wake is directly proportional to the seventh power of the boat's speed s. How can you express this relationship?   *(Lesson 2-1)*
   (a) $P = 7s$
   (b) $s = kP^7$
   (c) $P = ks^7$
   (d) $P = k^7s$

20. Solve $5(3m + 5) = 4m - 8$.   *(Lesson 1-5)*

21. Write an algebraic expression which describes each situation.   *(Lesson 1-1)*
   a. One pencil costs c cents. How much do p pencils cost?
   b. Two pencils cost d cents. How much do r pencils cost?

22. *Skill sequence.*   Solve. Remember to find two values for x.   *(Previous course)*
   a. $x^2 = 49$
   b. $36x^2 = 49$
   c. $2x = \frac{49}{2x}$

23. a. Draw $\triangle ABC$ where $A = (-2, -5)$, $B = (-3, 1)$, and $C = (-8, 0)$.
   b. Draw the image of $\triangle ABC$ after reflection over the y-axis.
   (Previous course)

Exploration

24. Complete the table of values at the left for the four functions $y_1 = x^2$, $y_2 = x^3$, $y_3 = x^4$, and $y_4 = x^5$.
   a. Describe some patterns that can be explained by the Fundamental Theorem of Variation.
   b. Describe some other patterns in the table.

x	y_1	y_2	y_3	y_4
1	1	1	1	1
2	4	8	16	32
3				
4				
5				
6				

LESSON
2-4

The Graph of y = kx

This time-lapse photo shows lightning above Tucson, Arizona

Recall from Question 14 in Lesson 2-1 that the distance you are from a lightning strike varies directly with the time elapsed between your seeing the lightning and hearing the thunder. The formula $d = \frac{1}{5}t$ describes this situation. This direct-variation function can also be represented graphically. A table and a graph for the equation $d = \frac{1}{5}t$ are given below.

t = time (in seconds)	5	10	15	20	25	30
d = distance (in miles)	1	2	3	4	5	6

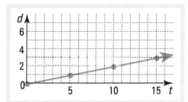

Note that neither distance nor time can be negative in this situation. Thus, the domain of this function is the set of nonnegative real numbers, and the range is also the set of nonnegative real numbers. So, when *all* real-world solutions to the equation $d = \frac{1}{5}t$ are plotted in the coordinate plane, the graph is a ray starting at the origin and passing through the first quadrant. There are no points on the graph in any other quadrants.

Slope of a Line

Recall that the steepness of a line is measured by a number called the *slope*. The slope of a line is the *rate of change* of y with respect to x determined by two points on the line. Let (x_1, y_1) and (x_2, y_2) be the two points. Then, as pictured below, the expression $y_2 - y_1$ is the vertical change (the change in the dependent variable) and $x_2 - x_1$ is the horizontal change (the change in the independent variable). The slope, or rate of change, is the quotient of these changes.

$$\text{slope} = \frac{\text{change in vertical distance}}{\text{change in horizontal distance}}$$
$$= \frac{\text{change in dependent variable}}{\text{change in independent variable}}$$
$$= \frac{\text{rise}}{\text{run}}$$

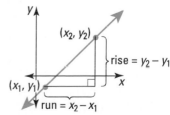

Example

Find the slope of the line with equation $d = \frac{1}{5}t$, where t is the independent variable (time in seconds) and d is the dependent variable (distance in miles).

Solution

Use the definition of slope. Because d is on the vertical axis and t is on the horizontal axis, the ordered pairs are of the form (t, d).

Find two points on the line; either point can be considered (t_2, d_2). Here we use $(t_1, d_1) = (10, 2)$ and $(t_2, d_2) = (15, 3)$.

$$\text{slope} = \frac{d_2 - d_1}{t_2 - t_1} = \frac{3 - 2}{15 - 10}\ \frac{mi}{sec} = \frac{1}{5}\ \frac{mi}{sec}$$

Refer back to the graph at the beginning of this lesson. Notice that for every change of 5 units to the right there is a change of 1 unit up. This is equivalent to saying that for every change of 1 horizontal unit, there is a change of $\frac{1}{5}$ of a vertical unit.

Awesome grade. *This road is in Quebec. A grade of 18% warns the driver that for every horizontal change of 100 meters there will be a vertical change of 18 meters.*

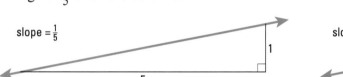

slope $= \frac{1}{5}$ slope $= \frac{\frac{1}{5}}{1} = \frac{1}{5}$

Notice also that the slope $\frac{1}{5}$ of the line $d = \frac{1}{5}t$ is the constant of variation of this direct-variation equation.

Properties of the Function with Equation $y = kx$

In general, the domain of the function with equation $y = kx$ is the set of real numbers. When $k \neq 0$ the range is the set of real numbers. Below are graphs of $y = kx$ for four values of k: 2, $\frac{3}{4}$, $-\frac{3}{4}$, and -2.

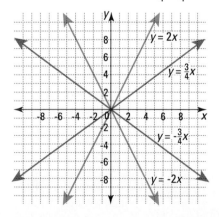

Observe that in each case the graph is a line through the origin with slope k. For example, the slope of $y = 2x$ is 2, and the slope of $y = -\frac{3}{4}x$ is $-\frac{3}{4}$. This is true for all values of k. When $k > 0$, the graph slants up as you read from left to right. When $k < 0$, the graph slants down as you read from left to right.

Theorem
The graph of the direct-variation function $y = kx$ has constant slope k.

Proof:
Let (x_1, y_1) and (x_2, y_2) be two distinct points on $y = kx$, with $k \neq 0$. Then, substitute these values into the variation function.

$$y_1 = kx_1$$
$$y_2 = kx_2$$

Subtract the equations. $\qquad y_2 - y_1 = kx_2 - kx_1$

Use the Distributive Property. $\quad y_2 - y_1 = k(x_2 - x_1)$

Solve for k. $\qquad\qquad\qquad \dfrac{y_2 - y_1}{x_2 - x_1} = k$

So k is the slope.

QUESTIONS

Covering the Reading

1. The slope of a line is found by dividing the change in __?__ distance by the change in __?__ distance between two points on the line.

2. Write a definition of slope in terms of independent and dependent variables.

3. By definition, $\frac{y_2 - y_1}{x_2 - x_1}$ is the slope of the line through which points?

4. What is the slope of the line $d = \frac{1}{5}t$?

5. A slope of $-\frac{3}{4}$ means that for every change of 4 units to the right there is a change of __?__ units __?__; it also means that for every change of 1 horizontal unit there is a vertical change of __?__ of a unit.

6. Use the functions $y = 3x$ and $y = \frac{1}{2}x$.
   a. Complete the table at the left.
   b. On a single set of axes, graph both lines using the values from the table.
   c. What is the slope of the line with equation $y = 3x$?
   d. What is the slope of the line with equation $y = \frac{1}{2}x$?

x	y = 3x	y = ½x
4		
3		
2		
1		
0		
-1		
-2		
-3		
-4		

7. The graph of every direct-variation function $y = kx$ is a __?__, with slope __?__, and passing through the point __?__.

8. Graphs that slant up as you read from left to right have __?__ slope; graphs that slant down as you read from left to right have __?__ slope.

Quick dive. *Pictured is the submarine* Santa Fe. *A submarine dives by flooding its ballast tanks with water. The added weight causes the ship to lose its positive buoyancy. A submarine can dive to a depth of over 100 feet in less than a minute.*

Applying the Mathematics

In 9–12, find the slope of

9. a mountain road which goes up 60 meters for each 1000 meters traveled horizontally.

10. a submarine dive if the submarine drops 2000 feet while moving forward 8000 feet.

11. the line through the points (6, 42) and (0, 0).

12. the line through the points (-2, 8) and (5, -40).

13. The cost c of gasoline varies directly with the number of gallons g bought.
    a. If 15 gallons cost $18.75, determine an equation for the variation function.
    b. Give the domain and range of this variation function.
    c. Find three ordered pairs of the function in part **a**.
    d. Graph these pairs. They should lie on the same ray.
    e. What is the slope of the ray in part **c**?
    f. What does the slope represent in this situation?

14. Match each graph with its equation. On each graph, the x-axis and the y-axis have the same scale.

   I: $y = 3x$ **II:** $y = -3x$ **III:** $y = \frac{1}{3}x$ **IV:** $y = -\frac{1}{3}x$

(a)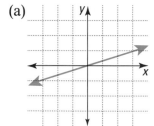

(b)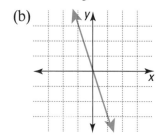

(c)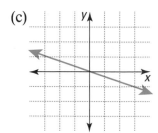

(d)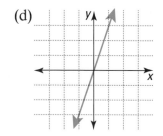

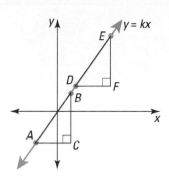

15. At the left is a graph of $y = kx$. Explain how similar triangles can be used to show that the slope of the line is the same no matter which points are chosen to find the slope.

Review

16. In the variation function $W = \frac{k}{d^2}$, what is the effect on W if
  a. d is tripled?
  b. d is halved? *(Lesson 2-3)*

17. Assume that the cost of a spherical ball bearing varies directly as the cube of its diameter. What is the ratio of the cost of a ball bearing 6 mm in diameter to the cost of a ball bearing 3 mm in diameter? *(Lesson 2-3)*

In 18–21, state whether the equation is a function of direct variation, a function of inverse variation, or neither. *(Lessons 2-1, 2-2)*

18. $y = -\frac{8}{x}$

19. $y = -\frac{x}{8}$

20. $y = x - 11$

21. the Law of the Lever

22. Consider the sequence defined recursively as follows. *(Lessons 1-8, 1-9)*
$$\begin{cases} t_1 = 6 \\ t_n = 3t_{n-1} + 1, \text{ for integers } n \geq 2 \end{cases}$$
  a. Describe the sequence in words.
  b. Find the first four terms of the sequence.

23. Ohm's Law, $I = \frac{V}{Z}$, relates current I (in amperes) to voltage V (in volts) and impedance Z (in ohms). *(Lesson 1-6)*
  a. Solve this formula for Z.
  b. Solve this formula for V.

24. *Skill sequence.* Solve for x. *(Lessons 1-5, 1-6)*
  a. $3x = 2$
  b. $3x = 2y$
  c. $3x = 2y + 6$
  d. $3(x + 5) = 2y + 6$

Exploration

25. Each of the following terms is a synonym for "slope." Find out who might use each term.
  a. marginal cost **b.** pitch **c.** grade

IN·CLASS

A C T I V I T Y

*Introduction
to Automatic
Graphers*

Work with a partner or in small groups. Each student needs an automatic grapher.

An **automatic grapher** is either a graphics calculator or function graphing software for use on a computer. Of course, no grapher is completely automatic. Here we discuss what you need to know in order to use any automatic grapher. Consult your calculator owner's manual or your function grapher's documentation for specific information about your grapher.

The part of the coordinate plane that shows on your grapher's screen is called a **window.** Every automatic grapher has a **default window,** that is, a window that is set by the manufacturer. The default window that is used on one automatic grapher is shown below.

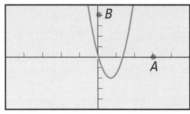

$-15 \leq x \leq 15, \quad x\text{-scale} = 3$
$-10 \leq y \leq 10, \quad y\text{-scale} = 2$

A description of the window is given below the display. Here, x goes from -15 to 15, and y goes from -10 to 10.

x-scale = 3 means that the tick marks along the x-axis are 3 units apart. Point A is (9, 0).

y-scale = 2 means that the tick marks along the y-axis are 2 units apart. Point B is (0, 8).

You should include a description like this whenever you sketch the output from your grapher.

1 **a.** Sketch the default window for your grapher.
   b. Describe one way to change the window on your grapher.

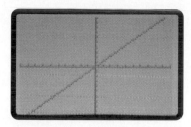
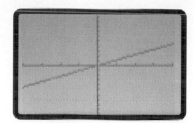

2 Find out how to enter equations on your grapher.
  a. Draw a graph of $y = -8x$ using each of the following windows.
  i. your default window
  ii. $-15 \leq x \leq 15$, x-scale = 3; $-10 \leq y \leq 10$, y-scale = 2
  iii. $-3 \leq x \leq 3$, x-scale = .5; $-30 \leq y \leq 30$, y-scale = 5
  iv. $-1.5 \leq x \leq 3.5$, x-scale = .25; $-5 \leq y \leq 5$, y-scale = 1
  b. What is the slope of each line in part **a?**
  c. Discuss the graphs with other members of your group. How do the dimensions of the viewing window affect your visual impression of the graph? Write a few sentences that describe your conclusion. Include sketches of the graphs you just made.
  d. Find out how to trace along the graph. Use the *trace* feature to determine the following values.
  i. Find y when $x = 1.7$.
  ii. Find x when $y = -12$.
  How close are your grapher's answers to those you get by calculating with paper and pencil?

3 Clear your screen. Use any convenient window. Graph $y = x$, $y = 3x$, and $y = 9x$ on one set of axes. Copy these onto another sheet of paper.

4 Clear your screen. Set it to the default window. Then do the following: enter an equation of the form $y = kx$, where $k \neq 0$, without letting the others in your group see the equation. Show your screen to the others, and have them identify what equation you used.

5 Is your automatic grapher capable of printing a *hard copy,* that is, a paper copy of your screen? If so, find out how to do this. Print a hard copy of one of the graphs in Questions 2–4.

The Graph of $y = kx^2$

Seeing red. *If the red car is traveling at a speed of 35 mph, has the driver put on the brakes soon enough to stop before the red light? One of the first things a student driver learns is how to use the brakes to stop a car safely and smoothly.*

Rate of Change

In Lesson 2-1, you learned that the distance it takes a car to stop after the brakes are applied varies directly with the square of the car's speed. The formula $d = \frac{1}{16}s^2$ describes this relation for a certain car. A table and a graph are given below.

s	d
0	0
10	6.25
20	25
30	56.25
40	100
50	156.25
60	225
70	306.25

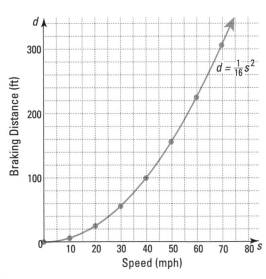

Notice that the points do not all lie on a straight line. You can verify this by calculating the rate of change between different points on the graph.

Example 1

Find the following rates of change, and explain what each means.

a. r_1, the rate of change from (20, 25) to (40, 100)

b. r_2, the rate of change from (40, 100) to (60, 225)

Solution

a. Use the definition of slope.

$$r_1 = \frac{100 \text{ ft} - 25 \text{ ft}}{40 \text{ mph} - 20 \text{ mph}} = \frac{75 \text{ ft}}{20 \text{ mph}} = \frac{3.75 \text{ ft}}{\text{mph}}$$

This means that on the average, when driving between 20 mph and 40 mph, for every increase of 1 mph in speed, you need 3.75 more feet to stop your car.

b. Similarly, $r_2 = \frac{225 \text{ ft} - 100 \text{ ft}}{60 \text{ mph} - 40 \text{ mph}} = \frac{125 \text{ ft}}{20 \text{ mph}} = 6.25 \frac{\text{ft}}{\text{mph}}$.

So on the average, between $s = 40$ and $s = 60$, for every change of 1 mph (the horizontal unit), there is a change of 6.25 feet of braking distance (the vertical unit).

Check

Look at the points on the graph. Let $A = (20, 25)$, $B = (40, 100)$, and $C = (60, 225)$. Is $\overline{BC}$ steeper than $\overline{AB}$? Yes, it is.

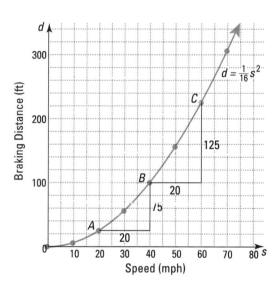

Because the rate of change determined by different pairs of points on the graph of $d = \frac{1}{16}s^2$ is not constant, two conclusions can be drawn:

1. The graph of $d = \frac{1}{16}s^2$ is not a line.

2. The steepness of the graph cannot be described by a single number.

Notice that the slope is larger where the graph is steeper; this means that the function is increasing faster.

The equation $d = \frac{1}{16}s^2$ is a direct-variation function of the form $y = kx^2$. All graphs of equations of this form share some properties.

The Graph of $y = kx^2$ when $k > 0$

Example 2

Graph solutions to $y = 2x^2$, $y = x^2$, and $y = \frac{1}{4}x^2$.

Solution 1

Use an automatic grapher with any window that allows you to see each graph clearly. Shown below is a good display and the window used to create it.

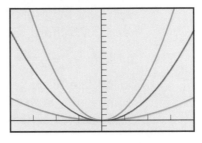

$$-4 \le x \le 4, \quad x\text{-scale} = 1$$
$$-2 \le x \le 20, \quad y\text{-scale} = 1$$

When you use an automatic grapher to draw a graph, you should also either print a hard copy or copy the graph by hand. In either case you should always label the axes, include a scale, and plot a few points. When you make two or more graphs on the same axes, be sure to label the graphs with the equations you used. Your paper should look something like either graph below.

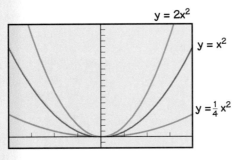

$$-4 \le x \le 4, \quad x\text{-scale} = 1$$
$$-2 \le x \le 20, \quad y\text{-scale} = 1$$

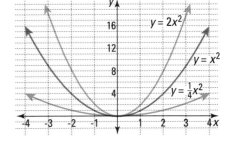

Solution 2

Make a table of values. To save space, the value of the independent variable is written only once.

x	0	1	2	3	-1	-2	-3
$y = x^2$	0	1	4	9	1	4	9
$y = 2x^2$	0	2	8	18	2	8	18
$y = \frac{1}{4}x^2$	0	$\frac{1}{4}$	1	$\frac{9}{4}$	$\frac{1}{4}$	1	$\frac{9}{4}$

Then plot the points. The graph should look something like the one drawn by hand in Solution 1.

The graphs of $y = x^2$, $y = 2x^2$, and $y = \frac{1}{4}x^2$ in Example 2 are curves called *parabolas*. Each parabola passes through the point (0, 0). Further, each parabola coincides with its reflection image over a line, specifically, the *y*-axis. So, each parabola is **reflection-symmetric,** and the *y*-axis is called the **line of symmetry.**

The Graph of $y = kx^2$ when $k < 0$

Example 3 shows graphs of $y = kx^2$ for three negative values of *k*.

Example 3

Graph $y = -2x^2$, $y = -x^2$, and $y = -\frac{1}{4}x^2$ on the same axes.

Solution

We use an automatic grapher with the window $-4 \leq x \leq 4$ and $-20 \leq y \leq 2$. On the left below is the grapher's display. On the right below is a graph drawn from the display.

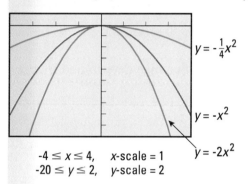

$-4 \leq x \leq 4$,   *x*-scale = 1
$-20 \leq y \leq 2$,   *y*-scale = 2

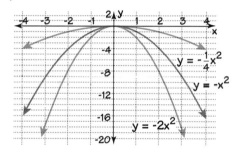

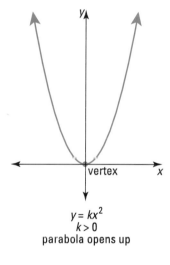

$y = kx^2$
$k > 0$
parabola opens up

$y = kx^2$
$k < 0$
parabola opens down

When you are asked to *copy* a graph from your automatic grapher, you must label the axes, include a scale, and label the equations used. However, you don't need to identify a large number of points. Usually it is enough to name those points that highlight key features of the graph. You can find those points by substitution or by using the TRACE feature of your grapher.

Domain and Range

In general, the domain of the function with equation $y = kx^2$ is the set of all real numbers. When $k > 0$, the range is the set of nonnegative real numbers, and the parabola *opens up*. That is, the vertex of the parabola is its *minimum* point. When $k < 0$, the range is the set of nonpositive real numbers and the parabola *opens down*. That is, the vertex of the parabola is its *maximum* point.

Example 4

How are the graphs in Example 3 like and unlike those in Example 2?

Solution

The graphs in Example 3 $\left(y = -2x^2, y = -x^2, \text{ and } y = -\frac{1}{4}x^2\right)$ are shaped like those in Example 2 $\left(y = 2x^2, y = x^2, \text{ and } y = \frac{1}{4}x^2\right)$. Each curve passes through the origin and is symmetric to the y-axis. In each case, the graph of $y = ax^2$ appears to be congruent to the graph of $y = -ax^2$. However, the curves in Example 3 open down, while those in Example 2 open up.

QUESTIONS

Covering the Reading

In 1 and 2, refer to the formula $d = \frac{1}{16}s^2$ relating speed and braking distance.

1. Find the rate of change determined by the pair of points and explain what it means.
   a. (10, 6.25) and (20, 25)
   b. (50, 156.25) and (60, 225)

2. *True or false.*   The rate of change on the graph is constant, regardless of the points used.

3. The graph of $y = kx^2$ ($k \neq 0$) is what type of curve?

4. Consider the equations $y = x^2$, $y = 3x^2$, and $y = -3x^2$.
   a. Make a table of values for $x = -2, -1, 0, 1,$ and 2.
   b. Graph the solutions on one set of axes. Use as the domain the set of real numbers between -2 and 2 inclusive. Do not use an automatic grapher.
   c. Compare and contrast the three graphs.

5. Explain what it means to say that the graph of $y = kx^2$ is symmetric to the y-axis.

6. Suppose $k < 0$. State the domain and range of the function $f(x) = kx^2$.

7. In general, for what values of k does the graph of $y = kx^2$
   a. open up?
   b. open down?

8. Match each graph with its equation. Each graph has the same scale.

$y = \frac{1}{2}x^2$ 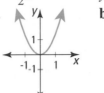 $y = -2x$ $y = -x^2$ $y = 1.3x^2$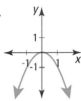

a. **b.** **c.** **d.**

9. Explain why the point $(0, 0)$ is on the graph of $y = kx^2$ for all values of k.

10. **a.** Use an automatic grapher to graph $y = \frac{1}{5}x^2$ on the window $-10 \le x \le 10$ and $-10 \le y \le 10$.
    b. Use the trace feature of your grapher to estimate the value of the function when $x = 3$.
    c. Compare the value obtained in part **b** with the value obtained by substituting $x = 3$ into $y = \frac{1}{5}x^2$.
    d. Use the trace feature to find the value(s) of x when $y = 10$.

11. Let N represent the number of houses that can be served by a water main of diameter d centimeters. Since the amount of water that can flow through a pipe is directly proportional to the area of a cross section of pipe, it is reasonable to assume $N = kd^2$. Suppose
$$N = \frac{1}{2}d^2.$$
    a. Make a table of solutions to this equation. For values of d use 0, 10, 20, 30, and 40.
    b. From your table, estimate the number of houses that can be served by a water main of diameter 35 cm.
    c. Graph $N = \frac{1}{2}d^2$.
    d. From your graph, estimate the number of homes that can be served by a water main of diameter 35 cm.
    e. How well do your estimates from parts **b** and **d** agree?
    f. From the variation equation, determine the actual number of homes that can be served by a water main of diameter 35 cm.

The pressure's on. *Most cities pump water into elevated storage tanks, such as this one in Stanton, Iowa, to help keep the water pressure high. When water is released from the tank, gravity pulls the water downward, giving it the pressure to rush through the water mains.*

12. Refer to the table and graph of $y = x^2$ in Example 2.
    a. Find the rate of change:
       i. from $(0, 0)$ to $(1, 1)$.   **ii.** from $(1, 1)$ to $(2, 4)$.
       iii. from $(2, 4)$ to $(3, 9)$.   **iv.** from $(3, 9)$ to $(4, 16)$.
    b. Use your results from **i–iv** in part **a** to make a conjecture about the rate of change between the points (n, n^2) and $(n + 1, (n + 1)^2)$.
    c. Prove your conjecture by calculating the rate of change for the points in part **b**.

Sound architecture. *The Sydney Opera House, internationally famous for its innovative design, contains a 1700-seat main hall, a 1550-seat auditorium, and a small theater.*

13. **a.** Graph the following four equations on one set of axes. *(Previous course, Lesson 2-4)*

$$y = 4x \qquad y = \tfrac{1}{4}x \qquad y = -\tfrac{1}{4}x \qquad y = -4x$$

   b. Find the slope of each graph.
   c. Give the equations of two lines that appear to be perpendicular.

14. What is another name for slope? *(Lesson 2-4)*

15. Architects designing auditoriums use the fact that sound intensity I is inversely proportional to the square of the distance d from the sound source. *(Lessons 2-2, 2-3)*
   a. Write the variation equation that represents this situation.
   b. A person moves to a seat 4 times as far from the source. How will the intensity of sound be affected?

16. The Fahrenheit and Celsius scales indicate temperature. Temperature can also be measured on the Kelvin scale, in kelvins. At a fixed pressure, the volume V of a fixed amount of air varies directly with its Kelvin temperature t. The lowest possible temperature occurs when t is zero, about -273°C. Suppose that a balloon contains 7.5 liters of air at 300 kelvins (about room temperature). *(Lesson 2-1)*
   a. Write a variation formula for V in terms of t.
   b. Use this formula to predict the volume of air in the balloon at temperatures of 400, 500, 600, and 1000 kelvins.

17. Consider a sequence defined explicitly as

$$t_n = 4n^2 + 3n - 2.$$

   a. What is the domain of the sequence?
   b. Find the first five terms of the sequence. *(Lessons 1-2, 1-7)*

18. Solve for x: $y = -\tfrac{1}{4}x$. *(Lesson 1-6)*

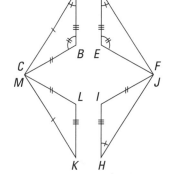

19. **a.** Which of the triangles pictured at the left can be proved congruent to $\triangle ABC$? Name them, with vertices in correct order.
   b. Give the justification for each triangle congruence you find. *(Previous course)*

20. Accurately graph the curve $y = x^2$ from $x = 0$ to $x = 3$. Estimate the area bounded by the curve, the x-axis, and the line $x = 3$. The area between the curve and the x-axis is an integer number of square units. What is this number?

IN·CLASS
ACTIVITY

*Automatic
Graphers
and Inverse
Variation*

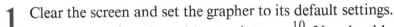

Work on this Activity with a partner. Each pair needs at least one automatic grapher.

1 Clear the screen and set the grapher to its default settings.
  a. Graph the function with equation $y = \frac{10}{x}$. You should see points in two quadrants. Sketch the graph.
  b. Set the cursor on the graph at some positive value of x by using the trace key. Trace along the curve to the right. Describe what happens to y as x gets larger and larger. What is the value of y when $x = 200$?
  c. Set the cursor on the graph at some negative value of x. Trace along the curve to the left. Describe what happens to y as x gets smaller and smaller. What is the value of y when $x = -200$?
  d. Does the graph of $y = \frac{10}{x}$ ever intersect the x-axis? If yes, give the coordinates of the point(s) of intersection. If no, explain why not.
  e. Does the graph of $y = \frac{10}{x}$ ever intersect the y-axis? If yes, give the coordinates of the point(s) of intersection. If no, explain why not.

2 Clear the screen.
  a. Graph the function with equation $y = \frac{10}{x^2}$. Sketch the graph.
  b. Place the cursor on the graph at some positive value of x. Trace along the curve to the right. Describe what happens to y as x gets larger and larger.
  c. Set the cursor on the graph at some negative value of x. Trace along the curve to the left. Describe what happens to y as x gets smaller and smaller.
  d. Does the graph of $y = \frac{10}{x^2}$ ever intersect the x-axis? If yes, give the coordinates of the point(s) of intersection. If no, explain why not.
  e. Does the graph of $y = \frac{10}{x^2}$ ever intersect the y-axis? If yes, give the coordinates of the point(s) of intersection. If no, explain why not.

3 In Questions 1 and 2 you investigated graphs of functions with equations of the forms $y = \frac{k}{x}$ and $y = \frac{k}{x^2}$. In those questions you set $k = 10$. Suppose $k = -10$.
  a. Make a conjecture about what you think the graphs will look like.
  b. Test your conjecture by graphing $y = -\frac{10}{x}$ and $y = -\frac{10}{x^2}$. Sketch the graphs, and write a sentence about some properties of each function.

LESSON 2-6

The Graphs of $y = \dfrac{k}{x}$ and $y = \dfrac{k}{x^2}$

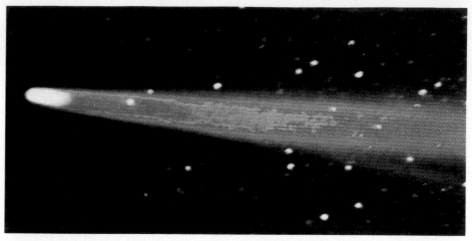

Energy deficit. *Periodic comets, like* Comet P/Halley *shown above, have negative total energy and so travel along an ellipse. Comets with positive total energy travel along a hyperbola and are seen only once, never to return.*

The Graph of $y = \dfrac{k}{x}$

The graph of every function with an equation of the form $y = \dfrac{k}{x}$, where $k \neq 0$, is a *hyperbola*. In the preceding In-class Activity you graphed the hyperbolas with equations $y = \dfrac{10}{x}$ and $y = -\dfrac{10}{x}$.

Example 1

a. Draw the graphs of $f(x) = \dfrac{16}{x}$ and $g(x) = -\dfrac{16}{x}$.

b. Identify the domain and range of the function f.

Solution

a. Use an automatic grapher.

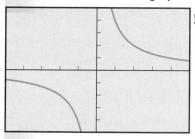
$y = \dfrac{16}{x}$

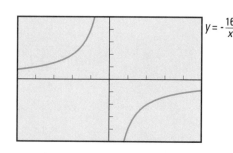
$y = -\dfrac{16}{x}$

$-10 \leqslant x \leqslant 10, \quad x\text{-scale} = 2$
$-10 \leqslant y \leqslant 10, \quad y\text{-scale} = 2$

$-10 \leqslant x \leqslant 10, \quad x\text{-scale} = 2$
$-10 \leqslant y \leqslant 10, \quad y\text{-scale} = 2$

b. To find the domain, think: what numbers can x be? In the equation $f(x) = \dfrac{16}{x}$, any real number except 0 can be substituted for x. So, For the function f, the domain = {x: x ≠ 0}. To find the range, think: what values can y have? From the graph, notice that y can be either positive or negative, large or small. But if $y = 0$, you would have $0 = \dfrac{16}{x}$ or $0 \cdot x = 16$ which is impossible. So, For the function f, the range = {y: y ≠ 0}.

Notice that each hyperbola consists of two separate parts, called **branches.** When $k > 0$, the branches of $y = \frac{k}{x}$ lie in the first and third quadrants; if $k < 0$, the branches lie in the second and fourth quadrants. In each case the domain and the range are the set of all nonzero real numbers.

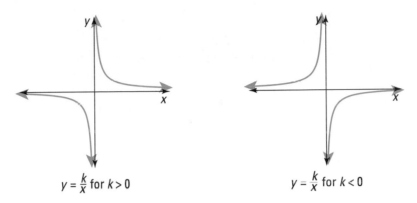

$y = \frac{k}{x}$ for $k > 0$ $y = \frac{k}{x}$ for $k < 0$

In some situations, only one branch of a hyperbola is relevant. For instance, recall the Metro Car Sales example of Lesson 2-2. The number of students s hired to wash cars and the number of hours t each will need to work are related by the equation $t = \frac{36}{s}$. A table of values for this equation and a graph are shown below.

s	1	2	3	4	5	6	7	8	9	10	12	18	24	30	36
t	36	18	12	9	$7\frac{1}{5}$	6	$5\frac{1}{7}$	$4\frac{1}{2}$	4	$3\frac{3}{5}$	3	2	$1\frac{1}{2}$	$1\frac{1}{5}$	1

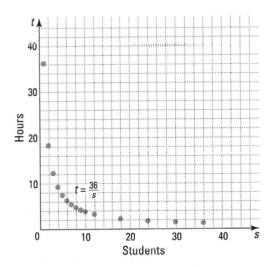

The domain is the set of natural numbers. Notice that it does not make sense to connect the points of this graph. Thus, the graph consists of a set of **discrete** (distinct or not connected) points on one branch of the hyperbola $y - \frac{36}{x}$.

The Graph of $y = \frac{k}{x^2}$

You have also studied inverse-square variation. The graph of an inverse-square variation function does not have a special name, so we just call it an **inverse-square curve.**

Example 2

a. Graph $y = \frac{16}{x^2}$ and $y = -\frac{16}{x^2}$.

b. Describe the symmetry in each graph.

Solution

a. Use an automatic grapher.

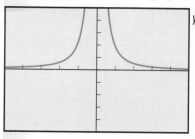

$y = \frac{16}{x^2}$

$-10 \leqslant x \leqslant 10, \quad x\text{-scale} = 2$
$-10 \leqslant y \leqslant 10, \quad y\text{-scale} = 2$

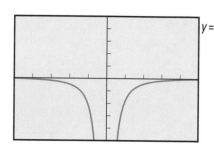

$y = -\frac{16}{x^2}$

$-10 \leqslant x \leqslant 10, \quad x\text{-scale} = 2$
$-10 \leqslant y \leqslant 10, \quad y\text{-scale} = 2$

b. Each graph is symmetric to the y-axis.

Notice that the inverse-square curve, like a hyperbola, has two distinct branches. These two branches, however, do not form a hyperbola because the shape and relative positions of the branches differ from those of the branches of a hyperbola. The domain of every inverse-square function is $\{x: x \neq 0\}$. The range depends on the value of k.

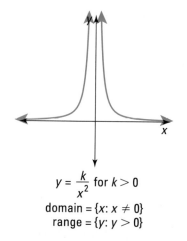

$y = \frac{k}{x^2}$ for $k > 0$
domain $= \{x: x \neq 0\}$
range $= \{y: y > 0\}$

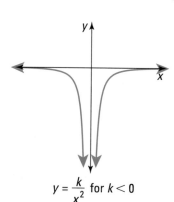

$y = \frac{k}{x^2}$ for $k < 0$
domain $= \{x: x \neq 0\}$
range $= \{y: y < 0\}$

Asymptotes

In general, when $x = 0$, $y = \frac{k}{x}$ and $y = \frac{k}{x^2}$ are undefined. So neither curve crosses the y-axis. However, when x is near 0, the functions are defined.

Activity

Graph $y = \frac{16}{x}$ on the window $-1 \le x \le 1$, $-100 \le y \le 100$. You should see something like the curve below.

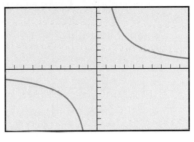

$-1 \le x \le 1$, x-scale = .1
$-100 \le y \le 100$, y-scale = 10

a. Set the cursor on the branch of the hyperbola in the first quadrant near $x = 1$. Trace to the left. Thus, the cursor will approach the line $x = 0$ from the right. Record the coordinates of three points on the hyperbola.

b. Set the cursor on the other branch of the hyperbola near the line $x = -1$. Trace so that the cursor approaches 0 from the left. Record the coordinates of three points on the curve.

You should have found in the Activity that, as the cursor gets closer and closer to $x = 0$ from the right, the y-value gets larger and larger without bound. Similarly, as the cursor gets closer and closer to $x = 0$ from the left, the y-value gets smaller and smaller without bound. If as the values of x get closer and closer to a vertical line, the values of the function get larger and larger without bound (or smaller and smaller without bound), that line is called a **vertical asymptote** of the function. The y-axis is a vertical asymptote to the graphs of $y = \frac{k}{x}$ and $y = \frac{k}{x^2}$, provided $k \ne 0$.

Similarly, the x-axis is a **horizontal asymptote** to the graph of $y = \frac{k}{x}$ and $y = \frac{k}{x^2}$. As x gets very very large (or very very small), the value of y gets closer and closer to the x-axis.

In general, asymptotes may be vertical, horizontal, or oblique. But not all graphs have asymptotes. For instance, the parabola $y = kx^2$ does not have any asymptotes. When a graph has an asymptote, that asymptote is *not* part of the graph.

Covering the Reading

1. Suppose $k \neq 0$. What is the graph of $y = \frac{k}{x}$ called?

In 2 and 3, consider the function f with equation $f(x) = \frac{12}{x}$.

2. **a.** Make a table of values. Include at least six ordered pairs in the first quadrant.
   b. Draw a graph of this function for $-12 \leq x \leq 12$ and $-12 \leq y \leq 12$.

3. State the domain and range of f.

In 4 and 5, refer to Example 1.

4. In which quadrants are the branches of the graph of $g(x) = -\frac{16}{x}$?

5. State the domain and range of g.

6. Refer to the graph of $t = \frac{36}{s}$ in this lesson.
   a. Why doesn't it make sense to connect the points on this graph?
   b. Why are there no points in the 2nd or 3rd quadrants?

7. Suppose $k \neq 0$. Tell whether the equation has a graph that is symmetric to the y-axis.
   (a) $y = \frac{k}{x}$ (b) $y = \frac{k}{x^2}$ (c) $y = kx$ (d) $y = kx^2$

8. In which quadrants are the branches of $y = \frac{k}{x^2}$
   a. if k is positive?   **b.** if k is negative?

9. **a.** Explain what happens to the y-coordinate of the graph of $y = \frac{16}{x}$ when x is negative and is getting closer and closer to 0. Use your work from the lesson's Activity.
   b. Identify two asymptotes of the graph of $y = \frac{16}{x}$.

10. Does the graph of $y = \frac{16}{x^2}$ have any asymptotes? If so, give an equation for each.

Applying the Mathematics

11. Ian is on a seesaw. He weighs 40 pounds and is sitting 5 feet from the fulcrum. $\left(\text{Remember, the Law of the Lever is } d = \frac{k}{w}.\right)$
    a. Find k and write a variation equation for this situation.
    b. Make a table of weights and distances from the pivot that would balance Ian.
    c. Plot your values from part **b.**
    d. Should you connect your points with a smooth curve? Explain why or why not.

12. Examine the graph of $y = \frac{16}{x}$ from Example 1.

    a. How many symmetry lines does the graph have?

    b. Write an equation for each symmetry line.

    c. Does the graph of $y = -\frac{16}{x}$ have the same symmetry lines? If not, what are the equations for its symmetry line(s)?

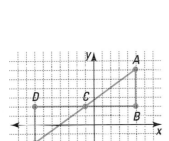

-10 ≤ x ≤ 10,   x-scale = 5
-10 ≤ y ≤ 10,   y-scale = 5

13. At the left is the graph of $f(x) = \frac{1}{x^2}$ on the window $-10 \le x \le 10$ and $-10 \le y \le 10$ as drawn by an automatic grapher. Is there a value of x with $f(x) = 0$? Justify your answer.

14. a. Use an automatic grapher to draw a graph of $y = \frac{24}{x}$.

    b. Find the rate of change from $x = 2$ to $x = 6$ for $y = \frac{24}{x}$.

    c. Use an automatic grapher to draw a graph of $y = \frac{24}{x^2}$.

    d. Find the rate of change from $x = 2$ to $x = 6$ for $y = \frac{24}{x^2}$.

    e. Which of the two graphs is falling faster from $x = 2$ to $x = 6$?

Review

15. In the figure at the left, parabolas P_1 and P_2 are congruent. If parabola P_1 has equation $y = 6x^2$, what is an equation for parabola P_2? *(Lesson 2-5)*

16. a. Draw a line that passes through the origin and has slope $\frac{5}{3}$.

    b. Write an equation for the line in part **a.** *(Lesson 2-4)*

17. Is line m graphed at the left an example of a direct variation? Justify your answer. *(Lesson 2-4)*

18. A small spherical balloon has a diameter of 6 inches when blown up. A larger spherical balloon has a diameter of 10 inches.

    a. How many more cubic inches of air are needed to blow up the large balloon than is needed for the small one?

    b. How many times the amount of air is in the larger balloon than is in the smaller balloon? *(Lesson 2-3)*

19. When she tried to solve the equation $\frac{1}{2}x + \frac{1}{3}x + 5 = 10$, Mikki's first step was $3x + 2x + 30 = 60$.

    a. Explain what Mikki did.

    b. Finish the solution to the equation. *(Lesson 1-5)*

20. In the graph at the left, the grid lines are 1 unit apart. Each labeled point is at the intersection of grid lines. Are triangles ABC and EDC congruent? Explain your answer. *(Previous course)*

Exploration

21. a. Use an automatic grapher to draw the graph of $y = 3x + \frac{1}{x}$.

    b. Identify the asymptote(s) of the graph in part **a.**

Under pressure. *Scuba divers, like this one at Poor Knights Island, New Zealand, can safely dive about 18 meters (59 feet).*

The following table and graph give the water pressure (in pounds per square inch, or psi) exerted on a diver at various depths (in feet).

depth of diver (ft)	0	10	25	40	55	75
pressure on diver (psi)	0	4.3	10.8	17.2	23.7	32.3

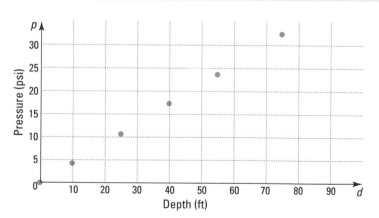

Because the pressure on the diver depends on his or her depth, pressure is the dependent variable and is placed on the vertical axis. Notice that the water pressure on a deep-sea diver increases as the diver goes deeper. How is the pressure related to the depth of water? The points seem to lie on a line through the origin. Therefore, it seems appropriate to describe the relation between these variables by saying the pressure varies directly as the depth. It makes sense that the origin is on this line because on the surface—that is, 0 feet under the water—the water pressure is 0 pounds per square inch (psi).

If p represents the pressure and d represents the depth, the formula for this variation is $p = kd$. The constant k can be determined from one of the data points. For instance, substitute $p = 4.3$ and $d = 10$ into the equation.

$$4.3 = k \cdot 10$$
$$k = 0.43 \left(\text{The unit is } \tfrac{\text{psi}}{\text{ft}} . \right)$$

So, the relation between p and d can be expressed as

$$p = 0.43d.$$

It is important to check that this formula holds for all the data in the table. One way to do this is to make a table for the function $p = 0.43d$. For instance, if $d = 25$, then $p = 0.43(25) = 10.75$.

Shown below is a table for $p = 0.43d$ using the values of d given in the table on page 110.

d	0	10	25	40	55	75
p	0	4.3	10.75	17.2	23.65	32.25

Notice that the ordered pairs in this table are very close to the experimental data given in the table on page 110. This suggests that this function describes or *models* the situation well.

The equation $p = 0.43d$ is a *mathematical model* of the real-world relation between pressure and depth. A **mathematical model** for a real situation is a description of that situation using the language and concepts of mathematics. A good model holds true for all the given information. The formula $p = 0.43d$ gives very close approximations for all the values in the table so it is a good model for this diving situation.

The model $p = 0.43d$ makes it possible to predict the pressure on a diver at depths other than those given in the table. At a depth of 125 ft, for instance, the model predicts that the pressure on a diver would be

$$p = (0.43)(125) = 53.75.$$

That is, the pressure would be about 54 psi.

Here is another situation whose mathematical model involves variation.

Example

X. Perri Menter was investigating the relation between the volume and pressure of a gas in a laboratory. While she held the temperature of the gas constant, she varied the pressure (the independent variable) and measured the volume (the dependent variable) to obtain the following data, which are graphed below.

Pressure P (in psi)	20	30	40	50	60	70	80
Volume V (in ft³)	83	55	42	33	28	24	21

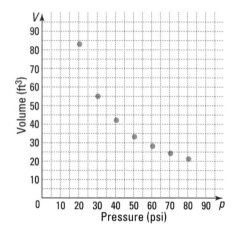

The shape of the graph suggests two possible models: V varies inversely as P or inversely as the square of P.

a. Does $V = \frac{k}{P^2}$ model the data? Justify your answer.

b. Does $V = \frac{k}{P}$ model the data? Justify your answer.

c. Predict the volume of gas if the pressure is 45 psi.

Solution

a. To test $V = \frac{k}{P^2}$, first substitute the coordinates of one data point to find k. For instance, when $P = 20$ psi, $V = 83$ ft³. So

$$83 = \frac{k}{(20)^2}$$
$$k = 33{,}200.$$

Next, decide whether the equation

$$V = \frac{33{,}200}{P^2}$$

is valid by making a table for this function.

P	$V = \frac{33{,}200}{P^2}$
20	83
30	36.889
40	20.75
50	13.28
60	9.2222
70	6.7755
80	5.1875

▶

Notice that Except for the pair (20, 83) in the table, none of the other values of V predicted by the equation $V = \frac{33,200}{P^2}$ fit the experimental data. So, $V = \frac{k}{P^2}$ is not a good model for these data.

b. To test if $V = \frac{k}{P}$ is a correct model, again substitute 20 psi for P and 83 ft³ for V. For these values, k = 1660, and the model is
$$V = \frac{1660}{P}.$$
Now check whether this equation is valid for all the data of the experiment by making a table for the function.

P	$V = \frac{1660}{P}$
20	83
30	55.33
40	41.5
50	33.2
60	27.67
70	23.71
80	20.75

Notice that The values of V are close to the experimental data. Thus, $V = \frac{1660}{P}$ is a good model for X. Perri Menter's data.

c. Substitute P = 45 psi into the model $V = \frac{1660}{P}$. Then
$$V = \frac{1660}{45} = 36.\overline{8}, \text{ or about 37. So the model predicts a}$$
volume of 37 ft³ at 45 psi.

Check

b. Use the Fundamental Theorem of Variation. If V varies inversely with P, then if V is doubled, P is divided by 2. Some ordered pairs are (20, 83), (40, 42), and (80, 21). As the P-coordinate doubles, the V-coordinate is halved. It checks.

c. Use the graph. The point (45, 37) appears to fit on the graph.

Masked man. *The Heliox Band mask, worn by this diver, was designed for deep dives. The divers must ascend to the surface slowly. The pressure in their lungs and other air spaces must equal the water pressure to avoid injury.*

QUESTIONS

Covering the Reading

In 1–3, refer to the data about deep-sea diving.

1. Describe in words the variation relationship between the pressure and the depth.

2. Use the model $p = .43d$ to predict the pressure on a diver who is 130 ft below the surface.

3. The maximum pressure that a deep-sea diver can withstand without using special equipment is about 65 psi. How deep can a diver go below the surface without special equipment?

4. Define *mathematical model*.

In 5–7, refer to the Example.

5. *True or false.* As the pressure on the gas increases, the volume of the gas increases.

6. Why did we conclude that $V = \frac{k}{P^2}$ is not a good model of this situation?

7. Use the correct model to predict the volume of a gas under a pressure of 18 psi.

Applying the Mathematics

8. Refer to the situation at the beginning of this lesson. Kwame found the constant of variation for the equation $p = kd$ as follows:

$$\frac{p_2 - p_1}{d_2 - d_1} = \frac{23.7 - 17.2}{55 - 40} = \frac{6.5}{15} \approx .43.$$

Explain what Kwame did and tell whether it is a valid method.

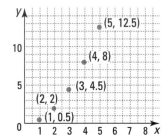

9. Refer to the graph at the left.
  a. *Multiple choice.* Which of the following equations could be a good model for this graph?
   I: $y = kx$ **II:** $y = kx^2$ **III:** $y = \frac{k}{x}$ **IV:** $y = \frac{k}{x^2}$
  b. Find the constant k for your model.
  c. Test your model to see if y is 8 when x is 4.
  d. Use your model to predict the value of y when x is 6.

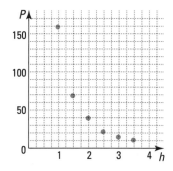

10. a. *Multiple choice.* Which formula best models the graph at the left?
   I: $P = kh$ **II:** $P = kh^2$
   III: $P = \frac{k}{h}$ **IV:** $P = \frac{k}{h^2}$
  b. Justify your answer.
  c. Predict the value of P when $h = 0.5$

11. A scientist dropped a ball from a cliff and used a slow motion film to determine the distance it fell over different periods of time. The data are summarized below.

t seconds	1	2	3	4	5
d meters	4.9	19.6	44.1	78.4	122.5

  a. Draw a graph to represent these data. Let t, the time in seconds, be the independent variable and d, the distance in meters, be the dependent variable. (Use a large enough scale on the d-axis to handle 122.5.)
  b. What variation equation appears to model this situation? Use one data point to calculate k and check the model with other data points. Revise your variation equation if necessary.
  c. Predict how far the ball would fall in 4.5 sec.

In 12–15, match each graph to its equation. The scales on the axes are the same for all four graphs. *(Lessons 2-4, 2-5, 2-6)*

(a) $y = 3x$ (b) $y = \frac{3}{x}$ (c) $y = \frac{3}{x^2}$ (d) $y = -\frac{3}{x}$ (e) $y = -\frac{1}{3}x^2$

12.

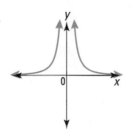

13.

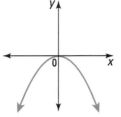

14.

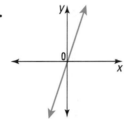

15.

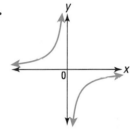

16. Find the rate of change determined by the points on $y = \frac{15}{x}$ where $x = 1$ and $x = 11$. *(Lessons 2-4, 2-6)*

17. Suppose that the value of x is halved. How is the value of y changed if y is directly proportional to x^4? *(Lesson 2-3)*

18. The table at the left gives the population of the United States from 1980 to 1992. *(Lessons 1-2, 1-3)*
   a. Let f be the function that maps the year y onto the population p. Find $f(1988)$.
   b. For what year y is $f(y) = 238,466,000$?

19. The formula $C = \frac{Wtc}{1000}$ gives the cost C of operating an electrical device that uses W watts and runs for t hours at a cost of c cents per kilowatt-hour. Find the cost of running a 550-watt microwave for 10 minutes if the electric company charges 10.819 cents per kilowatt-hour. *(Lesson 1-1)*

Year	Population (in thousands)
1980	227,726
1981	229,966
1982	232,188
1983	234,307
1984	236,348
1985	238,466
1986	240,651
1987	242,804
1988	245,021
1989	247,342
1990	249,900
1991	252,671
1992	255,462

Exploration

20. Look in a chemistry, physics, or general-science text. Find a formula relating water pressure to depth of an object, or pressure to volume of a gas. How does that formula relate to the ones in this lesson?

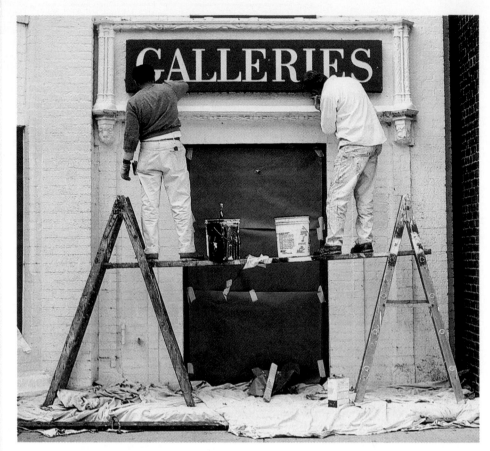

Don't get bored up here. *If this board were longer, more painters could fit on it. But would the board safely hold more painters? See Examples 1–3.*

All through this chapter you have seen situations in which two quantities vary. In many real-world situations there are more than two variables. Consider the situation presented on page 71, where the problem is to determine the maximum weight that can be supported by a board.

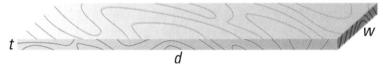

Three quantities which influence the maximum weight M are the board's width w, the board's thickness t, and the distance d between supports. The goal is to find an equation relating w, t, d, and M, the dependent variable.

The equation cannot be described by a single graph in two dimensions because there are four variables to be considered. However, by keeping all but one independent variable constant, we can investigate separately the relationship between the dependent variable M and that one independent variable.

To obtain these models, we use converses of the parts of the Fundamental Theorem of Variation. These converses can be proved.

> **Converse of the Fundamental Theorem of Variation**
> a. If multiplying every x-value of a function by c results in multiplying the corresponding y-value by c^n, then y varies directly as the nth power of x. That is, $y = kx^n$.
> b. If multiplying every x-value of a function by c results in dividing the corresponding y-value by c^n, then y varies inversely as the nth power of x. That is, $y = \frac{k}{x^n}$.

Maximum Weight as a Function of Width

We show how to find this model with a story. The data are made up, but the idea is not. Our heroine is again X. Perri Menter. She investigated the model as follows. First, Perri held the two independent variables d and t constant. She did this by choosing boards 10 feet long and 2 inches thick. Then she varied the widths of these boards and measured how much weight could be supported before the boards broke. Perri obtained the following data.

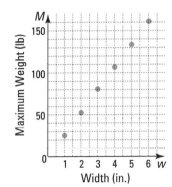

Width of board (in.) w	1	2	3	4	5	6
Maximum Weight (lb) M	27	53	80	107	133	160

The graph at the left shows how the maximum weight M depends on the width w of the board. Because the points seem to lie on a line through the origin, she concluded that M *varies directly as w*.

> ## Example 1
>
> Verify that M varies directly as w.
>
> **Solution**
>
> Suppose M = kw. Use a data point to find k. For instance, if w = 2 in. and M = 53 lb, then 53 = 2k
> and 26.5 = k.
> So the equation M = 26.5w should give values close to the experimental data. With k = 26.5, we get the data below.
>
> | w | 1 | 2 | 3 | 4 | 5 | 6 |
> |---|---|---|---|---|---|---|
> | M | 26.5 | 53 | 79.5 | 106 | 132.5 | 159 |
>
> These are almost identical to Perri's data.
> So M varies directly as w.

Maximum Weight as a Function of Thickness

Perri then investigated the relationship between M and the thickness t. She held the distance d between supports constant at 10 feet, and the width w constant at 3 inches. She varied the thickness and measured the maximum weight that could be supported. The table and graph at the top of page 118 present her findings.

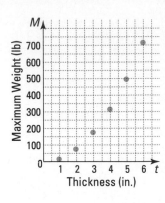

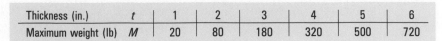

Thickness (in.)	t	1	2	3	4	5	6
Maximum weight (lb)	M	20	80	180	320	500	720

The points seem to lie on a parabola through the origin.

Example 2

Use the Converse of the Fundamental Theorem of Variation to determine how M varies with t.

Solution

Look at what happens to M when t doubles. Some ordered pairs are (1, 20), (2, 80), (4, 320). As t doubles, M is multiplied by $2^2 = 4$. When t triples, say from 2 to 6, the value of M increases from 80 to 720, that is, by a factor of $3^2 = 9$. Thus, M *varies as the square of t.*

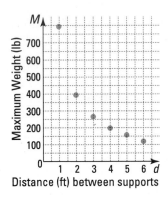

Distance (ft) between supports

Maximum Weight as a Function of Depth

Next, Perri investigated the relationship between M and d by holding t and w constant. She chose boards 2 inches thick and 3 inches wide, and measured the maximum weight boards of different lengths would hold. Perri obtained the following data.

Distance (ft)	d	1	2	3	4	5	6
Maximum weight (lb)	M	800	400	267	200	160	133

The graph at the left shows how M depends on d. It is not immediately clear whether M varies inversely as d or inversely as d^2.

Example 3

Explain why M varies inversely as d.

Solution 1

Look what happens to M as d doubles. When $d = 3$, $M = 267$, and when $d = 6$, $M = 133$. M is halved. This can be verified with other values of d and M. So M *varies inversely as d.*

Solution 2

Suppose M varies inversely as d. Then $M = \frac{k}{d}$. To find k, substitute a pair of values. If $d = 1$, $M = 800$. So $800 = \frac{k}{1}$ or $k = 800$. Make a table of values for $M = \frac{800}{d}$.

d	1	2	3	4	5	6
M	800	400	266.67	200	160	133.33

It agrees with the experimental data. So $M = \frac{800}{d}$ when $t = 2$ and $w = 3$ is a good model. Thus, M *varies inversely as d.*

Summary

Ms. Menter summarized her findings as follows:
M varies directly as w and the square of t;
M varies inversely as d.

These relations can be expressed in the single formula $M = \frac{kwt^2}{d}$, where M is in pounds, w and t are in inches, d is in feet, and k is the constant of variation. Notice that M varies directly as each independent variable in the numerator and inversely as each independent variable in the denominator. The formula tells you that the greater the width and the thickness and the shorter the distance between supports, the stronger the board will be.

In the next lesson you will calculate the constant of variation k in this equation.

QUESTIONS

Covering the Reading

1. How can one investigate the relationship between a dependent variable and more than one independent variable?

In 2–8, refer to the situation of this lesson.

2. What is the maximum weight that can be supported by a board 10 ft long, 2 in. wide, and 2 in. thick?

3. What is the maximum weight supported by a board 10 ft long, 3 in. wide, and 5 in. thick?

4. What is the shape of the graph of the relationship between M and d, when t and w are held constant?

5. Refer to the data for M as a function of t. Find an equation to model the data in the table when $d = 10$ ft and $w = 3$ in.

6. Use the Converse of the Fundamental Theorem of Variation to show that M varies inversely as d and not d^2.

7. If the thickness of a board is doubled and all other measurements are held constant, how is M affected? Explain your reasoning.

8. In the formula $M = \frac{kwt^2}{d}$, M varies directly as any variable in the _?_ of the expression.

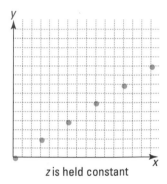

z is held constant

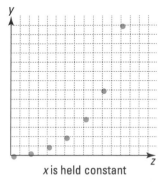

x is held constant

9. *Multiple choice.* The two graphs at the left show the relationships between the dependent variable y and the independent variables x and z.

Which equation best models this situation?

(a) $y = kx$ (b) $y = \frac{kx}{z}$ (c) $y = kxz^2$ (d) $y = kx^2z$

10. Cyrus Nathan Tist attempted to find how the volume of a gas (the dependent variable) depends on the temperature and pressure of the gas (the independent variables).
    a. When he held the pressure fixed at 250 millibars (abbreviated mbar), he obtained the following results. (The temperature is in kelvins, where Kelvin temperature = 273 + Celsius temperature.)

Temperature (kelvins) T	250	275	300	325	350
Volume (cm³) V	417	458	500	542	583

Graph these data points. On the V-axis, start at 400 and use an interval of length 20; that is, make tick marks at 400, 420, . . . , 600.
    b. How does V vary with T?
    c. When Cy held the temperature fixed at 300 kelvins, he obtained the following results.

Pressure (mbar) P	200	250	300	350	400
Volume (cm³) V	625	500	417	357	313

Graph these data points. On the V-axis, start at 300 and use an interval of length 50.
    d. How does V vary with P?
    e. Write an equation of variation to show how V depends on T and P. Do not solve for k.

11. Marc was trying to determine how the pressure exerted on the floor by the heel of a shoe depends on the width of the heel and the weight of the person wearing the shoe. He started by measuring the pressure (in psi) exerted by several people wearing a shoe with a heel width of 2.5 in. The data are summarized below.

Weight (lb) w	62	85	100	128	154	180
Pressure (psi) P	11.2	15.3	18.02	23.06	27.75	32.43

He then had his niece Mary, who weighs 142 lb, try on shoes with different heel widths, and he measured the pressure exerted. The data are summarized below.

Heel Width (in.) h	1.25	1.5	1.75	2	2.25	2.5
Pressure (psi) P	127.2	70.7	51.9	39.8	31.4	25.4

Assuming that P (the pressure), w (the weight), and h (the heel width) are related by a variation model, find an equation to describe that relationship. (Hint: Follow the steps in Question 10.) Do not solve for k.

Heel pressure. *Current fashion styles and trends usually dictate heel size, especially for women. Styles have included saddle shoes, platform shoes, spike heels, clogs, and earth shoes.*

12. Consider $y = \frac{-20}{x^2}$. *(Lesson 2-6)*
    a. What real number is excluded from the domain of x?
    b. *Multiple choice.* Which could be the graph of the equation?

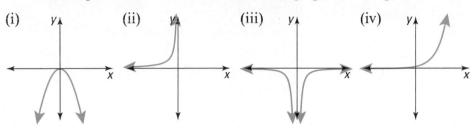

(i) (ii) (iii) (iv)

In 13–16, identify all asymptotes of the graph of the equation.
(Lessons 2-4, 2-5, 2-6)

13. $y = \frac{4x}{7}$ 14. $y = .08x^2$ 15. $y = \frac{-10}{x}$ 16. $y = \frac{\frac{1}{3}}{x^2}$

In 17 and 18, an equation for a function is given.
a. Give its domain. b. Give its range. *(Lessons 1-2, 2-5, 2-6)*

17. $y = -2x^2$ 18. $y = \frac{3}{x}$

19. The graph of the equation $y = \frac{3x}{4}$ is a __?__ with slope __?__. *(Lesson 2-4)*

20. Suppose y varies inversely as the cube of w. If y is 6 when w is 5, find y when w is 11. *(Lesson 2-1)*

21. Consider the sequence defined recursively as
    $$\begin{cases} s_1 = -3 \\ s_n = (s_{n-1})^2 + 2, \text{ for integers} \geq 2. \end{cases}$$
    (Lessons 1-8, 1-9)
    a. Describe the sequence in words.
    b. Find the first five terms of the sequence.

22. The perimeter of a rectangle is equal to 12.4 meters. Its length is 3.2 meters. What is its width? *(Lesson 1-5)*

23. The ability of a board to support a weight also depends on the type of wood. That is, the constant of variation k in the formula $M = \frac{kwt^2}{d}$ depends on the type of wood.
    a. For a stronger type of wood, is k larger or is k smaller?
    b. Which is strongest: oak, balsa, or pine?

2-9

Combined and Joint Variation

This photo is from the Chinese New Year parade in Chinatown in Los Angeles.

Combined Variation

When direct and inverse variations occur together, the situation is called **combined variation.** Perhaps the simplest equation of combined variation is

$$y = \frac{kx}{z}$$

where k is the constant of variation. The equation can be translated as "y varies directly as x and inversely as z."

A combined-variation equation can have more than two variables, and the independent variables can have any positive exponent. You saw an instance of this in Lesson 2-8 with the formula:

$$M = \frac{kwt^2}{d}.$$

This formula gives the maximum weight M in pounds that can be supported by a board of width w in., thickness t in., and distance d ft between supports.

You should be able to write a combined-variation situation expressed in words as an equation.

Example 1

The time T that it takes a parade to pass a reviewing stand varies directly as the length L of the parade and inversely as the speed s of the parade. Write a general equation to model this situation.

Solution

Because T is described in terms of L and s, T is the dependent variable. Because T varies directly as L, L will be in the numerator. Because T varies inversely as s, s will be in the denominator. The equation is

$$T = \frac{kL}{s}.$$

Finding the Constant of Variation in a Combined-Variation Model

To find k, the constant of variation in a combined-variation model, use the same idea you used to find the constant for direct and inverse variation. Find one instance that relates all the variables simultaneously, and substitute all values into the general variation equation. For instance, to find k in the equation $M = \frac{kwt^2}{d}$, refer to any one of the tables in Lesson 2-8. For example, the first table relating M and w gives six possible pairs of numbers for w and M. For each of these pairs, $t = 2$ in. and $d = 10$ ft.

w (in.)	1	2	3	4	5	6
M (lb)	27	53	80	107	133	160

Now choose a pair of values for w and M. If you use $M = 27$ lb and $w = 1$ in., and substitute into the formula, you get

$$27 = \frac{k(1)(2)^2}{10}.$$

Unit analysis

$$\text{lb} = \frac{k \cdot \text{in.} \cdot \text{in}^2}{\text{ft}}$$

$$k = \frac{\text{ft-lb}}{\text{in}^3}$$

The unit analysis shows the unit for k. So, the constant of variation is

$$k = 67.5 \frac{\text{ft-lb}}{\text{in}^3}.$$

This value for k should be checked by using other data points. You will do this in the Questions at the end of the lesson. Thus, the formula becomes

$$M = \frac{67.5 \, wt^2}{d}$$
$$= 67.5 \, \frac{wt^2}{d}.$$

Now it is possible to use this model.

Example 2

Find the maximum weight that can be supported by a board 3.75 in. wide and 1.75 in. thick with supports 8 ft apart.

Solution

Use the preceding formula with $d = 8$ ft, $w = 3.75$ in. and $t = 1.75$ in.

$$M = \frac{67.5 \, wt^2}{d}$$
$$= \frac{(67.5)(3.75)(1.75)^2}{8}$$
$$\approx 96.9$$

Unit analysis

$$\text{unit of } M = \frac{\frac{\text{ft-lb}}{\text{in}^3} \cdot \text{in.} \cdot \text{in}^2}{\text{ft}}$$
$$= \frac{\text{ft-lb}}{\text{ft}}$$
$$\text{unit of } M = \text{lb}$$

The board can support about 97 lb.

Joint Variation

Sometimes one quantity varies directly as the product of two or more independent variables, but not inversely as any variable. This is called **joint variation.** Perhaps the simplest equation of joint variation is

$$y = kxz,$$

where k is the constant of variation. The equation can be read as "y varies jointly as x and z" or "y varies directly as the product of x and z."

Recall from geometry the formula for the volume of a cone:

$$V = \tfrac{1}{3}\pi r^2 h.$$

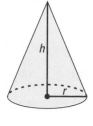

This can be expressed as "the volume varies jointly as the height and the square of the radius of the base." The constant of variation is $\tfrac{1}{3} \cdot \pi$, or $\tfrac{\pi}{3}$.

Solving joint-variation problems involves the same ideas as solving other variation problems.

Example 3

The wind force F on a vertical surface varies jointly as the area A of the surface and the square of the wind speed S. The force is 340 newtons on a vertical surface of area 1 m^2 when the wind blows at 18 m/sec. Find the force exerted by a wind of 35 m/sec on a vertical surface of area 2 m^2. $\left(\text{Note: One newton equals one } \frac{\text{kilogram-meter}}{\text{sec}^2}.\right)$

Solution

Use the same technique that you used when solving direct-variation and inverse-variation problems.

First, write the general equation relating F, A, and S.

$$F = kAS^2$$

Second, find k. If $A = 1$ and $S = 18$, then $F = 340$.

Unit analysis

Substitute. $\quad 340 = k \cdot 1 \cdot 18^2$	$\frac{\text{kg-m}}{\text{sec}^2} = k \cdot \text{m}^2 \cdot \frac{\text{m}^2}{\text{sec}^2}$
Solve for k. $\qquad k \approx 1.05$	$\frac{\text{kg}}{\text{m}^3} = k$

So, $\qquad\qquad\qquad k \approx 1.05 \, \frac{\text{kg}}{\text{m}^3}.$

Third, rewrite the formula with the calculated value of k.

$$F \approx 1.05AS^2$$

Fourth, substitute $A = 2$ and $S = 35$.

$$F \approx (1.05)(2)(35)^2$$
$$\approx 2572.5$$

The force exerted is about 2570 newtons.

Full of wind. *These two dhows (singular pronounced* dou*) are taking full advantage of the wind. They are shown on the Nile River.*

As in combined variation, a joint-variation equation can have more than two independent variables, and the independent variables can have any positive exponent.

QUESTIONS

Covering the Reading

1. What is combined variation?

2. Translate into a single formula: R varies directly as L and inversely as d^2.

3. Refer to the variation equation found in Example 1. Suppose a parade 4000 ft long walking at 1.5 mph needs 60 min to pass the reviewing stand. How long would it take a parade 1000 ft long walking at 2 mph to pass the reviewing stand?

In 4–7, refer to the situation about the maximum weight that can be supported by a board.

4. M varies directly as __?__ and __?__ and inversely as __?__.

5. Find k by using these data from Lesson 2-8: $M = 80$ lb, $w = 3$ in., $t = 2$ in., and $d = 10$ ft. (This checks that the value of k found in the lesson is reasonable.)

6. Find the maximum weight that can be supported by a board with supports 16 ft apart, 11.5 in. wide, and 1.5 in. thick. Use the value of k from the lesson.

7. Suppose that the maximum load that can be supported by a board is 2250 lb, and that the constant of variation is $67.5 \frac{\text{ft-lb}}{\text{in}^3}$. If the board is 10 in. thick and the supports are 12 ft apart, what is the width of the board?

8. Translate the formula $V = \frac{1}{3} \pi r^2 h$ into words, using the language of variation.

9. Refer to Example 3. Find the force exerted by a wind of 20 m/sec on a vertical surface of area 3.5 m^2.

Chicago parade. *The Bud Billiken Parade, honoring children, has been held every August in Chicago since 1929. Bud Billiken is a mythical character who protects children.*

Applying the Mathematics

10. Translate into a single formula: The time t it takes to finish algebra homework varies directly as the number n of questions assigned and inversely as the number d that can be solved with a calculator.

11. Consider the formula $V = \ell wh$ for the volume V of a rectangular solid with length ℓ, width w, and height h.
   a. Rewrite the formula as an English sentence using the language of variation.
   b. What is the constant of variation?

12. Suppose y varies directly as x and inversely as z. Describe how y changes when x and z are both doubled. Explain how you arrived at your conclusion.

13. The amount of heat H lost through a single pane of glass varies jointly as the area A of the pane and the difference $T_I - T_O$ in temperatures on either side of the window.
   a. Translate this statement into an equation of variation.
   b. Suppose when the indoor temperature is 70°F (T_I) and the outdoor temperature is 0°F (T_O), the heat lost through a 12-ft^2 window is 950 Btu (British thermal units). Find the constant of variation.
   c. Rewrite the equation of variation using the constant found in part b.
   d. Find the amount of heat lost through a 16-ft^2 window if the indoor temperature is 75°F and the outdoor temperature is -5°F.

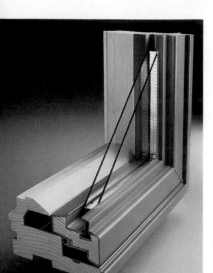

Energy efficient. *This is a double-pane thermal window. In cold weather the window restricts heat from getting out of the house, and in warm weather it restricts heat from getting into the house.*

Review

14. The resistance R in an electrical circuit is related to the diameter D of the wire and the length L of the wire. In an experiment, Dahlia obtained the following data using 50-ft lengths of wire. *(Lessons 2-7, 2-8)*

D (in.)	.05	.08	.11	.14	.17	3.20
R (ohms)	9.0	3.5	1.9	1.1	0.8	0.6

   a. Graph these data points.
   b. How does R vary with D?
   c. With wire of diameter .05 in., she obtained the following data.

L (ft)	25	50	75	100	125	150
R (ohms)	4.5	9	13.5	18	22.5	27

   Graph these data points.
   d. How does R vary with L?
   e. Write an equation that relates R, D, and L. You do not need to find the constant of variation.

In 15–18, use the function with equation $f(x) = \frac{20}{x^2}$. *(Lessons 1-2, 1-3, 2-6)*

15. Which is greater: $f(3)$ or $f(5)$?

16. Sketch a graph of this equation on the window $-10 \le x \le 10$, $-10 \le y \le 10$.

17. *True or false.* The graph of $y = \frac{20}{x^2}$ is a hyperbola.

18. State the domain and range of f.

19. Use the equation $y = 10x^2$. *(Lesson 2-5)*
   a. Make a table of values for $-3 \le x \le 3$.
   b. Use the values in your table to graph the equation.
   c. What is the name of this curve?
   d. Find the rate of change from $x = 1$ to $x = 2$.
   e. Would the answer to part **d** be the same for any two points on the graph? Explain your answer.

20. One general equation for a combined variation is $y = k\frac{xz}{w}$. Solve for k in terms of the other variables. *(Lesson 1-6)*

21. Given a function f defined by $f: x \rightarrow 4x^2 + 2$, find $f(\pi)$. *(Lesson 1-3)*

In 22 and 23, an instance of a general property is given. Name the property. *(Appendix A)*

22. $(4 + a) + z = (a + 4) + z$ **23.** $\frac{2}{3} + \frac{x}{3} = \frac{2 + x}{3}$

Exploration

24. Question 13 referred to heat loss through a window
   a. Find out what unit is used for heat in the metric system.
   b. How is this unit related to the Btu?

A project presents an opportunity for you to extend your knowledge of a topic related to the material of this chapter. You should allow more time for a project than you do for typical homework questions.

1 Pizza Prices

Consult several neighborhood restaurants that serve pizza in round pans. Choose a particular type of pizza, for instance, thin crust with mushrooms and onions. Find out the cost of this pizza, and the diameter of the pan.

a. For each restaurant, make a table of values, showing cost as a function of diameter for that type of pizza. Include diameters and prices for a small, medium, and large pizza.

b. Considering only area, price should vary with the square of the diameter. Do any of the restaurants you contacted follow this model? If yes, what equation describes their prices? If no, what other model seems to fit your data?

c. If the price structures are consistent, what should each of your restaurants charge for a giant pizza with a 30″ diameter?

2 The Maximum Load of a Balsa Board

Collect pieces of balsa wood of various lengths, widths, and thicknesses. Reconstruct *X. Perri Menter's* experiments regarding the maximum load *M* a board can hold described in Lessons 2-7, 2-8, and 2-9. Find an equation relating *d, w, t,* and *M* for balsa wood.

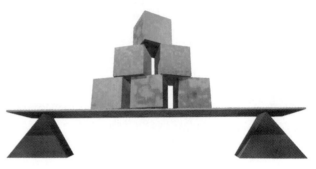

3 Sums or Differences of Variation Functions

Use an automatic grapher to explore graphs of sums of functions of variation equations.

a. For each function sketch a graph, state its domain and range, and identify any asymptotes of the graph.

(i) $y = x + \frac{1}{x}$

(ii) $y = x^2 + \frac{1}{x}$

(iii) $y = x^2 + \frac{1}{x^2}$

(iv) $y = x^2 + x$

b. Make up at least four other functions that can be expressed as sums or differences of variation functions. Sketch graphs of these functions.

c. Describe some patterns in the graphs.

4 Variation and Light

Use a small, bright, pocket-sized flashlight and a sheet of typing paper. Put the paper on a desk in a dimly lit room, and shine the light on the paper from various distances, for example, 1 in., 2 in., 4 in., 9 in., . . . , until the image becomes too dim to see.

a. At each step, measure the diameter of the circular image produced by the light beam. Create a table with the distance d of the flashlight from the paper in the first column, and the diameter D of the circular image in the second.

b. Graph the data in part **a.**

c. Write a formula showing how the diameter of the image depends on the distance. Does the diameter vary directly or inversely as d?

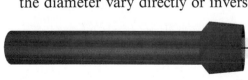

light

d. Borrow a light meter from a photographer or a science teacher. Repeat the above process, but rather than measure the diameters of the image, measure the *intensity,* or brightness, of the image at each step.

e. Theoretically the intensity I varies inversely as the square of d. How closely do your data fit this model?

5 The Law of the Lever

Does the Law of the Lever actually work? Get at least six people together of different weights and weigh each one of them. Choose one person to sit and remain on a see-saw a fixed distance from the fulcrum (This person determines the constant of variation in the law of the lever.) Have each of the remaining persons in turn sit on the see-saw and balance with the first person. Measure the distance from the fulcrum for each person.

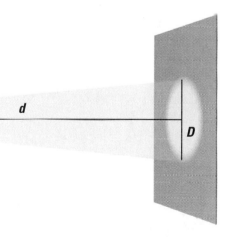

d

D

SUMMARY

Two types of functions studied in this chapter are direct variation and inverse variation. When $k \neq 0$ and $n > 0$, formulas of the form $y = kx^n$ represent direct-variation functions, and those of the form $y = \dfrac{k}{x^n}$ represent inverse-variation functions. Four special cases of direct and inverse variation occur frequently in real-world situations. The graph, domain D, and range R of each case is indicated below.

Direct-variation formulas

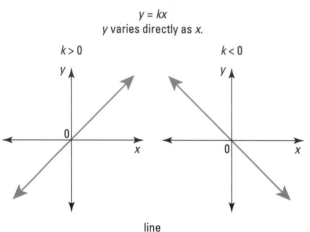

$y = kx$
y varies directly as x.

line
$D = R =$ the set of all real numbers

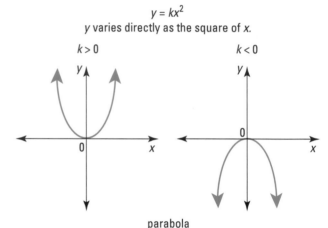

$y = kx^2$
y varies directly as the square of x.

parabola
$D =$ the set of all real numbers
$R = \{y: y \geq 0\}$

$D =$ the set of all real numbers
$R = \{y: y \leq 0\}$

Inverse-variation formulas

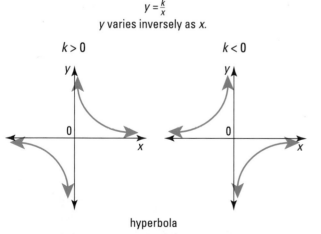

$y = \dfrac{k}{x}$
y varies inversely as x.

hyperbola
$D = R =$ the set of all nonzero real numbers

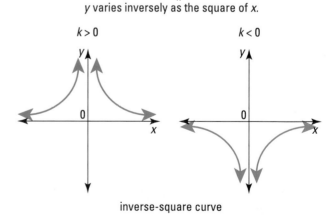

$y = \dfrac{k}{x^2}$
y varies inversely as the square of x.

inverse-square curve
$D =$ the set of all nonzero real numbers
$R = \{y: y > 0\}$

$D =$ the set of all nonzero real numbers
$R = \{y: y < 0\}$

In a formula where y is given in terms of x, it is natural to ask how changing x (the independent variable) affects the value of y (the dependent variable). In direct or inverse variation, when x is multiplied by a constant, the changes in y are predicted by the Fundamental Theorem of Variation. If y varies directly as x^n, then when x is multiplied by c, y is multiplied by c^n. If y varies inversely as x^n, then when x is multiplied by c, y is divided by c^n.

The rate of change from (x_1, y_1) to (x_2, y_2) is $\frac{y_2 - y_1}{x_2 - x_1}$, the slope of the line connecting them. For graphs of the form $y = kx$, the rate of change is the constant k. For other curves, the rate of change is not constant but varies depending on which points are used to calculate the rate of change.

Variation formulas may involve three or more variables. If all the independent variables are multiplied, then joint variation occurs. If they are not all multiplied, the situation is one of combined variation. Variation formulas can be derived from real data by examining two variables at a time and comparing their graphs with those given on page 130. We call this idea modeling, or forming a mathematical model of the data. Automatic graphers are useful tools to help in graphing or comparing functions.

The applications of slope, variation, and modeling are numerous. They include many perimeter, area, and volume formulas, the inverse-square laws of sound and gravity, and a variety of relationships among physical quantities such as distance, time, force, and pressure.

VOCABULARY

Below are the most important terms and phrases for this chapter.
You should be able to state each in words and give a specific example.
For the starred (*) terms, you should be able to supply a good definition.

Lesson 2-1
varies directly as
constant of variation
*direct variation
directly proportional to

Lesson 2-2
varies inversely as
*inverse variation
inversely proportional to
fulcrum
Law of the Lever
inverse-square variation
conjecture, prove

Lesson 2-3
Fundamental Theorem of Variation

Lesson 2-4
*slope
*rate of change

Lesson 2-5
automatic grapher
window
default window
parabola
reflection-symmetric
line of symmetry
copy
trace
opens up
opens down

Lesson 2-6
hyperbola
branches of a hyperbola
discrete
inverse-square curve
vertical asymptote
horizontal asymptote

Lesson 2-7
*mathematical model

Lesson 2-8
Converse of the Fundamental Theorem of Variation

Lesson 2-9
combined variation
joint variation

PROGRESS SELF-TEST

Take this test as you would take a test in class. Use graph paper and a ruler. Then check your work with the solutions in the Selected Answers section in the back of the book.

In 1 and 2, translate into a variation formula.

1. The number of trees n that can be planted per acre varies inversely as the square of the trees' distance d apart.

2. The weight w that a bridge column can support varies directly as the fourth power of its diameter d and inversely as the square of its length L.

3. Write the variation formula $y = kx^5$ in words.

4. If S varies directly as the square of p and $S = 10$ when $p = 3$, find S when $p = 8$.

5. For the variation equation $y = 3x^2$, what is the change in the y-value when an x-value is doubled? Give two specific pairs of (x, y) values that support your conclusion.

6. For the variation $y = \frac{6}{x}$, what is the change in the y-value when an x-value is multiplied by $c(c \neq 0)$?

7. Find the rate of change of $y = x^2$ between $x = 3$ and $x = 4$.

8. *True or false.* All graphs of variation equations pass through the origin.

9. The graph of $y = kx^2$ is called a __?__ and opens up if __?__.

10. Suppose $f(x) = \frac{17}{x^2}$. What is the domain of f?

11. Fill in the blank with "inversely," "directly," or "neither inversely nor directly."

   a. The surface area of a sphere varies __?__ as the cube of its radius.

   b. The number of different shares you can buy varies __?__ as the cost of each share, if you invest exactly $10,000.

   c. Explain your reasoning in part **a**.

12. a. Make a table of values for $y = -5x$. Include at least five pairs.

   b. Make a graph of the equation in part **a**.

   c. Find the slope of the graph in part **b**.

13. a. Use an automatic grapher to sketch a graph of $y = \frac{5}{x}$.

   b. Identify any asymptotes to the graph in part **a**.

14. *Multiple choice.* Find the equation whose graph looks the most like the graph shown below.

   (a) $y = -3x$ (b) $y = -\frac{3}{x}$

   (c) $y = -\frac{3}{x^2}$ (d) $y = -\frac{x}{3}$

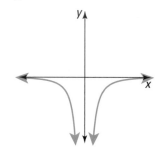

15. A student removing the bolts from the back of a large cabinet in a science lab knew that it was easier to turn a bolt with a long wrench than with a short one. The student decided to investigate the force required with various wrenches, and obtained the following data.

Length of wrench (in.) L	3	5	6	8	9
Force (lb) F	120	72	60	45	40

   a. Graph these data points.

   b. Which variation equation is a better model for this situation, $F = \frac{k}{L}$ or $F = \frac{k}{L^2}$? Justify your answer.

   c. How much force would be required to turn one of the bolts with a 12-in. wrench?

16. Give a real-world example of a direct-variation situation, other than one that is in this test.

17. Suppose the price of a single scoop of ice cream varies directly with the cube of the diameter of the scoop. If a scoop 2 inches in diameter costs 79¢, how much should a 3-inch-diameter scoop cost?

PROGRESS SELF-TEST

18. Here is a graph of $y = x^2$ on the window $-4 \leq x \leq 4$, $0 \leq y \leq 10$.

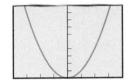

$-4 \leq x \leq 4$, x-scale = 1
$0 \leq y \leq 10$, y-scale = 1

Which of the graphs below cannot be a graph of this equation on some other window?

(a)

$-2 \leq x \leq 2$, x-scale = 1
$0 \leq y \leq 5$, y-scale = 1

(b)
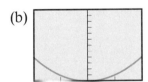

$-2 \leq x \leq 2$, x-scale = 1
$0 \leq y \leq 12$, y-scale = 1

(c)
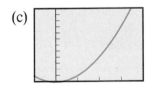

$-1 \leq x \leq 4$, x-scale = 1
$0 \leq y \leq 12$, y-scale = 1

(d)
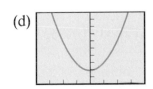

$-4 \leq x \leq 4$, x-scale = 1
$0 \leq y \leq 12$, y-scale = 1

19. Suppose that variables V, h, and g are related as illustrated in the graphs below. The points on the first graph lie on or near a parabola. The points on the second graph lie on a line through the origin.

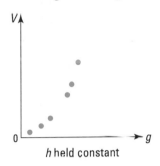

h held constant

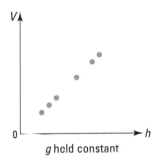

g held constant

Write a general equation approximating the relationship among V, h, and g.

20. Poiseuille's Law states that the speed S at which blood flows through arteries and veins varies directly with the blood pressure P and the fourth power of the radius r of the blood vessel. Suppose that blood flows at a rate of 0.09604 cm^3/sec through an artery of radius 0.065 cm when the blood pressure is 100 mmHg. What would be the blood pressure if plaque reduced the artery to 0.05 cm radius, and the speed stayed the same?

CHAPTER REVIEW

Questions on SPUR Objectives

SPUR stands for **S**kills, **P**roperties, **U**ses, and **R**epresentations. The Chapter Review questions are grouped according to the SPUR Objectives for this chapter.

SKILLS DEAL WITH THE PROCEDURES USED TO GET ANSWERS.

Objective A: *Translate variation language into formulas and formulas into variation language.*
(Lessons 2-1, 2-2, 2-9)

In 1–3, translate into a variation equation.

1. y varies directly as the square of x.

2. s varies inversely with p.

3. z varies jointly as x and t.

4. In the formula, $r = kstu$, r varies __?__ with __?__.

5. If $V = k\pi r^2$, then V varies __?__ as __?__.

In 6–8, write each variation equation in words.

6. $w = kh^4$ **7.** $y = \frac{k}{x^2}$ **8.** $t = \frac{kz^3}{w^2}$

Objective B: *Solve variation problems.*
(Lessons 2-1, 2-2, 2-9)

9. y varies directly as x. If $x = 4$, then $y = -12$. Find y when $x = -7$.

10. y varies directly as the square of x. When $x = -5$, $y = 75$. Find y when $x = 8$.

11. y varies inversely as the cube of x. If $x = 4$, $y = -\frac{1}{16}$. Find y when $x = \frac{1}{2}$.

12. z varies directly as the square of x and inversely as y. When $x = 3$ and $y = 5$, $z = 4.5$. Find z when $x = -2$ and $y = -1.5$.

Objective C: *Find slopes (rates of change).*
(Lessons 2-4, 2-5, 2-6)

13. Find the slope of the line through the points (15, 27) and (20, 36).

14. What is the slope of the line at the right?

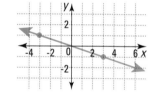

In 15 and 16, $y = 5x^2$.

15. Find the rate of change from $x = -2$ to $x = -1$.

16. Find the rate of change from $x = -3$ to $x = -2$.

In 17 and 18, find the rate of change from $x = 3$ to $x = 4$.

17. $y = \frac{9}{x}$ **18.** $y = \frac{9}{x^2}$

PROPERTIES DEAL WITH THE PRINCIPLES BEHIND THE MATHEMATICS.

Objective D: *Use the Fundamental Theorem of Variation.*   *(Lesson 2-3)*

In 19 and 20, suppose that the value of x is tripled. Tell how the value of y changes.

19. y varies directly as x.

20. y varies directly as x^3.

In 21 and 22, suppose that p varies inversely as the square of q. How does the value of p change if q is

21. doubled?   **22.** multiplied by ten?

23. If $y = \frac{k}{x^n}$ and x is multiplied by any nonzero constant c, then y is __?__.

24. If $y = kx^n$ and x is divided by any nonzero constant c, then y is __?__.

25. Suppose $y = \frac{kx^n}{z^n}$ and both x and z are multiplied by any nonzero constant c. What is the effect on y?

Objective E: *Identify the properties of variation functions.* *(Lessons 2-4, 2-5, 2-6)*

26. The graph of the equation $y = kx$ is a __?__ having slope __?__.

27. Graphs of all direct-variation formulas go through the point __?__.

In 28–30, refer to these four equations:

(a) $y = kx$
(b) $y = kx^2$
(c) $y = \frac{k}{x}$
(d) $y = \frac{k}{x^2}$.

28. Which equations have graphs that are symmetric to the y-axis?

29. The graph of which equation is a parabola?

30. *True or false.* When $k > 0$, all of the equations have points in quadrant I.

31. Identify the domain and range of $f(x) = kx^2$ for $k > 0$.

32. Identify the asymptotes to an equation of an inverse-square variation.

33. For what values of x is the inverse variation $y = \frac{k}{x}$ undefined?

34. Write a short paragraph explaining how the graphs of $y = -2x^2$ and $y = \frac{-2}{x^2}$ are alike and how they are different.

USES DEAL WITH APPLICATIONS OF MATHEMATICS IN REAL SITUATIONS.

Objective F: *Recognize variation situations.*
(Lessons 2-1, 2-2)

In 35–38, translate into a variation equation.

35. The number n of congruent marbles that fit into a box is inversely proportional to the cube of the radius r of each marble.

36. The area A of an image on a movie screen is directly proportional to the square of the distance d from the projector to the screen.

37. The gravitational pull p of a star on a planet with mass m varies directly as the mass and inversely as the square of the distance d of the planet from the star.

38. At a given speed, the distance a plane travels is directly proportional to the time traveled.

In 39–42, complete with "directly," "inversely," or "neither directly nor inversely."

39. The number of people invited to dinner varies __?__ as the amount of space each guest has at the table.

40. The temperature in a house varies __?__ as the number of hours the air conditioner has been on.

41. The volume of a cylinder of height 10 cm varies __?__ as the square of its radius.

42. Your distance above the ground while on a Ferris wheel varies __?__ as the number of minutes you have been riding on it.

Objective G: *Solve real-world variation problems.* *(Lessons 2-1, 2-2, 2-9)*

43. Suppose the price of a pizza varies directly with the square of its diameter. At Vic Yee's pizza parlor an 8-inch pizza costs $6.00. How much would a 12-inch pizza cost?

44. The refund r you get varies directly with n the number of cans you recycle. If you get a $7.50 refund for 150 cans, how much should you get for 400 cans?

45. One of Murphy's Laws is that the time t a committee spends debating a budget item is inversely proportional to d, the number of dollars involved. If a committee spends 10 minutes debating a $300 item, how much time is spent debating a $1000 item?

46. Recall that the weight of a body varies inversely with the square of its distance from the center of the Earth. If Daniel weighs 75 lb on the surface of the Earth, how much would he weigh in space 50,000 miles from the Earth's surface? (The radius of the Earth is approximately 4000 miles.)

47. The force needed to keep a car from skidding on a curve varies directly as the weight of the car and the square of the speed and inversely as the radius of the curve. It requires 266 lb of force to keep a 2200-lb car, traveling at 30 mph, from skidding on a curve of radius 500 ft. How much force is required to keep a 3000-lb car, traveling at 45 mph, from skidding on a curve of radius 400 ft?

48. An object is tied to a string and then twirled in a circular motion. The tension in the string varies directly as the square of the speed and inversely as the radius of the circle. When the radius is 5 ft and the speed is 4 ft/sec, then the tension in the string is 90 lb. If the radius is 3.5 ft and the speed is 4.4 ft/sec, find the tension in the string.

Objective H: *Fit an appropriate model to data.*
(Lessons 2-7, 2-8)

In 49–50, a situation and question are given.

    a. Draw a graph to represent the situation.

    b. Find a general variation equation to represent the situation.

    c. Find the value of the constant of variation and rewrite the variation equation.

    d. Answer the question asked in the problem.

49. Officer Friendly measured the length L of car skid marks when the brakes were applied at different speeds S. He obtained the following data.

S (mph)	20	30	40	50	60
L (ft)	18	41	72	113	162

How far would a car skid if the brakes are applied at 70 mph?

50. A man weighs 200 lb at sea level. The following table gives his weight W at various distances D from the center of the Earth.

D (miles)	4000	4500	5000	5500	6000
W (lb)	200	158	128	106	89

How much would the man weigh on the top of Mt. Everest, which is about 4005.5 miles from the Earth's center?

51. Cyrus N. Tist tried to discover how the power in an electric circuit is related to the strength of the current and the resistance of the wire.

    a. When he held the current I constant at 10 amps, he obtained the following data relating power P and resistance R.

R (ohms)	1	1.5	2	2.5	3	3.5
P (watts)	100	150	200	250	300	350

    i. Graph these data points.

    ii. How does P vary with R? Justify your answers.

    b. Then Cy held the resistance constant at 2 ohms. He obtained the data relating power P and current I shown below.

I (amps)	2	3	4	5	6	7
P (watts)	8	18	32	50	72	98

    i. Graph these data points.

    ii. How does P vary with I?

    c. Write a general equation of variation relating P, R, and I.

    d. Find the constant of variation, and write a variation equation relating P, R, and I.

    e. Predict the value of P when $I = 20$ amps and $R = 4$ ohms.

52. Erika performed an experiment to determine how the pressure P of a liquid on an object is related to the depth d of the object and the density D of the liquid. She obtained the graph on the left by keeping the depth constant and measuring the pressure on an object in solutions with different densities. She obtained the graph on the right by keeping the density constant and measuring the pressure on an object in a solution at various depths.

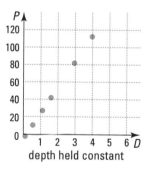
depth held constant

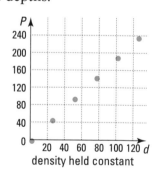
density held constant

Write a general equation relating P, d, and D. (Do not find the constant of variation.) Explain your reasoning.

REPRESENTATIONS DEAL WITH PICTURES, GRAPHS, OR OBJECTS THAT ILLUSTRATE CONCEPTS.

Objective I: *Graph variation equations.*
(Lessons 2-4, 2-5, 2-6)

In 53–56, an equation is given. **a.** Make a table of values. **b.** Graph the equation.

53. $y = \frac{1}{2}x$ **54.** $y = -2x$

55. $y = -2x^2$ **56.** $y = \frac{1}{2}x^2$

In 57 and 58, use an automatic grapher to sketch the graph of each equation.

57. $y = \frac{36}{x}$ **58.** $y = \frac{36}{x^2}$

63. In the graph of $y = kx^2$ shown at the right, what type of number must k be?

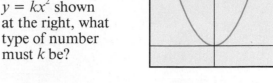

64. In the graph of $y = \frac{k}{x^2}$ shown at the right, what type of number is k?

Objective J: *Identify variation equations from graphs.* *(Lessons 2-4, 2-5, 2-6)*

Multiple choice. In 59–62, select the equation whose graph is most like the one shown. Assume the scales on the axes are equal.

59. (a) $y = 4x$
     (b) $y = -4x^2$
     (c) $y = -\frac{1}{4}x$
     (d) $y = -\frac{1}{4}$

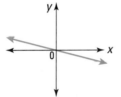

60. (a) $y = 10x^2$
     (b) $y = -x^2$
     (c) $y = -10x$
     (d) $y = -\frac{10}{x^2}$

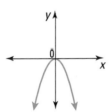

61. (a) $y = \frac{x^2}{6}$
     (b) $y = \frac{6}{x}$
     (c) $y = \frac{-6}{x}$
     (d) $y = \frac{-6}{x^2}$

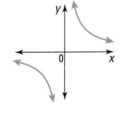

62. (a) $y = \frac{x^2}{6}$
     (b) $y = \frac{6}{x}$
     (c) $y = \frac{-6}{x}$
     (d) $y = \frac{-6}{x^2}$

Objective K: *Recognize the effects of a change in scale or viewing window on a graph of a variation equation.* *(Lesson 2-5)*

In 65 and 66, a graph of $y = 4x$ is drawn below using the window $-5 \le x \le 5$, $-25 \le y \le 25$.

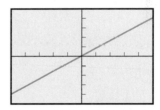

$-5 \le x \le 5$, x-scale = 1
$-25 \le y \le 25$, y-scale = 5

65. Sketch a graph of $y = 4x$ on the window shown at the right.

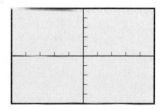

$-5 \le x \le 5$, x-scale = 1
$-5 \le y \le 5$, y-scale = 1

66. Does the slope of the line $y = 4x$ change when the viewing window is changed? Explain your answer.

67. Consider the function $f(x) = \frac{18}{x^2}$.

   a. Give the dimensions of a window that shows both branches of the graph.

   b. Give the dimensions of a window that shows only one branch of the graph.

CHAPTER

3

LINEAR FUNCTIONS

In Chapter 2 you studied direct-variation situations modeled by functions with equations of the form $y = kx$, or $f(x) = kx$. The graph of $f(x) = kx$ is a line, so f is called a *linear function*. Many other situations can be modeled by linear functions. Here are examples similar to the ones you will study in this chapter.

Constant Increase
A crate weighs 3 kilograms when empty. It is filled with oranges weighing 0.2 kilogram each. Then $W = 3 + .2n$ gives the weight W in kilograms of a crate containing n oranges.

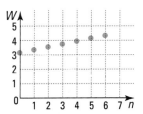

Linear Combination
A group bought A adult tickets at $7 each and S student tickets at $3 each. The group spent $42. The equation $7A + 3S = 42$ relates A, S, and the total amount spent.

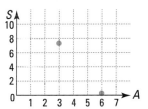

Point - Slope
Stuart Dent is conducting an experiment with a spring and a weight. The spring is 15 centimeters long with a 10-gram weight attached, and its length increases 0.8 centimeters for each additional gram weight. Then $L - 15 = .8(W - 10)$ relates spring length L and weight W.

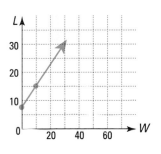

Step Functions
In 1995, first-class postage in the United States was 32¢ for the first ounce, and 23¢ for each additional ounce. The graph at the right pictures the relation between weight w in ounces and cost C in dollars.

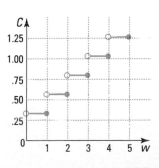

139

Temperature hike. *These hikers are prepared for changes in weather. As they climb higher the temperature tends to decrease; when they climb lower the temperature tends to increase.*

Constant-Increase Situations

In many situations there is an initial condition and a constant change applied to that condition. This kind of situation can always be modeled by a linear equation.

Example 1

The current temperature is 5°C and has been increasing 2°C an hour. If this trend continues, what will the temperature be *x* hours from now?

Solution

Write the temperature for several hours to find a general pattern. Let y = the temperature (in °C), x hours from now.

x	y
0	$5 + 0 \cdot 2 = 5$
1	$5 + 1 \cdot 2 = 7$
2	$5 + 2 \cdot 2 = 9$
3	$5 + 3 \cdot 2 = 11$
4	$5 + 4 \cdot 2 = 13$
5	$5 + 5 \cdot 2 = 15$

An equation relating temperature and hours is
$$5 + x \cdot 2 = y$$
or
$$y = 2x + 5.$$
The temperature x hours from now will be 2x + 5.

In $y = 2x + 5$, each value of x results in just one value for y. Thus, y is a function of x; y is the dependent variable and x is the independent variable. Solutions to the equation $y = 2x + 5$ are all the ordered pairs (x, y) whose values satisfy the equation.

Positive values of x represent times in the future, and negative values of x represent "hours ago," or times in the past. The graph of $y = 2x + 5$ is a *line* including points to the left of the vertical axis and below the horizontal axis, as shown at the right. The domain and range of the function are each the set of real numbers.

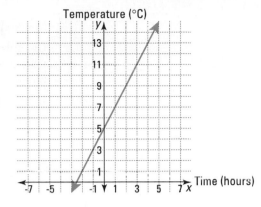
Temperature (°C)

Time (hours)

The graph of $y = 2x + 5$ crosses the y-axis at $(0, 5)$. We say that the *y-intercept* is 5. The y-intercept is the **initial condition** or the starting point in the situation, in this case, 5°. The slope of this line is the rate of change in the situation, 2°C per hour. To confirm this, we find the slope determined by the two points $(0, 5)$ and $(3, 11)$.

$$\frac{y_2 - y_1}{x_2 - x_1} = \frac{(11 - 5)°}{(3 - 0)\text{ hours}} = \frac{6°}{3\text{ hours}} = 2\frac{\text{degrees}}{\text{hour}}$$

The form $y = mx + b$, or $f(x) = mx + b$, is called the **slope-intercept form** of an equation for a line. Any function whose graph is a line is called a *linear function*. In general, a **linear function** is a function with an equation of the form $y = mx + b$.

Constant-Decrease Situations

Example 1 involves a *constant-increase situation*. Example 2 involves a *constant-decrease situation*. Every constant-increase or constant-decrease situation can be described by a linear function.

Feed the birds. *The house finch is a nonmigratory bird often seen at winter bird feeders, like the one shown here.*

Example 2

At the beginning of the month, Katie bought a 50-pound sack of wild-bird feed. She puts $\frac{2}{3}$ of a pound of feed in the bird feeder each morning.
a. Let y be the number of pounds left in the sack after x days. Write an equation in slope-intercept form relating y and x.
b. Graph the equation from part a.
c. How long will it take for the bird feed to run out?

Solution

a. This is an instance of constant decrease. So an equation is
$$y = mx + b,$$
where m and b need to be found. The rate of change m is $\frac{2}{3}$ of a pound per day. Because the amount of feed in the sack is decreasing, $m = -\frac{2}{3}$. The initial amount of food is 50 pounds, so 50 is the value of y when $x = 0$. Thus, the y-intercept is 50, and the equation is
$$y = -\frac{2}{3}x + 50.$$

▶

b. Make a table of values for the equation $y = -\frac{2}{3}x + 50$ from part **a.** Observe that we chose x-values that are multiples of 3 to make the computations easier. Part of the graph is shown below. Because the number of days is always an integer, the points are not connected.

x	0	3	6	9	12
y	50	48	46	44	42

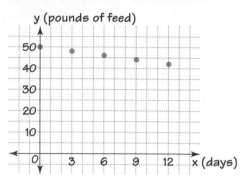

c. The supply runs out when $y = 0$. Substitute 0 for y in $y = -\frac{2}{3}x + 50$ and solve for x.

$$0 = -\frac{2}{3}x + 50 \qquad \text{substitution}$$
$$0 = -2x + 150 \qquad \text{Multiply by 3 to clear fractions.}$$
$$2x = 150 \qquad \text{Add } 2x \text{ to each side.}$$
$$x = 75 \qquad \text{Multiply each side by } \frac{1}{2}.$$

The supply will run out in 75 days.

In general, any constant-increase or constant-decrease situation can be modeled by an equation of the form $y = mx + b$. The graph of this equation is a line with slope m and y-intercept b. The slope m corresponds to the rate of change in the situation. The **y-intercept** b, which is the value of y when x is 0, corresponds to the initial value of the dependent variable.

In cases of constant increase, as in Example 1, the graph of the line slants up from left to right, indicating a *positive* slope. In cases of constant decrease, as in Example 2, the graph slants down from left to right, indicating a *negative* slope. In cases where the rate of change is 0, the graph of the line is horizontal.

Piecewise-Linear Functions

In some situations, the rate of change is constant for a while, but then changes to another constant rate. In such situations, the graph consists of two or more segments or rays. Such a graph is called **piecewise linear.** It is a union of segments or pieces of two or more linear functions.

Example 3

Horace made the graph below to describe his bicycle trip to a state park, 48 km from his home. It gives his distance *D*, in kilometers from home, as a function of *t*, the time in hours after leaving home. Write a story explaining the meaning of each segment of the piecewise linear graph.

Shown are bikers at the edge of a canyon near Moab, Utah.

Horace's Bicycle Trip

Solution

The graph shows that Horace started from home and bicycled at a constant rate for 2 hours, ending up 28 kilometers from home. So he traveled at a rate of 14 km/hr. After a half-hour stop, he traveled at a constant rate for another hour and a half, until he reached the state park 48 kilometers from home. During this part of the trip he traveled 20 km in 1.5 hr, so his rate was about 13.3 km/hr. He stayed at the park for two hours. Finally, he returned home traveling at a constant rate and reached home 9 hours after starting out.

QUESTIONS

Covering the Reading

1. Give three examples of situations which can be modeled by linear functions or piecewise-linear functions.

In 2–5, refer to Example 1.

2. **a.** Name the independent variable.
   b. Name the dependent variable.

3. What was the temperature after $3\frac{1}{2}$ hours?

4. In the equation $y = 2x + 5$, 5 represents the __?__ on the graph and the __?__ in the problem.

5. *True or false.*   The 2°C increase per hour is the slope of the line.

In 6 and 7, refer to the equation $y = mx + b$.

6. The coefficient of x tells you the __?__ of the line.

7. In cases of constant increase or constant decrease, the y-intercept b corresponds to __?__.

8. What is a linear function?

In 9 and 10, refer to Example 2.

9. How many pounds of bird feed are left after 21 days?

10. In how many days will there be 16 pounds of bird feed?

11. The graph at the left can represent a situation of constant __?__.

In 12 and 13, an equation for a line is given. **a.** Give its slope. **b.** Give its y-intercept.

12. $y = x + 3$ **13.** $y = kx$

In 14–17, refer to the bicycle trip in Example 3.

14. What does the slope of the segment from $(0, 0)$ to $(2, 28)$ represent?

15. *True or false.* On the trip from home to the state park, Horace traveled at a constant rate.

16. How fast (in kilometers per hour) was Horace going during the last three hours of his trip?

17. What is the total amount of time that Horace stopped during his trip?

Applying the Mathematics

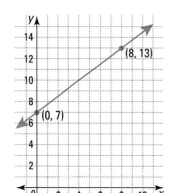

18. The line $y = \frac{3}{4}x + 7$ is graphed at the left.
   a. From the equation, what is its slope?
   b. Use the points $(0, 7)$ and $(8, 13)$ to verify your answer to part **a.**
   c. What is the y-intercept?
   d. Does this equation describe a function? Explain your answer.

19. An auto dealer is having a Fourth-of-July extravaganza. The dealership plans to be open for 72 hours straight. Suppose the dealer has 100 new cars on the lot and is able to sell 4 cars every 3 hours.
   a. Let h be the number of hours the dealer has been open and let C be the number of cars remaining on the lot. Find three other pairs of values that satisfy this relation.

h	0	3			
C	100	96			

   b. Write an equation that gives C as a function of h.
   c. After how many hours will there be only 60 cars left?
   d. If the dealership is able to maintain the pace of selling 4 cars every 3 hours, will the dealer sell all the cars on the lot during the sale? How can you tell?

20. Suppose y varies directly as x, and that y is 7 when x is 2.
   a. Find the constant of variation and write an equation describing the variation.
   b. Make a table of values and graph the equation.
   c. Verify that this direct-variation equation fits the $y = mx + b$ model by identifying the slope and y-intercept.
   d. Which situation does this variation represent: constant increase or constant decrease?

21. On the graph below, M = the number of miles Carmen is from home at time T. Carmen walks to school, then to her job after school, and then home.

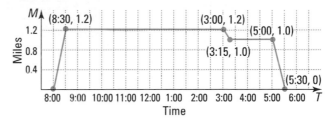

   a. How far is the school from Carmen's home?
   b. How long does it take Carmen to walk to school?
   c. Find the slope of the segment from (3:00, 1.2) to (3:15, 1.0) using hours as the unit for time.
   d. Write a question which can be answered by using the graph. Answer your question.

Review

22. a. Find the area of a circle inscribed in a square whose sides are 6 cm long.
   b. Find the area of a circle inscribed in a square whose sides are x cm long. *(Previous course)*

Dream vacation.
Shown is Waikiki Beach in Honolulu, Hawaii. Honolulu is a popular vacation spot.

23. a. Draw the line with slope $\frac{1}{4}$ that contains the point (0, 0).
   b. Write an equation for this line. *(Lesson 2-4)*

24. Solve for y: $x + 2y = 5$. *(Lesson 1-6)*

25. Solve for x: $1.5 + 0.5x = 0.45(x + 5)$. *(Lesson 1-5)*

26. Suppose B ounces of blended fruit juice contain 10% apple juice. How many ounces of juices other than apple juice are in the blend? *(Previous course)*

Exploration

27. What place in the world would you most like to visit? Find out how much it would cost to go there by air, and estimate how much your average daily expenses would be. Write an equation that can be used to calculate the total cost T of your visit if you stay for n days.

Equation solver. *Shown are methods commonly used to find the solutions to an equation.*

As you saw in the last lesson, the solutions to an equation of the form $y = mx + b$ lie on a line with slope m and y-intercept b. To graph an equation of this form you can: (1) make a table of values and plot points; (2) use an automatic grapher; or (3) use the slope and y-intercept to draw the line. You should be able to use all three methods.

Graphing Using Slope and Intercept

Example 1

Graph the line $y = 4x + 7$ using its slope and y-intercept.

Solution

The y-intercept is 7, so the line contains $(0, 7)$. Use the slope to locate another point. A slope of 4 means that every horizontal change of one unit to the right corresponds to a vertical change of four units up. Starting at $(0, 7)$, count 1 unit right and 4 up. This gives the new point $(0 + 1, 7 + 4) = (1, 11)$. Plot $(1, 11)$ and draw the line. Label it $y = 4x + 7$.

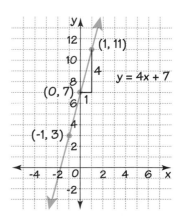

Check

The point (1, 11) satisfies the equation $y = 4x + 7$, because $11 = 4 \cdot 1 + 7$. The two points (0, 7) and (1, 11) determine a line, so the graph must be correct.

It is usually faster to graph a line using its slope and y-intercept than to construct a table of solutions. Often, it is even faster than using an automatic grapher. In fact, most automatic graphers require you to rewrite an equation in slope-intercept form before it can be graphed. Example 2 illustrates how this can be done.

Rewriting an Equation in Slope-Intercept Form

Example 2

Graph the line $2y = -3x + 16$.

Solution 1

To solve for y, divide each side by 2. $y = -\frac{3}{2}x + 8$

In this form, you can see that the slope is $-\frac{3}{2}$ and the y-intercept is 8.

First plot (0, 8). A slope of $-\frac{3}{2}$ means a vertical change of $-\frac{3}{2}$ units for every horizontal change of 1 unit. This is the same as 3 units down for every 2 units to the right. So, add 2 to the x-coordinate and -3 to the y-coordinate of (0, 8). $(0 + 2, 8 - 3) = (2, 5)$. Draw the line through the y-intercept and this point, and label the line $y = -\frac{3}{2}x + 5$.

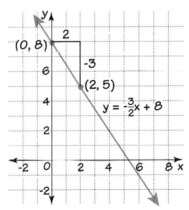

Check

Substitute the coordinates of the two points you plotted into the original equation.
Check (0, 8). Does $2(8) = -3(0) + 16$? Yes.
Check (2, 5). Does $2(5) = -3(2) + 16$? Yes.

Solution 2

Solve for y as in Solution 1.
Graph $y = -1.5x + 8$ using an automatic grapher. Copy and label what you see on the window of your grapher.

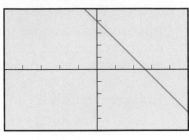

-10 ≤ x ≤ 10   x-scale = 2
-10 ≤ y ≤ 10   y-scale = 2

Check

Trace to find the coordinates of two points on the line. Our grapher shows (0, 8) and (1.06, 6.40). We checked (0, 8) in solution 1. Checking (1.06, 6.40) in the original equation, we see that $2(6.40) = 12.80$ and $-3(1.06) + 16 = 12.82$. It checks.

Horizontal Lines

Recall that lines with negative slope go down to the right. Lines with positive slope go up to the right. Lines with 0 slope are horizontal. (Vertical lines are a different matter; they are discussed in Lesson 3-4.)

Example 3

a. Graph the line $y = 2$.
b. Identify the domain and range of the function represented by this graph.

Solution

a. The equation $y = 2$ is equivalent to $y = 0x + 2$. This shows that the y-intercept is 2. So the line contains (0, 2) and the slope is 0. A slope of 0 means that for a horizontal change of 1 unit there is a vertical change of 0 units. In other words, there is no vertical change and the graph is a horizontal line.

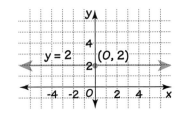

b. Because x does not appear in the equation, x can be any real number. However, y has the single value 2. Thus, **The domain is the set of all real numbers and the range is {2}.**

In general, a line is horizontal if and only if it has an equation of the form $y = b$. Its slope is 0 and its y-intercept is b.

Parallel Lines and Slope

Consider the graphs of $y = 3x - 7$ and $y = 3x + 4$ shown at the right. Both lines have slope 3. On each line, as you move 1 unit to the right, the line rises 3 units. Right triangles ABC and DEF are congruent by SAS Congruence, so these lines form congruent corresponding angles with the y-axis at A and D. Consequently, $\overline{AB}$ and $\overline{DE}$ are parallel.

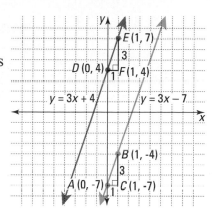

This argument can be repeated with any two lines that have the same slope. Thus, the following theorem can be proved.

> **Theorem**
> If two lines have the same slope, then they are parallel.

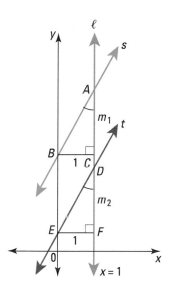

The converse of the theorem above is: If two lines are parallel, then they have the same slope. This can also be proved. Suppose lines s and t are parallel and not vertical. Let m_1 be the slope of s, and m_2 be the slope of t. We want to show $m_1 = m_2$. Draw line ℓ with equation $x = 1$. Note that ℓ is a transversal to the parallel lines. Draw horizontal segments from the y-intercepts of lines s and t to line ℓ as shown at the left. Then $m_1 = \text{slope of } s = \frac{AC}{BC} = \frac{AC}{1} = AC$, and $m_2 = \text{slope of } t = \frac{DF}{EF} = \frac{DF}{1} = DF$. Now recall that corresponding angles formed by parallel lines and a transversal are congruent. So $\angle BAC \cong \angle EDF$. By AAS Triangle Congruence, then, right triangles ABC and DEF are congruent, so $AC = DF$. Since $m_1 = AC$ and $DF = m_2$, the slopes are equal. We have proved the theorem below.

> **Theorem**
> If two non-vertical lines are parallel, then they have the same slope.

QUESTIONS

Covering the Reading

1. In the equation $y = mx + b$, the slope is __?__ and the y-intercept is __?__.

2. A slope of 7 means a __?__ change of __?__ units for every horizontal change of one unit.

3. Tell whether or not the rate of change means a slope of $-\frac{5}{6}$.
   a. a vertical change of -6 units for a horizontal change of 5 units
   b. a vertical change of $-\frac{5}{6}$ unit for every horizontal change of 1 unit
   c. a vertical change of 6 units for a horizontal change of -5 units
   d. a vertical change of 1 unit for a horizontal change of $-\frac{5}{6}$ unit

4. Refer to the line graphed in Example 1. Start at the point (1, 11).
   a. Going 1 unit to the right and 4 units up puts you at what point?
   b. Verify that your answer to part **a** lies on the line.

5. Refer to the line graphed in Example 2. It appears that (4, 2) lies on the line. Verify that this is true using the given equation for the line.

6. Consider the equation $4y = -7x - 20$.
   a. Rewrite the equation in slope-intercept form.
   b. Identify the slope and y-intercept.

7. Graph the line whose equation is $y = 1$.

8. The equation $y = b$ represents a __?__ line with slope __?__.

9. If two lines are parallel, what can be said about their slopes?

In 10 and 11, refer to the proofs dealing with parallel lines and slopes.

10. In the proof that lines with the same slope are parallel, name the corresponding sides and angles which were used to show $\triangle ABC$ to be congruent to $\triangle DEF$.

11. In the proof that parallel lines have the same slope, which corresponding sides and angles were used to show $\triangle ABC \cong \triangle DEF$?

12. Suppose line ℓ is parallel to $y = \frac{1}{3}x - 2$.
    a. What is the slope of ℓ?   b. What is a possible y-intercept of ℓ?

Applying the Mathematics

13. a. Draw the line with y-intercept -6 and slope $\frac{2}{5}$.
    b. Write the equation of this line in slope-intercept form.
    c. Use the equation to find x when y is 3. Check your answer by showing that the point is on the line.

14. a. Graph the lines $y = 2$ and $y = 2x$ on the same set of axes.
    b. Where do the lines intersect?

15. The Washington family has 20 newspapers in their recycling bin and is recycling 2 more newspapers each day. The Ebert family also recycles 2 newspapers a day; they now have only 6 papers.
    a. Graph the number of newspapers n each family has after d days.
    b. Relate the graphs to the content of this lesson.

16. A line does not cross the x-axis but goes through the point (17, -68).
    a. Give an equation for the line.   b. What is the slope of the line?

17. Recall Katie's bird feed from Lesson 3-1.
    a. Graph the amount of bird feed y that will be left in the sack after x days if Katie puts 2.5 pounds of feed into the feeder each day.
    b. Write an equation relating x and y.

18. Consider the equation $5x + 2y = 24$.
    a. Rewrite the equation in slope-intercept form.
    b. Identify the slope and the y-intercept.
    c. Graph the line using the slope and the y-intercept.

Review

In 19–21, tell whether the slope of the line is positive, negative, or zero. *(Lesson 3-1)*

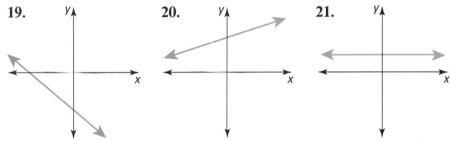

19.   20.   21.

22. A tank has a slow leak. The water level starts at 100 inches and falls $\frac{1}{2}$ inch per day.   *(Lesson 3-1)*
    a. What kind of situation is this: constant increase or constant decrease?
    b. Write an equation relating day d and the water level L.
    c. After how many days will the tank be empty?

23. Find S_4 if $S_1 = 2$ and $S_n = 3 \cdot S_{n-1}$.   *(Lesson 1-9)*

24. a. How much alcohol is in a 9-oz solution of water and alcohol that is 20% alcohol?
    b. How much alcohol is in an x-oz solution of water and alcohol that is 20% alcohol?
    c. How much water is in an x-oz solution of water and alcohol that is 20% alcohol?   *(Previous course, Lesson 1-1)*

Exploration

25. Consider the lines with equations $y = \frac{1}{5}x$ and $y = 5x$.
    a. Graph both lines on the same pair of coordinate axes.
    b. Find the slope of each line.
    c. Determine an equation for the bisector of the acute angles formed by these lines.
    d. Repeat parts **a–c**, using lines with equations $y = \frac{1}{3}x$ and $y = 3x$.
    e. Make a conjecture generalizing this problem and its results.
    f. Test your conjecture with an example where the lines have negative slopes.

Bird care. *House finch nestlings are cared for by both the male and the female. The male serenades and feeds his mate as she incubates the eggs.*

A tie is not pointless. *Shown are members of the New York Islanders and the New Jersey Devils of the National Hockey League. See Example 1. Football and soccer are other major team sports where a game can end in a tie.*

What Is a Linear Combination?

Consider the following problem.

> June has P 20¢ stamps for postcards and L 32¢ stamps for letters. Find the total value of her stamps.

The value of her 20¢ stamps is $20P$ cents, and the value of her 32¢ stamps is $32L$ cents. So the total value, in cents, of her stamps is

$$20P + 32L.$$

This expression is called a **linear combination** of P and L. In a linear combination, all variables are to the first power and are not multiplied or divided by each other.

Linear combinations occur in a wide variety of real-world situations.

An Example with a Discrete Domain

Example 1

In professional hockey, a win is worth 2 points, a tie is worth 1 point, and a loss is worth 0 points. Early in the season a team had 11 points.
a. Write an equation to express the relationship among the number of wins W, the number of ties T, and the number of losses L.
b. Make a table and a graph of all possible (W, T) that could have occurred.

▶

Solution

a. Each win is worth 2 points, so W wins are worth $2W$ points. A tie is worth 1 point, so T ties are worth $1T$ points. Because losses are worth 0 points, L losses add $0L$ to the total. This total is 11, so

$$2W + 1T + 0L = 11$$

that is $\qquad\qquad 2W + T = 11.$

b. Because of the situation, W and T must be nonnegative integers. Substitution and trial and error show that the only possible solutions are those shown in the table and the graph below.

W	T
0	11
1	9
2	7
3	5
4	3
5	1

Check

Check each ordered pair. For instance, 2 wins are worth 4 points, and 7 ties are worth 7 points. Altogether, this is 11 points. So (2, 7) checks.

An Example with a Continuous Domain

Linear-combination situations occur whenever quantities are mixed, as in chemicals, foods, diets, and medicines.

Example 2

A chemist mixes x ounces of a 20% acid solution with y ounces of a 30% acid solution. The final mixture contains 9 ounces of acid.
a. Write an equation relating x, y, and the total number of ounces of acid.
b. How many ounces of the 30% acid solution must be added to 2.7 ounces of the 20% acid solution to get 9 ounces of acid in the final mixture?

Solution

a. Find the amount of acid in each solution, and set the sum equal to 9. Let x = the number of ounces of 20% acid solution, and let y = the number of ounces of 30% acid solution.

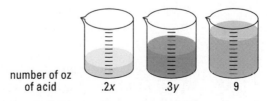

number of oz of acid .2x   .3y   9

Since 20% of the x ounces are acid, the amount of acid is 20% of x, or $0.2x$. Similarly, 30% of the y ounces are acid, so the amount of acid is $0.3y$. The total number of ounces of acid is thus $0.2x + 0.3y$. There are 9 ounces of acid, so an equation is $0.2x + 0.3y = 9$. ▶

b. Substitute 2.7 for x.

$$0.2(2.7) + 0.3y = 9$$
$$0.54 + 0.3y = 9$$
$$0.3y = 8.46$$
$$y = 28.2$$

So 28.2 ounces of the 30% acid solution must be added to 2.7 ounces of the 30% acid solution to get 9 ounces of acid in the final solution.

The equation $0.2x + 0.3y = 9$ from Example 2 can be graphed by solving for y and recognizing the slope-intercept form.

$$0.2x + 0.3y = 9$$
$$2x + 3y = 90 \qquad \text{Multiply each side by 10 to clear decimals.}$$
$$3y = -2x + 90 \qquad \text{Subtract } 2x \text{ from each side.}$$
$$y = -\tfrac{2}{3}x + 30 \qquad \text{Divide each side by 3.}$$

Thus, the slope is $-\tfrac{2}{3}$ and the y-intercept is 30. The graph is shown below.

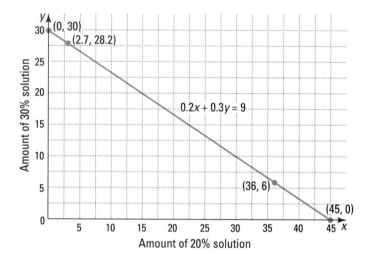

The graph lies entirely on the axes or in the first quadrant because the number of ounces of either solution must be a nonnegative real number. It is the segment connecting (0, 30) and (45, 0). Each point on the segment refers to a different mixture of the acid solutions. The point (2.7, 28.2) stands for 2.7 oz of 20% solution and 28.2 oz of 30% solution, as was found in Example 2b. The point (36, 6) means that 36 oz of the 20% solution could be mixed with 6 oz of the 30% solution to yield 9 oz of acid.

Any linear-combination situation in two variables is modeled by an equation whose graph is a line or a part of a line. This is the origin of the phrase "linear combination."

Covering the Reading

1. The expression $20P + 32L$ is called a __?__ of P and L.

2. Suppose you buy P pizzas at $12 each and D drinks at 75¢ each. Write an expression that tells how much you spend.

3. At a sale, Greta bought B blouses at $15 each, S skirts at $24 each, and H pairs of shoes at $26 a pair. Write a linear combination to find the amount she spent at the sale.

In 4 and 5, refer to Example 1.

4. With T as the dependent variable, $(3, 5)$ is a solution to $2W + T = 11$. This solution means the team won __?__ games, tied __?__ games, and earned a total of __?__ points.

5. In 1993–94 the Montreal Canadiens earned 96 points. If the Canadiens won 41 games that season, how many ties did they have?

6. Suppose that x ounces of a solution that is 60% acid are combined with y ounces of a 90% acid solution.
   a. How many ounces of acid are in the 60% solution?
   b. How many ounces of acid are in the 90% solution?
   c. How many total ounces of acid are there in the combination?
   d. If Alex wants 18 ounces of acid in the final mixture, what equation relates x, y, and the 18 total ounces of acid?
   e. Write the equation from part d in slope-intercept form.
   f. Graph the solutions to the equation.
   g. How many ounces of the 90% solution must be added to 9 ounces of the 60% solution to get 18 ounces of acid in the final mixture?

Applying the Mathematics

7. Suppose zucchini sells for 49¢ a pound and tomatoes sell for 59¢ per pound.
   a. What will be the cost of 2.5 pounds of zucchini and 3.4 pounds of tomatoes?
   b. What will be the cost of Z pounds of zucchini and T pounds of tomatoes?
   c. Write an equation indicating the amounts of zucchini Z and tomatoes T you can buy for $5.00.

8. William spent Saturday mowing lawns. He charged $10 for small lawns and $20 for large lawns and earned $140. Let S be the number of small lawns and L be the number of large lawns.
   a. What type of numbers makes sense for S and L in this context?
   b. Write an equation relating S, L, and the amount of money earned.
   c. Give all possible numbers of large and small lawns William could have mowed.
   d. Graph the equation of part b.

9. The Ironman triathlon is a sporting event made up of a 2.4-mile swim, a 112-mile bicycle race, and a marathon run of 26.2 miles. If a competitor goes at a rate of S minutes per mile swimming, B minutes per mile biking, and R minutes per mile running, what will be the competitor's total time for the triathlon?

10. Make up a problem whose solution is $3H + 4T = 12$.

Review

In 11 and 12, graph on a coordinate plane.   *(Lesson 3-2)*

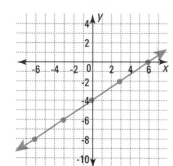

11. $y = 3$

12. $y = \frac{3}{4}x - 3$

13. What is the slope of a line parallel to $3y = 2x + 7$?   *(Lesson 3-2)*

14. For the line graphed at the left, determine each of the following.
    a. its slope
    b. its y-intercept
    c. an equation   *(Lessons 3-1, 3-2)*

15. The math department at a school has 100 reams of paper at the start of the school year. (A ream of paper contains 500 sheets.) Each day the department uses about $\frac{3}{4}$ of a ream.   *(Lesson 3-1)*
    a. Let d be the number of days and R be the number of reams remaining. Write a formula for R in terms of d.
    b. When the supply gets down to 10 reams, a new supply of paper needs to be ordered. After how many school days will paper need to be ordered?

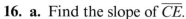

In 16 and 17, refer to the graph at the left, which shows the distance d in kilometers Terry is from a starting point S after t hours.   *(Lesson 3-1)*

16. a. Find the slope of $\overline{CE}$.
    b. Explain what the slope tells you about Terry's trip.

17. *True or false.*   During some part of the journey Terry traveled uphill. How can you tell?

In 18 and 19, find the distance between the given points.   *(Previous course)*

18. $(1, 3)$ and $(6, 15)$

19. (a, b) and (c, d)

Exploration

20. In many schools, a person's grade-point average is calculated using linear combinations. Some schools give 4 points for each A, 3 points for each B, 2 points for each C, and 1 point for each D. Suppose a person gets 7 As, 3 Bs, and 2 Cs.
    a. Calculate this person's total number of points.
    b. Divide your answer in part **a** by the total number of classes (12) to get the grade-point average.
    c. Calculate your own grade-point average for last year using this scheme.

The Graph of $Ax + By = C$

Chemists provide solutions. *Industrial chemists perform a variety of tasks. Some, like the one shown, do quality control work testing different items. Others conduct research, developing and testing new products.*

Recall the equation $0.2x + 0.3y = 9$ which describes the following linear combination situation from Lesson 3-3.

> A chemist mixes x ounces of a 20% acid solution with y ounces of a 30% acid solution. The final mixture contains 9 ounces of acid.

Because of the situation, both x and y must be positive. The graph was a segment. However, if you allow x and y to be any real numbers, then the graph of $0.2x + 0.3y = 9$ is the line shown below.

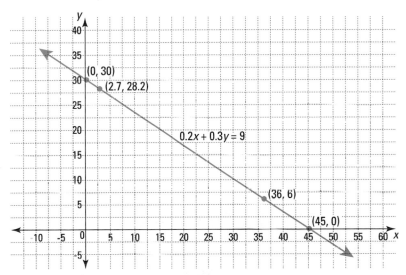

The equation $0.2x + 0.3y = 9$ is of the form $Ax + By = C$, with $A = 0.2$ and $B = 0.3$. When A and B are not both zero, the equation $Ax + By = C$ is called the **standard form of an equation for a line.**

An Equation for Any Line

The graph of $0.2x + 0.3y = 9$ is an instance of the following theorem.

Theorem
The graph of $Ax + By = C$, where A and B are not both 0, is a line.

Proof:
There are two cases to consider, depending on whether $B \neq 0$ or $B = 0$. When $B \neq 0$, the equation $Ax + By = C$ can be rewritten in slope-intercept form.

$$Ax + By = C$$
$$By = -Ax + C \qquad \text{Add } -Ax \text{ to each side.}$$
$$y = -\frac{A}{B}x + \frac{C}{B} \qquad \text{Multiply by } \frac{1}{B} \ (B \neq 0).$$

This is an equation of a line with slope $-\frac{A}{B}$ and y-intercept $\frac{C}{B}$. If $A \neq 0$, then the slope is not 0, and the line is oblique. If $A = 0$, then the slope is 0, and the line is horizontal. Then the line has equation $y = \frac{C}{B}$.
When $B = 0$, the equation $Ax + By = C$ can be written as follows.

$$Ax = C$$
$$x = \frac{C}{A} \qquad \text{Multiply by } \frac{1}{A} \ (A \neq 0).$$

This is an equation of a vertical line.

Vertical Lines

The equation $x + 0 \cdot y = 2$ is of the form $Ax + By = C$ with $B = 0$.

Example 1

a. Graph the line $x + 0 \cdot y = 2$.   **b.** Find the slope of the line.

Solution

a. In $x + 0 \cdot y = 2$, y can take on any value, but x is always 2. That is, $x = 2$. It is easy to make a table and plot the points.

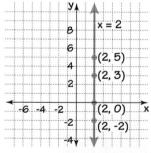

x	y
2	-2
2	0
2	3
2	5

b. Pick two points on the line. We use the points (2, 0) and (2, 3).

$$m = \frac{3 - 0}{2 - 2} = \frac{3}{0}$$

Since division by 0 is undefined, There is no slope. The slope is said to be *undefined*. By the same argument, the slope of any vertical line $x = h$ is undefined.

Only oblique and horizontal lines can be written in slope-intercept form. But any line in the coordinate plane has an equation that can be written in standard form.

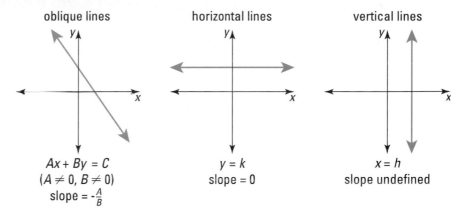

oblique lines

$Ax + By = C$
$(A \neq 0, B \neq 0)$
slope = $-\frac{A}{B}$

horizontal lines

$y = k$
slope = 0

vertical lines

$x = h$
slope undefined

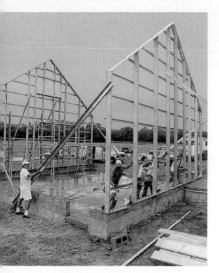

It's a frame-up. *Frames for most houses are made of wood, but steel frames like the one shown here are becoming more common. Can you find oblique, horizontal, and vertical lines in this picture?*

Notice that all oblique lines and all horizontal lines can be graphs of functions. A vertical line cannot be the graph of a function.

Graphing a Line Using Intercepts

Although you could rewrite the standard form of a linear equation in slope-intercept form in order to make a graph, it is often much quicker to graph such equations by hand using the x- and y-intercepts. The **x-intercept** is the value of x at the point where a graph crosses the x-axis. This point has second coordinate 0.

Example 2

Graph the equation $6x - 3y = 12$ using its intercepts.

Solution

To find the x-intercept, substitute 0 for y, and solve for x.

$$6x - 3(0) = 12$$
$$x = 2$$

The x-intercept is 2.

To find the y-intercept, substitute 0 for x, and solve for y.

$$6(0) - 3y = 12$$
$$y = -4$$

The y-intercept is -4.

Plot (2, 0) and (0, -4). Draw the line containing them, as shown at the right.

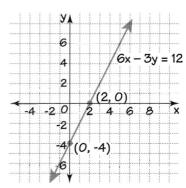

Check

The point (1, -2) appears to be on the graph. Substitute to see if its coordinates satisfy the equation.

Does $6(1) - 3(-2) = 12$?

Does $6 + 6 = 12$? Yes, it checks.

QUESTIONS

Covering the Reading

1. **a.** What is the standard form of an equation for a line?
   b. Given the equation $5x - 2y = 24$, identify A and B.

2. **a.** Rewrite the equation $x = -6$ in standard form.
   b. Graph the equation from part **a.**
   c. Use the definition of slope to explain why the slope of $x = -6$ is undefined.

In 3–5, refer to the proof of the theorem in this lesson.

3. **a.** If neither A nor B is 0, the equation $Ax + By = C$ can be written in slope-intercept form as __?__.
   b. The slope of the line is __?__.
   c. The y-intercept of the line is __?__.

4. If $A = 0$ and $B \neq 0$, then the line is __?__ and has __?__ slope.

5. If $A \neq 0$ and $B = 0$, then the line is __?__.

6. Match the line with a description of its slope.
   a. horizontal line   (i) zero slope
   b. vertical line   (ii) nonzero slope
   c. oblique line   (iii) slope undefined

7. The line whose equation is of the form $y = mx + b$ is oblique when m __?__ 0.

8. How many intercepts does an oblique line have?

9. The graph of $y = 7$ is a __?__ line with y-intercept __?__.

10. *True or false.* Every equation of the form $Ax + By = C$, where A and B are not both 0, represents a function. Justify your answer.

11. To find the x-intercept of a graph of points (x, y), for which variable should you substitute 0?

In 12 and 13, an equation is given. **a.** Find its x-intercept. **b.** Find its y-intercept. **c.** Graph the line using the points from parts **a** and **b**.

12. $4x - 3y = 24$ 13. $5x + 6y = 15$

Applying the Mathematics

In 14–16, an equation for a line is given. **a.** Tell whether the line is vertical, horizontal, or oblique and justify your answer. **b.** Give all intercepts for each line. **c.** Graph each equation.

14. $y = 4$ 15. $2x - 3y = 18$ 16. $2x = 16$

17. Give an equation in standard form for the line with y-intercept $-\frac{1}{5}$ and slope 2.

18. A line has no y-intercept and goes through the point (15, 34). Give an equation for the line.

19. Meg combines N oz of a solution that is 10% chlorine with Y oz of a solution that is 20% chlorine. She ends up with a mixture that contains 1.2 oz of chlorine.
    a. Write an equation relating N, Y, and the amount of chlorine in the mixture.
    b. Graph the equation you obtained in part **a** by finding the N- and Y-intercepts. Consider N to be the independent variable.
    c. Use your graph to find out how many ounces of the 20% solution must be added to 8 oz of the 10% solution to get the final mixture.

In 20–22, the equation represents a relation in the (x, y) plane. **a.** Give the set of possible values of x. **b.** Give the set of possible values of y.

20. $3x + 2y = 6$ 21. $3x = 6$ 22. $2y = 6$

Review

23. It costs $10.20 to run a 3-line classified advertisement in a local paper for five days, and $12.35 to run the same ad for seven days. What is the cost of x ads for five days and y ads for seven days? *(Lesson 3-3)*

24. You have some money saved from your job. You invest S of it in a savings account that pays 4% interest and the rest R in a checking account that pays 2%. You earn $50 interest in one year.
    a. Write an equation relating S, R, and the total amount of interest.
    b. Give three possible pairs of values for R and S. *(Lesson 3-3)*

25. A city police department pays police officers $2500 per month and sergeants $3000 per month. The total payroll for the month for officers and sergeants is $310,000. *(Lesson 3-3)*
    a. Write an equation relating the number of police officers P, the number of sergeants S, and the total monthly payroll.
    b. If there are 10 sergeants, how many police officers are there?

26. Do the changes mean a slope of $-\frac{4}{3}$? *(Lessons 2-4, 3-2)*
    a. a vertical change of -3 units for a horizontal change of 4 units
    b. a vertical change of -4 units for a horizontal change of 3 units
    c. a vertical change of $-\frac{4}{3}$ units for a horizontal change of 1 unit
    d. a vertical change of 1 unit for a horizontal change of $-\frac{4}{3}$ units

Drug education. *Pictured is a police officer teaching young students in an anti-drug program.*

Exploration

27. a. Find the x- and y-intercepts of $\frac{x}{2} + \frac{y}{7} = 1$.
    b. Find the x- and y-intercepts of $\frac{x}{-5} + \frac{y}{6} = 1$.
    c. Based on parts **a** and **b** above, make a conjecture about the x- and y-intercepts of $\frac{x}{a} + \frac{y}{b} = 1$. Either prove or give a counterexample to your conjecture.

3-5

Finding an Equation of a Line

Two points determine a line. You use this idea every time you draw a line through two points with a ruler. It is a postulate from geometry. In algebra, this idea raises the question: What is an equation of the line through two given points? This lesson will show you how to get such an equation.

Finding a Linear Equation

Example 1

Suppose you remember that 32°F = 0°C and 212°F = 100°C, but you have forgotten the conversion formula. You know that the formula is linear. Reconstruct the formula with F as a function of C.

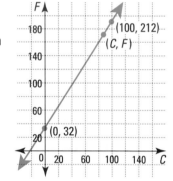

Solution 1

Because F is a function of C, the independent variable is C and the dependent variable is F. So, ordered pairs are of the form (C, F). Start with the slope-intercept form and substitute C for x and F for y.

$$F = mC + b$$

The values of m and b must be found. To find the slope m, use the given points $(0, 32)$ and $(100, 212)$.

$$m = \frac{212 - 32}{100 - 0} = \frac{180}{100} = 1.8$$

The F-intercept is 32, so $b = 32$. Substitute the values for the slope and the intercept into $F = mC + b$ to get

$$F = 1.8C + 32.$$

Solution 2

Using the points $(0, 32)$ and $(100, 212)$, find $m = 1.8$. Let (C, F) be any other point on the line. Substitute (C, F) and either $(0, 32)$ or $(100, 212)$ into the slope formula. We use $(100, 212)$.

$$\frac{F - 212}{C - 100} = 1.8$$

To put this equation in slope-intercept form, multiply both sides by $C - 100$.

$$F - 212 = 1.8(C - 100)$$

Now solve for F.

$$F - 212 = 1.8C - 180$$
$$F = 1.8C + 32$$

Each of the equations $F = 1.8C + 32$ and $F - 212 = 1.8(C - 100)$ describes the situation of Example 1. The slope-intercept form is useful for computing values of F quickly if you know the values of C. The form $F - 212 = 1.8(C - 100)$ shows the slope and a specific point on the graph, and it can be used to find the slope-intercept form.

Point-Slope Form of a Line

The method of Solution 2 of Example 1 can be generalized as follows.

Point-Slope Theorem

If a line contains (x_1, y_1) and has slope m, then it has equation $y - y_1 = m(x - x_1)$.

Proof:

Let ℓ be the line with slope m containing (x_1, y_1). If (x, y) is any other point on ℓ, then using the definition of slope,
$$m = \frac{y - y_1}{x - x_1}.$$
Multiplying both sides by $x - x_1$ gives:
$$y - y_1 = m(x - x_1).$$
This is the desired equation of the theorem.

The equation $y - y_1 = m(x - x_1)$ is called a **point-slope equation** for a line. The most convenient form to use depends on the information given. If you know the slope and the y-intercept, use $y = mx + b$. If you know the slope and some point other than the y-intercept, use $y - y_1 = m(x - x_1)$. If you know two points, find the slope and then use either point in the point-slope form. The following examples illustrate this procedure.

Example 2

Find an equation of the line through $(3, 5)$ and $(6, -1)$.

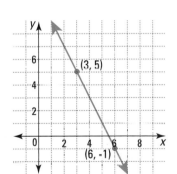

Solution

First, find the slope.
$$m = \frac{-1 - 5}{6 - 3} = \frac{-6}{3} = -2$$
Then, use the point-slope form with either point. We use $(3, 5)$. The equation is
$$y - 5 = -2(x - 3).$$
In standard form, this equation is $2x + y = 11$.

Check 1

Is the point $(3, 5)$ on the line? Does $5 - 5 = -2(3 - 3)$? Yes.
Is the point $(6, -1)$ on the line? Does $-1 - 5 = -2(6 - 3)$? Yes.

Check 2

Use the other point in the point-slope form. The point $(6, -1)$ gives
$$y - -1 = -2(x - 6) \text{ or } y + 1 = -2(x - 6).$$
In standard form, this becomes $2x + y = 11$, the same as before.

Example 3

In a physics experiment, a spring is 12 centimeters long with a 15-gram weight attached. Its length increases 0.5 centimeter with each additional gram of weight.
a. Write an equation relating the spring length L and weight W.
b. Find the length of the spring with a 20-gram weight attached.

Solution

a. The slope is 0.5 cm/gram. Because the units for slope are centimeters per gram, the length is the dependent variable and the weight is the independent variable. The point (15, 12) is on the line. Substituting this point into the point-slope form gives

$$L - 12 = 0.5(W - 15).$$

b. Substitute 20 for W.

$$L - 12 = 0.5(20 - 15)$$
$$L = 14.5$$

The spring is 14.5 centimeters long with a 20-gram weight.

Finding Equations for Piecewise-Linear Graphs

It is usually not possible to find a single equation to describe a piecewise-linear graph. However, each segment or ray can be described with an equation and the domain to which it applies.

Example 4

During the winter months, an electric company calculates bills for the kilowatt-hours (kwh) its residential customers use based on the following:

> $9.00 monthly service fee;
> $ .10 per kwh energy charge for the first 400 kwh;
> $ .07 per kwh energy charge for each kwh over 400.

a. Draw a graph showing how cost C (in dollars) is related to usage k (in kwh).
b. Express the relation between C and k in symbols.

Solution

a. For the first 400 kwh the increase is constant. This piece of the graph is a segment with one endpoint (0, 9) and a slope of $.10 per kwh. The other endpoint is the point where $k = 400$. When k = 400,
$C = 9 + .10(400) = 49.$
At (400, 49) a new constant rate of increase begins. Because the slope of $.07 per kwh is less for this piece of the graph, this piece increases more slowly. This part of the graph is a ray, since the rate of $.07 per kwh applies to all values of k greater than 400. The graph is shown at right.

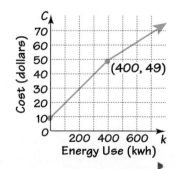

Water power. *Shown is a hydroelectric plant in Quebec, near James Bay. Quebec is a leading producer of hydroelectric power in North America. These plants supply about 95% of Quebec's electricity.*

b. The slope of the segment from the point (0, 9) to the point (400, 49) is 0.1. Thus,

$$C = .1k + 9 \text{ for } 0 \leq k \leq 400.$$

An equation for the ray beginning at (400, 49) with slope 0.07 is given by the point-slope form:

$$C - 49 = .07(k - 400)$$
$$C = .07(k - 400) + 49$$
$$C = .07k + 21, \text{ for } k > 400.$$

In summary, this situation can be described symbolically as follows:

$$C = \begin{cases} .1k + 9, & \text{for } 0 \leq k \leq 400 \\ .07k + 21, & \text{for } k > 400. \end{cases}$$

QUESTIONS

Covering the Reading

1. How many points determine a line?

2. A line contains the points (6, 4) and (2, 8).
   a. Find its slope.
   b. Use (6, 4) and the slope from part **a** to write an equation for the line.
   c. Does (2, 8) satisfy the equation of part **b?** Justify your answer.

3. *True or false.*   A line is determined by its slope and any point on it.

4. The point-slope form of the equation for a line with slope m and passing through point (x_1, y_1) is __?__ .

5. A line passes through the points (7, 12) and (5, 16).
   a. Find the slope of the line.
   b. Use the slope and the point (7, 12) in the point-slope form to find an equation of the line.
   c. Put your answer to part **b** in standard form.
   d. Check that (5, 16) satisfies the equation.

6. Refer to Example 3. Suppose a spring is 10 cm long with a 25-gram weight attached and its length increases 0.2 cm with each additional gram weight.
   a. Write an equation relating spring length L and weight W.
   b. Find the spring length with a 20-gram weight.

7. To find an equation for a line from the given information, which is generally easier to use, the point-slope form or the slope-intercept form?
   a. given the slope and a point other than the y-intercept
   b. given the y-intercept and the slope
   c. given two points

In 8 and 9, write an equation for the line with the given properties.

8. slope 6 and y-intercept -1

9. slope $\frac{2}{3}$ and passing through (7, 1)

In 10 and 11, refer to Example 4.

10. Find the monthly electric cost for a family that uses 500 kwh of electricity.

11. *Multiple choice.* The formula $C = .07k + 21$ gives the cost of k kwh of electricity for which values of k?
(a) $k > 0$ (b) $0 \leq k \leq 400$ (c) $k > 400$

Applying the Mathematics

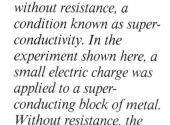

12. Scientists often use kelvins to measure temperature. On this scale, $32°F = 273.15$ kelvins and $212°F = 373.15$ kelvins. Let F represent the Fahrenheit temperature and K represent the Kelvin temperature. The relationship is linear. Find an equation giving temperature in kelvins as a function of temperature in degrees Fahrenheit.

13. A printer finds that it costs $1290 to print 30 books and $1335 to print 45 books. Let c be the cost of printing b books. Assume c can be written as a linear function of b.
a. Find an equation giving cost as a function of the number of books printed.
b. How much will it cost to print 0 books? (This is the set-up cost.)
c. How much will it cost to print 100 books?

14. Given the line with equation $y = \frac{4}{3}x - 7$, find an equation for the line that contains the point $(6, 4)$ and is parallel to the given line.

15. Use the following phone rates: A customer is allowed to make a maximum of 80 local calls per month for a $13.15 fee. Each local call after the 80th is billed at a rate of $0.035 per call.
a. Find the monthly phone bills for families making 70 local calls, 80 local calls, and 100 local calls.
b. Draw a graph of the relation between the cost C and the number of local calls n.
c. Describe this situation algebraically by finding an equation for each segment or ray.

Suspense. *At temperatures near 0 kelvins, some materials allow electricity to flow without resistance, a condition known as super-conductivity. In the experiment shown here, a small electric charge was applied to a super-conducting block of metal. Without resistance, the charge continues to flow, inducing a magnetic field in which the small metal cylinder is suspended.*

Review

16. Consider the equation $5x + 2y = 20$. *(Lesson 3-4)*
a. Find the x-intercept.
b. Find the y-intercept.
c. Graph the equation using the intercepts.

17. Let $P = (3, 4)$, $Q = (3, -5)$, $R = (-2, -5)$, and $S = (-2, 4)$.
a. Graph rectangle $PQRS$.
b. Give equations for the lines containing the four sides.
c. Find the area of $PQRS$. *(Previous course, Lessons 3-2, 3-4)*

18. As of November, 1994, the greatest combined number of points ever scored in a professional basketball game was 370 by the Denver Nuggets and the Detroit Pistons in December, 1983. A free throw is worth 1 point, a regular field goal is worth 2 points, and a three-point shot is worth 3 points. Suppose there were A free throws, B regular field goals, and C three-point shots made in that game. Write an equation which relates all these numbers.   *(Lesson 3-3)*

19. Recall that the weights of similarly-shaped people vary directly as the cube of their heights. Suppose a professional ballerina who is 5'8" tall weighs 120 pounds. How much would a 5'1" person with the same shape weigh?   *(Lesson 2-1)*

20. Write an inequality to describe the points graphed on the number line below.   *(Previous course)*

21. If $3x + 8 = 40$, find the value of $6x + 16$.   *(Previous course)*

22. Triangle MOP is shown below. Graph the reflection image of $\triangle MOP$ over the x-axis. Label the points M', O', and P', respectively, and give their coordinates.   *(Previous course)*

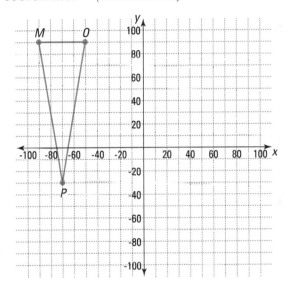

Dance career. *Ballet dancers usually become professional by age 20 and retire by 45.*

Exploration

23. In 1957, the world record in the 800-meter freestyle swimming event was about 10 min 30 sec for women and 9 min 15 sec for men. Between 1957 and 1980, these records decreased at a rate of about 4 sec/yr for women and 3 sec/yr for men.
   a. According to this information, what should the records have been in 1992?
   b. Check a book of records and see if these predictions for 1992 were true.
   c. According to this information, predict the records for the year 2000, and explain how you made your predictions.

*Using Linear
Models to
Approximate
Data*

IN-CLASS
ACTIVITY

Work with a partner.

Many real-world situations involve more than two pairs of
measurements. These pairs can be graphed, and, although the
points might not all lie on a line, it may be possible to find a line
that is close to many of the points. Finding an equation for such a
line is the purpose of this Activity.

The table and the graph below show the number of cars and the
number of people killed in traffic accidents in ten countries in 1990.

Country	Cars (millions)	Traffic deaths
Argentina	4.3	3,054
Australia	7.7	4,210
Belgium	3.8	1,937
Bulgaria	1.3	1,409
Canada	12.6	4,210
France	23.6	10,198
Israel	0.8	439
Italy	27.3	8,717
Japan	34.9	14,398
W. Germany	30.7	7,435

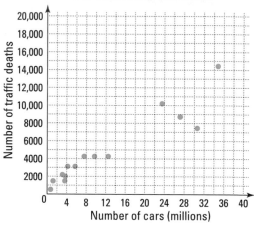

**Traffic Deaths and Number of Cars
for Selected Nations**

Source: U.S. Bureau of the Census,
Statistical Abstract of the United States: 1993

1 Graph the data in the table or copy the graph.

2 With your partner, draw a line on your graph that
both of you agree best describes the trend in the
points. This is called "fitting a line to the data by eye."

3 Estimate the coordinates of two points on your
line, and use a technique discussed in Lesson 3-5 to
find an equation for this line.

4 In 1990, there were about 143.5 million cars in the
United States. Use the equation for the line you
found in step 3 to estimate the number of traffic deaths
in the United States in 1990.

LESSON
3-6

Fitting a Line to Data

Traffic jam. *Shown is traffic that has come to a halt due to an accident on an autobahn in Germany. Autobahns, four-lane express highways in Germany, have no speed limits.*

In the Activity on page 168, you were first given a **scatterplot** of discrete points. If you used the points $A = (4, 2000)$ and $B = (38, 13,000)$ to find an equation for a line close to the points in the scatterplot, then your graph would look like the one below.

Example 1

Use the line through the points $A = (4, 2000)$ and $B = (38, 13,000)$ to fit a line to the data points about number of cars and traffic deaths.

Solution

First, find its slope.

$$\text{slope} = \frac{13,000 - 2000}{38 - 4}$$

$$= \frac{11,000}{34}$$

$$\approx 323.5$$

Then, use one of the points with this slope in the point-slope form. We use $(4, 2000)$.

$$y - 2000 = 323.5(x - 4)$$

Solve for y.

$$y - 2000 = 323.5x - 1294$$
$$y = 323.5x + 706$$

An equation of the line is $y = 323.5x + 706$.

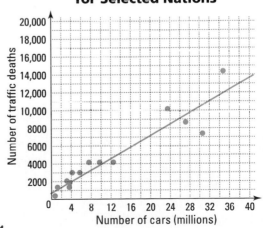

Traffic Deaths and Number of Cars for Selected Nations

This equation describes the relation between x and y in the data set reasonably well. The slope 323.5 tells you that for each increase of 1,000,000 in the number of cars in a country there is an increase of about 324 in the number of traffic deaths. The y-intercept 706 indicates that in a country with no cars, there would be 706 traffic deaths. Of course, this would not happen. Thus, this linear model does not apply to countries with a small number of cars.

Finding the Line of Best Fit

As you saw from the Activity, many possible equations could arise from fitting a line by eye. In fact, everybody who does the Activity could have a different, but reasonable, equation for a line through the points.

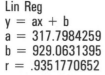

Lin Reg
$y = ax + b$
$a = 317.7984259$
$b = 929.0631395$
$r = .9351770652$

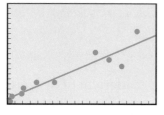

$0 \le x \le 40,$ x-scale $= 2$
$0 \le y \le 20,000,$ y-scale $= 1000$

Which line is the best fit for these data? Some computer software packages and most graphics calculators will find the equation of a line called the *line of best fit* or the *least squares line* or the *regression line*. Why this line is the "best" is explained in more advanced courses, but it is based on conventions that are accepted all over the world.

To find the regression line, use a calculator or computer to calculate the line of best fit. For details, check the manual or ask your teacher.

You may find that your calculator or computer displays only the slope and the y-intercept of the regression line; you must write the equation yourself. For instance, one calculator displays an equation of the regression line for the data set about traffic deaths as shown at the left. Thus, the regression equation is about $y = 317.8x + 929.1$. Notice that the slope and y-intercept of this line are close to those in Example 1.

Before. Prior to 1986, Chicago's Lake Shore Drive (shown between the tall, white Amoco Oil Building and the black building in front) contained two right angles.

Example 2

In 1990, there were approximately 143.5 million cars in the United States. Predict the number of traffic deaths using
a. the equation in Example 1.
b. the equation for the regression line.

Solution

a. Substitute 143.5 for x in the equation
$$y = 323.5x + 706.$$
$$y = 323.5 \cdot 143.5 + 706$$
$$y = 47{,}128.25$$
About 47,128 people would be predicted to die in traffic accidents if this line is accurate.

b. Substitute 143.5 for x in the equation
$$y = 317.8x + 929.1.$$
$$y = 317.8 \cdot 143.5 + 929.1$$
$$y = 46{,}533.4$$
About 46,533 people would have been predicted to die in traffic accidents using this model.

The actual number of traffic deaths in the U.S. in 1990 was 46,586. The prediction using the line of best fit is very accurate in this instance.

Correlation Coefficient

Most software packages and calculators display a third number, r, when finding the regression line. The number r is called the *correlation coefficient*. The correlation coefficient is a number from -1 to 1.

The sign of r indicates the direction of the relation between the variables. If r is positive, the slope of the regression line is positive. If r is negative, the slope of the regression line is negative. For the data in Example 2, $r \approx 0.935$. Because the correlation coefficient is positive, the slope of the regression line is also positive.

The absolute value of the correlation coefficient indicates the *strength* of the linear relationship. When $|r| = 1$, there is a perfect linear relationship; the points are collinear. Such perfect correlations are rare. The closer the magnitude of r is to 1, the stronger the linear relationship is. A correlation coefficient of 0 or close to 0 indicates there is no linear relationship between the variables. The value of r for the situation in this lesson, 0.935, is close to 1. It indicates a reasonably strong linear relationship between the variables. In other words, the regression equation is a reasonably accurate model for this set of data.

After. *To improve traffic safety, Lake Shore Drive was curved (shown under construction on landfill in the foreground).*

The Line of Best Fit Through Two Points

Because two points determine a line, the equation of the line of best fit determined by two points will have a perfect correlation of ±1. So, you can use a calculator or computer to find an equation for the line between two given points. Example 3 shows how this is done.

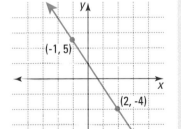

Example 3

Use linear regression to find an equation of the line through the points (-1, 5) and (2, -4).

Solution

Enter the points as data points. Then find the equation of the regression line. Our graphics calculator gives
$$y = -3x + 2$$
with $r = -1$.

Check

Substitute each ordered pair into this equation to see if the ordered pair satisfies the equation. For instance, if $x = -1$, then $y = -3 \cdot -1 + 2 = 3 + 2 = 5$, which checks. If $x = 2$, then $y = -3 \cdot 2 + 2 = -4$, which checks.

Notice that the correlation coefficient in Example 3 is -1. This was to be expected; this is a perfect linear fit, and the slope of the line is negative.

QUESTIONS

Covering the Reading

1. Refer to the scatterplot in the Activity on page 168. What does the ordered pair (34.9, 14,398) signify?

2. Briefly describe two ways to fit a line to data.

3. Refer to the graph in Example 1.
   a. How many data points are above the line determined by the equation $y = 323.5x + 706$?
   b. How many data points are below this line?

In 4–7, use the given equation to estimate the number of traffic deaths in a country where there are 50 million cars.

4. your equation from the In-class Activity on page 168

5. the equation $y = 323.5x + 706$ of Example 1

6. the equation $y = 317.8x + 929.1$ of the regression line from the lesson

7. *Multiple choice.* Which seems to be the most reasonable domain for x in the regression equation $y = 317.8x + 929.1$?
   (a) the set of all real numbers
   (b) $\{x: 0 \le x \le 40\}$
   (c) $\{x: 0.5 \le x \le 40\}$
   (d) $\{x: x \ge 0.5\}$

8. Explain what the correlation coefficient measures.

9. *True or false.* The correlation coefficient of a set of data points is the same as the slope of the equation of the regression line.

10. What does a correlation coefficient of 0 indicate?

In 11 and 12, refer to the owner's manual or other documentation for your technology.

11. What is the *line of best fit* called on your calculator or computer?

12. Does your technology give an equation for the line of best fit or merely the slope and y-intercept?

13. a. Use the regression line to find an equation of the line through the points (6, 7) and (-2, 13).
    b. Find the value of the correlation coefficient, and explain what it indicates about the equation you just formed.

In 14 and 15, use the figure at the right. It shows the line of best fit for survey data of the scores on a test and the number of hours spent studying for that test. An equation for the line is $y = 3.510x + 51.642$, and $r = 0.89$.

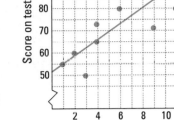

14. How many students were surveyed?

15. Consider the student who studied ten hours.
    a. What did this student score on the test?
    b. What test score was predicted by the regression equation?

In 16–19, multiple choice. What is the best description of the relation between the variables in the scatterplot?
(a) strong negative correlation
(b) weak positive correlation
(c) strong positive correlation
(d) correlation approximately 0

16. 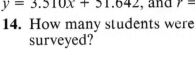 17. 18. 19.

20. The average weight in kilograms for girls of various ages in the United States is given below.

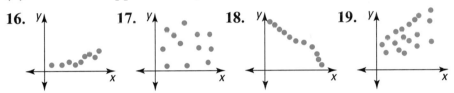

age in years	.5	1	2	3	4	6	8	10
weight in kg	7.2	9.1	11.3	13.6	15	20.4	25.4	31.3

a. Draw a scatterplot of these data or print a hard copy from your technology.
b. Find an equation of the regression line.
c. What is the unit for the slope of the regression line?
d. Write a sentence about the meaning of the y-intercept in this situation.
e. Find the correlation coefficient.
f. What does the correlation coefficient suggest about the relationship between the variables?
g. Draw the graph of the regression line on your scatterplot.
h. Use your equation for the regression line to predict the average weight of a seven-year-old girl.
i. Over what domain do you expect the regression equation to be a good model for the relation between the age and weight of a female in the United States? Explain your reasoning.

21. The sum of the measures of the angles of a triangle is 180°. In a convex quadrilateral the sum is 360°. Let n be the number of sides in a convex polygon and S be the sum of its angle measures. Use these two points to find an equation relating n and S, assuming there is a linear relationship. Let S be the dependent variable. *(Lesson 3-5)*

22. Find an equation of the line whose graph is shown below. *(Lesson 3-5)*

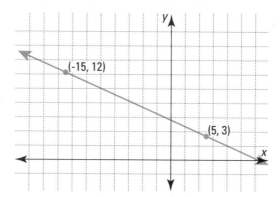

23. Graph f when $f(x) = \begin{cases} -8, & \text{for } x < 1 \\ -8x, & \text{for } x \geq 1. \end{cases}$ *(Lessons 3-2, 3-5)*

24. Graph $90x - 75y = 4500$. *(Lesson 3-4)*

25. Suppose x ounces of a 40% acid solution are mixed with y ounces of a 20% acid solution, and the result contains 10 ounces of acid. *(Lesson 3-3)*
    a. Write an equation relating x, y, and the total ounces of acid.
    b. How many ounces of the 40% acid solution must be mixed with 5.1 ounces of the 20% acid solution to get 10 ounces of acid?

26. Consider the sequence defined as follows. *(Lesson 1-9)*
    $$\begin{cases} t_1 = 20 \\ t_n = t_{n-1} - 3, \text{ for integers } n \geq 2 \end{cases}$$
    a. Write the first four terms of the sequence.
    b. If $t_{12} = -13$, what does t_{13} equal?

27. It costs $1.25 to park for up to an hour in a hospital parking lot plus $.75 for each additional half hour or fraction thereof. How much does it cost to park in this lot for 4 hours and 15 minutes? *(Previous course)*

Exploration

28. Refer to the regression line given in this lesson.
    a. Find the average of the x-values of the data. Call it $\bar{x}$.
    b. Find the average of the y-values of the same data. Call it $\bar{y}$.
    c. $(\bar{x}, \bar{y})$ is called the *center of gravity* of this data set. Verify that the center of gravity is on the regression line.

LESSON 3-7

Recursive Formulas for Arithmetic Sequences

Mechanically inclined. *Auto mechanics receive training from experienced mechanics, factory service instructors, or formal schools.*

What Is an Arithmetic Sequence?

If you buy a new car, you may be advised to have an oil change after driving 1000 miles and then every 3000 miles thereafter. Then the following sequence gives the mileage when oil changes are required:

$$1000, 4000, 7000, 10{,}000, 13{,}000, 16{,}000, \ldots$$

This is a constant-increase situation. There is a constant difference of 3000 between successive terms. A sequence with a *constant difference* between successive terms is called an **arithmetic sequence.** (Here the word *arithmetic* is used as an adjective; it is pronounced *arith<u>met</u>ic*.)

A recursive formula for the sequence above is

$$\begin{cases} a_1 = 1000 \\ a_n = a_{n-1} + 3000, \text{ for integers } n \geq 2. \end{cases}$$

The second line of this formula can be rewritten as

$$a_n - a_{n-1} = 3000.$$

This shows that the difference between the nth term and the $(n-1)$st term is a constant. So this recursive formula defines an arithmetic sequence.

This result can be generalized.

Theorem

The sequence defined by the recursive formula

$$\begin{cases} a_1 \\ a_n = a_{n-1} + d, \text{ for integers } n \geq 2 \end{cases}$$

is the arithmetic sequence with first term a_1 and constant difference d.

Proof:
When $n \geq 2$, we can rewrite the second line as $a_n - a_{n-1} = d$. This means that the difference between consecutive terms is the constant d. By definition, the sequence is arithmetic.

Example 1

Consider the sequence generated by

$$\begin{cases} a_1 = 2000 \\ a_n = a_{n-1} + 40, \text{ for integers } n \geq 2. \end{cases}$$

a. Describe this sequence in words.
b. Write the first five terms of the sequence.

Solution

a. This is an arithmetic sequence with first term 2000 and constant difference 40. A constant difference of 40 means that each term is 40 more than the previous term.
b. To generate terms after the first, add 40 to the previous term. The first five terms are 2000, 2040, 2080, 2120, 2160.

The constant difference of an arithmetic sequence can also be negative.

Example 2

Briana borrowed $370 from her parents for airfare to Europe. She will pay them back at the rate of $30 per month. Let a_n be the amount she still owes after n months. Find a recursive formula for this sequence.

Solution

After 1 month she owes $340, so $a_1 = \$340$. Each month she pays back $30, so the difference between successive terms is $d = -30$. A recursive formula for this sequence is

$$\begin{cases} a_1 = 340 \\ a_n = a_{n-1} - 30, \text{ for integers } n \geq 2. \end{cases}$$

Examples 1 and 2 illustrate that, in a recursive formula for an arithmetic sequence, the initial condition is the first term; the constant increase or decrease is the constant difference.

Graphs of Arithmetic Sequences

Because an arithmetic sequence represents a constant-increase or constant-decrease situation, its graph is a set of collinear points. Here is the graph of the sequence that began this lesson.

n	a_n
1	1000
2	4000
3	7000
4	10,000
5	13,000
$\vdots$	$\vdots$

$$\begin{cases} a_1 = 1000 \\ a_n = a_{n-1} + 3000, \\ \text{for integers } n \geq 2 \end{cases}$$

For this reason, arithmetic sequences are also called **linear sequences**.

An eyeful. *The Eiffel Tower, designed by Gustave Eiffel for the 1889 World's Fair in Paris, rises 300 meters. The tower has restaurants, a weather station, and is used to transmit television programs.*

Generating Sequences Recursively Using Programs

Calculator and computer programs can generate arithmetic sequences using recursive formulas. The following programs generate the first six terms of the sequence graphed on page 176. The program on the right below is for a graphics calculator; the one on the left is in the BASIC computer language.

BASIC
```
10 LET A = 1000
20 FOR N = 1 TO 6
30 PRINT A              loop
40 A = A + 3000
50 NEXT N
60 END
```

CALCULATOR
```
: 1000→A
: For(N,1,6)
: Disp A                loop
: A + 3000→A
: End
```

The first line of each program defines the first term of the sequence and stores it in a memory location called A. Each program contains a *loop*. The For line tells the calculator or computer to execute the commands six times. Disp A tells the calculator to display the current value of A on the screen. PRINT A tells the computer to print the current value of A. The line A = A + 3000 tells the computer to add 3000 to the previous term and store the result in A. Both programs produce the output below.

```
 1000
 4000
 7000
10000
13000
16000
```

These are the same six terms generated in the opening paragraph of this lesson.

QUESTIONS

Covering the Reading

1. Give a recursive formula for the arithmetic sequence with first term a_1 and constant difference d.

2. Explain why arithmetic sequences are sometimes called linear sequences.

3. Consider the sequence $\begin{cases} a_1 = 7 \\ a_n = a_{n-1} - 2, \text{ for integers } n \geq 2. \end{cases}$
   a. Find the first five terms of the sequence.
   b. Graph the sequence.
   c. The points from your graph in part b should lie on a line. What is the slope of this line?

4. What is the relation between the slope of the graph of an arithmetic sequence and the constant difference of the sequence?

5. Refer to Example 2. How much money will Briana owe her parents after 10 months?

In 6 and 7, refer to the sequence $\begin{cases} a_1 = 1 \\ a_n = a_{n-1} + 6, \text{ for integers } n \geq 2. \end{cases}$

6. Write the first four terms of this sequence.

7. a. Write a calculator or computer program to generate the first 25 terms of this sequence.
   b. Run the program and identify the 25th term.

Applying the Mathematics

8. *Multiple choice.* Which sequence is *not* an arithmetic sequence?
   (a) 5, 9, 13, 17, . . .
   (b) 3, 6, 12, 24, . . .
   (c) $\frac{1}{2}, 1, \frac{3}{2}, 2, \frac{5}{2}, \ldots$
   (d) 0, -1, -2, -3, -4, . . .

9. Write a recursive formula for the arithmetic sequence 13, 19, 25, 31, ...

10. Mr. Jefferson has a tax-sheltered annuity in which he is saving money for retirement. He deposits $400 a month into the account. Suppose he started with a zero balance.
    a. Write a recursive formula for the total amount he has deposited after each month.
    b. How much has he deposited after 5 months?

11. In a set of steps leading to a ship, the third step is 52 inches above the water and the fifth step is 63 inches above the water. Assuming all steps have equal rise, how far is the ninth step above the water?

12. What will the following BASIC program print when run?

```
10 LET A = 625
20 FOR N = 1 TO 15
30 PRINT A
40 A = A + 15
50 NEXT N
60 END
```

In 13 and 14, write a calculator or computer program to generate:

13. the first 1000 positive odd numbers.

14. the first 500 positive multiples of 11.

year after 1900	reported number of child abuse cases (in millions)
76	.67
78	.80
80	1.15
82	1.22
84	1.65
86	2.05
88	2.20
90	2.55
92	2.86
93	2.99

Review

15. The table at the left shows the number of cases of child abuse from 1976 to 1993 as reported by the National Committee for the Prevention of Child Abuse.
    a. Make a scatterplot of these data.
    b. Find an equation for the line of best fit.
    c. Draw the regression line on the scatterplot.
    d. Use the regression equation to predict the number of child abuse cases reported in the year 1990.
    e. Use the regression equation to predict the number of child abuse cases that will be reported in the year 2002.
    f. Describe at least two factors that might cause your estimate in part e to be too high.   *(Lesson 3-6)*

16. Find an equation for the line which passes through (75, 90) and is parallel to the line with equation $y = -.4x - 20$.   *(Lessons 3-2, 3-5)*

17. How are lines ℓ and m with the equations below related?
    $$\ell: 3x + 4y = 12$$
    $$m: y = -\frac{3}{4}x + 3$$
    Justify your answer.   *(Lessons 3-2, 3-4)*

18. Suppose z varies jointly as x and y. If z is 25 when x is 40 and y is 60, find z when x is 15 and y is 75.   *(Lesson 2-9)*

19. The number n of square tiles needed to cover a floor varies inversely as the square of the length s of a side of a tile. How is the number of tiles affected if the length of a side of a tile is doubled?   *(Lesson 2-3)*

20. The function A defined by $A(e) = 6e^2$ gives the surface area of a cube with edge length e. Find $A(4.?)$ *(Lesson 1-3)*

21. Match each inequality to its graph.   *(Previous course)*

    a. $-2 < x < 5$ (i)

    b. $-2 \leq x < 5$ (ii)

    c. $2 < x \leq 5$ (iii)

    d. $-2 \leq x \leq 5$ (iv)

Exploration

22. Describe all the positive integers that meet all three of these conditions at the same time:
    a. They leave a remainder of 1 when divided by 2.
    b. They leave a remainder of 3 when divided by 4.
    c. They leave a remainder of 5 when divided by 6.
    (Hint: 35 is one such number.)

LESSON

3-8

*Explicit
Formulas
for
Arithmetic
Sequences*

Consider again the sequence from Lesson 3-7 that gives the mileage when oil changes are required:

$$1000, 4000, 7000, 10{,}000, 13{,}000, 16{,}000, \ldots$$

It has the recursive formula

$$\begin{cases} a_1 = 1000 \\ a_n = a_{n-1} + 3000, \text{ for integers } n \geq 2. \end{cases}$$

Suppose you wanted the 50th term of the sequence. To use the recursive formula, you would need to find the first 49 terms. This is rather inefficient. It would be much easier to find the 50th term if you had an *explicit* formula for the sequence. The next example shows how an explicit formula can be found.

Developing an Explicit Formula for an Arithmetic Sequence

Example 1

a. Find an explicit formula for the arithmetic sequence of oil-change mileages 1000, 4000, 7000, 10,000, 13,000, 16,000, ...
b. Find a_{50} and tell what it represents.

Solution 1

a. To develop an explicit formula, use the constant difference of 3000 to write each term after the first. Consider the pattern in this table.

number of term	term
1	1000
2	$1000 + 1 \cdot 3000 = 4000$
3	$1000 + 2 \cdot 3000 = 7000$
4	$1000 + 3 \cdot 3000 = 10{,}000$
5	$1000 + 4 \cdot 3000 = 13{,}000$
$\vdots$	$\vdots$
n	$1000 + (n-1) \cdot 3000$

So, an explicit formula for the sequence is

$$a_n = 1000 + (n-1) \cdot 3000$$

or

$$a_n = 3000n - 2000.$$

Solution 2

a. The sequence can be thought of as (1, 1000), (2, 4000), (3, 7000), and so on. These points lie on a line with slope 3000. Use this slope and (1, 1000) in the point-slope form.

$$y - 1000 = 3000(x - 1)$$

Substitute. $\quad a_n - 1000 = 3000(n - 1)$

Solve for a_n. $\quad a_n = 1000 + 3000(n - 1) = 3000n - 2000$

▶

b. a_{50} is the 50th term. Substitute 50 for n.
$$a_{50} = 3000\,(50) - 2000$$
$$a_{50} = 148{,}000$$
If you kept the car long enough, the car would need its 50th oil change at 148,000 miles.

Example 1 suggests the following theorem.

Theorem (nth Term of an Arithmetic Sequence)
The nth term a_n of an arithmetic sequence with first term a_1 and constant difference d is given by the explicit formula
$$a_n = a_1 + (n - 1)d.$$

Proof:
Each term of the arithmetic sequence is of the form $(x, y) = (n, a_n)$. The first term is the ordered pair $(1, a_1)$. The slope is the constant difference which we call d. Use these values in the point-slope form of a linear equation.

$$y - y_1 = m(x - x_1)$$
$$a_n - a_1 = d(n - 1)$$
Solve for a_n. $\qquad a_n = a_1 + (n - 1)d$

Example 2

Find the 40th term of the arithmetic sequence 100, 97, 94, 91, ...

Solution 1

The first term $a_1 = 100$, and the constant difference $d = $ -3. So $a_n = 100 + (n - 1) \cdot$ -3. Because the 40th term is to be found, substitute $n = 40$ into the formula of the theorem:
$$a_{40} = 100 + (40 - 1) \cdot \text{-}3 = 100 + 39 \cdot \text{-}3 = \text{-}17.$$
The 40th term is -17.

Solution 2

The arithmetic sequence with first term equal to 100 and constant difference equal to -3 can be described recursively as follows.
$$\begin{cases} a_1 = 100 \\ a_n = a_{n-1} - 3 \end{cases}$$
Write a program to generate the first 40 terms of this sequence. When you run the program, you will see that the last term is -17.

You should be able to translate from a recursive to an explicit formula for an arithmetic sequence, and vice versa. To go from recursive to explicit form, use the method of Example 1, or substitute the known values of a and d into the result of the theorem above. To go from explicit to recursive form, you should substitute the known values of a and d into the recursive pattern for an arithmetic sequence, as in the check to Example 2.

Example 3

In a concert hall the first row has 20 seats in it, and each subsequent row has two more seats than the row in front of it. If the last row has 64 seats, how many rows are in the concert hall?

Solution

Because each succeeding row has two additional seats, the number of seats in each row generates the sequence

$$20, 22, 24, 26, \ldots, 64.$$

You know that $a_1 = 20$ and $d = 2$.

So $a_n = 20 + (n - 1)2,$

where n = the number of the row and a_n = the number of seats in the nth row. Here $a_n = 64$, and we wish to find n.

$64 = 20 + (n - 1)2$	
$44 = 2(n - 1)$	Subtract 20 from both sides.
$22 = n - 1$	Divide both sides by 2.
$23 = n$	Add 1 to both sides.

There are 23 rows of seats in the concert hall.

Check

Substitute $a_1 = 20$, $d = 2$, and $n = 23$ into $a_n = a_1 + (n - 1)d$.

$$a_{23} = 20 + (23 - 1)2 = 20 + 44 = 64$$

The last row has the correct number of seats.

Pictured is the Rialto Square Theater in Joliet, Illinois.

Programs Using Explicit Formulas

Calculators and computers can also generate arithmetic sequences using explicit formulas. The following programs will generate the first six terms of the sequence in Example 1.

```
BASIC
10 FOR N = 1 TO 6
20 LET A = 3000 * N − 2000
30 PRINT A
40 NEXT N
50 END
```

```
CALCULATOR
: For (N, 1, 6)
: 3000N − 2000→A
: Disp A
: End
```

Again the loop tells the calculator or computer to execute the instructions within the loop six times. The second line in each program defines the nth term. The third line tells the calculator or computer to display it. When the program is run the following will appear. These are the same values as those of the sequence in Example 1.

```
1000
4000
7000
10000
13000
16000
```

QUESTIONS

Covering the Reading

1. What is the 20th term of the sequence of Example 1?

2. What is the nth term of an arithmetic sequence with first term a_1 and constant difference d?

3. **a.** Find an explicit formula for the nth term of the arithmetic sequence
$$13, 15, 17, 19, 21, \ldots .$$
   b. Calculate the 51st term of this sequence.

In 4 and 5, an arithmetic sequence is given. **a.** Find an explicit formula for the nth term. **b.** Find the 100th term

4. 6, 15, 24, 33, . . .

5. 16, 14.5, 13, 11.5, . . .

6. Find the 400th term in the arithmetic sequence 9, 19, 29, 39, ...

7. Refer to Example 3. Suppose that in some other concert hall the first row has 40 seats, each subsequent row has two more seats than the row in front of it, and the last row has 70 seats. How many rows of seats are there?

8. Modify one of the programs at the end of the lesson so that it shows the first ten terms of the sequence defined explicitly as $a_n = 40 + 3n$.

Applying the Mathematics

9. Suppose $t_n = 10 + 7(n - 1)$.
   a. Is this formula explicit, recursive, or neither? How can you tell?
   b. Find t_{89}.

10. Write a recursive formula for the arithmetic sequence in which $a_n = 10.8 + 2.4n$.

11. Consider the sequence
$$\begin{cases} a_1 = 8.1 \\ a_n = a_{n-1} + 1.7, \text{ for integers } n \geq 2. \end{cases}$$
   a. Write its first three terms.
   b. Write an explicit formula for the sequence.

12. The BASIC program below generates several terms of a sequence using a recursive formula.

```
10 LET A = 15
20 FOR N = 1 TO 10
30 PRINT A
40 A = A + 3.5
50 NEXT N
```

a. What sequence is printed when the program is run?
b. Modify this program for a calculator or computer so that the sequence is defined explicitly.

13. Stu and Penelope decided to include bicycling as part of their exercise program. They start by biking 14 miles the first week. By the twenty-fifth week they want to bike 74 miles a week. If the number of miles biked in successive weeks forms an arithmetic sequence, what should be their weekly increase?

In 14 and 15, a local radio station is holding a contest to give away cash. The announcer makes a telephone call, and if the person who answers guesses the correct amount of money to be given away, he or she wins the money. If the person misses, $20 is added to the money to be given away.

14. On the 12th call, a contestant won $675. How much was to be given away on the 1st call?

15. Suppose the initial amount is $140. On what call will the winner receive $1100?

16. In a contest, the first-place winner gets $100,000. The tenth-place winner gets $23,500. If the winning amounts form an arithmetic sequence, find the cash difference between prizes.

Review

In 17 and 18, use the equation $5y = 7x - 4$.

17. a. Rewrite the equation in standard form.
   b. Identify the x- and y-intercepts.   *(Lesson 3-4)*

18. a. Rewrite the equation in slope-intercept form.
   b. Identify the slope and y-intercept. (This provides a check of Question 17b.)
   c. Graph the equation.   *(Lesson 3-2)*

19. Suppose that the variables H, z, and w are related as illustrated in the graphs at the left. The points on the top graph lie on a line through the origin. The points on the bottom graph lie on or near a hyperbola. Write an equation approximating the relationship between H, z, and w.   *(Lessons 2-8, 2-9)*

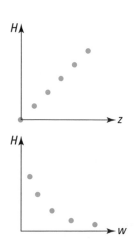

20. The following table gives the percent of the population in the United States aged 65 or over for each census year from 1860 to 1990.

year	1860	1870	1880	1890	1900	1910	1920
% aged 65 or over	2.7	3.0	3.4	3.9	4.1	4.3	4.7
year	1930	1940	1950	1960	1970	1980	1990
% aged 65 or over	5.4	6.8	8.1	9.2	9.8	11.3	12.5

  a. Make a scatterplot of these data.
  b. Find an equation for the regression line. Let x = the number of years after 1860.
  c. Separate the data set into two parts:
    (i) the years from 1860 to 1930; and
    (ii) the years from 1940 to 1990.
    Find a linear regression model for each of these data sets.
  d. Use the result of part **c** to write a piecewise-linear function to model the given data.
  e. Which of the functions, that in part **b** or that in part **d,** do you think is the better predictor of the percent of the U.S. population that will be 65 or over in the year 2010? Explain your reasoning. *(Lessons 3-5, 3-6)*

In 21 and 22, *multiple choice.* Refer to the graphs below.
(Lessons 2-4, 2-5, 2-6)

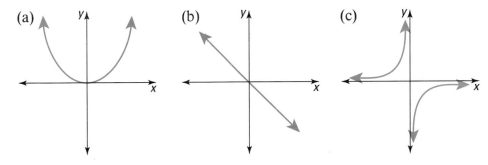

(a)   (b)   (c)

21. Which graph does not represent a direct-variation function?

22. Which graph represents a situation in which the constant of variation is positive?

23. Consider $S = \left\{ x: -8 \le x \le -\frac{21}{4} \right\}$.
  a. Name all integers that are members of S.
  b. Graph S on a number line. *(Previous course, Lesson 1-2)*

Exploration

---

24. Here is a sequence that is *not* an arithmetic sequence.
      3, 5, 9, 17, 33, 65, 129, 257, 513, . . .
  a. What is a possible next term?
  b. Find a recursive formula for the nth term.
  c. Find an explicit formula for the nth term.

Fit for life. *Exercise classes designed for senior citizens help combat heart disease, high blood pressure, and osteoporosis. More and more of the 31 million people over the age of 65 are taking such classes.*

Upper-level parking. *This is a multi-level parking garage located in
Tokyo. To make the best use of space, hydraulic lifts are used to move the
buses and cars into their spaces.*

The daily charge to park a car at a city lot is $3.00 for the first half
hour and $2.50 for each additional hour or portion of an hour. Thus,
the cost of parking is a function of time. The table below gives the
cost if a car is parked at 7:00 P.M. and picked up before 1:30 A.M.

time	n = number of minutes since 7:00 P.M.	c = cost (in dollars)
7:00 – 7:30	$0 < n \leq 30$	3.00
7:30 – 8:30	$30 < n \leq 90$	3.00 + 2.50 = 5.50
8:30 – 9:30	$90 < n \leq 150$	5.50 + 2.50 = 8.00
9:30 – 10:30	$150 < n \leq 210$	8.00 + 2.50 = 10.50
10:30 – 11:30	$210 < n \leq 270$	10.50 + 2.50 = 13.00
11:30 – 12:30	$270 < n \leq 330$	13.00 + 2.50 = 15.50
12:30 – 1:30	$330 < n \leq 390$	15.50 + 2.50 = 18.00

The graph at the right shows the cost of
parking a car for a given number of minutes
from 0 to 390. Notice that the graph
represents a function. You can verify this
with the Vertical-Line Test. The domain
consists of any positive real number of
minutes less than or equal to the 1440
minutes in a day. The range is the set of
costs: {$3.00, $5.50, $8.00, $10.50, . . . ,
$63.00}.

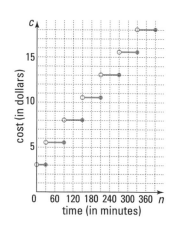

The function is not a linear function, but the
graph is piecewise linear. Because the graph
looks like a series of steps, this function is
called a **step function.** Each step is part of a
horizontal line.

The Greatest-Integer Function

The **greatest-integer symbol** $\lfloor \ \rfloor$ is defined as follows.

Definition

$\lfloor x \rfloor$ = the greatest integer less than or equal to x.

Example 1

Evaluate each of the following.

a. $\left\lfloor 2\frac{3}{4} \right\rfloor$ **b.** $\lfloor 7 \rfloor$ **c.** $\lfloor -3 \rfloor$ **d.** $\lfloor -2.1 \rfloor$

Solution

a. $\left\lfloor 2\frac{3}{4} \right\rfloor$ is the greatest integer less than or equal to $2\frac{3}{4}$. So $\left\lfloor 2\frac{3}{4} \right\rfloor = 2$.

b. $\lfloor 7 \rfloor$ is the greatest integer less than or equal to 7. $\lfloor 7 \rfloor = 7$

c. $\lfloor -3 \rfloor$ is the greatest integer less than or equal to -3. $\lfloor -3 \rfloor = -3$

d. $\lfloor -2.1 \rfloor$ is the greatest integer less than or equal to -2.1. Because $-3 < -2.1 < -2$, The greatest integer less than or equal to -2.1 is $\lfloor -2.1 \rfloor = -3$.

The **greatest-integer function** is the function f with $f(x) = \lfloor x \rfloor$, for all real numbers x. It is sometimes called the **rounding-down function** or **floor function.**

The Graph of the Greatest-Integer Function

Example 2

Graph the function defined by $f(x) = \lfloor x \rfloor$.

Solution

Make a table of values. For all x greater than or equal to 0 but less than 1, the greatest integer less than or equal to x is 0. For all x greater than or equal to 1 but less than 2, the greatest integer less than or equal to x is 1. In a similar manner you can get the other values in the table below. The graph is on the right below.

x	$f(x) = \lfloor x \rfloor$
$-3 \leq x < -2$	-3
$-2 \leq x < -1$	-2
$-1 \leq x < 0$	-1
$0 \leq x < 1$	0
$1 \leq x < 2$	1
$2 \leq x < 3$	2
$3 \leq x < 4$	3

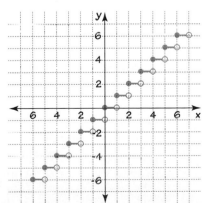

On the graph, the open circles at (1, 0), (2, 1), (3, 2), and so on, indicate that these values are *not* solutions to $f(x) = \lfloor x \rfloor$. At these points, the function value is jumping to the next step. Notice that the domain of the greatest-integer function is the set of real numbers, but the range is the set of integers.

On calculators and computers, the greatest-integer function is often labeled INT. For instance, INT(-6.1) = -7. If your automatic grapher has the INT function, it will graph the greatest-integer function for you. The graph from one automatic grapher is shown below. Because the grapher is programmed to connect successive pixels, it joins successive steps. This makes it appear as if the graph does not represent a function. On some graphers you can get the correct graph by switching from the connected mode to the dot mode.

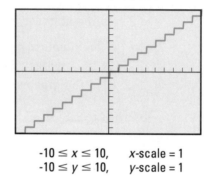

$-10 \le x \le 10, \quad x\text{-scale} = 1$
$-10 \le y \le 10, \quad y\text{-scale} = 1$

Applications of the Greatest-Integer Function

Formulas with the greatest-integer symbol are often found when function values must be integers but other formulas would give noninteger values.

Example 3

A bottling company prepares flavored water. The bottles are packaged in cartons of 8. Write an equation that gives the number of complete cartons c packaged each day, if the company prepares b bottles each day.

Solution

The number of cartons must be a nonnegative integer because the company cannot package part of a carton. The bottles are packaged in groups of 8, so it seems reasonable to divide the number of bottles by 8. However, $\frac{b}{8}$ may not be an integer. Since we are interested in the number of complete cartons, round the answer down to the nearest integer. Using the greatest-integer function, you can write the equation as

$$c = \left\lfloor \frac{b}{8} \right\rfloor.$$

Check

Substitute values of b. For instance, $\left\lfloor \frac{16}{8} \right\rfloor = \lfloor 2 \rfloor = 2$. Two complete cartons are needed for 16 bottles. Also $\left\lfloor \frac{21}{8} \right\rfloor = \left\lfloor 2\frac{5}{8} \right\rfloor = 2$. When there are 21 bottles, 2 complete cartons will be packaged.

Step Functions in Calculator or Computer Programs

The greatest-integer function may be used in calculator or computer programs. Consider the two below.

```
BASIC
10 INPUT N
20 PRINT INT(N + .5)
30 GOTO 10
40 END
```

```
CALCULATOR
: Lbl A
: Input N
: Disp int (N + .5)
: Goto A
```

In each program the statement INPUT N prints a ? on the screen, asking the user for a number for N. Then it computes $\lfloor N + .5 \rfloor$ and prints the result. The program then loops back to the beginning and asks the user for another input.

Example 4

Describe what either program above displays for any number N.

Solution

Pick several values of N and evaluate the expression $\lfloor N + .5 \rfloor$.

N	$\lfloor N + .5 \rfloor$
4	$\lfloor 4 + .5 \rfloor = \lfloor 4.5 \rfloor = 4$
4.5	$\lfloor 4.5 + .5 \rfloor = \lfloor 5 \rfloor = 5$
-7	$\lfloor -7 + .5 \rfloor = \lfloor -6.5 \rfloor = -7$
7.49	$\lfloor 7.49 + .5 \rfloor = \lfloor 7.99 \rfloor = 7$
7.6	$\lfloor 7.6 + .5 \rfloor = \lfloor 8.1 \rfloor = 8$

This program seems to take any number, round it to the nearest integer (rounding all halves up), and prints the result.

QUESTIONS

Covering the Reading

In 1 and 2, refer to the parking example on page 186.

1. What would be the cost to park a car for 4 hours and 45 minutes?

2. What is the domain of the function?

3. In your own words, write the meaning of $\lfloor x \rfloor$.

In 4–7, evaluate.

4. $\left\lfloor 3\frac{1}{2} \right\rfloor$ **5.** $\lfloor 11.9 \rfloor$ **6.** $\lfloor -11.7 \rfloor$ **7.** $\lfloor 8 + .5 \rfloor$

8. a. The function f defined by $f(x) = \lfloor x \rfloor$ is called the __?__ or __?__ function.
   b. The range of $f: x \rightarrow \lfloor x \rfloor$ is __?__.
   c. Why are there open circles at (1, 0), (2, 1), (3, 2), and so on in the graph of f?

9. Explain why some automatic graphers connect the steps when graphing $y = \lfloor x \rfloor$.

10. Give the domain and range of the function with equation $c = \lfloor \frac{b}{8} \rfloor$ for the situation in the Example 3.

11. Suppose a company packaged bottles in cartons of 6. Write an equation that relates the number of bottles b to the number of complete cartons c that can be packaged.

12. Refer to the expression INT(N + .5) in Example 4. What will the program display if each of the following is input for N?
   a. 17.3   **b.** 9.99   **c.** -3.29

Applying the Mathematics

13. *Multiple choice.* In 1995, the cost to mail a letter first class in the U.S. was 32¢ for up to one ounce, and 23¢ for each additional ounce or fraction thereof. Which is a correct graph for this function?

(a)
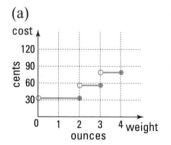

(b)

(c)
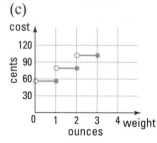

(d)
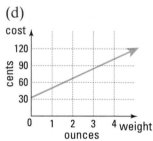

14. *Multiple choice.* An auditorium has 2500 seats available for graduation. There are g graduates. Which of the following represents the number of tickets each graduate may have if each graduate is given the same number of tickets?
   (a) $\frac{2500}{g}$ (b) $\lfloor \frac{2500}{g} \rfloor$ (c) $\lfloor 2500 \cdot g \rfloor$ (d) $\lfloor \frac{g}{2500} \rfloor$

15. Tyrone's salary is $275 a week plus $75 for each $300 he has in sales that week.
   a. Find his salary during a week in which he had $1000 in sales.
   b. When he has d dollars in sales, write an equation that gives his weekly earnings.

16. Consider the BASIC program in Example 4. If line 20 is changed to
   20 PRINT 10 * INT((N + 5)/10) a different type of rounding occurs.
   a. Make a table of at least five pairs of input and output for this program.
   b. What kind of rounding is done by this program?

17. The cost C in dollars of making a call lasting m minutes during the day from Chicago, Illinois, to Paris, France, is given by the formula $C = 1.71 - 1.08 \lfloor 1 - m \rfloor$.
   a. Evaluate this formula when $m = 2, 7.5,$ and 10.
   b. Graph this equation for $0 < m \le 10$.

18. Consider the sequence $\begin{cases} a_1 = -x \\ a_n = a_{n-1} + 3x, \end{cases}$ for integers $n \geq 2$.

    a. Write the first four terms of the sequence.

    b. Is the sequence arithmetic? How can you tell?

    c. Find an explicit formula for the sequence. *(Lesson 3-8)*

19. Consider the arithmetic sequence 46, 42.5, 39, 35.5, ...

    a. Write a recursive formula for the sequence.

    b. Write an explicit formula for the sequence. *(Lessons 3-7, 3-8)*

20. Suppose the first term of an arithmetic sequence is -4, and the 15th term is 38. Find the 30th term of the sequence. *(Lessons 3-7, 3-8)*

21. Write an equation of the line which passes through the points (3, -1) and (6, 5). *(Lesson 3-5)*

22. Graph $4x - 6y = 36$ using the intercepts. *(Lesson 3-4)*

23. Graph the function defined below. *(Lessons 3-1, 3-2, 3-5)*

$$f(x) = \begin{cases} 6, & \text{for } x > 2 \\ 3x, & \text{for } 0 \leq x \leq 2 \\ -\frac{1}{2}x, & \text{for } x < 0 \end{cases}$$

Exploration

24. The formula

$$W = d + 2m + \left\lfloor \frac{3(m+1)}{5} \right\rfloor + y + \left\lfloor \frac{y}{4} \right\rfloor - \left\lfloor \frac{y}{100} \right\rfloor + \left\lfloor \frac{y}{400} \right\rfloor + 2$$

gives the day of the week based on our current calendar where

    d = the day of the month of the given date.

    m = the number of the month in the year with January and February regarded as the 13th and 14th months of the previous year; that is, 2/22/1990 is 14/22/1989. The other months are numbered 3 to 12 as usual.

    y = the year.

Once W is computed, divide by 7 and the remainder is the day of the week, with Saturday = 0, Sunday = 1, . . . , Friday = 6.

    a. Use the formula to find the day of the week on which you were born.

    b. On what day of the week was the Declaration of Independence adopted?

Birth day. *Dolley Madison (top) was born on May 29, 1768. Eleanor Roosevelt (bottom) was born on Oct. 11, 1884. Both of these First Ladies played important roles in American history. On what day of the week was each woman born?*

A project presents an opportunity for you to extend your knowledge of a topic related to the material of this chapter. You should allow more time for a project than you do for typical homework questions.

1 A Graphical Investigation

Graphs of equations of the form $Ax + By = C$, where A, B, C are consecutive terms in an arithmetic sequence, share something in common.

a. Graph five equations of this type, for instance,

$$x + 2y = 3$$
$$3x + 5y = 7$$
$$8x + 6y = 4$$
$$-2x - 3y = -4$$
$$3x - y = -5.$$

b. Make a conjecture based on these five graphs.

c. Test your conjecture with a few more graphs.

d. Use the definitions or theorems about arithmetic sequences to verify your conjecture.

2 Taxi Meter

If you have ever been in a taxi during a traffic jam, you might have noticed that the meter will "keep running" if there is a charge for miles and waiting time. For instance, a taxi might charge $1.20 as a base rate for the first $\frac{1}{10}$ of a mile, $.20 for each additional $\frac{1}{10}$ of a mile, $0.20 for each full minute waiting time when the taxi isn't moving, and $0.50 for each additional passenger. Find out the taxi rates near where you live. Write a program in which the inputs are the distance traveled, the minutes of waiting time, and the number of additional passengers. Have the output be the amount due to the taxi driver.

3 Fines for Speeding

The amount a driver is fined for speeding is a function of the amount by which the speed exceeds the legal speed limit. In some states these functions are linear functions; in other states they are piecewise linear.

a. Find out what the fines are for speeding in your state. If they are not the same for cars and trucks, find out both sets of fines. Describe the speeding-fine function using a table and a graph. What are the domain and the range of this function? Does this function belong to any of the types you have studied in this course? If so, what kind of function is it? If possible, find an equation for the speeding fine function.

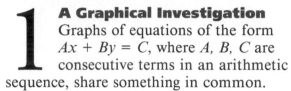

SPEED LIMIT 55

b. Find a state or province that has a different set of fines for speeding than your state. Describe those fines with a table, a graph, and an equation. Describe some ways the two speeding fine functions are alike and the ways they are different.

4 Time-Series Data

When the value of a dependent variable, such as a world's record in a particular sport, changes over time the data are called *time-series data*.

a. Find an example of time-series data in which the dependent variable appears to vary linearly with time. Make a scatterplot of the data.

b. Use a computer or calculator to find a line of best fit and draw it on the graph with the scatterplot.

c. According to your model, what will the value of the variable be in the years 2000, 2025, 2075, and 3000? Do your predictions seem reasonable? Why or why not?

5 Linear Combinations

Combinations of 23¢ and 32¢ stamps can be used for 55¢ postage, 78¢ postage, and many other amounts. But no combination of 23¢ and 32¢ stamps can be used for 10¢ postage, 30¢ postage, and so on.

a. What is the greatest amount of postage you *cannot* form from 23¢ and 32¢ stamps? Explain how you got your answer.

b. What is the greatest amount of postage you *cannot* form from stamps with denominations x¢ and y¢, if x and y have no common factors? Explain how you got your answer.

6 Graphing Piecewise Functions

Some automatic graphers allow you to use logical statements whose value is 1 if the statement is true and 0 if it is false. These statements can be used to graph an equation over a particular domain. For instance, to graph

$$y = x - 1 \text{ for } x > 1,$$

input $y = (x - 1) \div (x > 1)$. When $x > 1$, the logical statement is true and has a value of 1; so the grapher reads $y = (x - 1) \div (x > 1)$ as $y = x - 1$ and graphs a line. When $x \le 1$, the logical statement $x > 1$ is false and has a value of 0. Because division by 0 is undefined, no graph is displayed.

a. Graph $y = (x - 5) \div (x > 5)$. Describe what is displayed.

b. To graph a function defined by different equations over different domains, enter each equation as a separate function divided by its domain. Make up a piecewise-linear function. Show how to use your grapher to draw this graph.

SUMMARY

A linear equation in two variables is one that is equivalent to an equation of the form $Ax + By = C$ where A and B are not both zero. The graph of every linear equation is a line. The form $Ax + By = C$ where $A \neq 0$ and $B \neq 0$ is called the standard form of the equation for a line. If the line is not vertical, then its equation can be put into the form $y = mx + b$, with slope m and y-intercept b. Horizontal lines have slope 0; they have equations of the form $y = b$. Slope is not defined for vertical lines, which have equations of the form $x = h$.

If the slope and one point on a line are known, then the point-slope form of a linear equation can be used to obtain an equation for the line. In many real-world situations, a set of data points is roughly linear. In such cases, a regression line can be used to describe the data and make predictions. The correlation coefficient describes the strength of the linear relationship.

Linear equations result from two basic kinds of situations: constant increase or decrease, and linear combinations. Sequences with a constant increase or decrease have a constant difference between terms. Their graphs consist of collinear points, and they are called linear or arithmetic sequences. If a_n is the nth term of an arithmetic sequence with constant difference d, then the sequence can be described explicitly as $a_n = a_1 + (n - 1)d$, for integers $n \geq 1$, or recursively as

$$\begin{cases} a_1 \\ a_n = a_{n-1} + d, \text{ for integers } n \geq 2. \end{cases}$$

Some calculators and computers can be programmed to display terms of a sequence defined either explicitly or recursively.

A function whose graph is the union of segments and rays is called piecewise linear. Piecewise-linear functions result from situations in which rates are constant for a while but change at known points. Step functions, in particular the greatest-integer function, are special instances of piecewise-linear functions.

VOCABULARY

Below are the most important terms and phrases for this chapter.
You should be able to give a definition for those terms marked with *.
For all other terms you should be able to give a general description or a specific example.

Lesson 3-1
*y-intercept
initial condition
*slope-intercept form
linear function
constant-increase situation
constant-decrease situation
slope
piecewise-linear graph

Lesson 3-3
linear-combination
 situation

Lesson 3-4
standard form
*x-intercept
vertical line
oblique line
horizontal line

Lesson 3-5
*point-slope form

Lesson 3-6
scatterplot
regression line, line of best
 fit, least squares line
correlation coefficient

Lesson 3-7
constant difference
*arithmetic sequence, linear
 sequence
recursive formula for an
 arithmetic sequence

Lesson 3-8
*explicit formula for an
 arithmetic sequence

Lesson 3-9
step function
*greatest-integer function, $\lfloor x \rfloor$
rounding-down function,
 floor function, INT (x)

PROGRESS SELF-TEST

Take this test as you would take a test in class. Use graph paper and a ruler. Then check your work with the solutions in the Selected Answers section in the back of the book.

1. Graph the line with equation $y = 3x - 5$.

2. State whether the line determined by $x = -7$ is vertical, horizontal, or oblique.

3. Consider the line with equation $4x - 5y = 12$.
 a. What is its slope?
 b. What are its x- and y-intercepts?

4. For what values of m does the equation $y = mx + b$ model a constant-decrease situation?

5. Give an equation for the line through $(4, 2)$ and $(-5, 3)$.

6. Give an equation for the line that contains $(5, -1)$ and is parallel to $y = \frac{5}{3}x + 4$.

7. a. For what kind of lines is slope not defined?
 b. Which lines have a slope of zero?

8. A company makes 36″ and 48″ shoelaces by cutting off lengths from a spool of cord. Let S be the number of 36″ laces and L be the number of 48″ laces made. How much cord will be used in making S short and L long laces?

9. A scuba diver is 40 m below the surface. If the diver ascends at a constant rate of 0.8 meters per second, how long will it take to reach a depth of 10 meters?

10. Consider the following computer program.

```
10  LET A = 1
20  FOR N = 1 TO 12
30  PRINT A
40  A = A + 5
50  NEXT N
60  END
```

 a. What sequence is generated?
 b. Is the sequence arithmetic? Explain your answer.

11. In 1994, the cost of making a phone call from some airplanes was $7.50 for the first three minutes and $1.75 for each additional minute or portion of a minute.
 a. Graph the function for any call lasting up to 8 minutes.
 b. How much would it cost Isaiah Rich to make a $6\frac{1}{3}$ minute call?

12. Tell whether 8, 5, 2, -3, . . . could be the first four terms of an arithmetic sequence. Justify your answer.

13. Evaluate the expression $10 \cdot \lfloor (x + 5)/10 \rfloor$ when $x = 365$.

14. Celsius temperature and Réaumur temperature are related by a linear equation. Two pairs of corresponding temperatures are $0°C = 0°R$ and $100°C = 80°R$. Write a linear equation relating R and C, and solve it for R.

In 15 and 16, use the fact that -7, -10, -13, -16, . . . is an arithmetic sequence.

15. Write an explicit formula for the sequence.

16. Write a recursive formula for the sequence.

17. *Multiple choice.*   A store charges for copies as follows:

For		
1–50 copies		5¢ each
51–200 copies		4¢ each
more than 200 copies		3¢ each.

Which graph most closely describes the total cost C for printing n copies? (The scales on all four sets of axes are the same.)

(a)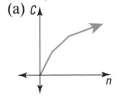

(b)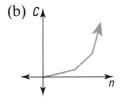

(c)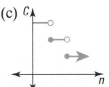

(d)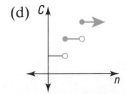

PROGRESS SELF-TEST

18. The table and the scatterplot below give the expected years of life left for people in the U.S. at selected ages in 1991.

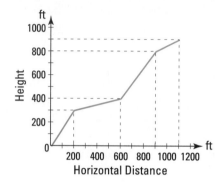

Age	Expected Years of Life Left
0	75.4
15	61.4
25	52.0
35	42.6
45	33.5
55	24.9
65	17.3
75	10.9
85	6.1

Source: *The World Almanac and Book of Facts 1993*

a. Do the data seem to be linearly related? Explain your answer.

b. Find an equation of the regression line for these data.

c. Use the answer to part **c** to estimate the expected years of life left of someone who is currently 42.

19. The graph below represents the height and horizontal distance moved by a ski lift.

a. What is the slope of the section whose horizontal distance goes from 600 to 900 feet?

b. Where is the slope of the lift steepest?

CHAPTER REVIEW

Questions on SPUR Objectives

SPUR stands for **S**kills, **P**roperties, **U**ses, and **R**epresentations. The Chapter Review questions are grouped according to the SPUR Objectives for this chapter.

SKILLS DEAL WITH THE PROCEDURES USED TO GET ANSWERS.

Objective A: *Determine the slope and intercepts of a line given its equation.*
(Lessons 3-1, 3-2, 3-4)

In 1–3, an equation for a line is given.
a. Give its slope. **b.** Give its *y*-intercept.

1. $y = 7x - 2$
2. $500x + 700y = 1200$
3. $y = 4$

In 4–6, an equation for a line is given.
a. Find its *x*-intercept. **b.** Find its *y*-intercept.

4. $3x + 5y = 45$
5. $x = -4.7$
6. $6y = 8x$

Objective B: *Find an equation for a line given two points on it or given a point on it and its slope.* *(Lessons 3-2, 3-5)*

In 7–12, find an equation for the line satisfying the given conditions.

7. The line has slope 8 and contains (40, 75).
8. The line has slope -0.25 and goes through the origin.
9. The line contains (2, 4) and (-1, 6).
10. The line contains (5, -9) and (5, 14).
11. The line is parallel to $3x + 2y = 9$ and contains (-1, 2).
12. The line is parallel to $y = 4x$ and contains (11, 0).

Objective C: *Evaluate expressions based on step functions.* *(Lesson 3-9)*

In 13 and 14, evaluate the expression.
13. **a.** $\lfloor 16.2 \rfloor$ **b.** $\lfloor -16.2 \rfloor$

14. **a.** $\lfloor 47.9 \rfloor$ **b.** $\lfloor -47.9 \rfloor$
15. Evaluate $\lfloor N + .5 \rfloor$ when $N = -17$.
16. Evaluate 10 INT $((X + 5)/10)$ when $X = 36$.

Objective D: *Evaluate or find explicit and recursive formulas for arithmetic sequences.*
(Lessons 3-7, 3-8)

In 17 and 18, an arithmetic sequence is given.
a. Find an explicit formula for the *n*th term.
b. Find a formula for the sequence in recursive form.
c. Find the 75th term of the sequence.

17. 7, 12, 17, 22, . . .
18. 8, 2, -4, -10, . . .
19. Give a recursive description of the sequence whose *n*th term is $a_n = 2n - 11$.
20. Find an explicit formula for the nth term of the sequence defined below.
$$\begin{cases} a_1 = \frac{1}{2} \\ a_n = a_{n-1} + 4 \text{ for integers } n \ge 2 \end{cases}$$

21. **a.** Write the first five terms of an arithmetic sequence.
    b. Describe the sequence recursively.
    c. Find an explicit formula for the *n*th term of your sequence.

In 22 and 23, a program is given. **a.** Tell how many terms will be printed; **b.** Give the first five terms that will be displayed.

22.
```
20 LET A = 100
30 FOR N = 1 TO 1000
40 PRINT A
50 A = A + 1
60 NEXT N
70 END
```

23.
```
For (N, 1, 10)
  Disp 2 * N + 3
End
```

PROPERTIES DEAL WITH THE PRINCIPLES BEHIND THE MATHEMATICS.

Objective E: *Recognize properties of linear functions.* *(Lessons 3-1, 3-2, 3-4, 3-5)*

In 24–26, consider the equation $y = mx + b$.

24. When does this equation model a constant-increase situation?

25. What constant represents the initial amount?

26. What is this form of a linear equation called?

27. Give the domain and range of $f(x) = 3x - 7$.

28. *Multiple choice.* Which of the following does *not* mean a slope of $-\frac{4}{3}$?

   (a)   a vertical change of -3 units for a horizontal change of 4 units

   (b)   a vertical change of -4 units for a horizontal change of 3 units

   (c)   a vertical change of $-\frac{4}{3}$ units for a horizontal change of 1 unit

   (d)   a vertical change of $\frac{4}{3}$ units for a horizontal change of -1 unit

29. What is true about the slopes of parallel lines?

In 30–32, refer to the equation $Ax + By = C$.

30. To find the x-intercept, substitute __?__ for __?__.

31. If $A = 0$ and $B \neq 0$, the graph of this equation is a __?__ line.

32. For what values of A and B does the equation *not* represent a function?

33. State whether the line is vertical, horizontal, or oblique.

   a. $y = -4$ **b.** $2x - 3y = 8$ **c.** $4x = 12$

34. Give the point-slope equation for a line.

35. *True or false.* The line with the equation $y - 5 = 3(x - 2)$ goes through the point $(5, 2)$.

Objective F: *Recognize properties of arithmetic sequences.* *(Lessons 3-7, 3-8)*

36. Describe how arithmetic sequences are formed.

37. Describe the graph of an arithmetic sequence.

In 38–41, tell whether the numbers could be the first four terms of an arithmetic sequence.

38. 1.2, 1.4, 1.6, 1.8

39. $\pi + 1, \pi + 2, \pi + 3, \pi + 4$

40. -9, -11, -13, -15   **41.** $4, 2, \frac{1}{2}, \frac{1}{4}$

In 42–44, does the formula generate an arithmetic sequence?

42. $\begin{cases} a_1 = 1.5 \\ a_n = a_{n-1} + 13, \text{ for integers } n \geq 2 \end{cases}$

43. $\begin{cases} a_1 = 9 \\ a_n = 2a_{n-1}, \text{ for integers } n \geq 2 \end{cases}$

44. $a_n = 3n^2 + 2$

USES DEAL WITH APPLICATIONS OF MATHEMATICS IN REAL SITUATIONS.

Objective G: *Model constant-increase or constant-decrease situations or situations involving arithmetic sequences.* *(Lessons 3-1, 3-7, 3-8)*

45. A crate weighs 3 kg when empty. It is filled with grapefruit weighing 0.2 kg each.
   a. Write an equation relating the weight w and the number n of grapefruit.
   b. Find the weight when there are 22 grapefruit in the crate.

46. A math teacher has a ream of 500 sheets of graph paper. Each week the advanced algebra class uses about 90 sheets.
   a. About how many sheets are left after w weeks?

   b. After how many weeks will there be 50 sheets left?

47. The number of feet traveled during each second of free fall is given by the formula $a_n = 16 + 32(n - 1)$. What distance is traveled during the eighth second?

48. When Florence Flask joined a laboratory, she was given a $26,000 salary and promised at least an $1800 raise each year. What is the longest time it could take for her salary to reach $35,000?

Objective H: *Model situations leading to linear combinations.* *(Lesson 3-3)*

49. Lubbock Lumber sells 6-foot long 2-by-4 boards for $1.70 each and 8-foot long 2-by-6 boards for $2.50 each. Last week they sold $250 worth of these boards. Let F be the number of 2-by-4s and S be the number of 2-by-6s.
   a. Write an equation to model this situation.
   b. If 100 2-by-4s were sold, how many 2-by-6s were sold?

50. A maintenance engineer of a swimming pool combines A gallons of solution that is 6% chlorine and B gallons of solution that is 8% chlorine.
   a. How much solution is there altogether?
   b. How much chlorine is there altogether?
   c. Two gallons of chlorine are needed in the pool. Write an equation that describes this situation.
   d. List three ordered pairs that are solutions to the equation in part c.

Objective I: *In a real-world context, find an equation for a line containing two points.* *(Lesson 3-5)*

51. Woody Bench finds that it costs his business $7,600 to make 30 desks and $16,000 to make 100 desks. Assuming a linear relationship exists between the cost and the number of desks, how much will it cost to make 1000 desks?

52. Charlotte finds that the cost of making shoes is linearly related to the number of shoes her company makes. It costs $1450 to make 150 pairs of shoes and $1675 to make 225 pairs of shoes.
   a. Let C = the cost of making p pairs of shoes. Write a formula relating C to p.
   b. Rewrite the equation in part a in slope-intercept form, and explain what the slope represents in this situation.
   c. How much will it cost to make 500 pairs of shoes?

Objective J: *Fit lines to data.* *(Lesson 3-6)*

53. The table below gives the percent of the population in 1981–83 in 16 countries who said more emphasis on family life would be good and the percent who said greater respect for authority would be good.
   a. Make a scatterplot of these data.
   b. Does it appear that a linear equation would be a good model for this data set? Explain your answer.
   c. Find an equation of the linear regression line.
   d. What is the correlation coefficient? Is it consistent with your answer to part b? Explain your answer.

Country	Family emphasis	Respect authority
United States	94.9	85.0
Finland	94.8	29.0
Norway	92.2	36.3
Ireland	90.5	84.4
Australia	90.2	68.3
Canada	90.2	75.6
Denmark	88.0	35.9
France	87.9	56.8
Italy	87.9	56.8
Belgium	84.0	60.0
Sweden	83.7	38.9
Germany	83.5	76.0
Spain	83.5	76.0
United Kingdom	82.1	70.0
Japan	79.7	6.5
Netherlands	70.4	56.9

Source: *We're Number One!* by Andrew L. Shapiro, Vintage Books, 1992

54. The table gives the number of people who belong to Health Maintenance Organizations (HMOs).

Year	Enrollment (in thousands)
1976	6,016
1980	9,100
1984	16,743
1985	18,894
1986	23,664
1987	28,587
1988	31,848
1989	32,557
1990	34,663
1991	35,263
1992	38,842

Source: *Statistical Abstract of the United States:* 1993

a. Make a scatterplot of these data.
b. Does it seem reasonable to fit a line to these data? Explain.
c. Find an equation of the regression line.
d. Does the correlation coefficient suggest that a linear equation is a reasonable way to model these data? Explain.
e. According to this model, what will the HMO enrollment be in 1995?

55. Use the table below.

Year	Number of Physicians (thousands)
1970	334.0
1975	393.7
1980	467.7
1985	552.7
1986	569.2
1988	585.6
1989	600.8
1990	615.4

Source: *Statistical Abstract of the United States:* 1993

a. Make a scatterplot of these data.
b. Fit a line by eye to these data points and determine the equation of this line.
c. Use the line you drew in part b to estimate the number of physicians in the U.S. in 1984.
d. Find an equation of the line of best fit.
e. Explain what the slope of the regression line means.
f. Use your equation from part d to estimate the number of physicians in the U.S. in 1984.
g. Explain how you know that a linear equation models these data well.

56. The display shows the number of public school students (in millions) in the U.S. since 1955. The equation of the regression line is $y = .458x - 856.75$ with $r = .74$.

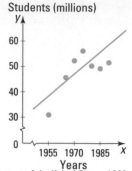

Students (millions)

Years
Source: *Statistical Abstract of the United States:* 1992

a. There is no data point for 1960. Estimate the number of public school students in 1960.
b. For 1970, **(i)** what is the observed enrollment? **(ii)** what is the predicted enrollment?
c. Estimate the public school enrollment in 2000.

Objective K: *Model situations leading to piecewise-linear functions or to step functions.* *(Lessons 3-1, 3-5, 3-9)*

In 57 and 58, a gas company charges $.4287 per therm for the first 50 therms used and $0.3432 per therm for each therm over 50.

57. Find the gas bill for a customer who used **a.** 45.76 therms. **b.** 232.99 therms.

58. Describe this situation algebraically.

59. Refer to the graph below. Cory traveled from her cousin's house to her grandmother's and then back home.

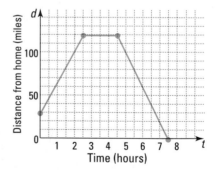

a. How far from home did Cory start?
b. How fast did Cory travel during the first two hours?
c. What was the total distance Cory traveled?

REPRESENTATIONS DEAL WITH PICTURES, GRAPHS, OR OBJECTS THAT ILLUSTRATE CONCEPTS.

60. A cheetah trots along at 5 mph for a minute, spies a small deer and accelerates to 60 mph in just 6 seconds. After chasing the deer at this speed for 30 seconds, the cheetah gives up, and over the next 20 seconds, slows to a stop. Graph this situation plotting time on the horizontal axis and speed on the vertical axis.

61. *Multiple choice.* Suppose you earn $5 per hour and the time you work is rounded down to the nearest hour. What is a rule for the function that relates time t in hours to wages w in dollars?

(a) $w = \lfloor 5t \rfloor$

(b) $w = \lfloor \frac{t}{5} \rfloor$

(c) $w = 5\lfloor t \rfloor$

(d) $w = 5\lfloor t - \frac{1}{2} \rfloor$

62. Evening phone rates, not including taxes, between two cities are 47¢ for the first minute and 33¢ for each additional minute or portion of a minute. How much will it cost to make a 12 minute 10 second phone call between these cities?

Objective L: *Graph or interpret graphs of linear equations.* (Lessons 3-2, 3-6, 3-7)

63. Graph the line with slope 5 and y-intercept 7.

64. Graph the line $4x - 6y = 36$ using its intercepts.

65. Graph $x = 3$ in the coordinate plane.

66. Graph $y = -1$ in the coordinate plane.

In 67–69, tell whether the slope of the line is positive, negative, zero, or undefined.

67.   **68.**   **69.**

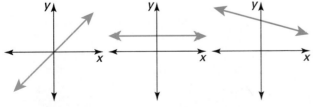

70. What is an equation for the line graphed here?

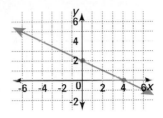

Objective M: *Graph or interpret graphs of piecewise-linear functions or step functions.* (Lessons 3-1, 3-9)

71. Use the function f, where
$$f(x) = \begin{cases} x + 3, & \text{for } x \le 0 \\ 3, & \text{for } x > 0 \end{cases}.$$
a. Draw a graph of $y = f(x)$.
b. Find the domain and range of f.

72. Consider the function $g(x) = \lfloor x \rfloor + 4$.
a. Draw a graph of $y = g(x)$.
b. Give the domain and range of g.

73. The evening phone rate between two cities is 21¢ for the first minute and 18¢ for each additional minute or fraction thereof. Graph the cost C as a function of the time t for the domain $\{t: 0 \le t \le 10\}$.

74. A salesperson earns a bonus of $270 for every $1000 of merchandise sold. Draw a graph of the bonuses as a function of the value of the merchandise sold.

75. The New York City subway system opened in 1904 with the cost of a one-way ticket being 5¢. In 1948 the fare increased (for the first time) to 10¢. Below is a table of the fares charged and the years these fares took effect.
a. Make a graph of the fares on the New York City Subway from 1904 to 1994. Let y be the independent variable.
b. *True or false.* The set of ordered pairs (y, f) for $1904 \le y \le 1992$ is a linear function.

year (y)	1904	1948	1953	1966	1970	1972	1975	1981	1984	1986	1990	1992
fare (f)	.05	.10	.15	.20	.30	.35	.50	.75	.90	1.00	1.15	1.25

CHAPTER

4

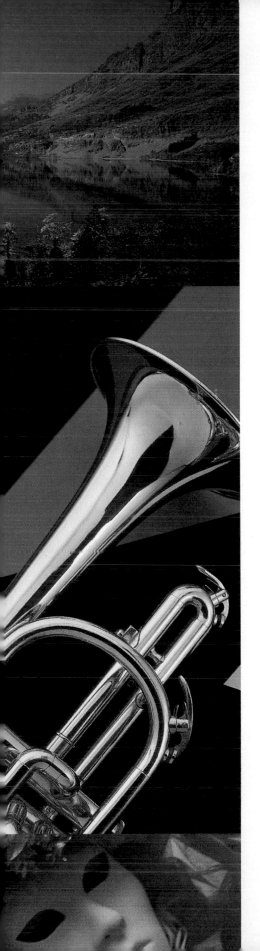

MATRICES

A **matrix** is a rectangular arrangement of objects, each of which is called an **element** of the matrix. The plural of "matrix" is "matrices." One use of matrices is to store data. In the matrix below, the elements represent data describing the nine planets in our solar system.

	Mean Distance from Sun (millions of km)	Period of Revolution	Equatorial Diameter (km)
Mercury	57.9	88 days	4,880
Venus	108.2	224.7 days	12,100
Earth	149.6	365.2 days	12,756
Mars	227.9	687 days	6,794
Jupiter	778.3	11.86 years	142,800
Saturn	1,427	29.46 years	120,660
Uranus	2,870	84 years	51,810
Neptune	4,497	165 years	49,528
Pluto	5,900	248 years	2,290

Source: *The 1994 Information Please Almanac*

The brackets [ ] identify the matrix. The titles of the rows and columns are not part of the matrix.

A second use of matrices is to describe transformations of geometric figures. On the graph below, *QUAD* has been reflected over the *x*-axis. The vertices of *QUAD* may be described by a matrix. You will learn that the coordinates of the image *Q′U′A′D′* can be found by multiplying two matrices.

$$\begin{array}{c} \quad\;\; Q \;\; U \;\; A \;\; D \\ \begin{array}{c} x \\ y \end{array} \left[\begin{array}{cccc} -4 & 0 & 5 & 4 \\ 3 & 6 & 5 & 1 \end{array}\right] \end{array}$$

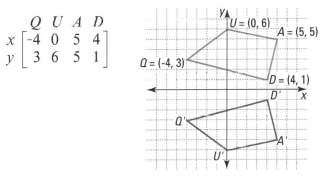

In this chapter you will study various operations using matrices, and how those operations can be applied to geometric transformations and real-life situations.

Brass sounds. *Most horns are made of brass. The differences in tubing produce a wide variety of sounds, from the brilliant sound of the trumpet to the mellow sound of the French horn. See Example 1.*

Information is often stored in matrices. The inventory of athletic clothing owned by a high school cross-country team is shown in the matrix below.

	sweat pants	sweat shirts	shorts	
small	9	10	8	row 1
medium	18	20	19	row 2
large	20	24	23	row 3
x-large	11	11	(12)	row 4
	column 1	column 2	column 3	

the element in the 4th row and the 3rd column

Dimensions of a Matrix

The elements of the matrix above are enclosed by large square brackets. (Sometimes, large parentheses are used in place of brackets.) This matrix has 4 *rows* and 3 *columns*. It is said to have the *dimensions* 4 by 3, written 4×3. In general, a matrix with m rows and n columns has **dimensions** $m \times n$.

The entries in spreadsheets also constitute a matrix. Recall that the entries in a spreadsheet are called cells, and the location of a cell is identified by giving its column first (a letter) and row second (a number). Matrices use the reverse order for identifying an element—row first and column second. Matrices, like spreadsheets, can have headings to identify their rows and columns. In matrices, the headings are placed outside the matrix.

Example 1

The Matterhorn Company produced 1500 trumpets and 1200 French horns in September; 2000 trumpets and 1400 French horns in October; and 900 trumpets and 700 French horns in November.
a. Store the company's production in a matrix.
b. What are the dimensions of the matrix?

Solution

a. Two matrices can be written. Matrix M_1 has the months as rows, and matrix M_2 has the months as columns. Either matrix is an acceptable way to store the data.

Matrix M_1

$$\begin{array}{c c c} & \text{trumpets} & \text{French horns} \\ \text{Sept.} & 1500 & 1200 \\ \text{Oct.} & 2000 & 1400 \\ \text{Nov.} & 900 & 700 \end{array}$$

Matrix M_2

$$\begin{array}{c c c c} & \text{Sept.} & \text{Oct.} & \text{Nov.} \\ \text{trumpets} & 1500 & 2000 & 900 \\ \text{French horns} & 1200 & 1400 & 700 \end{array}$$

b. Matrix M_1 has 3 rows and 2 columns. The dimensions of M_1 are 3 × 2. Matrix M_2 has 2 rows and 3 columns. The dimensions of M_2 are 2 × 3.

Although matrices M_1 and M_2 are both acceptable ways to store the data, the two matrices are not considered equal. Two matrices are **equal matrices** if and only if they have the same dimensions and their corresponding elements are equal.

Matrices and Geometry

Points and polygons can also be represented by matrices. The ordered pair (x, y) is generally represented by the matrix

$$\begin{bmatrix} x \\ y \end{bmatrix}.$$

This 2 × 1 matrix is called a **point matrix.** Notice that the element in the first row is the x-coordinate and the element in the second row is the y-coordinate. Thus, the point $(5, -1)$ is represented by the matrix

$$\begin{bmatrix} 5 \\ -1 \end{bmatrix}.$$

Similarly, polygons can be written as matrices. The first row of the matrix contains the x-coordinates of the vertices in the order in which the polygon is named. The second row contains the corresponding y-coordinates. Example 2 illustrates this.

Example 2

a. Write pentagon *PENTA* as a matrix.
b. Write pentagon *NEPAT* as a matrix.
c. Are the two matrices equal? Why or why not?

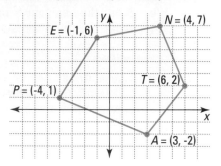

Solution

a. Starting with the coordinates of *P*, write the *x*-coordinates in the first row and the *y*-coordinates in the second row.

$$
\begin{array}{c c}
 & \begin{array}{ccccc} \text{P} & \text{E} & \text{N} & \text{T} & \text{A} \end{array} \\
\begin{array}{c} x \\ y \end{array} & \left[\begin{array}{rrrrr} -4 & -1 & 4 & 6 & 3 \\ 1 & 6 & 7 & 2 & -2 \end{array} \right]
\end{array}
$$

b. Start with the coordinates of *N* and use the same procedure that was used in part **a.**

$$
\begin{array}{c c}
 & \begin{array}{ccccc} \text{N} & \text{E} & \text{P} & \text{A} & \text{T} \end{array} \\
\begin{array}{c} x \\ y \end{array} & \left[\begin{array}{rrrrr} 4 & -1 & -4 & 3 & 6 \\ 7 & 6 & 1 & -2 & 2 \end{array} \right]
\end{array}
$$

c. The two matrices are not equal because not all of the corresponding elements are equal. However, both matrices are valid ways to represent the polygon.

QUESTIONS

Covering the Reading

1. What is a matrix?

2. *True or false.*   Each element in a matrix must be a number.

In 3 and 4, **a.** Give the number of rows in the indicated matrix. **b.** Give the number of columns. **c.** Give its dimensions.

3. the matrix of basic planetary data from page 203

4. the matrix for pentagon *PENTA* in Example 2

In 5–7, refer to the clothing matrix at the start of this lesson.

5. How many large sweatshirts did the cross-country team have?

6. What does the element in the 2nd row, 3rd column represent?

7. What does the sum of the elements in the 3rd column represent?

8. Refer to Example 1. Suppose the Matterhorn Company produces 2500 trumpets and 3800 French horns in December. Construct a 4 × 2 matrix that gives the company's production through December.

9. The ordered pair (a, b) can be represented by the matrix __?__. This matrix is called a __?__ matrix.

10. *Multiple choice.* Which matrix represents the point (-136, 4.9)?
(a) $\begin{bmatrix} -136 & 4.9 \end{bmatrix}$
(b) $\begin{bmatrix} 4.9 & -136 \end{bmatrix}$
(c) $\begin{bmatrix} -136 \\ 4.9 \end{bmatrix}$
(d) $\begin{bmatrix} 4.9 \\ -136 \end{bmatrix}$

11. Refer to Example 2.
a. Write pentagon *EPATN* as a matrix.
b. Are the matrices for *EPATN* and *ENTAP* equal? Why or why not?
c. Do *EPATN* and *ENTAP* both represent the pentagon?

12. Write △*TRI* as a matrix.

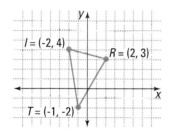
$I = (-2, 4)$
$R = (2, 3)$
$T = (-1, -2)$

13. The matrix below describes a hexagon. Graph this hexagon.

$$\begin{bmatrix} 7 & 8 & 4 & 5 & 1 & 0 \\ 11 & -1 & 3 & 9 & 2 & 6 \end{bmatrix}$$

Applying the Mathematics

14. The matrix below gives the numbers of active-duty U.S. military personnel in 1992.

	Commissioned Officers	Enlisted Personnel
Army	85,953	561,104
Navy	71,826	500,459
Marines	19,132	165,397
Air Force	92,000	394,800

Sources: *1993 and 1994 World Almanacs*

a. What are the dimensions of this matrix?
b. What does the sum of the elements in row 2 represent?
c. What does the sum of the elements in column 1 represent?

15. If $\begin{bmatrix} 7 & 4 \\ y & 2 \end{bmatrix} = \begin{bmatrix} x & 4 \\ 8 & 2 \end{bmatrix}$, then $x =$ __?__ and $y =$ __?__.

16. If $\begin{bmatrix} 3a + 1 \\ b + 4 \end{bmatrix} = \begin{bmatrix} 7 \\ 4 \end{bmatrix}$, then $a =$ __?__ and $b =$ __?__.

Heralding the President. *The U.S. Army Herald Trumpets perform at many events—such as this one during the Inaugural festivities in 1993.*

17. Adam, Barbara, and Clem write to each other from time to time. Last year, Adam received 2 letters from Barbara and 5 from Clem. Barbara received 3 from Adam and 3 from Clem. Clem received 1 from Adam and 4 from Barbara.
    a. Organize this information in a 3 × 3 matrix. (Hint: There are three zeros in the matrix.)
    b. How many letters did each person write?

Review

18. Which three of the following describe the graph of $y = -4x^2$?
    (Lesson 2-5)
    (a) hyperbola   (b) direct variation
    (c) parabola   (d) inverse variation
    (e) symmetric to y-axis   (f) symmetric to x-axis

19. Write a short paragraph describing the changes in the graph of the line $y = kx$ as k increases from -100 to 100. Include sketches.
    (Lesson 2-4)

20. Find two more pairs (x, y) in the following function of inverse variation. *(Lesson 2-1)*

x	1	2	3	4	5
y	36	18	12	9	7.2

21. Find the area of rectangle *ABCD*. *(Previous course)*

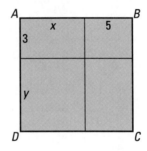

In 22 and 23, state the property. *(Appendix B)*

22. Commutative Property of Multiplication

23. Associative Property of Addition

Exploration

24. Information in newspapers is often organized in matrices. However, newspapers usually don't write square brackets to identify matrices. Find an example of a matrix from a newspaper.

LESSON

4-2

Matrix Addition

How Are Matrices Added?

There are many situations which require adding the information stored in matrices. For instance, suppose matrix C represents the current inventory of the Chic Boutique.

$$
\begin{array}{c}
 & \text{sizes} \\
 & \begin{array}{ccccc} 8 & 10 & 12 & 14 & 16 \end{array} \\
\begin{array}{l} \text{dresses} \\ \text{suits} \\ \text{skirts} \\ \text{blouses} \end{array} &
\left[\begin{array}{ccccc}
5 & 7 & 8 & 10 & 9 \\
3 & 4 & 6 & 2 & 2 \\
15 & 20 & 18 & ㉓ & 7 \\
12 & 18 & 14 & 21 & 11
\end{array}\right]
\end{array} \quad \begin{array}{l} \text{Current Inventory} \\ = C \end{array}
$$

The quantities of new items received by the boutique are represented by the numbers stored in matrix D.

$$
\begin{array}{c}
 & \text{sizes} \\
 & \begin{array}{ccccc} 8 & 10 & 12 & 14 & 16 \end{array} \\
\begin{array}{l} \text{dresses} \\ \text{suits} \\ \text{skirts} \\ \text{blouses} \end{array} &
\left[\begin{array}{ccccc}
3 & 2 & 4 & 3 & 1 \\
1 & 2 & 3 & 4 & 2 \\
5 & 6 & 4 & ③ & 5 \\
4 & 3 & 5 & 7 & 6
\end{array}\right]
\end{array} \quad \begin{array}{l} \text{Delivery} \\ = D \end{array}
$$

The new inventory is found by taking the sum of matrices C and D. This **matrix addition** is performed according to the following definition.

Definition

If two matrices A and B have the same dimensions, their **sum** $A + B$ is the matrix in which each element is the sum of the corresponding elements in A and B.

For the previously mentioned matrices, the sum $C + D$ is a 4×5 matrix. Add corresponding elements of C and D to find the elements of $C + D$. We have circled one set of corresponding elements: $23 + 3 = 26$.

$$
\begin{array}{c}
 & \text{sizes} \\
 & \begin{array}{ccccc} 8 & 10 & 12 & 14 & 16 \end{array} \\
\begin{array}{l} \text{dresses} \\ \text{suits} \\ \text{skirts} \\ \text{blouses} \end{array} &
\left[\begin{array}{ccccc}
8 & 9 & 12 & 13 & 10 \\
4 & 6 & 9 & 6 & 4 \\
20 & 26 & 22 & ㉖ & 12 \\
16 & 21 & 19 & 28 & 17
\end{array}\right]
\end{array} \quad \begin{array}{l} \text{New Inventory} \\ = C + D \end{array}
$$

Because addition of real numbers is commutative, *addition of matrices is commutative.* Thus, if two matrices A and B can be added, then $A + B = B + A$. Also, for all matrices A, B, and C, *addition of matrices is associative;* that is, $(A + B) + C = A + (B + C)$.

How Are Matrices Subtracted?

Subtraction of matrices is defined like matrix addition is. Given two matrices A and B with the same dimensions, their **difference** $A - B$ is the matrix whose element in each position is the difference of the corresponding elements in A and B.

Example 1

The matrix $W1$ below represents the costs of 1 dozen each of eggs and oranges in three different markets during one week. The matrix $W2$ represents the cost of these same items in the same markets during another week.

$$W1 = \begin{bmatrix} .97 & .90 & .95 \\ 1.99 & 1.79 & 1.59 \end{bmatrix} \begin{matrix} \text{eggs} \\ \text{oranges} \end{matrix} \qquad W2 = \begin{bmatrix} .97 & .85 & 1.05 \\ 1.49 & 1.79 & 1.89 \end{bmatrix} \begin{matrix} \text{eggs} \\ \text{oranges} \end{matrix}$$

market
1 2 3

a. Find $W2 - W1$.

b. For which of the markets was the change in the price of oranges from Week 1 to Week 2 greatest? What was the change in price?

Solution

a.

$$\underset{W2}{\begin{bmatrix} .97 & .85 & 1.05 \\ 1.49 & 1.79 & 1.89 \end{bmatrix}} - \underset{W1}{\begin{bmatrix} .97 & .90 & .95 \\ 1.99 & 1.79 & 1.59 \end{bmatrix}} = \underset{W2-W1}{\begin{bmatrix} 0 & -.05 & .10 \\ -.50 & 0 & .30 \end{bmatrix}}$$

b. The changes in prices of oranges are given in row 2 of the matrix $W2 - W1$. The change in price of oranges from week 1 to week 2 was greatest in market 1. The price of oranges decreased $.50 per dozen in this period.

Egg-xamination.
Eggs are collected onto a moving belt that carries them to be washed, rinsed, and dried. They are then candled—rolled over bright lights to be graded and inspected for flaws. Once inspected they are weighed, separated according to size, and packed in cartons for shipping.

Scalar Multiplication

Matrix addition is related to a special multiplication involving matrices called *scalar multiplication.* Consider this repeated addition:

$$\begin{bmatrix} 7 & 8 \\ 4 & 2 \end{bmatrix} + \begin{bmatrix} 7 & 8 \\ 4 & 2 \end{bmatrix} + \begin{bmatrix} 7 & 8 \\ 4 & 2 \end{bmatrix} = \begin{bmatrix} 21 & 24 \\ 12 & 6 \end{bmatrix}$$

Notice that in the final result, every element of the original matrix has been multiplied by 3. With real numbers, we use multiplication as a shorthand for repeated addition; for example, $17 + 17 + 17$ can be written as 3×17. Similarly, we can rewrite the above sum as

$$3 \begin{bmatrix} 7 & 8 \\ 4 & 2 \end{bmatrix}.$$

The constant 3 is called a *scalar.* **Scalar multiplication** is defined below.

> **Definition**
> The product of a scalar k and a matrix A is the matrix kA in which each element is k times the corresponding element in A.

Example 2

Find the product $5 \begin{bmatrix} 7 & 2 & -1 \\ 4 & 9 & 11 \end{bmatrix}$.

Solution

Multiply each element in the matrix by 5.

$$5 \begin{bmatrix} 7 & 2 & -1 \\ 4 & 9 & 11 \end{bmatrix} = \begin{bmatrix} 5 \cdot 7 & 5 \cdot 2 & 5 \cdot (-1) \\ 5 \cdot 4 & 5 \cdot 9 & 5 \cdot 11 \end{bmatrix} = \begin{bmatrix} 35 & 10 & -5 \\ 20 & 45 & 55 \end{bmatrix}$$

QUESTIONS

Covering the Reading

1. What must be true about the dimensions of two matrices in order for addition or subtraction to be possible?

In 2 and 3, refer to the clothing matrices C and D at the beginning of this lesson.

2. Does $C + D = D + C$?

3. Suppose the shop gets another delivery described by matrix P below.

$$P = \begin{array}{c} \\ \\ \\ \text{sizes} \\ \begin{array}{ccccc} 8 & 10 & 12 & 14 & 16 \end{array} \\ \begin{bmatrix} 5 & 2 & 1 & 0 & 3 \\ 4 & 1 & 1 & 1 & 2 \\ 3 & 6 & 4 & 10 & 5 \\ 4 & 2 & 5 & 11 & 12 \end{bmatrix} \begin{array}{l} \text{dresses} \\ \text{suits} \\ \text{skirts} \\ \text{blouses} \end{array} \end{array}$$

Find the new inventory matrix $P + C + D$.

4. Refer to Example 1.
   a. In which market did the price of a dozen eggs change the most from Week 1 to Week 2?
   b. Was the change an increase or a decrease?

In 5 and 6, let $A = \begin{bmatrix} 3 & 5 \\ 0 & -3 \end{bmatrix}$, $B = \begin{bmatrix} 4 & -5 \\ -2 & 1 \end{bmatrix}$, and $C = \begin{bmatrix} 1 & -1 \\ -6 & 3 \end{bmatrix}$.

5. a. Find $A - B$.
   b. Find $B - A$.
   c. *True or false.* Subtraction of matrices is commutative.

6. a. Find $(A + B) + C$.
   b. Find $A + (B + C)$.
   c. The results of parts **a** and **b** show an instance of which property?

7. If $M = \begin{bmatrix} 3 & 2 & 5 \\ 4 & 9 & 1 \end{bmatrix}$, what does $10M$ equal?

8. The matrices *N, C,* and *S* give the enrollments by sex and grade at North, Central, and South High Schools. In each matrix, Row 1 gives the number of boys and Row 2 the number of girls. Columns 1 to 4 give the number of students in grades 9 to 12, respectively. Determine the matrix *T* that shows the total enrollment by sex and grade in the three schools.

$$N = \begin{matrix} & 9 & 10 & 11 & 12 & \\ & \begin{bmatrix} 250 & 245 & 240 & 235 \\ 260 & 250 & 240 & 230 \end{bmatrix} & \begin{matrix} \text{boys} \\ \text{girls} \end{matrix} \end{matrix}$$

$$C = \begin{bmatrix} 200 & 190 & 180 & 170 \\ 200 & 195 & 190 & 185 \end{bmatrix} \begin{matrix} \text{boys} \\ \text{girls} \end{matrix}$$

$$S = \begin{bmatrix} 140 & 135 & 130 & 125 \\ 130 & 130 & 125 & 120 \end{bmatrix} \begin{matrix} \text{boys} \\ \text{girls} \end{matrix}$$

9. Final standings from the National Hockey League Northeast Division (formerly the Adams Division) for 1992–1993 and 1993–1994 are given in the matrices below.

1992–1993

	W	L	T	Pts
Boston	51	26	7	109
Quebec	47	27	10	104
Montreal	48	30	6	102
Buffalo	38	36	10	86
Hartford	26	52	6	58
Ottawa	10	70	4	24

1993–1994

	W	L	T	Pts
Boston	42	29	13	97
Quebec	34	42	8	76
Montreal	41	29	14	96
Buffalo	43	32	9	95
Hartford	27	48	9	63
Ottawa	14	61	9	37

a. Subtract the left matrix from the right matrix. Call the difference *M*.
b. What is the meaning of the 4th column of *M*?
c. What is the meaning of the 1st row of *M*?

10. A toymaker makes handcrafted toys for children. His output last year is represented by the matrix *M* below.

$$\begin{matrix} & \text{sm} & \text{med} & \text{lg} & \\ \text{dolls} & \begin{bmatrix} 5 & 10 & 18 \\ 12 & 22 & 9 \end{bmatrix} & & = M \\ \text{stuffed animals} \end{matrix}$$

a. Suppose he wants to increase his output by 30%. Write the matrix that represents the needed output.
b. Find 2*M* and explain what the matrix represents.

11. Suppose $M = \begin{bmatrix} a & b \\ c & d \end{bmatrix}$. Find kM.

In 12 and 13, solve for *a, b, c,* and *d.*

12. $\begin{bmatrix} 15 & 20 \\ 25 & 30 \end{bmatrix} + \begin{bmatrix} a & b \\ c & d \end{bmatrix} = \begin{bmatrix} 8 & 18 \\ 28 & 38 \end{bmatrix}$

13. $3\begin{bmatrix} a & -1 \\ c & 4 \end{bmatrix} - 5\begin{bmatrix} 3 & b \\ 11 & -2.5 \end{bmatrix} = \begin{bmatrix} 9 & 0 \\ 8 & d \end{bmatrix}$

14. In June, The Faucets & Fixtures Company produced 50 porcelain sinks, 40 stainless steel sinks, and 17 molded plastic sinks. In July, they produced 100 porcelain, 80 stainless steel, and 3 molded plastic sinks. In August they produced 42 porcelain, 58 stainless steel, and 5 molded plastic sinks. Write two different 3×3 matrices to store these data. *(Lesson 4-1)*

15. Use the matrix $\begin{bmatrix} 0 & -1 & 0 & 1 \\ 3 & 0 & -3 & 0 \end{bmatrix}$.
    a. Graph the polygon represented by the matrix.
    b. What kind of polygon is it?
    c. Graph the image of this polygon under a size change of magnitude 2. *(Previous course, Lesson 4-1)*

16. Write an expression to describe the cost of C concert tickets at $20 each, P programs at $3.50 each, and D drinks at $1.50 each. *(Lesson 3-3)*

17. Consider the function with equation $f(x) = \frac{-3}{x^2}$.
    a. Graph the function.
    b. Identify its domain and range. *(Lesson 2-6)*

18. Write a paragraph explaining some differences and similarities between the functions with equations $y = kx$ and $y = kx^2$. Include some sketches. *(Lessons 2-4, 2-5)*

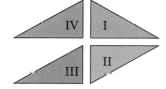

In 19–22, use the figures at the left. Triangles I, II, III, and IV are congruent. Segments that look parallel are. Fill in the blank with one of the words *translation, rotation,* or *reflection.* *(Previous course)*

19. IV is a __?__ image of I.

20. III is a __?__ image of IV.

21. II is a __?__ image of IV.

22. II is a __?__ image of III.

23. Let $P = (-3, 5)$ and $Q = (12, -3)$. Use the Pythagorean Theorem or the distance formula to find PQ. *(Previous course)*

Exploration

24. With most graphics calculators you can store data in matrices, add and subtract matrices, and perform scalar multiplication.
    a. Store $A = \begin{bmatrix} 3 & \frac{1}{3} \\ 0 & -2 \end{bmatrix}$, $B = \begin{bmatrix} 1 & 2 \\ 3 & -4 \end{bmatrix}$, and $C = \begin{bmatrix} -60 \\ 5 \end{bmatrix}$ in a calculator.
    b. Find $A + B$ using the calculator.
    c. What happens when you try to find $A - C$ using the calculator?
    d. Find $7.34B$ using the calculator.

Reel attractions. *Most movie theaters charge varying admission fees based on age and time of day to attract as many people as possible.*

Row-by-Column Multiplication

In linear-combination applications, it is quite useful to store data in matrices which can then be multiplied. Consider a movie theater which charges $6.00 for adults over 17, $4.00 for students 13–17 years old, and $2.50 for children 12 or under. How much does it cost for a family with 2 adults, 1 student, and 3 children to enter the theater?

The answer is $6.00 · 2 + $4.00 · 1 + $2.50 · 3 = $23.50. This is the same arithmetic needed to calculate the *product* of these matrices:

$$\begin{bmatrix} 6 & 4 & 2.50 \end{bmatrix} \cdot \begin{bmatrix} 2 \\ 1 \\ 3 \end{bmatrix}$$

cost per category · number of people in each category

Multiplying Two Matrices

More generally, the product $A \cdot B$ or AB of two matrices A and B is found by multiplying rows of A by columns of B. Multiply the first element in the row by the first element in the column, the second element in the row by the second element in the column, and the third element in the row by the third element in the column. In general, multiply the nth element in the row by the nth element in the column. Finally, add the resulting products.

$$\begin{bmatrix} 6 & 4 & 2.50 \end{bmatrix} \cdot \begin{bmatrix} 2 \\ 1 \\ 3 \end{bmatrix} = \begin{bmatrix} 6 \cdot 2 + 4 \cdot 1 + 2.50 \cdot 3 \end{bmatrix} = \begin{bmatrix} 23.50 \end{bmatrix}$$

The product is the 1×1 matrix $\begin{bmatrix} 23.50 \end{bmatrix}$, corresponding to $23.50. If matrix A has 2 rows and matrix B has 3 columns, then there are

6 ways to multiply a row by a column. The six products of these rows and columns are the 6 elements of the product matrix AB.

Example 1

Let $A = \begin{bmatrix} 8 & -2 \\ 4 & 1 \end{bmatrix}$ and $B = \begin{bmatrix} 1 & 3 & 5 \\ 0 & 4 & 2 \end{bmatrix}$. Find the product AB.

Solution

The product of row 1 of A and column 1 of B is $8 \cdot 1 + -2 \cdot 0 = 8$. This is put in the 1st row and 1st column of the answer. Thus far we have

$$\begin{bmatrix} 8 & -2 \\ 4 & 1 \end{bmatrix}\begin{bmatrix} 1 & 3 & 5 \\ 0 & 4 & 2 \end{bmatrix} = \begin{bmatrix} 8 & - & - \\ - & - & - \end{bmatrix}.$$

The products from row 1 of A and column 2 of B yield $8 \cdot 3 + -2 \cdot 4 = 16$. Now you know the element in the 1st row, 2nd column of the answer.

$$\begin{bmatrix} 8 & -2 \\ 4 & 1 \end{bmatrix}\begin{bmatrix} 1 & 3 & 5 \\ 0 & 4 & 2 \end{bmatrix} = \begin{bmatrix} 8 & 16 & - \\ - & - & - \end{bmatrix}$$

The other four elements of AB are found using this pattern.

row ▨ by column ▨

For instance, the element in the 2nd row, 3rd column of AB is found by multiplying the 2nd row of A by the 3rd column of B, as shown here, along with the final result.

$$\begin{bmatrix} 8 & -2 \\ 4 & 1 \end{bmatrix}\begin{bmatrix} 1 & 3 & 5 \\ 0 & 4 & 2 \end{bmatrix} = \begin{bmatrix} 8 & 16 & 36 \\ 4 & 16 & 22 \end{bmatrix}$$

To multiply matrices with other dimensions, do all possible products using rows from matrix A and columns from matrix B.

> **Definition of Matrix Multiplication**
> Suppose A is an $m \times n$ matrix and B is an $n \times p$ matrix. Then the product $A \cdot B$ (or AB) is the $m \times p$ matrix whose element in row i and column j is the product of row i of A and column j of B.

From the definition, *the product of two matrices A and B exists only when the number of columns of A equals the number of rows of B.* So if A is $m \times n$, B must be $n \times p$ in order for AB to exist.

These matrices can be multiplied.

$$\begin{bmatrix} 8 & -2 \\ 4 & 1 \end{bmatrix}\begin{bmatrix} 1 & 3 & 5 \\ 0 & 4 & 2 \end{bmatrix} = \begin{bmatrix} 8 & 16 & 36 \\ 4 & 16 & 22 \end{bmatrix}$$

$2 \times 2 \quad 2 \times 3 \qquad 2 \times 3$

equal

dimensions of product

These matrices cannot be multiplied.

$$\begin{bmatrix} 1 & 3 & 5 \\ 0 & 4 & 2 \end{bmatrix}\begin{bmatrix} 8 & -2 \\ 4 & 1 \end{bmatrix}$$

$2 \times 3 \quad 2 \times 2$

not equal

These two cases indicate that, in general, *multiplication of matrices is not commutative.*

Setting Up Matrices to Be Multiplied

When matrices arise from real situations you often have several choices for arranging your data. However, when two matrices are to be multiplied, the *number* of columns of the first matrix must match the *number* of rows of the second matrix. In order to set up matrices that can be multiplied, it may help to think about the units involved. That is, the *headings* of the columns of the left matrix must match the *headings* of the rows of the right matrix.

Example 2

Costumes have been designed for the school play. Each boy's costume requires 5 yards of fabric, 4 yards of ribbon, and 3 packets of sequins. Each girl's costume requires 6 yards of fabric, 5 yards of ribbon, and 2 packets of sequins. Fabric costs $4 per yard, ribbon costs $2 per yard, and sequins cost $.50 per packet. Use matrix multiplication to find the total cost of the materials for each costume.

Solution

The information about the needed materials for each type of costume can be put in a 2 × 3 matrix. (In the Questions you are asked to start with a 3 × 2 matrix to arrive at the same result.) Then we put the information about the unit cost of each material in a matrix. So that the headings "match," we write this as a 3 × 1 matrix and position it to the right of the original matrix.

$$
\begin{array}{c}
\\
\text{Boys} \\
\text{Girls}
\end{array}
\begin{array}{ccc}
\text{Fab.} & \text{Rib.} & \text{Seq.}
\end{array}
\begin{bmatrix}
5 & 4 & 3 \\
6 & 5 & 2
\end{bmatrix}
\qquad
\begin{array}{c}
\\
\text{Fabric} \\
\text{Ribbon} \\
\text{Sequins}
\end{array}
\begin{array}{c}
cost
\end{array}
\begin{bmatrix}
4 \\
2 \\
.50
\end{bmatrix}
$$

materials for each costume   unit cost of materials

When we multiply the "materials for each costume" matrix by the "unit cost of materials" matrix we get the total cost of the materials for each costume.

$$
\begin{bmatrix}
5 & 4 & 3 \\
6 & 5 & 2
\end{bmatrix}
\begin{bmatrix}
4 \\
2 \\
.50
\end{bmatrix}
=
\begin{bmatrix}
5 \cdot 4 + 4 \cdot 2 + 3 \cdot .50 \\
6 \cdot 4 + 5 \cdot 2 + 2 \cdot .50
\end{bmatrix}
=
\begin{bmatrix}
29.50 \\
35
\end{bmatrix}
$$

Notice that units work with matrix multiplication as they do with normal multiplication. In Example 2, the units of the original matrices are

$$
\begin{array}{c}
\\
\text{Boys} \\
\text{Girls}
\end{array}
\begin{array}{ccc}
\text{Fab.} & \text{Rib.} & \text{Seq.}
\end{array}
\begin{bmatrix}
 & & \\
& &
\end{bmatrix}
\qquad
\begin{array}{c}
\\
\text{Fabric} \\
\text{Ribbon} \\
\text{Sequins}
\end{array}
\begin{array}{c}
cost
\end{array}
\begin{bmatrix}
 \\
\\
\end{bmatrix},
$$

so the product has units $\begin{array}{c} \\ \text{Boys} \\ \text{Girls} \end{array} \begin{array}{c} cost \end{array} \begin{bmatrix} \\ \\ \end{bmatrix}$. Thus the cost of the materials for a boy's costume is $29.50, and for a girl's costume it is $35.00.

Multiplying More Than Two Matrices

To multiply more than two matrices, multiply two at a time.

Example 3

Find the product below.

$$\left(\begin{bmatrix} 2 & 5 \end{bmatrix} \begin{bmatrix} 3 & 1 & -4 \\ 2 & 0 & 1 \end{bmatrix} \right) \begin{bmatrix} 5 \\ 2 \\ 1 \end{bmatrix}$$

Solution

$$\left(\begin{bmatrix} 2 & 5 \end{bmatrix} \begin{bmatrix} 3 & 1 & -4 \\ 2 & 0 & 1 \end{bmatrix} \right) \begin{bmatrix} 5 \\ 2 \\ 1 \end{bmatrix} = \begin{bmatrix} 16 & 2 & -3 \end{bmatrix} \begin{bmatrix} 5 \\ 2 \\ 1 \end{bmatrix} = \begin{bmatrix} 81 \end{bmatrix}$$

In the Questions, you are asked to verify that changing the way the matrices are grouped does not affect the answer. You are verifying that *matrix multiplication is associative.*

Using Technology to Multiply Matrices

Most graphics calculators and some computer software can be used to multiply matrices. Consult a manual for the specific key sequences needed to input and multiply matrices with your technology.

Activity

Use a graphics calculator or computer software to find AB, where

$$A = \begin{bmatrix} 8 & -2 \\ 4 & 1 \end{bmatrix} \text{ and } B = \begin{bmatrix} 1 & 3 & 5 \\ 0 & 4 & 2 \end{bmatrix}.$$

This should check Example 1.

QUESTIONS

Covering the Reading

In 1 and 2, multiply the matrices.

1. $\begin{bmatrix} 3 & 5 & 7 \end{bmatrix} \begin{bmatrix} 1 \\ 0 \\ -2 \end{bmatrix}$

2. $\begin{bmatrix} 1 & -1 & 1 & -1 \end{bmatrix} \begin{bmatrix} 10 \\ 9 \\ 8 \\ 7 \end{bmatrix}$

3. Let $M = \begin{bmatrix} 6 & 2 \\ 0 & 3 \end{bmatrix}$ and $N = \begin{bmatrix} 5 & 8 & -2 \\ -4 & 1 & 0 \end{bmatrix}$. Find the product MN.

4. If A has dimensions 11×15 and B has dimensions 15×19, what are the dimensions of AB?

5. If A is $m \times n$ and B is $p \times q$, when does AB exist?

In 6 and 7, two matrices are given. **a.** Determine the dimensions of each matrix. **b.** Decide whether the product can or cannot be found. **c.** If the product exists, calculate it; if it does not exist, explain why not.

6. $\begin{bmatrix} 8 & 1 & 0 \\ 6 & 3 & -4 \end{bmatrix} \begin{bmatrix} 2 & 8 \\ 5 & 4 \end{bmatrix}$

7. $\begin{bmatrix} 9 & 4 & 8 & 6 \\ 2 & 0 & 3 & 1 \\ 1 & -2 & 5 & 0 \end{bmatrix} \begin{bmatrix} 12 & 2 \\ 15 & 1 \\ 3 & 9 \\ 8 & 11 \end{bmatrix}$

8. Answer the question of Example 2 starting with a 3 × 2 matrix.

9. Refer to Example 2. How much will it cost to make 8 boys' and 10 girls' costumes?

10. Suppose $X = \begin{bmatrix} 3 & 0 & 5 \\ -1 & 4 & 2 \end{bmatrix}$ and $Y = \begin{bmatrix} 2 & -2 \\ 0 & 1 \\ -3 & 4 \end{bmatrix}$.

    a. Calculate XY.   **b.** Calculate YX.
    c. *True or false.* Matrix multiplication is commutative.

11. a. Find the product $\begin{bmatrix} 2 & 5 \end{bmatrix} \left(\begin{bmatrix} 3 & 1 & -4 \\ 2 & 0 & 1 \end{bmatrix} \begin{bmatrix} 5 \\ 2 \\ 1 \end{bmatrix} \right)$.

    b. What property is verified in Example 3 and part **a?**

Applying the Mathematics

From dough to bread.
Shown above is a Native American making fry bread. Fry bread is much like pita bread and is often used to make sandwiches. Below are Mexican sweet breads known as pan dulce. *Sweet breads are generally eaten during the* merienda—*a light supper—with coffee and milk.*

12. The matrix D gives the daily delivery of cases of four bakery products to two restaurants, Pierre's and Pauline's. The matrix C gives the cost per case for each product.

Matrix D:

	wheat bread	white bread	rye bread	English muffins
Pierre's	5	10	3	5
Pauline's	0	15	8	10

Matrix C:

	cost per case
wheat bread	7.00
white bread	7.00
rye bread	6.50
English muffins	8.00

    a. Find DC.
    b. What is the daily cost of these bakery products at Pierre's?
    c. Pierre's restaurant is open 20 days this month and Pauline's is open 25 days. Let $M = \begin{bmatrix} 20 & 25 \end{bmatrix}$. Find the total cost of these bakery items for the month at these two restaurants.

13. East, Central, and West schools need some new uniforms next year. East needs 10 band uniforms, 20 basketball uniforms, 15 track uniforms, and 10 swimsuits. Central needs 20 band uniforms and 10 track uniforms. West needs 15 band uniforms, 20 basketball uniforms, and 10 swimsuits. A band uniform costs $90, a basketball uniform costs $40, a track uniform costs $50, and a swimsuit costs $25.
    a. Set up a matrix that gives the number of each type of uniform needed by each school.
    b. Use matrix multiplication to find the total cost of uniforms for each school.

14. The matrix $\begin{bmatrix} 1 & 0 \\ 0 & 1 \end{bmatrix}$ is called **the 2 × 2 identity matrix** for multiplication. To see why, calculate the products in parts **a** and **b**.

a. $\begin{bmatrix} 1 & 0 \\ 0 & 1 \end{bmatrix}\begin{bmatrix} a & b \\ c & d \end{bmatrix}$
b. $\begin{bmatrix} a & b \\ c & d \end{bmatrix}\begin{bmatrix} 1 & 0 \\ 0 & 1 \end{bmatrix}$

c. *True or false.* Matrix multiplication with the identity matrix is commutative.

15. Solve for x. $\begin{bmatrix} 3 & 1 \\ 0 & 2 \end{bmatrix}\begin{bmatrix} x \\ 9 \end{bmatrix} = \begin{bmatrix} 10 \\ 18 \end{bmatrix}$

Review

16. a. Find the matrix M such that $M - \begin{bmatrix} 3 & 7 \\ -1 & 2 \end{bmatrix} = \begin{bmatrix} 4 & 5 \\ 9 & 8 \end{bmatrix}$.

  b. Check your answer.   *(Lesson 4-2)*

17. A matrix contains the vertices of an *n*-gon. What are the dimensions of the matrix?   *(Lesson 4-1)*

18. The matrices below represent U.S. exports and imports by region in millions of dollars.

	1992 Exports	1992 Imports		1991 Exports	1991 Imports
North Atlantic area	89.2	130.6		90.6	123.6
South Atlantic area	38.7	37.8		34.5	33.6
Gulf area	70.9	70.4		66.0	67.6
Pacific area	129.5	139.5		120.0	127.3
Great Lakes area	85.7	121.3		80.6	108.1
other	33.3	25.3		30.2	22.8

  a. Find the difference in exports and imports between 1992 and 1991.
  b. From which area did the imports increase the most?   *(Lesson 4-2)*

19. The matrix $\begin{bmatrix} 2 & 7 & -5 \\ -1 & 6 & 4 \end{bmatrix}$ can represent a triangle. Use the distance formula to show that this triangle is isosceles.   *(Previous course, Lesson 4-1)*

20. Solve $3(m + 1) + 5(m + 1) = 40$.   *(Lesson 1-5)*

21. Suppose $f: x \rightarrow \frac{10}{x}$. What is $f(20)$?   *(Lesson 1-3)*

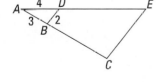

22. In the figure at the left, $\triangle ADB \sim \triangle AEC$. If $DE = 12$, find BC and CE.   *(Previous course)*

23. Simplify $\frac{\sqrt{81}}{\sqrt{9}}$.

24. Simplify $\frac{\sqrt{5}}{\sqrt{80}}$.   *(Previous course)*

Exploration

25. In Question 14 you used the 2 × 2 identity matrix for multiplication. Find the matrix that is the 3 × 3 identity matrix for multiplication.

*Size
Changes*

IN·CLASS

A C T I V I T Y

Materials: Each person in the group needs graph paper and a straightedge. Work on this Activity in a small group.

1
  a. Draw quadrilateral *QUAD* with $Q = (-1, 3)$, $U = (1, 1)$, $A = (3, 1)$, and $D = (4, 4)$.
  b. Complete the matrix below to represent *QUAD*.

$$\begin{bmatrix} -1 & 1 & \text{—} & \text{—} \\ 3 & \text{—} & \text{—} & \text{—} \end{bmatrix}$$

c. Multiply the matrix for *QUAD* from part **b** by $\begin{bmatrix} 3 & 0 \\ 0 & 3 \end{bmatrix}$.

d. The answer to part **c** represents a quadrilateral. Draw this quadrilateral. How is it related to the quadrilaterals you drew in parts **a** and **c** above?

2
  a. Draw a polygon of your own choice on graph paper.
  b. Write a matrix for your polygon.

c. Multiply the matrix in part **b** by $\begin{bmatrix} k & 0 \\ 0 & k \end{bmatrix}$, where you pick the value of k. Don't pick $k = 0$ or $k = 1$.
d. The answer to part **c** represents a polygon. Draw this polygon.

3
  Draw conclusions. Look back at the work you and the others in your group did in Questions 1 and 2. What generalization(s) can you make? When you multiply a

matrix for a polygon by the matrix $\begin{bmatrix} k & 0 \\ 0 & k \end{bmatrix}$, what happens?

Matrices for Size Changes

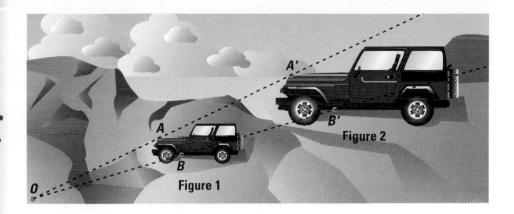

Figure 2

Figure 1

In the drawing above, we began with Figure I and point O. For point A on Figure I, a point A' (read "A prime") was found on Figure II. The position of A' was determined by the following rules: (1) A' is on $\overrightarrow{OA}$, and (2) $\frac{OA'}{OA} = 2$. This procedure was repeated with point B to find B' and with all other points of the smaller car.

Recall from geometry that a **transformation** is a one-to-one correspondence between the points of a preimage and the points of an image. Every point in Figure I above corresponds to exactly one point in Figure II, and vice versa, so a transformation has taken place. Specifically, the above transformation is a size change with center O and magnitude 2. We call Figure I the **preimage** and Figure II the **image** under the transformation. The two figures are similar with the ratio of similitude being 2.

What Is a Size Change?

Consider $\triangle PQR$ with $P = (3, 1)$, $Q = (-4, 0)$, and $R = (-3, -2)$. The size change with center $(0, 0)$ and magnitude 3, denoted S_3, can be performed by multiplying each x- and y-coordinate on $\triangle PQR$ by 3. We write

$$S_3(3, 1) = (9, 3)$$
$$S_3(-4, 0) = (-12, 0)$$
$$S_3(-3, -2) = (-9, -6)$$

and, in general, $\qquad S_3(x, y) = (3x, 3y).$

We read the top line as "A size change of magnitude 3 maps (3, 1) onto (9, 3)." Recall that the symbol $\rightarrow$ is often used in mathematics to denote "maps onto"; so the sentence "a size change of magnitude 3 maps (x, y) onto $(3x, 3y)$" can also be written in mapping notation as $S_3: (x, y) \rightarrow (3x, 3y)$.

Definition

For any $k \neq 0$, the transformation that maps (x, y) onto (kx, ky) is called the **size change** with **center** (0, 0) and **magnitude** k, and is denoted S_k.

$$S_k(x, y) = (kx, ky)$$

S_3 transforms $\triangle PQR$. $\triangle PQR$ is the preimage; $\triangle P'Q'R'$ is its image. We also say that S_3 maps $\triangle PQR$ onto $\triangle P'Q'R'$.

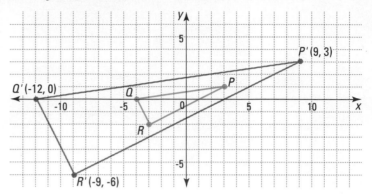

Notice that $\triangle PQR$ and $\triangle P'Q'R'$ are similar with ratio of similitude equal to 3. In general, when two figures are images of each other under a size change, they are similar.

Using Matrices to Perform Size Changes

Size-change images can also be found by multiplying matrices. As you should have seen in the Activity preceding this lesson, when the matrix

for the point (x, y) is multiplied by $\begin{bmatrix} k & 0 \\ 0 & k \end{bmatrix}$, the matrix for the point (kx, ky) results.

$$\begin{bmatrix} k & 0 \\ 0 & k \end{bmatrix} \begin{bmatrix} x \\ y \end{bmatrix} = \begin{bmatrix} kx + 0y \\ 0x + ky \end{bmatrix}$$

$$= \begin{bmatrix} kx \\ ky \end{bmatrix}$$

This proves the following theorem.

> **Theorem**
>
> $\begin{bmatrix} k & 0 \\ 0 & k \end{bmatrix}$ is the matrix for S_k.

Example 1

Given $ABCD$ with $A = (0, 3)$, $B = (-2, -4)$, $C = (-6, -4)$, and $D = (-6, 4)$, find the image $A'B'C'D'$ under S_4.

Solution

Write $ABCD$ and S_4 in matrix form and multiply.

$$\overset{S_4}{\begin{bmatrix} 4 & 0 \\ 0 & 4 \end{bmatrix}} \overset{\begin{matrix} A & B & C & D \end{matrix}}{\begin{bmatrix} 0 & -2 & -6 & -6 \\ 3 & -4 & -4 & 4 \end{bmatrix}} = \overset{\begin{matrix} A' & B' & C' & D' \end{matrix}}{\begin{bmatrix} 0 & -8 & -24 & -24 \\ 12 & -16 & -16 & 16 \end{bmatrix}}$$

Thus $A'B'C'D'$ has vertices $A' = (0, 12)$, $B' = (-8, -16)$, $C' = (-24, -16)$, and $D' = (-24, 16)$. ▶

Check

Graph the preimage and image. They should look similar, and they do.

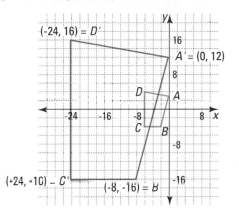

The transformation that maps each point (x, y) onto itself is called the **identity transformation.** When a point matrix $\begin{bmatrix} x \\ y \end{bmatrix}$ is multiplied on the left by $\begin{bmatrix} 1 & 0 \\ 0 & 1 \end{bmatrix}$ each point (x, y) coincides with its image. Thus $\begin{bmatrix} 1 & 0 \\ 0 & 1 \end{bmatrix}$ represents the identity transformation.

$$\begin{bmatrix} 1 & 0 \\ 0 & 1 \end{bmatrix} \begin{bmatrix} x \\ y \end{bmatrix} = \begin{bmatrix} 1 \cdot x + 0 \cdot y \\ 0 \cdot x + 1 \cdot y \end{bmatrix} = \begin{bmatrix} x \\ y \end{bmatrix}$$

The size change of magnitude 1 is the identity transformation.

Properties of Size-Change Images

Since size-change images are similar to preimages, corresponding angles are congruent and ratios of corresponding segments equal the ratio of similitude. This can be verified using the distance formula.

Standard size car. *Shown is a standard size car. Notice its size with respect to the person.*

Compact car. *Again notice the size of the car with respect to the person. What transformation do you think took place?*

Example 2

Refer to the quadrilaterals in Example 1. Calculate each ratio.

a. $\frac{D'C'}{DC}$

b. $\frac{A'B'}{AB}$

Solution

a. $D'C' = 32$ and $DC = 8$. So $\frac{D'C'}{DC} = \frac{32}{8} = 4$.

b. Using the distance formula, we see that

$$AB = \sqrt{(-2 - 0)^2 + (-4 - 3)^2} \quad = \sqrt{53}$$

and $A'B' = \sqrt{(-8 - 0)^2 + (-16 - 12)^2} = \sqrt{848}$.

So $\frac{A'B'}{AB} = \frac{\sqrt{848}}{\sqrt{53}} = 4$, which is the magnitude of the size change.

In general, the size change of magnitude k multiplies distance by $|k|$, so that the ratio of image lengths to preimage lengths is $|k|$. You will prove this in Chapter 7.

QUESTIONS

Covering the Reading

In 1–3, how is the expression read?

1. $S_3: (3, 1) \rightarrow (9, 3)$ 2. $S_5(2, 1) = (10, 5)$ 3. $\triangle P'Q'R'$

4. a. Draw $\triangle ABC$ where $A = (0, 0)$, $B = (\text{-}3, 1)$, and $C = (4, 1)$.
   b. Draw the image of $\triangle ABC$ under S_2.
   c. What is the matrix for the transformation that maps $\triangle ABC$ onto $\triangle A'B'C'$?

5. The size-change transformation with center $(0, 0)$ and magnitude k maps (x, y) onto __?__.

6. The matrix $\begin{bmatrix} k & 0 \\ 0 & k \end{bmatrix}$ is associated with a(n) __?__ change with center __?__ of magnitude __?__.

In 7–9, refer to Example 1.

7. What are the coordinates of the *image* of point B?

8. What is the magnitude of the size change?

9. Verify that $\frac{A'D'}{AD} = 4$.

10. The identity transformation maps each point onto __?__.

In 11–13, answer *true or false*.

11. Under every size change an angle and its image are congruent.

12. Under every size change a segment and its image are congruent.

13. Under every size change a figure and its image are similar.

Applying the Mathematics

14. $\triangle ABC$ has matrix $\begin{bmatrix} 6 & \text{-}4 & 2 \\ 8 & 2 & \text{-}2 \end{bmatrix}$. Graph $\triangle ABC$ and its image $\triangle A'B'C'$ under $S_{\frac{1}{2}}$.

15. Suppose $P = (3, 4)$. Let P' be the image of P under the size change with center $O = (0, 0)$ and magnitude 2.5.
   a. Verify that $\frac{OP'}{OP} = 2.5$.
   b. Give an equation for the line containing O, P, and P'.

16. A 3×5 drawing is enlarged to 6×10 by using a size change.
  a. What is the matrix for the size change?
  b. Suppose $A = (1, 3.5)$, $B = (1.5, 3.1)$, and $C = (2, 4.1)$ are three points located on the smaller drawing. Write a matrix for the location of these points on the enlargement.
  c. The skateboard in the enlargement is how many times as long as the skateboard in the original drawing?

Bart Simpson

17. Refer to Example 1.
  a. Find the slope of $\overline{AB}$.
  b. Find the slope of $\overline{A'B'}$.
  c. Is $\overline{AB}$ parallel to $\overline{A'B'}$? Why or why not?

18. a. Draw $\triangle PQR$ represented by the matrix $\begin{bmatrix} 6 & 2 & -8 \\ 12 & -7 & -5 \end{bmatrix}$.
  b. Find the product $\begin{bmatrix} -\frac{1}{2} & 0 \\ 0 & -\frac{1}{2} \end{bmatrix} \begin{bmatrix} 6 & 2 & -8 \\ 12 & -7 & -5 \end{bmatrix}$.
  c. The matrix in part **b** represents $\triangle P'Q'R'$, the image of $\triangle PQR$ under a size change of magnitude $-\frac{1}{2}$. Draw $\triangle P'Q'R'$.
  d. How are the lengths of the sides of $\triangle P'Q'R'$ related to the lengths of the sides of $\triangle PQR$?
  e. How are the areas of $\triangle PQR$ and $\triangle P'Q'R'$ related?

Review

19. Solve for a and b: $\begin{bmatrix} 2 & a \\ 3 & b \end{bmatrix} \begin{bmatrix} 5 \\ 6 \end{bmatrix} = \begin{bmatrix} 7 \\ 8 \end{bmatrix}$. *(Lesson 4-3)*

20. A German clothing manufacturer has factories in Berlin, Hamburg, and Munich. Sales (in thousands of items) can be summarized by the following matrix S.

	Berlin	Hamburg	Munich	
Blouses	9	5	3	
Dresses	14	7	3	$= S$
Skirts	12	7	2	
Slacks	18	10	4	

  a. What are the dimensions of S?
  b. The selling price of a blouse is DM38. A dress sells for DM105. A skirt sells for DM45, and a pair of slacks sells for DM44. (DM is the symbol for Deutsche mark, the currency of Germany.) Write a 1×4 matrix representing the selling prices of the items.
  c. Use matrix multiplication to determine the total revenue of each factory. *(Lessons 4-1, 4-2)*

In 21 and 22, refer to the graph below, which shows the location of an elevator over a one-minute period. *(Lessons 1-4, 2-4, 3-1)*

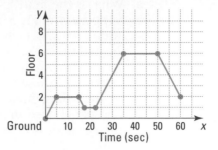

21. When is the elevator on the sixth floor?

22. a. At what rate does the elevator ascend?
   b. Does the elevator descend at the same speed that it ascends? Justify your answer.

23. Two figures, *F* and *G,* are similar. The perimeter of *F* is 20 cm, and the perimeter of *G* is 15 cm. If the area of *F* is 100 cm^2, what is the area of *G?* *(Previous course)*

Up-lifting. *Pictured are three glass elevators on the cruise ship* Sovereign of the Seas.

Exploration

24. *ABCD* is the square defined by the matrix $\begin{bmatrix} 0 & 2 & 2 & 0 \\ 0 & 0 & 2 & 2 \end{bmatrix}$.

Transform *ABCD* by multiplying its matrix by each of the following size-change matrices (and by some others of your own choice).

a. $\begin{bmatrix} 2 & 0 \\ 0 & 2 \end{bmatrix}$ **b.** $\begin{bmatrix} 3 & 0 \\ 0 & 3 \end{bmatrix}$ **c.** $\begin{bmatrix} 4 & 0 \\ 0 & 4 \end{bmatrix}$ **d.** $\begin{bmatrix} 5 & 0 \\ 0 & 5 \end{bmatrix}$

e. Find the area of each image. Enter your results in a table like this one:

Area of Preimage	Matrix	Area of Image
4 units^2	$\begin{bmatrix} 2 & 0 \\ 0 & 2 \end{bmatrix}$	_?_ units^2
4 units^2	$\begin{bmatrix} 3 & 0 \\ 0 & 3 \end{bmatrix}$	_?_ units^2
_?_	$\begin{bmatrix} 4 & 0 \\ 0 & 4 \end{bmatrix}$	_?_
_?_	$\begin{bmatrix} 5 & 0 \\ 0 & 5 \end{bmatrix}$	_?_

f. There is a connection between the matrix associated with a size change and the effect the matrix has on the area of a figure. What is this connection?

25. Use the procedure from the beginning of this lesson to make an enlargement of a photo or of a picture from a magazine.

LESSON 4-5

Matrices for Scale Changes

Original	Horizontal scale change of magnitude 2 (a stretch)	Vertical scale change of magnitude $\frac{1}{3}$ (a shrink)	Horizontal scale change of magnitude 2 and vertical scale change of magnitude $\frac{1}{3}$

What Is a Scale Change?

In contrast to a size change, which you studied in the previous lesson, a *scale change* can transform a figure by stretching or shrinking it in either a horizontal direction, a vertical direction, or both directions at once.

> **Definition**
> For any nonzero numbers a and b, the transformation that maps (x, y) onto (ax, by) is called the **scale change** with **horizontal magnitude** a and **vertical magnitude** b, and is denoted $S_{a,b}$.

When $|a| > 1$ (or $|b| > 1$), the scale change is a *stretch* in the horizontal (or vertical) direction. When $|a| < 1$ (or $|b| < 1$), the scale change is a *shrink* in the horizontal (or vertical) direction.

In mapping notation, we write $S_{a,b}: (x, y) \rightarrow (ax, by)$. In $f(x)$ notation, $S_{a,b}(x, y) = (ax, by)$.

Example 1

Consider quadrilateral $ABCD$ with $A = (0, 3)$, $B = (-2, -4)$, $C = (-6, -4)$, and $D = (-6, 4)$. Find its image $A'B'C'D'$ under $S_{2,5}$.

Solution

$S_{2,5}(x, y) = (2x, 5y)$. That is, to find each image point, multiply the x-coordinate of the preimage by 2 and the y-coordinate of the preimage by 5. So,

$$S_{2,5}(0, 3) = (0, 15),$$
$$S_{2,5}(-2, -4) = (-4, -20),$$
$$S_{2,5}(-6, -4) = (-12, -20), \text{ and}$$
$$S_{2,5}(-6, 4) = (-12, 20).$$

Quadrilateral A'B'C'D' has vertices
A' = (0, 15), B' = (-4, -20), C' = (-12, -20), and D' = (-12, 20). ▶

Graph the preimage and image.

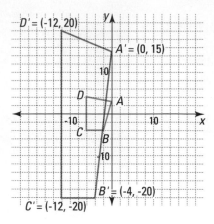

The image of a preimage point on the *y*-axis should be 5 times as far from the origin as the preimage, and the image of a preimage point on the *x*-axis should be 2 times as far from the origin as the preimage. The preimage point $A = (0, 3)$ is 3 units from the origin. The image of $(0, 3)$ is $(0, 15)$, which is 15 units from the origin. For the preimage point $(-6, 0)$, the image is the point $(-12, 0)$, which is twice as far from the origin as its preimage. It checks.

Example 1 shows that a scale-change image is not necessarily similar to its preimage. The ratios of the lengths of corresponding sides in *ABCD* and *A′B′C′D′* are *not* equal. For instance,

$$\frac{D'C'}{DC} = \frac{40}{8} = 5 \qquad \text{(the vertical stretch)}$$

$$\frac{B'C'}{BC} = \frac{8}{4} = 2 \qquad \text{(the horizontal stretch)}$$

and $\dfrac{A'B'}{AB} = \dfrac{\sqrt{(-20-15)^2 + (-4-0)^2}}{\sqrt{(-4-3)^2 + (-2-0)^2}} = \dfrac{\sqrt{1241}}{\sqrt{53}} \approx 4.84$ (a result of both horizontal and vertical stretches)

Because the ratios of corresponding sides are different, the two quadrilaterals are not similar.

Matrices for Scale Changes

Because a size change has a 2×2 matrix, it is reasonable to expect that a scale change also has one. Suppose that $S_{a,b}$ has the matrix

$$\begin{bmatrix} e & f \\ g & h \end{bmatrix}$$

where *e*, *f*, *g*, and *h* are real numbers. Because $S_{a,b}$ maps (x, y) onto (ax, by), we want to find *e*, *f*, *g*, and *h* so that

$$\begin{bmatrix} e & f \\ g & h \end{bmatrix} \begin{bmatrix} x \\ y \end{bmatrix} = \begin{bmatrix} ax \\ by \end{bmatrix}.$$

Notice that by matrix multiplication, $\begin{bmatrix} a & 0 \\ 0 & b \end{bmatrix} \begin{bmatrix} x \\ y \end{bmatrix} = \begin{bmatrix} ax \\ by \end{bmatrix}$. Thus $e = a$, $f = 0$, $g = 0$, and $h = b$. We have proved the following theorem.

Example 2

Refer to quadrilateral *ABCD* from Example 1. Use matrix multiplication to find its image *A'B'C'D'* under $S_{2,5}$.

Solution

Write $S_{2,5}$ and *ABCD* in matrix form.

$$
\begin{array}{cc}
S_{2,5} & \\
\begin{bmatrix} 2 & 0 \\ 0 & 5 \end{bmatrix} &
\begin{array}{cccc} A & B & C & D \\ \end{array} \\
\end{array}
$$

$$
\begin{array}{c} S_{2,5} \\ \begin{bmatrix} 2 & 0 \\ 0 & 5 \end{bmatrix} \end{array}
\begin{array}{c} \begin{array}{cccc} A & B & C & D \end{array} \\ \begin{bmatrix} 0 & -2 & -6 & -6 \\ 3 & -4 & -4 & 4 \end{bmatrix} \end{array}
=
\begin{array}{c} \begin{array}{cccc} A' & B' & C' & D' \end{array} \\ \begin{bmatrix} 0 & -4 & -12 & -12 \\ 15 & -20 & -20 & 20 \end{bmatrix} \end{array}
$$

Check

The product matrix gives the same result for quadrilateral *A'B'C'D'* that was found in Example 1.

Notice that a scale change may stretch or shrink by different factors in the horizontal and vertical directions. If the factors are the same in both directions, then the scale-change matrix has the form $\begin{bmatrix} k & 0 \\ 0 & k \end{bmatrix}$ and is really just a size change. Conversely, a size change with magnitude k is a scale change with horizontal magnitude k and vertical magnitude k. Thus *size changes are special types of scale changes.* In symbols, $S_k = S_{k,k}$.

A stretch of the imagination. *The curved mirror shown is performing a horizontal stretch.*

QUESTIONS

Covering the Reading

1. $S_{a,b}$ maps (x, y) onto ___?___.

2. **a.** What is the image of $(6, 5)$ under $S_{3,2}$?
   b. Describe $S_{3,2}$ in words.

3. If $S_{2,5}(1, -2) = (m, n)$, find m and n.

4. If the horizontal and vertical scale changes in the final picture of the clown at the beginning of this lesson were to be done by applying $S_{a,b}$ to the first drawing, what are the values of a and b?

5. **a.** Draw $\triangle ABC$ with $A = (-3, 0)$, $B = (0, -3)$, and $C = (5, 5)$.
   b. Find its image $\triangle A'B'C'$ under $S_{2,4}$.
   c. *True or false.* $\triangle ABC$ and $\triangle A'B'C'$ are similar. Justify your answer.

6. **a.** Find the matrix product $\begin{bmatrix} 100 & 0 \\ 0 & 200 \end{bmatrix}\begin{bmatrix} 7 \\ 9 \end{bmatrix}$.
   b. You have found the image of ___?___ under ___?___.

7. The scale change with matrix $\begin{bmatrix} 0.5 & 0 \\ 0 & 1.5 \end{bmatrix}$ is a horizontal __?__ and a vertical __?__.

8. The size change S_4 can be thought of as the scale change $S_{a,b}$. What are the values of a and b?

Applying the Mathematics

9. Refer to Example 1. Prove or disprove: $\overline{AB} \ /\!/ \ \overline{A'B'}$.

10. I.   II.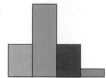

Describe the scale change under which bar graph II is the image of bar graph I.

11. **a.** Quadrilateral *TOPS* is represented by the matrix $\begin{bmatrix} 0 & 4 & 4 & 0 \\ 0 & 0 & 6 & 6 \end{bmatrix}$. What type of quadrilateral is *TOPS*?
    b. Find the matrix of the image $T'O'P'S'$ of quadrilateral *TOPS* under the scale change represented by $\begin{bmatrix} 3 & 0 \\ 0 & 2 \end{bmatrix}$.
    c. What type of quadrilateral is $T'O'P'S'$?
    d. Graph the preimage *TOPS* and the image $T'O'P'S'$.

12. Consider the matrix equation below.
$$\begin{array}{c} S_{a,b} \\ \begin{bmatrix} a & 0 \\ 0 & b \end{bmatrix} \end{array} \begin{array}{c} T\ \ R\ \ Y \\ \begin{bmatrix} 1 & 2 & 3 \\ 4 & -1 & 3 \end{bmatrix} \end{array} = \begin{array}{c} T'\ \ \ R'\ \ \ Y' \\ \begin{bmatrix} 5 & 10 & 15 \\ 16 & -4 & 12 \end{bmatrix} \end{array}$$

An apple of a pear. *The Japanese pear-apple combines the tastes and textures of both fruits.*

   a. What scale change is represented by this equation?
   b. Draw the preimage $\triangle TRY$ and the image $\triangle T'R'Y'$.
   c. Find $\dfrac{T'R'}{TR}$ and $\dfrac{T'Y'}{TY}$.
   d. Should the ratios be the same? Why or why not?

Review

13. The matrix below gives the daily delivery of boxes of apples and pears to two markets.

$$\begin{array}{c} \ \ \ \ \text{Troy's} \ \ \text{Abby's} \\ \begin{array}{c} \text{apples} \\ \text{pears} \end{array} \begin{bmatrix} 5 & 4 \\ 1 & 2 \end{bmatrix} \end{array}$$

During peak season, the demand for fruit at each market triples.
   a. What size change is needed to meet the increased demand? Represent the size change by a matrix.
   b. Multiply the original matrix by the size-change matrix to find the new matrix which meets the increased demand. *(Lesson 4-4)*

14. Find AX when $X = \begin{bmatrix} 2 & -1 & -2 \\ 0 & 1 & -3 \\ 3 & 4 & 0 \end{bmatrix}$ and $A = \begin{bmatrix} 3 & 2 & 1 \\ 1 & 0 & 1 \\ -3 & -2 & 0 \end{bmatrix}$. *(Lesson 4-3)*

15. In 1990, about 35 million people in the U.S. did not have health insurance. The four states with the highest percent of uninsured residents were Louisiana (19.7%), Mississippi (19.9%), Texas (21.1%) and New Mexico (22.2%). Their populations in 1990 were as follows: Louisiana, 4,219,973; Mississippi, 2,573,216; Texas, 16,986,510; and New Mexico, 1,515,069. Source: *The 1993 Information Please Almanac* *(Lessons 4-1, 4-3)*
   a. Use matrix multiplication to determine how many people in these four states had no health insurance.
   b. From this information, which state do you think has the biggest health-insurance problem? Why?

16. Evaluate $P_n = 500(2)^{n-1}$ when $n = 1$. *(Lesson 1-7)*

17. *Skill sequence.* Solve the equation. *(Lesson 1-5)*
   a. $\dfrac{\frac{x}{3}}{5} = 15$
   b. $\dfrac{\frac{y}{3}}{\frac{5}{9}} = 15$

18. If $A = (-2, 8)$, $B = (8, -2)$, and $C = (7, 7)$, prove that $AC = BC$. *(Previous course)*

Exploration

19. Let $ABCD$ be the square defined by the matrix $\begin{bmatrix} 0 & 2 & 2 & 0 \\ 0 & 0 & 2 & 2 \end{bmatrix}$.

Transform $ABCD$ by multiplying its matrix by each of the following matrices (and by some others of your own choice).
   a. $\begin{bmatrix} 3 & 0 \\ 0 & 4 \end{bmatrix}$
   b. $\begin{bmatrix} 3 & 0 \\ 0 & 1 \end{bmatrix}$
   c. $\begin{bmatrix} 3 & 0 \\ 0 & 2 \end{bmatrix}$
   d. $\begin{bmatrix} 2 & 0 \\ 0 & 1 \end{bmatrix}$
   e. Complete the table.

Area of Preimage	Matrix	Area of Image
4 units^2	$\begin{bmatrix} 3 & 0 \\ 0 & 4 \end{bmatrix}$	__?__ units^2
4 units^2	$\begin{bmatrix} 3 & 0 \\ 0 & 1 \end{bmatrix}$	__?__ units^2
__?__	$\begin{bmatrix} 3 & 0 \\ 0 & 2 \end{bmatrix}$	__?__
__?__	$\begin{bmatrix} 2 & 0 \\ 0 & 1 \end{bmatrix}$	__?__

   f. What is the connection between the elements a and b of the scale-change matrix $\begin{bmatrix} a & 0 \\ 0 & b \end{bmatrix}$ and the effect the scale change has on area?

LESSON
4-6

Matrices for Reflections

Mountain reflection. *Shown is Mt. McKinley and its reflection on a lake in Denali National Park in Alaska. The Athabaskan Indians called the mountain* Denali, *The Great One, or The High One.*

What Is a Reflection?

Recall from geometry that the **reflection image of a point *A* over a line *m*** is:

1. the point *A*, if *A* is on *m;*

2. the point *A'* such that *m* is the perpendicular bisector of $\overline{AA'}$, if *A* is not on *m.*

The line *m* is called the **reflecting line** or **line of reflection.**

A **reflection** is a transformation that maps a figure to its reflection image. The figure below shows the *reflection image* of a leaf and the point *A* over the line *m.*

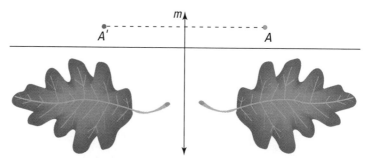

Reflection over the *y*-axis

Suppose that $A = (x, y)$ is reflected over the *y*-axis, as shown below.

Since $\overline{AA'}$ must be perpendicular to the y-axis, it is a horizontal segment. Therefore, the y-coordinate of A' is the same as the y-coordinate of A. Because A' and A are equidistant from the y-axis, and A is x units from the y-axis, the point A' must also be x units from the y-axis. Since A and A' are on opposite sides of the y-axis, this means that the x-coordinate of A' is $-x$. So the reflection image of (x, y) over the y-axis is $(-x, y)$.

Reflection over the y-axis can be denoted $r_{y\text{-axis}}$ or r_y. In this book we use r_y. We write

$$r_y: (x, y) \rightarrow (-x, y)$$

This is read "the reflection over the y-axis maps (x, y) onto $(-x, y)$."
Or we write

$$r_y(x, y) = (-x, y).$$

This is read "the reflection image of (x, y) over the y-axis is $(-x, y)$."

Notice that $\begin{bmatrix} -1 & 0 \\ 0 & 1 \end{bmatrix} \begin{bmatrix} x \\ y \end{bmatrix} = \begin{bmatrix} -1 \cdot x + 0 \cdot y \\ 0 \cdot x + 1 \cdot y \end{bmatrix} = \begin{bmatrix} -x \\ y \end{bmatrix}.$

This means that there is a matrix for r_y and proves the next theorem.

Theorem

$\begin{bmatrix} -1 & 0 \\ 0 & 1 \end{bmatrix}$ is a matrix for r_y.

Example 1

If $A = (1, 2)$, $B = (1, 4)$, and $C = (2, 4)$, find the image of $\triangle ABC$ under r_y.

Solution

Represent r_y and $\triangle ABC$ as matrices and multiply.

$$\overset{r_y}{\begin{bmatrix} -1 & 0 \\ 0 & 1 \end{bmatrix}} \overset{\triangle ABC}{\begin{bmatrix} 1 & 1 & 2 \\ 2 & 4 & 4 \end{bmatrix}} = \overset{\triangle A'B'C'}{\begin{bmatrix} -1 & -1 & -2 \\ 2 & 4 & 4 \end{bmatrix}}$$

The image $\triangle A'B'C'$ has $A' = (-1, 2)$, $B' = (-1, 4)$, and $C' = (-2, 4)$.

Check

Graph the preimage and image. It checks.

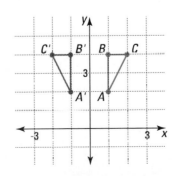

Reflections over Other Lines

Two other important reflecting lines are the x-axis and the line with equation $y = x$. Reflection over the x-axis is denoted by r_x, and reflection over the line $y = x$ is denoted by $r_{y=x}$.

The following graphs show the effects of r_x and $r_{y=x}$ on $\triangle ABC$ of Example 1.

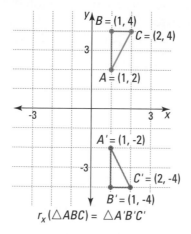

$r_x(\triangle ABC) = \triangle A'B'C'$

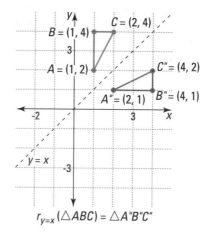

$r_{y=x}(\triangle ABC) = \triangle A''B''C''$

You can verify the following results.

$$r_x: (x, y) \rightarrow (x, -y)$$
$$r_{y=x}: (x, y) \rightarrow (y, x)$$

The matrices for r_x and $r_{y=x}$ also involve only 0s, 1s, and -1s.

Theorem

$\begin{bmatrix} 1 & 0 \\ 0 & -1 \end{bmatrix}$ is the matrix for r_x.

Proof:

$$\begin{bmatrix} 1 & 0 \\ 0 & -1 \end{bmatrix} \begin{bmatrix} x \\ y \end{bmatrix} = \begin{bmatrix} 1 \cdot x + 0 \cdot y \\ 0 \cdot x + -1 \cdot y \end{bmatrix} = \begin{bmatrix} x \\ -y \end{bmatrix}$$

Activity

Use $\triangle ABC$ from Example 1. Give a matrix multiplication that could be used to find the image of $\triangle ABC$ under r_x.

Theorem

$\begin{bmatrix} 0 & 1 \\ 1 & 0 \end{bmatrix}$ is the matrix for $r_{y=x}$.

Proof:
You are asked to do the proof in Question 13.

Example 2

Find the reflection image of pentagon *WEIRD* over the line $y = x$ if
$W = (-1, -1)$, $E = (3, -2)$, $I = (6, 0)$, $R = (6, 5)$ and $D = (-1, 7)$.

Solution

Represent *WEIRD* and $r_{y=x}$ by matrices and multiply.

$$\begin{array}{c} \\ \overset{r_{y=x}}{\begin{bmatrix} 0 & 1 \\ 1 & 0 \end{bmatrix}} \end{array} \overset{\begin{array}{ccccc} W & E & I & R & D \end{array}}{\begin{bmatrix} -1 & 3 & 6 & 6 & -1 \\ -1 & -2 & 0 & 5 & 7 \end{bmatrix}} = \overset{\begin{array}{ccccc} W' & E' & I' & R' & D' \end{array}}{\begin{bmatrix} -1 & -2 & 0 & 5 & 7 \\ -1 & 3 & 6 & 6 & -1 \end{bmatrix}}$$

W'E'I'R'D' is represented by the product matrix.

Check

WEIRD and *W′E′I′R′D′* are graphed at the left.

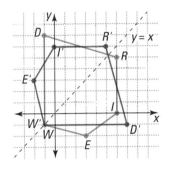

It is important to note one significant way that reflections, size changes, and scale changes differ. All reflections preserve shape and size, so reflection images are always congruent to their preimages. All size changes preserve shape, but only S_1 and S_{-1} yield congruent images. However, size-change images are similar to their preimages. In general, scale-change images are neither congruent nor similar to their preimages.

Remembering Matrices

At this point, you have seen matrices for some size changes, some scale changes, and three reflections. You may wonder: How do I remember them? Here is one way. To write the 2×2 matrix for a particular transformation T, use this rule: The first column is the image of $(1, 0)$ under T. The second column is the image of $(0, 1)$ under T. So, for example, to remember the matrix for r_y, use the picture below. Arrows show the images of $(1, 0)$ and $(0, 1)$ under the transformation r_y.

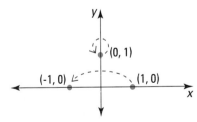

The image of $(1, 0)$ under r_y is $(-1, 0)$.

$$\begin{bmatrix} 1 & 0 \\ 0 & 1 \end{bmatrix} \qquad \begin{bmatrix} -1 & 0 \\ 0 & 1 \end{bmatrix}$$

The image of $(0, 1)$ under r_y is $(0, 1)$.

This rule gives us a way to write matrices for transformations quickly without memorizing them. The rule is called the *Matrix Basis Theorem* because the matrix is *based* on the images of points (1, 0) and (0, 1).

Matrix Basis Theorem
Suppose T is a transformation represented by a 2 $\times$ 2 matrix.
If $T: (1, 0) \rightarrow (x_1, y_1)$ and $T: (0, 1) \rightarrow (x_2, y_2)$ then T has the matrix
$$\begin{bmatrix} x_1 & x_2 \\ y_1 & y_2 \end{bmatrix}.$$

Proof:
Let $\begin{bmatrix} a & b \\ c & d \end{bmatrix}$ be the matrix representing T. Then, since $T: (1, 0) \rightarrow (x_1, y_1)$,
$$\begin{bmatrix} a & b \\ c & d \end{bmatrix} \begin{bmatrix} 1 \\ 0 \end{bmatrix} = \begin{bmatrix} x_1 & x_2 \\ y_1 & y_2 \end{bmatrix}$$
or
$$\begin{bmatrix} a \\ c \end{bmatrix} = \begin{bmatrix} x_1 \\ y_1 \end{bmatrix}.$$
Therefore, $a = x_1$ and $c = y_1$. Similarly, since $T: (0, 1) \rightarrow (x_2, y_2)$,
$$\begin{bmatrix} a & b \\ c & d \end{bmatrix} \begin{bmatrix} 0 \\ 1 \end{bmatrix} = \begin{bmatrix} x_2 \\ y_2 \end{bmatrix}$$
or
$$\begin{bmatrix} b \\ d \end{bmatrix} = \begin{bmatrix} x_2 \\ y_2 \end{bmatrix}.$$
Therefore, $b = x_2$ and $d = y_2$. So the matrix for T is $\begin{bmatrix} x_1 & x_2 \\ y_1 & y_2 \end{bmatrix}$.

QUESTIONS

Covering the Reading

1. Suppose that A is not on line m and that A' is the reflection image of A over m. Then m is the __?__ of $\overline{AA'}$.

2. **a.** What is the reflection image of a point A over a line m if A is on m?
   b. Which vertex of pentagon *WEIRD* in Example 2 shows this?

3. How can the following sentence be read? $r_x(x, y) = (x, -y)$

4. In Example 1, what matrix multiplication could be used to find the image of $\triangle ABC$ under $r_{y=x}$?

In 5–7, *multiple choice.* Choose the matrix which corresponds to the given reflection.

(a) $\begin{bmatrix} 1 & 0 \\ 0 & -1 \end{bmatrix}$　(b) $\begin{bmatrix} -1 & 0 \\ 0 & 1 \end{bmatrix}$　(c) $\begin{bmatrix} -1 & 0 \\ 0 & -1 \end{bmatrix}$　(d) $\begin{bmatrix} 0 & 1 \\ 1 & 0 \end{bmatrix}$　(e) $\begin{bmatrix} 0 & -1 \\ -1 & 0 \end{bmatrix}$

5. r_x　　　　　　　　6. r_y　　　　　　　　7. $r_{y=x}$

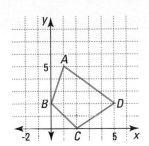

8. **a.** Write a matrix for quadrilateral $ABCD$ shown at the left.
   b. Use matrix multiplication to draw $A'B'C'D'$, its reflection image over the x-axis.

9. Refer to Example 2. Find the matrix for $W''E''I''R''D''$, the reflection image of $WEIRD$ over the y-axis. Graph $WEIRD$ and $W''E''I''R''D''$.

10. *True or false.* Reflection images are congruent to their preimages.

11. The matrix equation below shows that the reflection image of the point __?__ over the line __?__ is the point __?__.

$$\begin{bmatrix} -1 & 0 \\ 0 & 1 \end{bmatrix}\begin{bmatrix} 2 \\ 3 \end{bmatrix} = \begin{bmatrix} -2 \\ 3 \end{bmatrix}$$

Applying the Mathematics

---

12. Explain how the Matrix Basis Theorem enables a person to remember the matrix for $r_{y=x}$.

13. Prove that $\begin{bmatrix} 0 & 1 \\ 1 & 0 \end{bmatrix}$ is the matrix for the reflection over the line $y = x$.

14. Suppose $P = (x, y)$ and $Q = (y, x)$. Let $R = (a, a)$ be any point on the line $y = x$.

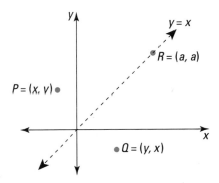

   a. Verify that $PR = QR$.
   b. From part **a,** what theorem from geometry allows you to conclude that the line $y = x$ is the perpendicular bisector of $\overline{PQ}$?

Review

---

15. A polygon P is represented by the matrix $\begin{bmatrix} 3 & 2 & 5 & -3 \\ 4 & -1 & 0 & 1 \end{bmatrix}$. Find the change in area when P is multiplied by $\begin{bmatrix} 3 & 0 \\ 0 & 3 \end{bmatrix}$. *(Lesson 4-4)*

16. Find an equation for the line parallel to $y = 3x - 4$ and containing the point $(2, -5)$. *(Lessons 3-2, 3-5)*

17. A company makes three kinds of skis: trick skis, slalom skis, and cross-country skis. The matrix A represents the time in hours required to make one of each type of ski. The company has two manufacturing plants X and Y in different parts of the country. The hourly rates for each department are given in matrix B.

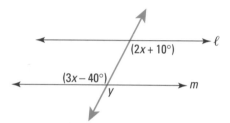

Find the total labor costs for each ski at each factory.   *(Lesson 4-3)*

18. Refer to the diagram below. Find y given $\ell \; // \; m$.   *(Previous course)*

Trick ski. *Snowboards, a type of trick ski, are used to perform tricks and acrobatics.*

19. Trace the figure below. Point H on pentagon *HOUSE* has been rotated counterclockwise 90° about point C to the position of H'. Rotate the other vertices of *HOUSE* 90° about C and draw $H'O'U'S'E'$.   *(Previous course)*

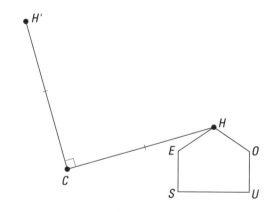

Exploration

20. Let $r_{y=-x}$ denote reflection over the line $y = -x$.
   a. By graphing and testing points, complete this sentence.
   $r_{y=-x}(x, y) = $ _____.
   b. Find the 2×2 matrix for $r_{y=-x}$.

*Composites of
Transformations*

IN-CLASS

ACTIVITY

Materials: Each member of the group should use graph paper with the same scales on the axes throughout this activity. Work on this Activity in groups of three.

Each person in the group will have a different task to complete. Decide who will be person *X*, who will be person *Y*, and who will be person *Z*. Then, follow the directions accordingly.

1 **Person X**
a. Draw quadrilateral *ABCD* with $A = (2, 1)$, $B = (3, 3)$, $C = (5, 2)$, and $D = (4, -2)$.
b. Draw the reflection image of *ABCD* over the *x*-axis. Label the image *A'B'C'D'*.
c. Draw the reflection image of *A'B'C'D'* over the line $y = x$. Label the new image *A"B"C"D"*.
d. What single transformation could you apply to *ABCD* to get the image *A"B"C"D"*?

Person Y
a. Draw quadrilateral *ABCD* with $A = (2, 1)$, $B = (3, 3)$, $C = (5, 2)$, and $D = (4, -2)$.
b. Draw the reflection image of *ABCD* over the line $y = x$. Label the image *A'B'C'D'*.
c. Draw the reflection image of *A'B'C'D'* over the *x*-axis. Label the new image *A"B"C"D"*.
d. What single transformation could you apply to *ABCD* to get the image *A"B"C"D"*?

Person Z
a. Draw quadrilateral *ABCD* with $A = (2, 1)$, $B = (3, 3)$, $C = (5, 2)$, and $D = (4, -2)$.
b. Write a matrix to describe this quadrilateral. Call this matrix *Q*.
c. Multiply *Q* by the matrix $\begin{bmatrix} 0 & -1 \\ 1 & 0 \end{bmatrix}$ and call the image *Q'*.
d. Graph *Q'*.

2 As a group, summarize what you found.

Transformations and Matrices

Singapore parade. *Shown are the rotating flags of a flag team in a parade in Singapore. More than 75% of the 2.8 million people who live in Singapore are Chinese. See the Example on page 242.*

Recall that a function is a correspondence between variables such that each value of the first variable corresponds to exactly one value of the second variable. Variables can represent points, so there can be functions that map points to points.

Because transformations are correspondences between sets of points, they are functions. Like other functions, transformations can be described by rules. These rules may be algebraic (a formula), geometric (giving the location of image points), or arithmetic (a matrix). In the preceding three lessons, you have encountered 2×2 matrices for size changes, scale changes, and reflections. Multiplying each of these matrices by a matrix for a polygon results in a transformation image of that polygon. Here we summarize some of the properties of multiplication of 2×2 matrices, and show other ways that matrices are related to transformations.

Properties of Matrix Multiplication

Multiplication of 2×2 matrices has some of the same properties as multiplication of real numbers.

1. Closure: *The set of 2×2 matrices is closed under multiplication. Closure* means: If you multiply two 2×2 matrices, the result is a 2×2 matrix. This property follows from the definition of multiplication of matrices.

2. Associativity: *Multiplication of 2×2 matrices is associative.* In fact, for any three matrices which can be multiplied, it can be shown that $(AB)C = A(BC)$.

▶

3. **Identity:** *The matrix* $\begin{bmatrix} 1 & 0 \\ 0 & 1 \end{bmatrix}$ *is the multiplicative identity for* 2×2 *matrices.*

We prove the Identity property below.

Proof:
Recall that for the real numbers, 1 is the multiplicative identity because for all real numbers a, $1 \cdot a = a \cdot 1 = a$. So we need to show that for all 2×2 matrices M,

$$\begin{bmatrix} 1 & 0 \\ 0 & 1 \end{bmatrix} \cdot M = M \cdot \begin{bmatrix} 1 & 0 \\ 0 & 1 \end{bmatrix} = M.$$

Let $M = \begin{bmatrix} a & b \\ c & d \end{bmatrix}$.

Then, $\begin{bmatrix} 1 & 0 \\ 0 & 1 \end{bmatrix}\begin{bmatrix} a & b \\ c & d \end{bmatrix} = \begin{bmatrix} 1 \cdot a + 0 \cdot c & 1 \cdot b + 0 \cdot d \\ 0 \cdot a + 1 \cdot c & 0 \cdot b + 1 \cdot d \end{bmatrix} = \begin{bmatrix} a & b \\ c & d \end{bmatrix}$ and

$\begin{bmatrix} a & b \\ c & d \end{bmatrix}\begin{bmatrix} 1 & 0 \\ 0 & 1 \end{bmatrix} = \begin{bmatrix} a \cdot 1 + b \cdot 0 & a \cdot 0 + b \cdot 1 \\ c \cdot 1 + d \cdot 0 & c \cdot 0 + d \cdot 1 \end{bmatrix} = \begin{bmatrix} a & b \\ c & d \end{bmatrix}$.

(Both multiplications are needed because multiplication is not always commutative.) Thus $\begin{bmatrix} 1 & 0 \\ 0 & 1 \end{bmatrix}$ is the multiplicative identity for 2×2 matrices.

Remember that $\begin{bmatrix} 1 & 0 \\ 0 & 1 \end{bmatrix}$, the transformation that maps each point (x, y) onto itself, is the identity transformation. When a point matrix $\begin{bmatrix} x \\ y \end{bmatrix}$ is multiplied on the left by $\begin{bmatrix} 1 & 0 \\ 0 & 1 \end{bmatrix}$, each point (x, y) coincides with its image. Thus $\begin{bmatrix} 1 & 0 \\ 0 & 1 \end{bmatrix}$ represents the identity transformation.

$$\begin{bmatrix} 1 & 0 \\ 0 & 1 \end{bmatrix}\begin{bmatrix} x \\ y \end{bmatrix} = \begin{bmatrix} 1 \cdot x + 0 \cdot y \\ 0 \cdot x + 1 \cdot y \end{bmatrix} = \begin{bmatrix} x \\ y \end{bmatrix}$$

The size change of magnitude 1 is the identity transformation.

Thus the identity for multiplication of matrices represents an identity for transformations as well. One property of multiplication of real numbers that is not true for 2×2 matrices is commutativity. That is, there are 2×2 matrices A and B with $AB \neq BA$. In the questions, you are asked to find an example of such 2×2 matrices.

Multiplying the Matrices of Two Transformations

The Example below shows how the product of two transformation matrices is related to the corresponding transformations.

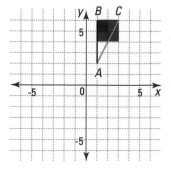

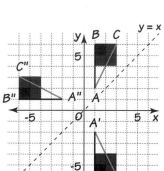

Example

A flag has key points at $A = (1, 2)$, $B = (1, 6)$, and $C = (3, 6)$, as shown at the left. Use matrices to reflect this flag over the x-axis. Then reflect its image over the line $y = x$.

Solution

Represent A, B, and C by the matrix $\begin{bmatrix} 1 & 1 & 3 \\ 2 & 6 & 6 \end{bmatrix}$.

To find the first image, multiply this matrix by the matrix for r_x.

$$\overset{r_x}{\begin{bmatrix} 1 & 0 \\ 0 & -1 \end{bmatrix}} \overset{A \quad B \quad C}{\begin{bmatrix} 1 & 1 & 3 \\ 2 & 6 & 6 \end{bmatrix}} = \overset{A' \quad B' \quad C'}{\begin{bmatrix} 1 & 1 & 3 \\ -2 & -6 & -6 \end{bmatrix}}$$

To find the second image, multiply the matrix for A', B', and C' by the matrix for $r_{y=x}$.

$$\overset{r_{y=x}}{\begin{bmatrix} 0 & 1 \\ 1 & 0 \end{bmatrix}} \overset{A' \quad B' \quad C'}{\begin{bmatrix} 1 & 1 & 3 \\ -2 & -6 & -6 \end{bmatrix}} = \overset{A'' \quad B'' \quad C''}{\begin{bmatrix} -2 & -6 & -6 \\ 1 & 1 & 3 \end{bmatrix}}$$

The points $A'' = (-2, 1)$, $B'' = (-6, 1)$ and $C'' = (-6, 3)$ enable you to draw the final flag shown at the left.

Composites of Transformations

We call the final flag the image of the original flag under the *composite* of the reflections r_x and $r_{y=x}$. In the In-class Activity on page 239, you found the image of another figure under the composite of these two reflections. In general, any two transformations can be *composed*.

> ### Definition
> Suppose transformation T_1 maps figure F onto figure F', and transformation T_2 maps figure F' onto figure F''. The transformation that maps F onto F'' is called the **composite** of T_1 and T_2, written $T_2 \circ T_1$.

The symbol $\circ$ is read "following." In the Example, r_x came first and then $r_{y=x}$, so we write $r_{y=x} \circ r_x$ and say "$r_{y=x}$ following r_x" or "the composite of $r_{y=x}$ and r_x."

The Composite of $r_{y=x}$ and r_x

To describe $r_{y=x} \circ r_x$ as one transformation, again consider the flags in the Example. The flag containing points A, B, and C is the preimage. The flag containing A', B', and C' is the first image, and the flag containing

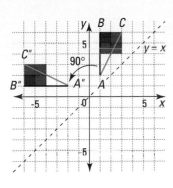

A″, B″, C″ is the final image. Look only at the preimage and the final image as shown at the left. How are the two flags related?

The graph shows that the composite is neither a reflection nor a size or scale change. The composite is the turn, or **rotation** 90° counterclockwise about (0, 0). We denote this rotation by R_{90}. This rotation is the composite of the two reflections:

$$R_{90} = r_{y=x} \circ r_x.$$

Persons *X* and *Z* should have found that in the In-class Activity preceding this lesson, the final figure is the image of the original figure under a rotation of 90° counterclockwise.

Matrices and Composites of Transformations

How do we find the matrix associated with R_{90}? Notice that the composite of transformations of the Example led to the following matrix multiplication.

$$\overset{r_{y=x}}{\begin{bmatrix} 0 & 1 \\ 1 & 0 \end{bmatrix}} \overset{r_x}{\begin{bmatrix} 1 & 0 \\ 0 & -1 \end{bmatrix}} \overset{A\ B\ C}{\begin{bmatrix} 1 & 1 & 3 \\ 2 & 6 & 6 \end{bmatrix}} = \overset{A'\ B'\ C'}{\begin{bmatrix} -2 & -6 & -6 \\ 1 & 1 & 3 \end{bmatrix}}$$

Because matrix multiplication is associative, this product can also be computed as follows:

$$\overset{r_{y=x}}{\begin{bmatrix} 0 & 1 \\ 1 & 0 \end{bmatrix}} \overset{r_x}{\begin{bmatrix} 1 & 0 \\ 0 & -1 \end{bmatrix}} \overset{A\ B\ C}{\begin{bmatrix} 1 & 1 & 3 \\ 2 & 6 & 6 \end{bmatrix}} = \begin{bmatrix} 0 & -1 \\ 1 & 0 \end{bmatrix} \overset{A\ B\ C}{\begin{bmatrix} 1 & 1 & 3 \\ 2 & 6 & 6 \end{bmatrix}} = \overset{A''\ B''\ C''}{\begin{bmatrix} -2 & -6 & -6 \\ 1 & 1 & 3 \end{bmatrix}}$$

Observe that multiplying the two reflection matrices gives the single matrix $\begin{bmatrix} 0 & -1 \\ 1 & 0 \end{bmatrix}$. Applying this matrix to *A, B,* and *C* results in the final images *A″, B″,* and *C″*. Thus, the single matrix $\begin{bmatrix} 0 & -1 \\ 1 & 0 \end{bmatrix}$ is the matrix of the composite $r_{y=x} \circ r_x$. The general idea is summarized below.

Theorem
If M_1 is the matrix for transformation T_1 and M_2 is the matrix for transformation T_2, then $M_2 M_1$ is the matrix for $T_2 \circ T_1$.

QUESTIONS

Covering the Reading

1. Explain why every transformation is a function.

2. When a 2 × 2 matrix is multiplied by a 2 × 2 matrix, what are the dimensions of the product matrix?

3. Let $X = \begin{bmatrix} 1 & -2 \\ 3 & 4 \end{bmatrix}$, $Y = \begin{bmatrix} 0 & 1 \\ 4 & -2 \end{bmatrix}$, and $Z = \begin{bmatrix} \frac{1}{2} & 1 \\ 0 & 1 \end{bmatrix}$.

    a. Show that $(XY)Z = X(YZ)$.

    b. The answer to part **a** is an instance of what property of matrices?

4. **a.** Multiply $\begin{bmatrix} 1 & 0 \\ 0 & 1 \end{bmatrix}$ by $\begin{bmatrix} \pi & \sqrt{2} \\ -3 & \frac{3}{4} \end{bmatrix}$.

    b. What property of matrix multiplication is illustrated in part **a?**

5. Find two 2×2 matrices A and B such that $AB = BA$.

6. Find two 2×2 matrices A and B such that $AB \neq BA$.

7. The multiplicative-identity matrix for 2×2 matrices is the matrix for what transformation?

8. What property of matrix multiplication justifies this equation?

$$\begin{bmatrix} 0 & 1 \\ 1 & 0 \end{bmatrix} \left(\begin{bmatrix} 1 & 0 \\ 0 & -1 \end{bmatrix} \begin{bmatrix} 1 & 1 & 3 \\ 2 & 6 & 6 \end{bmatrix} \right) = \left(\begin{bmatrix} 0 & 1 \\ 1 & 0 \end{bmatrix} \begin{bmatrix} 1 & 0 \\ 0 & -1 \end{bmatrix} \right) \begin{bmatrix} 1 & 1 & 3 \\ 2 & 6 & 6 \end{bmatrix}?$$

9. What does the symbol $\circ$ mean?

10. In $r_{y=x} \circ r_x$, which reflection is done first, $r_{y=x}$ or r_x?

11. R_{90} represents a rotation of _?_ degrees around _?_ in a(n) _?_ direction.

12. If T_1 has matrix $\begin{bmatrix} -2 & 0 \\ 0 & 2 \end{bmatrix}$ and T_2 has matrix $\begin{bmatrix} 0 & 1 \\ -1 & 0 \end{bmatrix}$, what is a matrix for $T_2 \circ T_1$?

Applying the Mathematics

13. **a.** Find the matrix for $r_x \circ r_{y=x}$.

    b. To what single transformation is $r_x \circ r_{y=x}$ equivalent?

    c. How does your answer to part **b** compare with $r_{y=x} \circ r_x$?

14. Graph the image of the flag of this lesson under the transformation $S_2 \circ r_y$.

15. **a.** To what single transformation is each of the following equivalent?

       (i) $r_x \circ r_x$ (ii) $r_y \circ r_y$ (iii) $r_{y=x} \circ r_{y=x}$

    b. Explain the geometric meaning of the results of part **a**.

16. Use $A = \begin{bmatrix} a & b \\ c & d \end{bmatrix}$, $B = \begin{bmatrix} e & f \\ g & h \end{bmatrix}$, and $C = \begin{bmatrix} i & j \\ k & l \end{bmatrix}$.

    Prove that multiplication of 2×2 matrices is associative by calculating the following.

    a. $(AB)C$ **b.** $A(BC)$

17. Prove that if C is any 2×2 matrix and M_k is the matrix for S_k, then $M_k \cdot C = C \cdot M_k$.

18. $\triangle BAT$ can be represented by $\begin{bmatrix} 0 & 3 & -1 \\ 5 & -2 & -1 \end{bmatrix}$.   *(Lesson 4-6)*

    a. Find a matrix to represent the image $\triangle B'A'T'$ under r_x.
    b. *True or false.* $\triangle BAT \cong \triangle B'A'T'$.

19. **a.** Graph the triangle represented by $\begin{bmatrix} -7 & 7 & 0 \\ 0 & 0 & 7 \end{bmatrix}$.

    b. What kind of triangle is this?
    c. What matrix describes the image of the triangle in part **a** under

      the transformation given by $\begin{bmatrix} 4 & 0 \\ 0 & 1 \end{bmatrix}$?

    d. Graph the image.
    e. What special kind of triangle is the image?
    f. Are the triangles in parts **a** and **c** similar? How can you tell?
    (Lessons 4-1, 4-5)

20. Some communication trends in Kenya and New Zealand for the years 1970 and 1990 are given below. "Newspapers" refers to the number of copies of daily newspapers circulated per 1000 people. "TV" refers to the number of televisions owned per 1000 people.

	1970 Newspapers	TV	1990 Newspapers	TV
Kenya	14	1.4	15	9
New Zealand	375	235	324	442

    a. Which country had the greatest change in newspaper circulation?
    b. Name a trend common to both countries.   *(Lesson 4-2)*

21. During the 1990s, the population of Guatemala has been increasing by about 250,000 per year. In contrast, the population of Hungary has been decreasing by about 11,000 per year. The population of Guatemala in 1990 was about 9,300,000. The population of Hungary in 1990 was about 10,500,000.   *(Lessons 3-1, 3-7)*
    a. In what year will the two populations be about the same?
    b. What will the populations be then?

22. A line has the equation $y - 2 = \frac{1}{2}(x + 3)$. Give the slope of the line and a point it contains.   *(Lessons 3-4, 3-5)*

23. *Multiple choice.*   Which expression equals $-(x_1 - x_2)$?
    (Previous course)
    (a) $x_2 - x_1$ (b) $x_1 - x_2$ (c) $x_1 + x_2$
    (d) $-x_{-1}$ (e) none of these

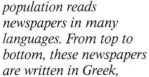

International news.
*America's diverse
population reads
newspapers in many
languages. From top to
bottom, these newspapers
are written in Greek,
Chinese, Russian, and
Arabic.*

Exploration

24. Three properties of multiplication of 2×2 matrices are given in this lesson. Explore whether multiplication of 3×3 matrices has properties identical or similar to these.

Human rotation. *Pictured are members of the Jessie White Tumblers performing the Human Chain. A tumbler rotates 360° while soaring over the backs of up to 20 others. The team, based in Chicago, performs worldwide.*

Rotations are closely related to angles. The arcs used to denote angles suggest turns. Angles with larger measures indicate greater turns. The amount and direction of the turn determine the magnitude of the **rotation.** Counterclockwise turns have positive magnitudes. Clockwise turns have negative magnitudes. The rotation of magnitude x around the origin is denoted by R_x.

Caution: A rotation is denoted with a capital R, while a reflection is denoted with a lowercase r.

The Composite of Two Rotations

Rotations often occur one after the other, as when going from one frame to another in animated cartoons or in computer generated images.

In the frames on page 246, the character undergoes a series of 60° counterclockwise rotations. Notice that figures two frames apart are turned 180°. This is a result of a fundamental property of rotations, which is itself derived from the Angle Addition Postulate in geometry.

> **Theorem**
> A rotation of $b°$ following a rotation of $a°$ with the same center results in a rotation of $(a + b)°$. In symbols: $R_b \circ R_a = R_{a+b}$.

Matrices for Rotations

Can rotations have 2×2 matrices? Notice that $\begin{bmatrix} a & b \\ c & d \end{bmatrix} \begin{bmatrix} 0 \\ 0 \end{bmatrix} = \begin{bmatrix} 0 \\ 0 \end{bmatrix}$.

So any transformation represented by a 2×2 matrix must map $(0, 0)$ onto itself. Thus, the only rotations that can have 2×2 matrices are those with center $(0, 0)$.

In the previous lesson, you learned that $\begin{bmatrix} 0 & -1 \\ 1 & 0 \end{bmatrix}$ is the matrix for the composite $r_{y=x} \circ r_x = R_{90}$. We state this result as a theorem.

> **Theorem**
> $\begin{bmatrix} 0 & -1 \\ 1 & 0 \end{bmatrix}$ is the matrix for R_{90}.

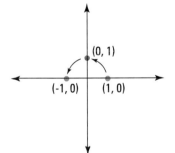

We can verify this theorem using the Matrix Basis Theorem.

The image of $(1, 0)$ is $(0, 1)$.

$$\begin{bmatrix} 1 & 0 \\ 0 & 1 \end{bmatrix} \qquad \begin{bmatrix} 0 & -1 \\ 1 & 0 \end{bmatrix}$$

The image of $(0, 1)$ is $(-1, 0)$.

Matrices for Rotations of Multiples of 90°

By composing two 90° rotations, a matrix for $\boldsymbol{R_{180}}$ can be found.

> **Example**
>
> Find the matrix for R_{180}.
>
> **Solution**
>
> A rotation of 180° can be considered as a 90° rotation followed by another 90° rotation. That is, $R_{90} \circ R_{90} = R_{180}$. In matrix form,
> $$\begin{bmatrix} 0 & -1 \\ 1 & 0 \end{bmatrix} \begin{bmatrix} 0 & -1 \\ 1 & 0 \end{bmatrix} = \begin{bmatrix} -1 & 0 \\ 0 & -1 \end{bmatrix}. \text{ So } \begin{bmatrix} -1 & 0 \\ 0 & -1 \end{bmatrix} \text{ is the matrix for } R_{180}.$$

▶

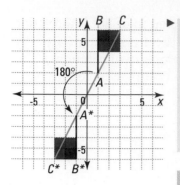

Check

Apply this matrix to a figure. We use *A, B,* and *C* from the last lesson.

$$R_{180} \quad \begin{matrix} A & B & C \end{matrix} \qquad \begin{matrix} A^* & B^* & C^* \end{matrix}$$
$$\begin{bmatrix} -1 & 0 \\ 0 & -1 \end{bmatrix} \begin{bmatrix} 1 & 1 & 3 \\ 2 & 6 & 6 \end{bmatrix} = \begin{bmatrix} -1 & -1 & -3 \\ -2 & -6 & -6 \end{bmatrix}.$$

The graph verifies that *A, B,* and *C* have been rotated 180°.

Activity

Prove that the matrix for R_{270} is $\begin{bmatrix} 0 & 1 \\ -1 & 0 \end{bmatrix}$.

The image of any figure under a rotation with a negative magnitude can also be found by a clockwise rotation. For instance, R_{-90} represents a 90° turn clockwise. Because a rotation of -90° has the same images for every point as a rotation of 270°, R_{-90} equals R_{270}.

Here is a summary of the rotations of this lesson:

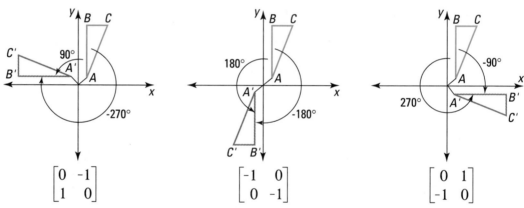

Rotation of 90° or -270° Rotation of 180° or -180° Rotation of 270° or -90°

$$\begin{bmatrix} 0 & -1 \\ 1 & 0 \end{bmatrix} \qquad \begin{bmatrix} -1 & 0 \\ 0 & -1 \end{bmatrix} \qquad \begin{bmatrix} 0 & 1 \\ -1 & 0 \end{bmatrix}$$

These matrices make it possible to describe rotation images algebraically. For instance, for R_{90},

$$\begin{bmatrix} 0 & -1 \\ 1 & 0 \end{bmatrix} \begin{bmatrix} x \\ y \end{bmatrix} = \begin{bmatrix} 0 \cdot x + -1 \cdot y \\ 1 \cdot x + 0 \cdot y \end{bmatrix} = \begin{bmatrix} -y \\ x \end{bmatrix}. \text{ Thus, } R_{90}(x, y) = (-y, x).$$

Rotations, being composites of reflections, are transformations that yield images congruent to their preimages.

QUESTIONS

Covering the Reading

1. A rotation of negative magnitude is a turn in which direction?

2. How much of a turn does the character pictured at the bottom of page 246 undergo from the first frame to the fourth frame?

3. The composite of a rotation of 45° and a rotation of 90° is a rotation of __?__.

4. In general, $R_c \circ R_d = $ __?__.

5. A rotation of -270° is the same as a rotation with what positive magnitude?

In 6–8, *multiple choice.* Identify the matrix for the given rotation. (Hint: Use the Matrix Basis Theorem.)

(a) $\begin{bmatrix} 0 & -1 \\ 1 & 0 \end{bmatrix}$ (b) $\begin{bmatrix} 0 & -1 \\ -1 & 0 \end{bmatrix}$ (c) $\begin{bmatrix} -1 & 0 \\ 0 & -1 \end{bmatrix}$ (d) $\begin{bmatrix} 0 & 1 \\ -1 & 0 \end{bmatrix}$ (e) $\begin{bmatrix} -1 & 0 \\ 0 & 1 \end{bmatrix}$

6. R_{90} **7.** R_{-90} **8.** R_{180}

9. Write your proof from the Activity that $\begin{bmatrix} 0 & 1 \\ -1 & 0 \end{bmatrix}$ is the matrix for R_{270}.

10. Find the image of (x, y) under R_{270}.

Applying the Mathematics

11. Use matrix multiplication to determine $R_{180} \circ R_{180}$. Write a sentence or two explaining your results.

12. Consider $\triangle DEF$ with $D = (1, -2)$, $E = (3, 5)$, and $F = (-3, -1)$. Use matrix multiplication to find the image of $\triangle DEF$ under R_{90}.

13. Quadrilateral $MATH$ has coordinates $M = (0, 0)$, $A = (5, 0)$, $T = (5, 7)$, and $H = (-1, 3)$. Graph $MATH$ and its image under R_{180}.

14. a. Calculate a matrix for $R_{180} \circ r_y$.
     b. To what transformation does the matrix in part **a** correspond?

15. The point $(3, 4)$ lies on the circle with center $(0, 0)$ and radius 5.
     a. Rotate this point 90°, 180°, and 270° around $(0, 0)$ to find the coordinates of three other points on this circle.
     b. Graph all four points.

No time for needling.
Shown is a group using rotations of a compass to navigate. They use the angles of the compass to find each checkpoint.

Review

In 16–19, write the matrix for the transformation.   *(Lessons 4-4, 4-5, 4-6, 4-7)*

16. the size change of magnitude 3, center $(0, 0)$

17. the identity transformation

18. $r_{y=x}$ **19.** $S_{1,3}: (x, y) \rightarrow (x, 3y)$

20. Use the figure at the left. Describe the transformation which maps $ABCDE$ onto $A'B'C'D'E'$ *(Lesson 4-6)*
     a. in words.
     b. with a matrix.
     c. using mapping notation.

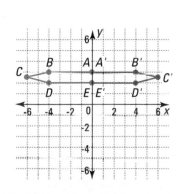

21. Let $T = (-2, 3)$, $R = (5, 3)$, $A = (5, 0)$, and $P = (3, 0)$.
   a. Graph polygon *TRAP*.
   b. Graph the image of *TRAP* under the transformation with matrix $\begin{bmatrix} 2 & 0 \\ 0 & -2 \end{bmatrix}$.
   c. Describe the transformation.   *(Lessons 4-1, 4-5)*

22. Let $M = \begin{bmatrix} 2 & 1 \\ 0 & -2 \end{bmatrix}$, $N = \begin{bmatrix} -2 & 3 \\ -5 & 0 \end{bmatrix}$, and $P = \begin{bmatrix} 1 & -4 \\ 1 & 2 \end{bmatrix}$.
   a. Compute $M(N + P)$.
   b. Compute $MN + MP$.
   c. Is matrix multiplication distributive over matrix addition in this case?   *(Lessons 4-2, 4-3)*

23. What algebraic sentence does $\begin{bmatrix} a & b \end{bmatrix} \cdot \begin{bmatrix} a \\ b \end{bmatrix} = \begin{bmatrix} c^2 \end{bmatrix}$ represent?
   (Lessons 4-1, 4-3)

In 24–26, two figures are given. **a.** Fill in the blank with one of the following: similar, similar and congruent, or neither similar nor congruent. **b.** Identify the type of transformation that maps the figure on the left to the one on the right.   *(Previous course, Lessons 4-3, 4-4, 4-5)*

24.  and   are __?__.

25.  and   are __?__.

26.  and   are __?__.

27. Calculate the slope of the line joining $(-4, 6)$ and $(2, 3)$.   *(Lesson 2-4)*

28. Give a definition of *perpendicular lines*.   *(Previous course)*

Exploration

29. The matrix for R_{30} is $\begin{bmatrix} \dfrac{\sqrt{3}}{2} & \dfrac{-1}{2} \\ \dfrac{1}{2} & \dfrac{\sqrt{3}}{2} \end{bmatrix}$.

   Use this information to determine matrices for some other rotations.

30. $\begin{bmatrix} 0.6 & -0.8 \\ 0.8 & 0.6 \end{bmatrix}$ is a matrix for a rotation R_x about the point $(0, 0)$. By carefully plotting points and their images, estimate x.

LESSON

4-9

Rotations and Perpendicular Lines

Two lines are perpendicular if and only if they form a 90° angle. Thus, if a rotation of magnitude 90° is applied to a line, then the image line is perpendicular to the preimage.

Consider a line passing through points (x_1, y_1) and (x_2, y_2). The line can be represented by the 2×2 matrix $\begin{bmatrix} x_1 & x_2 \\ y_1 & y_2 \end{bmatrix}$. Its slope is $\frac{y_2 - y_1}{x_2 - x_1}$.

Example 1

Consider the line $\overleftrightarrow{AB}$ containing points $A = (-5, -3)$ and $B = (4, 1)$.
a. Represent $\overleftrightarrow{AB}$ as a 2×2 matrix.
b. Calculate the slope of $\overleftrightarrow{AB}$.
c. Find $\overleftrightarrow{A'B'}$, the image of $\overleftrightarrow{AB}$ under R_{90}.
d. Graph $\overleftrightarrow{AB}$ and $\overleftrightarrow{A'B'}$.
e. Calculate the slope of $\overleftrightarrow{A'B'}$.

Solution

a. $\overleftrightarrow{AB}$ can be represented by the matrix $\begin{bmatrix} -5 & 4 \\ -3 & 1 \end{bmatrix}$.

b. slope of $\overleftrightarrow{AB}$ = $\frac{y_2 - y_1}{x_2 - x_1}$ = $\frac{1 - -3}{4 - -5}$ = $\frac{4}{9}$

c. To find $\overleftrightarrow{A'B'}$, multiply the matrix for R_{90} by the matrix for $\overleftrightarrow{AB}$.

$$\overset{R_{90}}{\begin{bmatrix} 0 & -1 \\ 1 & 0 \end{bmatrix}} \overset{(\overleftrightarrow{AB})}{\begin{bmatrix} -5 & 4 \\ -3 & 1 \end{bmatrix}} = \overset{\overleftrightarrow{A'B'}}{\begin{bmatrix} 3 & -1 \\ -5 & 4 \end{bmatrix}}$$

So $\overleftrightarrow{A'B'}$ is the line containing the points $(3, -5)$ and $(-1, 4)$.

d. The preimage $\overleftrightarrow{AB}$ and image $\overleftrightarrow{A'B'}$ are graphed at the right.

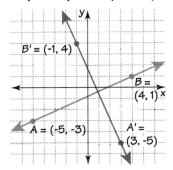

e. slope of $\overleftrightarrow{A'B'}$ = $\frac{4 - -5}{-1 - 3}$ = $\frac{9}{-4}$ = $-\frac{9}{4}$

Notice that the slopes of the lines in Example 1 are negative reciprocals of each other. Their product is -1. The steps of Example 1 can be generalized to prove the following theorem.

Theorem

If two lines with slopes m_1 and m_2 are perpendicular, then $m_1m_2 = -1$.

Proof:

We are given lines with slopes m_1 and m_2. We must show that the product m_1m_2 is -1. The given lines either contain the origin or they are parallel to lines with the same slopes that contain the origin. We prove the theorem for two lines through the origin; this proves the property for perpendicular lines elsewhere.

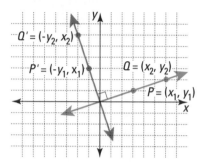

Let $P = (x_1, y_1)$ and $Q = (x_2, y_2)$ be two points on line $\overleftrightarrow{PQ}$ that contains the origin. Then $\overleftrightarrow{PQ}$ can be represented by $\begin{bmatrix} x_1 & x_2 \\ y_1 & y_2 \end{bmatrix}$.
Let $P' = R_{90}(P)$ and $Q' = R_{90}(Q)$. Then

$$R_{90}\left(\overleftrightarrow{PQ}\right) = \overleftrightarrow{P'Q'}.$$

In matrix form, $\begin{bmatrix} 0 & -1 \\ 1 & 0 \end{bmatrix}\begin{bmatrix} x_1 & x_2 \\ y_1 & y_2 \end{bmatrix} = \begin{bmatrix} -y_1 & -y_2 \\ x_1 & x_2 \end{bmatrix}$.

From the matrix, $P' = (-y_1, x_1)$ and $Q' = (-y_2, x_2)$. Let the slope of the preimage be m_1 and the slope of the image be m_2.

$$m_1 = \text{slope of } \overleftrightarrow{PQ} = \frac{y_2 - y_1}{x_2 - x_1}$$

$$m_2 = \text{slope of } \overleftrightarrow{P'Q'} = \frac{x_2 - x_1}{-y_2 - (-y_1)} = \frac{x_2 - x_1}{-(y_2 - y_1)} = -\frac{x_2 - x_1}{y_2 - y_1}$$

The product of the slopes is $m_1m_2 = \left(\frac{y_2 - y_1}{x_2 - x_1}\right)\left(-\frac{x_2 - x_1}{y_2 - y_1}\right) = -1$.

Example 2

Line n contains $(-4, 1)$ and is perpendicular to line ℓ whose equation is $y = -\frac{3}{2}x + 2$. Find an equation for line n.

Solution

From the equation for line ℓ, we see that its slope is $-\frac{3}{2}$. Since $n \perp \ell$, the slope of n is $\frac{2}{3}$ because the negative reciprocal of $-\frac{3}{2}$ is $\frac{2}{3}$. Now use the point-slope equation for a line. Since line n contains (-4, 1), an equation for n is $y - 1 = \frac{2}{3}(x + 4)$.

▶

Check

Rewrite $y - 1 = \frac{2}{3}(x + 4)$ in slope-intercept form.

$$y - 1 = \frac{2}{3}x + \frac{8}{3}$$
$$y = \frac{2}{3}x + \frac{11}{3}$$

Graph the equations on the same set of axes. The lines are perpendicular, and line n passes through the point $(-4, 1)$.

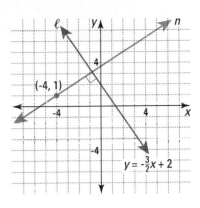

What if line ℓ_1 has slope m_1 and line ℓ_2 has slope m_2, and $m_1m_2 = -1$? Are ℓ_1 and ℓ_2 always perpendicular? The answer is yes, for the following reason. Any line ℓ_3 perpendicular to ℓ_1 has slope m_3, where $m_1m_3 = -1$. So $m_1m_3 = m_1m_2$, and $m_3 = m_2$. Thus ℓ_3 and ℓ_2 have the same slope. So $\ell_3 // \ell_2$. But $\ell_1 \perp \ell_3$. If a line is perpendicular to one of two parallel lines, it must be perpendicular to the other. So $\ell_1 \perp \ell_2$. We have proved the converse of the previous theorem.

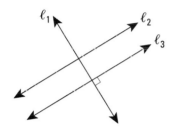

> **Theorem**
> If two lines have slopes m_1 and m_2 and $m_1m_2 = -1$, then the lines are perpendicular.

QUESTIONS

Covering the Reading

1. The line which contains the points (x_1, y_1) and (x_2, y_2) can be represented by the 2×2 matrix __?__.

2. Let $\overleftrightarrow{AB}$ contain points $A = (3, 5)$ and $B = (-1, -6)$.
   a. Find two points on the image of $\overleftrightarrow{AB}$ under R_{90}.
   b. Graph $\overleftrightarrow{AB}$ and its image $\overrightarrow{A'B'}$.
   c. Find the slopes of $\overleftrightarrow{AB}$ and of $\overleftrightarrow{A'B'}$.
   d. What is the product of the slopes?

In 3 and 4, *true or false.*

3. If two lines have slopes m_1 and m_2, and $m_1m_2 = -1$, then the lines are perpendicular.

4. If two lines are perpendicular, and they have slopes m_1 and m_2, then $m_1m_2 = -1$.

5. Find an equation of the line through (6, 1) and perpendicular to the line $y = \frac{4}{3}x - 2$.

6. Find an equation of the line perpendicular to $y = -3x + 7$ passing through the point (-21, 20).

Applying the Mathematics

7. Consider the line with equation $2x + 6y = 1$ and the point $P = (7, -2)$.
   a. Find the slope of a line perpendicular to the given line.
   b. Find an equation for the line through P and perpendicular to the given line.

8. Consider the line with equation $y = \frac{1}{2}x - 3$.
   a. Find an equation of a line perpendicular to the given line.
   b. Graph the two lines.

9. Why do the statements of the theorems in this lesson apply only to lines with nonzero slopes?

10. *Multiple choice.* What is the slope of a line perpendicular to the line with equation $x = 7$?
    (a) 0   (b) undefined slope   (c) $\frac{-1}{7}$ (d) 7

11. Find an equation for the line through (6, 2) and perpendicular to $y = 4$.

12. Let $A = (6, 0)$ and $B = (12, 5)$. Find an equation for the perpendicular bisector of $\overline{AB}$. (Hint: First find the midpoint of $\overline{AB}$.)

13. Let $A = (7, 3)$ and $B = (-4, 1)$.
    a. Find the coordinates of A' and B' under R_{270}.
    b. Graph both $\overleftrightarrow{AB}$ and $\overleftrightarrow{A'B'}$.
    c. Find the slopes of $\overleftrightarrow{AB}$ and $\overleftrightarrow{A'B'}$.
    d. What relationship exists between the slopes? What does this tell you about the lines?
    e. A counterclockwise rotation of 270° is the same as a clockwise rotation of __?__.

In 14–17, assume all lines lie in the same plane.
a. Fill each blank with // or ⊥.
b. Draw a picture to illustrate each situation.

14. If ℓ // m and m // n, then ℓ __?__ n.

15. If ℓ // m and $m \perp n$, then ℓ __?__ n.

16. If $\ell \perp m$ and m // n, then ℓ __?__ n.

17. If $\ell \perp m$ and $m \perp n$, then ℓ __?__ n.

Air *lines.* *Many airports, like San Francisco's International Airport shown below, have parallel or perpendicular runways and taxiways.*

18. a. Calculate the matrix of $r_y \circ R_{180}$.
   b. In $r_y \circ R_{180}$, which transformation is performed first, R_{180} or r_y?
   c. To what transformation does this matrix correspond?
   (Lessons 4-6, 4-7)

In 19 and 20, consider the rectangle $R = \begin{bmatrix} 3 & 3 & 6 & 6 \\ 8 & 0 & 0 & 8 \end{bmatrix}$. *(Lesson 4-5)*

19. Suppose M is a matrix for a scale change and $MR = \begin{bmatrix} 4 & 4 & 8 & 8 \\ 6 & 0 & 0 & 6 \end{bmatrix}$. Find M.

20. Find the matrix for a scale change that maps the rectangle R onto a square with sides of length 2.

21. Matrix D gives the daily delivery of fish to two markets. Matrix C gives the unit cost for each item in the market.

$$\begin{array}{c} \\ \\ \text{Albert's} \\ \text{Carlita's} \end{array} \begin{array}{ccc} \text{cod} & \text{perch} & \text{grouper} \\ \end{array}$$

$$\begin{array}{c} \text{Albert's} \\ \text{Carlita's} \end{array} \begin{bmatrix} 12 & 6 & 20 \\ 8 & 5 & 32 \end{bmatrix} = D \qquad \begin{array}{c} \\ \text{cod} \\ \text{perch} \\ \text{grouper} \end{array} \begin{array}{c} \text{unit cost} \\ \begin{bmatrix} 2.89 \\ 2.59 \\ 1.98 \end{bmatrix} = C \end{array}$$

   a. Calculate DC.
   b. What is the cost of the daily delivery of fish at Carlita's?

22. Find an equation for the line passing through the point (16, 50) and parallel to the line $y = -x$.   *(Lessons 3-2, 3-5)*

23. a. Solve for x: $u + vx = w + yx$.
   b. When does the equation in part **a** have no solution?
   (Lessons 1-2, 1-5)

24. *Skill sequence.*   Use the Distributive Property to rewrite each expression.   *(Previous course, Lesson 1-5)*
   a. $x(x + 9)$ **b.** $5(x + 9)$ **c.** $(x + 5)(x + 9)$

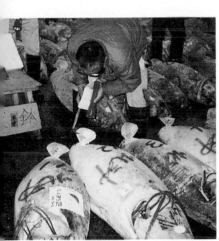

Going once, going twice, sold! *Wholesalers, like this one inspecting whole frozen tuna at the Tsukiji fish market in Tokyo, buy their fish at auction. This market supplies 90% of the fish consumed in Tokyo.*

Exploration

25. Rework Example 1, but use R_{180} in place of R_{90}. What general relationship do you find between the slope of the preimage line connecting the two points and the slope of the image line?

4-10

Translations and Parallel Lines

In this chapter, you have found transformation images by multiplying 2×2 matrices. There is one transformation for which images can be found by *adding* matrices.

Translations

Consider $\triangle ABC$ and $\triangle A'B'C'$ below.

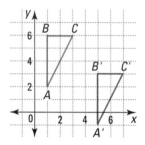

$\triangle A'B'C'$ is a *slide* or *translation image* of the preimage $\triangle ABC$. The matrices M and M' for these triangles are given below.

$\triangle ABC$ has matrix $M = \begin{bmatrix} 1 & 1 & 3 \\ 2 & 6 & 6 \end{bmatrix}$.

$\triangle A'B'C'$ has matrix $M' = \begin{bmatrix} 5 & 5 & 7 \\ -1 & 3 & 3 \end{bmatrix}$.

Calculate $M' - M$.

$$M' - M = \begin{bmatrix} 5 & 5 & 7 \\ -1 & 3 & 3 \end{bmatrix} - \begin{bmatrix} 1 & 1 & 3 \\ 2 & 6 & 6 \end{bmatrix} = \begin{bmatrix} 4 & 4 & 4 \\ -3 & -3 & -3 \end{bmatrix}$$

In $M' - M$, all the elements in the first row are equal, and all the elements in the second row are equal. Thus, to get the image $\triangle A'B'C'$, add 4 to every x-coordinate and -3 to every y-coordinate of the preimage.

$$\begin{bmatrix} 4 & 4 & 4 \\ -3 & -3 & -3 \end{bmatrix} + \overset{M}{\begin{bmatrix} 1 & 1 & 3 \\ 2 & 6 & 6 \end{bmatrix}} = \overset{M'}{\begin{bmatrix} 5 & 5 & 7 \\ -1 & 3 & 3 \end{bmatrix}}$$

This leads to an algebraic definition of *translation*.

Definition
The transformation that maps (x, y) onto $(x + h, y + k)$ is a translation of h units horizontally and k units vertically, and is denoted by $T_{h,k}$.

Using mapping or $f(x)$ notation we can write

$$T_{h,k}: (x, y) \rightarrow (x + h, y + k), \text{ or } T_{h,k}(x, y) = (x + h, y + k).$$

In the figure on page 256, $\triangle A'B'C'$ is the image of $\triangle ABC$ under the translation $T_{4,-3}$.

Matrices for Translations

A given translation cannot be represented by a single matrix because the dimensions of the translation matrix depend on the figure being translated.

For instance, in the Example below, to translate the quadrilateral, you must add a matrix with dimensions 2 by 4.

Example

A quadrilateral has vertices $Q = (-4, 2)$, $U = (-2, 6)$, $A = (0, 5)$ and $D = (0, 3)$.
a. Find its image under the transformation $T_{3,5}$.
b. Graph the image and preimage on the same set of axes.

Solution

a.
$$T_{3,5}(x, y) = (x + 3, y + 5)$$
$$Q' = T_{3,5}(-4, 2) = (-1, 7)$$
$$U' = T_{3,5}(-2, 6) = (1, 11)$$
$$A' = T_{3,5}(0, 5) = (3, 10)$$
$$D' = T_{3,5}(0, 3) = (3, 8)$$

In matrix form:

$$\overset{T_{3,5}}{\begin{bmatrix} 3 & 3 & 3 & 3 \\ 5 & 5 & 5 & 5 \end{bmatrix}} + \overset{Q \quad U \quad A \quad D}{\begin{bmatrix} -4 & -2 & 0 & 0 \\ 2 & 6 & 5 & 3 \end{bmatrix}} = \overset{Q' \quad U' \quad A' \quad D'}{\begin{bmatrix} -1 & 1 & 3 & 3 \\ 7 & 11 & 10 & 8 \end{bmatrix}}$$

The image is quadrilateral $Q'U'A'D'$.

b. $Q'U'A'D'$ is the image under a translation 3 units to the right and 5 units up, as expected. The graph is shown below.

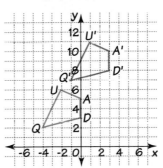

Activity

Consider $\overline{UA}$ and $\overline{U'A'}$ from the Example. It appears that $\overleftrightarrow{UA} \parallel \overleftrightarrow{U'A'}$. Calculate slopes to show that they are parallel.

Properties of Translation Images

The Activity provides a special case of the following more general result.

> **Theorem**
> Under a translation, a line is parallel to its image.

Proof:
Let $P = (x_1, y_1)$ and $Q = (x_2, y_2)$ be two different points on the line $\overleftrightarrow{PQ}$. The image of the line under $T_{h,k}$ contains $P' = (x_1 + h, y_1 + k)$ and $Q' = (x_2 + h, y_2 + k)$.

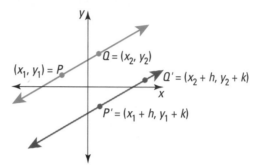

If $\overleftrightarrow{PQ}$ is a vertical line, then $x_1 = x_2$, and so $x_1 + h = x_2 + h$. Then, $\overleftrightarrow{P'Q'}$ is also a vertical line. Thus, in this case $\overleftrightarrow{PQ} \parallel \overleftrightarrow{P'Q'}$. If $\overleftrightarrow{PQ}$ is not vertical, then $x_1 \neq x_2$, and both $\overleftrightarrow{PQ}$ and $\overleftrightarrow{P'Q'}$ have a slope. Let m_1 be the slope of the preimage and let m_2 be the slope of the image.

$$m_1 = \text{slope of } \overleftrightarrow{PQ} = \frac{y_2 - y_1}{x_2 - x_1}$$

$$m_2 = \text{slope of } \overleftrightarrow{P'Q'} = \frac{(y_2 + k) - (y_1 + k)}{(x_2 + h) - (x_1 + h)} = \frac{y_2 - y_1}{x_2 - x_1}$$

The slopes are equal. So $\overleftrightarrow{PQ} \parallel \overleftrightarrow{P'Q'}$.

QUESTIONS

Covering the Reading

1. Refer to $\triangle ABC$ and $\triangle A'B'C'$ at the beginning of the lesson.
   a. What translation maps $\triangle ABC$ onto $\triangle A'B'C'$?
   b. What translation maps $\triangle A'B'C'$ onto $\triangle ABC$?

2. A translation of the plane is a transformation mapping (x, y) to __?__.

3. $T_{h,k}$ is a translation of __?__ units horizontally and __?__ units vertically.

4. Find the image of the point under $T_{-2, 6}$.
   a. $(0, 0)$ b. $(100, -98)$ c. (a, b)

5. Consider $\triangle PQR$ represented by the matrix $\begin{bmatrix} 3 & 18 & 12 \\ -5 & 0 & 0 \end{bmatrix}$. Use matrix addition to find $\triangle P'Q'R'$, the image of $\triangle PQR$ under a translation 5 units to the left and 6 units down.

6. The matrix $\begin{bmatrix} 9 & 4 & 3 & 1 & 6 \\ -1 & 10 & 5 & 1 & 8 \end{bmatrix}$ represents pentagon *FAITH*.

    a. Apply the translation $T_{2,-7}$ to the pentagon.

    b. Graph the preimage and the image on the same set of axes.

In 7 and 8, refer to the Example.

7. a. What is the slope of $\overline{QU}$?   **b.** What is the slope of $\overline{Q'U'}$?

8. Show that $\overline{UA} \parallel \overline{U'A'}$.

Applying the Mathematics

---

9. Suppose lines ℓ_1 and ℓ_2 are not parallel. Can they be translation images of each other? Explain your reasoning.

10. Refer to the graph at the left.

    a. What translation maps *ABCDE* onto *A'B'C'D'E'*?

    b. Verify that $\overline{BB'} = \overline{EE'}$.

    c. Verify that $\overline{BC} \parallel \overline{B'C'}$.

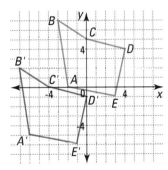

11. $\triangle CUB$ is translated under $T_{4,9}$ to get $\triangle C'U'B'$. $\triangle C'U'B'$ is then translated under $T_{6,5}$ to get $\triangle C''U''B''$. What single translation will give the same result as $T_{6,5} \circ T_{4,9}$?

12. What is the image of the line with equation $y = -3$ under the indicated translation?

    a. $T_{0,5}$ **b.** $T_{5,0}$ **c.** $T_{5,5}$

13. Consider the line whose equation is $y = -2x + 3$. Find an equation for the image of this line under $T_{5,-1}$.

14. Line ℓ has the equation $y = 3x - 7$. Line ℓ' is the image of ℓ under a translation. If ℓ' contains the point $(0, 5)$, find an equation for ℓ'.

Review

---

15. Suppose line ℓ_1 has slope $\frac{2}{3}$, and $\ell_1 \perp \ell_2$. What is the slope of ℓ_2?
(Lesson 4-9)

16. Find an equation of the perpendicular bisector of $\overline{PQ}$, where $P = (1, 4)$ and $Q = (10, -6)$.   *(Lesson 4-9)*

17. $H = (5, 1)$ and $I = (-3, -1)$.   *(Lessons 4-6, 4-9)*

    a. Find the image $\overline{H'I'}$ under r_y.

    b. Find HI and $H'I'$ and compare the two lengths.

    c. Are $\overline{HI}$ and $\overline{H'I'}$ perpendicular? Justify your answer.

In 18–23, a transformation is given. **a.** Give the matrix for the transformation. **b.** Give the image of (a, b).   *(Lessons 4-4, 4-5, 4-6, 4-8)*

18. R_{270} **19.** r_x

20. a size change of magnitude 4   **21.** $S_{1,6}$

22. $(x, y) \rightarrow (-x, -y)$ **23.** reflection over the line $y = x$

24. By what matrix must you multiply $\begin{bmatrix} 5 & 0 & -1 \\ 2 & 6 & -4 \end{bmatrix}$ to get $\begin{bmatrix} -10 & 0 & 2 \\ 1 & 3 & -2 \end{bmatrix}$?
(Lesson 4-5)

25. Find a single matrix equal to the following.   *(Lessons 4-2, 4-3)*

$$\begin{bmatrix} 1 & -1 & 2 \\ 0 & 2 & 1 \end{bmatrix} \begin{bmatrix} 2 & 8 \\ -1 & 0 \\ 1 & -2 \end{bmatrix} - \begin{bmatrix} 5 & 5 \\ 4 & -4 \end{bmatrix}$$

26. A housing contractor builds four model houses—I, II, III, and IV—in three housing developments—Hill, Plain, and Dale. Matrix A gives the number of doors and windows in each model, matrix B gives the number of each model built last year, and matrix C gives the unit cost in dollars of each door and window.

Matrix A

	Doors	Windows
I	2	12
II	2	20
III	3	15
IV	3	20

Matrix B

	I	II	III	IV
Hill	10	5	1	2
Plain	5	10	2	5
Dale	6	4	5	3

Matrix C

	cost
Doors	150
Windows	90

a. Calculate BA and tell what it represents.
b. Set up a matrix product that gives the contractor's total cost for doors and windows last year.
c. Calculate the product in part **b.**   *(Lesson 4-3)*

Exploration

---

27. A transformation has the following rule: The image of (x, y) is $(3x, y + 2)$. Find images of a figure of your own choosing. Geometrically describe what the transformation does to a figure.

A project presents an opportunity for you to extend your knowledge of a topic related to the material of this chapter. You should allow more time for a project than you do for typical homework questions.

1 Predicting the Weather

A weather forecaster has collected data to predict whether tomorrow will be sunny (S), cloudy (C), or rainy (R), given today's weather conditions. If today is sunny, then tomorrow the probabilities are 75% for S, 15% for C, and 10% for R. If today is cloudy, then the probabilities for tomorrow are 30% for S, 60% for C, and 10% for R. If today is rainy, then the probabilities for tomorrow are 25% for S, 20% for C, and 55% for R. Represent this information as a 3×3 matrix. Multiply this matrix by itself to calculate predictions of what the weather will be 2 days from now, 3 days from now, 4 days from now, . . . , n days from now, based on today's weather. In general, what can you tell about the weather in the forecaster's area?

2 History of Matrices

Investigate the development of matrices and the early work of Arthur Cayley and James Sylvester in the mid 1800s.

a. What were "matrices" used for before they were given that name?

b. When were matrices first used to describe transformations?

c. What other mathematicians and terms are associated with the history and use of matrices?

Arthur Cayley   *James Sylvester*

▶

3 Overhead Projectors as Size Changers

Overhead projectors are used because they project images of figures under a size change. A light source passes through a transparency (preimage), reflects off the mirror and projects a picture onto the screen (image). As you move the projector away from

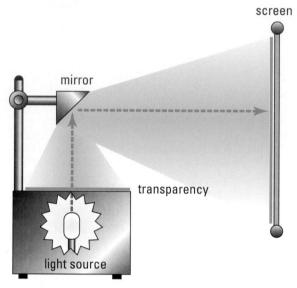

Overhead Projector

the screen, the image gets larger, and as you move the projector closer to the screen, the image gets smaller. Investigate whether the following is true.

$$\frac{\text{length of a line segment on transparency}}{\text{corresponding length of a line segment on screen}} = \frac{\text{distance from transparency to mirror}}{\text{distance from mirror to screen.}}$$

Abbreviated, $\frac{P}{I} = \frac{V}{H}$. (Here, P stands for preimage, I for image, V for vertical, and H for horizontal). Experiment with an overhead projector in your school. Make a table with columns P, I, V, H, $\frac{P}{I}$ and $\frac{V}{H}$ for your data. Does the proportion hold for your data? Write a report explaining your results.

4 Translations using Matrix Multiplication

The translations in this chapter are done using matrix addition. All other transformations of points here are performed using matrix multiplication and a 2×2 matrix particularly chosen for that transformation. A translation cannot be done using multiplication of 2×2 matrices. However, to translate a point using matrix multiplication, first write the point $\begin{bmatrix} x \\ y \end{bmatrix}$ in *homogeneous form* as $\begin{bmatrix} x \\ y \\ 1 \end{bmatrix}$. In homogeneous form, the translation

$$T_{h,k}: \begin{bmatrix} x \\ y \end{bmatrix} \rightarrow \begin{bmatrix} x + h \\ y + k \end{bmatrix}$$

becomes $T_{h,k}: \begin{bmatrix} x \\ y \\ 1 \end{bmatrix} \rightarrow \begin{bmatrix} x + h \\ y + k \\ 1 \end{bmatrix}$.

a. Multiply $\begin{bmatrix} 1 & 0 & h \\ 0 & 1 & k \\ 0 & 0 & 1 \end{bmatrix} \cdot \begin{bmatrix} x \\ y \\ 1 \end{bmatrix}$ to see how matrix multiplication can be used for translations.

b. When a point is written in homogeneous form, the 2×2 matrices for transformations for size changes, scale changes, reflections, and rotations need to be written as 3×3 matrices in this form. For example,

$$\begin{array}{ccc} r_x & (x, y) & (x, -y) \end{array}$$
$$\begin{bmatrix} 1 & 0 \\ 0 & -1 \end{bmatrix} \begin{bmatrix} x \\ y \end{bmatrix} = \begin{bmatrix} x \\ -y \end{bmatrix}$$

in homogeneous form becomes

$$\begin{bmatrix} 1 & 0 & h \\ 0 & -1 & k \\ 0 & 0 & 1 \end{bmatrix} \begin{bmatrix} x \\ y \\ 1 \end{bmatrix} = \begin{bmatrix} x + h \\ -y + k \\ 1 \end{bmatrix}.$$

What are the 3×3 matrices for S_k, $S_{a,b}$, R_{90}, and r_y?

c. When translations are represented by 3×3 matrices, you can compose translations with size changes, scale changes, reflections, and rotations. What is the 3×3 matrix for $T_{2,3} \circ R_{90}$?

SUMMARY

A matrix is a rectangular array of objects. Matrices are frequently used to store data and to represent transformations. Matrices can be added if they have the same dimensions. Addition of matrices can be used to obtain translation images of figures. A matrix can be multiplied by a single number, called a scalar.

The product of two matrices exists only if the number of columns of the left matrix equals the number of rows of the right matrix. The element in the nth row and nth column of AB is the product of the nth row of A and the nth column of B. Matrix multiplication is associative but not commutative.

Matrices with 2 rows can represent points, lines, and other figures in the coordinate plane. Multiplying such a matrix by a 2×2 matrix on its left may yield a transformation image of the figure. Transformations for which 2×2 matrices are given in this chapter include reflections,

rotations, size changes, and scale changes. They are summarized below. The rotation of 90° about the origin is a particularly important transformation. It helps in the proof that two nonvertical lines are perpendicular if and only if the product of their slopes is -1.

The set of 2×2 matrices is closed under multiplication. The identity matrix for multiplying 2×2 matrices is $\begin{bmatrix} 1 & 0 \\ 0 & 1 \end{bmatrix}$. The identity transformation maps any figure onto itself. It can be considered as the size change S_1, the rotation R_0, or the translation $T_{0,0}$.

The Matrix Basis Theorem provides a way for you to remember matrices for transformations. If a transformation T can be represented by a 2×2 matrix, with $T(1, 0) = (x_1, y_1)$ and $T(0, 1) = (x_2, y_2)$, then T has the matrix $\begin{bmatrix} x_1 & x_2 \\ y_1 & y_2 \end{bmatrix}$.

Transformations Yielding Images Congruent to Preimages

Reflections:

over x-axis
$$\begin{bmatrix} 1 & 0 \\ 0 & -1 \end{bmatrix}$$
$r_x\colon (x, y) \rightarrow (x, -y)$

over y-axis
$$\begin{bmatrix} 1 & 0 \\ 0 & 1 \end{bmatrix}$$
$r_y\colon (x, y) \rightarrow (-x, y)$

over the line $y = x$
$$\begin{bmatrix} 0 & 1 \\ 1 & 0 \end{bmatrix}$$
$r_{y=x}\colon (x, y) \rightarrow (y, x)$

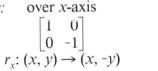

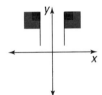

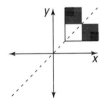

Rotations with center (0, 0):

magnitude 90°
$$\begin{bmatrix} 0 & -1 \\ 1 & 0 \end{bmatrix}$$
$R_{90}\colon (x, y) \rightarrow (-y, x)$

magnitude 180°
$$\begin{bmatrix} -1 & 0 \\ 0 & -1 \end{bmatrix}$$
$R_{180}\colon (x, y) \rightarrow (-x, -y)$

magnitude 270°
$$\begin{bmatrix} 0 & 1 \\ -1 & 0 \end{bmatrix}$$
$R_{270}\colon (x, y) \rightarrow (y, -x)$

Translations:

No general matrix

$T_{h,k}\colon (x, y) \rightarrow (x + h, y + k)$

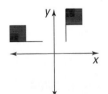

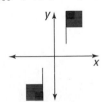

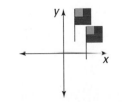

SUMMARY

Transformations Yielding Images Similar to Preimages

Size changes with center (0, 0), magnitude k:

$$\begin{bmatrix} k & 0 \\ 0 & k \end{bmatrix}$$

$S_k: (x, y) \rightarrow (kx, ky)$

Other Transformations

Scale change with horizontal magnitude a and vertical magnitude b:

$$\begin{bmatrix} a & 0 \\ 0 & b \end{bmatrix}$$

$S_{a,b}: (x, y) \rightarrow (ax, by)$

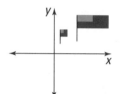

VOCABULARY

Below are the most important terms and phrases for this chapter.
You should be able to give a definition for those terms marked with *.
For all other terms you should be able to give a general description or a specific example.

Lesson 4-1
matrix; matrices
element of a matrix
dimensions $m \times n$
row
column
equal matrices
point matrix

Lesson 4-2
*matrix addition
sum of matrices
difference of matrices
scalar multiplication

Lesson 4-3
matrix multiplication
headings
2×2 identity matrix

Lesson 4-4
*transformation
*size change, S_k
preimage
image
similar
ratio of similitude
center
magnitude of size change
*identity transformation

Lesson 4-5
*scale change, $S_{a,b}$
horizontal magnitude
vertical magnitude
stretch, shrink

Lesson 4-6
reflection image of a point
over a line
*reflection image
reflecting line
line of reflection
reflection
$r_x, r_y, r_{y=x}$
Matrix Basis Theorem

Lesson 4-7
closure
*composite of transformations
composed
rotation
R_{90}

Lesson 4-8
rotation
$R_x, R_{90}, R_{180}, R_{270}$

Lesson 4-10
*translation, $T_{h,k}$
slide or translation image

PROGRESS SELF-TEST

Take this test as you would take a test in class. You will need graph paper. Then check your work with the solutions in the Selected Answers section in the back of the book.

1. Graph the polygon described by the matrix
$$\begin{bmatrix} 3 & -5 & -6 & -5 & -1 & 5 \\ 4 & 2 & 0 & -2 & -3 & -4 \end{bmatrix}.$$

2. One day on Fruitair flying from Appleburg, there were 14 first-class and 120 economy passengers going to Peachport; 3 first-class and 190 economy passengers bound for Bananasville; and 8 first-class and 250 economy passengers flying to Grapetown. Write a 2 × 3 matrix for this information.

In 3–6, use matrices A, B, and C below.

$$A = \begin{bmatrix} 5 & -2 \\ 4 & -2 \\ -1 & 0 \end{bmatrix} \quad B = \begin{bmatrix} 2 & 0 \\ 1 & 5 \end{bmatrix} \quad C = \begin{bmatrix} 8 & 6 \\ -2 & 2 \end{bmatrix}$$

3. Which product exists, AB or BA?

4. Find BC. 　　　　**5.** Find $B - C$.

6. Calculate $7B$.

7. Why is $\begin{bmatrix} 1 & 0 \\ 0 & 1 \end{bmatrix}$ called the identity matrix?

8. Find an equation for the line through $(3, -2.5)$ that is perpendicular to $y = 5x - 3$.

9. Calculate the matrix for $r_x \circ R_{270}$.

In 10–12, refer to the graph at the right.

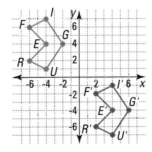

10. What translation maps *FIGURE* onto $F'I'G'U'R'E'$?

11. What is true about $\overleftrightarrow{FI}$ and $\overleftrightarrow{F'I'}$?

12. Find the image of *FIGURE* under the transformation r_y.

13. A shoe manufacturer has factories in Los Angeles, Tucson, and Santa Fe. One year's sales (in thousands) can be summarized by the following matrix.

	Deck Shoes	Pumps	Sandals	Boots
Los Angeles	23	8	10	5
Tucson	11	5	10	15
Santa Fe	2	3	15	15

The selling prices of the deck shoes, pumps, sandals, and boots are $18, $58, $12, and $76, respectively. Use matrix multiplication to find the total revenue for each factory.

14. If the matrices below represent two pet store's inventories before they merge, what will be their inventory immediately after the merger?

	Lois's Pet Shop Males	Females		Doug's Pet Shop Males	Females
Dogs	8	11		10	14
Cats	5	4		11	13
Birds	15	16		7	9
Monkeys	2	0		0	3

In 15 and 16, identify the matrix you might use to perform the given transformation.

15. a horizontal stretch of magnitude 2 and a vertical shrink of magnitude $\frac{1}{4}$

16. $T: (x, y) \rightarrow (y, x)$

In 17 and 18, $\triangle ABC$ has vertices $A = (7, 6)$, $B = (-1, 2)$, and $C = (3, -4)$.

17. Describe the matrix multiplication.
$$\begin{bmatrix} 1 & 0 \\ 0 & -1 \end{bmatrix}\begin{bmatrix} 7 & -1 & 3 \\ 6 & 2 & -4 \end{bmatrix} = \begin{bmatrix} 7 & -1 & 3 \\ -6 & -2 & 4 \end{bmatrix}$$

18. Graph $\triangle ABC$ and $R_{90}(\triangle ABC)$.

19. The translation with matrix $\begin{bmatrix} 3 & 3 \\ 2 & 2 \end{bmatrix}$ is applied to the line $x + 2y = 5$. Find an equation for its image. (Hint: Find two points on the line.)

20. Find the matrix of the image of $\begin{bmatrix} 5 & 2 & -3 \\ -3 & 2 & -5 \end{bmatrix}$

　　a. under S_3. 　　　**b.** under $S_4 \circ S_3$.

CHAPTER REVIEW

Questions on SPUR Objectives

SPUR stands for **S**kills, **P**roperties, **U**ses, and **R**epresentations. The Chapter Review questions are grouped according to the SPUR Objectives for this chapter.

SKILLS DEAL WITH THE PROCEDURES USED TO GET ANSWERS.

Objective A: *Add, subtract, and find scalar multiples of matrices.*   *(Lesson 4-2)*

1. Find a single matrix for $\begin{bmatrix} 8 & 6 \\ 3 & -2 \\ 4 & -1 \end{bmatrix} - \begin{bmatrix} -3 & 0 \\ -1 & 6 \\ -4 & -3 \end{bmatrix}$.

In 2–4, let $A = \begin{bmatrix} 2 & 3 & 4 \\ 7 & 5 & -1 \\ 1 & 2 & 0 \end{bmatrix}$ and $B = \begin{bmatrix} 1 & -6 & 0 \\ 2 & 3 & 1 \\ 4 & 9 & 2 \end{bmatrix}$.

Find the following.

2. $A + B$ 3. $2A + B$ 4. $3A - 4B$

In 5 and 6, solve for a and b.

5. $\begin{bmatrix} a & 16 \\ 10 & b \end{bmatrix} + \begin{bmatrix} .4 & -1 \\ -10 & 3.1 \end{bmatrix} = \begin{bmatrix} 2 & 15 \\ 0 & -7 \end{bmatrix}$

6. $2\begin{bmatrix} -1 & 9 \\ b & -.5 \end{bmatrix} - \begin{bmatrix} a & 7 \\ -3 & 3 \end{bmatrix} = \begin{bmatrix} 6 & 11 \\ 13 & -4 \end{bmatrix}$

Objective B: *Multiply matrices.*   *(Lesson 4-3)*

In 7–10, calculate the product.

7. $\begin{bmatrix} 6 & -1 & -4 \end{bmatrix} \begin{bmatrix} 8 \\ -3 \\ -2 \end{bmatrix}$ 8. $\begin{bmatrix} 3 & 5 \\ 4 & 2 \end{bmatrix} \begin{bmatrix} 3 & 5 \\ 4 & 2 \end{bmatrix}$

9. $\begin{bmatrix} -3 & -2 & -1 \\ 0 & 1 & 1 \\ 5 & 0 & 5 \end{bmatrix} \begin{bmatrix} 1 & 2 & 3 \\ 4 & 5 & 6 \\ 7 & 8 & 9 \end{bmatrix}$

10. $\begin{bmatrix} 1 & 2 & 3 \end{bmatrix} \begin{bmatrix} 4 & 7 \\ 5 & 8 \\ 6 & 9 \end{bmatrix} \begin{bmatrix} 16 & 0 \\ 0 & 4 \end{bmatrix}$

In 11 and 12, solve for a and b.

11. $\begin{bmatrix} a & 0 \\ 0 & b \end{bmatrix} \begin{bmatrix} 2 \\ -9 \end{bmatrix} = \begin{bmatrix} 10 \\ 27 \end{bmatrix}$

12. $\begin{bmatrix} 0 & -1 \\ 1 & 0 \end{bmatrix} \begin{bmatrix} a \\ b \end{bmatrix} = \begin{bmatrix} -5 \\ 8 \end{bmatrix}$

Objective C: *Determine equations of lines perpendicular to given lines.*   *(Lesson 4-9)*

13. Find an equation for the line through $(3, -1)$ and perpendicular to $y = -\frac{1}{2}x + 4$.

14. Find an equation for the line through $(7, 8)$ and perpendicular to $x = -4$.

15. Given $A = (6, 1)$ and $B = (-2, 3)$, find an equation for the perpendicular bisector of $\overline{AB}$.

16. Let $\triangle SMR$ be represented by the matrix
$\begin{bmatrix} 123 & -13 & 43 \\ 65 & 432 & -105 \end{bmatrix}$.
Let $\triangle S'M'R' = R_{90}(\triangle SMR)$.
   a. What is the slope of $\overleftrightarrow{SR}$?
   b. What is the product of the slopes of $\overleftrightarrow{SR}$ and $\overleftrightarrow{S'R'}$?
   c. Use your answers to parts **a** and **b** to find the slope of $\overleftrightarrow{S'R'}$.

PROPERTIES DEAL WITH THE PRINCIPLES BEHIND THE MATHEMATICS.

Objective D: *Recognize properties of matrix operations.* *(Lessons 4-2, 4-3, 4-7)*

In 17 and 18, a statement is given. **a.** Is the statement true or false? **b.** Give an example to support your answer.

17. Matrix addition is commutative.

18. Matrix multiplication is associative.

19. Determine whether the product exists.

a. $\begin{bmatrix} 1 & 6 & 4 \end{bmatrix} \begin{bmatrix} 2 \\ 8 \end{bmatrix}$

b. $\begin{bmatrix} 3 & 1 & 6 \\ 5 & 8 & -2 \end{bmatrix} \begin{bmatrix} 1 & -1 & 0 & 7 \\ 1 & 0 & 0 & 0 \\ 0 & 1 & 5 & 2 \end{bmatrix}$

20. N and T are matrices. N has dimensions $r \times p$ and T has dimensions $q \times r$.

    a. Which product must exist, NT or TN?

    b. What are the dimensions of your answer in part **a?**

21. What matrix is the identity for multiplication of 2×2 matrices?

Objective E: *Recognize relationships between figures and their transformation images.* *(Lessons 4-4, 4-5, 4-6, 4-8, 4-9, 4-10)*

In 22–25, *multiple choice.*
(a) not necessarily similar or congruent
(b) similar, but not necessarily congruent
(c) congruent
(d) parallel

22. A figure and its size-change image are __?__.

23. A figure and its scale-change image are __?__.

24. A figure and its reflection image are __?__.

25. A line and its translation image are __?__.

26. Consider two lines. One is the image of the other under R_{90}. The slope of one of the lines is -0.2. What is the slope of the other line?

27. Repeat Question 26 if the transformation is the translation $T_{1,2}$.

Objective F: *Relate transformations to matrices, and vice versa.* *(Lessons 4-4, 4-5, 4-6, 4-7, 4-8, 4-10)*

28. Translate the following matrix equation into English by filling in the blanks.

$$\begin{bmatrix} 0 & 1 \\ 1 & 0 \end{bmatrix} \begin{bmatrix} 5 \\ -2 \end{bmatrix} = \begin{bmatrix} -2 \\ 5 \end{bmatrix}$$

The reflection image of the point __?__ over the line __?__ is the point __?__.

29. Multiply the matrix for r_y by itself, and tell what transformation the product represents.

30. The matrix $\begin{bmatrix} 6 & 0 \\ 0 & 6 \end{bmatrix}$ is associated with a __?__ change with center __?__ and magnitude __?__.

31. **a.** Calculate a matrix for $r_x \circ R_{180}$.

    b. What single transformation corresponds to your answer?

32. Find two reflections whose composite is R_{180}.

33. **a.** What translation maps $PEAR$ onto $P'E'A'R'$, as shown below.

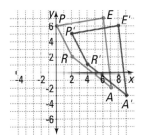

    b. Explain how to use a matrix operation to transform $PEAR$ to $P'E'A'R'$.

In 34–36, give a matrix for the transformation.

34. $r_{y=x}$ **35.** $S_{4,6}$ **36.** R_{90}

37. Find the image of $\begin{bmatrix} -1 & 0 & 4 & 0 \\ 3 & .5 & -1 & 5 \end{bmatrix}$ under r_y.

38. $GOLD$ has coordinates $G(0, 0)$, $O(4, 1)$, $L(3, 5)$, and $D(-1, 4)$. Find the matrix of the image of $GOLD$ under R_{270}.

39. Find the matrix of the image of $\begin{bmatrix} 6 & 8 & 2 \\ 0 & 4 & 0 \end{bmatrix}$ under $S_{\frac{1}{2}}$.

USES DEAL WITH APPLICATIONS OF MATHEMATICS IN REAL SITUATIONS.

Objective G: *Use matrices to store data.*
(Lesson 4-1)

40. Chuck makes handcrafted furniture. Last year he made 5 oak tables, 10 oak chairs, 3 pine tables, 12 pine chairs, 1 maple table, and 6 maple chairs. Store this data in a 2×3 matrix.

41. A high school has 490 freshman boys, 487 freshman girls, 402 sophomore boys, 416 sophomore girls, 358 junior boys, 344 junior girls, 293 senior boys, and 300 senior girls. Write a 4×2 matrix to describe the school's enrollment.

In 42 and 43, the matrix below gives the cost of several items at three different markets.

42. Which element gives the cost of plums in Market I?

43. What does the sum of the numbers in the second column represent?

	Market I	Market II	Market III
eggs (12)	.89	.95	.99
plums (lb)	.90	.79	.82
peaches (lb)	1.49	1.50	1.59
bananas (lb)	.33	.28	.25

Objective H: *Use matrix addition, matrix multiplication, and scalar multiplication to solve real-world problems.* *(Lessons 4-2, 4-3)*

44. A large pizza costs $12.50, a medium pizza costs $8.90, and a small pizza costs $5.20. An order for a Journalism Club party consists of 7 large pizzas, 2 medium pizzas, and 4 small pizzas. Write matrices C and N for the cost and number ordered, then calculate CN to find the total cost of the order.

45. An electronics manufacturer has two factories. Sales (in thousands) can be summarized by the matrix below. The selling price of a VHS recorder is $270, a TV is $320, and a compact disc player is $210. Use matrix multiplication to determine the total revenue for each factory.

	Factory 1	Factory 2
VHS	15	6
TV	10	8
CD	2	1

46. A printing company has two presses. Print runs for two years are given in the matrices below.

1993
	Textbooks	Novels	Nonfiction
Press 1	250,000	125,000	312,000
Press 2	60,000	48,000	90,000

1994
	Textbooks	Novels	Nonfiction
Press 1	190,000	100,000	140,000
Press 2	45,000	60,000	72,000

a. Calculate the matrix that represents the growth in production of each press from 1993 to 1994.

b. Which type of book decreased the most in production?

Making books. *Publishing companies, such as ScottForesman (top), develop and edit manuscripts. Printing companies, such as Donnelley (bottom), print the books.*

REPRESENTATIONS DEAL WITH PICTURES, GRAPHS, OR OBJECTS THAT ILLUSTRATE CONCEPTS.

47. Normal fares (in $) of an airline to three cities are given in the matrix below. To increase air travel, the airline plans to reduce fares by 40%.

$$\begin{array}{c} \text{city 1} \quad \text{city 2} \quad \text{city 3} \end{array}$$
$$\begin{array}{c} \text{first class} \\ \text{economy} \end{array} \begin{bmatrix} 415 & 672 & 258 \\ 198 & 394 & 109 \end{bmatrix}$$

a. What scalar multiplication will yield the new fares?

b. Find the new fares for travel to these three cities.

Objective I: *Graph figures and their transformation images.*
(Lessons 4-1, 4-4, 4-5, 4-6, 4-7, 4-8, 4-10)

48. Draw the polygon described by the matrix
$$\begin{bmatrix} 3 & 0 & 3 \\ -3 & -3 & 0 \end{bmatrix}.$$

49. Refer to the graph below.
Write quadrilateral *HOPE* as a matrix.

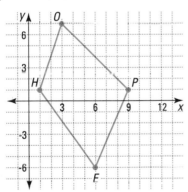

50. Draw the polygon $\begin{bmatrix} 1 & 5 & 3 & -2 \\ 4 & 6 & -2 & -2 \end{bmatrix}$ and its image under $S_{\frac{1}{2}}$.

51. Trapezoid *ABCD* is represented by
$\begin{bmatrix} -1 & 6 & 5 & 0 \\ 0 & 0 & 4 & 4 \end{bmatrix}$. Graph the preimage and image under r_y.

Ideal location. *Pictured is a beach in Cancun, Mexico, a popular destination for airlines.*

52. Consider the quadrilateral defined by the matrix $\begin{bmatrix} 0 & -1 & 0 & 1 \\ 1 & 0 & -1 & 0 \end{bmatrix}$.

a. Graph the quadrilateral and its image under $\begin{bmatrix} 3 & 0 \\ 0 & 3 \end{bmatrix}$.

b. Are the image and preimage similar?

c. Are they congruent?

CHAPTER 5

SYSTEMS

The 1995 postal regulations describing those post cards that could be sent through the mail at a rate of $.20 are given below.

"Each piece must be rectangular. Additionally each piece

$$
\text{must be }
\begin{cases}
\text{at least } 0.007 \text{ inches thick,} \\
\text{no more than } 0.25 \text{ inches thick,} \\
\text{between } 3.5 \text{ and } 4.25 \text{ inches high, and} \\
\text{between } 5 \text{ and } 6 \text{ inches long."}
\end{cases}
$$

Mathematically, if T represents thickness, h represents height, and ℓ stands for length, we could say the *constraints* for a rectangular piece of mail are as follows:

$$
\begin{aligned}
& T \geq 0.007'' \\
\text{and } \ & T \leq 0.25'' \\
\text{and } \ & 3.5'' \leq h \leq 4.25'' \\
\text{and } \ & 5'' \leq \ell \leq 6''.
\end{aligned}
$$

Notice that each constraint in the system of inequalities involves just one variable. In other situations, constraints may involve more than one variable.

When mathematical conditions are joined by the word *and,* the set of conditions or sentences is called a *system.* Many methods for solving systems of equations were developed in the 19th century. This early work on systems involved some of the greatest mathematicians of all time, including Karl Friedrich Gauss from Germany, Jean Baptiste Joseph Fourier from France, and Arthur Cayley from Great Britain.

Systems have many applications. For instance, in the early 19th century Gauss used systems of equations to calculate the orbits of asteroids from sightings made by a few astronomers. In 1939, the Russian mathematician Leonid Kantorovich was the first to announce that large systems might have applications for production planning in industry. In 1945, the American economist George Stigler used systems to determine a best diet for the least cost. Kantorovich and Stigler each received a Nobel Prize for his work, Kantorovich in 1975 and Stigler in 1982. (Both prizes were in economics; there is no Nobel Prize in mathematics.) In this chapter you will study systems of equations and inequalities, including simplified examples of the kinds of problems studied by Kantorovich and Stigler.

Leaping ahead. *The long jump requires an athlete to run down a runway and leap from a take-off board into a sand pit. The jump is measured from the edge of the take-off board to the place in the sand where the athlete lands.*

Suppose that in order to qualify to compete in the long jump in a track meet, you must jump at least 5 meters. If J is the length of your jump, then you qualify if $J \geq 5$. The *open sentence* $J \geq 5$ is called an **inequality** because it contains one of the symbols $<$, $>$, $\leq$, $\geq$, or $\neq$. There are infinitely many solutions to the sentence $J \geq 5$. Some solutions are 5, 5.2, 5.9, 6, and 6.5. Not all solutions can be listed. One way to describe all possible solutions is to graph them on a number line. A graph of the solutions to $J \geq 5$ is shown below.

The shaded circle means that 5 is included in the solution set. Shaded circles are used in the graphs of inequalities of the form $\leq$ or $\geq$. An open circle at the endpoint indicates that the endpoint is not included in the solution set. Open circles are used for inequalities involving $<$, $>$, or $\neq$. The graph below shows all solutions of $m < {-}0.2$.

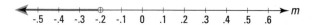

The sets $\{J: J \geq 5\}$ and $\{m: m < {-}0.2\}$ are *intervals.* The graph of the interval $\{J: J \geq 5\}$ is a ray. The graph of $\{m: m < {-}0.2\}$ is a ray without its endpoint.

Compound Sentences Using the Word *and*

In English, you can use the words *and* and *or* as conjunctions to join two or more clauses. In mathematics, these two words are used in a similar way. A sentence in which two clauses are connected by the word *and* or by the word *or* is called a **compound sentence.** The sentences below are compound sentences.

$$4 < x \text{ and } x < 8.$$
$$n \text{ is a negative integer and } n > \text{-}5.$$
$$A < 12 \text{ or } A < 65.$$
$$x = 9 \text{ or } x = \text{-}3.$$

The solution set for a compound sentence using *and* is the *intersection* of the solution sets to the individual sentences. Recall that the **intersection** of two sets is the set consisting of those values common to both sets. The graph of the intersection consists of the points common to the graphs of the individual sets.

Example 1

On page 271, the post-office regulations for post cards are given. Among these are that a post card must be "at least 0.007 inches thick, no more than 0.25 inches thick." Graph all possible thicknesses for a post card.

Solution
Translate each regulation into an inequality. We let T be the thickness. The first regulation is that $T \geq 0.007$. Here is its graph.

The second regulation is that $T \leq 0.25$. Here is its graph, on the same scale.

The intersection of these two graphs is the graph of allowable thicknesses.

The intersection can be described as an *and* statement, such as "$T \geq 0.007$ and $T \leq 0.25$," or by the single compound sentence $0.007 \leq T \leq 0.25$.

... wish you were here.
These leather post cards were mailed in the early 1900s. Cards like these could be mailed today because their dimensions are within the postal regulations.

Recall that the symbol used for intersection is ∩. So the intersection of sets A and B is written $A \cap B$. In set notation,

$$\{x: 0.007 \leq x \leq 0.25\} = \{x: x \geq 0.007\} \cap \{x: x \leq 0.25\}.$$

This can be read "the set of numbers from 0.007 to 0.25 equals the intersection of the set of numbers greater than or equal to 0.007 and the set of numbers less than or equal to 0.25."

When describing an interval in words, it is sometimes difficult to know if the endpoints are included. In this book, we use the following language:

"x is from 3 to 4"   means $3 \le x \le 4$.   The endpoints are included.

"x is between 3 and 4"   means $3 < x < 4$.   The endpoints are not included.

This is consistent with the use of the word "between" in geometry. When just one endpoint is included, as in "$3 \le x < 4$," you can say "x is 3 or between 3 and 4."

Compound Sentences Using the Word *or*

The solution set for a compound sentence using *or* is the *union* of the solution sets to the individual sentences. Recall that the **union** of two sets is the set consisting of those values in either *one or both* sets. This meaning of *or* is somewhat different from the everyday meaning of *either, but not both*. The symbol often used for union is ∪. The union of sets A and B is written $A \cup B$.

The Americans with Disabilities Act of 1990 requires mass transportation systems to be accessible to people with disabilities.

Example 2

People who are either younger than 12 or older than 65 do not pay full fare on a bus.
a. Write this information as a compound inequality.
b. Graph the solution set on a number line.

Solution

a. Let A = the age of the passenger. Then the set of ages A which do not pay full fare on buses is {A: A < 12 or A > 65}.
b. The graph is the union of the graphs of the individual parts.

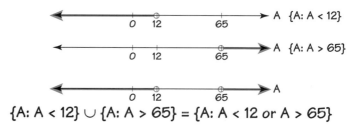

$$\{A: A < 12\} \cup \{A: A > 65\} = \{A: A < 12 \text{ or } A > 65\}$$

Solving Inequalities

The properties of inequality ensure that solving an inequality is very much like solving an equation.

Properties of Inequality
For all real numbers a, b, and c:
If $a < b$, then $a + c < b + c$.   **Addition Property of Inequality**
If $a < b$ and $c > 0$, then $ac < bc$.   **Multiplication Properties of**
If $a < b$ and $c < 0$, then $ac > bc$.   **Inequality**

Example 3

Solve $2m + 57 > 113$ and graph the solution set.

Solution

$$2m + 57 > 113$$
$$2m > 56 \quad \text{Add } -57 \text{ to each side.}$$
$$m > 28 \quad \text{Multiply each side by } \tfrac{1}{2}.$$

The set of solutions is $\{m: m > 28\}$. The graph of the solution set is a ray without its endpoint.

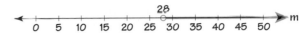

Check

First check the endpoint. Substitute 28 for m in the original sentence. The two sides should be equal. Does $2 \cdot 28 + 57 = 113$? Yes.

Now check the direction of the inequality. Pick a value of m in the solution set. We pick $m = 40$. Is $2 \cdot 40 + 57 > 113$? Yes.

Notice that there are two Multiplication Properties of Inequality. Solving inequalities is different from solving equations only when you multiply or divide each side of an inequality by a negative number. Then you must *reverse* the inequality sign.

There are many applications for inequalities. The words "or less" are clue words for setting up the inequality in Example 4.

Example 4

A ticket agency has 275 tickets to a playoff game. Each caller receives 2 tickets. When there are 50 tickets or less remaining, the agency tries to obtain more tickets. How many callers can be served before more tickets are needed?

Solution

Let c = the number of callers that can be served. Then, after c callers, there will be $275 - 2c$ tickets left. The ticket agency needs to obtain more tickets when c satisfies $275 - 2c \le 50$.

$$275 - 2c \le 50$$
$$-2c \le -225 \quad \text{Subtract 275 from both sides.}$$
$$\frac{-2c}{-2} \ge \frac{-225}{-2} \quad \text{Divide each side by } -2, \text{ so reverse the inequality.}$$
$$c \ge 112.5$$

The agency can serve 112 callers before the agency needs to obtain more tickets.

Covering the Reading

In 1 and 2, could a rectangular-shaped piece of mail with the given dimensions have been mailed for 20¢ in 1995?

1. $\frac{1}{10}''$ thick, 4" high, and 5" long

2. $\frac{1}{3}''$ thick, 3" high, and $5\frac{1}{2}''$ long

3. *Multiple choice.*   Which inequality is the sentence "x is greater than or equal to 5"?
   (a) $x \le 5$ (b) $5 < x$ (c) $5 > x$ (d) $x \ge 5$

4. Write an inequality for the set of numbers graphed below.

5. The solution set to a compound sentence using *and* is the __?__ of the solution sets to the individual sentences.

6. The solution set to a compound sentence using *or* is the __?__ of the solution sets to the individual sentences.

7. On a number line, graph the solution set to the sentence $n \ge 2.95$ *and* $n \le 3.005$.

In 8 and 9, graph the solution set on a number line.

8. **a.** $x \le -7$ or $x > -2$ **b.** $x \le -7$ and $x > -2$

9. **a.** $x \ge -7$ or $x < -2$ **b.** $x \ge -7$ and $x < -2$

10. Match each set at the left with its graph at the right.
    a. $\{x: x > 1 \text{ and } x < 4\}$
    b. $\{x: x > 1 \text{ or } x < 4\}$
    c. $\{x: x < 1 \text{ or } x > 4\}$
    d. $\{x: x < 1 \text{ and } x > 4\}$

    (i)
    (ii)
    (iii)
    (iv)
    (v)

In 11 and 12, solve and graph all solutions.

11. $2m < 1.4$

12. $-4n - 5 > 1$

13. Suppose the ticket agency of Example 4 had 300 tickets and sells 4 tickets to each customer. If more tickets are ordered when the supply falls below 75, how many customers can be served?

14. When you multiply or divide both sides of an inequality by a negative number, you must __?__ the inequality sign.

15. To check that $b \geq 9$ is the simplest sentence equivalent to $5 - \frac{b}{3} \leq 2$, what two values might you pick for b? Why must you pick two values?

Applying the Mathematics

16. Louise wants to buy a car. She will spend more than $8000 but less than $11,000 on a new car, or she will buy a good used car for no more than $5000. Let c represent the cost of the car she will buy.
   a. Write a sentence using set notation describing the amount she may spend.
   b. Graph the possible values of c on a number line.

17. Some parts of interstate highways have a maximum speed limit of 55 mph, while the minimum speed is 45 mph.
   a. Graph the possible legal speeds on a number line.
   b. Write the set of possible legal speeds in set notation.

18. A truck weighs 5000 kg when empty. It is used for carrying sacks of pistachio nuts, each sack weighing 50 kg.
   a. Write a sentence for the total weight T of the truck loaded with s sacks of pistachios.
   b. How many sacks of pistachios can the truck carry over a bridge with a weight limit of 8000 kg? (Hint: T must be less than or equal to the weight limit.)

19. Will solved $x^2 = 4$ and wrote "$x = 2$ *and* $x = -2$." What is wrong with Will's answer?

20. Solve $4x - 3 < 5$ and $8 - 2x < 9$. Then graph the solution set.

21. Cheap Rentals rents cars at $10 per day plus 12¢ per mile. Ruby needs a car for four days. How many miles can she drive if the total cost of renting the car is not to exceed $100?

22. Most programming languages for computers and calculators use the words *and* and *or*. Use either of the programs below.

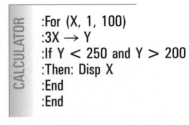

```
BASIC
10 FOR X = 1 TO 100
20 LET Y = 3*X
30 IF Y < 250 AND Y > 200
     THEN PRINT Y
40 NEXT X
50 END
```

```
CALCULATOR
:For (X, 1, 100)
:3X → Y
:If Y < 250 and Y > 200
:Then: Disp X
:End
:End
```

   a. What numbers will be displayed when these programs are run?
   b. What numbers will be displayed if the word AND in the programs is changed to OR?

Nutty information. *The pistachio nut is the small seed of the pistachio tree. The tree grows in the eastern Mediterranean, southwest Asia, and the southwestern U.S. Often the shells of pistachio nuts are dyed red.*

23. Write equations for the two lines graphed below.   *(Lessons 3-5, 3-6)*

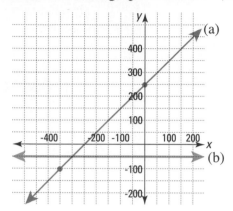

24. Graph the line with equation $8x - 4y = 16$.   *(Lessons 3-2, 3-4)*

25. If y varies inversely as x^3, how is the value of y changed if x is
   a. quadrupled?   **b.** halved?   *(Lesson 2-3)*

Exploration

26. Normal weights are often given in a table as a range of values depending on height. Find such a table in a health book or almanac. Graph the interval of normal weights for your height.

27. In ordinary usage, replacing *or* by *and* can dramatically change the meaning of a sentence. For instance, "Give me liberty and give me death" differs from Patrick Henry's famous saying by only that one word. Find examples of other sayings whose meanings are changed by replacing "and" with "or," or vice versa.

Outspoken. *This drawing depicts Patrick Henry (1736–1799) who was an American Revolutionary patriot, orator, and statesman. His famous "Give me liberty" speech was given in 1775 before the Virginia Provincial Convention. In 1776 he was elected as the first governor of Virginia.*

LESSON

5-2

Solving Systems Using Tables or Graphs

Remember that a **system** is a set of conditions joined by the word *and*. It is a special kind of compound sentence. A system is often denoted by a brace. Thus, the compound sentence

$$y = 7x - 3 \text{ and } y = 6x + 2$$

can be written as the system $\begin{cases} y = 7x - 3 \\ y = 6x + 2. \end{cases}$

Systems with one or two variables can be represented algebraically (as shown above), graphically, or in a table.

The **solution set for a system** is the intersection of the solution sets for the individual sentences. When the system involves two equations each with two variables, the solution can often be found by making a table of values or a graph.

Finding Solutions with Tables or Graphs

Example 1

Solve the system $\begin{cases} y = 7x - 3 \\ y = 6x + 2. \end{cases}$

Solution 1

Make a table of values for each sentence.

x	y = 7x − 3	x	y = 6x + 2
0	-3	0	2
1	4	1	8
2	11	2	14
3	18	3	20
4	25	4	26
5	32	5	32

The pair of values $x = 5$ and $y = 32$ is in both tables. So $x = 5$, $y = 32$ is a solution to the system. However, we do not know if this is the only solution.

▶

Solution 2

Graph the system $\begin{cases} y = 7x - 3 \\ y = 6x + 2 \end{cases}$ by hand.

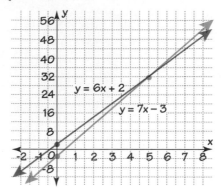

The solution to the system consists of all points of intersection of the two lines. The two lines intersect in only one point, so there is only one solution. The lines *seem to intersect* when x = 5 and y = 32.

Check

Substitute the point (5, 32) into both sentences.
Is $32 = 7 \cdot 5 - 3$? Yes.
Is $32 = 6 \cdot 5 + 2$? Yes. This tells us that the solution is exact.
The single solution is the point (5, 32).

Solutions to systems of equations can be described in three ways. For example, the solution to the system of Example 1 could be expressed in any of the following ways.

1. listing the solution: (5, 32)

2. writing the solution set: {(5, 32)}

3. writing a simplified equivalent system: $\begin{cases} x = 5 \\ y = 32 \end{cases}$

Using an Automatic Grapher to Estimate Solutions

Graphing a system can quickly indicate the number of solutions, but if the solutions are not integers, graphing may not give an exact answer. As Example 2 shows, using an automatic grapher gives more precise approximations.

Example 2

A family owns land on a straight stretch of the Oldman River. They plan to fence in a rectangular piece of land along the river. They have 80 meters of fencing material, and want to enclose an area of 500 square meters. What can the dimensions of this region be?

Solution

Draw a picture. Let W and S be the width and sides of the rectangle.

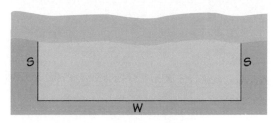

The length of the fence is one width plus two sides. So, $W + 2S = 80$. Since the area is 500 m^2, $WS = 500$. The system is

$$\begin{cases} W + 2S = 80 \\ \qquad WS = 500. \end{cases}$$

To graph, solve each equation for W.

$$\begin{cases} W = 80 - 2S \\ W = \dfrac{500}{S} \end{cases}$$

On an automatic grapher, if we let S be the independent variable, these equations become $y_1 = 80 - 2x$ and $y_2 = \dfrac{500}{x}$. Since W and S represent lengths, they must be positive. Therefore, it is not necessary to draw the third-quadrant branch of the hyperbola.

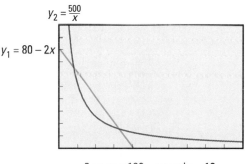

$$y_2 = \frac{500}{x}$$
$$y_1 = 80 - 2x$$

$0 \le x \le 100,\quad x\text{-scale} = 10$
$0 \le y \le 100,\quad y\text{-scale} = 10$

The graphs intersect at two points, so there are two solutions. Tracing and rounding to the nearest integer gives intersections near (8, 65) and near (32, 15). The family can make the width about 65 m and the other two sides about 8 m, or they can make the width about 15 m and the other two sides 32 m.

Check

Substitute $S = 8$ and $W = 65$ into both sentences.

$$65 + 2 \cdot 8 = 81 \; \approx 80$$
$$65 \cdot 8 = 520 \approx 500$$

The check shows that (8, 65) is an approximate solution. The check of (32, 15) is left to you in the Questions. The solution set has two ordered pairs, and is approximately {(8, 65), (32, 15)}.

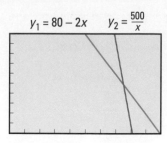

$y_1 = 80 - 2x$ $y_2 = \frac{500}{x}$

$7 \leq x \leq 8,$ x-scale = .1
$64 \leq y \leq 65,$ y-scale = .1

On an automatic grapher, you can estimate the solution of a system to a high degree of accuracy by *rescaling* or *zooming* in on each point of intersection. With a different window, our automatic grapher shows that a more accurate solution to the system in Example 2 is $S \approx 7.8$ and $W \approx 64.5$.

$$64.5 + 2(7.8) = 80.1 \approx 80$$
$$64.5 \cdot 7.8 = 503.1 \approx 500$$

To get a solution accurate to the nearest hundredth, you can zoom or rescale again.

QUESTIONS

Covering the Reading

1. *Multiple choice.* The solution to a system is
   (a) the union of the solution sets of the individual sentences.
   (b) the intersection of the solution sets of the individual sentences.
   (c) all points satisfying at least one of the individual sentences.

In 2 and 3, consider the system $\begin{cases} y = 3x + 2 \\ y = 20. \end{cases}$

2. *Multiple choice.* The system means
   (a) $y = 3x + 2$ and $y = 20$.
   (b) $y = 3x + 2$ or $y = 20$.

3. **a.** The solution to the system is ___?___.
   b. Verify that the ordered pair you found in part a is the solution.

4. Find a solution to the system $\begin{cases} y = 3x + 6 \\ y = -2x + 16 \end{cases}$ by making a table for each equation for integer values of x from 0 to 5.

5. Solve the system $\begin{cases} y = \frac{1}{2}x - 5 \\ y = 2x - 1 \end{cases}$ by making a graph.

In 6–8, refer to Example 2.

6. *Multiple choice.* $WS = 500$ represents a relationship involving
   (a) perimeter.   (b) area.   (c) volume.

7. Why do you not need to draw the third-quadrant branch of the hyperbola?

8. **a.** Show that (32, 15) is an approximate solution to the system.
   b. Show that (32.25, 15.51) is a closer approximate solution to the system than (32, 15) is.

9. Describe one advantage and one disadvantage of graphing to find a solution to a system.

In 10–12, use the given systems and their graphs.
   a. Tell how many solutions the system has.
   b. Estimate the solutions, if there are any.
   c. Verify that your solutions satisfy all equations of the system.

10. $\begin{cases} y = x^2 \\ y = x - 5 \end{cases}$
   11. $\begin{cases} xy = 2 \\ 2x - y = 3 \end{cases}$
   12. $\begin{cases} y = x^2 \\ xy = 8 \end{cases}$

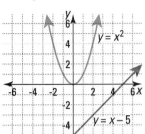
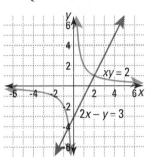
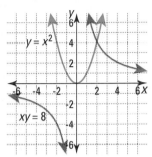

In 13 and 14 a system is given. **a.** Graph each system. **b.** Tell how many solutions the system has. **c.** Estimate any solutions to the nearest tenth.

13. $\begin{cases} y = \frac{1}{2}x^2 \\ x + y - 5 \end{cases}$
   14. $\begin{cases} y = \frac{3}{x^2} \\ y - x + 1 \end{cases}$

15. The system $\begin{cases} y = 5x + 40 \\ y = 9x \end{cases}$ could represent the following situation:

A child challenged her father to a race. The father gave her a head start of 40 m. He ran at 9 meters per second. She ran at 5 meters per second. Let y be distance and x be time in seconds, after the father started.
   a. What equation represents the father's distance from the start after x seconds?
   b. After 1 second, how far was the father from the start? How far was the daughter?
   c. When did the father catch up to his daughter?
   d. When the father caught up to her, how far from the start were they?

16. Use a graph to show that there is no pair of real numbers x and y whose product is 30 and whose sum is 10.

17. Graphing can be the first step in helping to solve more complicated systems by search procedures. Consider the system below.

$$\begin{cases} y = 3x^2 \\ y = 4x + 10 \end{cases}$$

A rough graph shows a solution between $x = -2$ and $x = 0$.
   a. Generate a table or use the zoom features of your automatic grapher to approximate this solution to two decimal places.
   b. The other solution is between $x = 2$ and $x = 3$. Find this solution, correct to two decimal places.

$y_1 = 3x^2$

$y_2 = 4x + 10$

$-4 \le x \le 4, \quad x\text{-scale} = 1$
$-20 \le y \le 40, \quad y\text{-scale} = 10$

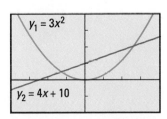

18. Graph $\{x: -0.5 \le x < 4\}$ on a number line. *(Lesson 5-1)*

19. **a.** Graph $\{x: x > 4 \ or \ x > -2\}$.
b. Graph $\{x: x > 4 \ and \ x > -2\}$. *(Lesson 5-1)*

20. Given $a_n = 3 - 5n$, find all n for which $a_n \le -102$.
(Lessons 1-7, 3-8, 5-1)

21. Children in northeastern Tanzania sometimes play a game called *tarumbeta* involving stones or beans arranged in rows like those shown below. Each row has one more stone than the row above it. Older children play the game using more stones.

a. How many stones are needed to form a tarumbeta game with 9 rows?
b. Find a formula, either explicit or recursive, for the number S_n of stones in a game with n rows. *(Lessons 1-7, 1-9, 3-8)*

22. An equation relating the total surface area T of a cylinder with height h and radius r is $T = 2\pi r^2 + 2\pi rh$. Solve this equation for h. *(Lesson 1-6)*

23. Suppose two figures of the indicated type have equal area. Must they be congruent? Explain your reasoning. *(Previous course)*
a. squares **b.** triangles

24. Solve $x^2 = 121$. *(Previous course)*

Shown are school children in Tanzania, a country in Africa with a population of about 31 million.

Exploration

25. Consider both branches of the hyperbola with equation $y = \frac{1}{x}$. Is there a line that intersects this hyperbola in exactly one point? If there is, give an equation for one such line. If not, explain why not.

Solving Systems by Substitution

Making tables and graphing enables you to find solutions to systems, but sometimes these methods do not give exact solutions. To find exact solutions, you usually need to use algebraic techniques. The next several lessons discuss algebraic techniques for solving systems of equations.

The first technique uses the Substitution Property of Equality, which states that if $a = b$, then a may be substituted for b in any arithmetic or algebraic expression. So, if $y = 2 - 10x$, for example, you can substitute $2 - 10x$ for y in any other expression. In this lesson we use the Substitution Property to solve systems with two and three equations.

Solving Systems with Two Linear Equations

Example 1

Solve the system $\begin{cases} 6x + 12y = 5 \\ \qquad y = 2 - 10x. \end{cases}$

Solution

Notice that the second equation is solved for y. If you substitute $2 - 10x$ for y in the first equation, the resulting equation will have only one variable. So you can solve it for x.

$6x + 12(2 - 10x) = 5$ Substitute $(2 - 10x)$ for y.

$6x + 24 - 120x = 5$ Apply the Distributive Property.

$-114x = -19$ Add -24 to both sides and add like terms.

$x = \dfrac{1}{6}$

Now substitute this value for x back into either equation to find y. We use the second equation because it is already solved for y.

$$y = 2 - 10\left(\dfrac{1}{6}\right)$$

$$y = 2 - \dfrac{10}{6}$$

$$y = \dfrac{2}{6} = \dfrac{1}{3}$$

So, the solution set is $\left\{\left(\dfrac{1}{6}, \dfrac{1}{3}\right)\right\}$.   ▶

Check

Substitute $\frac{1}{6}$ for x and $\frac{1}{3}$ for y in the first equation.

Does $6\left(\frac{1}{6}\right) + 12\left(\frac{1}{3}\right) = 5$? Yes, $1 + 4 = 5$.

Solving Systems with Three or More Linear Equations

Substitution can also be used when there are more than two variables and two conditions, as Example 2 illustrates.

Example 2

Suppose a part of a stadium has a seating capacity of 4216. There are four times as many lower-level seats as there are upper-level seats. Also, there are three times as many mezzanine seats as there are upper-level seats. How many seats of each type are there?

Solution

Let L = the number of lower-level seats,
    M = the number of mezzanine seats, and
    U = the number of upper-level seats.

Then the system to be solved is

$$\begin{cases} L + M + U = 4216 \\ \quad\quad\quad L = 4U \\ \quad\quad\quad M = 3U. \end{cases}$$

Substitute the expressions for L and M given in the last two equations into the first equation.

$$4U + 3U + U = 4216$$
$$8U = 4216$$
$$U = 527$$

Substitute to find L. $L = 4 \cdot 527 = 2108$

Substitute to find M. $M = 3 \cdot 527 = 1581$

There are 527 upper-level seats, 2108 lower-level seats, and 1581 mezzanine seats.

Check

There are 4216 seats. $527 + 2108 + 1581 = 4216$. It checks.

Solving Nonlinear Systems

You may also use the Substitution Property of Equality with a system that has one or more nonlinear equations. Write one equation in terms of a single variable, and substitute the expression into the other equation.

Example 3

Solve the system $\begin{cases} y = 3x \\ xy = 48. \end{cases}$

World's largest soccer stadium. *Maracaña Municipal Stadium in Rio de Janeiro, Brazil has a capacity of 205,000. In 1950, the World Cup final between Brazil and Uruguay drew a record crowd of 199,854. A dry moat, 3 meters wide and more than 1.5 meters deep, separates the players from the spectators.*

▶

▶ **Solution**

The first equation is already solved for y, so substitute $3x$ for y in the second equation.

$$x(3x) = 48$$
$$3x^2 = 48$$
$$x^2 = 16$$
$$x = 4 \text{ or } x = -4$$

(Note the word *or*. The solution set is the union of all possible answers.) Each value of x yields a value of y. Substitute each value of x into either of the original equations. We substitute into $y = 3x$. If $x = 4$, then $y = 3(4) = 12$. If $x = -4$, then $y = 3(-4) = -12$. The solution set is $\{(4, 12), (-4, -12)\}$.

Check 1

Substitute the coordinates of each point into each equation. For (4, 12), does $12 = 3 \cdot 4$? Yes. Does $4 \cdot 12 = 48$? Yes. In the Questions, you are asked to check the other point.

Check 2

Graph the equations. The graph below shows that there are two solutions. One solution seems near (4, 12); the other near (-4, -12).

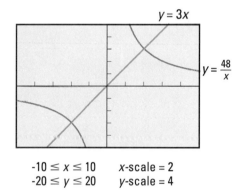

$-10 \leq x \leq 10$ x-scale = 2
$-20 \leq y \leq 20$ y-scale = 4

Examples 1 to 3 illustrate that substitution may be an appropriate method when at least one of the following applies.

1. At least one of the equations has been or can easily be solved for one of the variables.
2. There are two or more linear equations and two or more variables.
3. The system has one or more nonlinear equations.

Consistent and Inconsistent Systems

Systems are classified into two groups depending on whether or not solutions exist. A system which has *one or more solutions* is called a **consistent** system. Examples 1, 2, and 3 involve consistent systems. A system which has no solutions is called an **inconsistent** system. Example 4 illustrates an inconsistent system.

Example 4

Solve the system $\begin{cases} x = 3 - 2y \\ 3x + 6y = 6. \end{cases}$

Solution 1

Substitute $3 - 2y$ for x in the second equation.

$$3(3 - 2y) + 6y = 6$$
$$9 - 6y + 6y = 6 \quad \text{Use the Distributive Property.}$$
$$9 = 6 \quad \text{Add like terms.}$$

What's going on here? We came up with a statement that is never true! This statement indicates that **The system has no solutions.**

Solution 2

Graph this system. We use the intercepts of each line. Notice that the lines are parallel. (If we check the slopes, we see that they are both $-\frac{1}{2}$). Therefore, there are no intersections, and so **The system has no solutions.**

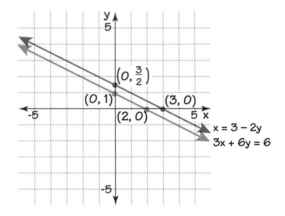

Example 5 illustrates a consistent system with infinitely many solutions.

Example 5

Solve the system $\begin{cases} y = 6 - \frac{x}{6} \\ \frac{x}{3} + 2y = 12. \end{cases}$

Solution 1

Substitute $6 - \frac{x}{6}$ for y in the second equation.

$$\frac{x}{3} + 2\left(6 - \frac{x}{6}\right) = 12$$
$$\frac{x}{3} + 12 - \frac{x}{3} = 12$$
$$12 = 12$$

This statement is *always* true. Thus, this system has an *infinite number of solutions.* **The solutions are all ordered pairs satisfying either equation.**

▶

Solution 2

Graph the system. Notice that the two graphs are the same line.

Graphs of $y = 6 - \frac{x}{6}$ and $\frac{x}{3} + 2y = 12$

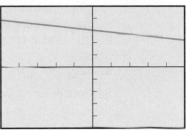

| $-10 \le x \le 10$ | x-scale = 2 |
| $-10 \le y \le 10$ | y-scale = 2 |

QUESTIONS

Covering the Reading

1. Write the Substitution Property of Equality.

2. Solve the following system using substitution.
$$\begin{cases} y = 3x + 5 \\ 4x - 3y = 12 \end{cases}$$

In 3 and 4, refer to Example 2.

3. After the expressions in the second and third equations were substituted into the first equation, how many variables were in this new equation?

4. A second part of the stadium was built to the same specifications but has a seating capacity of 6904. How many upper-level seats are there?

5. Solve $\begin{cases} 3x + 2y + z = 24 \\ \quad x = 5z - 20 \\ \quad y = -2z. \end{cases}$

6. Verify that $(-4, -12)$ is a solution to the system of Example 3.

7. **a.** Solve the system $\begin{cases} y = 5x \\ xy = 500. \end{cases}$

   b. Is the system consistent?

In 8–10, a system is given. **a.** Graph the system. **b.** Tell whether the system is consistent or inconsistent. **c.** Solve the system using substitution.

8. $\begin{cases} y = 4 - 2x \\ y = \frac{1}{2}(8x - 12) \end{cases}$

9. $\begin{cases} y = 3x - 1 \\ y = \frac{1}{2}(6x - 2) \end{cases}$

10. $\begin{cases} y = 3x - 1 \\ y = \frac{1}{2}(6x - 8) \end{cases}$

11. Follow the directions of Questions 8–10 for the system $\begin{cases} y = x^2 \\ y = x. \end{cases}$

12. A sports stadium seats 60,000 people. The home team gets 4 times as many tickets as the visiting team. Let H be the number of tickets for the home team and V be the number of tickets for the visiting team.

    a. *Multiple choice.* Which system represents the given conditions?

    (i) $\begin{cases} 4H + 4V = 60,000 \\ H = 4V \end{cases}$ (ii) $\begin{cases} H + V = 60,000 \\ V = 4H \end{cases}$ (iii) $\begin{cases} H = 4V \\ H + V = 60,000 \end{cases}$

    b. Solve the correct system for H and V.

13. FASTPIC offers to process a roll of film for 30¢ per print with free developing. A competitor, QUALIPRINT, will process a roll for 25¢ per print plus a $2.00 developing charge. Let x = the number of prints made and y = the cost of making x prints.

    a. Set up a system of two equations to describe this situation.

    b. For what number of prints will the cost be the same at FASTPIC and QUALIPRINT?

    c. What is the cost for this number of prints?

14. A recipe which makes 7 cups of French dressing uses tomato juice, vinegar, and olive oil. It calls for 3 times as much vinegar as tomato juice and $4\frac{1}{2}$ times as much olive oil as vinegar.

    a. Set up a system of three equations to describe this situation.

    b. Solve the system to determine how much of each ingredient should be used.

15. The Buy-A-Gadget company pays sales representatives on a commission basis. The weekly salary P (in dollars) depends on the number n of gadgets sold. When Matt started with the company, the pay scale was $P = 200 + 2n$. In order to encourage higher sales, management decided to change to an incentive scale where $P = 100 + 3n$.

    a. Graph the original pay scale and the incentive scale on the same axes.

    b. At what number of gadgets sold do the two scales pay the same wage?

    c. Interpret the graph in terms of which sales representatives benefit from the incentive plan and which don't. Is anyone hurt by the new plan?

    d. If Matt sells 150 gadgets the first week of June, what would his pay be under the old plan and under the incentive plan?

    e. Explain why the new pay scale is called an incentive plan.

16. Graph on a number line.

    a. $x > 2$

    b. $-2 \le x$ and $2 > x$

    c. $-2 \le x$ and $2 < x$ *(Lesson 5-1)*

In 17 and 18, multiply.   *(Lesson 4-3)*

17. $\begin{bmatrix} 2 & 4 \\ -3 & 1 \end{bmatrix}\begin{bmatrix} 7 & -1 & 0 \\ 11 & 4 & -1 \end{bmatrix}$

18. $\begin{bmatrix} a & b \\ c & d \end{bmatrix}\begin{bmatrix} x \\ y \end{bmatrix}$

19. Kathy has $5 to buy apples and pears for her volleyball team. Let a = number of apples and p = the number of pears.
   a. If apples are 40¢ each and pears are 50¢ each, write an equation that expresses the combinations of fruits she can get for $5.
   b. List three possible ways for Kathy to spend the entire $5.00.
   (Lesson 3-3)

20. Philip David Bayusun fills the bathtub slowly at a constant rate. He turns off the water, then gets in the tub and bathes. After a few minutes he gets out of the tub and pulls the plug. The water drains quickly. Which of the graphs below shows the relation between the height h of water in the tub and time t?   *(Lesson 3-1)*

(a)

(b)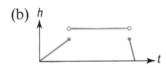

(c)

(d)

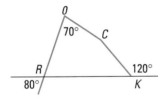

21. Find the measures of all four angles of quadrilateral *ROCK* drawn at the left.   *(Previous course)*

Exploration

22. **a.** Choose an inconsistent system, such as the one in Example 4. Set up a table with columns for *x*- and *y*-values and the differences of the two *y*-values, such as the one started below. Does the difference column surprise you? Why or why not?

x	$y_1 = -.5x + 1.5$	$y_2 = -.5x + 1$	$y_1 - y_2$
0	1.5	1	.5
1	1	.5	.5
2			

   b. Make a similar table for the system in Example 5 of this lesson. Explain how the difference column relates to the number of solutions of the system.
   c. Make a similar table for a consistent system that has exactly one solution. How does the difference column tell you what the solution to the system is?

5-4

Solving Systems Using Linear Combinations

In the last lesson, you solved systems using the Substitution Property. To use this method, one of the variables must be solved in terms of the other variables. However, when linear equations are written in standard form, it may be more efficient to solve the system using the Addition and Multiplication Properties of Equality.

Linear Combinations with Systems of Two Equations

Example 1

Solve the system $\begin{cases} x + y = 9 \\ 2x - y = 2. \end{cases}$

Solution

Notice that the coefficients of y are 1 and -1, which add to zero. Adding the sides of the two equations gives an equation in one variable.

$$\begin{array}{rl} x + y = 9 & \\ \underline{2x - y = 2} & \\ 3x = 11 & \text{Addition Property of Equality} \end{array}$$

Thus $\quad x = \frac{11}{3}$ or $3\frac{2}{3}$.

Substitute the value of x into either of the two original equations, and solve for y. We choose the first equation.

$$\frac{11}{3} + y = 9$$

$$y = \frac{16}{3}$$

The solution is $\left(\frac{11}{3}, \frac{16}{3}\right)$.

Check 1

Graph the lines $x + y = 9$ and $2x - y = 2$. Use an automatic grapher's trace feature to determine whether the points of intersection are close to $\left(\frac{11}{3}, \frac{16}{3}\right) = (3.\overline{6}, 5.\overline{3})$. They are.

Check 2

Verify that $\left(\frac{11}{3}, \frac{16}{3}\right)$ satisfies each of the given sentences.

Does $\frac{11}{3} + \frac{16}{3} = 9$? Yes.

Does $2 \cdot \frac{11}{3} - \frac{16}{3} = 2$? Yes, so it checks.

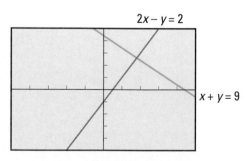

$2x - y = 2$

$x + y = 9$

-10 ≤ x ≤ 10, x-scale = 2
-10 ≤ y ≤ 10, y-scale = 2

Systems arise in important practical endeavors. During World War II, the United States and many other countries had to ration certain foods. Mathematicians were involved in determining diets that met the minimum requirements of protein, vitamins, and minerals. Here is an example. Notice that we use both the Multiplication Property of Equality and the Addition Property of Equality in its solution.

Example 2

The table shows the protein and calcium contents for a dinner of roast beef and mashed potatoes. How many servings of each are needed to get 29 grams of protein and 61 milligrams of calcium?

	Roast Beef	Mashed Potatoes
protein (g) per serving	25	2
calcium (mg) per serving	11	25

Solution

Let r = the number of servings of roast beef and m = the number of servings of mashed potatoes.
Then for protein, the servings must satisfy $25r + 2m = 29$ and for calcium, the servings must satisfy $11r + 25m = 61$.
The system to be solved is $\begin{cases} 25r + 2m = 29 \\ 11r + 25m = 61. \end{cases}$

To use the addition method of Example 1, the coefficients of the variables must be opposites. Notice that the least common multiple of 2 and 25 (the coefficients of m) is 50. Using the Multiplication Property of Equality, multiply the first equation by 25 and the second equation by -2 to get equivalent equations with opposite coefficients for m.

$25r + 2m = 29$ Multiply by 25 to get: $625r + 50m = 725$
$11r + 25m = 61$ Multiply by -2 to get: $\underline{-22r - 50m = -122}$
Then add the resulting equations. $603r \qquad\quad = 603$
Solve for r. $r \qquad = 1$
Now substitute $r = 1$ into either equation and solve for m.
We substitute into $25r + 2m = 29$.

$$25(1) + 2m = 29$$
$$m = 2$$

So you need 1 serving of roast beef and 2 servings of potatoes to get the required amounts of protein and calcium.

Check

Substitute $r = 1$ and $m = 2$ into the other sentence, $11r + 25m = 61$.
Does $11(1) + 25(2) = 61$? Yes.

In Example 2, we also could have found the least common multiple for the coefficients of r, which is 275. Then the first equation would be multiplied by 11, and the second equation would be multiplied by -25. Solving the system in this way would result in the same solution.

A rational plan. *Shown are ration coupons from World War II. During the war, the Office of Price Administration set up a rationing program to ensure that scarce goods were distributed fairly. Each family received a book of coupons to be used to purchase such goods as sugar, meat, butter, and gasoline.*

Recall that an expression of the form $Ax + By$ is called a *linear combination* of the two variables x and y. The method used in Examples 1 and 2 is often called the **Linear Combination Method** of solving systems because it involves adding multiples of the given equations.

Linear Combinations with Systems of Three Equations

The Linear Combination Method can also be used to solve a system with three linear equations. First, take any pair of equations and eliminate one of the variables; then take another pair of equations and eliminate the same variable. This process results in a system of two equations with two variables. Then the methods of Examples 1 and 2 can be used.

Example 3

Solve the system $\begin{cases} 2x + 3y + z = 13 \\ 5x - 2y - 4z = 7 \\ 4x + 5y + 3z = 25. \end{cases}$

Solution

We choose to eliminate z first. Consider the first two equations. Notice that the coefficients of z in these equations are 1 and -4. To eliminate z from these equations, multiply the first equation by 4 and add the result to the second equation.

$2x + 3y + z = 13$ Multiply by 4 to get: $8x + 12y + 4z = 52.$
$5x - 2y - 4z = 7$

Add. $\underline{5x - 2y - 4z = 7}$
$13x + 10y = 59$

Now eliminate z from another pair of equations. We choose to use the first and third equations. The least common multiple of 1 and 3, the coefficients of z, is 3.

$2x + 3y + z = 13$ Multiply by -3 to get: $-6x - 9y - 3z = -39.$
$4x + 5y + 3z = 25$

Add. $\underline{4x + 5y + 3z = 25}$
$-2x - 4y = -14$

We are now left with the following system to solve.

$$\begin{cases} 13x + 10y = 59 \\ -2x - 4y = -14 \end{cases}$$

Use the Linear Combination Method on this system. We eliminate y first.

$13x + 10y = 59$ Multiply by 2 to get: $26x + 20y = 118.$
$-2x - 4y = -14$ Multiply by 5 to get: $-10x - 20y = -70.$

Add. $\underline{}$
$16x = 48$
$x = 3$

Substitute $x = 3$ into one of the two equations involving just two variables. We substitute into $13x + 10y = 59$.

$$13(3) + 10y = 59$$
$$y = 2$$

Now substitute $x = 3$ and $y = 2$ into one of the original equations of the system. We use $2x + 3y + z = 13$.

$$2(3) + 3(2) + z = 13$$
$$z = 1$$

So The solution is $x = 3$, $y = 2$, and $z = 1$. This can be written as the ordered triple $(3, 2, 1)$.   ▶

► **Check**

Substitute $x = 3$, $y = 2$, and $z = 1$ into each equation.
Does $2(3) + 3(2) + 1 = 13$? Yes.
Does $5(3) - 2(2) - 4(1) = 7$? Yes.
Does $4(3) + 5(2) + 3(1) = 25$? Yes.

The equations in the examples are consistent, and each has a unique solution. However, if you use the Linear Combination Method and get a statement such as $0 = -36$, which is always false, the original system is inconsistent and has no solutions.

If, after applying the Linear Combination Method with all the given equations, you get a statement such as $0 = 0$, which is always true, the original system is consistent, and there are infinitely many solutions.

QUESTIONS

Covering the Reading

1. Refer to Example 1. The equation $3x = 11$ is the result of __?__ the two original equations.

2. Solve the system $\begin{cases} a + b = 14 \\ 3a - b = 50. \end{cases}$

3. Refer to Example 2. Multiply the first equation by 11 and the second equation by -25. Solve the resulting system.

4. The table below gives the number of grams of protein and fat in one serving of two foods.

	Beef Stew with Vegetables	Bread
protein (g) per serving	16	2
fat (g) per serving	11	1

Lynne wants to get 26 grams of protein and 17.5 grams of fat from one meal of beef stew and bread.
a. Let s = the number of servings of stew and b = the number of servings of bread. Write a system of equations that describes these conditions.
b. How many servings of each does Lynne need to eat?

In 5–8, use the Linear Combination Method to solve the system.

5. $\begin{cases} 5x - 6y = 3 \\ 2x + 12y = 12 \end{cases}$

6. $\begin{cases} a + b = \frac{1}{3} \\ a - b = \frac{1}{4} \end{cases}$

7. $\begin{cases} p + q = 4 \\ 5p + 6q = 7 \end{cases}$

8. $\begin{cases} 2x - y + 3z = 9 \\ 4x + 2y - 2z = 10 \\ x - y - z = -5 \end{cases}$

9. Refer to Example 3.

   a. Suppose you want to use the first and second equations to eliminate y. If you multiply the first equation by __?__ and the second equation by __?__, then when you add the two equations, you get an equation in x and z only.

   b. Suppose you want to use the second and third equations to eliminate y. If you multiply the second equation by __?__ and the third equation by __?__, then when you add the two equations, you get an equation in x and z only.

   c. The equations resulting from parts **a** and **b** form a system of two equations in two variables. Solve this system.

   d. Substitute the values of x and z from part **c** into one of the original three equations, and solve for y.

   e. *True or false.*   The method used in parts **a** to **d** gives the same solution as in Example 3.

10. Morris was solving a system of two equations and got $0 = 0$ after adding the equations together. This result means that Morris has what kind of system?

In 11 and 12, use the Linear Combination Method to determine whether the system is inconsistent or consistent.

11. $\begin{cases} 2x + 3y = 4 \\ 4x + 6y = 9 \end{cases}$

12. $\begin{cases} 2m + 3n = 4 \\ 4m + 6n = 8 \end{cases}$

Applying the Mathematics

13. At the zoo, Jay bought 3 slices of vegetable pizza and 1 small lemonade for $5.40. Terri paid $4.80 for 2 slices of vegetable pizza and 2 small lemonades. What is the cost of a small lemonade?

14. N mL of a 60% salt solution are mixed with S mL of an 80% salt solution. The result is 35 mL of a 72% salt solution.

   a. Write an equation relating N, S, and the total number of mL.

   b. The amount of salt in the 72% solution is 0.72(35), or 25.2 mL. Write an equation relating the amount of salt in the 60%, 80%, and 72% solutions.

   c. Solve the system represented by your answers in parts **a** and **b**. How many mL of the 60% and the 80% solutions are needed?

Worth its weight in salt. *Shown is a salt-harvesting machine in France collecting salt obtained from evaporated seawater.*

In 15 and 16, solve by the Linear Combination Method.

15. $\begin{cases} 5u + \quad 4v = \text{-}18 \\ 0.04u - 0.12v = 0.16 \end{cases}$

16. $\begin{cases} 9x^2 - 6y^2 = 291 \\ 3x^2 + 2y^2 = 197 \end{cases}$

17. While checking its records, a publishing company noted the following costs for publishing books.

Job 1: 60 hr for design, 100 hr for editing, 200 hr for production, total cost = $23,000

Job 2: 30 hr for design, 300 hr for editing, 400 hr for production, total cost = $49,500

Job 3: 40 hr for design, 80 hr for editing, 150 hr for production, total cost = $17,400

How much did the company charge per hour for design, how much for editing, and how much for production?

Designing books. *Shown is Steve Curtis (center), designer of this book, meeting with some of the ScottForesman editorial staff. Designers deal with such issues as how copy is positioned on a page, artwork, and photos.*

Review

18. **a.** Solve the system $\begin{cases} y = 200 - 2x \\ y = 150 + 1.5x \end{cases}$ by substitution.

   b. Make up a question that could be answered by solving this system. *(Lesson 5-3)*

19. *Multiple choice.* A system whose graph is two different parallel lines ___?___ has a solution. *(Lesson 5-3)*
   (a) always   (b) sometimes   (c) never

20. Consider the system $\begin{cases} y = \frac{1}{2}x^2 \\ y = \frac{1}{3}x + 1. \end{cases}$

   a. Graph the solution set of each sentence.
   b. Approximate the solutions to the system. *(Lesson 5-2)*

21. Consider the arithmetic sequence 112, 120, 128, 136, ...
   a. Find a recursive formula.
   b. Find an explicit formula. *(Lessons 3-7, 3-8)*

In 22 and 23, find the product. *(Lesson 4-3)*

22. $\begin{bmatrix} 4 & 7 \\ 2 & 1 \end{bmatrix}\begin{bmatrix} -.1 & .7 \\ .2 & -.4 \end{bmatrix}$

23. $\begin{bmatrix} 1 & 0 & 0 \\ 0 & 1 & 0 \\ 0 & 0 & 1 \end{bmatrix}\begin{bmatrix} 7 & 8 & 9 \\ \sqrt{5} & \sqrt{6} & \sqrt{7} \\ -1 & 0 & 1 \end{bmatrix}$

24. Find the multiplicative inverse of each number. *(Previous course)*
   a. 4   **b.** $-\frac{2}{3}$ **c.** a (when $a \neq 0$)   **d.** 0

Exploration

25. This problem was made up by the Indian mathematician Mahavira and dates from about 850 A.D. "The price of nine citrons and seven fragrant wood apples is 107; again, the mixed price of seven citrons and nine fragrant wood apples is 101. Oh you arithmetician, tell me quickly the price of a citron and a wood apple here, having distinctly separated these prices well." At this time algebra had not yet been developed. How could this question be answered by someone without using algebra?

*Matrices
and
Inverses*

IN-CLASS
ACTIVITY

Materials: You will need technology that can perform matrix operations.
Work on this Activity with a partner.

1 **a.** Plot quadrilateral *ABCD* with *A* = (-2, 1), *B* = (1, 1),
     C = (1, -2), and *D* = (-2, -2).
b. Plot *A'B'C'D'*, the image of *ABCD* under $S_{2,5}$.
c. Plot *A"B"C"D"*, the image of *A'B'C'D'* under $S_{\frac{1}{2},\frac{1}{5}}$.
d. Describe how *A"B"C"D"* and *ABCD* are related.

2 **a.** Write the matrix for $S_{2,5}$.
     b. Write the matrix for $S_{\frac{1}{2},\frac{1}{5}}$.
c. Find the product of the matrices in parts **a** and **b**.
d. What is the matrix in part **c** called?

3 Two matrices whose product is the identity matrix are called
     inverse matrices. Most graphics calculators and some computer
programs can find the inverse of some matrices. Check with your
teacher, a friend, or the manual for your technology.
a. Enter the matrix for $S_{2,5}$ on a calculator.
b. Use the calculator to find the inverse of the matrix in part **a**.
c. What transformation does the matrix in part **b** represent?

4 **a.** Write the matrix for some scale change other than $S_{2,5}$ and $S_{\frac{1}{2},\frac{1}{5}}$.
     b. Use the calculator to find the inverse of the matrix in part **a**.
c. What transformation does the matrix in part **b** represent?
d. Do the transformations represented by the matrices in parts **a** and **b**
undo each other? How can you tell?

So far in this chapter, you have seen three methods for solving systems: graphing, substitution, and linear combination. A fourth method involves matrices and their inverses.

In the Activity on page 298, you saw that the scale-change matrix $\begin{bmatrix} 2 & 0 \\ 0 & 5 \end{bmatrix}$ has an inverse $\begin{bmatrix} \frac{1}{2} & 0 \\ 0 & \frac{1}{5} \end{bmatrix}$ because

$$\begin{bmatrix} 2 & 0 \\ 0 & 5 \end{bmatrix} \begin{bmatrix} \frac{1}{2} & 0 \\ 0 & \frac{1}{5} \end{bmatrix} = \begin{bmatrix} 1 & 0 \\ 0 & 1 \end{bmatrix} \text{ and}$$

$$\begin{bmatrix} \frac{1}{2} & 0 \\ 0 & \frac{1}{5} \end{bmatrix} \begin{bmatrix} 2 & 0 \\ 0 & 5 \end{bmatrix} = \begin{bmatrix} 1 & 0 \\ 0 & 1 \end{bmatrix}.$$

What Are Inverse Matrices?

In general, 2×2 matrices M and N are **inverse matrices** if and only if their product is the 2×2 identity matrix for multiplication, that is, if and only if

$$MN = NM = \begin{bmatrix} 1 & 0 \\ 0 & 1 \end{bmatrix}.$$

All the transformation matrices you saw in Chapter 4 have inverses. But not all matrices represent transformations, and not all matrices have inverses. In general, only **square matrices,** those with the same number of rows and columns, can have inverses.

When a matrix M represents a transformation, its inverse represents a transformation that undoes the effect of M. For instance, $\begin{bmatrix} 4 & 0 \\ 0 & 3 \end{bmatrix}$ is the matrix for the scale change $S_{4,3}$, and $\begin{bmatrix} \frac{1}{4} & 0 \\ 0 & \frac{1}{3} \end{bmatrix}$ is the matrix for the scale change $S_{1/4,1/3}$. As shown below, $S_{1/4,1/3}$ undoes the effect of $S_{4,3}$ on $\triangle ABC$. The product of the matrices for $S_{4,3}$ and $S_{1/4,1/3}$ is the identity matrix. The composite of the transformations is the identity transformation, and so the final image is identical to the preimage.

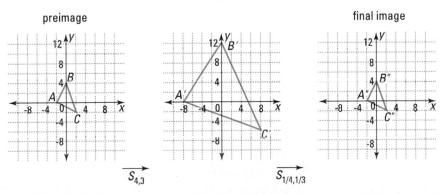

In the Activity you saw two other pairs of inverse scale-change matrices.

Example 1

Verify that the inverse of $\begin{bmatrix} 2 & 5 \\ 3 & 8 \end{bmatrix}$ is $\begin{bmatrix} 8 & -5 \\ -3 & 2 \end{bmatrix}$.

Solution

We multiply the matrices. The product is the identity matrix.

$$\begin{bmatrix} 8 & -5 \\ -3 & 2 \end{bmatrix}\begin{bmatrix} 2 & 5 \\ 3 & 8 \end{bmatrix} = \begin{bmatrix} 8(2) + -5(3) & 8(5) + -5(8) \\ -3(2) + 2(3) & -3(5) + 2(8) \end{bmatrix} = \begin{bmatrix} 1 & 0 \\ 0 & 1 \end{bmatrix}$$

Activity 1

Multiply the matrices in reverse order. Do you get the identity matrix?

Recall that the multiplicative inverse of a real number x is sometimes written as x^{-1}. Similarly, the multiplicative inverse of a matrix M can be written as M^{-1}. Thus, from Example 1 we can write

$$\begin{bmatrix} 2 & 5 \\ 3 & 8 \end{bmatrix}^{-1} = \begin{bmatrix} 8 & -5 \\ -3 & 2 \end{bmatrix}.$$

Finding the Inverse of a 2 × 2 Matrix

The following theorem tells when an inverse matrix exists, and gives you a formula to find it.

Inverse Matrix Theorem

If $ad - bc \neq 0$, the inverse of $\begin{bmatrix} a & b \\ c & d \end{bmatrix}$ is $\begin{bmatrix} \frac{d}{ad - bc} & \frac{-b}{ad - bc} \\ \frac{-c}{ad - bc} & \frac{a}{ad - bc} \end{bmatrix}$.

Proof:

We need to show that the product of the two matrices in either order is the identity matrix. Below, we show one order. In the Questions, you are asked to verify the multiplication in the reverse order.

$$\begin{bmatrix} a & b \\ c & d \end{bmatrix}\begin{bmatrix} \frac{d}{ad - bc} & \frac{-b}{ad - bc} \\ \frac{-c}{ad - bc} & \frac{a}{ad - bc} \end{bmatrix} = \begin{bmatrix} \frac{ad}{ad - bc} + \frac{-bc}{ad - bc} & \frac{-ab}{ad - bc} + \frac{ab}{ad - bc} \\ \frac{cd}{ad - bc} + \frac{-cd}{ad - bc} & \frac{-bc}{ad - bc} + \frac{ad}{ad - bc} \end{bmatrix}$$

$$= \begin{bmatrix} \frac{ad - bc}{ad - bc} & \frac{0}{ad - bc} \\ \frac{0}{ad - bc} & \frac{ad - bc}{ad - bc} \end{bmatrix} = \begin{bmatrix} 1 & 0 \\ 0 & 1 \end{bmatrix}$$

The formula in the Inverse Matrix Theorem is programmed into most graphics calculators. However, when the numbers in a 2 × 2 matrix are integers, it is often quicker to calculate the inverse matrix by hand than to find it with a calculator.

Example 2

Find the inverse of the matrix $\begin{bmatrix} 0 & -2 \\ 2 & 1 \end{bmatrix}$.

Solution

Use the Inverse Matrix Theorem. In $\begin{bmatrix} 0 & -2 \\ 2 & 1 \end{bmatrix}$, $a = 0$, $b = -2$, $c = 2$, and $d = 1$. So $ad - bc = 0(1) - (-2)(2) = 4$.

Substitute into the formula to get

$\begin{bmatrix} \frac{1}{4} & \frac{2}{4} \\ \frac{-2}{4} & \frac{0}{4} \end{bmatrix}$, which can be written as $\begin{bmatrix} \frac{1}{4} & \frac{1}{2} \\ -\frac{1}{2} & 0 \end{bmatrix}$.

Check

Does $\begin{bmatrix} 0 & -2 \\ 2 & 1 \end{bmatrix} \begin{bmatrix} .25 & .5 \\ -.5 & 0 \end{bmatrix} = \begin{bmatrix} 1 & 0 \\ 0 & 1 \end{bmatrix} = \begin{bmatrix} .25 & .5 \\ -.5 & 0 \end{bmatrix} \begin{bmatrix} 0 & -2 \\ 2 & 1 \end{bmatrix}$?

Yes, it checks.

Activity 2

Verify Example 2 by storing the matrix $A = \begin{bmatrix} 0 & -2 \\ 2 & 1 \end{bmatrix}$ on a computer or calculator and using the technology to find A^{-1}.

Determinants and Inverses

The formula in the Inverse Matrix Theorem can be simplified by using scalar multiplication. When $ad - bc \neq 0$,

$$\begin{bmatrix} \dfrac{d}{ad - bc} & \dfrac{-b}{ad - bc} \\ \dfrac{-c}{ad - bc} & \dfrac{a}{ad - bc} \end{bmatrix} = \frac{1}{ad - bc} \begin{bmatrix} d & -b \\ -c & a \end{bmatrix}.$$

Not all square matrices have inverses. You can see that the inverse of $\begin{bmatrix} a & b \\ c & d \end{bmatrix}$ exists only if $ad - bc \neq 0$. Because the number $ad - bc$ determines whether or not a matrix has an inverse, it is called the **determinant** of the matrix. We abbreviate the word *determinant* as *det*. Thus, the Inverse Matrix Theorem can be written:

Let $M = \begin{bmatrix} a & b \\ c & d \end{bmatrix}$. $M^{-1} = \dfrac{1}{\det M} \begin{bmatrix} d & -b \\ -c & a \end{bmatrix}$ if and only if $\det M \neq 0$.

Example 3

Determine whether $\begin{bmatrix} 3 & 2 \\ -5 & 4 \end{bmatrix}$ has an inverse.

Solution

Evaluate the determinant of the matrix.

$$\det \begin{bmatrix} 3 & 2 \\ -5 & 4 \end{bmatrix} = 3 \cdot 4 - 2 \cdot -5 = 22 \neq 0.$$

So the matrix has an inverse.

Example 4

Explain why $\begin{bmatrix} 3 & 1 \\ 6 & 2 \end{bmatrix}$ does not have an inverse.

Solution 1

Suppose $\begin{bmatrix} 3 & 1 \\ 6 & 2 \end{bmatrix}$ had an inverse $\begin{bmatrix} e & f \\ g & h \end{bmatrix}$. Then by the definition of

inverses, $\begin{bmatrix} 3 & 1 \\ 6 & 2 \end{bmatrix}\begin{bmatrix} e & f \\ g & h \end{bmatrix} = \begin{bmatrix} 3e + 9 & 3f + h \\ 6e + 2g & 6f + 2h \end{bmatrix} = \begin{bmatrix} 1 & 0 \\ 0 & 1 \end{bmatrix}.$

This leads to two systems:

$\begin{cases} 3e + g = 1 \\ 6e + 2g = 0 \end{cases}$ and $\begin{cases} 3f + h = 0 \\ 6f + 2h = 1 \end{cases}.$

Both of these systems are inconsistent.
Therefore, the matrix can have no inverse.

Solution 2

$\det \begin{bmatrix} 3 & 1 \\ 6 & 2 \end{bmatrix} = 3 \cdot 2 - 1 \cdot 6 = 0$, so the matrix has no inverse.

Determinants were first used by the Japanese mathematician Seki Kowa in 1683, and independently ten years later by the German mathematician Gottfried Leibniz (1646–1716). Formulas for inverses of 3×3 or larger matrices exist, but are quite complicated. Calculators and computers can often find inverses of square matrices larger than 2×2.

Gottfried Leibniz (1646–1716)

QUESTIONS

Covering the Reading

1. a. If a and b are real numbers and multiplicative inverses of each other, what does ab equal?
   b. If M and N are 2×2 matrices and multiplicative inverses of each other, what does MN equal?

2. a. Find the inverse of $M = \begin{bmatrix} 5 & 0 \\ 0 & 3 \end{bmatrix}$.

b. What transformation does M represent?

c. What transformation does the inverse of M represent?

3. *True or false.* $\begin{bmatrix} 12 & 13 \\ 2 & 3 \end{bmatrix}$ and $\begin{bmatrix} .3 & -1.3 \\ -.2 & 1.2 \end{bmatrix}$ are inverses of each other. Justify your answer.

4. Verify the second part of the proof of the Inverse-Matrix Theorem. That is, show that the identity matrix is the product of the two matrices in the reverse order.

In 5 and 6, *true or false.*

5. Only square matrices have inverses.

6. All square matrices have inverses.

7. Give an expression for $\det \begin{bmatrix} a & b \\ c & d \end{bmatrix}$.

8. M is a matrix with nonzero determinant. What does M^{-1} denote?

9. If A is a 2×2 matrix and A^{-1} exists, what does $A^{-1}A$ equal?

10. Show the result you obtained for Activity 1 in this lesson.

11. What did you need to do to complete Activity 2 in this lesson?

In 12–14, a matrix is given. **a.** Find its determinant. **b.** Find its inverse, if it has one. **c.** Check your answer to part **b** by multiplying.

12. $\begin{bmatrix} 5 & 4 \\ 2 & 2 \end{bmatrix}$

13. $\begin{bmatrix} -1 & -3 \\ 4 & -8 \end{bmatrix}$

14. $\begin{bmatrix} \frac{1}{2} & \frac{1}{2} \\ \frac{1}{2} & \frac{1}{2} \end{bmatrix}$

Applying the Mathematics

15. Give an example of a matrix not mentioned in this lesson that does not have an inverse.

16. The inverse of a 2×2 matrix can also be found by solving a pair of systems. Here is how. If the inverse of $\begin{bmatrix} 0 & -2 \\ 3 & 1 \end{bmatrix}$ is $\begin{bmatrix} e & f \\ g & h \end{bmatrix}$, then $\begin{bmatrix} 0 & -2 \\ 3 & 1 \end{bmatrix}\begin{bmatrix} e & f \\ g & h \end{bmatrix} = \begin{bmatrix} 1 & 0 \\ 0 & 1 \end{bmatrix}$.

This yields the systems $\begin{cases} 0e - 2g = 1 \\ 3e + g = 0 \end{cases}$ and $\begin{cases} 0f - 2h = 0 \\ 3f + h = 1. \end{cases}$

a. Solve these two systems and determine the inverse matrix.

b. Check your answer to part **a** by finding the inverse matrix using the Inverse Matrix Theorem.

17. a. Find the inverse of the matrix for R_{90}.
   b. Explain the result to part a geometrically.

Review

18. Solve the system $\begin{cases} 5x + y + 6z = 3 \\ x - y + 10z = 9 \\ 5x + y - 2z = -9. \end{cases}$ *(Lesson 5-4)*

19. A chef reports that with 5 kilograms of flour, 12 loaves of bread and 6 pizza crusts can be made. With 2 kilograms of flour, 1 loaf of bread and 10 pizza crusts can be made. How much flour is needed to make one loaf of bread? How much is needed for one pizza crust?   *(Lesson 5-4)*

20. Alan Aska wanted to purchase a new air conditioner. One brand costs $540 to purchase and $20 a month to operate. A less efficient brand costs $320 to purchase and $24 a month to operate.
   a. Plot the costs over time of both brands on a single graph.
   b. What does the point of intersection of the graphs denote?
      (Lessons 3-1, 5-2, 5-3)

21. Solve $\begin{cases} y = 4x \\ 3x + 2y = 22. \end{cases}$ *(Lessons 5-2, 5-3)*

22. Multiply $\begin{bmatrix} 1 & -2 \\ 4 & 5 \end{bmatrix} \begin{bmatrix} x \\ y \end{bmatrix}$.   *(Lesson 4-3)*

23. During a flood watch, a group of volunteers filled sandbags at a rate of 500 sandbags an hour. In how many hours will the group have prepared at least 10,000 sandbags?   *(Lesson 5-1)*

Shown are volunteers filling sandbags during the Midwest floods of 1993.

Exploration

24. a. If $A = \begin{bmatrix} -7 & 4 \\ -9 & -4 \end{bmatrix}$ and $B = \begin{bmatrix} 3 & 3 \\ 0 & 3 \end{bmatrix}$, find det A, det B, and det (AB).
   b. Pick two other 2×2 matrices A and B with nonzero determinants. Find det A, det B, and det (AB).
   c. Generalize the results of parts **a** and **b**.

25. a. Find the area of the triangle with vertices $(0, 0)$, $(-3, 0)$, and $(-7, 8)$.
   b. Calculate $\frac{1}{2}$ det $\begin{bmatrix} -3 & -7 \\ 0 & 8 \end{bmatrix}$.
   c. Find the area of the triangle with vertices $(0, 0)$, $(5, 2)$, and $(4, 0)$.
   d. Calculate $\frac{1}{2}$ det $\begin{bmatrix} 5 & 4 \\ 2 & 0 \end{bmatrix}$.
   e. Make a conjecture based on parts **a–d**.
   f. Test your conjecture with another example.

Solving Systems Using Matrices

Taking stock. *Pictured are parts of the stock certificates an investor receives when purchasing stocks. Superimposed are stock prices. A large system of equations could be used to plan an investment portfolio. See Example 2.*

Using Matrices to Solve 2 × 2 Linear Systems

In the middle of the nineteenth century, the British mathematician Arthur Caylcy developed a way to solve systems of linear equations by using matrices. To see how this method works, we begin with an example. Notice that

$$\begin{bmatrix} 1 & 3 \\ 2 & -1 \end{bmatrix}\begin{bmatrix} x \\ y \end{bmatrix} = \begin{bmatrix} x + 3y \\ 2x - y \end{bmatrix}.$$

Thus, we can represent the system $\begin{cases} x + 3y = 22 \\ 2x - \ y = 2 \end{cases}$

by the matrix equation $\begin{bmatrix} 1 & 3 \\ 2 & -1 \end{bmatrix}\begin{bmatrix} x \\ y \end{bmatrix} = \begin{bmatrix} 22 \\ 2 \end{bmatrix}.$

This is called the **matrix form of a system.** The matrix $\begin{bmatrix} 1 & 3 \\ 2 & -1 \end{bmatrix}$ contains the coefficients of the variables, so it is called the **coefficient matrix.** The matrix $\begin{bmatrix} 22 \\ 2 \end{bmatrix}$ contains the constants on the right sides of the equations. It is called the **constant matrix.**

To solve a system in matrix form, multiply each side of the matrix equation by the inverse of the coefficient matrix.

Example 1

Use matrices to solve $\begin{cases} x + 3y = 22 \\ 2x - y = 2. \end{cases}$

Solution

Write the matrix form of the system.

$$\begin{bmatrix} 1 & 3 \\ 2 & -1 \end{bmatrix} \begin{bmatrix} x \\ y \end{bmatrix} = \begin{bmatrix} 22 \\ 2 \end{bmatrix}$$

Use the Inverse Matrix Theorem to find the inverse of the coefficient matrix.

The inverse of $\begin{bmatrix} 1 & 3 \\ 2 & -1 \end{bmatrix}$ is $\begin{bmatrix} \frac{-1}{-7} & \frac{-3}{-7} \\ \frac{-2}{-7} & \frac{1}{-7} \end{bmatrix} = \begin{bmatrix} \frac{1}{7} & \frac{3}{7} \\ \frac{2}{7} & \frac{-1}{7} \end{bmatrix}$.

Multiply both sides of the matrix equation by the inverse. Because matrix multiplication is not commutative, the inverse matrix must be *at the left on each side* of the equation.

$$\begin{bmatrix} \frac{1}{7} & \frac{3}{7} \\ \frac{2}{7} & \frac{-1}{7} \end{bmatrix} \begin{bmatrix} 1 & 3 \\ 2 & -1 \end{bmatrix} \begin{bmatrix} x \\ y \end{bmatrix} = \begin{bmatrix} \frac{1}{7} & \frac{3}{7} \\ \frac{2}{7} & \frac{-1}{7} \end{bmatrix} \begin{bmatrix} 22 \\ 2 \end{bmatrix}$$

Multiply the matrices.

$$\begin{bmatrix} 1 & 0 \\ 0 & 1 \end{bmatrix} \begin{bmatrix} x \\ y \end{bmatrix} = \begin{bmatrix} 4 \\ 6 \end{bmatrix}$$

The presence of the identity matrix verifies that the inverse matrix was calculated correctly. Thus

$$\begin{bmatrix} x \\ y \end{bmatrix} = \begin{bmatrix} 4 \\ 6 \end{bmatrix}.$$

So, x = 4 and y = 6.

Check

Substitute $x = 4$ and $y = 6$ into the original equations.
Does $4 + 3(6) = 22$? Does $2(4) - 6 = 2$? Yes, it checks.

In general, to solve the system $\begin{cases} ax + by = e \\ cx + dy = f \end{cases}$ using matrices, first rewrite the system as the matrix equation

$$\begin{bmatrix} a & b \\ c & d \end{bmatrix} \begin{bmatrix} x \\ y \end{bmatrix} = \begin{bmatrix} e \\ f \end{bmatrix}.$$

This equation is of the form $A \begin{bmatrix} x \\ y \end{bmatrix} = B$, where A is the coefficient matrix and B is the constant matrix. As long as $ad - bc \neq 0$, A^{-1} exists. To solve this equation, multiply each side by A^{-1}.

$$A^{-1} A \begin{bmatrix} x \\ y \end{bmatrix} = A^{-1} B$$

$$\begin{bmatrix} 1 & 0 \\ 0 & 1 \end{bmatrix} \begin{bmatrix} x \\ y \end{bmatrix} = A^{-1} B$$

▶

► So,
$$\begin{bmatrix} x \\ y \end{bmatrix} = A^{-1} B$$

The last equation shows that the solution to the system is the product of the inverse of the coefficient matrix and the constant matrix.

Example 2

Ku invested $25,000, some in a savings account and the rest in bonds. If the return on his savings account was 4% last year and the return on his bonds was 6%, how did Ku divide his investments if the total interest was $1300?

Solution 1

Let s be the amount invested in savings and b be the amount invested in bonds. Because the total invested was $25,000, one equation to be satisfied is

$$s + b = 25,000.$$

The interest on savings was $.04s$ and the interest on the bonds was $.06b$. Because the total interest was $1300, a second equation to be satisfied is

$$.04s + .06b = 1300.$$

Now write the matrix equation for this system.

$$\begin{bmatrix} 1 & 1 \\ .04 & .06 \end{bmatrix} \begin{bmatrix} s \\ b \end{bmatrix} = \begin{bmatrix} 25,000 \\ 1300 \end{bmatrix}$$

Find the inverse of the coefficient matrix. It is $\begin{bmatrix} \frac{.06}{.02} & \frac{-1}{.02} \\ \frac{-.04}{.02} & \frac{1}{.02} \end{bmatrix}$ or $\begin{bmatrix} 3 & -50 \\ -2 & 50 \end{bmatrix}$.

Multiply both sides of the matrix equation *on the left* by this matrix.

$$\begin{bmatrix} 3 & -50 \\ -2 & 50 \end{bmatrix} \begin{bmatrix} 1 & 1 \\ .04 & .06 \end{bmatrix} \begin{bmatrix} s \\ b \end{bmatrix} = \begin{bmatrix} 3 & -50 \\ -2 & 50 \end{bmatrix} \begin{bmatrix} 25,000 \\ 1300 \end{bmatrix}$$

$$\begin{bmatrix} 1 & 0 \\ 0 & 1 \end{bmatrix} \begin{bmatrix} s \\ b \end{bmatrix} = \begin{bmatrix} 10,000 \\ 15,000 \end{bmatrix}$$

So the solution is $s = 10,000$ and $b = 15,000$. Ku invested $10,000 in a savings account and $15,000 in bonds.

Solution 2

Set up the system $\begin{cases} s + b = 25,000 \\ .04s + .06b = 1300 \end{cases}$ as in Solution 1.

Use the Linear Combination Method to solve the system. (You are asked to do this in the Questions.)

Check

Does $10,000 + 15,000 = 25,000$? Yes.
Does $.04(10,000) + .06(15,000) - 1300$? Yes.

The Number of Solutions to a 2 × 2 Linear System

Matrices can be used to determine the number of solutions to a 2 × 2 linear system. Consider the system $\begin{cases} ax + by = e \\ cx + dy = f \end{cases}$. When the determinant $ad - bc$ of the coefficient matrix is not 0, the inverse of the coefficient matrix exists, and the system has exactly one solution.

When $ad - bc = 0$, the coefficient matrix has no inverse. Thus, when the determinant of the coefficient matrix is 0, the system has infinitely many solutions or none at all. This is summarized in the next theorem.

> **System-Determinant Theorem**
> A 2 × 2 system has exactly one solution if and only if the determinant of the coefficient matrix is *not* zero.

To determine whether there are no solutions or infinitely many solutions, find an ordered pair that satisfies one of the equations and test it in the other one. If it satisfies the other equation, the two equations represent the same line and there are infinitely many solutions. If the given ordered pair does not satisfy the other equation, the two lines are parallel and there are no solutions.

Example 3

How many solutions does the system $\begin{cases} 6x - 9y = 12 \\ 2x - 3y = 4 \end{cases}$ have?

Solution 1

The determinant is $ad - bc = 6 \cdot (-3) - (-9) \cdot (2) = 0$.
So either the two equations describe the same line, or they describe two parallel lines. Find an ordered pair that satisfies one equation. The point (2, 0) satisfies the first equation. Does it satisfy the second? Does $2 \cdot 2 - 3 \cdot 0 = 4$? Yes. Thus, the system has infinitely many solutions. The two equations describe the same line.

Solution 2

Use the Linear Combination Method. Multiply each side of the second equation by -3, and add the resulting equations.

$$6x - 9y = 12 \qquad\qquad 6x - 9y = 12$$
$$-3(2x - 3y) = -3 \cdot 4 \qquad \underline{-6x + 9y = -12}$$
$$0 = 0$$

There are infinitely many solutions.

Solution 3

Multiplying both sides of the first equation by $\frac{1}{3}$ gives the second equation. Thus, the two given equations are equivalent. So their graphs are identical, and the system has infinitely many solutions.

Using Matrices to Solve 3 × 3 Linear Systems

Matrices can be used to solve $n \times n$ linear systems whenever the inverse of the coefficient matrix exists. In the next example, we solve a system of three equations with three unknowns using 3×3 matrices.

The identity matrix for 3×3 matrices is $\begin{bmatrix} 1 & 0 & 0 \\ 0 & 1 & 0 \\ 0 & 0 & 1 \end{bmatrix}$. The calculation of the inverse of a 3×3 matrix by hand is complicated, so use a graphics calculator or computer software to find the inverse of a 3×3 matrix whenever it is needed.

Example 4

Solve the system $\begin{cases} x - y + 3z = 9 \\ x + 2z = 3 \\ 2x + 2y + z = 10. \end{cases}$

Solution

First, rewrite each equation in standard form if necessary. The second equation becomes $1 \cdot x + 0 \cdot y + 2 \cdot z = 3$. Then rewrite the system as a matrix equation. To the right of the solution steps below, we have given the general form of the steps.

$$\begin{bmatrix} 1 & -1 & 3 \\ 1 & 0 & 2 \\ 2 & 2 & 1 \end{bmatrix} \begin{bmatrix} x \\ y \\ z \end{bmatrix} = \begin{bmatrix} 9 \\ 3 \\ 10 \end{bmatrix} \qquad\qquad A \begin{bmatrix} x \\ y \\ z \end{bmatrix} = B$$

Multiply each side by the inverse of the coefficient matrix, then simplify. Use technology to obtain the inverse.

$$\begin{bmatrix} 1 & 0 & 0 \\ 0 & 1 & 0 \\ 0 & 0 & 1 \end{bmatrix} \begin{bmatrix} x \\ y \\ z \end{bmatrix} = \begin{bmatrix} 4 & -7 & 2 \\ 3 & 5 & -1 \\ -2 & 4 & -1 \end{bmatrix} \begin{bmatrix} 9 \\ 3 \\ 10 \end{bmatrix} \qquad A^{-1}A \begin{bmatrix} x \\ y \\ z \end{bmatrix} = A^{-1}B$$

You should find that

$$\begin{bmatrix} x \\ y \\ z \end{bmatrix} = \begin{bmatrix} 35 \\ -22 \\ -16 \end{bmatrix} \qquad\qquad \begin{bmatrix} x \\ y \\ z \end{bmatrix} = A^{-1}B$$

So the solution is x = 35, y = -22, and z = -16.

Check

Substitute $x = 35$, $y = -22$, and $z = -16$ into each equation of the system. You are asked to do this in the Questions.

Note that on many calculators and computers you can compute $A^{-1}B$ without actually finding the elements in matrix A^{-1}.

In general, to solve a linear system using matrices, first rewrite the system so that all equations are in standard form. Then set up the coefficient matrix as matrix A and the constant matrix as matrix B. The solution is given by $A^{-1}B$. If the system is 2×2 you can compute $A^{-1}B$ by hand. For larger systems you will probably need to use a graphics calculator or computer software.

Covering the Reading

In 1 and 2, refer to the system in Example 1.

1. What does the matrix $\begin{bmatrix} 1 & 3 \\ 2 & -1 \end{bmatrix}$ represent?

2. Noel found the inverse $\begin{bmatrix} \frac{1}{7} & \frac{3}{7} \\ \frac{2}{7} & \frac{-1}{7} \end{bmatrix}$ and multiplied as shown here.

$$\begin{bmatrix} 1 & 3 \\ 2 & -1 \end{bmatrix}\begin{bmatrix} \frac{1}{7} & \frac{3}{7} \\ \frac{2}{7} & \frac{-1}{7} \end{bmatrix}\begin{bmatrix} x \\ y \end{bmatrix} = \begin{bmatrix} 22 \\ 2 \end{bmatrix}\begin{bmatrix} \frac{1}{7} & \frac{3}{7} \\ \frac{2}{7} & \frac{-1}{7} \end{bmatrix}$$

Did Noel find a solution? Explain.

3. **a.** Rewrite the system $\begin{cases} ax + by = c \\ dx + ey = f \end{cases}$ in matrix form.

   b. What matrix in part **a** is the coefficient matrix for the system?

In 4 and 5, refer to Example 2.

4. Use the Linear Combination Method to finish Solution 2.

5. Suppose the return on Ku's savings account was 3% and the return on his bonds was 5%. How did Ku invest his money if he received a total of $1200 interest?

In 6 and 7, solve each system using matrices.

6. $\begin{cases} -8x - 3y = 10 \\ 4x + 6y = 5 \end{cases}$

7. $\begin{cases} 4x + y = 2 \\ 9x - 2y = 4 \end{cases}$

In 8 and 9, how many solutions does the system have? Justify your answer.

8. $\begin{cases} -10x + 15y = 30 \\ 4x + 6y = 12 \end{cases}$

9. $\begin{cases} -30p - 18q = 42 \\ 35p - 21q = 76 \end{cases}$

10. Complete the check of Example 4.

11. **a.** Solve the system $\begin{cases} 4x - 2y + 3z = 1 \\ 8x - 3y + 5z = 4 \\ 7x - 2y + 4z = 5 \end{cases}$ using matrices.

    b. Check your answer.

Applying the Mathematics

12. **a.** Solve this matrix equation: $\begin{bmatrix} 1 & 2 \\ 3 & 4 \end{bmatrix}\begin{bmatrix} w & x \\ y & z \end{bmatrix} = \begin{bmatrix} 5 & 6 \\ 7 & 8 \end{bmatrix}$

    b. What two systems does part **a** simultaneously solve?

13. A bicycle, three tricycles, and an unicycle cost $561. Seven bicycles and a tricycle cost $906. Five unicycles, two bicycles and seven tricycles cost $1758.
   a. Set up a system of equations that can be used to find the cost of each item.
   b. Use matrices to solve the system.

In 14 and 15, determine all values of n that satisfy the condition.

14. $\begin{cases} 2x + 4y = n \\ x + 2y = 7 \end{cases}$ has infinitely many solutions.

15. $\begin{cases} 4x - 6y = 5 \\ 2x + ny = 2 \end{cases}$ has no solution.

Review

Who needs two wheels?
Unicycles first appeared in the late 1800s as part of circus acts and similar entertainment.

16. If the determinant of a 2×2 matrix is 0, what can you conclude about the inverse of the matrix? *(Lesson 5-5)*

17. a. Find a 2×2 matrix whose determinant equals 1.
   b. Find the inverse of the matrix in part **a.** *(Lesson 5-5)*

In 18–20, solve using any method. *(Lessons 5-2, 5-3, 5-4, 5-5)*

18. $\begin{cases} y = |x| \\ y = 9 \end{cases}$ **19.** $\begin{cases} x + y = 10 \\ x + 2y = 3 \end{cases}$ **20.** $\begin{cases} 3x - 2y = z \\ x = y + 9 \\ z = 8x \end{cases}$

In 21 and 22, use the facts that whole milk is 4% butterfat, and some low fat milk is 1% butterfat.

21. If you mix one quart (32 oz) of each of these two types of milk, how many ounces of butterfat are in the result? *(Previous course, Lesson 3-3)*

22. How much of each type of milk must you mix to get one quart of milk that is 2% butterfat? *(Lessons 5-4, 5-5)*

23. Graph on a number line: $\{x: x > 11\} \cap \{x: x > 12\}$. *(Lesson 5-1)*

24. In the coordinate plane, graph the line with equation $x = -3$. *(Lesson 3-6)*

Exploration

25. a. Show how the *two* systems
   $\begin{cases} -3w + 4y = 5 \\ w + 2y = 0 \end{cases}$ $\begin{cases} -3x + 4z = 1 \\ x + 2z = 3 \end{cases}$
   can be rewritten as a single matrix equation with three 2×2 matrices.
   b. When can two 2×2 systems be rewritten as in part **a?**
   c. Can three 2×2 systems ever be rewritten as a single matrix equation? Explain your answer.

LESSON

5-7

Graphing Inequalities in the Coordinate Plane

Can you ride? *Ferris wheels, roller coasters, and other rides may have height restrictions for riders. See Example 2.*

As you learned in Lesson 5-1, solutions to compound sentences with inequalities involving only one variable can be graphed on a number line. Before studying compound sentences with inequalities with two or more variables, we must study the graphs of inequalities in the plane.

When a line is drawn in a plane, the line separates the plane into two distinct regions called **half-planes.** The line itself is the **boundary** of the two regions. The boundary does not belong to either half plane.

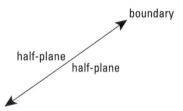

Inequalities with Horizontal or Vertical Boundaries

Inequalities with one variable can be graphed either on a number line or in the coordinate plane.

Example 1

Graph the solutions to $x > 5$
a. on a number line.   **b.** in the coordinate plane.

Solution

a. Plot all points on a number line with x-coordinate greater than 5. These are the points to the right of $x = 5$. Use an open circle to show that 5 is not included in the solution set.

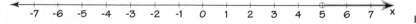

312

b. Plot all points in the coordinate plane with *x*-coordinate greater than 5. The line with equation *x* = 5 is the boundary for the half-plane that is the solution set. Draw a dashed line because the points on this line are not part of the solution set. The solution set consists of all points to the right of *x* = 5. It is shaded below.

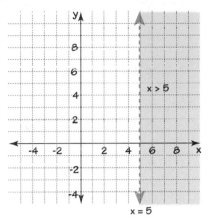

Check

b. Pick a point in the shaded region. We pick (7, 2). Its coordinates should satisfy *x* > 5. Is 7 > 5? Yes. So (7, 2) is in the solution set. If you pick a point in the unshaded region, its coordinates should not satisfy the inequality. We pick (1, 4). Is 1 > 5? No. So (1, 4) is not in the solution set.

Riding tall. Shown is a sign at a children's ride at an amusement park in Australia.

Some applications use these kinds of simple inequalities in the plane.

Example 2

A person's safety on many amusement park rides depends on the person's size rather than age. So someone may need to be 4 feet tall to go on a ride, regardless of age. Graph the constraint ''A person must be 4 feet tall or taller, regardless of age.'' Let *H* = the height of a person and *A* = the person's age.

Solution

Let *A* be the independent variable. Notice that age is always positive. So the solution is the set of points (*A*, *H*) such that *H* ≥ 4 and *A* > 0. These are the points in the first quadrant on or above the line *H* = 4. Since points on the line are in the solution set, draw the part of the line in the first quadrant with a solid line. Mark the points in the solution set as shown by the shaded region below.

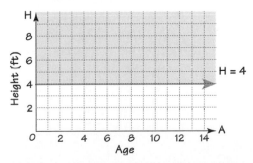

Half-Planes with Oblique Boundaries

All half-planes are easily described by inequalities. The half-planes determined by the line $y = mx + b$ are described by $y > mx + b$ and $y < mx + b$.

Example 3

Graph the linear inequality $y > 2x + 7$.

Solution

Step 1: Graph the boundary $y = 2x + 7$. These points must be connected with a dashed line because the boundary points do not satisfy the inequality. This line is shown below at the left.

Step 2: To determine which half-plane contains the solutions to the inequality, test a point not on the boundary. Usually $(0, 0)$ is an easy point to test. Does $(0, 0)$ satisfy $y > 2x + 7$? To find out, substitute $(0, 0)$ into $y > 2x + 7$. Is $0 > 2(0) + 7$? No. So $(0, 0)$ is not on the graph of $y > 2x + 7$. Thus the solutions to the inequality are all ordered pairs in the half-plane on the side of the line *not* containing $(0, 0)$. Shade the solution set as shown below at the right.

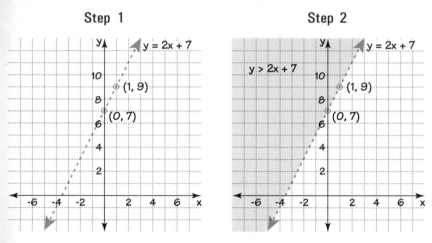

Check

Pick a point in the shaded region. We pick $(-3, 6)$. Do the coordinates satisfy $y > 2x + 7$? Is $6 > 2(-3) + 7$? Yes.

Notice that in Examples 1 to 3, the domain of all variables is the set of real numbers or the set of real numbers greater than a certain constant. In such cases, the graph of an inequality consists of all points in a region, which we indicate by shading. However, in some situations the domain of the variables consists of only integers. Then, the solution set consists of points whose coordinates are integers, called **lattice points.** If there are not too many lattice point solutions, you should indicate each with a dot on the plane, rather than with a shaded region.

Example 4

A ferryboat transports cars and buses across a river. It has space for 12 cars, and a bus takes up the space of 3 cars. Draw a graph showing all possible combinations of cars and buses that can be taken in one crossing.

Solution

This is a linear-combination situation. A car occupies 1 space, so x cars need x spaces. A bus occupies 3 spaces, so y buses need 3y spaces. The ferry has only 12 spaces, so a sentence describing the situation is

$$x + 3y \leq 12.$$

First note that $x \geq 0$ and $y \geq 0$. So only points in the first quadrant or its boundary rays need to be considered. Since x and y are integers, only lattice points are possible. Next, graph the boundary $x + 3y = 12$. Five points on this line, including (0, 4) and (12, 0), are lattice points. They satisfy the inequality.

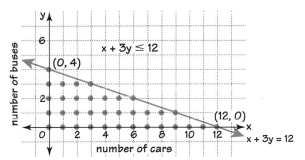

To determine the half-plane in which the solutions lie, try an ordered pair not on the boundary. Substituting (0, 0) gives $0 + 3(0) \leq 12$. This is true; so all solutions lie below $x + 3y = 12$. Thus the graph consists of only those points on or below the line $x + 3y = 12$ whose coordinates are whole numbers. The solutions are represented by the dots in the graph above. Thirty-five combinations of cars and buses can be taken by the ferry in one crossing.

Check

To check that the correct half-plane was chosen, pick a point on the *other* side of the boundary. It should not satisfy the inequality. We choose (10, 5). Substitute into the inequality. Is $10 + 3(5) \leq 12$? No. So the solution checks.

In summary, to graph a linear inequality in the coordinate plane:

1. Graph the appropriate boundary line, either a dashed or solid line, to the corresponding linear equation.
2. Test a point in one half-plane to see if the point satisfies the inequality. If it does not lie on the boundary, the point (0, 0) is often used.
3. Shade the half-plane that satisfies the inequality, or plot points if the situation is discrete.

Beats swimming.
Pictured is a large ferry at Horseshoe Bay in Vancouver, British Columbia.

Covering the Reading

1. **a.** A line separates a plane into two distinct regions called __?__.
   b. The line itself is called the __?__ of these regions.

2. Graph the solutions to $x \le 6$: **a.** on a number line. **b.** in the coordinate plane.

3. Could a 42-year-old man, whose height in inches matches his age in years, go on the ride in Example 2?

4. The graph in the plane of all solutions to $x < \text{-}2$ consists of points to one side of the line $x = \text{-}2$. Which side?

5. Graph the set of all ordered pairs (x, y) that satisfy $y \ge 3$.

In 6–8, refer to Example 3. Justify your answer.

6. *True or false.* The ordered pair $(\text{-}4, 3)$ is a solution to the inequality.

7. Why is the boundary line dashed rather than solid?

8. How would the graph change if the inequality were $y < 2x + 7$?

In 9 and 10, graph the inequality.

9. $y \ge \text{-}2x - 3$

10. $x + y < 5$

11. What is a lattice point?

In 12–14, refer to Example 4.

12. Why are there no points in the solution set in the second, third, or fourth quadrants?

13. **a.** In how many ways can all the spaces be filled for one ferry crossing?
    b. List them.
    c. On which part of the graph are these solutions found?

14. **a.** In how many ways can less than the twelve spaces be filled?
    b. List four of them.
    c. Where are these solutions found on the graph?

In 15 and 16, write an inequality that describes the shaded region.

15.

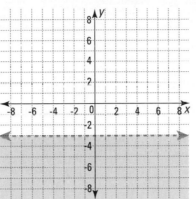

16.

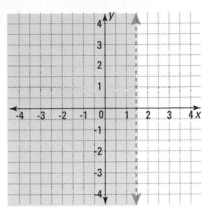

17. A person wants to buy x pencils at 10¢ each and y erasers at 15¢ each and spend less than 75¢.
  a. Write an inequality with x and y describing this situation.
  b. Graph all solutions.
  c. How many solutions are there?

18. In order to make the playoffs, the Dunkers basketball team must win at least twice as many of its 38 games as it loses. Graph the set of points (L, W) which satisfy these conditions. Plot L on the horizontal axis.

19. Write an inequality which describes the shaded region below.

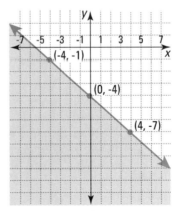

20. Explain why the sentence $y > 2x + 7$ does not describe a function.

21. a. What linear system is represented by the matrix equation
$$\begin{bmatrix} 3 & 4 \\ 1 & 2 \end{bmatrix} \begin{bmatrix} x \\ y \end{bmatrix} = \begin{bmatrix} 0 \\ -2 \end{bmatrix}?$$
  b. Solve this system. *(Lesson 5-6)*

Helping hands. *Shown are American Red Cross workers aiding flood victims in Iowa. More than 135 nations have Red Cross societies. The name* Red Cross *comes from the organization's flag—honoring Switzerland, where the Red Cross was founded in 1863. Most Muslim countries use a red crescent for their symbol. In Israel, the symbol is a red Star of David.*

22. A relief organization supplies cots, tables, and chairs to victims of floods, fires, or other natural disasters. Recently the organization purchased the following items.

	Number of cots	Number of tables	Number of chairs	Total cost ($)
January	10	10	40	1950
February	20	0	20	1800
March	10	5	20	1350

The unit cost of cots, tables, and chairs was constant during this period.
   a. Write a system of equations that can be used to determine the cost of one cot, one table, or one chair.
   b. Solve the system.   *(Lessons 5-4, 5-6)*

23. Let $M = \begin{bmatrix} 3 & 15 \\ -20 & x \end{bmatrix}$. For what value(s) of x does M^{-1} not exist?
   (Lesson 5-5)

In 24 and 25, refer to the following systems.   *(Lessons 5-3, 5-4)*

(a) $\begin{cases} y = 3x + 1 \\ 4x - 3y = 12 \end{cases}$ (b) $\begin{cases} 5x - 7y = 12 \\ -12x + 8y = 19 \end{cases}$

(c) $\begin{cases} 3x + 2y + z = -22 \\ x = 15z \\ y = -12z \end{cases}$ (d) $\begin{cases} x + 5y = 12 \\ 2x + 5y = 8 \end{cases}$

24. Which systems are written in a form which is convenient to be solved using linear combinations?

25. Which systems can be solved conveniently using substitution?

26. Consider the translation $T_{3,1}$ applied to the point $(6, -10)$ on the line $y = -2x + 2$.   *(Lessons 3-5, 4-10)*
   a. What is the image of $(6, -10)$ under the translation?
   b. What is the equation of the line through this image, parallel to $y = -2x + 2$?
   c. Repeat parts **a** and **b** with the point $(0, 2)$.
   d. Show that the lines in **b** and **c** are the same line.
   e. What is the image of the line $y = -2x + 2$ under the translation $T_{3,1}$?

Exploration

27. Earlier in this chapter you encountered many lines that arise from real situations.
   a. Pick one of the situations and modify it so that it leads to an inequality.
   b. Graph the inequality.

LESSON

5-8

Systems of Linear Inequalities

Because the graph of a linear inequality in two variables is a half-plane, the graph of the solution to a system of linear inequalities is the intersection of half-planes. The set of solutions to a system of linear inequalities is often called the **feasible set** or **feasible region** for that system. The boundaries are always parts of lines. The intersections of the boundaries are called **vertices** of the feasible set.

Example 1

Graph the feasible set for the system $\begin{cases} y > -2x + 6 \\ y \le \frac{1}{4}x - 3. \end{cases}$

Solution

Graph each inequality on the same set of axes. The graph of $y > -2x + 6$ is the set of points above and to the right of the line with equation $y = -2x + 6$. In the graphs below, this set is indicated by the shading ▭. The graph of $y \le \frac{1}{4}x - 3$ consists of points on or below the line with equation $y = \frac{1}{4}x - 3$, indicated by the shading ▭. The part of the plane shaded ▭ is the feasible set for this system. As shown in the graph below to the right, the feasible set for this system is the union of the interior of an angle and one of the rays of the angle except for the vertex.

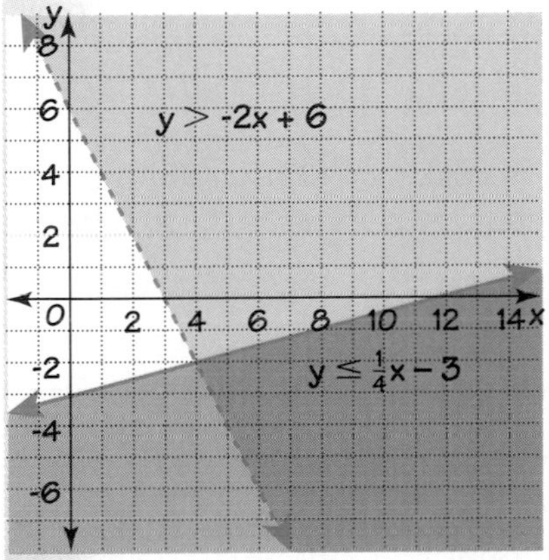

Feasible Set

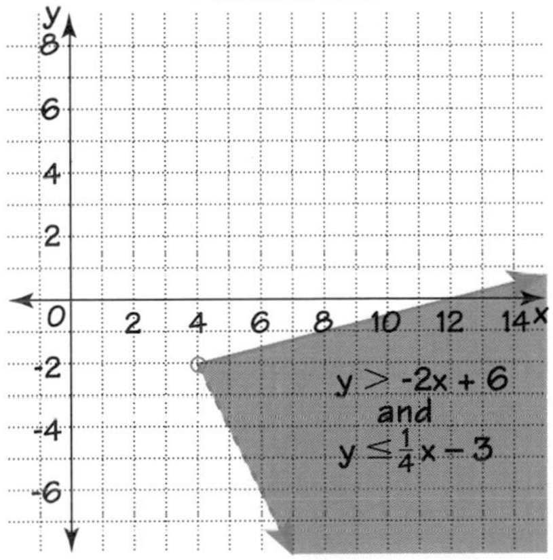

Check

First find the vertex of the feasible set by solving the system of equations

$$\begin{cases} y = -2x + 6 \\ y = \frac{1}{4}x - 3. \end{cases}$$

▶

This vertex is (4, −2), which checks with the graph. Second, pick a point in the shaded region, such as (8, −5), and substitute it into each inequality.
Is −5 > −2(8) + 6? Yes.
Is −5 ≤ $\frac{1}{4}$(8) − 3? Yes. So the solution checks.
You should also try points outside the region to show that they do not satisfy both inequalities.

Many automatic graphers can graph inequalities and systems of inequalities. If your grapher has this capability, you should repeat Example 1 using your grapher.

The applications of systems to production planning in industry, discovered by Kantorovich in 1939, may involve systems with thousands of variables. Computers are needed to solve these problems. But simple examples can be done by hand. Here is a simplified example of a business application.

Hand-crafted. *Shown is a woodworker in a furniture plant in Winchester, Virginia.*

Example 2

The Biltrite Furniture Company makes wooden desks and chairs. Carpenters and finishers work on each item. On the average, the carpenters spend four hours working on each chair and eight hours on each desk. There are enough carpenters for up to 8000 worker-hours per week. The finishers spend about two hours on each chair and one hour on each desk. There are enough finishers for a maximum of 1300 worker-hours per week. Given these constraints, find the feasible region for the number of chairs and desks that can be made per week.

Solution

Make a table to organize this information.

	chairs	desks	total hours available
hrs. of carpentry per piece	4	8	8000
hrs. of finishing per piece	2	1	1300

Let x = the number of chairs that can be made per week and y = the number of desks that can be made per week.

Write sentences for each constraint. Because x and y represent pieces of furniture, they must be nonnegative integers; that is,

x ≥ 0 and y ≥ 0, and (x, y) is a lattice point.

The carpentry hours must satisfy 4x + 8y ≤ 8000.

The finishing hours must satisfy   2x + y ≤ 1300.

Thus, the feasible region for this situation is the set of lattice-point solutions of the system

$$\begin{cases} x \geq 0 \\ y \geq 0 \\ 4x + 8y \leq 8000 \\ 2x + y \leq 1300. \end{cases}$$

▶ Now graph the solution to the system. The first two inequalities indicate all solutions are in the first quadrant or on the positive axes. So it is sufficient to examine solutions to the last two inequalities only in the first quadrant. The graph is shown below.

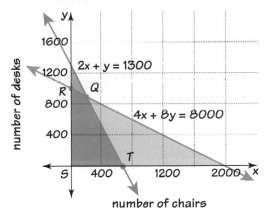

The feasible region is the set of lattice points in the quadrilateral QRST and its interior.

You can find the coordinates of each vertex of *QRST* by solving a system of equations. For instance, *Q* is the solution to the system

$$\begin{cases} 4x + 8y = 8000 \\ 2x + \ y = 1300. \end{cases}$$

Notice that even though only integer values are solutions, shading is used because it would be too difficult to show all the dots.

Check

Choose a point in the feasible region and see if it is a solution of each inequality of the system. We choose (400, 200).
Is $400 \geq 0$? Yes.
Is $200 \geq 0$? Yes.
Is $4(400) + 8(200) \leq 8000$? Yes.
Is $2(400) + 200 \leq 1300$? Yes.
So one option available to the Biltrite Furniture Company is to make 400 chairs and 200 desks per week.

As a further check, choose a point just outside the feasible region, such as (700, 400), and show that there is at least one inequality which is not satisfied by this point.

Covering the Reading

1. The solution to a system of linear inequalities can be represented by the ___?___ of half-planes.

2. The solution to a system of linear inequalities is often called the ___?___.

In 3–5, refer to Example 1. Verify the following statements.

3. (10, -6) is a solution to the system.

4. (5, 0) is *not* a solution to the system.

5. (4, -2) is the vertex of the feasible set for the system.

In 6–10, refer to Example 2.

6. Which inequality expresses the amount of time that the company can have its finishers working?

7. Why is it sufficient to consider only the first quadrant in graphing the feasible set?

8. Find the coordinates of vertex Q.

9. What system of equations gives vertex R as its solution?

10. Could the company manufacture 600 chairs and 200 desks per week under the given operating conditions? Justify your answer.

In 11 and 12, a system of inequalities is given. **a.** Graph the feasible region. **b.** Find the coordinates of each vertex of the region.

11. $\begin{cases} y \geq 3x + 1 \\ y \leq -2x + 4 \end{cases}$

12. $\begin{cases} x + 3y \leq 18 \\ 2x + y \leq 16 \\ x \geq 0 \\ y \geq 2 \end{cases}$

Applying the Mathematics

13. Refer to the graph at the left.
    a. Write a system of inequalities which is represented by this feasible set.
    b. *True or false.* The point (3, 4) is in the feasible region.
    c. Find the coordinates of vertex C.

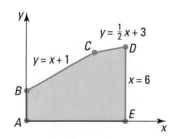

14. A clothier makes women's suits and coats from nylon lining and wool tweed. Each suit requires 2 yards of nylon lining and 3 yards of wool tweed. Each coat requires 3 yards of nylon lining and 4 yards of wool tweed. A total of 420 yards of nylon lining and 580 yards of wool are in stock.

a. Let s be the number of suits and c be the number of coats. Complete the translation of this situation into a system of inequalities.

$$\begin{cases} s \geq \underline{\ ?\ } \\ c \geq \underline{\ ?\ } \\ 2s + 3c \leq \underline{\ ?\ } \\ \underline{\ ?\ } \leq 580 \end{cases}$$

b. Graph the feasible region of points (s, c) for this system and label the vertices.

15. An electronics firm makes two kinds of televisions: black-and-white and color. The firm has enough equipment to make as many as 1000 black-and-white sets per month or 600 color sets per month. It takes 20 worker-hours to make a black-and-white set and 30 worker-hours to make a color set. The firm has up to 24,000 worker-hours of labor available each month. Let x be the number of black-and-white TVs and y be the number of color TVs made in a month.

a. Translate this situation into a system of inequalities.

b. Graph the feasible set for this system, and label the vertices.

Review

16. Write an inequality to describe the region shaded below. *(Lesson 5-7)*

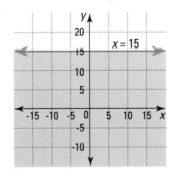

17. Jack has a savings account from which to pay his share of the rent at college for 9 months. The account must also cover a security deposit of $200.

a. If the account contains D dollars, write an inequality that expresses the monthly rent R he can afford without running out of money.

b. If Jack's account has $1500 dollars in it, what is the maximum monthly rent he can afford? *(Lesson 5-7)*

18. Solve the following matrix equation. *(Lesson 5-6)*

$$\begin{bmatrix} -1 & 2 \\ 3 & 4 \end{bmatrix} \begin{bmatrix} x \\ y \end{bmatrix} = \begin{bmatrix} -6 \\ 8 \end{bmatrix}$$

In 19 and 20, solve each system.   *(Lessons 5-2, 5-3, 5-4, 5-5)*

19. $\begin{cases} y = \dfrac{8}{x} \\ y = x^2 \end{cases}$

20. $\begin{cases} x + 2y - z = -11 \\ 4x + 3y + 2z = -39 \\ 2x - y + 3z = -17 \end{cases}$

21. If a toy manufacturer makes \$5 profit on each board game and \$8 profit on each video game, how much profit will be earned from x board games and y video games?   *(Lesson 3-3)*

22. What is the slope of the line with equation $5x + 8y = 1000$?
(Lesson 3-2)

23. *Skill sequence.*   Multiply and simplify where possible.
(Previous course)
    a. $8(n + 8)$ **b.** $n(n + 8)$
    c. $(n + 8)(n + 8)$ **d.** $(n + 8)^2$

24. A region of the plane is said to be **convex** if and only if any two points of the region can be connected by a line segment which is itself entirely within the region. The pentagon below is convex but the quadrilateral is not.

Convex   Not convex

Tell whether or not the shaded region is convex.   *(Previous course)*

a.   **b.**   **c.**   **d.**

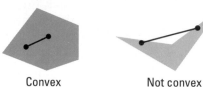

Exploration

25. Refer to Question 24.
    a. At most how many pieces can you get out of a circular pie with 4 straight cuts? How many of these pieces are convex? Draw a picture to illustrate this situation.
    b. At most, how many pieces can you get out of a circular pie with n straight cuts? How many of these pieces are convex?

Long live the king. *It is believed that chess originated in India around 600 A.D. It soon spread to Persia and later to Europe. The term "checkmate" derives from the Persian* shah mat, *meaning "the king is dead."*

In Example 2 of Lesson 5-8, a system of linear inequalities describes constraints of the manufacturing operations of the Biltrite Furniture Company. The feasible set describes various combinations of chairs and desks that the company can make with the given constraints. Now suppose that the company also knows that it earns a profit of $15 on each chair and $20 on each desk it makes. Given these known constraints, how can the production schedule be set to maximize the profit?

If x chairs and y desks are sold, the profit P is given by the formula
$$15x + 20y = P.$$
For instance, the solutions to
$$15x + 20y = 3000$$
are ordered pairs that will yield a $3000 profit.

The shaded region in the figure below again shows the feasible set for Biltrite's system of inequalities. The figure also shows the graphs of some lines that result from substituting different values of P into the profit formula.

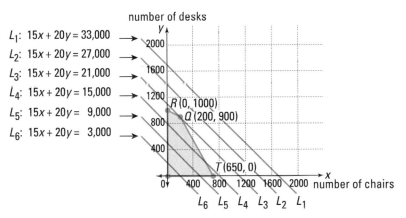

L_1: $15x + 20y = 33{,}000$

L_2: $15x + 20y = 27{,}000$

L_3: $15x + 20y = 21{,}000$

L_4: $15x + 20y = 15{,}000$

L_5: $15x + 20y = 9{,}000$

L_6: $15x + 20y = 3{,}000$

All lines with equations of the form $15x + 20y = P$ are parallel because each has slope $-\frac{3}{4}$. Some of these lines intersect the feasible region and some do not. In the figure above, the lines above L_3 do not intersect the feasible set. Lines such as L_3 and L_4 that do intersect the feasible region indicate possible profits. The greatest profit will occur when the line with equation $15x + 20y = P$ is as high as possible, but still intersects the feasible region. This will happen when the profit line passes through vertex $Q = (200, 900)$. This is the line L_3. Thus to maximize profits, the company should manufacture 200 chairs and 900 desks per week. The maximum profit under these conditions is $21,000.

Problems such as this one, which lead to systems of linear inequalities, are called **linear-programming problems.** The word "programming" does not refer to a computer; it means that the solution gives a "program," or course of action, to follow. The most profitable program for Biltrite Furniture Company is to make 200 chairs and 900 desks per week.

In 1826, the French mathematician Jean Baptiste Joseph Fourier proved the following theorem.

> **Linear-Programming Theorem**
> The feasible region of every linear-programming problem is convex, and the maximum or minimum quantity is determined at one of the vertices of this feasible region.

The Linear-Programming Theorem tells you where to look for the greatest or least value of a linear combination expression in a linear-programming situation, without having to draw many lines through the feasible region.

In 1945, George Stigler (then at Columbia University, later at the University of Chicago) was looking for the least expensive diet that would provide a person's daily needs of calories, protein, and various vitamins and minerals. Here is a simplified diet problem of the type first considered by Stigler.

An amazing building.
Shown is Mitchell Corn Palace in Mitchell, South Dakota. The building is decorated with murals made of different colors of ears of corn. The decorations are replaced every September. Concerts, dances, and many other events are held there.

Example

Here is a table with some of the nutritional value of fried chicken and corn on the cob.

	Vitamin A (I.U.)	Potassium (mg)	Iron (mg)	Calories
Fried Chicken (one small piece)	100	0	1.2	122
Corn (one ear)	310	151	1.0	70

Suppose you want at least 1000 units of vitamin A, 200 mg of potassium, 6 mg of iron, and at least 600 calories of energy from these foods. Let n = the number of pieces of chicken and e = the number of ears of corn.

$$\begin{cases} n \geq 0 & \\ e \geq 0 & \\ 100n + 310e \geq 1000 & \text{(at least 1000 units vitamin A)} \\ 151e \geq 200 & \text{(at least 200 mg potassium)} \\ 1.2n + e \geq 6 & \text{(at least 6 mg iron)} \\ 122n + 70e \geq 600 & \text{(at least 600 calories)} \end{cases}$$

A graph of this system is shown on the next page.

▶

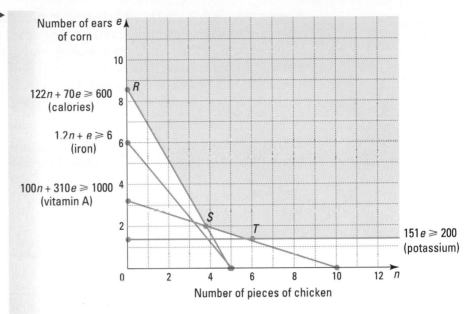

Number of ears *e* of corn

$122n + 70e \geqslant 600$
(calories)

$1.2n + e \geqslant 6$
(iron)

$100n + 310e \geqslant 1000$
(vitamin A)

$151e \geqslant 200$
(potassium)

Number of pieces of chicken

a. Find the vertices of the feasible set.

b. Now suppose that each piece of chicken costs $.80 and each ear of corn costs $.50. Determine the values of *n* and *e* which will meet the nutritional requirements above at the lowest cost *C*.

Solution

a. There are three vertices *R, S,* and *T.* To find *R,* let $n = 0$ in the equation $122n + 70e = 600$.

$$122(0) + 70e = 600$$
$$e \approx 8.6$$

So the coordinates of R are about (0, 8.6).

Vertex *S* is the intersection of lines with equations $122n + 70e = 600$ and $100n + 310e = 1000$. So solve this system. We find $e \approx 2$ and $n \approx 3.8$. So the coordinates of *S* are about (3.8, 2).

Finally, *T* is the intersection of $100n + 310e = 1000$ and $151e = 200$. Solve the second equation for *e* and substitute into the first. The coordinates of *T* are about (6, 1.3).

b. The cost formula is $C = .80n + .50e$. The Linear-Programming Theorem says that the minimum value of *C* occurs at a vertex of the feasible region. So evaluate *C* at each vertex to see which combination of *n* and *e* gives the minimum cost. Because *n* and *e* are integers and the feasible set has no values less than those found at the vertices, the coordinates of each vertex must be rounded up to the nearest integer. Thus we use (0, 9) for *R,* (4, 2) for *S,* and (6, 2) for *T.*

$$at \ (0, 9): \quad C = .80(0) + .50(9) = 4.50$$
$$at \ (4, 2): \quad C = .80(4) + .50(2) = 4.20$$
$$at \ (6, 2): \quad C = .80(6) + .50(2) = 5.80$$

Thus the minimum cost satisfying the given constraints is $4.20. To satisfy this nutritional need for the lowest cost, you should eat 4 pieces of chicken and 2 ears of corn.

Stigler originally considered 70 possible foods and found that the lowest-cost diet satisfying an adult's need for calories, protein, calcium, vitamin A, thiamine, riboflavin, niacin, and ascorbic acid was a combination of wheat flour, cabbage, and pork liver. By eating just these three foods, a person was thought to be able to live in good health in 1945 for $59.88 a year. Costs today are higher, so it might now cost about $500 a year for that diet.

Of course many other vitamins, minerals, and foods are now taken into account by dietitians planning well-balanced meals. Stigler could not consider a greater number of possible foods nor consider more health needs because computers were not available in 1945. Today it is possible to consider hundreds of foods and many more daily needs.

QUESTIONS

Covering the Reading

In 1–5, refer to the discussion of the Biltrite Furniture Company at the beginning of this lesson.

1. What is the company trying to maximize?

2. What do the numbers 15 and 20 represent in the profit equation?

3. *True or false.* If a line with equation $15x + 20y = P$ intersects the feasible region for the system of inequalities, it is possible for Biltrite to make a profit of P dollars.

4. **a.** What is the maximum weekly profit Biltrite can earn?
   b. How many chairs and how many desks must the company produce to make this profit?

5. Find the profit if 199 chairs and 899 desks are made.

6. To what does the word "program" refer in a linear-programming problem?

7. In a linear-programming problem, why is it necessary to find the vertices of the feasible region?

In 8 and 9, refer to the Example in this lesson.

8. **a.** Which linear combination must be minimized?
   b. What is the minimum value of C satisfying the constraints of the problem?
   c. At which vertex does the maximum cost occur?

9. If 10 mg of iron were needed in this diet, the constraint for iron would be $1.2n + e \geq 10$.
   a. Regraph the feasible region of the system with this new iron requirement.
   b. In the new feasible region, which vertex yields the minimum cost?

10. a. A diet problem like the one in the Example was first modeled mathematically by whom and in what year?

   b. Why can more variables be dealt with now than could be considered when these diet problems were first studied?

Applying the Mathematics

11. Use the feasible set graphed at the right. Which vertex maximizes the profit equation $P = 3x + 4y + 250$? Justify your answer.

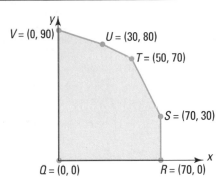

$V = (0, 90)$
$U = (30, 80)$
$T = (50, 70)$
$S = (70, 30)$
$Q = (0, 0)$
$R = (70, 0)$

12. Refer to Question 15 in Lesson 5-8. Suppose the electronics firm earns a profit of $25 on each black-and-white TV and $40 on each color TV.

   a. Write a formula for the monthly profit earned.

   b. What combination of black-and-white and color TVs will maximize profit?

In 13–15, suppose that a farmer has no more than 50 acres for planting either corn or soybeans or both and has a maximum of $12,000 to spend on the planting. Further suppose that it costs $250 per acre to plant corn and $200 per acre to plant soybeans. If C is the number of acres of corn and S is the number of acres of soybeans that the farmer plants, the system for this problem is given below.

Harvest time. *Shown is a combine harvesting a soybean crop in Iowa. The combine cuts the plants, and threshes and cleans the seeds in one operation. Soybeans are usually harvested in late summer or early fall.*

$$\begin{cases} C + S \leq 50 \\ 250C + 200S \leq 12,000 \\ C \geq 0 \\ S \geq 0 \end{cases}$$

(Source: USDA, 1991)

13. Match each inequality in the system with its meaning.

   a. $C + S \leq 50$

   b. $250C + 200S \leq 12,000$

   c. $C \geq 0$

   d. $S \geq 0$

   (i) The number of acres of soybeans is not negative.

   (ii) The total number of acres is not more than 50.

   (iii) The cost of planting must be no more than $12,000.

   (iv) The least number of acres of corn is zero.

14. Graph the feasible region of points (C, S).

15. Suppose the profit per acre for corn is $75 and for soybeans is $85.

   a. Find the vertices of the feasible region.

   b. State the profit formula.

   c. At which vertex is P maximized?

16. A landscaping contractor uses a combination of two brands of fertilizers, each containing different amounts of phosphates and nitrates, as shown in the table below.

	Phosphate Content per Package	Nitrate Content per Package
Brand *A*	4 lb	2 lb
Brand *B*	6 lb	5 lb

A certain lawn requires a mixture of at least 24 lb of phosphates and at least 16 lb of nitrates. If *a* is the number of packages of Brand A and *b* is the number of packages of Brand B, then the constraints of the problem are given by the following system of inequalities.

$$\begin{cases} a \geq 0 \\ b \geq 0 \\ 4a + 6b \geq 24 \\ 2a + 5b \geq 16 \end{cases}$$

a. Graph the feasible region.

b. If a package of Brand A costs $6.99 and a package of Brand B costs $17.99, which pair (a, b) in the feasible region gives the lowest cost?

Review

17. *Multiple choice.* Which of the following systems describes the graph at the left? *(Lesson 5-8)*

(a) $\begin{cases} 2y + x \geq 6 \\ y - 3x \leq 4 \end{cases}$
 (b) $\begin{cases} 2y + x \leq 6 \\ y - 3x \geq 4 \end{cases}$
 (c) $\begin{cases} 2y + x < 6 \\ y - 3x > 4 \end{cases}$

18. Evaluate $\det \begin{bmatrix} x & -3 \\ 2 & y \end{bmatrix}$. *(Lesson 5-5)*

19. Find the inverse of the matrix for S_{-3}. *(Lessons 4-4, 5-5)*

20. Find an equation for the line through the points $(-2, -4)$ and $(5, 7)$. *(Lesson 3-5)*

21. At the left is a graph of the function $f(x) = x^2$. Write an inequality to describe the shaded region. *(Lesson 5-7)*

22. The strength S of a rectangular beam varies directly as its width w and the square of its depth d, and varies inversely as its length L. Suppose a beam can support 1750 pounds, and its dimensions are $w = 4''$, $d = 8''$, and $L = 20$ feet. What is the strength of a beam of the same material with $w = 4''$, $d = 8''$, and $L = 25$ feet? *(Lesson 2-9)*

Exploration

23. Suppose $a + b + c + d = 100$, and a, b, c, and d are all nonnegative. What are the largest and smallest possible values of $abcd$?

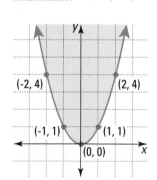

LESSON

5-10

Linear Programming II

Golden touch. *Although most jewelry today is made by machine, many people make jewelry as a hobby. This woman is working with brass wires to make jewelry to be sold at a local craft fair. See the Example on page 332.*

In Lesson 5-9, you practiced using the Linear-Programming Theorem for a given feasible region and linear combination expression to be maximized or minimized. In this lesson you will learn to solve linear-programming problems from scratch. To solve a linear-programming problem, follow these steps.

1. Identify the variables.
2. Identify the constraints, and translate them into a system of inequalities relating the variables. If necessary, make a table.
3. Graph the system of inequalities; find the vertices of the feasible set.
4. Write a formula or an expression to be maximized or minimized.
5. Apply the Linear-Programming Theorem.
6. Interpret the results.

Because linear-programming problems are long and involved, you must organize your work and write neatly. We illustrate the entire process in the following Example.

Example

Some students make necklaces and bracelets in their spare time and sell all that they make. Every week they have available 10,000 g of metal and 20 hours to work. It takes 50 g of metal to make a necklace and 200 g to make a bracelet. Each necklace takes 30 minutes to make and each bracelet takes 20 minutes. The profit on each necklace is $3.50, and the profit on each bracelet is $2.50. The students want to earn as much money as possible. Because you are taking this course, they ask you to give them advice. What numbers of necklaces and bracelets should they make each week? How much profit can they make?

Solution

1. Identify the variables.
   Let x = the number of necklaces to be made per week and y = the number of bracelets to be made per week.
2. Identify the constraints, and translate them into a system of inequalities. Negative numbers cannot be used; thus

   $$x \geq 0$$
   $$y \geq 0.$$

The following table summarizes the information about production.

	Metal Used	Time to Make
for each necklace	50 g	30 min
for each bracelet	200 g	20 min
Total available	10,000 g	20 hours (or 1200 min)

The amount of metal (in grams) used satisfies

$$50x + 200y \leq 10,000.$$

The amount of time (in minutes) needed satisfies

$$30x + 20y \leq 1200.$$

3. Graph the system and find the vertices. Only the feasible set is shown.

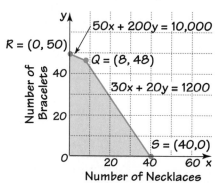

The vertices $R = (0, 50)$, $Q = (8, 48)$, and $S = (40, 0)$ are found by solving systems of equations. For instance, Q is found by solving $50x + 200y = 10,000$ and $30x + 20y = 1200$. The vertices are labeled on the graph.

4. Write a formula to be maximized or minimized.
   The profit formula is $P = 3.50x + 2.50y$. P is to be maximized.

▶

5. Apply the Linear-Programming Theorem. Substitute the coordinates of each vertex into the profit formula.
For (0, 0): P = 3.5(0) + 2.5(0) = 0.
For (40, 0): P = 3.5(40) + 2.5(0) = 140.
For (8, 48): P = 3.5(8) + 2.5(48) = 148.
For (0, 50): P = 3.5(0) + 2.5(50) = 125.

6. Interpret the results. The maximum profit of $148 occurs at vertex Q = (8, 48). To earn the maximum weekly profit of $148, the students should make 8 necklaces and 48 bracelets each week.

Linear programming is often used in industries in which all the competitors make the same product (such as gasoline, paper, appliances, clothing, and so on). Efficiency in the use of labor and materials determines the amount of profit. These situations can involve as many as 5000 variables and 10,000 inequalities. Although we use graphing to solve linear-programming problems in this book, more efficient methods of solution are used by computers and calculators. The most efficient procedure for most linear-programming problems is the *simplex algorithm* invented in 1947 by the econometrician Leonid Hurwicz and the mathematicians George Dantzig and Tjalling Koopmans, all from the United States. It is for this work that Koopmans shared the Nobel Prize with Kantorovich in 1975.

Noble prize. Shown are Tjalling Koopmans (top) and Leonid Kantorovich. The Nobel Prize, named for Alfred Nobel, was first awarded in 1901. There are six categories— physics, chemistry, physiology (medicine), literature, economic sciences, and international peace.

QUESTIONS

Covering the Reading

In 1–5, refer to the Example in this lesson.

1. What are the students trying to find?

2. What is the *x*-intercept of the equation which limits the amount of metal to be used? Is it in the feasible region?

3. How much more profit do the students make with the linear combination at (8, 48) than at the next best vertex?

4. a. What would be the profit if the students made 9 necklaces and 47 bracelets?
b. Why can they not do this?

5. Suppose the students in the Example decide to put semiprecious gems in their jewelry: six in each necklace and one in each bracelet. They can use 150 gems each week.
a. Translate this constraint into an inequality.
b. The entire system, including this new constraint, is graphed at the left. Find the new vertices, *T* and *U*.
c. If the students will now earn $6.50 profit on each necklace and $3.00 profit on each bracelet, do they need to change their program to keep profits at a maximum? Justify your answer.

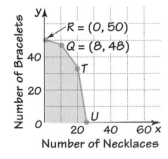

6. a. Name a method for solving linear-programming problems without graphing.
b. Who developed this method, and when?

Applying the Mathematics

7. Some parents shopping for their family want to know how much hamburger and how many potatoes to buy. From a nutrition table, they find that one ounce of hamburger has 8 mg of iron, 10 units of Vitamin A, and 6.5 grams of protein. One medium potato has 1.1 mg of iron, 0 units of Vitamin A, and 4 grams of protein.

For this meal the parents want each member of the family to have at least 5 mg of iron, 30 units of Vitamin A, and 35 grams of protein. One potato costs $0.10 and 1 ounce of hamburger costs $0.15. The parents want to be economical (minimize their costs), yet meet daily requirements. They need a program for the quantities of hamburger and potatoes to buy for the family.
a. Identify the variables for this problem.
b. Translate the constraints of the problem into a system of inequalities. (You should have five inequalities; a table may help.)
c. Graph the system of inequalities in part **b,** and find the vertices of the feasible set. (Note: Usually the variable that comes first in the alphabet is plotted on the horizontal axis.)
d. Write an expression for the cost to be minimized.
e. Apply the Linear-Programming Theorem to determine which vertex minimizes the cost expression of part **d.**
f. Interpret your answer to part **e.** What is the best program for this family?

8. A company makes two kinds of tires: model R (regular) and model S (snow). Each tire is processed on three machines, A, B, and C. To make one model R tire requires $\frac{1}{2}$ hour on machine A, 2 hours on B, and 1 hour on C. To make one model S tire requires 1 hour on A, 1 hour on B, and 4 hours on C. During the next week, machine A will be available for at most 20 hours, machine B for at most 60 hours, and machine C for at most 60 hours. If the company makes a $10 profit on each model R tire and a $15 profit on each model S tire, how many of each tire should be made to maximize the company's profit? (Hint: Use the six steps suggested in this lesson.)

Review

9. a. Graph the feasible set of the system $\begin{cases} x + 2y \geq 12 \\ 2x + y \leq 1 \\ x \leq 4 \\ y \geq 0. \end{cases}$

b. What are the coordinates of its vertices? *(Lesson 5-9)*

10. a. Graph all solutions to the system $\begin{cases} x > -4 \\ y < 6 \end{cases}$.

    b. Is the point $(-7, 2)$ in the solution set? How can you tell?
    (Lessons 5-7, 5-8)

11. Solve the following system by using matrices.   *(Lessons 5-5, 5-6)*
$$\begin{cases} 5x + 3y = 7 \\ 2x - 7y = -30 \end{cases}$$

12. Does the matrix $\begin{bmatrix} 4 & -6 \\ -6 & 9 \end{bmatrix}$ have an inverse? How can you tell?
(Lesson 5-5)

13. An alloy containing 65% aluminum is made by melting together two alloys that are 25% aluminum and 75% aluminum. How many kilograms of each alloy must be used to produce 160 kilograms of the 65% alloy?
(Lesson 5-4)

14. Solve $12 - 5x < 16$.   *(Lesson 5-1)*

15. A formula for the nth term of a sequence is $a_n = 10 - .5(n - 1)$.
    a. Write the first three terms.
    b. Write a recursive formula for this sequence.
    c. Solve the explicit formula for n.
    d. One term of the sequence is -39. Which term is it?
    (Lessons 1-5, 1-7, 3-3, 3-4)

16. *Skill sequence.*   Solve.   *(Previous course)*
    a. $x^2 = 49$ **b.** $x^2 + 2 = 51$ **c.** $x^2 + 2 = 49$

Plainly aluminum. *Shown is a McDonnell-Douglas MD-80 series jet. This jet contains approximately 18,000 kilograms of aluminum.*

Exploration

17. In this lesson and in the last lesson six situations are described that lead to linear-programming problems. For instance, one situation is found in the Example of this lesson and another is in Question 7 of this lesson. Make up another situation that would lead to this kind of problem.

A project presents an opportunity for you to extend your knowledge of a topic related to the material of this chapter. You should allow more time for a project than you do for typical homework questions.

1 Nutritious and Cheap?

a. Look up the recommended daily allowance (RDA) of protein, vitamin A, vitamin C, and calcium for someone your age.

b. Name some foods that contain these nutrients, and find the cost of one serving of each of these foods.

c. Use linear programming to find out how many servings of these foods you would need to eat each day in order to get the RDA of these nutrients at the lowest possible cost.

d. Make up some other problem about nutrients that can be solved using a system of equations or inequalities. Solve your problem.

2 Inverses of 3 × 3 Matrices

Find a book that tells how to find the inverse of a 3 × 3 matrix by hand.

a. Write an explanation of the method using a matrix of your own choosing.

b. Apply the method to solve a 3 × 3 system.

c. Tell what 3 × 3 matrices do not have inverses.

3 Systems Involving a Hyperbola and a Line

In Lesson 5-2 you found the intersection points of a hyperbola and line. Use an automatic grapher to examine the possible intersections of all systems written in the form below.

$$\begin{cases} y = mx + b \\ y = \dfrac{n}{x} \end{cases}$$

Make conjectures, if you can, about the following situations.

a. If m and n are both positive and constant, how does varying b affect the number of intersections? What are the numbers of intersections possible? (Try positive and negative values for b.)

b. Repeat part **a** with m and n both negative.

c. Repeat part **a** with m and n having opposite signs. Does it make any difference which is positive and which negative? How do the graphs compare?

4 History of Linear Programming

Throughout the chapter, there have been brief references to the history of linear programming. Research the work of the mathematicians mentioned in this chapter and their contributions to the development of linear-programming techniques. Also, find out more about recent developments using computers to solve such problems, such as work done by Karmarkar in 1984. Write a report that summarizes your findings.

5

Using Matrices to Code and Decode Messages

Between 1929 and 1931, the mathematician Lester Hill devised a method of encoding messages using matrices. Every integer is assigned a letter according to the scheme:

$1 = A, 2 = B, 3 = C, \ldots, 25 = Y, 26 = Z,$
$27 = A, 28 = B, \ldots,$ and $0 = Z,$
$-1 = Y, \ldots, -24 = B, -25 = A, -26 = Z \ldots.$

a. To code or encipher the word FOUR, follow these steps.

Step 1 Put the letters into a 2×2 matrix four at a time. With $6 = F$, $15 = O$, $21 = U$, $18 = R$, use the matrix

$$\begin{bmatrix} 6 & 15 \\ 21 & 18 \end{bmatrix}.$$

Step 2 Choose a 2×2 coding or key matrix, such as $\begin{bmatrix} 0 & 1 \\ 1 & 2 \end{bmatrix}$. Multiply each 2×2 matrix by the coding matrix. For instance,

$$\begin{bmatrix} 0 & 1 \\ 1 & 2 \end{bmatrix}\begin{bmatrix} 6 & 15 \\ 21 & 18 \end{bmatrix} = \begin{bmatrix} 21 & 18 \\ 48 & 51 \end{bmatrix}.$$

Step 3 Change the resulting matrix $\begin{bmatrix} 21 & 18 \\ 48 & 51 \end{bmatrix}$ back to letters to write the coded message: URVY.

Step 4 Repeat this as many times as necessary to encode a longer message. Code MEET ME AT NOON using the key $\begin{bmatrix} 0 & 1 \\ 1 & 2 \end{bmatrix}$.

b. To decode or decipher a message, follow these steps.

Step 1 Break the message up into groups of four letters and write as matrices using the corresponding numbers. Each letter-group matrix should be:

$$\begin{bmatrix} \text{1st letter} & \text{2nd letter} \\ \text{3rd letter} & \text{4th letter} \end{bmatrix}.$$

Step 2 Find the inverse of the key matrix and multiply each letter-group matrix by the inverse.

The following message was also encoded using the key $\begin{bmatrix} 0 & 1 \\ 1 & 2 \end{bmatrix}$:

YTKOFOTISBGVITWKOULO.

What is the original message?

c. Make up a coding matrix and a coded message of your own. (Note: In order for Hill's method to work the determinant of your coding matrix must be 1 or -1.)

SUMMARY

When two or more sentences are joined by the words *and* or *or*, a compound sentence results. The solution set to *A or B* is the union of the solution sets of *A* and *B*. If the word joining them is *and*, the compound sentence is called a system. The solution set to *A and B* is the intersection of the solution sets of *A* and *B*.

Systems have many applications and may contain any number of variables. If the system contains one variable, then its solutions may be graphed on a number line or in the plane, depending on the situation. If the system contains two variables, then its solutions may be graphed in the plane. Graphing in the plane often tells you the number of solutions but may not yield the exact solutions.

In using algebra to solve systems of linear equations you may use linear combinations, substitutions, or matrices. The matrix method converts a system of *n* equations in *n* unknowns to a single matrix equation. To get the solution to the system, both sides of the equation are multiplied by the inverse of the coefficient matrix.

The graph of a single linear inequality in two variables is a half-plane or a half-plane with its boundary. For a system of two linear inequalities, if the boundary lines intersect, then the graph is the interior of an angle plus perhaps one or both of its sides.

Systems with two variables but more than two inequalities arise in linear-programming problems. In a linear-programming problem, you look for a solution to the system that maximizes or minimizes the value of a particular expression or formula. To solve such a problem, first find the set of solutions to the system. This feasible set is always a convex region. The Linear-Programming Theorem states that the desired point must be a vertex of the feasible set, so all vertices must be tested. Applications of linear programming are a relatively recent development in mathematics and are quite important in industry.

VOCABULARY

Below are the most important terms and phrases for this chapter. You should be able to give a definition for those terms marked with *. For all other terms you should be able to give a general description and a specific example.

Lesson 5-1
constraint
system
open sentence
interval
compound sentence
* union of sets, *or*
* intersection of sets, *and*
inequality
Addition Property of Inequality
Multiplication Properties
   of Inequality

Lesson 5-2
system
* solution for a system
rescale, zoom

Lesson 5-3
consistent system
inconsistent system

Lesson 5-4
Linear Combination Method

Lesson 5-5
* inverse of a matrix M, M^{-1}
Square matrix
Inverse Matrix Theorem
* determinant of a 2×2
   matrix M
det M

Lesson 5-6
matrix form of a system
coefficient matrix
constant matrix
System-Determinant Theorem
3×3 identity matrix

Lesson 5-7
half-plane
boundary
lattice point

Lesson 5-8
* feasible set, feasible region
* vertices of feasible region
convex region

Lesson 5-9
linear-programming problem
Linear-Programming Theorem

PROGRESS SELF-TEST

Take this test as you would take a test in class. Use graph paper and a calculator. Then check your work with the solutions shown in the Selected Answers section in the back of the book.

1. Solve and graph the solution set to $-3n + 16 < 22$ on a number line.

2. On a number line, graph $\{x: x \le -5 \text{ or } x \ge 7\}$.

3. A graph of the system $\begin{cases} y = .5x - 2 \\ y = -x^2 \end{cases}$
   is shown below. Estimate the solutions to the system, to the nearest tenth.

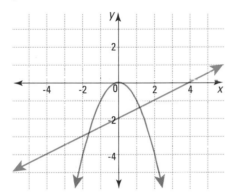

4. Consider the system $\begin{cases} 2x - 9 = 8 \\ 2x - 9 = -7. \end{cases}$

   a. Is this system consistent or inconsistent?

   b. Explain how you can tell.

In 5 and 6, solve each system. Show your work, or explain how you used technology to find the solution.

5. $\begin{cases} s = 4t \\ r = t + 11 \\ 3r - 8s = 4 \end{cases}$ 6. $\begin{cases} -3x + 3y = 2 \\ -4x - 2y = 3 \end{cases}$

7. At Eggs-N-Links Restaurant you can get a Double Duo Breakfast of 2 eggs with 2 sausage links for $2.78 and a Triple Quad Breakfast of 3 eggs with 4 sausage links for $4.99. From this information, use a system of equations to determine what Eggs-N-Links might charge for 1 egg. Be sure to identify what each variable represents.

8. Consider the system $\begin{cases} 8x + 3y = 41 \\ 6x + 5y = 39. \end{cases}$

   a. What is the coefficient matrix?

   b. Find the inverse of the coefficient matrix.

   c. Use a matrix equation to solve the system.

9. a. Give an example of a 2×2 matrix that does not have an inverse.

   b. How can you tell that the inverse does not exist?

10. Graph the solution set of $y < -2x + 6$.

11. *Multiple choice.* The graph at the right shows the feasible set for which system?

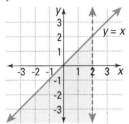

   (a) $\begin{cases} y \le 3x \\ x \ge 2 \end{cases}$ (b) $\begin{cases} y > x \\ x \le 2 \end{cases}$

   (c) $\begin{cases} y \le x \\ x < 2 \end{cases}$ (d) $\begin{cases} y < x \\ x < 2 \end{cases}$

12. A furniture manufacturer makes upholstered chairs and sofas. On the average it takes carpenters 7 hours to build a chair and 4 hours to build a sofa. There are enough carpenters for no more than 133 worker-hours per day. Upholsterers average 2 hours per chair and 6 hours per sofa. There are enough upholsterers for no more than 72 worker-hours per day. The profit per chair is $80 and the profit per sofa is $70. How many sofas and chairs should be made per day to maximize the profit?

   a. Translate the constraints into a system of linear inequalities. Call the variables c and s.

   b. Graph the system of inequalities and find the vertices of the feasible set.

   c. Apply the Linear-Programming Theorem and interpret the results.

CHAPTER REVIEW

Questions on SPUR Objectives

SPUR stands for **S**kills, **P**roperties, **U**ses, and **R**epresentations. The Chapter Review questions are grouped according to the SPUR Objectives for this chapter.

SKILLS DEAL WITH THE PROCEDURES USED TO GET ANSWERS.

Objective A: *Solve 2 × 2 and 3 × 3 systems using the Linear Combination Method or substitution.* *(Lessons 5-3, 5-4)*

1. *Multiple choice.* After which of the following does the system $\begin{cases} 2x + 3y = 19 \\ 4x - y = 17 \end{cases}$ yield $-7y = -21$?
   a. Multiply the first equation by -2 and add.
   b. Multiply the second equation by 3 and add.
   c. Multiply the first equation by 2, the second equation by -1, and then add.
   d. Multiply the second equation by 3 and subtract.

2. Consider the system $\begin{cases} y = 5x \\ -3x + 2y = -28. \end{cases}$
   a. Which method do you prefer to use to solve this system?
   b. Solve and check the system using the method you prefer.

In 3–8, solve and check.

3. $\begin{cases} 2a - 4b = 18 \\ .5a - b = 22 \end{cases}$

4. $\begin{cases} 3m + 10n = 16 \\ m = -6n \end{cases}$

5. $\begin{cases} y = x - 4 \\ 2x - y = -2.5 \end{cases}$

6. $\begin{cases} 3x + 6y = -3 \\ -5x - 8y + 22 = 0 \end{cases}$

7. $\begin{cases} 2r + 15t = 6 \\ r = 3s \\ t = \frac{2}{5}s \end{cases}$

8. $\begin{cases} a = 3b - 2 \\ b = 4c + 5 \\ c = 5a + 1 \end{cases}$

Objective B: *Find the determinant and inverse of a square matrix.* *(Lesson 5-5)*

In 9–12, a matrix is given. **a.** Calculate its determinant. **b.** Write the inverse, if it exists.

9. $\begin{bmatrix} 2 & 0 \\ 0 & 1 \end{bmatrix}$

10. $\begin{bmatrix} 6 & 4 \\ -3 & 2 \end{bmatrix}$

11. $\begin{bmatrix} 2 & -4 \\ 5 & -10 \end{bmatrix}$

12. $\begin{bmatrix} a & b \\ c & d \end{bmatrix}$

13. Suppose $M = \begin{bmatrix} 1 & 9 \\ -7 & 6 \end{bmatrix}$. Find M^{-1}.

14. If the inverse of $\begin{bmatrix} p & q \\ r & s \end{bmatrix}$ does not exist, what must be true about its determinant?

15. Explain why the matrix $\begin{bmatrix} 3 & 2 \\ 6 & 4 \end{bmatrix}$ does not have an inverse.

16. **a.** Find $\begin{bmatrix} -1 & 2 & 3 \\ 4 & 5 & 6 \\ 7 & 8 & 9 \end{bmatrix}^{-1}$ using a calculator or a computer.
    b. Check your result.

Objective C: *Use matrices to solve systems of two or three linear equations.* *(Lesson 5-6)*

In 17–20, solve each system using matrices.

17. $\begin{cases} 2x - 9y = 14 \\ 6x - y = 42 \end{cases}$

18. $\begin{cases} 4a - 5b = -19 \\ 3a + 7b = 18 \end{cases}$

19. $\begin{cases} 3m = 4n + 5 \\ 2m = 3n - 6 \end{cases}$

20. $\begin{cases} 36 = 3x - 4y + 2z \\ 3 = x + 8y \\ 20 = 2x - y + 6z \end{cases}$

PROPERTIES DEAL WITH THE PRINCIPLES BEHIND THE MATHEMATICS.

Objective D: *Recognize properties of systems of equations.*   *(Lessons 5-2, 5-3, 5-4, 5-6)*

21. Are the systems $\begin{cases} 3x - y = 19 \\ 5x + 2y = 39 \end{cases}$ and

$\begin{cases} x = 7 \\ x + y = 9 \end{cases}$ equivalent? Why or why not?

22. Give the simplest system equivalent to $3x = 6$ and $x + y = 10$.

23. What is a system with no solutions called?

In 24–27, a system is given. **a.** Identify the system as inconsistent or consistent.
b. Determine the number of solutions.

24. $\begin{cases} 3x + 5y = 15 \\ 3x + 5y = 45 \end{cases}$ **25.** $\begin{cases} 6m - 4n = 9 \\ -3m = -2n - \frac{9}{2} \end{cases}$

26. $\begin{cases} 8a - 5b = 40 \\ 2a + b = -6 \end{cases}$ **27.** $\begin{cases} y = -x^2 \\ y = x - 5 \end{cases}$

28. For what value of k does $\begin{cases} 2x + ky = 6 \\ 14x + 7y = 42 \end{cases}$ have infinitely many solutions?

29. For what value of t does $\begin{cases} 3x + 9y = t \\ 4x + 12y = 7 \end{cases}$ have infinitely many solutions?

30. Suppose the determinant of the coefficient matrix of a system of equations is not zero. What can you conclude about the system?

Objective E: *Recognize properties of systems of inequalities.*   *(Lessons 5-8, 5-9)*

31. *True or false.* The boundaries are included in the graph of the solution set of $\begin{cases} y > 2 \\ y < 4 - x. \end{cases}$

32. A system of inequalities was graphed as shown below. Tell whether the point is a solution to the system. Justify your answer.
  a. (3, 2)   **b.** (6, 3)

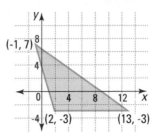

33. Which two of the shaded regions could be feasible sets in a linear-programming situation?

  (a)   (b)   (c)   (d)

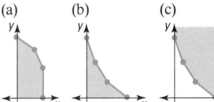
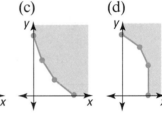

34. Where in a feasible set are the possible solutions to a linear programming problem?

35. Does the point M in the region at the right represent a possible solution to a linear programming problem? Why or why not?

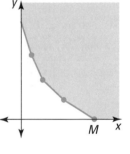

USES DEAL WITH APPLICATIONS OF MATHEMATICS IN REAL SITUATIONS.

Objective F: *Use systems of two or three linear equations to solve real-world problems.*
(Lessons 5-3, 5-4, 5-6)

36. At Kit's Kitchen, it costs $4.20 for two hamburgers and one order of fries. It costs $14.40 for six hamburgers and six orders of fries. At these prices, how much should one hamburger cost?

37. Billy likes to mix two cereals for breakfast. He wants to reduce his sugar intake without giving up Sugar-O's, his favorite cereal. Sugar-O's contains 20% sugar while Health-Nut contains 5% sugar. How much of each cereal does he need to fill a small bowl with a 25 g mixture that is 10% sugar?

38. One night at the circus, the big top attraction sold out, selling all 3,050 seats. There are four times as many lower-level seats as upper-level seats. How many upper-level seats are there?

39. A farm worker can earn $5.00 per hour plus $.30 per bag picking cucumbers at one farm and $5.40 per hour and $.24 per bag at another farm. Suppose a worker picks cucumbers for 8 hours each day.

   a. How many bags of cucumbers must the worker pick in order to earn the same daily wage at each farm?

   b. How many bags of cucumbers must the worker pick daily in order to earn more at the first farm?

40. After two tests Barbara's average in math was 76. After three tests it was 83. If the teacher drops Barbara's lowest test score, Barbara's average will be 88. What are Barbara's three test scores?

41. A travel agent books three charter groups to go on a weekend cruise. A group of surgeons reserves 9 first-class, 22 business-class, and 15 tourist-class rooms for $30,760. A group of journalists reserves 4 first-class, 13 business-class, and 8 tourist-class rooms for $16,660. A teacher's association reserves 5 first-class, 7 business-class, and 25 tourist-class rooms for $22,600. What is the weekend charge for each class of room?

Objective G: *Use linear programming to solve real-world problems.* *(Lessons 5-9, 5-10)*

42. Jocelyn's Jewelry Store makes rings and pendants. Every week the staff uses at most 500 g of metal and spends at most 80 hours making jewelry. It takes 5 g of metal to make a ring and 20 g to make a pendant. Each ring takes 1.5 hours to make and each pendant takes 1 hour. The profit on each ring is $90 and the profit on each pendant $40. The store wants to earn as much profit as possible.

   a. Identify the variables and translate the constraints into a system of inequalities.

   b. Graph the system and find the vertices of the feasible set.

   c. Write an expression to be maximized.

   d. Apply the Linear-Programming Theorem and interpret the results.

43. Surehold Shelving company produces two types of decorative shelves. The Olde English style takes 20 minutes to assemble and 10 minutes to finish. The Cool Contemporary style takes 10 minutes to assemble and 20 minutes to finish. Each day, there are 48 worker-hours of labor available in the assembly department and 64 worker-hours of labor available in the finishing department. To fulfill its commitments, the company must produce at least 200 shelving units per day. The cost of materials for the Olde English shelf is $2.00 each. The cost of materials for the Cool Contemporary shelf is $2.50 each. How many of each type of shelf should the company produce to minimize the cost of materials and still meet its production commitments?

REPRESENTATIONS DEAL WITH PICTURES, GRAPHS, OR OBJECTS THAT ILLUSTRATE CONCEPTS.

Objective H: *Solve and graph linear inequalities in one variable.* *(Lesson 5-1)*

44. Graph all solutions to $y \leq 6$ on a number line.

45. *Multiple choice.* Which inequality is graphed below?

(a) $x > \text{-}7$ (b) $x < \text{-}7$ (c) $x \geq \text{-}7$ (d) $x \leq \text{-}7$

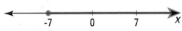

46. Write an inequality that describes the graph below.

In 47 and 48, solve the inequality and graph its solution set.

47. $\text{-}4x + 12 > 22$

48. $3n + 6(n - 12) \geq 9$

In 49–52, graph.

49. $\{x: x > 9 \text{ and } x < 14\}$

50. $\{t: \text{-}2 \leq t < 7\} \cap \{t: t \geq 0\}$

51. $\{n: n > 5\} \cup \{n: n > 3\}$

52. $\{y: y \leq 4 \text{ or } 5 \leq y \leq 6\}$

53. Write the compound sentence graphed below.

Objective I: *Estimate solutions to systems by graphing.* *(Lesson 5-2)*

In 54–56, estimate all solutions by graphing.

54. $\begin{cases} y = 3x - 5 \\ y = 1.5x + 3 \end{cases}$ **55.** $\begin{cases} 4x - y = \text{-}6 \\ y = x^2 \end{cases}$

56. $\begin{cases} 3x + 5y = \text{-}20 \\ \quad xy = 6 \end{cases}$

Objective J: *Graph linear inequalities in two variables.* *(Lesson 5-7)*

In 57–60, graph on a coordinate plane.

57. $x < \text{-}2 \text{ or } y \geq 0$ **58.** $x \geq 5 \text{ and } y \geq 12$

59. $y \geq \text{-}3x + 1$ **60.** $3x - 4y < 6$

61. Write an inequality to describe the shaded region at the right.

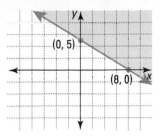

Objective K: *Solve systems of inequalities by graphing.* *(Lessons 5-1, 5-8)*

In 62–64, graph the solution set.

62. $\begin{cases} x \geq 2 \\ y \leq \text{-}3 \end{cases}$ **63.** $\begin{cases} 7c + 3d < 21 \\ 7c - 3d > \text{-}1 \end{cases}$

64. $\begin{cases} 5x \geq \text{-}10 \\ 3(x + y) \leq 6 \\ 6 > y - 4x \end{cases}$

65. Use a compound sentence to describe the region below.

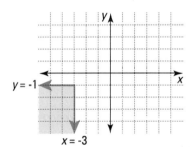

66. *Multiple choice.* Which of the following systems describes the region below?

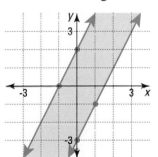

(a) $\begin{cases} y < \frac{1}{2}x + 2 \\ y < \frac{1}{2}x - 3 \end{cases}$ (b) $\begin{cases} y > 2x + 2 \\ y < 2x - 3 \end{cases}$

(c) $\begin{cases} y \leq 2x + 2 \\ y \geq 2x - 3 \end{cases}$ (d) $\begin{cases} y \leq 2x - 1 \\ y \geq 2x + 1.5 \end{cases}$

CHAPTER

6

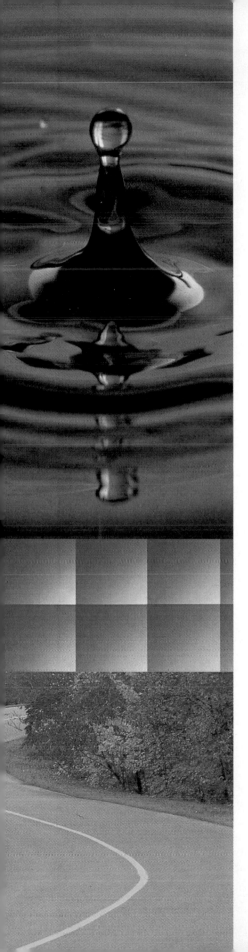

QUADRATIC FUNCTIONS

The word quadratic comes from the Latin word *quadratus,* which means "to make square."

Many situations lead to quadratic functions. You studied direct-variation quadratic functions of the form $f(x) = kx^2$ in Chapter 2. The area formulas $A = s^2$ (for a square) and $A = \pi r^2$ (for a circle) are equations for quadratic functions. The area of a rectangle with length x and width y is xy, a quadratic expression in the two variables x and y.

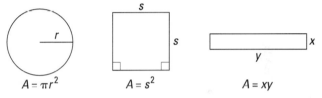

$A = \pi r^2$ $A = s^2$ $A = xy$

Quadratic functions also arise from studying the paths of objects. The path traveled by a baseball or a basketball closely follows the path of a parabola. The paths of the jets of a fountain also outline parabolas. All parabolas can be described by quadratic equations.

In this chapter, you will study many uses of quadratic expressions and functions. You also will learn how to solve all quadratic equations, including those whose solutions involve square roots of negative numbers.

Pooling resources. *International swim meets are held in pools that are 50 m long and are divided into 6, 8, or 10 lanes, each 2.1 or 2.4 m wide. The water must be at least 1.2 m deep and have a temperature of about 26°C.*

The word *quadratic* in today's mathematics refers to expressions, equations, and functions that involve sums of constants and first and second powers of variables and no higher powers. That is, they are of degree 2. Specifically, when $a \neq 0$:

$ax^2 + bx + c$ is the general **quadratic expression** in the variable x,
$ax^2 + bx + c = 0$ is the general **quadratic equation** in the variable x, and
$f: x \rightarrow ax^2 + bx + c$ is the general **quadratic function** in the variable x.

We call $ax^2 + bx + c$ the **standard form of a quadratic.** Some expressions, equations, and functions of degree 2 are not in standard form, but they can be rewritten in standard form. We still call them quadratic.

There can also be quadratics in two or more variables. The general quadratic expression in two variables is $Ax^2 + Bxy + Cy^2 + Dx + Ey + F$, and there are corresponding equations and functions of two variables. These are the subject of Chapter 12.

The simplest quadratic expression, x^2, equals the product of the simplest linear expressions x and x. More generally, the product of any two linear expressions $ax + b$ and $cx + d$ is a quadratic expression. Since all area formulas involve the product of two lengths, they all involve quadratic expressions.

Quadratic Expressions from Rectangles

Example 1

Suppose a rectangular swimming pool 50 m by 20 m is to be built with a walkway around it. If the walkway is w meters wide, write the total area of the pool and walkway in standard form.

▶

▶

Solution

Draw a picture. The pool with walkway occupies a rectangle with length 50 + 2w meters and width 20 + 2w meters. The area of this rectangle is (50 + 2w)(20 + 2w) square meters.

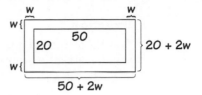

Apply the Distributive Property several times to find the product of $(50 + 2w)(20 + 2w)$.

$$(50 + 2w)(20 + 2w) = (50 + 2w) \cdot 20 + (50 + 2w) \cdot 2w$$
$$= 1000 + 40w + 100w + 4w^2$$
$$= 1000 + 140w + 4w^2$$

The total area of the pool and the walkway is $4w^2 + 140w + 1000$ square meters.

Check

Suppose the walkway is 3 meters wide. The width of the pool with the walkway is 20 + 2(3) = 26 m and the length is 50 + 2(3) = 56 m. So the area is 26 · 56 = 1456 m^2. Does 1456 = 1000 + 140(3) + 4(3^2)? 1456 = 1000 + 420 + 36. Yes, it checks.

Notice in Example 1 that the expression $4w^2 + 140w + 1000$ is of the form $ax^2 + bx + c$, with $w = x$, $a = 4$, $b = 140$, and $c = 1000$.

Quadratic Expressions from Squares

When a linear expression is multiplied by itself, the result, its square, is a quadratic expression. In Example 2, the linear expression $(x + 7)$ is squared. It is taken to the 2nd power. Writing this power as a quadratic expression is called *expanding* the power.

Example 2

Expand $(x + 7)^2$.

Solution

Use the definition of the second power. Then apply the Distributive Property several times.

$$(x + 7)^2 = (x + 7)(x + 7)$$
$$= (x + 7)x + (x + 7)7$$
$$= x^2 + 7x + 7x + 49$$
$$= x^2 + 14x + 49$$

Check

Let $x = 5$. Then $(x + 7)^2 = (5 + 7)^2 = 12^2 = 144$. Also, $x^2 + 14x + 49 = 5^2 + 14 \cdot 5 + 49 = 144$. It checks.

The square of a binomial can be thought of as the area of a square whose side is the binomial.

Example 3

Write the area of the square with sides of length $x + y$ in standard form.

Solution 1

Draw a picture of the square. Notice that its area is the sum of the four smaller areas: a square of area x^2, two rectangles, each with area xy, and a square of area y^2. So the area of the square is $x^2 + 2xy + y^2$.

Solution 2

The area of a square with side $x + y$ is $(x + y)^2$.
Expand $(x + y)^2$.

$$
\begin{aligned}
(x + y)^2 &= (x + y)(x + y) && \text{definition of 2nd power} \\
&= (x + y)x + (x + y)y && \text{Distributive Property} \\
&= x^2 + yx + xy + y^2 && \text{Distributive Property} \\
&= x^2 + 2xy + y^2 && \text{Commutative Property} \\
& && \text{of Multiplication and} \\
& && \text{Distributive Property}
\end{aligned}
$$

The area of the square is $x^2 + 2xy + y^2$.

Squares of binomials occur so often that their expansions are identified as a theorem.

Binomial Square Theorem
For all real numbers x and y,
$$(x + y)^2 = x^2 + 2xy + y^2 \text{ and}$$
$$(x - y)^2 = x^2 - 2xy + y^2.$$

The proof of the second part of the theorem is left to you. Of course, the theorem holds for *any* numbers or expressions. It is important for you to be able to apply it automatically.

Example 4

Expand $\left(3m - \frac{k}{4}\right)^2$.

Solution

Use the Binomial Square Theorem with $x = 3m$ and $y = \frac{k}{4}$.

$$
\begin{aligned}
(x - y)^2 &= x^2 - 2 \cdot x \cdot y + y^2 \\
\left(3m - \frac{k}{4}\right)^2 &= (3m)^2 - 2(3m)\left(\frac{k}{4}\right) + \left(\frac{k}{4}\right)^2 \\
&= 9m^2 - \frac{3k}{2}m + \frac{k^2}{16}
\end{aligned}
$$

▶

Check

Let $m = 3$ and $k = 8$. The left side is then 7^2 or 49. The right side is

$$9 \cdot 3^2 - \tfrac{3 \cdot 8}{2} \cdot 3 + \tfrac{8^2}{16} = 81 - 36 + 4 = 49.$$

QUESTIONS

Covering the Reading

1. *Multiple choice.* Which is *not* a quadratic equation?
   (a) $y = \tfrac{1}{2}x^2$ (b) $xy = 4$ (c) $x^2 + y^2 = 10$ (d) $y = 2x$

2. Is $x^2 - \sqrt{3}x + 4$ a quadratic expression? Explain your answer.

3. What kind of path does a thrown baseball follow?

4. A swimming pool 50 m by 25 m is to be built with a walkway w meters wide around it. Write the total area of the pool and walkway in the form $ax^2 + bx + c$.

In 5–8, multiply and simplify.

5. $(4x + y)(2x + 3y)$

6. $(x + 1)(x - 2)$

7. $(2 - y)(3 - y)$

8. $(12 + a)(5 + a)$

9. Expand.
   a. $(10 + 2)^2$ b. $(n + 2)^2$ c. $(n + q)^2$

10. Expand $(x - 5)^2$ and check your work.

11. Describe two different ways to find the area of a square with side of length $p + 3$.

12. Prove the second part of the Binomial Square Theorem.

In 13–16, rewrite the expression in the form $ax^2 + bx + c$.

13. $(5a + b)^2$

14. $(a - 8)^2$

15. $\left(2w - \tfrac{1}{2}\right)^2$

16. $(7e + 3f)^2$

Applying the Mathematics

17. Refer to the walkway around the swimming pool mentioned in Example 1. What is the area of the walkway?

18. Refer to the rectangles at the left. What is the area of the shaded region?

In 19–21, expand and simplify.

19. $\tfrac{1}{2}n(n + 1)^2$ 20. $(x + y)^2 - (x - y)^2$ 21. $-3(x - 2)^2$

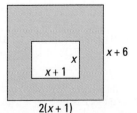

$x + 6$

x

$x + 1$

$2(x + 1)$

In 22 and 23, solve for h.

22. $x^2 + 50x + 625 = (x + h)^2$ **23.** $x^2 - 2x + 1 = (x + h)^2$

Located in Crater Lake National Park in Oregon, Crater Lake is the deepest lake in the U.S. with a depth of 1932 feet. The lake was formed when the top of Mount Mazama, an inactive volcano, collapsed and left a huge bowl that gradually filled with water.

Review

24. a. Draw $\triangle ABC$ with vertices $A = (0, 0)$, $B = (1, 1)$, $C = (2, 4)$.
   b. Draw $\triangle A'B'C'$, its image under the transformation
   $(x, y) \rightarrow (x - 5, y + 2)$.
   c. Describe the effect of this transformation on $\triangle ABC$.   *(Lesson 4-10)*

In 25 and 26, use the following data about the United States National Parks.   *(Lessons 2-5, 3-7)*

Year	Number of Visits (in millions)	Federal Appropriations (in billions of $)
1988	244.8	.97
1989	250.9	1.12
1990	258.7	1.14
1991	267.7	1.42
1992	274.7	1.49

Source: U.S. Department of the Interior

25. Find the rate of change from 1988 to 1992 for each of the following.
   a. number of visits   **b.** federal appropriations

26. a. Find an equation for the line of best fit describing the number of visits as a function of the year using only the years 1988, 1990, 1992. Let x be the number of years after 1988.
   b. How well does the line of best fit predict the values for 1991?

27. a. Graph $y = \frac{1}{2}x^2$ and $y = -2x^2$ on the same set of axes.
   b. Describe two properties that apply to both graphs.   *(Lesson 2-5)*

28. Which numbers between 100 and 200 are perfect squares?
   (Previous course)

Exploration

29. If you know the square of an integer, you can find the square of the next higher integer in your head. For instance, since 70^2 is known to be 4900, add 70 and 71 to 4900 to find 71^2.
   a. How could you find 101^2 in your head?
   b. Explain why this procedure works for any integer x.

LESSON 6-2

Absolute Value, Square Roots, and Quadratic Equations

Alge-robics. *These students are doing function "exercises." The graphs formed by their arms represent absolute value functions. See page 352.*

The Absolute Value Function

Geometrically, the absolute value of a number n, written $|n|$, is the distance of n from 0 on the number line. For instance, $|27| = 27$ and $|-27| = 27$. Both 27 and -27 are 27 units away from zero.

Algebraically, the **absolute value** of a number can be defined piecewise as

$$|x| = \begin{cases} x, \text{ for } x \geq 0 \\ -x, \text{ for } x < 0. \end{cases}$$

Examine the definition carefully. Because $-x$ is the opposite of x, $-x$ is positive when x is negative. For instance, $|-18| = -(-18) = 18$. Thus, $|x|$ and $|-x|$ are never negative, and in fact, $|x| = |-x|$.

Example 1

Solve for x: $|x - 2| = 5.3$

Solution 1

Use the algebraic definition of absolute value.
Either   x – 2 = 5.3   *or*   x – 2 = -5.3.
So   x = 7.3   *or*   x = -3.3.

Solution 2

Think of distance on the number line. $|x - 2|$ represents the distance between the points with coordinates x and 2. This distance equals 5.3. So measure 5.3 units from 2 in each direction.

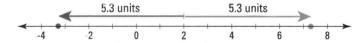

Check

The solutions are –3.3 or 7.3. $|7.3 - 2| = |5.3| = 5.3$ and $|-3.3 - 2| = |-5.3| = -(-5.3) = 5.3$. It checks.

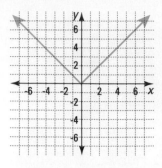

Because every real number has exactly *one* absolute value, $f: x \rightarrow |x|$ is a function.

The graph of $f(x) = |x|$ is shown at the left. When $x \geq 0$ the graph is a ray with slope 1 and endpoint (0, 0). This is the ray in the first quadrant. When $x < 0$, the graph is the ray with slope -1 and endpoint (0, 0). This is the ray in the second quadrant. The graph of $f(x) = |x|$ is the union of two rays, so the graph of $f(x) = |x|$ is an angle.

This function is called the **absolute value function.** Its domain is the set of real numbers, and its range is the set of nonnegative real numbers.

On many automatic graphers and spreadsheets and in many calculator and computer languages, the absolute value function is denoted ABS or abs. That is, $\text{ABS}(x) = |x|$. Parentheses must usually be used to indicate the argument of the absolute value function. For example, to evaluate $|x - 3|$, you need to enter $\text{ABS}(x - 3)$, not $\text{ABS } x - 3$.

Absolute Value and Square Roots

Recall that the square root or radical sign $\sqrt{}$ stands for only the *nonnegative* square root of a real number. Thus, $\sqrt{16} = 4$. To write the negative square root, you need to write $-\sqrt{16}$.

Activity 1

1. Evaluate each of the following. Use your calculator if necessary. $\sqrt{4^2}$, $\sqrt{(-4)^2}$, $\sqrt{9.3^2}$, $\sqrt{(-9.3)^2}$
2. Find a value of x that is a solution to $\sqrt{x^2} = x$.
3. Find a value of x that is not a solution to $\sqrt{x^2} = x$.

As you should have noticed from Activity 1, $\sqrt{x^2}$ cannot always be simplified to x.

If x is positive, then $\sqrt{x^2} = x$.
If $x = 0$, then $\sqrt{x^2} = 0$.
If x is negative, then $\sqrt{x^2} = -x$, which is a positive number.

This proves a surprising relationship between square roots and absolute value.

x	x^2	$\sqrt{x^2}$	$|x|$
-17	289	17	17
-3.14	9.8596	3.14	3.14
-1	1	1	1
$-\frac{2}{3}$	$\frac{4}{9}$	$\frac{2}{3}$	$\frac{2}{3}$
0	0	0	0
$\frac{2}{3}$	$\frac{4}{9}$	$\frac{2}{3}$	$\frac{2}{3}$
1	1	1	1
3.14	9.8596	3.14	3.14
10	100	10	10

> **Absolute Value – Square Root Theorem**
> For all real numbers x, $\sqrt{x^2} = |x|$.

You can verify this theorem by making a table of values like the one shown at the left. Notice how the values in the last two columns are always equal.

Solving Some Quadratic Equations

The simplest quadratic equations are of the form $x^2 = k$. As you know, if $k \geq 0$, the solutions to $x^2 = k$ are the **square roots** of k, namely $\sqrt{k}$ and $-\sqrt{k}$.

The Absolute Value-Square Root Theorem can be used to solve some quadratic equations.

Example 2

Solve $x^2 = 40$.

Solution

Take the positive square root of each side.

$$\sqrt{x^2} = \sqrt{40}.$$

Use the Absolute Value-Square Root Theorem.

$$|x| = \sqrt{40}$$

So either $x = \sqrt{40}$ or $x = -\sqrt{40}$.

Check

Use your calculator to evaluate $(\sqrt{40})^2$ and $(-\sqrt{40})^2$. Each equals 40.

Questions about squares intrigued the ancient Greek mathematicians. They wondered: What should be the radius of a circle if it is to have the same area as a given square? With algebra, it is possible to find it.

Example 3

A square and a circle have the same area. The square has side 10. What is the radius of the circle?

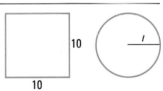

Solution

The area of the square is 100. Let r be the radius of the circle.

$$\pi r^2 = 100$$

$$r^2 = \frac{100}{\pi} \qquad \text{Divide by } \pi.$$

$$|r| = \sqrt{\frac{100}{\pi}} \qquad \text{Take the square root of each side.}$$

$$r = \pm \sqrt{\frac{100}{\pi}}$$

The $\pm$ sign here means there are two solutions, one with the $+$ sign, one with the $-$ sign. In this situation we can ignore the negative solution because the radius can never be negative. Using a calculator, $\sqrt{\frac{100}{\pi}} \approx 5.64$. The radius of the circle is approximately 5.64 units.

Check

The diameter of the circle should be greater than the side of the square. (Do you see why?) Since $d = 2r$, $d \approx 11.28$, which is greater than 10.

The Greek mathematicians also wondered if square roots could be expressed as simple fractions. For instance, they knew that $\sqrt{40}$ was between 6 and 7, because $\sqrt{36} = 6$ and $\sqrt{49} = 7$. But was it some number like $\frac{13}{2}$, which as a mixed number is $6\frac{1}{2}$? No, $\left(6\frac{1}{2}\right)^2 = 42\frac{1}{4}$, so $6\frac{1}{2} > \sqrt{40}$.

Activity 2

a. Square $6\frac{8}{25}$.

b. Is $6\frac{8}{25}$ more or less than $\sqrt{40}$?

Recall from Lesson 1-2 that a *rational number* is a number that can be written in the form $\frac{a}{b}$, where a and b are integers and $b \neq 0$. So the Greeks were asking: Is $\sqrt{40}$ a rational number? Around 430 B.C., they proved that unless an integer was a perfect square (like 49, or 625, or 10,000), its square root was an *irrational number*. An **irrational number** is a real number that cannot be written as a simple fraction. You may have seen proofs that certain square roots are irrational in an earlier course. In general, irrational numbers are exactly those numbers which have infinite non-repeating decimal expansions.

QUESTIONS

Covering the Reading

1. Evaluate $|17.8|$.

2. Evaluate $|-11|$.

3. **a.** Evaluate $|x + 2|$ when $x = -3$.
   b. Compare your answer in part **a** to the value of $|x| + 2$ when $x = -3$.

4. Solve $|x - 6| = 4.5$.

5. Solve $|n + 1.8| = 5$.

6. State the domain and range of the function f with $f(x) = |x|$.

7. Is the graph of $f(x) = |x|$ piecewise linear? Justify your answer.

8. The square roots of 25 are __?__ and __?__.

9. Give your answers to the questions in Activity 1.

10. Give your answers to the questions in Activity 2.

11. When $x < 0$, what does $\sqrt{x^2}$ equal?

12. When $x < 0$, what does $|x|$ equal?

13. Solve $t^2 = 720$.

14. A circle has the same area as a square of side 6. What is the radius of the circle?

15. About how many years ago was it first shown that certain numbers are irrational?

In 16–21, tell whether the number is rational or irrational. If rational, write the number as a simple fraction.

16. $\sqrt{10}$

17. $\sqrt{100}$

18. -5

19. $\frac{203}{317}$

20. $\sqrt{2} - \sqrt{2}$

21. π

Applying the Mathematics

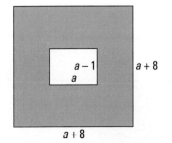

Shown is the beginning of a 100-meter race with U.S. record holder Mary T. Meagher in the foreground. International swimming competitions are held annually.

22. Suppose the ideal length of an item is I units. If the item is actually p units long, then the formula $e = |p - I|$ gives the error e in the length of the item. When constructing a swimming pool for international competitions the contractor aims for a length of 50.015 meters with an acceptable value of e of no more than .015 m.
   a. Write a mathematical sentence satisfied by the possible lengths p of pools in this situation.
   b. What are the possible solutions to this sentence?

23. a. Graph on the same set of axes: $f(x) = \sqrt{x^2}$ and $g(x) = |x|$.
   b. How are the two graphs related?

24. a. Graph the function d with $d(x) = -|x|$.
   b. State the domain and the range of d.

25. If $\sqrt{(x - 3)^2} = |k|$, then $k = \underline{\ ?\ }$.

26. On a brand-name pizza box, the directions read: "Spread dough to edges of pizza pan or onto a 10″ by 14″ rectangular cookie sheet." How big a circular pizza could you make with this dough, assuming it is spread the same thickness as for the rectangular pizza?

27. Solve.
   a. $x^2 + 36 = 49$
   b. $x^2 + 36 = 50$

Review

28. *Skill sequence.* Expand and simplify. *(Lesson 6-1)*
   a. $(x - 10)^2$
   b. $(3x - 10)^2$
   c. $3(x - 10)^2$

29. *True or false.* $(n + 11)^2 = n^2 + 11^2$. Justify your answer. *(Lesson 6-1)*

30. Find the area of the shaded region determined by the rectangles at the left. *(Lesson 6-1)*

31. Consider the graph of $y = 3x^2$. Find an equation for its image under r_x and under r_y. *(Lesson 4-6)*

$a - 1$

a

$a + 8$

$a + 8$

Exploration

32. a. Find a simple fraction within .000001 of $\sqrt{40}$.
   b. How close to 40 is the square of your number?

IN · CLASS

ACTIVITY

Materials: Use an automatic grapher.
Work in small groups. Clear the screen of your grapher after you have answered each question.

In 1 and 2, equations for two functions are given. **a.** Draw the graphs of both functions on the same set of axes. **b.** What transformation maps the graph of the first function onto the graph of the second?

1 $y_1 = x^2$, $y_2 = (x - 8)^2$

2 $y_1 = x^2$, $y_2 = (x + 4)^2$

3 **a.** Make a conjecture describing the transformation that maps the graph of $y_1 = x^2$ onto the graph of $y_2 = (x - h)^2$.
b. Test your conjecture with another value of h.

In 4 and 5, follow the directions of Questions 1 and 2.

4 $y_1 = x^2$, $y_2 = x^2 + 3$

5 $y_1 = x^2$, $y_2 = x^2 - 6$

6 **a.** Make a conjecture describing the transformation that maps the graph of $y_1 = x^2$ onto the graph of $y_2 = x^2 + k$.
b. Test your conjecture with another value of k.

7 Find the vertex of the parabola $y = (x - 2)^2 + 4$.

8 Find an equation for a parabola congruent to $y = x^2$ with vertex at (-6, 3).

Shown is a multi-exposure print which illustrates the Graph-Translation Theorem.

What Is the Graph-Translation Theorem?

Consider the graphs of the equations $y_1 = x^2$ and $y_2 = (x - 8)^2$ that you made in the preceding In-class Activity.

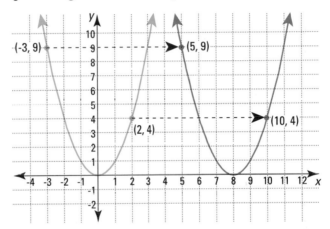

As the arrows indicate, the graph of $y = (x - 8)^2$ can be obtained from the graph of $y = x^2$ by a translation 8 units to the right. Thus, replacing x by $x - 8$ in the equation for the function translates its graph 8 units to the right.

In general, replacing x by $x - h$ in a sentence translates its graph h units to the right.

Similarly, vertical translations of graphs result from replacing y by $y - k$. For example, you saw that the graph of the equation $y = x^2 + 3$ is 3 units above the graph of $y = x^2$. Note that we can rewrite this equation as $y - 3 = x^2$. Thus, replacing y by $y - 3$ in the equation for a function translates its graph 3 units up. In general, replacing y by $y - k$ in a sentence translates its graph k units up.

Recall that the translation $T_{h,k}$ slides a figure h units to the right and k units up at the same time. Thus we can translate horizontally and vertically at the same time. We can summarize these examples as follows.

> **Graph-Translation Theorem**
> In a relation described by a sentence in x and y, the following two processes yield the same graph:
> (1) replacing x by $x - h$ and y by $y - k$;
> (2) applying the translation $T_{h,k}$ to the graph of the original relation.

The Graph-Translation Theorem applies to all relations that can be described by a sentence in x and y.

Example 1

Find an equation for the image of the graph of $y = |x|$ under the translation $T_{5,-3}$.

Solution

Applying $T_{5,-3}$ is equivalent to replacing x by $x - 5$ and y by $y - -3$ or $y + 3$ in the equation for the preimage. An equation for the image is $y + 3 = |x - 5|$ or $y = |x - 5| - 3$.

Check

Graph $y = |x|$ and $y = |x - 5| - 3$ on the same set of axes. As shown at the left, the graph of the second equation is the image of the graph of the first equation under a translation 5 units to the right and 3 units down.

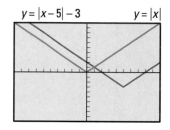

$y = |x - 5| - 3$ $\qquad$ $y = |x|$

$-10 \leqslant x \leqslant 10$, x-scale $= 1$
$-10 \leqslant y \leqslant 10$, y-scale $= 1$

Applying the Graph-Translation Theorem to Graph Parabolas

Recall from Chapter 2 that the graph of $y = ax^2$ is a parabola. If we replace x by $x - h$ and y by $y - k$ in the equation $y = ax^2$, we obtain $y - k = a(x - h)^2$. Since any figure is congruent to its translation image, the image is also a parabola.

This argument proves the following *corollary* to the Graph-Translation Theorem. (Recall that a **corollary** is a theorem that follows immediately from another theorem.)

> **Corollary**
> The image of the parabola $y = ax^2$ under the translation $T_{h,k}$ is the parabola with the equation
> $$y - k = a(x - h)^2.$$

Example 2

Sketch the graph of $y - 7 = 3(x - 6)^2$.

Solution
The equation is the result of replacing x by $x - 6$ and y by $y - 7$ in $y = 3x^2$. So its graph is the image of $y = 3x^2$ under the translation $T_{6,7}$. Thus, the graph is a parabola with vertex $(6, 7)$. Because $y = 3x^2$ opens up, so does the graph of $y - 7 = 3(x - 6)^2$. To find some other points on the graph, find some solutions to $y = 3x^2$ and then find their images under $T_{6,7}$. For example,

$$\begin{array}{c} \text{preimage} \rightarrow \text{image} \\ (-2, 12) \rightarrow (4, 19) \\ (-1, 3) \rightarrow (5, 10) \\ (1, 3) \rightarrow (7, 10) \\ (2, 12) \rightarrow (8, 19) \end{array}$$

The graph is shown below.

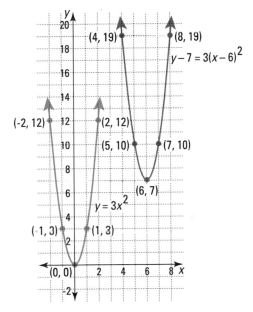

Graphic display. *These students are illustrating the Graph-Translation Theorem for a parabola.*

Check
Use an automatic grapher to draw the graph. Trace to verify that $(6, 7)$ is the vertex.

You know that $(0, 0)$ is the vertex of the parabola $y = ax^2$. The translation image of $(0, 0)$ is $T_{h,k}(0, 0) = (0 + h, 0 + k) = (h, k)$. So, the vertex of the parabola with equation $y - k = a(x - h)^2$ is (h, k). For this reason, the equation $y - k = a(x - h)^2$ is called the **vertex form of an equation for a parabola.** The line with equation $x = h$ is the line of symmetry or **axis of symmetry** of the parabola. If $a > 0$, then the parabola opens up and the y-coordinate of the vertex is the **minimum** y-value. If $a < 0$, then the parabola opens down and the graph has a **maximum** y-value. When the equation for a parabola is in vertex form, the parabola can be graphed quickly even if you do not have an automatic grapher.

Example 3

a. Sketch the graph of $y = -\frac{1}{2}(x + 3)^2$.

b. Give an equation for the axis of symmetry of the parabola.

Solution

a. Rewrite the equation in vertex form: $y - 0 = -\frac{1}{2}(x - -3)^2$. The graph is the image of the parabola $y = -\frac{1}{2}x^2$ under $T_{-3,0}$. Thus it opens down and its vertex is $(-3, 0)$. Find other points by substituting values into $y = -\frac{1}{2}(x + 3)^2$. Use some values of x less than -3 and some values greater than -3. A table and a graph are shown below.

x	y
-7	-8
-5	-2
-3	0
-1	-2
1	-8

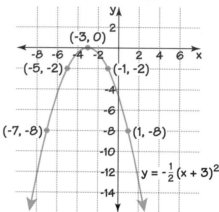

b. The axis of symmetry is the vertical line through the vertex. The axis of symmetry has equation $x = -3$.

Finding Equations for Parabolas

You can apply the Graph-Translation Theorem to a known parabola to find an equation for its image under a given translation.

Example 4

Consider the parabolas at the left. The one that passes through the origin has equation $y = 2x^2$. The other is its image under a translation. Find an equation for the image.

-5 ≤ x ≤ 5, x-scale = 1
-5 ≤ y ≤ 10, y-scale = 1

Solution

The translation image appears to be 3 units to the left and 4 units up from the preimage. So the translation is $T_{-3,4}$. Applying $T_{-3,4}$ is equivalent to replacing x by $x + 3$ and y by $y - 4$ in the equation for the preimage. An equation for the image is $y - 4 = 2(x + 3)^2$ or $y = 2(x + 3)^2 + 4$.

Check

Use an automatic grapher. Plot $y = 2x^2$ and $y = 2(x + 3)^2 + 4$ on the same set of axes. You should see that the graph of the second equation is the image of the graph of the first under a translation 3 units left and 4 units up.

Covering the Reading

In 1 and 2, tell how the graphs of the two equations are related.

1. $y_1 = x^2$ and $y = (x - 8)^2$

2. $y_1 = x^2$ and $y - 3 = x^2$

3. a. What is the image of (x, y) under $T_{6,0}$?
   b. Under $T_{6,0}$, what is an equation for the image of the graph of $y = x^2$?

4. a. On the same axes, draw the graphs of $y = |x|$ and $y + 3 = |x - 5|$.
   b. Describe how the two graphs are related.

5. Suppose the parabola with equation $y = \frac{4}{7}x^2$ undergoes the translation $T_{2,-3}$. Find an equation for its image.

6. The graph of $y - k = a(x - h)^2$ is __?__ units above and __?__ units to the right of the graph of $y = ax^2$.

7. *True or false.* The graphs of $y = ax^2$ and $y = a(x - h)^2 + k$ are congruent.

8. What is the vertex of the parabola with equation $y - k = a(x - h)^2$?

9. Refer to Example 2.
   a. Give an equation of the axis of symmetry of $y = 3x^2$.
   b. Give an equation of the axis of symmetry of $y - 7 = 3(x - 6)^2$.

In 10 and 11, an equation for a parabola is given.
a. Give the coordinates of the vertex of the parabola.
b. Give an equation for the axis of symmetry.
c. Tell whether the parabola opens up or down.
d. Graph the solution set to the equation.

10. $y + 2 = -3(x + 7)^2$ **11.** $y = (x - 2)^2 + 6$

12. Find an equation for the translation image of $y = |x|$ graphed at the left below.

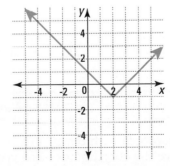   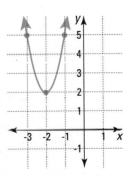

13. The parabola graphed at the right above is a translation image of $y = 3x^2$. What is an equation for this parabola?

Applying the Mathematics

14. A parabola congruent to $y = 7x^2$ has vertex (2, -5) and opens down. What is an equation for the parabola?

15. Consider the graph of $y = -1.4x^2$. Write an equation for its translation image. **a.** with vertex (0, 3). **b.** with vertex (3, 0).

16. a. Solve $x^2 = 100$.
 b. Solve $(x - 3)^2 = 100$.
 c. How are the solutions in parts **a** and **b** related to the Graph-Translation Theorem?

17. One solution to $x^2 + 5x + 3 = 87$ is 7. Use this information to find a solution to $(x - 4)^2 + 5(x - 4) + 3 = 87$.

18. The point-slope form of a line, $y - y_1 = m(x - x_1)$, can be thought of as the image of the line with equation __?__ under the translation $T_{h,k}$, where $h = $ __?__ and $k = $ __?__.

Review

In 19 and 20, solve and check. *(Lesson 6-2)*

19. $|2n + 1| = 0.5$ **20.** $x^2 - 17 = 22$

21. A 100'-by-60' rectangular lot is located in a town that allows no building closer than 5 feet to the edge of a lot.
 a. How much room is there to build?
 b. If no building were allowed to be closer than x feet to the edge of the lot, how much room would there be to build?   *(Lesson 6-1)*

22. A student claims that the graphs of $y_1 = (x - 9)^2$ and $y_2 = x^2 - 18x + 81$ coincide. How can you tell without graphing whether the student is right?   *(Lesson 6-1)*

23. Expand and simplify.
 a. $2(x + 4)^2$ **b.** $2(x + 4)^2 + 3$ *(Lesson 6-1)*

24. a. Find an equation of the line through the points (-2, 8) and (1, 17).
 b. Find a point on this line which lies in the third quadrant.
 (Lesson 3-5)

25. A school has buses that hold 40 students each. If s students are to be transported, how many buses are needed?   *(Lessons 1-1, 3-9)*

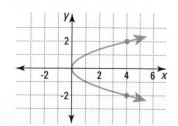

Exploration

26. The parabola with equation $x = y^2$ is graphed at the left.
 a. Give the coordinates of five points on this parabola.
 b. Graph its image under the translation $T_{3,-1}$.
 c. Write an equation for the image.
 d. Does the Graph-Translation Theorem hold for parabolas that open to the side?

Natural curve. *Shown are Manabezho Falls at the Porcupine Mountains State Wilderness Park in the Upper Peninsula of Michigan. The path of the water forms part of a parabola.*

Standard Form for the Equation of a Parabola

In general, any parabola whose equation can be written in vertex form $y - k = a(x - h)^2$ can be rewritten in the standard form $y = ax^2 + bx + c$. Here we show how.

Example 1

Show that the two formulas $y = 2(x + 3)^2 - 8$ and $y = 2x^2 + 12x + 10$ are equivalent.

Solution

Begin with $y = 2(x + 3)^2 - 8$. Expand the binomial, and simplify the right side of the equation.

$$y = 2(x^2 + 6x + 9) - 8$$
$$y = 2x^2 + 12x + 18 - 8$$
$$y = 2x^2 + 12x + 10$$

So the two formulas are equivalent.

Check

Graph $y = 2(x + 3)^2 - 8$ and $y = 2x^2 + 12x + 10$.
The graphs are the same.
Each looks like the graph shown here.

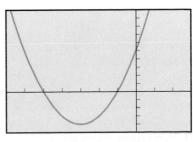

$-7 \leq x \leq 3, \quad x\text{-scale} = 1$
$-10 \leq y \leq 20, \quad y\text{-scale} = 2$

In general, to change vertex form to standard form, proceed as follows.

$$y - k = a(x - h)^2$$
$$y = a(x - h)^2 + k \qquad \text{Add } k \text{ to each side.}$$
$$y = a(x^2 - 2hx + h^2) + k \qquad \text{Square the binomial}$$
$$y = ax^2 - 2ahx + ah^2 + k \qquad \text{Use the Distributive Property}$$

This is in standard form, with $b = -2ah$ and $c = ah^2 + k$. With these substitutions, the equation becomes

$$y = ax^2 + bx + c.$$

Because the parabola determined by the equation $y - k = a(x - h)^2$ is a translation image of the parabola determined by the equation $y = ax^2$, the two parabolas are congruent.

> **Theorem**
> The graph of the equation $y = ax^2 + bx + c$ is a parabola congruent to the graph of $y = ax^2$.

Recall that any function f with an equation that can be put in the form $f(x) = ax^2 + bx + c$, where $a \neq 0$, is a quadratic function. Thus the graph of every quadratic function is a parabola, with y-intercept at $f(0) = c$. Unless otherwise specified, the domain of a quadratic function is the set of real numbers. The range is determined by examining the graph of the function.

The vertex form of a parabola is useful because it provides a quick way to visualize or sketch its graph, and it tells you the axis of symmetry and the vertex of the parabola. The standard form of a parabola is important because it has many applications in the real world, and because it tells you the y-intercept of the graph.

Applications of Quadratic Functions

Some applications of quadratic functions have been known for centuries. In the 16th century, Galileo described the motion of objects in free fall using mathematics. In the 17th century, Isaac Newton formulated his laws of motion and the law of universal gravitation. According to these laws, a ball thrown straight up at a velocity of 44 feet per second (30 mph) would go up

$$44t \text{ feet}$$

t seconds after it was released if there were no force acting to pull the ball downward. However, objects in free fall near the surface of the earth are acted upon by a force which he called *gravity* that acts to pull the object downward. After the first t seconds, gravity decreases the height of the ball by $16t^2$ feet. Thus, after t seconds the height h of the ball would be

$$44t - 16t^2 \text{ feet.}$$

Galileo (1564–1642), an Italian astronomer and physicist, conducted research on motion consistent with a moving Earth. From this research he developed the law of falling bodies and the law of the pendulum.

If the ball is released from a point 5 feet above ground level, then the height of the ball above the ground after the first t seconds is

$$44t - 16t^2 + 5 \text{ feet.}$$

Letting h be the height after t seconds, we can write $h = -16t^2 + 44t + 5$. By substituting values for t, the height h can be found after any number of seconds. The pairs (t, h) can be graphed.

Example 2

Suppose $h = -16t^2 + 44t + 5$.
a. Find h when $t = 0, 1, 2,$ and 3.
b. Explain what each pair (t, h) tells you about the height of the ball.
c. Graph the pairs (t, h) over the domain of the function.

Solution

a. At $t = 0$, $h = -16 \cdot 0^2 + 44 \cdot 0 + 5 = 5$
   At $t = 1$, $h = -16 \cdot 1^2 + 44 \cdot 1 + 5 = 33$
   At $t = 2$, $h = -16 \cdot 2^2 + 44 \cdot 2 + 5 = 29$
   At $t = 3$, $h = -16 \cdot 3^2 + 44 \cdot 3 + 5 = -7$

b. The pair $(0, 5)$ means that at 0 seconds, the time of release, the ball is 5 feet above the ground. The pair $(1, 33)$ means the ball is 33 feet high after 1 second. The pair $(2, 29)$ means the ball is 29 feet high after 2 seconds. (It is already on its way down.) The pair $(3, -7)$ means that after 3 seconds, the ball is 7 feet below ground level. Unless the ground is not level, it has already hit the ground.

c. The points in part a are plotted below on the left. The points do not tell much about the shape of the graph. More points are needed to show the parabola.

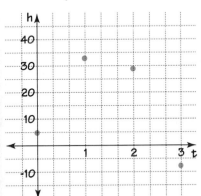
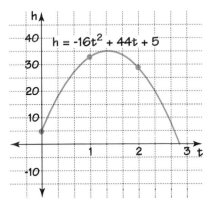

By calculating $h(t)$ for other values of t, or by using an automatic grapher, you can obtain a graph similar to the one above on the right.

In fact, because of the theorem on page 364, we know that, if $h = -16t^2 + 44t + 5$ is graphed for all real numbers t, it is a translation image of the graph of $y = -16x^2$.

Sir Isaac Newton (1642–1727) published his discoveries on the laws of motion and theories of gravitation in 1687 in Philosophiae Naturalis Principia Mathematica. *This publication is considered one of the greatest single contributions in the history of science. Queen Anne of Great Britain knighted him in 1705.*

Two natural questions about the thrown ball are related to questions about this parabola:

1. **How high does the ball get?** That asks for the largest possible value of h. From the graph, it seems to be about 35 feet.
2. **When does the ball hit the ground?** That asks for the larger t-intercept of the graph. It is between 2 and 3 seconds, nearer 3.

In later lessons, you will learn how to determine more precise answers to these questions.

The equation in Example 2 is a special case of a general formula for the height h of an object at time t with an initial upward velocity v_0 and initial height h_0 that was discovered by Newton. That formula is

$$h = -\tfrac{1}{2}gt^2 + v_0t + h_0$$

where g is a constant measuring the **acceleration due to gravity.** Recall that *velocity* is the rate of change of position with respect to time. Velocity is measured in units such as miles per hour, feet per second, or meters per second. Acceleration is a measure of how fast the velocity changes. This "rate of a rate" is measured in units like feet per second per second (or feet per second^2). The acceleration due to gravity varies depending on how close the object is to the center of a massive object. Near the surface of Earth, g is about $32\frac{\text{ft}}{\text{sec}^2}$, or $9.8\frac{\text{m}}{\text{sec}^2}$. In Example 2, $v_0 = 44$ and the height $h_0 = 5$.

Caution! The equation

$$h = -\tfrac{1}{2}gt^2 + v_0t + h_0$$

represents the height h of the ball off the ground at time t. It *does not* describe the path of the ball. However, the actual path of a ball thrown up into the air at any angle except straight up or straight down is almost parabolic, and an equation for its path is quadratic.

The paths of some objects in free fall can be described with simpler equations. For instance, when a ball is dropped (not thrown downward), its initial velocity is 0. So $v_0 = 0$ in the formula $h = -\tfrac{1}{2}gt^2 + v_0t + h_0$ and there is one less term. Example 3 illustrates such a situation.

Example 3

A ball is dropped from the top of a 20 meter tall building.
a. Find an equation describing the relation between h, the ball's height above the ground, and time t.
b. Graph its height h after t seconds.
c. Estimate how much time it takes the ball to fall to the ground. Explain your reasoning.

Solution

a. Because the unit of height is meters, use $g = 9.8$ m/sec^2. The ball is dropped, so $v_0 = 0$. Because the ball started 20 meters up, $h_0 = 20$. So the height in this situation is determined by the equation at the top of page 367.

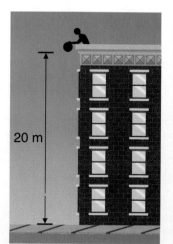

20 m

$$h = -\frac{1}{2}(9.8)t^2 + (0)t + 20$$

This equation is equivalent to $h = -4.9t^2 + 20$.

b. Negative values of t are not in the domain of t because a negative value of t would refer to something that had happened before the ball is dropped. Thus, the graph is entirely to the right of the y-axis. Negative values of h are not in the range because the ball is never below the ground, so the graph is entirely above the x-axis.

t	$h = -4.9t^2 + 20$
0	20
0.5	18.775
1	15.1
1.5	8.975
2	0.4
2.5	-10.625

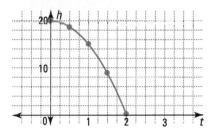

c. At $t = 2$ the ball is 0.4 m above the ground; at $t = 2.5$, according to the equation, the ball will be about 10.6 m below ground. Therefore, the ball hits the ground between 2 and 2.5 seconds after the ball is dropped.

Notice in Example 3 that the curve becomes steeper and steeper as t increases from 0 to 2. This reflects the increasing speed of the ball as it falls.

QUESTIONS

Covering the Reading

1. Give the standard form for the equation of a parabola.

In 2 and 3, rewrite the equation in standard form.

2. $y = (x + 6)^2$ **3.** $y = -2(x + 3)^2 + 4$

4. *True or false.* For any values of a, b, and c, the graph of $y = ax^2$ is congruent to the graph of $y = ax^2 + bx + c$.

In 5–7, use the equation $h = -\frac{1}{2}gt^2 + v_0t + h_0$ for the motion of a body in free fall.

5. Give the meaning of each of the following variables.
   a. h **b.** h_0 **c.** v_0 **d.** t **e.** g

6. If v_0 is measured in meters per second, what value of g should be used?

7. What is the value of v_0 if a ball is dropped?

In 8–10, refer to the graph of Example 2.

8. About how high is the ball after 2.5 seconds?

9. When the ball hits the ground, the value of h is __?__.

10. About when will the ball in Example 2 be 10 feet from the ground? (There are two answers.)

In 11 and 12, refer to Example 3.

11. What point corresponds to the time the ball is dropped?

12. Tell whether the ball is above or below ground at $t = 2.1$. Justify your answer.

Applying the Mathematics

In 13 and 14, graph the given equation for $-4 \le x \le 6$. On your sketch of the graph, label the vertex and the x- and y-intercepts.

13. $y = x^2 - 4x + 3$

14. $y = -2x^2 + 10x$

15. Consider the function defined by the equation $f(x) = x^2 - 5x - 6$.
   a. Sketch the graph of the function.
   b. Give an equation for the line of symmetry of the graph.
   c. Estimate the coordinates of the lowest point on the graph.

16. Suppose a juggler throws an object from his hand at a height of 1 meter with an initial upward velocity of 10 meters per second.
   a. Write an equation to describe the height of the object after t seconds.
   b. How high is the object after 1 second?
   c. Graph the equation from part a.
   d. Estimate the maximum height the object reaches.

17. I.M. Chisov of the USSR set a record in January 1942 for the highest altitude from which someone survived after bailing out of an airplane without a parachute. He bailed out at about 6700 meters.
   a. Write an equation describing his height at t seconds.
   b. Graph the equation in part a.
   c. About how long did his fall take?

18. Consider the parabolas with equations $y = \frac{1}{2}x^2 + \frac{x}{2}$ and $y = \frac{1}{2}(x + 1)^2$.
   a. Are the parabolas congruent? Why or why not?
   b. Do the parabolas coincide? Why or why not?

19. Find an equation in standard form for the image of the graph of $y = -.5x^2$ under the translation $T_{-2,4}$.

Up in the air. *Juggling is throwing and catching more than one object in one hand, or three or more objects in two hands. Juggling helps to develop hand-eye coordination and is often taught in elementary school.*

Review

In 20 and 21, **a.** Draw the graph of both equations on the same set of axes. **b.** Describe in a sentence or two how the graphs are related. *(Lessons 6-1, 6-3)*

20. $y = x^2$ and $y = (x - 2)^2$

21. $y = |x|$ and $y + 4 = |x - 5|$

368

22. A half-gallon of paint is supposed to cover an area of 450 square feet. Find the diameter of the largest circle that can be painted with this paint. *(Lesson 6-2)*

23. A mat w inches wide is to surround a picture that is 8″ by 12″. A thin frame surrounds the mat.
   a. What is the area of the picture with its mat?
   b. What is the inner perimeter of the frame surrounding the mat?
   (Lesson 6-1)

24. Evaluate when $x = -0.5$. *(Lessons 1-2, 6-1, 6-2)*
   a. $|x|$ **b.** $|-x|$ **c.** $-|x|$ **d.** x^2
   e. $-x^2$ **f.** $(-x)^2$ **g.** $\sqrt{x^2}$ **h.** $-\sqrt{x^2}$

25. Solve for a: $x^2 + 24x + 144 = (x + a)^2$. *(Lesson 6-1)*

26. Given the feasible region at the right, find the vertex at which $80x + 120y = P$ is maximized. *(Lesson 5-9)*

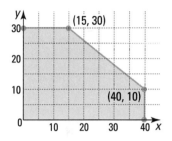

27. The table below gives the book value of a 1986 Chevrolet S-10 pickup truck from the year of its purchase until 1993.

Year	1987	1988	1990	1993
Value	5825	5450	4225	3425

   a. Find an equation for the line of best fit for the data.
   b. Use the equation to predict the value of a 1986 S-10 pickup in 1996.
   c. Why is it not a good idea to use this equation to predict the value of the truck in the year 2006? *(Lesson 3-8)*

Exploration

28. Draw a dot and a line on a large piece of plain paper. Now fold the paper so that the dot falls on a point of the line. Then unfold the paper. Do this 25 times, each time with a different point on the line, and describe what happens.

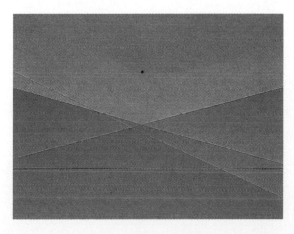

Completing the Square

You have now seen two forms for an equation of a parabola.

$$y = ax^2 + bx + c \qquad \text{standard form}$$
$$y - k = a(x - h)^2 \qquad \text{vertex form}$$

Because each form is useful, converting from one form to the other is helpful. Lesson 6-4 covered how to convert from vertex form to standard form. In this lesson, you will learn to convert from standard form to vertex form.

Completing the Square Geometrically

One method for converting from standard form to vertex form is called *completing the square*. Remember that $(x + h)^2 = x^2 + 2hx + h^2$. The trinomial $x^2 + 2hx + h^2$ is called a **perfect-square trinomial** because it is the square of a binomial. You can picture the equation

$$(x + h)^2 = x^2 + 2hx + h^2$$

as shown below.

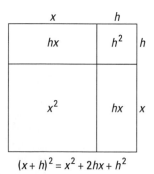

$(x + h)^2 = x^2 + 2hx + h^2$

Example 1

What number should be added to $x^2 + 10x$ to make a perfect-square trinomial?

Solution 1

Draw a picture to represent $x^2 + 10x$.

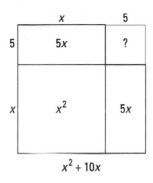

$$x^2 + 10x$$

Think: What is the area of the missing square in the upper right corner that allows you to complete the larger square? (This is the reason this process is called "completing the square.") A square with area 25 would complete the larger square. So, 25 must be added to $x^2 + 10x$ to make a perfect square.

Solution 2

Compare $x^2 + 10x + \underline{\ ?\ }$ with the perfect-square trinomial $x^2 + 2hx + h^2$. The first terms, x^2, are identical. To make the second terms equal, set
$$10x = 2hx.$$
So
$$h = 5.$$
The term added should be h^2 or 25.

Check

Apply the Binomial Square Theorem: $(x + 5)^2 = x^2 + 10x + 25$.

To generalize Example 1, consider the expression
$$x^2 + bx + \underline{\ ?\ }.$$

What must be put in the blank so the result is a perfect-square trinomial?
$$x^2 + bx + \underline{\ ?\ } = x^2 + 2hx + h^2$$

Because $b = 2h$, $h = \frac{1}{2}b$. Then $h^2 = \left(\frac{1}{2}b\right)^2$. This proves the following theorem.

Theorem

To complete the square on $x^2 + bx$, add $\left(\frac{1}{2}b\right)^2$.
$$x^2 + bx + \left(\frac{1}{2}b\right)^2 = \left(x + \frac{1}{2}b\right)^2 = \left(x + \frac{b}{2}\right)^2$$

The theorem on page 371 can be used to transform an equation of a parabola from standard form into vertex form.

Example 2

a. Rewrite the equation $y = x^2 + 10x + 8$ in vertex form.
b. Locate the vertex of the parabola.

Solution

a. Rewrite the equation so that only terms with x are on one side.

$$y - 8 = x^2 + 10x + \underline{\ ?\ }$$

$$\text{Here } b = 10, \text{ so } \left(\tfrac{1}{2}b\right)^2 = 25. \qquad \text{Complete the square on } x.$$

$$y - 8 + 25 = x^2 + 10x + 25 \qquad \text{Add 25 to both sides.}$$

$$y + 17 = x^2 + 10x + 25 \qquad \text{Simplify the left side.}$$

$$y + 17 = (x + 5)^2 \qquad \text{Apply the Binomial Square Theorem.}$$

b. The vertex of the parabola is (-5, -17).

Check 1

Use an automatic grapher. Draw graphs of $y = x^2 + 10x + 8$ and $y = (x + 5)^2 - 17$ on the same set of axes. They appear to be identical. Tracing verifies that the vertex is near (-5, -17).

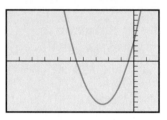

$$-20 \le x \le 4, \quad x\text{-scale} = 2$$
$$-20 \le y \le 20, \quad y\text{-scale} = 2$$

Check 2

To check the vertex, try x-values on either side of -5. When $x = -4$, $y = -16$ and when $x = -6$, $y = -16$. Because those values are equal, and because of the symmetry of the parabola, the vertex must be midway between $x = -4$ and $x = -6$.

Example 2 involves a parabola in which the coefficient of x^2 is 1. Example 3 shows how to complete the square on a quadratic expression if the coefficient of x^2 is not 1. This kind of expression occurs in describing paths of projectiles.

Example 3

Suppose a ball is thrown straight up from a height of 5 ft with an initial velocity of 60 ft/sec. The height h after t seconds is given by the equation

$$h = -16t^2 + 60t + 5.$$

Find the maximum height of the ball.

A toss up. *Shown is Lindsey Davenport at the 1994 U.S. Open. In a tennis serve, the player throws the ball straight up before hitting it.*

Solution

We need to find the vertex of the parabola that describes the height. Subtract 5 from each side to remove the constant from the right side.

$$h - 5 = -16t^2 + 60t$$

Divide each side by -16, the coefficient of t^2.

$$\frac{h - 5}{-16} = t^2 - \frac{15}{4}t$$

Now we can complete the square on the right side. Here $b = -\frac{15}{4}$. So

$$\left(\frac{b}{2}\right)^2 = \left(\frac{-15}{8}\right)^2 = \frac{225}{64}.$$

Add $\frac{225}{64}$ to each side. $\frac{h - 5}{-16} + \frac{225}{64} = t^2 - \frac{5}{4}t + \frac{225}{64}$

Rewrite the perfect-square trinomial as the square of a binomial.

$$\frac{h - 5}{-16} + \frac{225}{64} = \left(t - \frac{15}{8}\right)^2$$

Thus the maximum height occurs when $t = \frac{15}{8}$. To put the equation in vertex form, multiply each side by -16, the original coefficient of t^2.

$$h - 5 - \frac{225}{4} = -16\left(t - \frac{15}{8}\right)^2$$

Simplify the left side. $\quad h - \frac{245}{4} = -16\left(t - \frac{15}{8}\right)^2.$

The vertex of this parabola is $\left(\frac{15}{8}, \frac{245}{4}\right)$. So the maximum height of the ball is $\frac{245}{4} = 61.25$ feet.

Check 1

Graph $y = -16x^2 + 60x + 5$. Trace to locate the vertex. Using the window at the left, our grapher shows the vertex to be between the points (1.80, 61.1) and (1.89, 61.2) on the curve. It checks.

Check 2

Make a table. We checked values of x from 1.80 to 1.90 in the table below. The values are symmetric to $x = 1.875$. So the axis of symmetry must be $x = 1.875$, and hence the vertex is (1.875, 61.25). It checks.

x	$y = -16x^2 + 60x + 5$
1.80	61.16
1.81	61.1824
1.82	61.2016
1.83	61.2176
1.84	61.2304
1.85	61.24
1.86	61.2464
1.87	61.2496
1.88	61.2496
1.89	61.2464
1.90	61.24

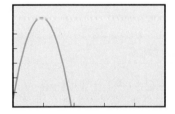

$0 \le x \le 10, \quad x\text{-scale} = 2$
$0 \le y \le 70, \quad y\text{-scale} = 10$

Completing the square helps to find key points on graphs involving quadratic expressions. But perhaps its most important application is in the proof of the *Quadratic Formula,* which you shall study in Lesson 6-7.

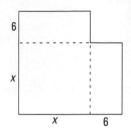

Covering the Reading

1. **a.** Give the sum of the areas of the three rectangles at the left.
   b. What number must be added to this sum to complete the square?
   c. Interpret your answer to part **b** geometrically.

In 2–5, find a number to write in the blank to make the expression a perfect square trinomial.

2. $x^2 + 18x +$ __?__

3. $x^2 - 6x +$ __?__

4. $z^2 - 3z +$ __?__

5. $x^2 + bx +$ __?__

6. Find an equation in vertex form equivalent to $y = x^2 + 40x + 10$.

In 7–10, an equation in standard form is given. **a.** Rewrite the equation in vertex form. **b.** Find the vertex of the parabola represented by each equation.

7. $y = x^2 + 18x + 6$

8. $y = x^2 - 4x + 5$

9. $y = 4x^2 + 64x + 251$

10. $y = -2x^2 - 12x - 13$

11. **a.** Find the vertex of the parabola $h = -16t^2 + 44t + 5$ graphed in Lesson 6-4.
    b. Write a sentence describing what the vertex tells you about the height of the ball.

Applying the Mathematics

12. Suppose a ball is thrown straight up from a height of 4 ft with an initial upward velocity of 22 ft/sec.
    a. Write an equation to describe the height h of the ball after t seconds.
    b. How high is the ball after 1 second?
    c. Graph your equation from part **a.**
    d. Determine the maximum height attained by the ball by completing the square.

13. Suppose $f(x) = 3x^2 + 12x + 16$.
    a. What is the domain of f?
    b. What is the vertex of its graph?
    c. What is the range of f?

14. What term must be added to the expression $x^2 + \frac{b}{a}x +$ __?__ to make a perfect square trinomial?

15. The KTHI-TV transmitting tower between Fargo and Blanchard, North Dakota is about 629 meters tall.
   a. If a tennis ball were dropped from the top, what would its height be after t seconds?
   b. In about how many seconds would it hit the ground?   *(Lesson 6-4)*

16. A student stated that $x^2 - 6x + 9 = (x - 3)^2$. Another student stated that $x^2 - 6x + 9 = (3 - x)^2$. Who is right? Justify your answer.   *(Lessons 6-1, 6-4)*

17. Consider the function with equation $\frac{y}{6} = \left(x + \frac{1}{2}\right)^2$.
   a. Convert the equation to standard form.
   b. Graph the function.
   c. What is the y-intercept of the graph?   *(Lessons 6-3, 6-4)*

18. Find an equation for the image of the graph of $y = (x + 1)^2$ under the translation $T_{4,0}$.   *(Lesson 6-3)*

19. Solve $y^2 = 20$.   *(Lesson 6-2)*

20. Refer to the graph of a bird's flight shown below.   *(Lessons 1-4, 2-4, 3-1)*
   a. When was the bird on the ground?
   b. What was the bird's average speed during its first descent?

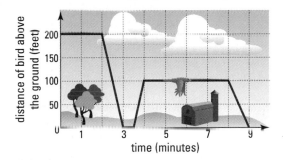

21. Simplify $\dfrac{-11 + \sqrt{121 - 96}}{6}$.   *(Previous course)*

22. *PONM* is a square of sides $a + b + c$. The areas of three regions inside have been given.
   a. Find the areas of the other six rectangles.
   b. Use the drawing to expand $(a + b + c)^2$.
   c. Show a drawing to expand $(a + b + c + d)^2$.

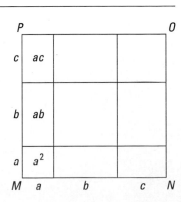

Fitting a Quadratic Model to Data

Using diagonals. *Sticks placed along the diagonals of a kite provide its support. Kites are the oldest form of aircraft, probably originating in China almost 3000 years ago.*

In Chapter 3, you studied how to find an equation for the line through two points, and how to find a linear model for data that lie approximately on a straight line. In this lesson you will learn how to find an equation for a parabola that passes through three points. You will apply the techniques of solving systems of equations you learned in the last chapter.

An Example from Geometry

Some students could not remember the formula for the number of diagonals in a polygon. So, they decided to try to rediscover it. They drew some polygons and their diagonals. In the figures below, n is the number of sides and d is the number of diagonals.

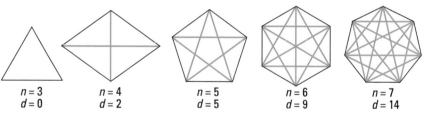

$n = 3$	$n = 4$	$n = 5$	$n = 6$	$n = 7$
$d = 0$	$d = 2$	$d = 5$	$d = 9$	$d = 14$

Using n as the independent variable, they plotted the ordered pairs (n, d) to get the graph at the right. From the graph they could see that the points do not lie on a line. The points look as if they might lie on a parabola. So they decided to see if a *quadratic model* would fit these data.

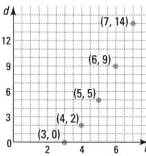

Example 1

Show that the ordered pairs graphed on page 376 satisfy an equation of the form

$$d = an^2 + bn + c.$$

Solution

Because the ordered pairs (n, d) are solutions of the equation $d = an^2 + bn + c$, they can be substituted to produce equations with a, b, and c. To find the three numbers, three equations must be used. Substitute the first three data points to get the system below.

When $n = 3$, $d = 0$: $0 = a(3)^2 + b(3) + c.$
When $n = 4$, $d = 2$: $2 = a(4)^2 + b(4) + c.$
When $n = 5$, $d = 5$: $5 = a(5)^2 + b(5) + c.$

So a, b, and c are solutions to the following system.

$$\begin{cases} 0 = 9a + 3b + c \\ 2 = 16a + 4b + c \\ 5 = 25a + 5b + c \end{cases}$$

This 3-by-3 system can be solved using linear combinations, substitution, or matrices. We use linear combinations. Subtract each equation from the one below it to eliminate c. This results in the 2-by-2 system below.

$$\begin{cases} 2 = 7a + b \\ 3 = 9a + b \end{cases}$$

Solve the 2-by-2 system by subtracting the top equation from the one below it to solve for a.

$$1 = 2a \quad \Rightarrow \quad a = \frac{1}{2}$$

Substitute $\frac{1}{2}$ for a in the top equation of the 2-by-2 system to find b.

$$2 = 7\left(\frac{1}{2}\right) + b \quad \Rightarrow \quad b = -\frac{3}{2}$$

Substitute $\frac{1}{2}$ for a and $-\frac{3}{2}$ for b into $0 = 9a + 3b + c$ to get

$0 = 9\left(\frac{1}{2}\right) + 3\left(-\frac{3}{2}\right) + c$. Thus, $c = 0$. So, a quadratic equation

which models the students' data is $d = \frac{1}{2}n^2 - \frac{3}{2}n$.

Check

Compare the actual number of diagonals with the number of diagonals predicted by the formula.

n	d (actual)	$d = \frac{1}{2}n^2 - \frac{3}{2}n$ (predicted)
3	0	0
4	2	2
5	5	5
6	9	9
7	14	14

The predicted and actual values for d are the same. The model here turns out to be exact.

An Example from Physics

Many physical relationships, such as that between car speed and braking distance, or between time elapsed and distance fallen in a free fall, can be modeled by quadratic functions. The following example shows how.

Example 2

Based on tests, here are the distances in feet it takes to stop a certain car in minimum time under emergency conditions.

Speed (mph)	10	20	30	40	50	60	70
Distance (ft)	19	42	73	116	173	248	343

a. Construct a scatterplot of these data.
b. Fit a quadratic model to these data using the points (10, 19), (20, 42), and (30, 73).
c. Graph the equation found in part b on top of the scatterplot of part a.

Solution

a. Let s be the speed in mph, and let d be the distance in ft. Use s as the independent variable. A scatterplot is given below. The scatterplot suggests that the data may be quadratic.

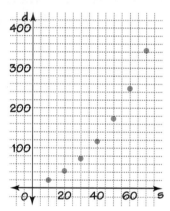

b. A quadratic model will be of the form $d = as^2 + bs + c$. To solve for a, b, and c, you need three equations. Substitute.
When $s = 10$, $d = 19$: $19 = a(10)^2 + b(10) + c$.
When $s = 20$, $d = 42$: $42 = a(20)^2 + b(20) + c$.
When $s = 30$, $d = 73$: $73 = a(30)^2 + b(30) + c$.
Solve the system.

$$\begin{cases} 19 = 100a + 10b + c \\ 42 = 400a + 20b + c \\ 73 = 900a + 30b + c \end{cases} \Rightarrow \begin{cases} 23 = 300a + 10b \\ 31 = 500a + 10b \end{cases} \Rightarrow 8 = 200a$$

From the equation $8 = 200a$, $a = .04$.
Now substitute .04 for a in $23 = 300a + 10b$ to get $23 = 300(.04) + 10b$. Thus, $b = 1.1$. Now substitute .04 for a and 1.1 for b in $19 = 100a + 10b + c$ to get $19 = 100(.04) + 10(1.1) + c$. Thus, $c = 4$. Therefore, the equation

$$d = .04s^2 + 1.1s + 4$$

models these data. ▶

On the road again.
Shown is Cascade Lakes Highway Scenic Byway in the Deschutes National Forest in Cascade Range, Oregon.

c. Below, a graph of $d = .04s^2 + 1.1s + 4$ is superimposed on the scatterplot.

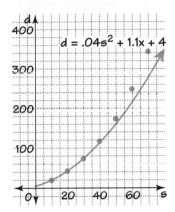

Notice that although the model goes through the points (10, 19), (20, 42), and (30, 73), it does not fit the other data exactly. However, because it passes close to the other data points, it seems to be a reasonably good model.

Two points not on a vertical line determine a linear function. Analogously, three noncollinear points, no two on the same vertical line, determine a quadratic function. If the data you are modeling can be described exactly with a quadratic function as in Example 1, any three points you use will lead to the same equation. However, if the data are only approximately quadratic, as in Example 2, the model will change. In either case, if you use three points whose x-values are equally spaced, for instance $x = 1, 2, 3$, or $x = 10, 20, 30$, the computation for the solution of the system of equations will be simpler.

*Give me a **brake**. Braking distances should be increased in poor weather conditions, such as snow, rain, or fog.*

Activity

Find the model for the braking-distance data obtained by using the data points (10, 19), (40, 116), and (70, 343). Draw a graph of the parabola determined by your equation. Do you think this is a better model than the one in the example? Why or why not?

Some computer software and some graphics calculators will find an equation of the parabola which in some sense is a best fit for the data set. This equation uses all of the data points. Our technology gave us

$$d \approx .072s^2 - .490s + 19.71$$

for the best quadratic fit to the braking-distance data. If you have technology with a *quadratic regression* feature, you may wish to learn how to use it.

Covering the Reading

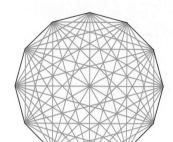

In 1–5, refer to Example 1.

1. How many diagonals does a heptagon (a polygon with seven sides) have?

2. Explain why the students thought it inappropriate to fit a linear model to their data.

3. Use three points different from those the students used to show that these points also lead to the model $d = \frac{1}{2}n^2 - \frac{3}{2}n$.

4. How many diagonals does the dodecagon pictured at the left have?

5. Some geometry books give the formula $d = \frac{n(n-3)}{2}$ for the number of diagonals of an n-gon. Is this formula equivalent to the one developed by the students? Explain why or why not.

In 6 and 7, refer to Example 2 about braking distance.

6. a. Use the equation $d = .04s^2 + 1.1s + 4$ found in Example 2 to determine the stopping distance of a car traveling 60 mph.
   b. What equation did you obtain in the Activity?
   c. Use your equation to determine the stopping distance of a car traveling 60 mph.
   d. Use the quadratic regression equation $d = .072s^2 - .490s + 19.71$ to determine the stopping distance of a car traveling 60 mph.
   e. Which model comes closest to the observed distance?

7. Find the predicted stopping distance for a car traveling at 0 mph using
   a. the equation of the Activity.
   b. the equation of Example 2.
   c. the quadratic regression equation.

Applying the Mathematics

8. Refer to Example 2.
   a. Explain why (0, 0) should be a point in this data set.
   b. Add (0, 0) to the data set. Fit a new quadratic model to this enlarged data set.

t (sec)	h (ft)
1	364
2	371
3	346
4	289
5	200
6	79

9. A ball is thrown off the top of a tall building. The table at the left shows the height h in feet of the ball above ground.
   a. Fit a quadratic model to these data.
   b. How tall is the building?
   c. When will the ball be 100 feet above the ground?
   d. When will the ball hit the ground?

10. Consider the sums of the integers from 1 to n.

n	$S = 1 + 2 + 3 + \ldots + n$
2	3
3	6
4	10
5	?
6	?

$$1 + 2 = 3$$
$$1 + 2 + 3 = 6$$
$$1 + 2 + 3 + 4 = 10$$

a. Complete the table of the sum S as a function of n.
b. Fit a quadratic model to the data in part **a**.
c. Use your model to find the sum of the first 1000 positive integers.

Review

11. Express the area of the largest rectangle at the left **a.** as a product of binomials, and **b.** in the form $ax^2 + bx + c$.   *(Lessons 6-1, 6-4)*

12. Rewrite $y - 18 = -2(x - 2)^2$ in the standard form of a parabola. *(Lesson 6-4)*

13. Consider the equation $y = 2x^2 + 24x - 20$.   *(Lessons 6-3, 6-4, 6-5)*
a. Without making a graph, identify the y-intercept of the graph.
b. Rewrite the equation in vertex form.
c. Without making a graph, identify the vertex of the graph.

14. Solve $\sqrt{(x + 2)^2} = 5$.   *(Lesson 6-2)*

15. Use matrices to solve the 3-by-3 system in Example 2.   *(Lesson 5-7)*

16. Consider the sequence 7, 10, 13, 16, ...
a. Write a recursive formula for this sequence.
b. Write an explicit formula for this sequence.
c. What is the 200th term of the sequence?
d. Which term of the sequence is 1966?   *(Lessons 1-7, 1-9, 3-3, 3-4)*

In 17–19, recall that for all nonnegative real numbers a and b, $\sqrt{a}\,\sqrt{b} = \sqrt{(ab)}$. Simplify each expression.   *(Previous course)*

17. $\sqrt{3}\,\sqrt{12}$ **18.** $\sqrt{20}\,\sqrt{50}$ **19.** $(\sqrt{17})^2$

In 20 and 21, *true or false*.   Justify your answer.   *(Previous course)*

20. $\sqrt{50} = 5\sqrt{10}$ **21.** $\dfrac{4 \pm \sqrt{12}}{6} = \dfrac{2 \pm \sqrt{3}}{3}$

Coming to your census.
Shown is interviewer Marie Cioffi collecting data from Margaret Napolitana in New York City for the 1930 census. Since 1960 the Census Bureau has used a combination of interviewers and individuals filling out the census forms themselves.

Exploration

22. The U.S. Constitution states that a census must be taken every ten years.
a. Why did the framers of the Constitution include such a requirement?
b. The first census was in 1790. Find the U.S. population for each census from 1790 to 1990.
c. Draw a scatterplot of these data and fit a quadratic model to the data.
d. Use your model to estimate the U.S. population in
(i) 1975. (ii) 2000. (iii) 2090.

6-7

The Quadratic Formula

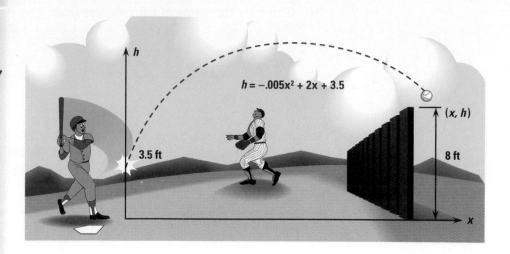

$$h = -.005x^2 + 2x + 3.5$$

3.5 ft   (x, h)   8 ft

Pop Fligh, the famous baseball player, hit a pitch that was 3.5 ft high. The ball traveled towards the outfield along a nearly parabolic path. Let x be the distance along the ground (in feet) of the ball from home plate, and let $h(x)$ be the height (in feet) of the ball at that distance. Then the path of his ball can be described by the function h, where

$$h(x) = -.005x^2 + 2x + 3.5.$$

Suppose we want to know where the ball was 8 feet high. Because we wish to know the horizontal distance x when the height was 8, we substitute 8 for $h(x)$ in the equation above.

$$8 = -.005x^2 + 2x + 3.5$$

By adding -8 to each side, you can put the equation into the standard form $ax^2 + bx + c = 0$.

$$0 = -.005x^2 + 2x - 4.5.$$

This equation can be solved by rewriting it in vertex form. But the arithmetic is messy. It is much easier to solve the equation by using the **Quadratic Formula.** This formula is very important—*if you have not learned it before, you should memorize it now.* The Quadratic Formula is a theorem; that is, it can be proved using the basic properties of algebra.

What Is the Quadratic Formula?

Quadratic Formula Theorem
If $ax^2 + bx + c = 0$ and $a \neq 0$, then
$$x = \frac{-b \pm \sqrt{b^2 - 4ac}}{2a}.$$

Our proof of the Quadratic Formula relies on completing the square.

Proof

Given: the equation $ax^2 + bx + c = 0$, where $a \neq 0$.

1. $x^2 + \frac{b}{a}x + \frac{c}{a} = \frac{0}{a}$

First divide both sides by a so the coefficient of x^2 is 1.

2. $x^2 + \frac{b}{a}x = -\frac{c}{a}$

Add $-\frac{c}{a}$ to each side.

3. $x^2 + \frac{b}{a}x + \frac{b^2}{4a^2} = \frac{b^2}{4a^2} - \frac{c}{a}$

Complete the square by adding $\left(\frac{1}{2} \cdot \frac{b}{a}\right)^2$ to both sides.

4. $\left(x + \frac{b}{2a}\right)^2 = \frac{b^2}{4a^2} - \frac{c}{a}$

Write the left side as a binomial squared.

5. $\left(x + \frac{b}{2a}\right)^2 = \frac{b^2 - 4ac}{4a^2}$

Add the fractions on the right side.

6. $x + \frac{b}{2a} = \pm\sqrt{\frac{b^2 - 4ac}{4a^2}}$

Take the square roots of both sides.

7. $x + \frac{b}{2a} = \frac{\pm\sqrt{b^2 - 4ac}}{2a}$

Use the definition of square root.

8. $x = \frac{-b \pm \sqrt{b^2 - 4ac}}{2a}$

Add $-\frac{b}{2a}$ to both sides.

Using the Quadratic Formula

Example 1

Solve $3x^2 + 11x - 4 = 0$.

Solution

To use the Quadratic Formula, you need to know the values of a, b, and c. Here $a = 3$, $b = 11$, and $c = -4$. Write the formula. Then substitute for a, b, and c.

$$x = \frac{-b \pm \sqrt{b^2 - 4ac}}{2a}$$

$$= \frac{-11 \pm \sqrt{11^2 - 4 \cdot 3 \cdot -4}}{2 \cdot 3}$$

$$= \frac{-11 \pm \sqrt{121 - -48}}{6}$$

$$= \frac{-11 \pm \sqrt{169}}{6}$$

$$= \frac{-11 \pm 13}{6}$$

So $x = \frac{-11 + 13}{6}$ or $x = \frac{-11 - 13}{6}$

$x = \frac{1}{3}$ or $x = -4$

▶

Check

Each solution should be checked.

Does $3 \cdot \left(\frac{1}{3}\right)^2 + 11 \cdot \frac{1}{3} - 4 = 0$? Yes, $\frac{1}{3} + \frac{11}{3} - 4 = 0$.

Does $3 \cdot (-4)^2 + 11 \cdot -4 - 4 = 0$? Yes, $48 - 44 - 4 = 0$.

In Example 1, since the number $b^2 - 4ac$ under the radical sign is a perfect square, the solutions are integers. In applications, however, the numbers are not always so nice. The Quadratic Formula still works, but you may need a calculator to estimate the solutions. Here is the problem posed at the beginning of this lesson.

Example 2

Find out when the ball hit by Pop Fligh was 8 feet high.

Solution

We need to solve $-.005x^2 + 2x - 4.5 = 0$. In this situation $a = -.005$, $b = 2$, and $c = -4.5$. Substitute into the formula.

$$x = \frac{-2 \pm \sqrt{2^2 - 4 \cdot (-.005) \cdot (-4.5)}}{2 \cdot (-.005)}$$

$$= \frac{-2 \pm \sqrt{4 - .09}}{-.01}$$

$$= \frac{-2 \pm \sqrt{3.91}}{-.01}$$

Use a calculator to estimate the square root. Separate the two solutions.

$$x \approx \frac{-2 + 1.977}{-.01} \text{ or } x \approx \frac{-2 - 1.977}{-.01}$$

So $x \approx 2.3$ or $x \approx 397.7$.

As you might expect, There are two places where the ball is 8 ft high. The first is when the ball is about 2.3 ft away from home plate and on the way up. The second is when the ball is about 398 ft away from home plate and on the way down.

Check

Graph $h(x) = -.005x^2 + 2x + 3.5$ with an automatic grapher. Use the trace feature to find the value of y when x is about 2.3 or 397.7. The value of y should be close to 8.

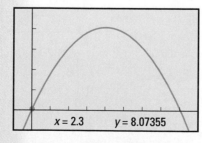

$x = 2.3$ $\quad$ $y = 8.07355$

$-50 \leqslant x \leqslant 450$, x-scale $= 50$
$-50 \leqslant y \leqslant 250$, y-scale $= 50$

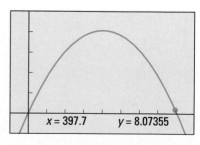

$x = 397.7$ $\quad$ $y = 8.07355$

$-50 \leqslant x \leqslant 450$, x-scale $= 50$
$-50 \leqslant y \leqslant 250$, y-scale $= 50$

An equation of the form $ax^2 + bx + c = 0$, with $a \neq 0$, is said to be in the **standard form of a quadratic equation.** If a quadratic equation is not in standard form, use the properties of algebra to rewrite it in that form before applying the Quadratic Formula. Example 3 illustrates this.

Example 3

The 3-4-5 right triangle has sides which are consecutive integers. Are there any other right triangles with this property?

Solution

If n is an integer, then n, n + 1, and n + 2 are consecutive integers. By the Pythagorean Theorem,

$$n^2 + (n + 1)^2 = (n + 2)^2.$$

Therefore, $n^2 + n^2 + 2n + 1 = n^2 + 4n + 4.$

$$2n^2 + 2n + 1 = n^2 + 4n + 4$$

Rewrite this equation in standard form by adding $-n^2 - 4n - 4$ to each side.

$$n^2 - 2n - 3 = 0$$

Use the Quadratic Formula with $a = 1$, $b = -2$, and $c = -3$.

$$n = \frac{-(-2) \pm \sqrt{(-2)^2 - 4(1)(-3)}}{2(1)}$$

$$= \frac{2 \pm \sqrt{16}}{2}$$

$$= \frac{2 \pm 4}{2}$$

So $n = \dfrac{2 + 4}{2} = 3$ or $n = \dfrac{2 - 4}{2} = -1$.

When $n = 3$, then $n + 1 = 4$ and $n + 2 = 5$. This is the 3-4-5 triangle. The value $n = -1$ must be rejected because a side of a triangle must have positive length. Thus the only right triangle with consecutive integer sides is the 3-4-5 right triangle.

QUESTIONS

Covering the Reading

1. If $ax^2 + bx + c = 0$, and $a \neq 0$, give the two values of x in terms of a, b, and c.

2. *Multiple choice.*   The Quadratic Formula is a
   (a) theorem.   (b) postulate.   (c) definition.

In 3–6, refer to the proof of the Quadratic Formula.

3. Why is it necessary to divide both sides by a in Step 1?

4. Write $x^2 + \frac{b}{a}x + \frac{b^2}{4a^2}$ as the square of a binomial.

5. Write $\frac{b^2}{4a^2} - \frac{c}{a}$ as a single fraction.

6. Why can't a equal 0 in the Quadratic Formula?

In 7–10, a quadratic equation is given. Solve the equation using the Quadratic Formula.

7. $10x^2 + 13x + 3 = 0$ **8.** $6v^2 - 5v - 3 = 0$

9. $x^2 + 2x - 15 = 0$ **10.** $2n^2 - 11n = 0$

In 11–13, consider the equation $h(x) = -.005x^2 + 2x + 3.5$ and the situation at the start of this Lesson.

11. What do x and $h(x)$ represent?

12. Manny Walker pitched the ball that Pop hit. How high was the ball when it passed over Manny's head, 60 ft from home plate?

13. Where was Pop's hit 100 ft high?

14. In Example 2, we could have multiplied both sides of the equation by -1 and solved $.005x^2 - 2x + 4.5 = 0$. Find the solutions to this equation to the nearest tenth.

15. Refer to Example 3. If n is the first of three consecutive integers, what are the other two?

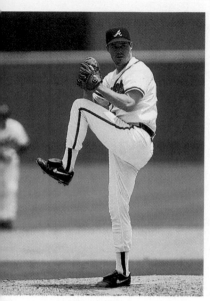

Stingy on runs. *Shown is Greg Maddux of the Atlanta Braves. In 1994, Maddux had a league low 1.56 earned run average.*

Applying the Mathematics

16. Find all right triangles whose sides are consecutive *even* integers n, $n + 2$, and $n + 4$.

In 17 and 18, solve.

17. $n^2 + 9 = 6n$ **18.** $4(m^2 - 3m) = -9$

19. Consider the parabola with equation $y = x^2 + 6x - 1$.
   a. Find the values of x if $y = 0$. What are these points called?
   b. Find the vertex of this parabola.
   c. Graph the parabola.
   d. Give an equation for its axis of symmetry.

20. As shown below, the graphs of $y = -\frac{1}{2}x^2 + 4x$ and $y = 4$ intersect at two points.
   a. Find the coordinates of the points of intersection without graphing.
   b. Check by using an automatic grapher.

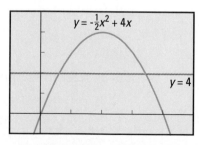

$-2 \leqslant x \leqslant 10$, x-scale $= 2$
$-2 \leqslant y \leqslant 10$, y-scale $= 2$

386

21. When a beam of light in air strikes the surface of water it is *refracted,* or bent. Below are the earliest known data on the relation between *i,* the angle of incidence in degrees, and *r,* the angle of refraction in degrees. The measurements are recorded in the *Optics of Ptolemy,* a Greek scientist who lived in the second century A.D.

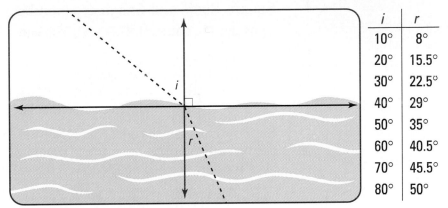

i	r
10°	8°
20°	15.5°
30°	22.5°
40°	29°
50°	35°
60°	40.5°
70°	45.5°
80°	50°

   a. Draw a scatterplot of these data.
   b. Fit a quadratic model to these data.
   c. Fit a linear model to these data.
   d. Decide which model seems more appropriate. Explain why you made that decision.   *(Lessons 3-8, 6-6)*

22. Explain how you can tell without graphing whether or not the equations $y = 3x^2 + 24x + 50$ and $y - 2 = 3(x + 4)^2$ have the same graph.   *(Lesson 6-5)*

23. a. Graph the solution sets to $y = \frac{36}{x}$ and $y + 2 = \frac{36}{x}$ on the same axes.
   b. Find equations for the lines of symmetry for the image and preimage.   *(Lessons 2-6, 6-3)*

24. For what values of x is it true that $\sqrt{x^2} = x$?   *(Lesson 6-2)*

25. A door 36″ × 78″ is surrounded by a wooden frame of width w inches on three sides.
   a. What is the total area of the door and frame?
   b. If $w = 4$, find the total area of the door and frame.   *(Lesson 6-1)*

26. Find the slope of the line through $(1, -4)$ and parallel to the *x*-axis.
   (Lessons 3-1, 3-2)

Exploration

27. For Question 14, we multiplied both sides of the equation $-.005x^2 + 2x - 4.5 = 0$ by -1 and asked you to solve it. Multiply both sides of this equation by 1000 and solve it. Compare your results with the solutions in Example 2 and Question 14. Write a few sentences explaining your results.

6-8

Imaginary Numbers

Square Roots and Imaginary Numbers

Consider the equation $t^2 = 400$. You can solve it for t as follows.

$$t^2 = 400$$
$$\sqrt{t^2} = \sqrt{400} \quad \text{Take square roots.}$$
$$|t| = 20 \quad \text{Use the Absolute Value-Square Root Theorem}$$
$$t = \pm 20. \quad \text{Solve the absolute value equation.}$$

But now consider the equation $t^2 = -400$. You know that this equation has no real solutions because the square of a real number is never negative. However, if you followed the solution above you might write

$$t^2 = -400$$
$$\sqrt{t^2} = \sqrt{-400}$$
$$|t| = ?$$

So far, $\sqrt{x}$ has been defined only for $x \geq 0$. If you try to evaluate $\sqrt{-400}$ on most calculators, an error message will be displayed.

Until the 1500s, mathematicians were also puzzled by square roots of negative numbers. They knew that if they solved certain quadratics, they would get negative numbers under the radical sign. They did not know, however, what to do with them.

One of the first to work with these numbers was Girolamo Cardano. In a book called *Ars Magna* ("Great Art") published in 1545, Cardano reasoned as follows: When k is positive, the equation $x^2 = k$ has two solutions, $\sqrt{k}$ and $-\sqrt{k}$. If we solve the equation $x^2 = -k$ in the same way, then the two solutions are $\sqrt{-k}$ and $-\sqrt{-k}$. These, then, are symbols for the square roots of negative numbers.

> **Definition**
> When $k > 0$, the two solutions to $x^2 = -k$ are denoted $\sqrt{-k}$ and $-\sqrt{-k}$.

By the definition, $(\sqrt{-k})^2 = -k$. This means that we can say, for *all* real numbers r,

$$\sqrt{r} \cdot \sqrt{r} = r.$$

Cardano called these square roots of negatives "fictitious numbers." In the 1600s, Descartes called them **imaginary numbers** in contrast to the numbers everyone understood, which he called "real numbers." In his book *De Formulis Differentialibus Angularibus,* written in 1777, Euler wrote, "In the following I shall denote the expression $\sqrt{-1}$ by the letter i so that $i\,i = -1$."

The number i is the **imaginary unit.**

Putting these two definitions together, we conclude that i and $-i$ are the two solutions to

$$x^2 = -1. \text{ That is, } i^2 = -1 \text{ and } (-i)^2 = -1.$$

Now consider multiples of i, such as $5i$. By the definition of i, $5i = 5\sqrt{-1}$. If we assume that multiplication of imaginary numbers is commutative and associative, then

$$
\begin{aligned}
(5i)^2 &= 5i \cdot 5i \\
&= 5^2 \cdot i^2 \\
&= 25 \cdot -1 \\
&= -25.
\end{aligned}
$$

So $5i$ is a square root of -25. We write $5i = \sqrt{-25}$ and then $-5i = -\sqrt{-25}$. The following theorem generalizes this result.

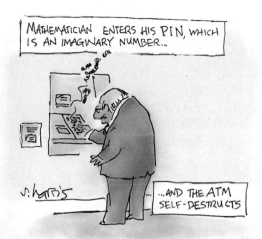

MATHEMATICIAN ENTERS HIS PIN, WHICH IS AN IMAGINARY NUMBER...

...AND THE ATM SELF-DESTRUCTS

S. Harris

Theorem
If $k > 0$, $\sqrt{-k} = i\sqrt{k}$.

Thus all square roots of negative numbers are multiples of i.

Example 1

Solve $t^2 = -400$.

Solution

Apply the first definition in this Lesson.
$$t = \sqrt{-400} \quad \text{or} \quad t = -\sqrt{-400}$$
Now use the theorem above.
$$t = i\sqrt{400} \quad \text{or} \quad t = -i\sqrt{400}$$
Simplify.
$$t = 20i \quad \text{or} \quad t = -20i$$

Example 2

Show that $i\sqrt{3}$ is a square root of -3.

Solution

Multiply $i\sqrt{3}$ by itself.
$$
\begin{aligned}
i\sqrt{3} \cdot i\sqrt{3} &= i \cdot i \cdot \sqrt{3} \cdot \sqrt{3} \\
&= i^2 \cdot 3 \\
&= -1 \cdot 3 \\
&= -3
\end{aligned}
$$

The other square root of -3 is $-i\sqrt{3}$.

Due to the long history of quadratic equations, solutions to them are described in different ways. The following all refer to the same numbers.

the solutions to $x^2 = -3$
the square roots of -3
$\sqrt{-3}$ and $-\sqrt{-3}$
$i\sqrt{3}$ and $-i\sqrt{3}$

Operations with Imaginary Numbers

All the Field Postulates listed in Appendix B, including the commutative, associative, and distributive properties of addition and multiplication, hold for imaginary numbers, as do all theorems based on these postulates. Consequently, you can use them when working with multiples of i, just as you would when working with multiples of any real numbers.

Example 3

Simplify the following.

a. $(2i)(5i)$ b. $\sqrt{-9} - \sqrt{-25}$ c. $\sqrt{-27} + \sqrt{-3}$ d. $\dfrac{\sqrt{-9}}{\sqrt{-25}}$

Solution

a. $(2i)(5i) = 10i^2 = 10 \cdot -1 = -10$

b. $\sqrt{-9} - \sqrt{-25} = 3i - 5i = -2i$

c. $\sqrt{-27} + \sqrt{-3} = 3i\sqrt{3} + i\sqrt{3} = 4i\sqrt{3}$

d. $\dfrac{\sqrt{-9}}{\sqrt{-25}} = \dfrac{3i}{5i} = \dfrac{3}{5}$

The next example shows how to multiply imaginary numbers expressed in radical form. Notice that the order of operations must be followed. Take square roots before multiplying.

Example 4

Simplify $\sqrt{-16}\,\sqrt{-25}$.

Solution

Convert to multiples of i.

$$\sqrt{-16} \cdot \sqrt{-25} = i\sqrt{16} \cdot i\sqrt{25}$$
$$= 4i \cdot 5i$$
$$= 20i^2$$
$$= -20$$

You are familiar with the property $\sqrt{ab} = \sqrt{a}\sqrt{b}$ for *nonnegative* real numbers a and b. Does this property hold when a and b are negative? Consider Example 4. If we assume that $\sqrt{a}\sqrt{b} = \sqrt{ab}$ for negative values of a and b, then

$$\sqrt{-16}\sqrt{-25} = \sqrt{(-16)(-25)}$$
$$= \sqrt{400}$$
$$= 20.$$

Clearly, this is different from the answer of Example 4. This counterexample shows that the property $\sqrt{a}\sqrt{b} = \sqrt{ab}$ does not hold when a and b are both negative numbers.

QUESTIONS

Covering the Reading

1. *Multiple choice.* About when did mathematicians begin to use roots of negative numbers as solutions to equations?
   (a) sixth century (b) twelfth century
   (c) sixteenth century (d) twentieth century

2. *True or false.* Not all negative numbers have square roots.

3. Write the solutions to $x^2 = -1$.

4. Who first used the term "imaginary number"?

5. Who was the first person to suggest using i for $\sqrt{-1}$?

6. *Multiple choice.* $\sqrt{-b} \neq i\sqrt{b}$ when $b \underline{\quad ? \quad} 0$.
   (a) > (b) = (c) <

7. Solve for x. Write the solutions to each equation with a radical sign, and without a radical sign.
   a. $x^2 + 16 = 0$ b. $x^2 - 16 = 0$

8. *True or false.* $i\sqrt{5}$ is a square root of -5. Justify your answer.

9. Show that $-3i$ is a square root of -9.

In 10–12, simplify.

10. $\sqrt{-7}$ 11. $\sqrt{-144}$ 12. $-\sqrt{96}$

In 13–18, perform the indicated operations. Give answers as real numbers or as multiples of i.

13. $3i + 4i$ 14. $8i - i$ 15. $(9i)(8i)$

16. $2\sqrt{-9} + \sqrt{-49}$ 17. $\dfrac{\sqrt{-16}}{\sqrt{-4}}$ 18. $-\sqrt{25} + \sqrt{-25}$

In 19–21, simplify the product.

19. $\sqrt{-3} \cdot \sqrt{-3}$ 20. $\sqrt{-6} \cdot \sqrt{-3}$ 21. $\sqrt{-2} \cdot \sqrt{2}$

22. When does $\sqrt{xy} \neq \sqrt{x}\sqrt{y}$?

Applying the Mathematics

In 23 and 24, simplify.

23. $\sqrt{-434{,}281}$

24. $\frac{2i + 3i}{i}$

In 25 and 26, *true or false.* If false, give a counterexample.

25. The sum of two imaginary numbers is imaginary.

26. The product of two imaginary numbers is imaginary.

27. Verify your solutions to $x^2 = -1$ in Question 3 by using the Quadratic Formula to solve $x^2 + 0x + 1 = 0$.

28. Solve $4x^2 + 25 = 0$ using the Quadratic Formula.

29. Solve for x.
  a. $x^2 + 15 = 6$ **b.** $(x - 3)^2 + 15 = 6$

Concentric circles in design. *Shown is a shell gorget with a snake motif, a Native American artifact from 4000–3000 B.C. The gorget is from a mound in Hamilton County, Tennessee.*

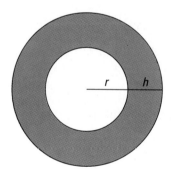

Review

In 30 and 31, solve.   *(Lesson 6-7)*

30. $4m^2 - 4m + 1 = 0$.   **31.** $2 - x^2 = 12x$.

32. A ball is thrown upwards from a height of 3 feet with an initial velocity of 28 feet per second.
  a. What is the height of the ball after t seconds?
  b. What is the maximum height of the ball?
  c. When does the ball hit the ground?   *(Lessons 6-3, 6-5, 6-7)*

In 33 and 34, use the two concentric circles shown at the left. The smaller circle has radius r; the larger circle has radius $r + h$. *(Previous course)*

33. Find the area of the shaded region.

34. **a.** How much greater is the circumference of the larger circle than the circumference of the smaller.
  b. The Earth's equator is nearly a circle with a radius of about 3963 miles. If a rope were held up along the equator 6 feet above ground level, how much longer would the rope be than the equator?

Exploration

35. By definition, $i^2 = -1$. So $i^3 = i^2 \cdot i = -1 \cdot i = -i$ and $i^4 = i^3 \cdot i = -i \cdot i = -i^2 = -(-1) = 1$.
  a. Continue this pattern to evaluate and simplify each power of i: i^5, i^6, i^7, and i^8.
  b. Generalize your result to predict the values of i^{1995}, i^{1996}, and i^{2000}.

LESSON 6-9
Complex Numbers

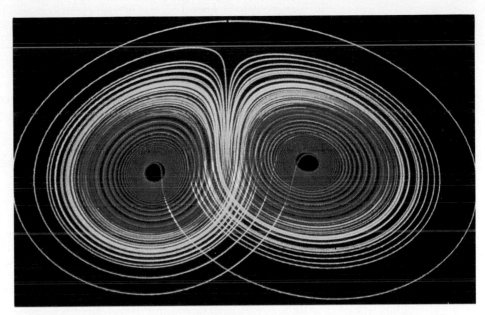

Strange Attractor. *Shown is the Lorenz Attractor, discovered in the early 1960s by MIT meteorologist Edward Lorenz. This shape shows that chaotic behavior can arise from simple dynamical systems. See page 395.*

What Are Complex Numbers?

The set of numbers of the form bi, where b is a real number, are the imaginary numbers. When a real number and an imaginary number are added, the sum is called a **complex number.**

> **Definition**
> A complex number is a number of the form $a + bi$, where a and b are real numbers and $i = \sqrt{-1}$; a is called the real part and b is called the imaginary part.

For example, $-3 + 4i$ is a complex number. The *real part* of $-3 + 4i$ is -3 and the *imaginary part* is 4 (not $4i$).

Two complex numbers $a + bi$ and $c + di$ are **equal** if and only if their real parts are equal and their imaginary parts are equal. That is, $a + bi = c + di$ if and only if $a = c$ and $b = d$. For example, if $x + yi = 2i - 3$, then $x = -3$ and $y = 2$.

Operations with Complex Numbers

> **Postulate**
> All postulates for real numbers except those for inequalities (see Appendix A) also hold for the set of complex numbers.

Thus, you can use the properties to operate with complex numbers in a manner consistent with the way you operate with real numbers.

Example 1

Add and simplify: $(3 + 4i) + (7 + 8i)$.

Solution

Use the Associative and the Commutative Properties of addition to regroup the real parts together and the imaginary parts together.

$(3 + 4i) + (7 + 8i) = (3 + 7) + (4i + 8i)$

Then use the Distributive Property to add the multiples of i.

$$= 10 + (4 + 8)i$$
$$= 10 + 12i$$

The Distributive Property can also be used to multiply a complex number by a real number or by an imaginary number.

Example 2

Simplify $2i(8 + 5i)$.

Solution

$$
\begin{aligned}
2i(8 + 5i) &= 2i(8) + 2i(5i) && \text{Distributive Property} \\
&= 16i + 10(i^2) && \text{Associative and Commutative} \\
& && \text{Properties of Multiplication} \\
&= 16i + 10(-1) && \text{Definition of } i \\
&= -10 + 16i && \text{Commutative Property} \\
& && \text{of Addition}
\end{aligned}
$$

In Example 2, notice that i^2 was simplified using the fact that $i^2 = -1$. Generally, all answers to complex-number computations should be put in the form $a + bi$. This makes it easy to identify the real and imaginary parts.

To multiply complex numbers, think of complex numbers as linear expressions in i, and multiply using the Distributive Property. Then use $i^2 = -1$ to simplify your answer.

Example 3

Multiply and simplify: $(8 - 3i)(4 + 5i)$.

Solution

$$
\begin{aligned}
(8 - 3i)(4 + 5i) &= 32 + 40i - 12i - 15i^2 && \text{Distributive Property} \\
&= 32 + 28i - 15i^2 && \text{Distributive Property} \\
& && \text{(like terms)} \\
&= 32 + 28i - 15(-1) && \text{Definition of } i \\
&= 47 + 28i && \text{Arithmetic}
\end{aligned}
$$

Applications of Complex Numbers

The first use of the term "complex number" is generally credited to Karl Friedrich Gauss. Gauss applied complex numbers to the study of electricity. Later in the 19th century, applications using complex numbers were found in geometry and acoustics. In the 1970s, a new field called *dynamical systems* arose in which complex numbers play a pivotal role.

Here is a situation arising from electricity. The *impedance* in an alternating-current (AC) circuit is the amount by which the circuit resists the flow of electricity. Impedance is described by a complex number. Two electrical circuits may be connected *in series* or *in parallel*.

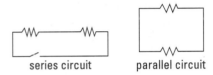

series circuit   parallel circuit

The total impedance Z_T of a circuit is a function of the impedances Z_1 and Z_2 of the individual circuits. In a series circuit, $Z_T = Z_1 + Z_2$. In a parallel circuit, $Z_T = \dfrac{Z_1 Z_2}{Z_1 + Z_2}$. Thus, to find the total impedance in a parallel circuit, division of complex numbers is needed.

Shown are electrical engineers aligning elements of the OPAL particle detector at CERN, the European centre for particle physics near Geneva, Switzerland. Electrical engineers deal with the development, production, and testing of electrical and electronic devices and equipment.

Example 4

Find the total impedance in a parallel circuit if $Z_1 = 3 - i$ ohms and $Z_2 = 3 + i$ ohms.

Solution

Substitute into the impedance formula for parallel circuits, and evaluate.

$$Z_T = \frac{Z_1 Z_2}{Z_1 + Z_2}$$
$$= \frac{(3 - i)(3 + i)}{(3 - i) + (3 + i)}$$
$$= \frac{9 + 3i - 3i - i^2}{6}$$
$$= \frac{9 - (-1)}{6}$$
$$= \frac{10}{6}$$
$$= \frac{5}{3}$$

The total impedance is $\frac{5}{3}$ ohms.

The complex numbers $3 - i$ and $3 + i$ given in Example 4 are *complex conjugates* of each other. Notice that the product $(3 - i)(3 + i)$ in the numerator of the fraction in Example 4 is a real number. In general, the **complex conjugate** of $a + bi$ is $a - bi$. In the Questions, you are asked to prove that the product of any two complex conjugates is a real number.

Complex conjugates are useful when dividing complex numbers. To divide two complex numbers, multiply both numerator and denominator by the conjugate of the denominator. When the denominator is multiplied by its complex conjugate, the result is a real number. Then divide each term in the numerator by this real number.

Example 5

Write $\frac{3-4i}{2+5i}$ in $a + bi$ form.

Solution

The complex conjugate of $2 + 5i$ is $2 - 5i$, so multiply both the numerator and the denominator by $\frac{2-5i}{2-5i}$, and divide the numerator by the real number that results.

$$\frac{3-4i}{2+5i} = \frac{3-4i}{2+5i} \cdot \frac{2-5i}{2-5i} \qquad \text{Identity Property of Multiplication}$$

$$= \frac{6-15i-8i+20i^2}{4-10i+10i-25i^2} \qquad \text{Distributive Property}$$

$$= \frac{6-23i+20i^2}{4-25i^2} \qquad \text{Distributive Property (like terms)}$$

$$= \frac{6-23i+20(-1)}{4-25(-1)} \qquad \text{Definition of } i$$

$$= \frac{-14-23i}{29}$$

$$= \frac{-14}{29} - \frac{23}{29}i \qquad \text{Distributive Property (adding fractions)}$$

The Various Kinds of Complex Numbers

Because $a + 0i = a$, every real number a is a complex number. Thus, the set of real numbers is a subset of the set of complex numbers. Likewise, every imaginary number bi equals $0 + bi$, so the set of imaginary numbers is also a subset of the set of complex numbers.

The diagram below shows how various kinds of complex numbers are related.

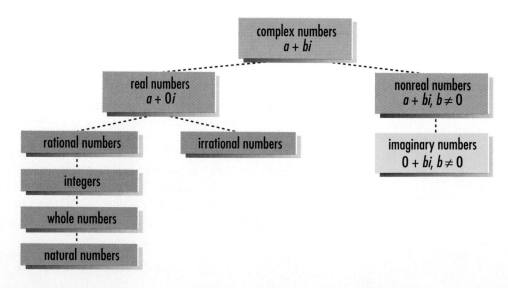

QUESTIONS

Covering the Reading

1. A complex number is a number of the form $a + bi$ where a and b are __?__ numbers.

In 2–4, give the real and imaginary parts of the complex number.

2. $14 + 5i$

3. $3 - i\sqrt{2}$

4. i

5. If $x + 12i = 7 - yi$, find x and y.

In 6–9, perform the operation and write your answer in $a + bi$ form.

6. $(11 - 4i) + (18 - 3i)$

7. $3i(2 + 9i)$

8. $(2 - i)(2 + i)$

9. $(-8 + i)(5 - 3i)$

10. Provide reasons for each step.

$$
\begin{aligned}
(12 + 7i)(8 + 4i) &= 96 + 48i + 56i + 28i^2 \qquad &\textbf{a. } \underline{\ ?\ } \\
&= 96 + 104i + 28i^2 \qquad &\textbf{b. } \underline{\ ?\ } \\
&= 96 + 104i + 28(-1) \qquad &\textbf{c. } \underline{\ ?\ } \\
&= 96 + -28 + 104i \qquad &\textbf{d. } \underline{\ ?\ } \\
&= 68 + 104i
\end{aligned}
$$

11. Who is thought to be the first to use the term "complex number"?

12. What are the complex numbers $3 + 7i$ and $3 - 7i$ called?

13. Given two electrical circuits with impedances $Z_1 = 10 + 5i$ ohms and $Z_2 = 10 - 5i$ ohms, find the total resistance if these two circuits are connected as described.
    a. in series
    b. in parallel

14. Name two fields in which complex numbers are applied.

15. Write in $a + bi$ form.
    a. $\dfrac{6 - 2i}{5}$
    b. $\dfrac{5}{6 - 2i}$

In 16–18, write in $a + bi$ form.

16. $\dfrac{3 + i}{4 - 7i}$

17. $\dfrac{4 + i}{-2 - 5i}$

18. $\dfrac{20 + 15i}{5i}$

19. *True or false.* Every real number is also a complex number.

Applying the Mathematics

20. Write the additive identity for the complex numbers in $a + bi$ form.

21. Write the multiplicative identity for the complex numbers in $a + bi$ form.

22. Write $\sqrt{-9}$ in $a + bi$ form.

In 23 and 24, consider the complex numbers $a - bi$ and $a + bi$.

23. Calculate their sum.

24. Find their product and explain why it is a real number.

25. **a.** Solve $x^2 - 2x + 5 = 0$ using the Quadratic Formula. Write your solutions in $a + bi$ form.
    b. Check your answer(s).

Review

26. *Multiple choice.* Which is not a square root of -9? *(Lesson 6-8)*
    (a) $-3i$ (b) $3i$ (c) -3 (d) $\sqrt{-9}$

In 27 and 28, suppose each letter of the English alphabet except for i is a variable representing a real number, and $i^2 = -1$. *(Lesson 6-8)*

27. Simplify Mississippi.

28. Name a state that represents an imaginary number.

29. Solve $4a^2 = 2 + 4a$. *(Lesson 6-7)*

30. The size of a human embryo is often measured by calculating its *crown-rump length,* as shown by segment $\overline{CR}$ at the left. The table below gives the average crown-rump length L in millimeters of a human embryo m months after conception.

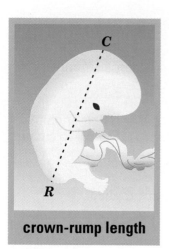

crown-rump length

m	.75	1	2	3	4	5
L	2	5	23	56	112	160

    a. Make a scatterplot of the data.
    b. Find a quadratic equation to model these data.
    c. According to your model, about how long is the average human embryo 2.5 months after conception?
    d. At full term (9 months after conception) the average crown-rump length of a human fetus is 350 mm. Does the model in part **b** overestimate or underestimate this length? *(Lesson 6-6)*

Exploration

31. A complex number $a + bi$ is graphed as the point (a, b) with the x-axis as the real axis and the y-axis as the imaginary axis.
    a. Graph $z = 1 + i$ as the point $(1, 1)$.
    b. Compute and graph z^2, z^3, and z^4.
    c. What pattern emerges? Can you predict where z^5 will be?

IN-CLASS
ACTIVITY

*Predicting
the Number
of Real
Solutions
to a
Quadratic
Equation*

Materials: Automatic grapher
Work on this activity in a small group.

Record your results in a table like the one below.

$y = ax^2 + bx + c$	number of x-intercepts of graph	solutions to $ax^2 + bx + c = 0$	number of real solutions to $ax^2 + bx + c = 0$	value of $b^2 - 4ac$
a. $y = 2x^2 - 12x + 18$				
b. $y = 2x^2 - 12x + 13$				
c. $y = 2x^2 - 12x + 23$				

1 Use an automatic grapher to sketch a graph of the quadratic function. Identify the number of x-intercepts.
a. $y = 2x^2 - 12x + 18$
b. $y - 2x^2 - 12x + 13$
c. $y = 2x^2 - 12x + 23$

2 Solve the equations using the Quadratic Formula, and tell how many of the solutions are real.
a. $2x^2 - 12x + 18 = 0$
b. $2x^2 - 12x + 13 = 0$
c. $2x^2 - 12x + 23 - 0$

3 Discuss with others in your group any patterns you notice in the table so far.

4 When $ax^2 + bx + c = 0$, the value of the expression $b^2 - 4ac$ can be used to predict the number of real solutions to the quadratic equation. Calculate the value of $b^2 - 4ac$ for each of the equations in Question 2, and record it in your table.

5 Each person in the group should do the following.
a. Make up an equation of the form $y = ax^2 + bx + c$, where a, b, and c are real numbers and $a \neq 0$.
b. Sketch a graph of $y = ax^2 + bx + c$.
c. Complete the table for your equation.

6 Look again at the entries in the table.
a. If $b^2 - 4ac > 0$, what conclusion(s) can you draw?
b. If $b^2 - 4ac = 0$, what seems to be true?
c. If $b^2 - 4ac < 0$, what conclusion(s) can you draw?

*Analyzing
Solutions to
Quadratic
Equations*

Stamps of approval. *Many countries have issued stamps to honor great mathematicians and their works. The mathematicians shown are Descartes (top), al-Khowarizmi (left), and Gauss (right).*

A Brief History of Quadratics

As early as 1700 B.C., ancient mathematicians considered problems that today we would solve using quadratic equations. The Babylonians even described solutions to these problems using words that indicate that they had general procedures for finding solutions that were like the Quadratic Formula. However, the ancients had neither our modern notation nor the notion of complex numbers. The history of the solving of quadratic equations helped lead to the acceptance of irrational numbers, negative numbers, and complex numbers.

The Pythagoreans in the 5th century B.C. thought of x^2 as the area of a square with side x. So if $x^2 = 2$, as in the square pictured below, then $x = \sqrt{2}$. These Greek mathematicians proved that $\sqrt{2}$ was an irrational number, and so they realized that irrational numbers have meaning. But they never considered the negative solution to the equation $x^2 = 2$, because lengths could not be negative.

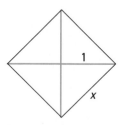

Writings of Indian and Arab mathematicians from the years 800 A.D. to 1200 A.D. indicate that they could solve quadratic equations. In 825 A.D., the Arab mathematician al-Khowarizmi, in his book *Hisab al-jabr w'al muqabala* (from which we get the word "algebra"), solved quadratics like the Babylonians did. His contribution is that he did not think of the unknown as having to stand for a length. Thus the unknown became an abstract quantity. Around 1200, al-Khowarizmi's book was translated into Latin by Fibonacci, and European mathematicians had a method for solving quadratics.

The first to use letters and coefficients the way we do was François Viète, a French lawyer and mathematician in the late 1500s. For $a^2 - 3a = 10$, Viète would write "1AQ–3A aequatur 10". (The Q stands for "quadratus.") Our modern notation with exponents is first found in a book by René Descartes published in 1637. These symbols forced European mathematicians to consider negative solutions to quadratic equations.

In the 16th century, mathematicians began using complex numbers because these numbers arose as solutions to equations. In the 19th century, Gauss brought both geometric and physical meaning to complex numbers. The geometric meaning, which you will encounter in later mathematics courses, uses the coordinate plane of Descartes. You have seen one of the physical meanings of complex numbers, the representation of impedance, in Lesson 6-9. In 1848, Gauss was the first to allow the *coefficients* in his equations to be complex numbers.

Finding Real Solutions by Graphing

Descartes' coordinate system provides a graphical way to find real-number solutions to quadratic equations, one that you have seen many times in this book. To solve $x^2 - 3x = 10$, notice that the graphs of $y = x^2 - 3x$ and $y = 10$ intersect at the two points (-2, 10) and (5, 10). This verifies that the solutions to $x^2 - 3x = 10$ are -2 and 5. This is shown on the graph below.

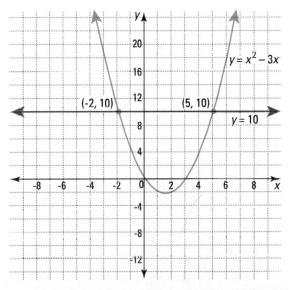

Another graphical way to solve the same equation is shown below. We have converted the equation $x^2 - 3x = 10$ to the standard form $x^2 - 3x - 10 = 0$. Now we look for intersections of the graphs of $y = x^2 - 3x - 10$ and $y = 0$. The graphs intersect at the two points $(-2, 0)$ and $(5, 0)$. This is shown below.

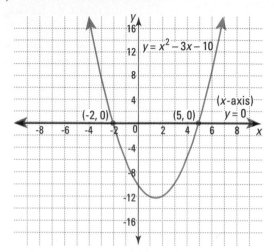

There is an advantage to the second graphical method. By converting the equation to standard form, all real-number solutions can be found by examining the x-axis. The x-intercepts of the graph of $y = ax^2 + bx + c$ indicate the real-number solutions to $ax^2 + bx + c = 0$. You should have seen in the In-class Activity that the number of real-number solutions to each quadratic equation was the same as the number of x-intercepts.

How Many Real Solutions Does a Quadratic Equation Have?

Now consider the general quadratic equation. When a, b, and c are real numbers and $a \neq 0$, the Quadratic Formula indicates that the two solutions to $ax^2 + bx + c = 0$ are

$$x = \frac{-b \pm \sqrt{b^2 - 4ac}}{2a}.$$

Because a and b are real numbers, the numbers $-b$ and $2a$ are real, so only $\sqrt{b^2 - 4ac}$ could possibly be nonreal. There are now three possibilities:

(1) If $b^2 - 4ac$ is positive, then $\sqrt{b^2 - 4ac}$ is a positive number. There will then be two real solutions.

(2) If $b^2 - 4ac$ is zero, then $\sqrt{b^2 - 4ac} = \sqrt{0} = 0$. Then $x = \frac{-b \pm 0}{2a} = \frac{-b}{2a}$, and there is only one real solution.

(3) If $b^2 - 4ac$ is negative, then $\sqrt{b^2 - 4ac}$ is an imaginary number. There will then be two nonreal solutions. Furthermore, since these solutions are of the form $m + ni$ and $m - ni$, they are complex conjugates.

Furthermore, if a, b, and c are rational numbers, and if $b^2 - 4ac$ is a perfect square, then the solutions to the equation will also be rational.

Because of the three properties on page 402, the number $b^2 - 4ac$ is called the **discriminant** of the quadratic equation. It determines the *nature of the solutions* to the equation.

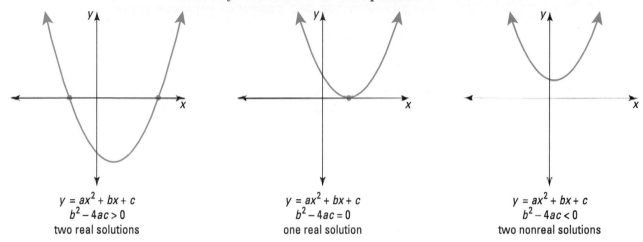

$y = ax^2 + bx + c$
$b^2 - 4ac > 0$
two real solutions

$y = ax^2 + bx + c$
$b^2 - 4ac = 0$
one real solution

$y = ax^2 + bx + c$
$b^2 - 4ac < 0$
two nonreal solutions

Geometrically, the graph of $y = ax^2 + bx + c$ intersects the x-axis in two points in situation (1), in one point in situation (2), and in no points in situation (3). Above are drawn these situations when a is positive, so the parabolas open up. The results you found from the In-class Activity should be consistent with this theorem.

> **Discriminant Theorem**
> Suppose a, b, and c are real numbers with $a \neq 0$.
> Then the equation $ax^2 + bx + c = 0$ has
> (i)   two real solutions if $b^2 - 4ac > 0$.
> (ii) one real solution if $b^2 - 4ac = 0$.
> (iii) two complex conjugate solutions if $b^2 - 4ac < 0$.

Solutions to quadratic (and other) equations are sometimes called **roots**. The number i allows square roots of negative numbers to be considered as complex solutions.

In Example 1, we let D stand for the discriminant. This shortens some of the writing.

Example 1

Determine the nature of the roots of the following equations. Then solve.
a. $4x^2 - 12x + 9 = 0$
b. $2x^2 + 3x + 4 = 0$
c. $2x^2 - 3x - 9 = 0$

Solution
a. Use the Discriminant Theorem. Here $a = 4$, $b = -12$, and $c = 9$.
$$D = b^2 - 4ac = (-12)^2 - 4(4)(9) = 0$$
Thus, the equation has one real solution or root.
By the quadratic formula, $x = \frac{12 \pm \sqrt{0}}{8} = \frac{3}{2} = 1.5$ ▶

b. Here $a = 2$, $b = 3$, and $c = 4$,
so $D = b^2 - 4ac = 9 - 4 \cdot 2 \cdot 4 = -23 < 0$.
The equation has two complex conjugate roots.

By the quadratic formula, $x = \dfrac{-3 \pm \sqrt{-23}}{4}$.

So $x = \dfrac{-3}{4} + \dfrac{\sqrt{23}}{4} i$ or $x = \dfrac{-3}{4} - \dfrac{\sqrt{23}}{4} i$.

c. Here $a = 2$, $b = -3$, and $c = -9$;
so $D = b^2 - 4ac = (-3)^2 - 4 \cdot 2 \,(-9) = 81 > 0$.
The equation has two real roots. Because D is a perfect square, the roots are rational.

By the quadratic formula, $x = \dfrac{3 \pm \sqrt{81}}{4} = \dfrac{3 \pm 9}{4}$.

So $x = 3$ or $x = -\dfrac{6}{4} = -\dfrac{3}{2} = -1.5$.

Applying the Discriminant Theorem

The number of real solutions to a quadratic equation can tell us something about the situation that led to the equation. For instance, recall the formula

$$h(x) = -.005x^2 + 2x + 3.5$$

describing the height of the baseball hit by Pop Fligh. When we let $h(x) = 8$, we solved

$$8 = -.005x^2 + 2x + 3.5$$

and found two solutions. These solutions indicated that the ball was 8 feet above the ground in two places. You can use the Discriminant Theorem to determine whether the ball ever reached a particular height.

Example 2

Does Pop Fligh's ball ever reach a height of 40 feet?

Solution

Pop Fligh's ball will reach a height of 40 feet if there are real values of x with
$$40 = -.005x^2 + 2x + 3.5.$$

To calculate the discriminant of this equation, first rewrite the equation in standard form:
$$0 = -.005x^2 + 2x - 36.5$$

The discriminant is
$$D = b^2 - 4ac = 2^2 - 4 \cdot (-.005) \cdot (-36.5) = 3.27.$$

Since D is positive, there are two real-number solutions to this equation. This means that the ball gets to a height of 40 feet twice. One of these times is on the way up, the other on the way down.

Covering the Reading

1. Match the idea at the left with the estimated length of time it has been known.
   a. today's notation for quadratics
   b. problems leading to quadratics

   (i)   about 3700 years
   (ii) about 2300 years
   (iii) about 1750 years
   (iv) about 1150 years
   (v)   about 350 years

2. The word *algebra* stems from a word in which language?

3. Why did the Pythagoreans think there was only one solution to $x^2 = 2$?

4. Consider the equation $ax^2 + bx + c = 0$, where a, b, and c are real.
   a. What is its discriminant?   b. What are its roots?

5. The discriminant of an equation $ax^2 + bx + c = 0$ is -1000. What does this indicate about the graph of $y = ax^2 + bx + c$?

In 6 and 7, use the discriminant to determine the nature of the roots to the equation.

6. $3x^2 - 4x + 5 = 0$ 7. $5y^2 - 10y + 5 = 0$

8. a. Solve $3x^2 + 4x + 5 = 0$, and write the solutions in $a + bi$ form.
   b. *True or false.*   The roots of this equation are complex conjugates.

In 9 and 10, the graph of a quadratic function $f(x) = ax^2 + bx + c$ is shown.
a. Tell whether the value of $b^2 - 4ac$ is positive, negative, or zero.
b. Tell how many real roots the equation $f(x) = 0$ has.

9.
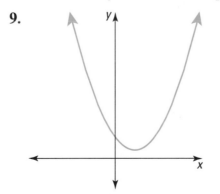

10.
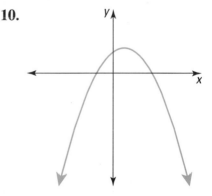

11. Without drawing a graph, tell whether the x-intercepts of $y = -4x^2 + 2x + 1$ are rational or irrational.

In the colorized medieval wood art shown, Pythagoras is producing music from carefully graduated bells and water glasses. Until they discovered irrational numbers, the Pythagoreans thought that all things could be described by whole numbers or ratios of them.

12. Pop Fligh wanted to know if the ball he hit went higher than the top of the stadium, 200 feet high. Was the ball ever 200 feet high? Justify your answer.

Applying the Mathematics

13. The three graphs on page 403 cover the possible cases when the parabolas open up. Draw the three graphs corresponding to $b^2 - 4ac > 0$, $b^2 - 4ac = 0$, and $b^2 - 4ac < 0$ when the parabolas open down.

In 14 and 15, *true or false.* If true, prove the statement; if false, give a counterexample.

14. Every parabola which has an equation of the form $y = ax^2 + bx + c$ has a *y*-intercept.

15. Whenever a parabola has one *x*-intercept, it also has a second *x*-intercept.

In 16 and 17, the following program in BASIC uses the discriminant to solve quadratic equations of the form $ax^2 + bx + c = 0$.

```
 90  PRINT "SOLVE A*X^2 + B*X + C = 0."
100  INPUT "COEFFICIENTS A =";A
120  INPUT "B ="; B
130  INPUT "C ="; C
200  D = B^2 - 4*A*C
250  IF D < 0 THEN 700
300  X1 = (-B + SQR(D))/(2*A)
400  X2 = (-B - SQR(D))/(2*A)
500  PRINT "THE ROOTS ARE";X1; "OR";X2
600  GO TO 999
700  PRINT "THERE ARE NO REAL ROOTS."
999  END
```

BASIC

16. Check to see that this BASIC program gives the correct solutions to the following equations. (Or translate this program to your calculator's language.)
   a. $-2x^2 + 40x = 0$ **b.** $5x^2 - 150x + 1185 = 0$

17. **a.** Describe what happens when you input $a = 0$.
    b. Modify the program so it tests whether a is 0, and prints "NOT A QUADRATIC EQUATION" when $a = 0$.

18. **a.** Find the value(s) of k for which the graph of the equation $y = x^2 + kx + 9$ will have exactly one x-intercept. (Hint: When is the discriminant zero?)
   b. Check your answers by graphing.

Review

In 19–24, perform the operation(s), and write the result in $a + bi$ form. *(Lessons 6-8, 6-9)*

19. $(i \cdot 3i)^2$

20. $\sqrt{-64} + \sqrt{-36}$

21. $\sqrt{-64} \cdot \sqrt{-36}$

22. $(6i + 1)(6i - 1)$

23. $\frac{3 + 5i}{i}$

24. $\frac{6 + i}{2 - 3i}$

25. For an electrical circuit, the formula $V = ZI$ gives the voltage V in volts in terms of the impedance Z in ohms and the current I in amps. Find the voltage if the current is $10 + 2i$ amps and the impedance is $5 - 3i$ ohms. *(Lesson 6-9)*

26. Solve $-7 - 3n^2 = 5n$. *(Lesson 6-7)*

27. The graph of a quadratic function f passes through the points $(-4, 3)$, $(2, 8)$, and $(1, 13)$.
   a. Find an equation for this graph.
   b. Check your work by graphing the equation in part **a.**
   (Lessons 6-4, 6-6)

28. Janice didn't want to deal with the fraction $\frac{25}{4}$ when solving $n^2 - 5n + \frac{25}{4} = 0$ so she multiplied each side by 4 and solved $4n^2 - 20n + 25 = 0$ instead. Do both equations have the same roots? Why or why not? *(Lessons 1-6, 6-6)*

Exploration

29. **a.** Find two nonreal complex numbers that are not complex conjugates and whose sum is a real number.
   b. Find two nonreal complex numbers that are not complex conjugates and whose product is a real number.
   c. Find two nonreal complex numbers that are not complex conjugates and whose quotient is a real number.

A project presents an opportunity for you to extend your knowledge of a topic related to the material of this chapter. You should allow more time for a project than you do for typical homework questions.

1 Projectile Motion

Paths of objects other than batted baseballs are almost parabolic. Find out about other objects that travel in parabolic paths. Make a poster or write a brief report illustrating and describing these paths. Find equations for some of the paths.

2 Sum and Product of Roots

Checking solutions to quadratic equations can be tedious, but there is an easier way to check than by substitution. The method checks the sum and the product of the roots. Look back over some quadratic equations you have solved. Let r_1 and r_2 be the roots of the equation $ax^2 + bx + c = 0$. For each equation calculate $r_1 + r_2$ and $r_1 \cdot r_2$. What patterns do you notice? Can you prove your generalizations hold for all quadratics? Use your results to check some of the solutions to quadratic equations in this chapter.

3 The Graph-Translation Theorem and Other Functions

The Graph-Translation Theorem applies to graphs of all functions, not just the quadratic and absolute value functions used in this chapter. Consider the family of inverse variation functions.

a. Graph the "parent" function $y = \frac{1}{x}$.

b. Graph 3 or 4 "offspring" whose equations have the form $y = \frac{1}{x - h}$.

c. Graph 3 or 4 "offspring" whose equations have the form $y = \frac{1}{x} + k$.

d. Graph 3 or 4 "offspring" whose equations have the form $y = \frac{1}{x - h} + k$.

e. Write a brief report summarizing the patterns you observe.

4 Quadratic Models

Look through an almanac or the *Statistical Abstract of the United States.* Find some current data for which x and y appear to be related by a quadratic model. Find an equation for a quadratic function that models the data. What seems to be a reasonable domain and range for this function? Why do you believe that a quadratic model fits the data set better than a linear model?

5 History of Quadratics

Throughout this chapter we have mentioned people who have contributed to our understanding of quadratic functions and equations. Make a time line marking important dates in this history. Consult a book on the history of mathematics and other references to find out about others who contributed to this history, and include their work on your time line.

6 Predicting the Areas of States or Countries

In a square with area A and side s, $s^2 = A$ exactly. The length of a side of shapes that are close to Squares can be a good predictor of area, even if the prediction is not exact. Use an atlas to find the lengths of sides

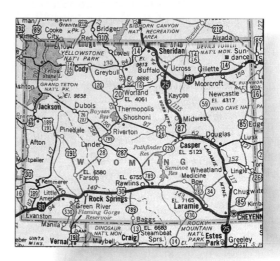

of several states in the United States or countries in the world that are fairly close to being square. On a map measure the length and width of the state or country with a ruler, and use the scale of the map to translate these distances into miles or kilometers. For each state or country you measure, look up its area in an atlas or encyclopedia. Fit a model to the data and compare the predicted and actual areas. Write a report which includes a table, graph, your model, an explanation of how you arrived at the model, and the conclusions you made.

SUMMARY

Quadratic expressions are those which involve one or more terms in x^2, y^2, or xy, but no higher powers of x or y. If $k > 0$, the equation $x^2 = k$ has two real solutions, $\sqrt{k}$ and $-\sqrt{k}$. When $k < 0$, the solutions are the imaginary numbers $i\sqrt{k}$ and $-i\sqrt{k}$, where, by definition, $\sqrt{-1} = i$. Any number of the form $a + bi$, where a and b are real, is a complex number. Complex numbers are added, subtracted, and multiplied using the field properties that apply to real numbers.

Work with quadratics requires some skill in manipulating squares and square roots. Among the theorems in this chapter are the Binomial Square Theorem: for all x and y, $(x + y)^2 = x^2 + 2xy + y^2$, and the Absolute Value-Square Root Theorem: for all real numbers x, $\sqrt{x^2} = |x|$. Additionally, for all nonnegative real numbers x and y, $\sqrt{xy} = \sqrt{x}\,\sqrt{y}$. This last theorem does not hold when x and y are both negative.

Areas, paths of projectiles, and relations between the initial velocity of an object and its height over time lead to problems involving quadratic equations and functions. If data involving two variables are graphed, and the scatterplot appears to be part of a parabola, you can use three points on the graph to set up a system of equations that will allow you to find a, b, and c in the equation $y = ax^2 + bx + c$.

When a, b, and c are real numbers and $a \neq 0$, the graph of the equation $y = ax^2 + bx + c$ is a parabola. Using the process known as completing the square, this equation can be rewritten in vertex form $y - k = a(x - h)^2$. This parabola is a translation image of the parabola $y = ax^2$ you studied in Chapter 2. Its vertex is (h, k), its line of symmetry is $x = h$, and it opens up if $a > 0$ and opens down if $a < 0$.

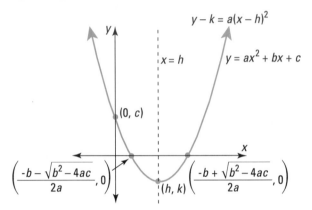

Its x-intercepts, the values of x for which $ax^2 + bx + c = 0$, can be found by the famous Quadratic Formula

$$x = \frac{-b \pm \sqrt{b^2 - 4ac}}{2a}$$

The expression $b^2 - 4ac$ is the discriminant of the quadratic equation, and reveals the nature of its roots. If $b^2 - 4ac > 0$, there are two real solutions. That is the situation pictured above. If $b^2 - 4ac = 0$, there is exactly one solution and the vertex of the parabola is on the x-axis. If a, b, and c are rational numbers and the discriminant is a perfect square, then the solutions are rational numbers. If $b^2 - 4ac < 0$, there are no real solutions and the parabola does not intersect the x-axis. The solutions are complex conjugates.

VOCABULARY

Below are the most important terms and phrases for this chapter.
You should be able to give a definition for those terms marked with a *.
For all other terms you should be able to give a general description or
a specific example.

Lesson 6-1
quadratic
quadratic expression
quadratic equation
quadratic function
standard form of a quadratic
Binomial Square Theorem

Lesson 6-2
absolute value
absolute value function
Absolute Value-Square Root Theorem
square root
simple fraction
irrational number

Lesson 6-3
*Graph-Translation Theorem
corollary
vertex form of an equation of a parabola
axis of symmetry
minimum, maximum

Lesson 6-4
standard form of an equation of a parabola
acceleration due to gravity
$h = -\frac{1}{2}gt^2 + v_0t + h_0$
velocity

Lesson 6-5
completing the square
perfect-square trinomial

Lesson 6-6
quadratic model
quadratic regression

Lesson 6-7
*Quadratic Formula
standard form of a quadratic equation

Lesson 6-8
*$\sqrt{-k}$
*$\sqrt{-1}$, i
imaginary number

Lesson 6-9
*complex number
*real part, imaginary part
*equal complex numbers
impedance
circuit in series, in parallel
*complex conjugate
hierarchy

Lesson 6-10
*discriminant of a quadratic equation
nature of the solutions
*root of an equation
Discriminant Theorem

PROGRESS SELF-TEST

Take this test as you would take a test in class. Use graph paper and a calculator. Then check your work with the solutions in the Selected Answers section in the back of the book.

In 1–3, consider the parabola with equation $y = x^2 - 8x + 12$.

1. Rewrite the equation in vertex form.

2. What is the vertex of this parabola?

3. What are the x-intercepts of this parabola?

In 4–7, perform the operations and simplify.

4. $2i \cdot i$

5. $\sqrt{-8} \cdot \sqrt{-2}$

6. $\dfrac{4 + \sqrt{-8}}{2}$

7. $(3i + 2)(6i - 4)$

8. If $z = 2 - 4i$ and $w = 1 + 5i$, what is $z - w$?

9. *Multiple choice.* How does the graph of $y - 2 = -(x + 1)^2$ compare with the graph of $y = -x^2$?
 (a) It is 1 unit to the right and 2 units below.
 (b) It is 1 unit to the right and 2 units above.
 (c) It is 1 unit to the left and 2 units above.
 (d) It is 1 unit to the left and 2 units below.

10. Graph the solution set to $y - 2 = -(x + 1)^2$.

In 11–13, find all solutions.

11. $3x^2 + 14x - 5 = 0$

12. $(m + 40)^2 = 2$

13. $3x^2 + 15 = 18x - 60$

14. *True or false.* $\sqrt{x^2} = |x|$ for all real values of x.

In 15 and 16, use the equation $y = ax^2 + bx + c$, assuming $a \neq 0$ and a, b, and c are real numbers.

15. How many x-intercepts does the graph have
 a. if its discriminant is 0?
 b. if its discriminant is 1?

16. If $y = 0$, describe the nature of the roots
 a. if its discriminant is 0.
 b. if its discriminant is -5.

In 17 and 18, expand.

17. $(a - 3)^2 + (a + 3)^2$

18. $(8v + 1)^2$

19. A ball is thrown upward from an initial height of 20 meters at an initial velocity of 10 meters per second. Write an equation for the height h of the ball t seconds after being thrown.

In 20 and 21, the height in feet h of a ball at time t is given by $h = -16t^2 + 12t + 4$.

20. How high is the ball .5 second after it is thrown?

21. When does the ball hit the ground?

22. A rectangular piece of metal is 40 cm by 30 cm. A strip s cm wide is cut parallel to each side of the metal. What is the area of the remaining rectangle in terms of s?

23. *Multiple choice.* Which parabola is not congruent to the others?
 (a) $y = 2x^2$ (b) $y = x^2 + 2$
 (c) $y = (x + 2)^2$ (d) $y + 2 = x^2$

24. There is a pattern to the number of handshakes needed in a group of people if everyone shakes hands with everybody else exactly once. Examine the table below.

Number of people n	1	2	3	4	5
Number of handshakes	0	1	3	6	10

 a. Draw a scatterplot of these data.
 b. Fit a quadratic model to these data.
 c. How many handshakes would be needed if 100 people shake hands with each other exactly once?

25. $y = 2x^2 - 6x + 4$ is written in standard form. Rewrite this equation in vertex form, and write a sentence detailing information about the graph you can obtain just by looking at the equation in this form.

CHAPTER REVIEW

Questions on SPUR Objectives

SPUR stands for **S**kills, **P**roperties, **U**ses, and **R**epresentations. The Chapter Review questions are grouped according to the SPUR Objectives for this chapter.

SKILLS DEAL WITH THE PROCEDURES USED TO GET ANSWERS.

Objective A: *Expand squares of binomials.*
(Lesson 6-1)

In 1–6, expand.

1. $(a + x)^2$ **2.** $(y - 11)^2$

3. $(3x + 4)^2$ **4.** $2(x - 2)^2$

5. $9(t - 5)^2$ **6.** $3(a + b)^2 - 4(a - b)^2$

Objective B: *Transform quadratic equations from vertex form to standard form, and vice versa.*
(Lessons 6-4, 6-5)

In 7 and 8, transform into standard form.

7. $y = 3(x + 2)^2$ **8.** $y + 8 = \frac{1}{2}(x - 4)^2$

In 9 and 10, transform each equation into vertex form.

9. $y = x^2 + 10x - 6$ **10.** $y = 2x^2 - 6x - 1$

11. Find an equation in vertex form equivalent to $y = 3x^2 - 30x + 12$.

12. *Multiple choice.* Which equation is equivalent to $y = 2x^2 - 4x + 3$?
(a) $y - 1 = 2(x + 2)^2$
(b) $y - 1 = 2(x - 1)^2$
(c) $y - 3 = 2(x + 1)^2$
(d) $y - 2 = 2(x - 1)^2$

Objective C: *Solve quadratic equations.*
(Lessons 6-2, 6-7, 6-8, 6-10)

In 13–26, solve.

13. $(x - 3)^2 = 0$ **14.** $d^2 - 48 = 0$

15. $z^2 = -8$ **16.** $w^2 = -9$

17. $x^2 + x - 1 = 0$ **18.** $1 - y^2 - 7y = 6$

19. $z^2 - 8z + 11 = -5$ **20.** $0 = 4a^2 + 3a + 2$

21. $k^2 = 4k + 2$

22. $3x^2 + 2x + 6 = 2x^2 + 4x - 3$

23. $x^2 + 25 = 0$ **24.** $x(x + 1) = 1$

25. $3 = 5p + 2p^2$ **26.** $2(3n^2 + 2) = 4(n - 9)$

Objective D: *Perform operations with complex numbers.* *(Lessons 6-8, 6-9)*

In 27–34, simplify.

27. $-i^2$ **28.** $\sqrt{-36}$

29. $\sqrt{-16} \cdot \sqrt{-49}$ **30.** $\sqrt{2} \cdot \sqrt{-2}$

31. $10\sqrt{-50}$ **32.** $3i \cdot i$

33. $\frac{4 \pm \sqrt{-80}}{2}$ **34.** $\frac{-5 \pm \sqrt{-25}}{10}$

In 35–40, perform the operations and write the answer in $a + bi$ form.

35. $(3 + 7i) + (-2 + 5i)$ **36.** $(8 + i) - (12 - 6i)$

37. $i(10 + 6i)$ **38.** $(4 + i)(9 - i)$

39. $\frac{-6 + 3i}{2i}$ **40.** $\frac{12}{4 - i}$

In 41–46, suppose $u = 3 - i$ and $v = 8i + 5$. Evaluate and simplify.

41. uv **42.** u^2

43. $3u - v$ **44.** $iu + v$

45. $\frac{u}{v}$ **46.** $\frac{v}{u}$

PROPERTIES DEAL WITH THE PRINCIPLES BEHIND THE MATHEMATICS.

Objective E: *Apply the definition of absolute value and the Absolute Value-Square Root Theorem.* (Lesson 6-2)

47. For what real numbers does $|x| = -x$?

48. For what real numbers does $|x| = -2$?

In 49–52, simplify.

49. $\sqrt{(6 + 3)^2}$

50. $-\sqrt{y^2}$

51. $-\sqrt{(-4)^2}$

52. $\sqrt{(-3x)^2}$

Objective F: *Use the discriminant of a quadratic equation to determine the nature of the solutions to the equation.* (Lesson 6-10)

In 53–56, an equation is given. **a.** Evaluate its discriminant. **b.** Give the number of real solutions. **c.** Tell whether the solutions are rational, irrational, or nonreal.

53. $9 + 4y^2 - 12y = 0$

54. $z^2 = 100z + 100$

55. $6 + t = t^2 - 5$

56. $8x^2 + 9x + 6 = 0$

57. How many real solutions does $2x^2 = 3x$ have? How can you tell?

USES DEAL WITH APPLICATIONS OF MATHEMATICS IN REAL SITUATIONS.

Objective G: *Use quadratic equations to solve area problems or problems dealing with velocity and acceleration.* (Lessons 6-1, 6-2, 6-4, 6-7)

58. A 20″-by-36″ picture is to be surrounded by a frame w inches wide.

   a. What is the total area of the picture and frame?

   b. If the total area is to be $\frac{4}{3}$ the area of the picture, how wide should the frame be?

59. Daniel wants to construct a rectangular pen alongside his house for his dog. He plans to use 22 meters of chicken wire. What should be the dimensions of the pen if Daniel wants his dog to have as much area as possible?

60. Suppose a ball is thrown upward from a height of 4.5 feet with an initial velocity of 21 feet per second.

   a. Write an equation relating the time t and height h of the ball.

   b. When will the ball hit the ground?

61. A package of supplies is dropped from a helicopter hovering 100 m above the ground. The attached parachute fails to open. After how many seconds will the package reach the ground? (Neglect air resistance.)

62. A ball is hit by a bat when 3 feet off the ground. It is caught at the same height 300 feet from the batter. How far from the batter did it reach its maximum height?

Objective H: *Fit a quadratic model to data.* (Lesson 6-6)

63. The following pictures show the first five pentagonal numbers.

1   5   12   22   35

   a. Find the next two pentagonal numbers.

   b. Fit a quadratic model to find a formula for $p(n)$, the nth pentagonal number.

   c. Find the 50th pentagonal number.

64. Jeremy was in charge of scheduling for the local Little League. Each team played each other team twice. He needed to know the total number of games played so that he could provide umpires. The first year, with 4 teams, he scheduled 12 games. The second year, with 5 teams, he scheduled 20 games. The third year with 6 teams, he scheduled 30 games. Find the number of games needed for a league with

   a. 2 teams.   **b.** 3 teams.

   c. With the number of teams as the independent variable, make a scatterplot of these data.

   d. Fit an appropriate model to these data.

   e. How many games would be necessary for a 10-team league?

Objective I: *Use the Graph-Translation Theorem to interpret equations and graphs.* (Lessons 6-3)

65. $y = x^2$ is translated 7 units to the left and 5 units down. What is an equation for its image?

66. Describe how the graphs of $y = |x|$ and $y = |x - 3|$ are related.

67. *Multiple choice.* Which of the following is *not* true for the graph of the parabola with equation $y - 5 = -2(x + 1)^2$?
(a) The vertex is (-1, 5).
(b) The maximum point is (-1, 5).
(c) The equation of the axis of symmetry is $x = -1$.
(d) The graph opens up.

68. Compare the solutions to $6 = (x - 2)^2$ with the solutions to $6 = x^2$.

69. Compare the solutions to $(k - 1)^2 = 36$ with the solutions to $k^2 = 36$.

70. Assume that parabola A is congruent to parabola B in the graph below.

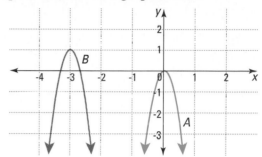

a. What translation maps parabola A onto B?
b. What is the equation of parabola B if parabola A has equation $y = -10x^2$?

71. If the graph of $y = (3x + 11)^2$ is congruent to the graph of $y = ax^2$, find a.

REPRESENTATIONS DEAL WITH PICTURES, GRAPHS, OR OBJECTS THAT ILLUSTRATE CONCEPTS.

Objective J: *Graph quadratic functions or absolute value functions and interpret them.* (Lessons 6-2, 6-3, 6-4)

In 72–75, graph the parabola, identifying its vertex and x-intercepts.

72. $y = 5x^2 - 20x$ **73.** $y - 4 = -\frac{1}{3}(x + 2)^2$

74. $y + 4 = 3(x - 1)^2$ **75.** $y = -2x^2 + 4x - 1$

In 76 and 77, refer to the parabolas below.

76. Tell whether the graph could represent solutions to an equation of the form $y - k = a(x - h)^2$.

a. 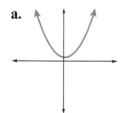   **b.**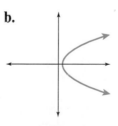

c. 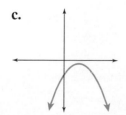   **d.**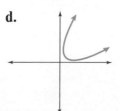

77. Given $y - k = a(x - h)^2$, for which of the graphs in Question 76 is a negative?

78. The height of a baseball thrown upward at time t is shown on the graph at the right.

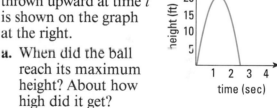

a. When did the ball reach its maximum height? About how high did it get?
b. When was the ball 10 feet high?

In 79 and 80, graph.

79. $y = |x + 3|$ **80.** $y - 2 = |x - 1|$

Objective K: *Use the discriminant of a quadratic equation to determine the number of x-intercepts of the graph.* (Lesson 6-10)

In 81 and 82, give the number of x-intercepts of the graph of the parabola.

81. $y = 3x^2 + 2x - 2$ **82.** $y = \frac{1}{2}(x + 5)^2 - 3$

83. Does the parabola $y = 6x^2 - 12x$ ever intersect the line $y = -5$? Justify your answer.

84. If the graph of $y = -\frac{1}{4}x^2$ has one x-intercept, how many x-intercepts does the graph of $y = \frac{1}{4}(x - a)^2$ have if $a \neq 0$?

POWERS

The table below compares typical weights with food consumption for five mammals of quite different sizes.

Mammal	Weight (kg)	Food per day (kg)	Ratio of food per day to weight
elephant	6400	145	.023
moose	450	19.5	.043
raccoon	10	2	.200
squirrel	0.45	0.33	.733
mouse	0.03	0.12	4.000

The ratio of F, the amount of food a mammal must eat per day, to m, its body mass, or weight on Earth, is not constant across species. For a mouse, $\frac{F}{m} \approx 4.000$; that is, a mouse must eat about 4 times its mass per day. On the other extreme, for an elephant $\frac{F}{m} \approx 0.023$, so an elephant needs to eat only about .02, or 2%, of its mass per day.

In general, scientists have shown that for each species of mammal, the ratio of amount of food eaten daily to body mass varies as the negative one-third power of the mass. This can be written as

$$\frac{F}{m} = km^{-\frac{1}{3}},$$

where the constant of variation k depends on the species.

In this chapter, you will study powers and related functions. You will learn how to interpret negative powers and fractional powers, and how to solve some equations involving powers. You will also learn about other applications of powers in the real world.

Power decisions. *Shown is Alex Trebek, host of the TV game show* Jeopardy. *If the probability of answering each question is p, what is the probability of answering an entire category of 5 questions? See Example 1.*

Recall that the expression x^n, read "*x* to the *n*th power" or the "*n*th power of *x*," is the result of an operation called **powering** or **exponentiation.** The variable *x* is called the **base,** *n* is called the **exponent,** and the expression x^n is called a **power.**

For any real number *b,* and any positive integer *n:*

$$b^n = \underbrace{b \cdot b \cdot b \cdot \ldots \cdot b}_{n \text{ factors}}.$$

This is the **repeated multiplication** meaning of a positive integer power. For instance, $x^7 = x \cdot x \cdot x \cdot x \cdot x \cdot x \cdot x$. This meaning enables you to use multiplication to calculate positive integer powers of any number. For example,

$$\left(-\tfrac{5}{4}\right)^3 = \left(-\tfrac{5}{4}\right) \cdot \left(-\tfrac{5}{4}\right) \cdot \left(-\tfrac{5}{4}\right)$$
$$= -\tfrac{125}{64}.$$

An Example of a Power Function

The *square* of *x*, x^2, is the area of a square with side *x*. The *cube* of *e*, e^3, is the volume of a cube with edge *e*. Powers with these and larger exponents arise in counting and probability situations. Here is a typical situation of this type.

Beth forgot to study for a true-false quiz with 5 questions. So she must guess at each answer. If each question is either true or false, the probability of guessing the correct answer to any particular question is $\frac{1}{2}$. If each answer were independent of the others, the probability of answering all five questions correctly would be

$$\frac{1}{2} \cdot \frac{1}{2} \cdot \frac{1}{2} \cdot \frac{1}{2} \cdot \frac{1}{2} = \left(\frac{1}{2}\right)^5 = .03125.$$

Now, if each question were multiple choice with four choices, then the probability of getting each question correct by guessing would be $\frac{1}{4}$, and the probability of getting all five answers correct would be

$$\frac{1}{4} \cdot \frac{1}{4} \cdot \frac{1}{4} \cdot \frac{1}{4} \cdot \frac{1}{4} = \left(\frac{1}{4}\right)^5 \approx .000977.$$

Example 1 generalizes Beth's situation.

Example 1

Suppose the probability of answering each question correctly on a five-question quiz is p, and that the questions are independent. Let A be the probability of answering all five questions correctly.
a. Find a formula for A in terms of p.
b. Make a table of values and a graph for values of p from 0 to 1 increasing by tenths.

Solution
a. $A = p^5$
b. The probability p must be a number from 0 to 1. A table and a graph are shown below.

p	$A = p^5$
0	0
0.1	0.00001
0.2	0.00032
0.3	0.00243
0.4	0.01024
0.5	0.03125
0.6	0.07776
0.7	0.16807
0.8	0.32768
0.9	0.59049
1	1

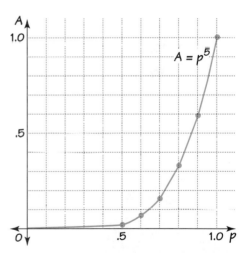

Notice that even if Beth had a 90% chance of answering each question correctly, the probability of her getting a perfect score on the quiz would be only $(.9)^5$, or about 59%.

Some Simple Power Functions

In general, the function defined by $f(x) = x^n$, where n is a positive integer, is called the **nth power function.** The function $y = x^5$ is the *5th power function.* The points graphed in Example 1 lie on the graph of the 5th power function.

The simplest power function is $f(x) = x^1$, which is called the **identity function.** The quadratic function, $f(x) = x^2$ is the **2nd power,** or **squaring, function.** Tables and graphs for the identity and squaring functions are shown below.

The Identity Function

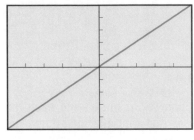

$-5 \le x \le 5, \quad x\text{-scale} = 1$
$-5 \le y \le 5, \quad y\text{-scale} = 1$

The Squaring Function

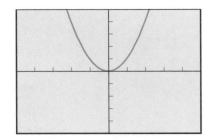

$-5 \le x \le 5, \quad x\text{-scale} = 1$
$-5 \le y \le 5, \quad y\text{-scale} = 1$

x	-5	-4	-3	-2	-1	0	1	2	3	4	5
$f(x) = x$	-5	-4	-3	-2	-1	0	1	2	3	4	5

x	-5	-4	-3	-2	-1	0	1	2	3	4	5
$f(x) = x^2$	25	16	9	4	1	0	1	4	9	16	25

Any real number can be raised to the first or second power. So the domain of each of the above functions is the set of all real numbers. The range of the identity function is also the set of real numbers. However, because the result of squaring a real number is always nonnegative, the range of the squaring function is the set of all nonnegative reals.

Caution: Some *scientific* calculators will give an error message when a negative number is raised to a power with the 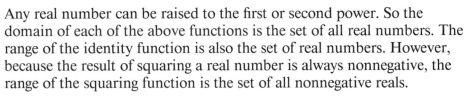 key. This means either that the calculator has not been programmed for this calculation, or that it cannot be done. Most graphics calculators do not give error messages when a negative number is raised to a positive *integer* power.

The function $f(x) = x^3$ is called the **cubing function.** The nth power functions where $n > 3$ do not have special names.

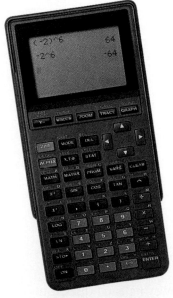

Caution must be taken when raising negative numbers to powers. Note, for example, that $-2^6 \neq (-2)^6$.

420

Example 2

Draw a graph and state the domain and the range for each function.

a. $f(x) = x^3$ **b.** $g(x) = x^4$

Solution

Make a table of values and plot points, or use an automatic grapher.

a.

x	-5	-4	-3	-2	-1	0	1	2	3	4	5
$f(x) = x^3$	-125	-64	-27	-8	-1	0	1	8	27	64	125

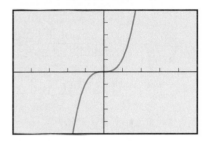

$-5 \le x \le 5$, x-scale = 1
$-5 \le y \le 5$, y-scale = 1

Both the domain and the range are the set of real numbers.

b.

x	-5	-4	-3	-2	-1	0	1	2	3	4	5
$g(x) = x^4$	625	256	81	16	1	0	1	16	81	256	625

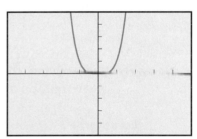

$-5 \le x \le 5$, x-scale = 1
$-5 \le y \le 5$, y-scale = 1

The domain is the set of real numbers; the range is the set of nonnegative real numbers.

Activity

a. Use an automatic grapher. Draw a graph of $f(x) = x^5$ on your default window.
b. What is the domain of f?
c. What is the range of f?
d. How is this graph like and unlike the other graphs in the lesson?

Properties of Power Functions

Several properties of the nth power functions $f(x) = x^n$, where n is a positive integer, can be deduced.

1. The graph of every nth power function $f(x) = x^n$ passes through the origin, because $0^n = 0$ for any positive integer value of n.

2. The domain of every nth power function is the set of real numbers, because you can raise any real number to a positive integer power.

3. To find the range, two cases must be considered for n.
    (a) *n is even:*
        If $x \geq 0$, then $x^n \geq 0$ because any nonnegative number raised to a power is nonnegative. If $x < 0$, then raising x to an even power results in a positive number. Thus, when n is even, the range of $f(x) = x^n$ is $\{y: y \geq 0\}$, and the graph of $f(x) = x^n$ is in quadrants I and II. You can check this with the graphs of $f(x) = x^2$ and $f(x) = x^4$ on the previous pages.
    (b) *n is odd:*
        If $x \geq 0$, then $x^n \geq 0$ because any nonnegative number raised to a power is nonnegative. If $x < 0$, then raising x to an odd power results in a negative number. Thus, when n is odd, the range of $f(x) = x^n$ is the set of all real numbers, and the graph of $f(x) = x^n$ is in quadrants I and III. This is illustrated by the graphs of $f(x) = x$, $f(x) = x^3$, and $f(x) = x^5$.

4. The graph of every power function has symmetry.
    (a) *n is even:*
        The even power functions have reflection symmetry. The graph of $f(x) = x^n$ is symmetric to the y-axis.
    (b) *n is odd:*
        The odd power functions have rotation symmetry. The graph of $f(x) = x^n$ when n is odd, can be mapped onto itself under a $180°$ rotation around the origin.

Example 3

Multiple choice. Which is the graph of the function with equation $y = x^7$? Justify your answer.

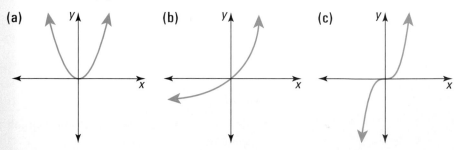

(a)   (b)   (c)

Solution

The graph of $y = x^7$ must have rotation symmetry about the origin; only graph (c) does. Thus (c) is the correct graph.

Masked mammal. *This is a raccoon. Why is it pictured here?*

QUESTIONS

Covering the Reading

1. What is the meaning of the number .200 in the table on page 417?

2. Cal forgot to study for a multiple-choice test with 8 independent questions.
   a. Let A be the probability of getting all questions correct. Find A if the probability of getting a single question correct is as follows.
   (i) .5   (ii) .2   (iii) p
   b. *True or false.* If the probability of getting each question correct is p, then A is a function of p. Justify your answer.

3. Give an equation describing the nth power function.

4. What is the function with equation $f(x) = x^3$ called?

5. Refer to the Activity in the lesson.
   a. Sketch your graph of $f(x) = x^5$.
   b. What is the domain of f?
   c. What is the range of f?
   d. How is this graph like and unlike other graphs in the lesson?

In 6 and 7, an equation for a power function is given. **a.** Sketch a graph without plotting points or doing any calculations. **b.** State the domain and the range of the function. **c.** Describe any symmetry the graph has.

6. $f(x) = x^8$

7. $f(x) = x^9$

8. If n is even, the range of $y = x^n$ is __?__ and the graph is in quadrants __?__ and __?__.

9. If n is odd, the range of $y = x^n$ is __?__ and the graph is in quadrants __?__ and __?__.

10. *Multiple choice.* Which of the following graphs could represent the function with equation $y = x^{10}$? Justify your answer.

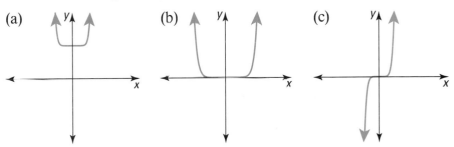

Historic windmills.
Shown are windmills in Holland de Zaanse Schans, a reconstructed historic village in Holland. Developed in the 1100s, these windmills were widely used to mill grain and drain water from the land.

11. *True or false.* The graphs of the odd power functions have no minimum or maximum values.

12. In Chapter 2, you read that the power P generated by a windmill is proportional to the cube of the wind speed w.
   a. Write a formula for P in terms of w.
   b. What is the domain of this function?
   c. Sketch a graph over the appropriate domain.

13. Phil learned to spell $\frac{p}{100}$ of the words on a long list when studying for a spelling contest. To qualify, he must spell correctly four words randomly chosen from the list. Let q = the probability of qualifying for the contest.
   a. Express q as a function of p.
   b. What is the domain of this function?

14. Graph $f: x \rightarrow x^{51}$ on the window $-5 \le x \le 5$, $-100 \le y \le 100$. On what interval(s) is the graph above the x-axis, and on what interval(s) is the graph below the x-axis?

In 15 and 16, an nth power function is graphed. Write an equation for each function.

15.

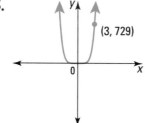

(3, 729)

16.

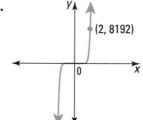

(2, 8192)

17. The point $(-2, 64)$ is on the graph of an nth power function where n is even. Must the point $(2, 64)$ also be on the graph? Justify your answer.

18. **a.** Use an automatic grapher to plot $f(x) = x^2$ and $g(x) = x^4$ on the same set of axes.
   b. For what value(s) of x is $f(x) = g(x)$?
   c. For what values of x is $x^2 > x^4$?

19. A manufacturer makes ribbons of cotton and silk. A yard of cotton ribbon costs $.40 to produce, and a yard of silk ribbon costs $.70 to produce. In one hour, the manufacturer can produce up to 4000 yards of ribbon, but no more than 2500 yards can be silk. The cost of producing these ribbons cannot exceed $2000 per hour.
 a. Translate the constraints of the problem into a system of inequalities. Let x = the number of yards of silk ribbon produced in an hour, and y = the number of yards of cotton ribbon produced in an hour.
 b. Graph the system of inequalities.
 c. The manufacturer makes a profit of $.15 on each yard of silk ribbon and $.08 profit for each yard of cotton. The manager thinks that the best way to maximize profit is to produce 2500 yards of silk ribbon and 625 yards of cotton ribbon. Do you agree with the manager? Why or why not? *(Lesson 5-10)*

20. Give an equation for the line perpendicular to $3x + 2y = 5$ at the point $(1, 1)$. *(Lesson 4-8)*

21. Name a transformation that maps $\begin{bmatrix} 2 & -3 & 0 \\ 1 & 4 & 6 \end{bmatrix}$ onto $\begin{bmatrix} 1 & 4 & 6 \\ -2 & 3 & 0 \end{bmatrix}$.
(Lesson 4-7)

In 22 and 23, solve for x. *(Previous course)*

22. $3^2 \cdot 3^4 = 3^x$ **23.** $(5^2)^3 = 5^x$

24. a. What is a postulate?
 b. What is a theorem? *(Previous course)*

25. You are at the hair stylist looking at a clock in a mirror. You see the figure at the left. *(Previous course)*
 a. What transformation do you need to apply in your mind to tell the time?
 b. What time is it?

26. Consider the table and the situation of Example 1. Suppose that Beth must get all 5 questions correct on the quiz to get an A. What probability must Beth have of getting each question correct in order to have a 75% chance of getting all 5 questions correct?

7-2

Properties of Powers

Earthrise. *Shown is the earthrise over the moon's horizon, taken from the Apollo 10 lunar module. Before the Apollo missions, there was no "earthrise" because no one had ever seen one. See Example 5.*

Using powers as a shortcut for multiplication when the factors are the same is often a person's first encounter with powers. However, powers can be calculated when the exponent is not a positive integer. In this chapter, you will learn how to calculate powers of any number with negative integer exponents, such as the following powers.

$$45^{-3} \qquad (-8)^{-2} \qquad (-.91)^{-12}$$

You will also learn how to calculate powers of positive numbers with rational exponents, such as those shown below.

$$2^{\frac{7}{2}} \qquad 81^{.25} \qquad 0.379^{\frac{2}{3}}$$

The properties of powers in this lesson apply not only to powers arising from repeated multiplication, but to all other powers as well.

Products of Powers

The repeated multiplication meaning of b^n enables you to simplify products of powers.

$$10^2 \cdot 10^3 = (10 \cdot 10) \cdot (10 \cdot 10 \cdot 10) = 10 \cdot 10 \cdot 10 \cdot 10 \cdot 10 = 10^5$$

$$x^4 \cdot x^2 = (x \cdot x \cdot x \cdot x) \cdot (x \cdot x) = x \cdot x \cdot x \cdot x \cdot x \cdot x = x^6$$

Each of these statements is an instance of a general pattern that we assume is true.

Product of Powers Postulate
For any nonnegative bases and nonzero real exponents, or any nonzero bases and integer exponents, $b^m \cdot b^n = b^{m+n}$.

You can use the Product of Powers Postulate to simplify expressions.

Example 1

A multiple-choice test is two pages long. Each question has c possible choices. Page 1 has 5 questions. Page 2 has 8 questions. How many different ways can students answer if no one leaves any questions blank?

Solution 1

There are c^5 ways to answer the questions on page 1 and c^8 ways to answer the questions on page 2. So there are $c^5 \cdot c^8 = c^{5+8} = c^{13}$ answer sheets possible.

Solution 2

The whole test has $5 + 8 = 13$ questions, each with c choices. So there are c^{13} answer sheets possible.

Powers of Powers

When a power is raised to a positive integer power, you can evaluate the expression using repeated multiplication.

$$(10^2)^3 = (10^2) \cdot (10^2) \cdot (10^2) = (10 \cdot 10) \cdot (10 \cdot 10) \cdot (10 \cdot 10) = 10^6$$

$$(x^4)^2 = x^4 \cdot x^4 = (x \cdot x \cdot x \cdot x) \cdot (x \cdot x \cdot x \cdot x) = x^8$$

These sentences are instances of the following property.

> **Power of a Power Postulate**
> For any nonnegative bases and nonzero real exponents, or any nonzero bases and integer exponents, $(b^m)^n = b^{m \cdot n}$.

Example 2

Simplify $(x^2)^5$.

Solution

Use the Power of a Power Postulate.
$$(x^2)^5 = x^{2 \cdot 5} = x^{10}$$

Check

Substitute any positive real number for x, such as 3. Does $(3^2)^5 = 3^{10}$?
Since $(3^2)^5 = 9^5 = 59{,}049$
and $\quad 3^{10} = 59{,}049$,
this case checks.

Power of a Product

Repeated multiplication also enables you to rewrite expressions involving positive-integer powers of products.

$$(3 \cdot 10)^4 = (3 \cdot 10) \cdot (3 \cdot 10) \cdot (3 \cdot 10) \cdot (3 \cdot 10) = 3^4 \cdot 10^4$$

$$(8y^5)^2 = (8y^5) \cdot (8y^5)$$
$$= (8 \cdot y \cdot y \cdot y \cdot y \cdot y) \cdot (8 \cdot y \cdot y \cdot y \cdot y \cdot y) = 8^2 y^{10} = 64y^{10}$$

These sentences are instances of the following property.

> **Power of a Product Postulate**
> For any positive nonnegative bases and nonzero real exponents, or any nonzero bases and integer exponents, $(ab)^m = a^m b^m$.

Example 3

Suppose $d \neq 0$ and $e \neq 0$. Simplify $\left(\frac{d}{e}\right)^4 \cdot \left(\frac{e}{d}\right)^4$.

Solution

Use the Power of a Product Postulate. Here $a = \frac{d}{e}$, $b = \frac{e}{d}$, and $m = 4$.

$$\left(\frac{d}{e}\right)^4 \cdot \left(\frac{e}{d}\right)^4 = \left(\frac{d}{e} \cdot \frac{e}{d}\right)^4 = 1^4 = 1$$

Check

Use repeated multiplication or a calculator to test a special case. The check is left to you.

Quotients Involving Powers

When two positive-integer powers with the same base are divided, you can rewrite the divisor and dividend using the repeated multiplication meaning of b^n.

Example 4

Verify that $\frac{2^{11}}{2^7} = 2^4$.

Solution 1

Rewrite the numerator and denominator using repeated multiplication.

$$\frac{2^{11}}{2^7} = \frac{\cancel{2} \cdot \cancel{2} \cdot \cancel{2} \cdot \cancel{2} \cdot \cancel{2} \cdot \cancel{2} \cdot \cancel{2} \cdot 2 \cdot 2 \cdot 2 \cdot 2}{\cancel{2} \cdot \cancel{2} \cdot \cancel{2} \cdot \cancel{2} \cdot \cancel{2} \cdot \cancel{2} \cdot \cancel{2}}$$
$$= \frac{2 \cdot 2 \cdot 2 \cdot 2}{1} = 2^4$$

Solution 2

Use a calculator. Two possible key sequences are

2 $\boxed{\wedge}$ 11 $\boxed{\div}$ 2 $\boxed{\wedge}$ 7 $\boxed{\text{ENTER}}$, or 2 $\boxed{y^x}$ 11 $\boxed{=}$ $\boxed{\div}$ 2 $\boxed{y^x}$ 7 $\boxed{=}$.

A calculator displays the answer 16, and $16 = 2^4$.

Quotients of certain powers can be simplified using these properties.

For any positive bases and real exponents, or any nonzero bases and integer exponents:

$$\frac{b^m}{b^n} = b^{m-n}.$$ **Quotient of Powers Postulate**

$$\left(\frac{a}{b}\right)^m = \frac{a^m}{b^m}.$$ **Power of a Quotient Postulate**

Properties of powers are often used when working with large numbers expressed in scientific notation.

Example 5

On average, Earth is about $150 \cdot 10^6$ kilometers from the sun and $390 \cdot 10^5$ kilometers from the moon. Light travels at about $3 \cdot 10^5 \frac{km}{sec}$. About how long does it take light from the sun to reach the moon when it is reflected off Earth?

Solution

Use the formula $t = \frac{d}{r}$ for each part of light's journey.

If d_1 = Earth's distance from the sun, d_2 = Earth's distance from the moon, and r = the speed of light, then the formula to use for the total time T is

$$T = \frac{d_1}{r} + \frac{d_2}{r}.$$

Substitute for d_1, d_2, and r. $T \approx \dfrac{150 \cdot 10^6 \text{ km}}{3 \cdot 10^5 \text{ km/sec}} + \dfrac{390 \cdot 10^5 \text{ km}}{3 \cdot 10^5 \text{ km/sec}}$

$$\approx \frac{150}{3} \cdot 10^1 \text{ sec} + 130 \text{ sec}$$

$$\approx 50 \cdot 10 \text{ sec} + 130 \text{ sec}$$

$$\approx 630 \text{ sec}$$

It takes about 630 seconds, or about 10.5 minutes, for light to travel from the sun to Earth and then to the moon.

This diamond ring is out of reach. *Shown is a 1983 solar eclipse with the "diamond ring" effect. This effect occurs when light from the sun shows through a crater or depression on the moon's surface.*

Division of Powers and Zero as an Exponent

When the Quotient of Powers Postulate is applied to two equal powers of the same base, the result is surprising to some people. For instance,

$$\frac{2^8}{2^8} = 2^{8-8} = 2^0.$$

But it is also true that $\dfrac{2^8}{2^8} = \dfrac{256}{256} = 1.$

The above statements and the Transitive Property of Equality prove that $2^0 = 1$.

In general, whenever b is a nonzero real number,

$$\frac{b^n}{b^n} = b^{n-n} = b^0.$$

Also, because any nonzero number divided by itself equals one, $\frac{b^n}{b^n} = 1$. We have proved the following theorem.

Zero Exponent Theorem
If b is a nonzero real number, $b^0 = 1$.

Example 6 applies several properties of powers, including the Zero Exponent Theorem.

Example 6

Three tennis balls stacked tightly as shown at the left just fill a can. What is the ratio of the volume of the balls to the volume of the can?

Solution

The balls may be considered as congruent spheres of radius r. The volume of each sphere is $\frac{4}{3}\pi r^3$. Let V_B equal the total volume of the three tennis balls. Then

$$V_B = 3\left(\frac{4}{3}\right)\pi r^3 = 4\pi r^3.$$

The can is a cylinder with radius r and height $6r$. Let V_C equal the volume of the can. Then

$$V_C = \pi r^2 h = \pi r^2(6r) = 6\pi r^3.$$

So the ratio of the volumes is

$$\frac{V_B}{V_C} = \frac{4\pi r^3}{6\pi r^3} = \frac{2}{3}.$$

That is, the tennis balls occupy $\frac{2}{3}$ of the volume of the can.

QUESTIONS

Covering the Reading

In 1–3, an expression is given. **a.** Write the expression as a single power using a postulate or a theorem from this lesson. **b.** Check your answer by applying repeated multiplication.

1. $6^2 \cdot 6^3$

2. $(4^2)^5$

3. $\frac{10^8}{10^2}$

4. A multiple-choice test has 8 questions on the first page and 7 questions on the second page. If each question can be answered in one of w ways, how many ways are there to answer all the multiple-choice questions?

5. Check Example 3.

6. Verify that $(2 \cdot 5)^4 = 2^4 \cdot 5^4$.

7. Indicate how you might explain to a friend how to evaluate the expression $\left(\frac{5}{2}\right)^4$.

In 8–13, name the property that justifies the statement.

8. $x^2 \cdot x^7 = x^9$

9. $(3a)^5 = 3^5 a^5$

10. For $y \neq 0$, $\frac{y^{12}}{y^3} = y^9$

11. $\left(\frac{x}{2}\right)^{10} = \frac{x^{10}}{2^{10}}$

12. $(b^3)^{13} = b^{39}$

13. For $x \neq 0$, $x^0 = 1$

In 14–19, simplify.

14. $(x^4)^3$

15. $(6x^7)^2$

16. $10x^7 \cdot 3x$

17. $\frac{n^{18}}{n^6}$

18. $\frac{n^{15}}{(n^3)^5}$

19. $\frac{\square^{100}}{z^0}$

20. Refer to Example 5. On average, Pluto is about $6 \cdot 10^9$ km from the sun. About how long does it take light to travel from the sun to Pluto?

In 21 and 22, refer to Example 6.

21. Which two properties of exponents are used in the solution?

22. Suppose a tennis can holds four balls stacked tightly on top of each other. What is the ratio of the volume of the balls to the volume of the can?

Applying the Mathematics

23. x^2 and x^6 are powers of x whose product is x^8. Find four more pairs of powers of x whose product is x^8.

24. **a.** Graph $y = (x^2)^4$.
    b. Find a number n such that the graph of $y = x^n$ coincides with the graph in part **a**.

In 25 and 26, solve by trial and error.

25. $(6 \cdot 10)^x = 216{,}000$

26. $0 < 10^y < 2$ if y is a nonnegative integer

In 27–29, simplify.

27. $\frac{w^5 \cdot w^6 \cdot z^7}{w^3 \cdot w^8 \cdot z}$

28. $\frac{(-8x^2)^3}{2x^4}$

29. $\left(\frac{4}{y}\right)^2 \left(\frac{y}{2}\right)^4$

30. To qualify for the Quiz Bowl, a student must answer all the questions in two qualifying rounds correctly. Each round consists of three history questions, three math questions, and three literature questions. The student estimates that the probability of correctly answering a history question is h, a math question is m, and a literature question is ℓ. Find the probability of qualifying for the Quiz Bowl. Assume the questions are independent.

31. a. Evaluate $F(r) = 4\pi r^2$, when $r = 4 \cdot 10^3$. Write your answer in scientific notation.

    b. The value of r given in part **a** is the approximate radius of Earth in miles. What does $F(r)$ represent?

In 32 and 33, use the fact that the population of the U.S. in 1992 was about $255 \cdot 10^6$.

32. If the land area of the U.S. was about $3.5 \cdot 10^6$ mi^2, what was the average number of people per square mile of land?

33. In 1992, people in the U.S. consumed about 24.26×10^9 lb of beef. How much beef was consumed per person in the U.S. that year?

Review

34. Consider the function $g(x) = x^{18}$.

    a. State its domain and its range.

    b. Sketch a graph.

    c. Describe any symmetries of the graph.   *(Lesson 7-1)*

35. Suppose the point (5, 125) is on the graph of an nth power function f.

    a. Write an equation for the function.

    b. What is $f(-5)$?   *(Lessons 1-3, 7-1)*

36. Solve the system $\begin{cases} y = x^2 \\ y = 3x + 4. \end{cases}$ *(Lessons 5-2, 6-6)*

37. What is the total cost of a bicycle that sells for b dollars, if you must pay a sales tax of 5% on the bicycle?   *(Previous course)*

Exploration

38. a. Below is a table of some powers of 2.

$$2^0 = 1$$
$$2^1 = 2$$
$$2^2 = 4$$
$$2^3 = 8$$
$$2^4 = 16$$
$$2^5 = 32$$
$$2^6 = 64$$
$$2^7 = 128$$

Look carefully at the last (units) digit of each numeral. *Predict* the last digit of 2^{13}. *Check* your prediction by calculating.

    b. What should be the last digit of 2^{20}? Justify your answer.

    c. Explain how to find the last digit of any positive integer power of 2.

    d. Explore powers of 3. Describe the patterns that occur in the last digits of these powers.

*Negative
Integer
Exponents*

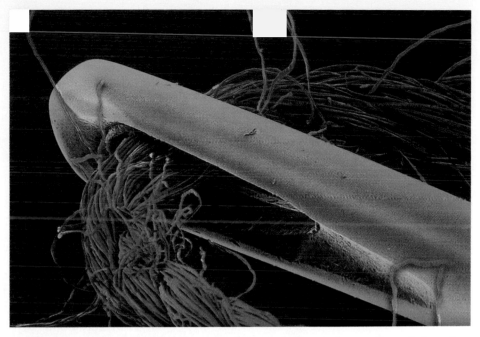

High-fiber content. *This electron micrograph photo of a needle and thread was shot at 16 times actual size. To find the size of the actual needle and thread, negative exponents can be used.*

The Meaning of Negative Integer Exponents

You have seen negative exponents when writing numbers in scientific notation. For example,

$$10^6 = 1,000,000 = \text{one million}$$
$$\text{and } 10^{-6} = .000001 = \text{one millionth.}$$

Note that $1,000,000 \cdot .000001 = 1$. This suggests that, in general, x^n and x^{-n} are reciprocals. This is easily proved.

Negative Exponent Theorem

For any positive base b and real exponent n, or any nonzero base b and integer exponent n, $b^{-n} = \frac{1}{b^n}$.

Proof

$$b^n \cdot b^{-n} = b^{n+-n} \qquad \text{Product of Powers Postulate}$$
$$= b^0 \qquad \text{Property of Opposites}$$
$$= 1 \qquad \text{Zero Exponent Theorem}$$

Dividing both sides by b^n (which can always be done because $b \neq 0$), gives

$$b^{-n} = \frac{1}{b^n}.$$

You should think of the Negative Exponent Theorem as stating b^n and b^{-n} are reciprocals. In particular, b^{-1} equals $\frac{1}{b^1}$, so b^{-1} is the reciprocal of b.

Example 1

Write 5^{-3} as a decimal.

Solution 1

5^{-3} is the reciprocal of 5^3. Since $5^3 = 125$, $5^{-3} = \frac{1}{125} = .008$.

Solution 2

$5^{-3} = \frac{1}{5^3} = \frac{1}{125} = .008$

Activity

Find a key sequence that gives 5^{-3} on your calculator.

Caution: A negative sign in an exponent does not make the expression negative. *All* the powers of a positive number are positive. Here are some powers of 9. Notice that as the value of n decreases, 9^n decreases, but never reaches 0.

$$9^0 = 1$$

$$9^1 = 9 \qquad\qquad 9^{-1} = \frac{1}{9^1} = \frac{1}{9}$$

$$9^2 = 81 \qquad\qquad 9^{-2} = \frac{1}{9^2} = \frac{1}{81}$$

$$9^3 = 729 \qquad\qquad 9^{-3} = \frac{1}{9^3} = \frac{1}{729}$$

$$9^4 = 6561 \qquad\qquad 9^{-4} = \frac{1}{9^4} = \frac{1}{6561}$$

The Negative Exponent Theorem allows expressions or sentences to be rewritten without fractions.

Example 2

Rewrite Newton's Law of Universal Gravitation,

$$W = \frac{Gm_1m_2}{r^2},$$

using negative exponents.

Solution

$$W = Gm_1m_2 \cdot \frac{1}{r^2} \qquad \text{Algebraic Definition of Division}$$
$$W = Gm_1m_2r^{-2} \qquad \text{Negative Exponent Theorem}$$

Properties of Negative Integer Exponents

Negative exponents satisfy the postulates involving powers stated in Lesson 7-2.

Example 3

Rewrite each expression as a single power.

a. $\dfrac{10^3}{10^7}$　　　　　　　　　　　　　　**b.** $t^5 \cdot t^{-1}$ (Assume $t > 0$.)

Solution

a. Use the Quotient of Powers Postulate.

$$\dfrac{10^3}{10^7} = 10^{3-7} = 10^{-4}$$

b. Use the Product of Powers Postulate.

$$t^5 \cdot t^{-1} = t^{5-1} = t^4$$

Check

a. Use the repeated multiplication meaning of 10^n.

$$\dfrac{10^3}{10^7} = \dfrac{\cancel{10} \cdot \cancel{10} \cdot \cancel{10}}{\cancel{10} \cdot \cancel{10} \cdot \cancel{10} \cdot 10 \cdot 10 \cdot 10 \cdot 10} = \dfrac{1}{10 \cdot 10 \cdot 10 \cdot 10} = \dfrac{1}{10^4} = 10^{-4}$$

b. From the Negative Exponent Theorem you know that

$$t^{-1} = \dfrac{1}{t}.$$

So $t^5 \cdot t^{-1} = t \cdot t \cdot t \cdot t \cdot t \cdot \dfrac{1}{t} = t^4.$

Some simplifications require more than one property of powers.

Example 4

Rewrite as a fraction: $(7k)^{-3} (2k)^5$.

Solution

$$
\begin{aligned}
(7k)^{-3} (2k)^5 &= 7^{-3} \cdot k^{-3} \cdot 2^5 \cdot k^5 && \text{Power of a Product Postulate} \\
&= 7^{-3} \cdot 2^5 \cdot k^{-3} \cdot k^5 && \text{Commutative and Associative} \\
& && \text{Properties of Multiplication} \\
&= 7^{-3} \cdot 2^5 \cdot k^2 && \text{Product of Powers Postulate} \\
&= \dfrac{1}{7^3} \cdot 2^5 \cdot k^2 && \text{Negative Exponent Theorem} \\
&= \dfrac{32k^2}{343}
\end{aligned}
$$

QUESTIONS

Covering the Reading

1. Suppose b is a positive real number. Simplify.

　a. $b^x \cdot b^y$　　　　　　　**b.** $b^x \cdot b^0$　　　　　　　**c.** $b^x \cdot b^{-x}$

2. *Multiple choice.*　If $a \neq 0$ and y is an integer, a^y and a^{-y} are
　(a) reciprocals.
　(b) opposites.
　(c) neither reciprocals nor opposites.

3. Solve for n: $5^{-4} = \frac{1}{5^n}$.

4. If $p^{-1} = \frac{1}{7}$, find p.

5. Write as a whole number or a fraction without an exponent.
   a. 8^0
   b. 8^{-1}
   c. 8^{-2}

6. Write 7^x as a fraction without an exponent for each value of x.
   a. -3
   b. -2
   c. -1

7. *Multiple choice.* If $b > 0$, for what values of n is $b^n < 0$?
   (a) $n < 0$
   (b) $0 < n < 1$
   (c) all values of n
   (d) no values of n

8. Write a key sequence that gives 5^{-3} on your calculator. (This is the Activity in the lesson.)

In 9 and 10, rewrite the right side of the formula using a negative exponent and no fraction.

9. $W = \frac{k}{d}$

10. $T = \frac{kL}{S}$

In 11–14, write without an exponent. Do not use a calculator.

11. $10^7 \cdot 10^{-6}$
12. $2^{-3} \cdot 13^0$
13. $(10^{-3})^2$
14. $(3^{-2})^{-4}$

In 15–18, simplify.

15. $\frac{8x^6}{6x^8}$
16. $(2p)^3(3p)^{-2}$
17. $\frac{4y^{-1}}{y^{-2}}$
18. $\frac{z^5}{10z^{-6}}$

Applying the Mathematics

19. If $x^3 = 5$, what is the value of x^{-3}?

In 20 and 21, *true or false.* Justify your answer.

20. $\frac{1}{2^{-5}} = \left(\frac{1}{2}\right)^{-5}$

21. $-3x^{-2} = \frac{1}{3x^2}$, for all $x \neq 0$

22. Write $(2.5 \cdot 10^{-2})^3$ in scientific notation.

23. Suppose $0 < x < 1$. Is x^{-2} smaller or larger than x? How can you tell?

24. In 1975, the U.S. Environmental Protection Agency set a standard of 50 parts per billion of lead in drinking water. In 1991, a new standard was set that safe water contains less than 15 parts per billion.
   a. Write each of these rates as fractions using positive powers of ten.
   b. Write each rate without fractions, using negative powers of ten.

25. The intensity I of light varies inversely as the square of the distance d from the observer. Let k be the constant of variation.
   a. Write this inverse variation equation with positive exponents.
   b. Rewrite this inverse variation equation with negative exponents.

26. Benjamin Franklin was one of the most famous scientists of his day. In one experiment he noticed that oil dropped on the surface of a lake would not spread out beyond a certain area. In modern units he found that 0.1 cm^3 of oil spread to cover about 40 m^2 of the lake. About how thick is such a layer of oil? Express your answer in scientific notation. (Nowadays we know that the layer of oil stops spreading when it is one molecule thick. Although in Franklin's time, no one knew about molecules, Franklin's experiment resulted in the first estimate of a molecule's size.)

Shown is a portrait of Benjamin Franklin painted in 1766 by David Martin. Franklin's diplomatic success was due in large part to his reputation as a scientist.

Review

27. Simplify $\dfrac{n^{1996} \cdot n^{-1997}}{(n^{1000})^2}$.   *(Lesson 7-2)*

28. Evaluate $\left(\frac{4}{3}\right)^3 \cdot \left(\frac{6}{4}\right)^3$ in your head. Explain what you did.   *(Lesson 7-2)*

29. Why can't c be zero in the statement $\dfrac{c^m}{c^n} = c^{m-n}$?   *(Lesson 7-2)*

30. A sphere of radius r fits tightly in a cube. About what percent of the volume of the cube is the volume of the sphere?
(Previous course, Lesson 7-2)

31. *Skill sequence.*   Rewrite the number in $a + bi$ form.   *(Lessons 6-8, 6-9)*

   a. $\sqrt{196}$ **b.** $\sqrt{-196}$ **c.** $\dfrac{10 + \sqrt{-196}}{2}$

32. Simplify $\dfrac{\frac{1}{10}}{9}$.   *(Previous course)*

Exploration

33. Examine these columns closely. Describe two patterns relating the powers of 5 at the left to the powers of 2 at the right.

5^6	$= 15{,}625$		2^6	$= 64$
5^5	$= 3{,}125$		2^5	$= 32$
5^4	$= 625$		2^4	$= 16$
5^3	$= 125$		2^3	$= 8$
5^2	$= 25$		2^2	$= 4$
5^1	$= 5$		2^1	$= 2$
5^0	$= 1$		2^0	$= 1$
5^{-1}	$= 0.2$		2^{-1}	$= 0.5$
5^{-2}	$= 0.04$		2^{-2}	$= 0.25$
5^{-3}	$= 0.008$		2^{-3}	$= 0.125$
5^{-4}	$= 0.0016$		2^{-4}	$= 0.0625$
5^{-5}	$= 0.00032$		2^{-5}	$= 0.03125$
5^{-6}	$= 0.000064$		2^{-6}	$= 0.015625$

34. a. Make a chart similar to the one in Question 33 using the powers of 4 and 2.5.
   b. Describe how the patterns in these charts are similar to the patterns in Question 33.
   c. Find another pair of numbers with the same properties.

Interest Compounded Annually

Suppose a person deposits $2000 in a saving institution that pays interest at an annual rate of 4%. If no money is added or withdrawn, after one year the account will have the original amount invested, plus 4% interest.

$$\text{amount after } \mathbf{1} \text{ year: } 2000 + .04(2000) = 2000(1 + .04)$$
$$= 2000(1.04) = 2080$$

There will be $2080 in the bank after one year.

Notice that to find the amount after 1 year, you do not have to add the interest; rather you can just multiply by 1.04. Similarly, at the end of the second year, there will be 1.04 times the amount after the first year.

$$\text{amount after } \mathbf{2} \text{ years: } 2000(1.04)(1.04) = 2000(1.04)^2 = 2163.20$$

There will be $2163.20 in the bank after two years.

$$\text{amount after } \mathbf{3} \text{ years: } 2000(1.04)^2(1.04) = 2000(1.04)^3 \approx 2249.73$$

There will be $2249.73 in the bank after three years.

Because the *interest* earns interest each year, the process is called **compounding.** Notice the general pattern.

$$\text{amount after } \mathbf{t} \text{ years: } \qquad\qquad 2000(1.04)^t$$

For example, after **18** years of *compounding annually* at 4% there will be

$$2000(1.04)^{18} \approx 4051.63.$$

After 18 years, the amount is more than double the original deposit.

There is a more general formula. Replace 4% by *r,* the annual interest rate, and 2000 by *P,* the **principal,** or original amount invested. Repeat the process shown above to find *A,* the amount the investment is worth after *t* years.

Annual Compound Interest Formula
Let *P* be the amount of money invested at an annual interest rate of *r* compounded annually. Let *A* be the total amount after *t* years. Then
$$A = P(1 + r)^t.$$

In the Compound Interest Formula, notice also that *A* varies directly as *P;* for example, doubling the principal doubles the amount at the end. However, *A* does not vary directly as *r;* doubling the rate does not necessarily double the amount earned.

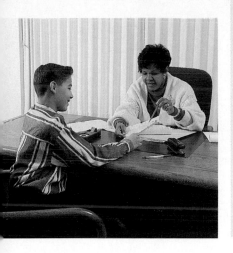

Example 1

Emilio invests $1000 in an account compounded at an annual rate of 3% and another $1000 in a second account compounded at an annual rate of 6%. How much interest will Emilio earn in each account after 4 years?

Solution

In Emilio's first account, P is $1000, r is .03, and t = 4.
$$P(1 + r)^t = A$$
$$1000(1.03)^4 \approx 1125.51$$
In his second account, he will have
$$1000(1.06)^4 \approx 1262.48.$$
Emilio will earn $125.51 interest in the 3% account and $262.48 in the 6% account.

Notice that the 6% account earns more than twice as much interest as the 3% account. This is because after the first compounding, the interest earns interest.

Example 2

In the situation of Example 1, how much interest will Emilio earn in the 5th year in the account paying 6% compounded annually?

Solution 1

One way to find the interest earned in the fifth year is to subtract the total in the fourth year from the total in the fifth. Let $F(t)$ be the amount in Emilio's second account at the end of t years.
$$F(5) = 1000(1.06)^5 \approx \$1338.23$$
$$F(4) = \underline{1000(1.06)^4 \approx \$1262.48}$$
difference $\approx \$\quad 75.75$
In the fifth year Emilio will earn $75.75 interest.

Solution 2

Another way to find the interest earned during a particular year in an account compounding interest annually is to multiply the balance from the previous year by the rate r. The fifth year's interest is
$(1262.48)(.06) \approx \$75.75.$

Interest Compounded More Than Once a Year

Most banks compound interest more than once a year. If a bank compounds **semi-annually,** the interest rate at each compounding is *half of the annual interest rate,* but there are *two compoundings each year* instead of just one. If an account pays 6% compounded semi-annually, you earn 3% on the balance every six months. In general, if r is the rate of interest, at the end of t years, interest paid semi-annually will have been paid $2t$ times. Therefore, the compound interest formula becomes

$$A = P(1 + \tfrac{r}{2})^{2t}.$$

A bank that compounds *quarterly* uses the compound interest formula

$$A = P\left(1 + \frac{r}{4}\right)^{4t}.$$

In this way, the compound interest formula can be generalized.

General Compound Interest Formula

Let P be the amount invested at an annual interest rate r compounded n times per year. Let A be the amount after t years. Then

$$A = P\left(1 + \frac{r}{n}\right)^{nt}.$$

Example 3

A bank is currently offering a certificate of deposit (CD) paying 5.25% interest compounded quarterly. Find the value of such a CD after two years if $1000 is invested.

Solution

Use the General Compound Interest Formula, with $P = 1000$, $r = 5.25\% = .0525$, $n = 4$ and $t = 2$.

$$A = P\left(1 + \frac{r}{n}\right)^{nt}$$

$$= 1000\left(1 + \frac{.0525}{4}\right)^{4\cdot2}$$

$$= 1000(1.013125)^8 \approx 1109.95$$

The CD will be worth about $1109.95 after two years.

Activity

Check the calculations in Example 3 on your calculator. Record the key sequence you used.

Banks are required by law in most states to advertise the **effective annual yield,** or **yield,** on every account. This is the rate of interest earned after all the compoundings have taken place in one year. To find the annual yield of an account paying 5.25% interest compounded quarterly, find the amount of interest $1 would earn in the account in one year.

$$\left(1 + \frac{.0525}{4}\right)^4 \approx 1.05354$$

So the interest earned is $1.05354 − $1 = $.05354. Thus, a rate of 5.25% compounded quarterly gives an annual yield of 5.354%.

Going Back in Time

In both compound interest formulas, you can think of P either as the principal, or as the *present amount*. In each of the previous examples, you can think of A as an amount that is determined after compounding. Then, the time t is represented by a positive number. But it is also possible to think of A as an amount some years ago that was compounded to get the present amount P. Then the time t is represented by a negative number.

Interest in bonds. *Bonds can be bought from brokers like the one shown. Governments and corporations issue bonds to raise money. Investors purchase bonds because they generally pay a steady rate of interest.*

Example 4

A bond paying 7% compounded annually has matured after 8 years, giving the owner $10,000. How much was invested 8 years ago?

Solution

Use the Annual Compound Interest Formula with $P = 10,000$, $r = .07$, and $t = -8$.

$$A = P(1 + r)^t$$
$$= 10000(1 + .07)^{-8}$$
$$= 10000(1.07)^{-8}$$
$$\approx 5820.09$$

The owner invested $5820.09, 8 years ago.

Check

Think of starting with 5820.09 and compounding for 8 years. Does $5820.09(1.07)^8 = 10,000$? Our calculator shows that $5820.09(1.07)^8 = 9999.998203$. It checks.

QUESTIONS

Covering the Reading

In 1 and 2, Lucy invests $3000 in a CD that pays interest at a rate of 5% compounded annually. The interest is left in the account and she makes no deposits or withdrawals.

1. To find next year's balance, you can multiply this year's balance by __?__.

2. How much will be in the account after four years?

3. A person deposits $2000 in a bank account that pays interest at an annual rate of 4%. If no money is added or withdrawn, find how much will be in the account after 1, 2, 3, 4, and 5 years.

In 4 and 5, refer to Examples 1 and 2.

4. *True or false.* Emilio earned more than $200 interest in each of his accounts in the first four years.

5. Using two different methods, find the interest earned in the fifth year for the account paying 3% compounded annually.

6. *True or false.* Justify your answer. Noel invests $1000 compounded annually at 4%. Chris invests $1000 compounded annually at 8%.
   a. In the first year, Chris's account will earn twice as much as Noel's.
   b. In the second year, Chris's account will earn twice as much as Noel's.

7. Write the key sequence for the calculator that you used in the Activity in this lesson to evaluate $1000\left(1 + \frac{.0525}{4}\right)^8$.

8. *Multiple choice.* In a compound interest situation, the total amount A is directly proportional to
(a) the rate r.
(b) the time t.
(c) the number n of compoundings in a year.
(d) the initial amount P.

9. Write the compound interest formula for an account that compounds interest
a. quarterly.   **b.** monthly.   **c.** daily.

10. Find the value of $5000 invested for 3 years at 6% compounded quarterly.

11. Find the effective annual yield of a 5% account which is compounded monthly.

12. Solve the equation $10,000 = P(1.05)^6$ for P. Round to the nearest hundredth.

13. A bond paying 6% compounded annually has matured after 5 years, giving the owner $5000. How much was invested 5 years ago?

Applying the Mathematics

14. Katie puts $10,000 in a 6-year 5.625% savings certificate in which interest is compounded daily.
a. How much interest will she earn during the entire 6-year period?
b. How much interest will she earn in the sixth year?

15. Gilda's parents bought airline tickets for $800 with a credit card which charges an annual rate of 18% and compounds the interest monthly. Find the amount of interest they must pay if they wait a month to pay the balance.

In 16 and 17, use the formula $I = Prt$ for **simple interest** where I is the interest, P is the principal, r is the annual rate, and t is the time in years. Simple interest is the interest paid on only the original principal P, not on the interest earned.

16. Suppose $1000 is invested at 6%.
a. How much simple interest is earned in 5 years?
b. How much interest would be earned in 5 years if the interest were compounded annually at 6% interest?
c. How much more does compound interest yield than simple interest for the situations in parts **a** and **b**?

17. A relative lends Jody $2000 to help her meet expenses for college. She insists that she be charged interest, so the relative makes the table at the left to determine how much Jody needs to pay back according to when she does pay it back.
a. Is her relative charging Jody simple or compound interest? Justify your answer.
b. What is the interest rate?

After:	Pay:
3 years	$2360
4 years	$2480
5 years	$2600
6 years	$2720
7 years	$2840

18. Refer to the BASIC program below.

```
10 PRINT "A PROGRAM TO CALCULATE BANK BALANCE"
20 INPUT "PRINCIPAL, ANNUAL RATE, NO. OF YEARS"; P, R, Y
30 PRINT "YEAR", "AMOUNT"
40 FOR C = 1 TO Y
50 A = P * (1 + R)
60 PRINT C, A
70 P = A
80 NEXT C
90 END
```

a. Lines 40 through 80 calculate A recursively. Modify the program so it calculates A explicitly from an interest formula for each year.
b. Use the given program or your modification to display the amount that will be in an account at the end of each of the first 10 years, if $250 is invested at a rate of 4% compounded annually.
c. In what year will the amount in the account first exceed $350?

Review

19. Write as a simple fraction: $5^{-3} \cdot 2^4$. Do not use a calculator.
(Lesson 7-3)

In 20–22, simplify.   *(Lessons 7-2, 7-3)*

20. $3x^{-2} \cdot 2x^3$ **21.** $(4z^2)^5$ **22.** $\dfrac{4a^5b^6}{(-2ab^2)^{-3}}$

23. Suppose a football team has a probability of .6 of winning each of 5 games it plays and a probability of .7 of winning each of the other 3 games it plays in a league. What is the probability that it will go through the season undefeated?   *(Lesson 7-1)*

24. The graph of $y = x^5$ is translated 3 units to the right and 6 units up. What is an equation for its image?   *(Lessons 6-2, 7-1)*

25. Consider the sequence t defined recursively as follows.
$$\begin{cases} t_1 = 4 \\ t_n = 3t_{n-1} \end{cases}$$
a. Find the first five terms of the sequence.
b. Is t an arithmetic sequence? Explain why or why not.
(Lessons 1-9, 3-7)

Exploration

26. a. Use either the computer program in Question 18 or your modification to find out how long it will take to double your money if it is invested annually at the given rate.
   i. 4%   **ii.** 6%   **iii.** 8%   **iv.** 10%
b. Generalize your results from part **a.**

LESSON
7-5

Geometric Sequences

That's the way a ball bounces. *This multiple-exposure photograph was made with a stroboscope, a device that makes a rapid series of equally-timed bright flashes. The camera shutter is kept opened, and an exposure is added every time the strobe flashes. See Example 3.*

Recursive Formulas for Geometric Sequences

In an arithmetic or linear sequence, each term after the first is found by adding a constant difference to the previous term. If, instead, each term after the first is found by *multiplying* the previous term by a constant, then a *geometric,* or *exponential, sequence* is formed. For instance, the sequence with first term 48 and *constant multiplier* 1.5 is

$$48, 72, 108, 162, 243, 364.5, \ldots.$$

> **Recursive Formula for a Geometric Sequence**
> The sequence defined by the recursive formula
> $$\begin{cases} g_1 \\ g_n = rg_{n-1}, \text{ for integers } n \geq 2, \end{cases}$$
> where r is a nonzero constant, is the **geometric,** or **exponential,** **sequence** with first term g_1 and constant multiplier r.

Solving the sentence $g_n = rg_{n-1}$ for r yields $\frac{g_n}{g_{n-1}} = r$. This indicates that in a geometric sequence, the ratio of successive terms is constant. For this reason, the constant multiplier r is also called the **constant ratio.**

You should be able to write the terms of any geometric sequence described by a recursive formula.

Example 1

Give the first six terms and the constant multiplier in the geometric sequence $\begin{cases} g_1 = 3 \\ g_n = 5g_{n-1}, \text{ for integers } n \geq 2. \end{cases}$

Solution

The value for g_1 is given. $g_1 = 3$. The rule for g_n tells you that each term after the first is found by multiplying the previous term by 5. The constant multiplier is 5.

$$g_2 = 5g_1 = 5 \cdot 3 \qquad = 15$$
$$g_3 = 5g_2 = 5 \cdot 15 \qquad = 75$$
$$g_4 = 5g_3 = 5 \cdot 75 \qquad = 375$$
$$g_5 = 5g_4 = 5 \cdot 375 \qquad = 1875$$
$$g_6 = 5g_5 = 5 \cdot 1875 = 9375$$

The first six terms of the sequence are 3, 15, 75, 375, 1875, 9375.

Explicit Formulas for Geometric Sequences

At the beginning of Lesson 7-4, a $2000 investment in an account at 4% interest compounded annually was discussed. The amounts in the account after successive years form a geometric sequence. In this case, the constant multiplier is 1.04, so $r = 1.04$. If we let the principal be $g_1 = 2000$, then g_2 is the amount the investment is worth at the end of the 1st year, and g_n is the amount the investment is worth at the end of the $(n - 1)$st year.

g_1	g_2	g_3	g_4
principal	*after 1 year*	*after 2 years*	*after 3 years* . . .
2000	$2000(1.04)^1$	$2000(1.04)^2$	$2000(1.04)^3$ . . .
2000	2080	2163.20	2249.73

Using the Annual Compound Interest Formula, we find that the amount in the account after $n - 1$ years is $2000(1.04)^{n-1}$. So $g_n = 2000(1.04)^{n-1}$. This process can be generalized to find an explicit formula for the nth term of a geometric sequence.

Explicit Formula for a Geometric Sequence
In the geometric sequence with first term g_1 and constant ratio r,
$g_n = g_1(r)^{n-1}$, for integers $n \geq 1$.

Notice that in the explicit formula, the exponent of the nth term is $n - 1$. When you substitute 1 for n to find the first term, the constant multiplier has an exponent of zero.

$$g_1 = g_1(r)^{1-1} = g_1 \, r^0$$

This is consistent with the Zero Exponent Theorem which states that if $r \neq 0$, $r^0 = 1$.

Constant multipliers in a geometric sequence can be negative. Then the terms of the sequence alternate between positive and negative values.

Example 2

Write the first five terms of the sequence defined by $g_n = 8(-5)^{n-1}$.

Solution

Substitute $n = 1, 2, 3, 4,$ and 5 into the formula for the sequence.

$$
\begin{aligned}
g_1 &= 8(-5)^0 = 8 \cdot 1 &&= 8 \\
g_2 &= 8(-5)^1 = 8 \cdot (-5) &&= -40 \\
g_3 &= 8(-5)^2 = 8 \cdot 25 &&= 200 \\
g_4 &= 8(-5)^3 = 8 \cdot (-125) &&= -1000 \\
g_5 &= 8(-5)^4 = 8 \cdot 625 &&= 5000
\end{aligned}
$$

The first five terms of the sequence are 8, -40, 200, -1000, 5000.

Example 3

Suppose a ball is dropped from a height of 5 meters, and it bounces up to 90% of its previous height after each bounce. (A "bounce" is counted when the ball hits the ground.) Let h_n be the greatest height of the ball after the nth bounce.

a. Find an explicit formula for h_n.

b. Find the greatest height of the ball after the tenth bounce.

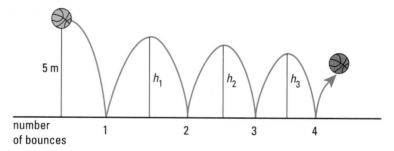

Solution

a. Because each term is the constant .9 times the previous term, the sequence is geometric. On the first bounce, the ball will bounce up to 5(.9) meters or 4.5 meters, so let $h_1 = 4.5$. Also, $r = .9$. Thus,
$$h_n = 4.5(.9)^{n-1}.$$

b. On the tenth bounce, $n = 10$ and the ball will bounce to
$$h_{10} = 4.5(.9)^{10-1} \approx 1.7.$$

So after the tenth bounce, the ball will rise to 1.7 meters.

Look back at the sequences generated in each of Examples 1 to 3. Notice that in Example 1, $r > 1$, and as n increases, g_n increases. In Example 3, $0 < r < 1$, and as n increases, g_n decreases. In Example 2, $r < 0$ and as n increases, g_n alternates between positive and negative values. These properties are true for all geometric sequences.

How Did *Geometric* Sequences Get Their Name?

In the figure below, the midpoints of the sides of each square are connected to form the next smaller square. When the side of one square has length g, the side of the next smaller square has length $\frac{\sqrt{2}}{2} g$. (This can be proved using the Pythagorean Theorem.)

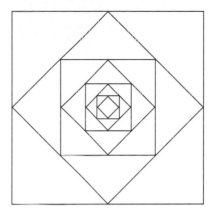

Consequently, the sides of the consecutively smaller and smaller squares form a sequence g in which $g_n = \frac{\sqrt{2}}{2} g_{n-1}$. From applications like this one, sequences with this property came to be known as geometric sequences.

QUESTIONS

Covering the Reading

1. In an arithmetic sequence, each term after the first is found by __?__ a constant to the previous term; in a geometric sequence, each term after the first is found by __?__ the previous term by a constant.

2. Could the sequence 5, 15, 45, 135 . . . be an exponential sequence? Justify your answer.

3. In a geometric sequence, the __?__ of successive terms is constant.

In 4 and 5, give the first five terms of the sequence defined by the given formula.

4. $g_n = 2 \cdot 3^{n-1}$, for integers $n \geq 1$

5. $\begin{cases} t_1 = 6 \\ t_n = \frac{2}{3} t_{n-1}, \text{ for integers } n \geq 2 \end{cases}$

6. Refer to Example 2.
   a. Find each ratio.
      i. $\frac{g_2}{g_1}$ ii. $\frac{g_3}{g_2}$ iii. $\frac{g_4}{g_3}$ iv. $\frac{g_5}{g_4}$
   b. What is true about the values in part **a**?

7. a. Write the first six terms of the geometric sequence whose first term is -3 and whose constant ratio is 4.
   b. Give a recursive formula for the sequence.
   c. Give an explicit formula for the sequence.

8. Consider the sequence whose first five terms are given below.

n	1	2	3	4	5
t_n	1000	$1000(1.04)$	$1000(1.04)^2$	$1000(1.04)^3$	$1000(1.04)^4$

Could this sequence be geometric? If yes, find an explicit formula for t_n. If no, explain why not.

9. Suppose a ball is dropped from a height of 5 meters and that it rises to 75% of its previous height each time it bounces.
   a. Find an explicit formula for the height on the nth bounce.
   b. Find the height of the tenth bounce.

10. In the figure below, the midpoints of the sides of the largest triangle have been joined to create the next smaller triangle, and this process has been continued. Each side of the next smaller triangle is half as long as the corresponding side of the previous triangle. Explain why the sequence of lengths of sides of the smaller and smaller triangles is a geometric sequence.

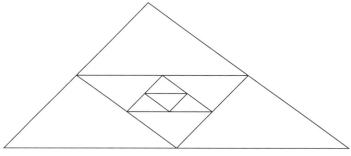

Applying the Mathematics

In 11 and 12, the first few terms of a geometric sequence are given.
a. Find the next term.
b. Find an explicit formula for the nth term of the geometric sequence.

11. 2, 6, 18, 54, 162, . . .   **12.** 100, 20, 4, .8, . . .

13. Consider the geometric sequence whose first four terms are 40, -40, 40, -40, . . .
   a. Write a recursive formula for the sequence.
   b. Write an explicit formula for the sequence.

14. The fifth term of a geometric sequence is 140. The constant multiplier is 2.
   a. What is the sixth term?   **b.** What is the first term?

15. A diamond was purchased for $2500. If its value increases 6% each year, give the value of the diamond after ten years.

Priceless crown. *Shown is the Russian Nuptial Crown, made in 1840. It is set with 1535 diamonds, weighing 283 carats. The crown was worn by three Imperial Russian brides, including the last Empress of Russia, Catherine the Great.*

16. Use the formula for simple interest, $I = Prt$, where
P = the principal sum invested,
r = the annual interest rate, and
t = the time in years.
  a. If you invest $1000 at 8% interest, write the sequence of your balances over the next 6 years.
  b. Use your answer to part **a.** Does simple interest lead to a geometric sequence? If so, what is the common ratio? If not, to what kind of sequence does simple interest lead?

Review

17. *Multiple choice.* Which of these accounts will have three times as much in it as an account where $1000 is compounded at an annual rate of 3% for 3 years? *(Lesson 7-4)*
  (a) an account where $3000 is compounded at 3% for 3 years
  (b) an account where $1000 is compounded at 9% for 3 years
  (c) an account where $1000 is compounded at 3% for 9 years

18. a. Solve for P: $853.25 = P(1.05)^3$.
  b. Make up a question that can be answered by solving the equation in part **a.** *(Lesson 7-4)*

19. Simplify $\dfrac{(6a^5b^4)^3\,c^2}{(a^{-3}\,b^{-2}\,c^{-1})^2}$. *(Lessons 7-2, 7-3)*

20. Let $f(x) = x^3 \cdot x^7$. Identify the domain and range of f. *(Lessons 7-1, 7-2)*

21. a. Solve $x^2 + 4 = 6x$ using the Quadratic Formula.
  b. What do the solutions tell you about the graph of $y = x^2 - 6x + 4$? *(Lessons 6-6, 6-10)*

22. a. Graph the function with equation $f(x) = \dfrac{3}{x}$.
  b. On the same coordinate axes as part **a,** sketch the image of $f(x) = \dfrac{3}{x}$ translated 2 units to the left.
  c. Write an equation for the image.
  d. Find the slope of the line(s) of symmetry of the graph of $f(x) = \dfrac{3}{x}$.
  (Lessons 2-4, 2-6, 6-3)

23. If $h: x \to 2(x - 1)^2$, evaluate $h(3) - h(0)$. *(Lesson 1-3)*

24. Prove: If the midpoints of the sides of a square are connected in order, then the figure formed is a square with sides $\dfrac{\sqrt{2}}{2}$ times as long as the sides of the original square. *(Previous course)*

Exploration

25. Find an example of how the term *geometric mean* is used in geometry, and explain what this term has to do with a geometric sequence.

LESSON
7-6

nth Roots

All that jazz. *Shown are jazz pianists Geri Allen and Hank Jones. Most pianos have 88 keys. There are 12 notes to an octave. So a piano's range is over 7 octaves.*

What Is an *n*th Root?

Recall that x is a square root of t if and only if $x^2 = t$. Similarly, x is a cube root of t if and only if $x^3 = t$. For instance, 4 is a cube root of 64 because $4^3 = 64$. Also, $-\frac{1}{3}$ is a 5th root of $-\frac{1}{243}$ because $\left(-\frac{1}{3}\right)^5 = -\frac{1}{243}$. Square roots and cube roots are special cases of the following more general idea.

> **Definition**
> Let n be an integer greater than 1.
> Then b is an **nth root** of x if and only if $b^n = x$.

There are no special names for nth roots other than *square roots* (when $n = 2$) and *cube roots* (when $n = 3$). Other nth roots are called *fourth roots, fifth roots,* and so on.

Musical Scales and *n*th Roots

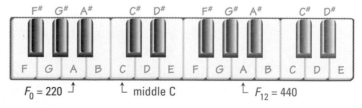

The purpose of tuning a piano or other musical instrument is so that the notes it plays have the proper frequencies. It is common to tune the A above middle C to 440 hertz. (The hertz is a unit of frequency equal to one cycle per second.) Notes that are one octave apart are tuned so that the note lower in pitch has exactly half the frequency of the note one octave higher. Thus, the A below middle C is tuned to a frequency of 220 hertz.

In most music today, an octave is divided into twelve steps. You can count the twelve steps of the octave beginning with the A below middle C on the keyboard on page 450. In order that a piece has the same sound in any key, notes are tuned so that the ratio of frequencies of consecutive notes are equal. To find these frequencies we let F_0 = the frequency of the A below middle C. Let F_n = the frequency of the note that is n notes above this note. Then F_{12} is the frequency of the A above middle C. Let r = the ratio of the frequencies of consecutive notes. Then for all integers $n \geq 1$,

$$\frac{F_n}{F_{n-1}} = r.$$

We multiply both sides of this equation by F_{n-1}. Then

$$F_n = rF_{n-1}, \text{ for all integers } n \geq 1,$$

This is the recursive formula for the nth term of a geometric sequence. It indicates that the frequencies of consecutive notes on a properly tuned piano are the elements of a geometric sequence. F_{12} is the 13th term of the musical sequence; the first term is F_0. From the explicit formula for the nth term of a geometric sequence,

$$F_{12} = r^{13-1}F_0 = r^{12}F_0.$$

To find r, we substitute the known values of F_{12} and F_0.

$$440 = r^{12} \cdot 220$$
$$2 = r^{12}$$

That is, the ratio of the frequencies of consecutive keys on a properly tuned piano is a 12th root of 2.

nth Roots and Graphs

The *real* nth roots of a number can be estimated from a graph.

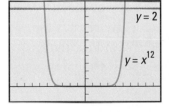

$-2 \leq x \leq 2, \quad x\text{-scale} = .2$
$-.4 \leq y \leq 2.2, \quad y\text{-scale} = .2$

Example 1

Estimate the real 12th roots of 2.

Solution
The real 12th roots of 2 are the real solutions to $x^{12} = 2$. So they are the x-coordinates of the points of intersection of the horizontal line $y = 2$ and the curve $y = x^{12}$. From the graph shown at the left, you can see that there are two real 12th roots of 2. Repeated zooming and tracing shows that The real 12th roots of 2 are near 1.0595 and -1.0595.

Check
$1.0595^{12} = (-1.0595)^{12} = 2.0008 \ldots$

Example 1 shows that there are two real 12th roots of 2. However, because frequencies of notes must be positive numbers, only the positive root is of relevance to music.

Example 2

Find the frequency of middle C, if the A below middle C is tuned to 220 hertz.

Solution

Middle C is three notes above this A. So its frequency is F_3, the 4th term of a geometric sequence with first term $F_0 = 220$. From Example 1, the common ratio of this sequence is about 1.0595. Now use the Explicit Formula for the nth term of a Geometric Sequence.

$F_3 = 220 \cdot 1.0595^{4-1} = 220 \cdot 1.0595^3 \approx 261.65$

The frequency of middle C is about 262 hertz.

A noteworthy profession.
When tuning a piano, the tension of the strings is adjusted. Loosening a string lowers the pitch; tightening a string raises the pitch. The man shown is a sixth-generation master craftsman of piano tuning and rebuilding.

Roots and Powers

One reason that powers are so important is that *the positive nth root of a positive number x is a power of x.* And the power is a simple one: the square root is the $\frac{1}{2}$ power; the cube root is the $\frac{1}{3}$ power; and so on. The general property is called the $\frac{1}{n}$ *Exponent Theorem.*

> $\frac{1}{n}$ **Exponent Theorem**
>
> When $x \geq 0$ and n is an integer greater than 1, $x^{\frac{1}{n}}$ is an nth root of x.

Proof
By the definition of nth root, b is an nth root of x if and only if $b^n = x$.
Suppose $b = x^{\frac{1}{n}}$.
Then
$$b^n = \left(x^{\frac{1}{n}}\right)^n \quad \text{Substitution}$$
$$= x^{\frac{1}{n} \cdot n} \quad \text{Power of a Power Property}$$
$$= x^1$$
$$= x.$$
Thus, $x^{\frac{1}{n}}$ is an nth root of x.

Mathematicians could decide to let the symbol $x^{\frac{1}{n}}$ be any of the nth roots of x. However, to ensure that $x^{\frac{1}{n}}$ has exactly one value, we restrict the base x to be a nonnegative real number, and to let $x^{\frac{1}{n}}$ stand for the *unique nonnegative nth root.* Specifically, $x^{\frac{1}{2}}$ is the positive square root of x, and $2^{\frac{1}{3}}$ is the positive cube root of 2.

Pay close attention to parentheses when applying the $\frac{1}{n}$ Exponent Theorem. We do not consider negative bases with these exponents because there are properties of powers that do not apply to them. While your calculator may give a value for $(-8)^{\frac{1}{3}}$, we do not consider such expressions in this book.

How Many Real *n*th Roots Does a Real Number Have?

The number of real *n*th roots of a real number k is the number of points of intersection of the line $y = k$ with the power function $y = x^n$. As the graphs of the power functions below show, the number of intersections is determined by whether the value of n is odd or even, and whether the real number k is positive or negative.

For instance, the negative number -4 has no real square roots, 4th roots, or 6th roots. It has one real cube root, one real 5th root, and one real 7th root.

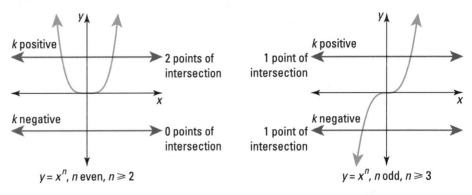

These properties of the *n*th power functions imply corresponding properties of *n*th roots.

> **Number of Real Roots Theorem**
> Every positive real number has: 2 real *n*th roots, when *n* is even.
>   1 real *n*th root, when *n* is odd.
> Every negative real number has: 0 real *n*th roots, when *n* is even.
>   1 real *n*th root, when *n* is odd.

Example 3

Evaluate the following in your head.

a. $81^{\frac{1}{4}}$ **b.** $-4^{\frac{1}{2}}$

Solution

a. $81^{\frac{1}{4}}$ is the positive real solution to $x^4 = 81$. $3^4 = 81$, so $81^{\frac{1}{4}} = 3$.

b. Follow the order of operations. Do the powers first. $4^{\frac{1}{2}}$ is the positive square root of 4, so $4^{\frac{1}{2}} = 2$. Thus $-4^{\frac{1}{2}} = -2$.

Because you can write roots as powers, you can use a calculator's powering key to evaluate the *n*th root of x as $x^{\frac{1}{n}}$.

Activity

Use the powering key on your calculator to evaluate $2^{\frac{1}{12}}$ to three decimal places. Record the key sequence you use. (Hint: You may need to use parentheses around the exponent.)

Example 4

Find all real solutions to $x^5 = 80$.

Solution

By the definition of nth root, the real solutions to $x^5 = 80$ are the real 5th roots of 80. So one solution is $x = 80^{\frac{1}{5}}$. By the Number of Real Roots Theorem, this is the only real 5th root of 80. A calculator shows that $80^{\frac{1}{5}} \approx 2.402$.

Check

$(2.402)^5 \approx 79.959$. It checks.

Nonreal nth Roots

Some of the nth roots of a real number are not real.

Example 5

Show that 2, -2, 2i, and -2i are fourth roots of 16.

Solution

To show that a number b is a fourth root of 16, verify that it satisfies $b^4 = 16$.

$$2^4 = 2 \cdot 2 \cdot 2 \cdot 2 = 16$$
$$(-2)^4 = -2 \cdot -2 \cdot -2 \cdot -2 = 16$$
$$(2i)^4 = 2^4 \cdot i^4 = 16 \cdot 1 = 16$$
$$(-2i)^4 = (-2i)(-2i)(-2i)(-2i) = (-2)^4 i^4 = 16 \cdot 1 = 16$$

So each of 2, -2, 2i, and -2i is a fourth root of 16.

At this time, you are not expected to find nonreal nth roots by yourself. Using mathematics beyond what you will learn in this course, it can be proved that every nonzero real number has n distinct nth roots. Thus, there are no other 4th roots of 16 than those given in Example 5.

QUESTIONS

Covering the Reading

In 1–3, refer to the discussion of musical frequencies in this lesson.

1. Into how many steps is an octave usually divided?

2. What is true about the ratios of the frequencies of consecutive notes on a properly tuned piano?

3. Find the frequency of the G above middle C.

4. Let n be an integer greater than 1. Then x is an nth root of t if and only if __?__.

In 5 and 6, without using a calculator, find the positive real root.

5. cube root of 8

6. 4th root of 81

7. Suppose the graphs of $y = x^n$ and $y = k$ are drawn on the same set of axes.
   a. If n is odd, in how many points do they intersect?
   b. How are the points of intersection related to the nth roots of k?

In 8 and 9, write all real roots of the equation as a decimal.

8. $x^3 = 125$

9. $x^4 = 10000$

10. a. For what values of n and x does $x^{\frac{1}{n}}$ have meaning?
   b. What is that meaning?

11. *True or false.*
   a. Both 5 and -5 are 6th roots of 15,625.
   b. $15,625^{\frac{1}{6}} = 5$
   c. $15,625^{\frac{1}{6}} = -5$
   d. $5i$ is a 6th root of 15,625.

In 12–14, evaluate without a calculator.

12. $125^{\frac{1}{3}}$

13. $144^{\frac{1}{2}}$

14. $-(49)^{\frac{1}{2}}$

In 15–17, use a calculator to approximate to the nearest thousandth.

15. $2^{\frac{1}{12}}$

16. $(1,419,857)^{\frac{1}{5}}$

17. $1000^{\frac{1}{10}}$

In 18 and 19, an equation is given.
a. Write the exact solution(s).
b. Approximate the solution(s) to the nearest thousandth.

18. $x^5 = 35$

19. $x^6 = 120$

20. Is the number a 6th root of 64? Justify your answer.
   a. 2   **b.** -2   **c.** $2i$ **d.** $-2i$

Stringing a song. *The sitar, which originated in India or Persia, is a stringed instrument with 7 main strings and 12 or more sympathetic strings. These strings vibrate when the main strings are played. The sitar is used in classical music of India, Pakistan, and Bangladesh.*

Applying the Mathematics

21. In some Asian Indian music, there are 22 steps between a note and the note an octave higher. What then is the ratio of the frequency of one note to the next step below?

22. According to Greek history about 400 B.C. the Delians (inhabitants of the island Delos) were suffering from a serious epidemic. When they consulted with an oracle (a person or object through whom a deity was believed to speak), they were told that if they *exactly* doubled the volume of their cubical altar to Apollo the epidemic would end. Suppose each edge of the original altar was 1 unit long.
   a. Find the volume of the original altar.
   b. Find the volume of the proposed altar.
   c. Find the length of an edge of new altar.

In 23–26, tell which symbol, <, =, or >, will give a true statement.

23. -7 __?__ $2401^{\frac{1}{4}}$

24. $-18^{\frac{1}{3}}$ __?__ $18^{\frac{1}{3}}$

25. $9^{\frac{1}{2}}$ __?__ $\left(\frac{1}{2}\right)^{9}$

26. $(27.1)^{\frac{1}{3}}$ __?__ 3

27. a. Evaluate $\left(\frac{16}{81}\right)^{\frac{1}{4}}$ mentally.

    b. Check your work.

28. a. Verify that $i\sqrt{2}$ is a 4th root of 4.

    b. Why then is $4^{\frac{1}{4}} \neq i\sqrt{2}$?

Review

In 29–31, the first six terms of a sequence are given. **a.** State whether the sequence could be arithmetic, geometric, or neither. **b.** If possible, write an explicit formula for the nth term of the sequence. **c.** If possible, write a recursive formula for the sequence. *(Lessons 3-7, 3-8, 7-5)*

29. 100, 90, 81, 72.9, 65.61, . . .

30. 100, 90, 81, 72, 63, 54, . . .

31. 100, 90, 80, 70, 60, 50, . . .

32. Suppose you invest \$400 in a bank that pays 6% interest compounded monthly. Your money is left in the bank for 18 months. *(Lesson 7-4)*

    a. Write an expression representing your final balance.

    b. Calculate the balance.

    c. Find the effective annual yield on this account.

33. Rewrite the expression $\frac{2x}{9x^{-2}} + \frac{-7}{9x^{-3}}$ without negative exponents.
(Lesson 7-3)

34. Evaluate $\left(\frac{5}{2}\right)^{-2}$ without using a calculator. Show the steps you use.
(Lessons 7-2, 7-3)

35. Write the reciprocal of i in $a + bi$ form. *(Lesson 6-9)*

Exploration

36. Consider the sequence $2^{\frac{1}{2}}, 2^{\frac{1}{3}}, 2^{\frac{1}{4}}, 2^{\frac{1}{5}}, . . .$

    a. As the sequence continues, what number do the terms approach?

    b. Suppose the base 2 is replaced by 4 in the sequence. What happens then?

    c. Generalize parts **a** and **b**.

IN·CLASS

ACTIVITY

*Noninteger
Power
Functions*

This activity shows some of the difficulties when working with noninteger rational exponents when the base is negative. Work in a group; if possible, use different calculators or computers to compare results. Use the standard default window when graphing.

1   **a.** Graph $y = x^{\frac{1}{3}}$.
   b. Clear the window from the graph in part **a**, and consider the equation $y = \left(x^{\frac{1}{6}}\right)^2$. Before graphing this equation, do you think the graph will be the same as the graph of $y = x^{\frac{1}{3}}$? Why or why not?
   c. Graph $y = \left(x^{\frac{1}{6}}\right)^2$ and tell whether or not your prediction was correct.

2   **a.** Graph $y = x^5$.
   b. Clear the window from the graph in part **a**, and consider the equation $y = \left(x^{10}\right)^{.5}$. Before graphing this equation, draw what you think the graph will be.
   c. Graph $y = \left(x^{10}\right)^{.5}$ and tell whether or not your prediction was correct.
   d. Clear the window from the graph in part **c**, and consider the equation $y = \left(x^{.5}\right)^{10}$. Before graphing this equation, draw what you think the graph will be. If you think the graph will be different from the graphs in parts **a** and **c**, explain why.
   e. Graph $y = \left(x^{10}\right)^{.5}$ and tell whether or not your prediction was correct.

3   **a.** Before graphing $y = x^{.5} \cdot x^{.5}$, predict what the graph will look like.
   b. Graph $y = x^{.5} \cdot x^{.5}$. Tell whether or not your prediction was correct.

4   From these examples, what properties of powers do not hold for noninteger rational exponents with negative bases?

From the In-class Activity, you should realize that many properties of powers do not hold when bases are allowed to be negative. For this reason, we define rational exponents only when the base is nonnegative.

The Meaning of Positive Rational Exponents

In Lesson 7-6, you learned that $x^{\frac{1}{n}}$ stands for the positive nth root of x. For instance, $16^{\frac{1}{4}}$ is the positive 4th root of 16. In this lesson, we look at $x^{\frac{m}{n}}$. For instance, what does $16^{\frac{3}{4}}$ signify? The answer can be found by examining the meaning of the fraction $\frac{3}{4}$ and using the Power of a Power Postulate.

$$16^{\frac{3}{4}} = 16^{\frac{1}{4} \cdot 3} \qquad \text{Rewrite } \tfrac{3}{4} \text{ as } \tfrac{1}{4} \cdot 3.$$
$$= \left(16^{\frac{1}{4}}\right)^3 \qquad \text{Use the Power of a Power Property.}$$

Thus, $16^{\frac{3}{4}}$ is the 3rd power of the positive 4th root of 16. With this interpretation, $16^{\frac{3}{4}}$ can be simplified.

$$16^{\frac{3}{4}} = \left(16^{\frac{1}{4}}\right)^3 = 2^3 = 8$$

Notice also that $16^{\frac{3}{4}} = 16^{3 \cdot \frac{1}{4}} = \left(16^3\right)^{\frac{1}{4}}$. So $16^{\frac{3}{4}}$ is also the 4th root of the 3rd power of 16. In general, with fractional exponents, the numerator is the power and the denominator is the root.

Rational Exponent Theorem

For any nonnegative real number x and positive integers m and n,

$x^{\frac{m}{n}} = \left(x^{\frac{1}{n}}\right)^{m}$, the mth power of the positive nth root of x, and

$x^{\frac{m}{n}} = \left(x^{m}\right)^{\frac{1}{n}}$, the positive nth root of the mth power of x.

Proof

$$x^{\frac{m}{n}} = x^{\frac{1}{n} \cdot m} \qquad \frac{m}{n} = \frac{1}{n} \cdot m$$
$$= \left(x^{\frac{1}{n}}\right)^{m} \qquad \text{Power of a Power Property}$$

Also,

$$x^{\frac{m}{n}} = x^{m \cdot \frac{1}{n}} \qquad \frac{m}{n} = m \cdot \frac{1}{n}$$
$$= \left(x^{m}\right)^{\frac{1}{n}}. \qquad \text{Power of a Power Property}$$

Because $x^{\frac{1}{n}}$ is defined only when $x \geq 0$, *the Rational Exponent Theorem applies only to a nonnegative base x.* Then, $x^{\frac{m}{n}} = \left(x^{m}\right)^{\frac{1}{n}} = \left(x^{\frac{1}{n}}\right)^{m}$.

An expression with a rational exponent can be simplified by finding either powers first or roots first. Usually, it is easier to find the root first, because you end up working with fewer digits.

Example 1

Simplify $25^{\frac{3}{2}}$.

Solution

Find the square root of 25 first, then cube it.
$$25^{\frac{3}{2}} = \left(25^{\frac{1}{2}}\right)^3 = 5^3 = 125$$

Check 1

Find the cube first: $25^{\frac{3}{2}} = \left(25^3\right)^{\frac{1}{2}} = 15{,}625^{\frac{1}{2}} = 125$. As expected, the result is the same in both cases.

Check 2

Use a calculator. You can use either $\frac{3}{2}$ or 1.5 for the exponent. The key sequence 25 $\boxed{y^x}$ 1.5 $\boxed{=}$ on a scientific calculator yields 125.

With practice, you should be able to simplify many expressions with fractional exponents mentally. You should also be able to estimate positive rational powers of numbers with a calculator.

Example 2

Approximate $25^{\frac{3}{5}}$ to the nearest thousandth.

Solution

Since $\frac{3}{5} = .6$, key in $25^{.6}$ on your calculator.
$$25^{\frac{3}{5}} \approx 6.899$$

Properties of Positive Rational Exponents

The answer 125 in Example 1 is larger than the base 25. In contrast, in Example 2 the answer 6.899 is less than the base. In general, when the base is larger than 1, the larger the exponent, the larger the power. This can be verified by calculating other rational powers of 25.

$$25^0 = 1$$
$$25^{\frac{1}{4}} = 2.236 \ldots$$
$$25^{\frac{1}{3}} = 2.924 \ldots$$
$$25^{\frac{1}{2}} = 5$$
$$25^{\frac{3}{4}} = 11.180 \ldots$$
$$25^1 = 25$$
$$25^{\frac{5}{4}} = 55.901 \ldots$$
$$25^{\frac{3}{2}} = 125$$
$$25^{\frac{7}{4}} = 279.508 \ldots$$
$$25^2 = 625$$

Thus, even without calculating, you should realize that $25^{\frac{5}{3}}$ is between 25 and 625, because $\frac{5}{3}$ is between 1 and 2.

Activity

a. Evaluate $36^{\frac{1}{2}}$, $36^{\frac{3}{4}}$, and $36^{\frac{5}{3}}$.

b. Evaluate 36^x for two other positive rational numbers x of your own choice.

c. Make a conjecture. For what positive rational values of x is $36^x > 36$? For what values of x is $36^x < 36$?

The properties of powers given in Lesson 7-2 hold for all positive rational powers. For instance, $25^{\frac{1}{2}} \cdot 25^1 = 25^{(\frac{1}{2})+1} = 25^{\frac{3}{2}}$ is an instance of the Product of Powers Property.

Example 3

Suppose $x > 0$. Simplify $\left(27x^6\right)^{\frac{4}{3}}$

Solution

$$\left(27x^6\right)^{\frac{4}{3}} = 27^{1\left(\frac{4}{3}\right)} \cdot x^{6\left(\frac{4}{3}\right)} \qquad \text{Power of a Product Postulate}$$

$$= 27^{\frac{4}{3}} \cdot x^{\frac{24}{3}} \qquad \text{Rational Exponent Theorem and arithmetic}$$

$$= \left(27^{\frac{1}{3}}\right)^4 \cdot x^8 \qquad \text{Product of Powers Postulate}$$

$$= 3^4 x^8 \qquad \frac{1}{n} \text{ Exponent Theorem}$$

$$= 81x^8 \qquad \text{arithmetic}$$

Solving Equations with Positive Rational Exponents

Properties of powers can be used to solve equations with positive rational exponents. To solve an equation of the form $x^{\frac{m}{n}} = k$, *raise each side of the equation to the $\frac{n}{m}$ power.* This can be done because any number can be substituted for its equal in an algebraic expression. In particular, if $a = b$, then $a^n = b^n$.

Example 4

Solve $x^{\frac{5}{4}} = 243$.

Solution

Recall that any number times its reciprocal equals 1. Thus to solve for x, raise both sides of the equation to the $\frac{4}{5}$ power.

$$x^{\frac{5}{4}} = 243$$

$$\left(x^{\frac{5}{4}}\right)^{\frac{4}{5}} = 243^{\frac{4}{5}} \qquad \text{Raise both sides to the } \frac{4}{5} \text{ power.}$$

$$x^1 = 243^{\frac{4}{5}} \qquad \text{Power of a Power Property}$$

$$x = \left(243^{\frac{1}{5}}\right)^4 \qquad \text{Rational Exponent Theorem}$$

$$x = (3)^4 = 81 \qquad \text{Simplify and calculate.}$$

Check

Does $81^{\frac{5}{4}} = 243$? Yes.

Rational exponents have many applications, including growth situations, investments, radioactive decay, and change-of-dimension situations (for example, area to volume and back). The following application was derived from observation of the planets by the astronomer Johannes Kepler (1571–1630). Remember that the *period* of a planet is the length of time it takes the planet to go around the sun. (The period of Earth is one year.)

Example 5

Kepler's third law states that the ratio of the squares of the periods of any two planets equals the ratio of the cubes of their mean distances from the sun. If the periods of two planets are t and T and their mean distances from the sun are d and D, respectively, then

$$\frac{T^2}{t^2} = \frac{D^3}{d^3}.$$

Find the ratio $\frac{T}{t}$ of the periods.

Solution

To solve the equation for $\frac{T}{t}$, note that the left side is the square of $\frac{T}{t}$.

$$\left(\frac{T}{t}\right)^2 = \left(\frac{D}{d}\right)^3$$

Now raise each side to the $\frac{1}{2}$ power.

$$\left(\left(\frac{T}{t}\right)^2\right)^{\frac{1}{2}} = \left(\left(\frac{D}{d}\right)^3\right)^{\frac{1}{2}}$$

Use the Power of a Power property.

$$\frac{T}{t} = \left(\frac{D}{d}\right)^{\frac{3}{2}}$$

Thus the ratio of the periods equals the $\frac{3}{2}$ power of the ratio of the planets' mean distances from the sun.

Shown is a telescope built by Galileo in the early 1600s. From his observations, Galileo became convinced of the truth of the Copernican theory that planets revolve around the sun. In 1597 he wrote a letter to Kepler stating his fear of ridicule if he declared his belief in the Copernican theory.

QUESTIONS

Covering the Reading

In 1 and 2, write as a power of x.

1. the 4th power of the 9th root of x

2. the seventh root of the cube of x

3. a. Rewrite $100{,}000^{\frac{4}{5}}$ in two ways as a power of a power of 100,000.
   b. Which way is easier to calculate mentally?
   c. Calculate $100{,}000^{\frac{4}{5}}$.

In 4–6, evaluate without a calculator.

4. $27^{\frac{2}{3}}$ **5.** $32^{\frac{3}{5}}$ **6.** $36^{\frac{3}{2}}$

In 7–9, evaluate with a calculator.

7. $729^{\frac{3}{2}}$ **8.** $729^{\frac{2}{3}}$ **9.** $1331^{\frac{5}{3}}$

10. Give the results you obtained for the Activity in the lesson.

In 11 and 12, suppose $x > 1$. Complete with $<$, $>$, or $=$.

11. $x^{\frac{5}{3}}$ _?_ $x^{\frac{3}{5}}$ **12.** $x^{\frac{3}{4}}$ _?_ $x^{\frac{3}{5}}$

In 13–15, suppose the value of each variable is positive. Simplify.

13. $\left(16x^8\right)^{\frac{3}{4}}$ **14.** $B^{\frac{2}{3}} \cdot B$ **15.** $\frac{2}{3}y^{\frac{2}{3}} \cdot \frac{3}{2}y^{\frac{3}{2}}$

In 16–18, solve and check.

16. $V^{\frac{5}{2}} = 100$ **17.** $k^{\frac{2}{3}} = 64$ **18.** $x^{\frac{4}{9}} = 12$

19. Solve Kepler's equation in Example 5 for $\frac{D}{d}$.

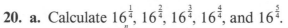

Applying the Mathematics

20. a. Calculate $16^{\frac{1}{4}}$, $16^{\frac{2}{4}}$, $16^{\frac{3}{4}}$, $16^{\frac{4}{4}}$, and $16^{\frac{5}{4}}$.
   b. Simplify $16^{\frac{n}{4}}$, where n is a positive integer.

21. This question gives another example to show why rational exponents are used only with nonnegative bases.
   a. If $(-8)^{\frac{1}{3}}$ were to equal the cube root of -8, then $(-8)^{\frac{1}{3}} = $ _?_ .
   b. If $(-8)^{\frac{2}{6}}$ follows the Rational Exponent Theorem, then
   $(-8)^{\frac{2}{6}} = \left((-8)^2\right)^{\frac{1}{6}} = $ _?_ .
   c. In this question, does $(-8)^{\frac{1}{3}} = (-8)^{\frac{2}{6}}$?

In 22–24, apply the Power of a Quotient Property, $\left(\frac{a}{b}\right)^m = \frac{a^m}{b^m}$, to simplify.

22. $\left(\frac{64}{27}\right)^{\frac{2}{3}}$ **23.** $\left(\frac{1000}{343}\right)^{\frac{4}{3}}$ **24.** $\left(\frac{16}{625}\right)^{\frac{3}{4}}$

25. Kepler used his third law (given in Example 5) to determine how far planets were from the Sun. He knew that for the Earth, $t \approx 365$ days and $d \approx 150,000,000$ km. He also knew that for Mars, $T = 687$ days. Use this information to find D, the mean distance of Mars from the Sun.

26. The diameter D of the base of a tree of a given species roughly varies directly with the $\frac{3}{2}$ power of its height h.
   a. Suppose a young sequoia 500 cm tall has a base diameter of 11.7 cm. Find the constant of variation.
   b. The most massive living tree is a California sequoia called "General Sherman." In 1991, its base diameter was about 807 cm. Approximately how tall was General Sherman then?
   c. One story of a modern office building is about 3 m high. General Sherman is about as tall as a _?_ story office building.

Treemendous. This is the "General Sherman," located in Sequoia National Park in California. It is one of the oldest living things on Earth, being between 2200 and 2500 years old.

27. a. Find $9^{\frac{1}{2}}$, $4^{\frac{1}{2}}$ and $(9 + 4)^{\frac{1}{2}}$.
   b. If $a > 0$ and $b > 0$, is $(a^2 + b^2)^{\frac{1}{2}} = a + b$?   *(Lesson 7-6)*

28. If $a^b = c$ then __?__ is a __?__ root of __?__.   *(Lesson 7-6)*

29. The first four terms of a geometric sequence are 5, -10, 20, -40.
   a. What is the constant ratio?
   b. What is the next term?
   c. Write an explicit formula for the nth term.
   d. What is the 15th term?   *(Lesson 7-5)*

30. a. Evaluate the expression $4000\left(1 + \frac{.04}{12}\right)^{36}$.
   b. Write a question about money and interest whose answer is the answer to part **a.**   *(Lesson 7-4)*

31. If x^t is the reciprocal of x, what is the value of t?   *(Lesson 7-3)*

32. Simplify: $\frac{2^{100}}{2^{-99}}$.   *(Lesson 7-3)*

33. Give the general property of powers that has been applied.
   $(x + 3)^2 \cdot (x + 3)^5 = (x + 3)^7$ *(Lesson 7-2)*

34. Find an equation for the line that goes through $(4, -3)$ and is perpendicular to $y = \frac{3}{2}x - 10$.   *(Lesson 4-9)*

35. Find the total surface area of the right circular cylinder with diameter 25 cm and height 40 cm shown at the left.   *(Previous course)*

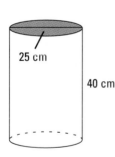

25 cm

40 cm

36. Choose a number x between 0 and 1.
   a. Using a calculator, estimate $x^{\frac{1}{10}}$, $x^{\frac{1}{4}}$, $x^{\frac{1}{3}}$, $x^{\frac{1}{2}}$, $x^{\frac{2}{3}}$, $x^{\frac{3}{4}}$ and $x^{\frac{9}{10}}$.
   b. Estimate x^2, $x^{\frac{5}{2}}$, x^3, $x^{\frac{7}{2}}$, and x^4.
   c. As the exponent increases, is there any pattern to the values of the powers?

Modern milking. *Shown is a computerized milking parlor. A computer records each cow's milk production and adjusts her feed ration accordingly. See Example 3.*

So far in this chapter you have learned meanings for positive rational exponents. For any positive number x and any positive integers m and n:

$$x^0 = 1 \qquad \qquad \text{(Zero Exponent Theorem, Lesson 7-2)}$$

$$x^{-n} = \frac{1}{x^n} \qquad \text{(Negative Exponent Theorem, Lesson 7-3)}$$

$$x^{\frac{1}{n}} = \text{positive } n\text{th root of } x \qquad \left(\tfrac{1}{n} \text{ Exponent Theorem, Lesson 7-6}\right)$$

$$x^{\frac{m}{n}} = \left(x^{\frac{1}{n}}\right)^m = \left(x^m\right)^{\frac{1}{n}} \qquad \text{(Rational Exponent Theorem, Lesson 7-7)}$$

Now we use the properties of powers to determine the meaning of negative rational exponents.

Evaluating Powers with Negative Rational Exponents

Consider the power $x^{-\frac{m}{n}}$. Because $x^{-\frac{m}{n}} = \left((x^{-1})^m\right)^{\frac{1}{n}} = \left((x^m)^{\frac{1}{n}}\right)^{-1} = \left((x^{\frac{1}{n}})^m\right)^{-1}$, and the exponents can be applied in any order, you have the choice of taking the reciprocal, the mth power, or the nth root first.

Example 1

Evaluate $81^{-\frac{1}{4}}$.

Solution

Here we take the reciprocal and then the 4th root.

$$81^{-\frac{1}{4}} = \frac{1}{81^{\frac{1}{4}}} = \frac{1}{3}$$

Check

Evaluate $81^{-\frac{1}{4}}$ on a calculator. Use parentheses as needed. Your calculator should display 0.3333333, which is approximately $\frac{1}{3}$. It checks.

Example 2

Simplify $\left(\frac{27}{1000}\right)^{-\frac{2}{3}}$.

Solution 1

Think $\left(\left(\left(\frac{27}{1000}\right)^{-1}\right)^{\frac{1}{3}}\right)^{2}$. This means take the reciprocal, cube root, and 2nd power in that order. Remember to work with the innermost parentheses first.

$$\left(\left(\left(\frac{27}{1000}\right)^{-1}\right)^{\frac{1}{3}}\right)^{2} = \left(\left(\frac{1000}{27}\right)^{\frac{1}{3}}\right)^{2} = \left(\frac{10}{3}\right)^{2} = \frac{100}{9}$$

Solution 2

Think $\left(\left(\left(\frac{27}{1000}\right)^{\frac{1}{3}}\right)^{2}\right)^{-1}$. This means take the cube root, 2nd power, and reciprocal in that order.

$$\left(\left(\left(\frac{27}{1000}\right)^{\frac{1}{3}}\right)^{2}\right)^{-1} = \left(\left(\frac{3}{10}\right)^{2}\right)^{-1} = \left(\frac{9}{100}\right)^{-1} = \frac{100}{9}$$

Check

Enter $\left(\frac{27}{1000}\right)^{-\frac{2}{3}}$ in your calculator. Ours gives 11.11111111, which is very close to $11.\overline{1}$ or $\frac{100}{9}$.

Example 3

The number of hours that pasteurized milk stays fresh is a function of the surrounding temperature T. Use the formula,

$$h(T) = 349 \cdot 10^{-.02T}$$

where the temperature T is in degrees Celsius, to predict how long newly pasteurized milk will stay fresh when stored at temperature 8°C.

Solution

Substitute $T = 8$ in the formula and evaluate using a calculator.

$$h(8) = 349 \cdot 10^{-.02(8)}$$
$$h(8) = 349 \cdot 10^{-.16}$$
$$h(8) \approx 241.45$$

When stored at 8°C, newly pasteurized milk will stay fresh about 241 hours, or a little more than 10 days.

Solving Equations Involving Negative Rational Exponents

The ideas used in Lesson 7-7 to solve equations with positive rational exponents can be used with negative rational exponents as well.

Example 4

Solve $x^{-\frac{2}{5}} = 9$.

Solution

The reciprocal of $-\frac{2}{5}$ is $-\frac{5}{2}$, so raise each side to the $-\frac{5}{2}$ power.

$$\left(x^{-\frac{2}{5}}\right)^{-\frac{5}{2}} = 9^{-\frac{5}{2}}$$

$$x = 9^{-\frac{5}{2}} = \left(\left(9^{\frac{1}{2}}\right)^5\right)^{-1} = \left(\left(3\right)^5\right)^{-1} = 243^{-1} = \frac{1}{243}$$

Check

Does $\left(\frac{1}{243}\right)^{-\frac{2}{5}} = 9$? $\left(\left(\left(\frac{1}{243}\right)^{-1}\right)^{\frac{1}{5}}\right)^2 = \left(243^{\frac{1}{5}}\right)^2 = 3^2 = 9$. Yes, it checks.

QUESTIONS

Covering the Reading

In 1–3 evaluate without using a calculator.

1. $125^{-\frac{1}{3}}$ **2.** $81^{-\frac{3}{4}}$ **3.** $\left(\frac{9}{4}\right)^{-\frac{5}{2}}$

4. Tell whether or not the expression equals $b^{-\frac{3}{4}}$ for $b > 0$.

  a. $\dfrac{1}{b^{\frac{3}{4}}}$ **b.** $\dfrac{1}{\left(b^3\right)^{\frac{1}{4}}}$ **c.** $\left(\left(b^{-1}\right)^3\right)^{\frac{1}{4}}$

  d. $-b^{\frac{3}{4}}$ **e.** $\left(b^{\frac{1}{4}}\right)^{-3}$ **f.** $\left(b^{-\frac{1}{4}}\right)^3$

In 5–7, estimate to the nearest thousandth.

5. $100^{-\frac{1}{2}}$ **6.** $50 \cdot 2.79^{-\frac{3}{5}}$ **7.** $5^{-.004}$

8. Refer to Example 3.
  a. About how long will newly pasteurized milk stay fresh if it is stored at 3°C?
  b. When milk is stored at 3°C it stays fresh about __?__ times as long as it will at 8°C.

In 9–11, solve.

9. $s^{-\frac{1}{4}} = 81$ **10.** $t^{-\frac{2}{3}} = 36$ **11.** $x^{-\frac{3}{2}} = \frac{1}{8}$

12. If $x^{-\frac{3}{5}} = 15$, find x to the nearest thousandth.

Applying the Mathematics

In 13–15, tell whether the number is positive, negative, or zero. Use a calculator only to check.

13. $(.98956)^{-\frac{3}{4}}$ **14.** $(1.0825)^0$ **15.** $(-.07)(3)^{-.4}$

16. Find n if $\left(\frac{99}{100}\right)^{-\frac{3}{4}} = \left(\frac{100}{99}\right)^n$.

17. **a.** Evaluate 64^x, where x increases by sixths from -1 to 1. (There are 13 values to evaluate: 64^{-1}, $64^{-\frac{5}{6}}$, $64^{-\frac{4}{6}} = 64^{-\frac{2}{3}}$, and so on, until 64^1.)
    b. Describe any patterns you observe in the answers to part **a.**

18. Let F be the amount of food a mammal with body mass m must eat daily to maintain its mass. On page 417, it was noted that $\frac{F}{m} = km^{-\frac{1}{3}}$ for a certain species of mammals. Solve this equation for F.

In 19–22, rewrite each expression in the form ax^n. Check your answer by substituting a value for x.

19. $\left(x^3\right)^{-\frac{1}{3}}$

20. $\left(25x^{-4}\right)^{-\frac{3}{2}}$

21. $\dfrac{x}{3x^{-\frac{2}{3}}} \cdot 6x^{\frac{1}{2}}$

22. $\dfrac{\frac{-3}{4}x^{-\frac{3}{4}}}{\frac{1}{4}x^{\frac{1}{4}}}$

23. Consider the expression $\dfrac{(a + b)^{\frac{2}{3}}}{(a + b)^{\frac{5}{3}}}$.
    a. Simplify as a power of $(a + b)$.
    b. Evaluate the expression when $a = 8$ and $b = 2$.

Review

In 24 and 25, use this information. The Galápagos Islands are a chain of islands in the Pacific Ocean that belong to Ecuador. They are famous for their unusual plant and animal life. A biologist has found that S, the number of different plant species on an island in the Galápagos, varies with the area A of the island in square kilometers according to the formula below.

$$S = 38.8A^{0.32}$$

Estimate the number of plant species for each of the following islands. *(Lesson 7-7)*

24. the largest island, Isabela, which has area of about 4588 square kilometers

25. the smallest major island in the Galápagos chain, Rábida, which has area of about 4.9 square kilometers

26. Recall that a sphere of radius r has volume $V = \frac{4}{3}\pi r^3$ and surface area $A = 4\pi r^2$.
    a. Solve the surface area formula for r.
    b. Substitute the expression you found for r in part **a** for r in the volume formula. Simplify.
    c. *True or false.* The volume of a sphere varies directly as its surface area. Justify your answer. *(Lessons 1-6, 2-1, 7-7)*

27. Suppose $f(x) = x^7$. For what value(s) of x is $f(x) = 14$? Explain how you got your answer. *(Lesson 7-6)*

28. **a.** Show that 5, -5, 5i, and -5i are 4th roots of 625.
    b. Use a graph to explain why the number 625 has only two real 4th roots. *(Lessons 6-7, 7-6)*

Shown is a giant land tortoise on the Galápagos Islands, 950 km west of Equador. During the 19th century, the islands abounded with the giant tortoises, whose Spanish name—galápagos—gave the islands their name.

29. A ball is dropped from a height of 20 ft. After each bounce it rebounds to 70% of the previous height it attained.
    a. How high does it get after the first bounce?
    b. How high does it get after the nth bounce? *(Lesson 7-5)*

30. Simplify $\left(\frac{2x}{5y}\right)^{-2}$. *(Lesson 7-3)*

31. The product of x^2 and x^3 is x^5. Find six more pairs of integer powers of x whose product is x^5. *(Lessons 7-2, 7-3)*

32. Solve this system $\begin{cases} 3x^{-1} + 2y^{-1} = 27 \\ 2x^{-1} - y^{-1} = 4. \end{cases}$
    (Hint: Let $a = x^{-1}$ and $b = y^{-1}$.) *(Lessons 5-3, 7-2)*

Exploration

---

33. The formula $P = 14.7 \cdot 10^{-.09h}$ estimates the atmospheric pressure P in pounds per square inch h miles above sea level. Find the atmospheric pressure:
    a. in Albuquerque, NM, where $h \approx .9$;
    b. in Miami, FL, which is approximately at sea level;
    c. on top of Mt. McKinley, AK, which is about 20,320 feet above sea level.
    d. Graph this relation, using the three points determined in parts **a–c** and two points of your choice.
    e. As h increases, does P increase or decrease?

Uphill battle. *Shown are some of the 16 women breast-cancer survivors who attempted to climb South America's tallest mountain, Aconcagua, in 1995. Three of the women made it to the summit of the 23,000-foot mountain.*

A project presents an opportunity for you to extend your knowledge of a topic related to the material of this chapter. You should allow more time for a project than you do for typical homework questions.

1 Musical Frequencies

In Lesson 7-6 you learned that frequencies of tones in music can be described using powers and roots. Find out about other musical scales and the mathematical relations they embody. For instance, find out about the discoveries of the Pythagoreans.

2 Noninteger Power Functions

Graph the functions with rules $y = x^5$, $y = x^{1.5}$, $y = x^{2.5}$, and so on. Give the domains and ranges of these functions, and compare their graphs with the graphs of $y = x^2$, $y = x^3$, and so on.

3 Financing Post-High School Education

a. Select a college or post-secondary school you have heard about or are considering attending. Find out its yearly tuition for each year of the past decade.

b. Based on the data in part **a**, estimate what tuition will cost during the years you will attend. Explain how you obtained this answer.

c. Suppose that 15 years ago, a benefactor set up an account for your education. This benefactor made a single deposit of $10,000 which has been earning 6% interest compounded quarterly. Since the deposit 15 years ago, no one has added money to the account or taken money from the account. Will this account be sufficient to cover tuition for all four years?

d. If it is not, find the smallest annual interest rate at which the account would have grown enough to pay for tuition for the years you will attend.

4 Fermat's Last Theorem

Recall that there are many right triangles with whole number solutions. That is, there are many triples of whole numbers that satisfy $x^2 + y^2 = z^2$. Some Pythagorean triples are 3-4-5, 5-12-13, and 8-15-17. In the 17th century, Pierre de Fermat investigated the possibility of whole number solutions to $x^3 + y^3 = z^3$, $x^4 + y^4 = z^4$, ..., $x^n + y^n = z^n$ where n is any integer greater than 2. He concluded that it is impossible to separate a cube into two cubes,

a fourth power into two fourth powers, or in general any power above the second into two powers of the same degree. In a book, he wrote, "I have found a truly marvelous proof of this theorem, but this margin is too narrow to contain it." For hundreds of years, mathematicians have searched for a proof to Fermat's Last Theorem. Investigate attempts to prove this theorem, including the widely publicized attempts in the 1990s by Andrew Wiles. Write a report or make an oral presentation about your findings.

5 Local Interest Rates

a. Conduct a survey of the interest rates on various savings accounts at several local savings institutions. Find out how often the interest is compounded at each place.

b. Suppose you had $5000 to invest as a 9th grader. Which account would yield the highest return four years later? Which would yield the lowest return on your investment after four years?

c. What are some other factors besides rate of interest and number of compounding periods that an investor should take into account?

6 Family of Equations

Graph at least 10 members of the family of equations of the form $ax + by = c$, where a, b, c, are consecutive terms in a geometric sequence. If you are using an automatic grapher, you may need to solve for y. Sketch the graphs. What pattern do you see? You may have to experiment to determine which values to use in the window you have chosen. How can you explain these results?

SUMMARY

When $x > 0$, the expression x^m is defined for any real number m. This chapter has covered the meanings and properties of x^m when m is a rational number. For any nonnegative bases and nonzero real exponents, or any nonzero bases and integer exponents:

Product of Powers Postulate: $x^m \cdot x^n = x^{m+n}$

Power of a Power Postulate: $(x^m)^n = x^{mn}$

Power of a Product Postulate: $(xy)^n = x^n y^n$

Quotient of Powers Postulate: For $x \neq 0$
$$\frac{x^m}{x^n} = x^{m-n}$$

Power of a Quotient Postulate: For $y \neq 0$
$$\left(\frac{x}{y}\right)^m = \frac{x^m}{y^m}.$$

From these postulates, the following theorems can be deduced. For any positive real number and any integers, m and n, $n \neq 0$:

Zero Exponent Theorem: $x^0 = 1$

Negative Exponent Theorem: $x^{-m} = \frac{1}{x^m}$

$\frac{1}{n}$ Exponent Theorem: $x^{\frac{1}{n}}$ is the positive solution to $b^n = x$.

Rational Exponent Theorem: $x^{\frac{m}{n}} = \left(x^m\right)^{\frac{1}{n}} = \left(x^{\frac{1}{n}}\right)^m$

These properties help in simplifying expressions and in solving equations of the form $x^n = b$. To solve such an equation, raise each side of the equation to the $\frac{1}{n}$ power. In the General Compound Interest Formula
$$A = P\left(1 + \frac{r}{n}\right)^{nt},$$

A is the value of an investment of $\$P$ earning interest at a rate r compounded n times per year for t years. A geometric sequence is a sequence in which each term is a constant multiple of the preceding term. That is, for all $n \geq 2$, $g_n = rg_{n-1}$. The nth term of a geometric sequence can be found explicitly using the formula $g_n = g_1 r^{n-1}$.

VOCABULARY

Below are the most important terms and phrases for this chapter. You should be able to give a definition or statement for those terms marked with a *. For all other terms you should be able to give a general description and a specific example of each.

Lesson 7-1
powering, exponentiation
*base, exponent, power
repeated multiplication
*nth power function
identity function
function
cubing function

Lesson 7-2
Product of Powers Postulate
Power of a Power Postulate
Power of a Product Postulate
Quotient of Powers Postulate
Power of a Quotient Postulate
Zero Exponent Theorem

Lesson 7-3
Negative Exponent Theorem

Lesson 7-4
compounded annually,
    semiannually, quarterly
principal
*Compound Interest Formula
General Compound Interest
    Formula
effective annual yield, yield
simple interest

Lesson 7-5
*geometric sequence,
    exponential sequence

constant multiplier,
    constant ratio
*recursive formula for a
    geometric sequence
*explicit formula for a
    geometric sequence

Lesson 7-6
*square root, cube root,
    nth root
$\frac{1}{n}$ Exponent Theorem
Number of Real Roots
    Theorem

Lesson 7-7
Rational Exponent Theorem

PROGRESS SELF-TEST

Take this test as you would take a test in class. You will need a calculator. Then check your work with the solutions in the Selected Answers section in the back of the book.

1. Order from largest to smallest: 3^{-4}, -3^4, $3^{\frac{1}{4}}$.

In 2–4, write as a whole number or simple fraction.

2. 9^{-2} **3.** $\left(\frac{1}{32}\right)^{-\frac{6}{5}}$ **4.** $(11,390,625)^{\frac{1}{6}}$

In 5–7, simplify. Assume $x > 0$ and $y > 0$.

5. $\left(625x^4y^8\right)^{\frac{1}{4}}$

6. $\frac{-96x^{15}y^3}{4x^3y^{-5}}$

7. $\frac{(2x^4)^5}{16x}$

In 8–10, find all real solutions.

8. $9x^4 = 144$ **9.** $c^{\frac{3}{2}} = 64$

10. $5^n \cdot 5^{21} = 5^{29}$

11. A bank account pays 3.75% interest compounded daily. If you deposit $200 in the account and leave it untouched for 5 years, how much will be in the account then?

In 12 and 13, *multiple choice.*

12. Which expression equals $a^{-\frac{4}{5}}$ for all $a > 0$?

(a) $\left(\frac{1}{a^4}\right)^{\frac{1}{5}}$ (b) $\left(-a^4\right)^{\frac{1}{5}}$

(c) $a^{\frac{5}{4}}$ (d) $\left(-a\right)^{\frac{4}{5}}$

13. Which equation could have the graph below?

(a) $y = x^3$
(b) $y = x$
(c) $y = x^4$
(d) $y = x^{-4}$

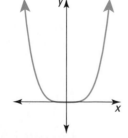

14. Find an explicit formula for the nth term in the geometric sequence 2, 8, 32, 128, . . .

15. Consider the sequence defined as follows.
$$\begin{cases} g_1 = 12 \\ g_n = \frac{1}{2}g_{n-1}, \text{ for integers } n \geq 2 \end{cases}$$
Write the first four terms of the sequence.

16. Write without an exponent: $\frac{2.1 \cdot 10^2}{10^{-3}}$.

17. If the frequency of the note A on a keyboard is 440 hertz, what is the frequency of B, the tone two steps above A?

18. To qualify for a quiz show a person must answer all questions correctly in three categories: literature, science, and current events. Suppose a person estimates that the probability of getting one question correct in literature is ℓ, in science is s, and in current events is c. If the person is asked 3 literature, 3 science, and 4 current events questions, what is the probability that person gets all the questions right?

19. Recall the formula $h(t) = 349 \cdot 10^{-.02t}$ for the number of hours h that milk stays fresh in a surrounding temperature t°C. How long might milk stay fresh when stored at 15°C?

20. Identify all real 4th roots of 81.

21. *True or false.* $64^{\frac{1}{6}} = -2$. Justify your answer.

22. A bond paying 5.4% interest compounded monthly for 8 years has matured giving the investor $5000. How much did the investor pay for the bond 8 years ago?

CHAPTER REVIEW

Questions on SPUR Objectives

SPUR stands for **S**kills, **P**roperties, **U**ses, and **R**epresentations. The Chapter Review questions are grouped according to the SPUR Objectives for this chapter.

SKILLS DEAL WITH THE PROCEDURES USED TO GET ANSWERS.

Objective A: *Evaluate b^n when $b > 0$ and n is a rational number.* (Lessons 7-2, 7-3, 7-6, 7-7, 7-8)

In 1–10, write as a simple fraction. Do not use a calculator.

1. 2^0

2. 6^{-2}

3. $3.4 \cdot 10^{-3}$

4. $(-2)^{-2}$

5. $\left(\frac{2}{3}\right)^{-1}$

6. $\left(\frac{1}{5}\right)^{-4}$

7. $1000^{\frac{1}{3}}$

8. $16^{\frac{3}{4}}$

9. $36^{-\frac{1}{2}}$

10. $\left(\frac{27}{216}\right)^{-\frac{2}{3}}$

In 11–14, estimate to the nearest hundredth.

11. $80^{\frac{2}{3}}$

12. $3 \cdot 27^{\frac{1}{8}}$

13. $\left(\frac{1}{64}\right)^{-\frac{3}{2}}$

14. $2^{1.5}$

In 15 and 16, *true or false.* Justify your answer.

15. $-7 = (117{,}649)^{\frac{1}{6}}$

16. $3^{-6.4} < 3^{-6.5}$

Objective B: *Simplify expressions or solve equations using properties of exponents.*
(Lessons 7-2, 7-3, 7-6, 7-7, 7-8)

In 17–20, solve.

17. $(9^5 \cdot 9^3) = 9^x$

18. $\frac{2^5}{2^{-1}} = 2^x$

19. $\left(7^{\frac{1}{2}}\right)^3 = 7^n$

20. $(2 \cdot 5)^{-3} = y^{-3}$

In 21–26, simplify. Assume all variables represent positive numbers.

21. $(-4x^2)^3$

22. $\frac{-8x^{10}y^{\frac{3}{2}}}{2xy^{\frac{1}{2}}}$

23. $\left(\frac{a}{b}\right)^3 \left(\frac{2b}{3a}\right)^4$

24. $\frac{15c}{(3c^{-6})(20c^6)}$

25. $\frac{12p^3q^{-2}}{16p^{-2}q}$

26. $\frac{\left(x^4y^2\right)^{\frac{1}{2}}}{xy^{\frac{3}{2}}}$

Objective C: *Describe geometric sequences explicitly and recursively.* (Lesson 7-5)

In 27 and 28, the first few terms of a geometric sequence are given.

a. Find an explicit formula for the nth term.

b. Find a recursive formula for the sequence.

c. Find the 12th term.

27. $-\frac{3}{8}, \frac{3}{4}, -\frac{3}{2}, 3, \ldots$ **28.** $10, 30, 90, \ldots$

29. Find the 50th term of a geometric sequence whose first term is 6 and whose constant multiplier is 1.05.

30. *Multiple choice.* Which of the following could be the first three terms of a geometric sequence?

(a) $16, 4, -8, \ldots$ (b) $3\frac{1}{3}, 33\frac{1}{3}, 333\frac{1}{3}, \ldots$

(c) $\frac{4}{5}, \frac{9}{5}, \frac{14}{5}, \ldots$ (d) $0.04, 0.16, 0.36, \ldots$

In 31–34, give the first four terms of the geometric sequence described.

31. constant ratio 4, first term 5

32. first term $\frac{1}{2}$, second term $\frac{3}{4}$

33. $\begin{cases} g_1 = 10 \\ g_n = -2g_{n-1}, \text{ for all integers } n \geq 2 \end{cases}$

34. $t_n = -2\left(\frac{3}{4}\right)^{n-1}$, for all integers $n \geq 1$

Objective D: *Solve equations of the form $x^n = b$, where n is a rational number.* (Lessons 7-6, 7-7, 7-8)

In 35–42, find all real solutions.

35. $3x^2 = 192$ **36.** $27 = a^4$

37. $x^3 = 12$ **38.** $x^{-2} = 9$

39. $5 = y^{\frac{1}{3}}$ **40.** $1.75 = m^{\frac{1}{5}}$

41. $m^{\frac{3}{2}} = \frac{1}{27}$ **42.** $4q^{-\frac{2}{5}} = 9$

PROPERTIES DEAL WITH THE PRINCIPLES BEHIND THE MATHEMATICS.

Objective E: *Recognize properties of nth powers and nth roots.* *(Lessons 7-2, 7-6, 7-7, 7-8)*

43. **a.** Identify all square roots of 225.
   b. Simplify $225^{\frac{1}{2}}$.

44. *True or false.* $2i$ is a 4th root of 16.

45. *True or false.* If $0 < x < 1$, $x^{\frac{1}{3}} > x$.

46. Suppose $x > 1$. Arrange from smallest to largest: $x, x^{\frac{1}{2}}, x^{-2}, x^{\frac{5}{4}}, x^{-\frac{2}{3}}$

In 47–50, use the properties below. Assume $Q > 0$, and $m \neq 0$ and $n \neq 0$. Identify the property or properties that justify the equality.

(a) $Q^0 = 1$ (b) $Q^{-n} = \frac{1}{Q^n}$

(c) $Q^{\frac{1}{n}}$ is the positive solution to $x^n = Q$.

(d) $Q^{\frac{m}{n}} = \left(Q^m\right)^{\frac{1}{n}} = \left(Q^{\frac{1}{n}}\right)^m$

47. $(6.789)^{5-5} = 1$

48. $\left(y^{\frac{1}{7}}\right)^7 = y$

49. $(25)^{-\frac{1}{2}} = \frac{1}{5}$

50. $\left(\frac{1}{y}\right)^{-\frac{3}{4}} = \left(y^{-3}\right)^{\frac{1}{4}}$

51. For what integer values of n does the equation $x^n = 11$ have exactly one real solution? How many solutions does it have for other integer values of n?

52. Explain why rational exponents are not defined for negative bases, using as examples $(-8)^{\frac{1}{3}}$ and $(-8)^{\frac{2}{6}}$.

53. The positive square root of a positive number equals the __?__ power of that number.

54. The positive cube root of a positive number equals the __?__ power of that number.

USES DEAL WITH APPLICATIONS OF MATHEMATICS IN REAL SITUATIONS.

Objective F: *Solve real-world problems which can be modeled by expressions with nth powers or nth roots.* *(Lessons 7-1, 7-2, 7-3, 7-4, 7-8)*

55. On part I of the test, there are 10 multiple-choice questions, each with c choices. On part II of the test are 5 multiple-choice questions, each with d choices. Assuming a person answers each question, how many answer sheets are possible?

56. The Pentagon has a floor space of about $6.2 \cdot 10^5$ square meters. This area is what percent of $1.6 \cdot 10^6$ square meters, which is the area of Monaco?

57. The power P of a radio signal varies inversely as the square of the distance d from the transmitter. Write a formula for P as a function of d using
   a. a positive exponent.
   b. a negative exponent.

58. A spherical balloon has a volume of 400 in^3. A second spherical balloon has a radius half as long.
   a. What is the radius of the second balloon?
   b. What is the volume of the second balloon?

59. A spherical raindrop has radius r millimeters. Through evaporation, the radius decreases by .05 millimeters. If the volume of the condensed drop is 7.2 mm^3, what was the original radius of the drop?

In 60 and 61, use this information about similar figures: If A_1 and A_2 are the surface areas of two similar figures and V_1 and V_2 are their volumes, then

$$\frac{A_1}{A_2} = \left(\frac{V_1}{V_2}\right)^{\frac{2}{3}}.$$

60. Two similar figurines have volumes 20 cm^3 and 25 cm^3. What is the ratio of the amounts of paint (surface area) they need?

61. Solve the formula for $\frac{V_1}{V_2}$.

62. The average pulse rate P for persons t cm tall is approximated by the formula $P = 940t^{-\frac{1}{2}}$
   a. Write this formula without a negative exponent.
   b. Find the average pulse rate for people 160 cm tall.

Objective G: *Apply the compound interest formula.* *(Lesson 7-4)*

63. Sue puts $150 in a savings account which pays 5.75% interest, compounded annually. How much money will be in the account if the $150 is left untouched for 6 years?

64. Investment A offers an annual interest rate of 6%, compounded daily. Investment B offers an annual interest rate of 4%, compounded daily. Leo is considering investing $200 in one of these accounts. Which will yield a higher amount: investment A for 3 years, or investment B for 4 years?

In 65 and 66, Caryn now has $6000 in an account earning interest at a rate of 5%, compounded quarterly.

65. Assuming she made no deposits or withdrawals in the past four years, how much money did she have four years ago?

66. How much interest will she have earned in the first two years?

67. What is the effective annual yield in an account paying 4.5% interest, compounded monthly?

68. A bond paying 6.8% interest compounded quarterly for 10 years has matured, giving the investor $10,000. How much did the investor pay for the bond 10 years ago?

Objective H: *Solve real-world problems involving geometric sequences.* *(Lessons 7-5, 7-6)*

69. The height reached by a bouncing ball on successive bounces generates a geometric sequence. Suppose a ball reaches heights in cm of 120, 96, and 76.8 on its first three bounces. How high will the ball reach on
  a. the next bounce?
  b. the 10th bounce?
  c. the nth bounce?

70. A copying machine is set to reduce linear dimensions to 95%.
  a. If an 8 in. by 10 in. original is reduced, what will be its dimensions?
  b. If each time a copy is made the resulting image is used as the preimage for the next copy, what will be the dimensions of the 5th image?

71. A vacuum pump removes 10% of the air from a chamber with each stroke.
  a. Find a formula for P_n, the percent of air that remains in the chamber after the nth stroke.
  b. How many strokes must be taken to remove 75% of the air in the chamber?

72. A ball on a pendulum moves 50 cm on its first swing. On each succeeding swing back or forth it moves 90% of the distance of the previous swing. Write the first four terms of the sequence of swing lengths.

50 cm

In 73 and 74, note A on a piano keyboard has been tuned to a frequency of 440 hertz.

73. What is the frequency of D, 5 notes above this A?

74. What is the frequency of G, 10 notes above this A?

REPRESENTATIONS DEAL WITH PICTURES, GRAPHS, OR OBJECTS THAT ILLUSTRATE CONCEPTS.

Objective I: *Graph nth power functions.*
(Lesson 7-1)

In 75 and 76, a function is given. **a.** Graph the function. **b.** Identify its domain and its range. **c.** Describe any symmetries of the graph.

75. $y = x^3$ **76.** $f(x) = x^6$

77. Use a graph to explain why the equation $x^8 = -10$ has no real solution.

78. A graph of $y = x^n$, where n is an integer, is shown at the right. What can you conclude about the value of n?

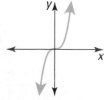

INVERSES AND RADICALS

If a real number is cubed, and then its real cube root is taken, the result of the two operations is the original number. For instance, begin with 2, cube it to get 8, and then take the real cube root of 8. The result is 2. The two processes, cubing and taking the cube root, are called *inverses* because they undo each other.

A table shows this arithmetically.

original number	cube	cube root of cube
-4	-64	-4
-3	-27	-3
-2	-8	-2
-1	-1	-1
0	0	0
1	1	1
2	8	2
3	27	3
4	64	4

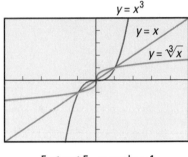

$-5 \leqslant x \leqslant 5, \quad x\text{-scale} = 1$
$-5 \leqslant y \leqslant 5, \quad y\text{-scale} = 1$

The three functions related to these processes are the identity function $y = x$, the cubing function $y = x^3$, and the cube root function $y = \sqrt[3]{x}$. You can see that the three graphs are also related. The graph of the cube root function is the reflection image of the graph of the cubing function over the line $y = x$.

In this chapter, you will study inverses of functions. We begin with the general idea of following one function by another. Then we study the properties and graphs of inverse functions. And finally, you will see how the work you did in Chapter 7 relates to the inverses of the power functions and to the *radical notation* $\sqrt[n]{}$. This notation is very common in the study of nth roots.

Composition of Functions

Price is a function of negotiations. *The price of a new or used car is often negotiated. The sticker price for a new car is the manufacturer's suggested retail price, but it is not necessarily what the customer actually pays.*

Suppose a car dealer offers a $1000 rebate and a 15% discount off the price of a new car. If the sticker price of the car is $13,000, how much will you pay?

If you are given the rebate first and then the discount, the selling price in dollars is

$$.85(13,000 - 1000) = 10,200.$$

(Recall that the price after a 15% discount is 85% of what it was before.) However, if you are given the discount first and then the rebate, the selling price in dollars is

$$.85(13,000) - 1000 = 10,050.$$

For a $13,000 car, calculating the 15% discount before the $1000 rebate results in a lower selling price.

Activity

Consider the situation above. Pick some other sticker price over $10,000. How much will you pay if
a. the rebate is given first, and then the discount?
b. the discount is given first, and then the rebate?

The Composite of Two Functions

Will calculating the discount before the rebate always result in a lower price? To analyze this situation for any sticker price, we use algebra. Let x be the sticker price, r the "rebate" function, and d the "discount" function. If we take the rebate first and then the discount, the final price is given by the expression $d(r(x))$, read "d of r of x".

Example 1

Suppose $r(x) = x - 1000$ and $d(x) = .85x$.
a. Find a formula for $d(r(x))$.
b. Check your answer to part a by letting $x = 13,000$.

Solution

a. $d(r(x)) = d(x - 1000)$ Apply the formula for r.

$ = .85(x - 1000)$ Apply the formula for d.

$ = .85x - 850$ Distributive Property

b. At the beginning of this lesson, $x = 13,000$, and when the rebate was taken before the discount, the selling price of the car was \$10,200. Using the formula in part a, $d(r(13,000)) = .85 \cdot 13,000 - 850 = 11,050 - 850 = 10,200$. It checks.

Example 1 indicates that when the rebate is given before the discount, you can find the selling price by multiplying the sticker price x by .85 and then subtracting \$850. In the Questions, you are asked to find a formula for $r(d(x))$, and to explain why it is always less than $d(r(x))$.

Notice how $d(r(x))$ is computed. First r is applied to x, and then d is applied to the result. This is the same idea you saw in Chapter 4 with transformations. Recall that the result of applying one transformation T after another S is called the composite of the two transformations, written $T \circ S$. Likewise, the function that maps x onto $d(r(x))$ is called the *composite* of the two functions r and d, and is written $d \circ r$.

The composite of two functions is a function. We can describe any function if we know its domain and a rule for obtaining its values. Thus we define the composite of two functions by indicating its rule and domain.

Definition

The **composite** $g \circ f$ of two functions f and g is the function that maps x onto $g(f(x))$, and whose domain is the set of all values in the domain of f for which $f(x)$ is in the domain of g.

Ways of Writing and Reading a Composite

The ways of writing a composite are the same as those used for transformations. Consider the composite $g \circ f$. We can describe the rule for a composite in two ways.

In mapping notation: $g \circ f: x \rightarrow g(f(x))$
In $f(x)$ notation: $g \circ f(x) = g(f(x))$

Either one of these can be read in any of the following ways:
"The composite f then g maps x onto g of f of x."
"The value of x under the composite f then g equals g of f of x."
"g composed with f of x equals g of f of x."

Is Function Composition Commutative?

The operation signified by the small circle ∘ is called **function composition,** or just **composition.** Example 2 shows that composition of functions is not necessarily commutative.

Example 2

Let $g(x) = |x|$ and $h(x) = -8x$. Evaluate
a. $g(h(-3))$.
b. $h(g(-3))$.

Solution

a.
$$
\begin{aligned}
g(h(-3)) &= g(-8 \cdot -3) \quad &&\text{Apply } h.\\
&= g(24) &&\text{Simplify.}\\
&= |24| &&\text{Apply } g.\\
&= 24 &&\text{Simplify.}
\end{aligned}
$$

b.
$$
\begin{aligned}
h(g(-3)) &= h(|-3|) \quad &&\text{Apply } g.\\
&= h(3) &&\text{Simplify.}\\
&= -8 \cdot 3 &&\text{Apply } h.\\
&= -24 &&\text{Simplify.}
\end{aligned}
$$

Notice in Example 2 that $g(h(-3)) \neq h(g(-3))$. This one example is enough to show that *composition of functions is not commutative.* Here is another example.

Example 3

Let $p(x) = x + 5$ and $s(x) = x^2$. Calculate $p \circ s(n)$ and $s \circ p(n)$.

Solution

$p \circ s$ means first square n, and then add 5 to the result.

$$p \circ s(n) = p(s(n)) = p(n^2) = n^2 + 5$$

To evaluate $s \circ p(n)$, first add 5 to n, and then square the result.

$$s \circ p(n) = s(p(n)) = s(n + 5) = (n + 5)^2 = n^2 + 10n + 25$$

Finding the Domain of a Composite of Functions

The domain for a composite function is the largest set for which the composite can be performed. That is, the domain includes exactly those values of x that allow the first function (f) to be performed and that give rise to values that are in the domain of the second function (g).

Example 4

Let x be a real number. Let r be the reciprocal function, $r(x) = \frac{1}{x}$, and let f be given by $f(x) = x^2 - 4$. What is the domain of $r \circ f$?

Solution 1

The domain of f is the set of all real numbers. However,

$$r \circ f(x) = r(f(x)) = r(x^2 - 4) = \frac{1}{x^2 - 4}.$$

Clearly x cannot be 2 or -2. So, the domain of $r \circ f$ is the set of real numbers other than 2 or -2.

Solution 2

There are no restrictions on the domain of f, but 0 is not in the domain of r. Thus, $f(x)$ cannot be 0.
Since $f(x) \neq 0$, $x^2 - 4 \neq 0$; thus $x^2 \neq 4$. So x cannot equal 2 or -2. The domain of $r \circ f$ is the set of real numbers other than 2 or -2.

QUESTIONS

Covering the Reading

In 1–4, refer to the rebate and discount functions in this lesson.

1. **a.** If the sticker price is \$18,500 and the rebate is given first, what is the selling price of the car?
   b. If the sticker price is \$18,500 and the discount is given first, what is the selling price of the car?

2. What answer did you get for the Activity at the start of the lesson?

3. **a.** Evaluate $r(d(20{,}000))$.
   b. Evaluate $r(d(x))$.
   c. Explain why $r(d(x))$ is less than $d(r(x))$ for any value of x.

4. Suppose the car dealer changes the rebate to \$2000 and the discount to 10%. Which results in a lower selling price—giving the rebate first or giving the discount first? Justify your answer.

In 5–7, let $f(x) = 3x^2$ and $g(x) = 4 - 5x$. Evaluate each expression.

5. **a.** $f(g(10))$ **b.** $f \circ g(10)$

6. **a.** $f \circ g(0)$ **b.** $f(g(0))$

7. **a.** $f \circ f(5)$ **b.** $g(g(5))$

8. Give an example other than the one given in the lesson that shows that composition of functions is not commutative.

In 9–11, let $f(n) = n^2 + n + 9$ and $g(n) = 7n$.

9. Evaluate each expression.
   a. $f(g(-5))$ **b.** $g(f(-5))$ **c.** $f(g(2))$ **d.** $g(f(2))$

10. Find an expression for each.
   a. $g(f(n))$ **b.** $f(g(n))$

11. Check Example 3.

12. The domain of $f \circ g$ consists of the set of values in the __?__ of __?__ for which __?__ is in the domain of __?__.

In 13 and 14, let $r(x) = \frac{1}{x}$ and $n(x) = x^2 - 9$.

13. Find the domain of $r \circ n$.

14. Find the domain of $n \circ r$.

Applying the Mathematics

15. Let $f(x) = \sqrt{x}$ and $g(x) = x^2$, where the domains of f and g are the set of real numbers.
   a. Calculate $f(g(-8))$.   **b.** Calculate $g(f(-8))$.

16. Let $c(t) = t^3$ and $m(t) = 3t$.
   a. Find a value of t for which $c \circ m(t) > m \circ c(t)$.
   b. Find a value of t for which $c \circ m(t) = m \circ c(t)$.
   c. Find a value of t for which $c \circ m(t) < m \circ c(t)$.

17. Consider $r(x) = \frac{1}{x}$.
   a. Simplify $r(r(x))$.   **b.** When is $r(r(x))$ undefined?

18. Composite functions can be used to describe relationships between functions of variation. Suppose w varies inversely as the square of z and z is proportional to the cube of x.
   a. Give an equation for w in terms of z.
   b. Give an equation for z in terms of x.
   c. Give an equation for w in terms of x.
   d. Use words to describe the function of part **c.**

Shown is a young barracuda. The barracuda, a swift, fearless, and destructive fish, is often called the "tiger of the sea." Barracudas live in the shallow waters of the Atlantic Ocean and can grow as long as 6 feet.

19. In the food chain, barracuda feed on sea bass and sea bass feed on shrimp. Suppose that the size of the barracuda population is estimated by the function $r(x) = 1000 + \sqrt{2x}$, where x is the size of the sea bass population. Also suppose that the size of the sea bass population is estimated by the function $s(x) = 2500 + \sqrt{x}$, where x is the size of the shrimp population.
   a. Find an equation of the composite which describes the size of the barracuda population in terms of the size of the shrimp population.
   b. About how many barracuda are there when the size of the shrimp population is 4,000,000?

3, 2, 1, liftoff! *Model rockets, like the two shown, fly the same way as giant space rockets. Model rockets like these weigh less than 3.5 pounds and measure from 8 to 24 inches long. They can rise as high as 2000 feet in a few seconds, traveling as fast as 300 mph.*

Review

20. The equation $h = -25(t - 2)^2 + 100$ gives the height h (in feet) of a model rocket t seconds after it is launched. What is the maximum height the rocket attains? *(Lesson 6-4)*

21. Find the inverse of the matrix $\begin{bmatrix} 0 & -2 \\ 3 & a \end{bmatrix}$. *(Lesson 5-5)*

22. Consider the transformation $T: (x, y) \rightarrow (y, x)$.
   a. Let $A = (0, -4)$, $B = (5, -1)$, and $C = (5, 2)$. Draw the image of $\triangle ABC$ under T.
   b. What transformation does T represent? *(Lesson 4-6)*

23. *True or false.* The line with equation $x = 7$ is the graph of a function. *(Lesson 3-4)*

24. The graph below shows the height of a flag on a 12-meter pole as a function of time. *(Lessons 1-2, 1-4)*
   a. Describe what is happening to the flag.
   b. Why are there some horizontal segments on the graph?
   c. What is the domain of this function?
   d. What is its range?

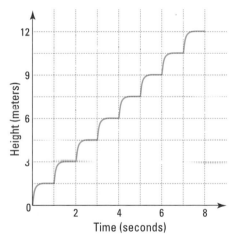

25. a. What number is the additive inverse of $\frac{4}{5}$?
   b. What number is the multiplicative inverse of $\frac{4}{5}$? *(Previous course)*

Exploration

26. Let $m(x)$ = the mother of x, let $b(x)$ = the brother of x, and let $h(x)$ = the husband of x. Then $m \circ h(x)$ is the mother-in-law of x. What relationship is defined by each of the following?
   a. $h \circ m(x)$ **b.** $b \circ m(x)$

27. Some graphics calculators allow function rules to be stated and then recalled for the purpose of evaluating composites. If you have such a calculator, learn how this works. Give the key sequence for doing Question **9a**.

Keeping current with currency. *The value of the Canadian dollar against the U.S. dollar changes daily—as is the case with all foreign currency. In March 1995, one Canadian dollar was worth $.705 U.S.*

In Canada, the unit of currency is the *Canadian dollar*. This dollar does not have the same value as one U.S. dollar; in recent times it has not been worth as much. This means that when you go to Canada, you will see prices (in Canadian dollars) that seem higher than prices in the United States. Here is a table of equivalent prices in October, 1994.

price in U.S. dollars x	price in Canadian dollars $y = \frac{4}{3}x$
$ 0	$ 0
15	20
30	40
45	60
60	80
75	100

In this table, the left column has values of the domain variable x and the right column has values of the range variable y. The function mapping the U.S. dollar price onto the Canadian dollar price can be described by the equation $y = \frac{4}{3}x$.

But if you are in Canada, you might want to relate the Canadian price to the U.S. price. You would switch the domain and range of the function. The Canadian price becomes x and the United States price becomes y. Now the function mapping the left column onto the right column can be described by the equation $x = \frac{4}{3}y$. Solving this equation for y, $y = \frac{3}{4}x$.

price in Canadian dollars	price in U.S. dollars
x	$y = \frac{3}{4}x$
$ 0	$ 0
20	15
40	30
60	45
80	60
100	75

The relations with equations $y = \frac{4}{3}x$ and $y = \frac{3}{4}x$ are obviously related. The ordered pairs of each one are found by switching the values of x and y in the other. They are called *inverse relations.*

Definition

The **inverse of a relation** is the relation obtained by reversing the order of the coordinates of each ordered pair in the relation.

Example 1

Let $f = \{(1, 4), (2, 8), (3, 8), (0, 0), (-1, -4)\}$. Find the inverse of f.

Solution

Switch the coordinates of each ordered pair. Let the inverse be called g. Then $g = \{(4, 1), (8, 2), (8, 3), (0, 0), (-4, -1)\}$.

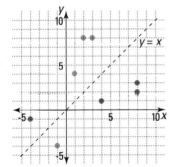

The blue dots at the left show the graph of the function f from Example 1. The orange dots show the graph of its inverse.

Recall that the points (x, y) and (y, x) are reflection images of each other over the *identity function,* that is, the line with equation $y = x$. That is, the reflection over the identity function switches the coordinates of the ordered pairs. So the graphs of any relation and its inverse are reflection images of each other over the line $y = x$. This is easily seen in the graph at the left.

The Domain and the Range of a Relation and Its Inverse

Recall that the domain of a relation is the set of possible values for the first coordinate and the range is the set of possible values for the second coordinate. Because the inverse is found by switching the coordinates, the domain and range of the inverse of a relation are found by switching the domain and range of the relation. For instance, in Example 1 above,

$$\text{domain of } f = \text{range of } g = \{1, 2, 3, 0, -1\}$$
$$\text{range of } f = \text{domain of } g = \{4, 8, 0, -4\}.$$

The following theorem summarizes the ideas you have read so far in this lesson.

> **Inverse Relation Theorem**
> Suppose f is a relation and g is the inverse of f. Then:
> (1) A rule for g can be found by switching x and y.
> (2) The graph of g is the reflection image of the graph of f over the line $y = x$.
> (3) The domain of g is the range of f, and the range of g is the domain of f.

Caution: the word *inverse*, when used in the term *inverse of a relation*, is different and unrelated to its use in the phrase *inverse variation*.

Determining Whether the Inverse of a Function Is a Function

The inverse of a relation is always a relation. But the inverse of a function is not always a function. In Example 1, f is a function but its inverse g is not a function because it contains the two pairs $(8, 2)$ and $(8, 3)$.

Example 2

Consider the function with domain the set of all real numbers and equation $y = x^2$.
a. What is an equation for the inverse?
b. Graph the function and its inverse on the same coordinate axes.
c. Is the inverse a function? Why or why not?

Solution

a. To find an equation for the inverse switch the values of x and y. Given the function with equation $y = x^2$, its inverse has equation $x = y^2$.
b. The graphs of $y = x^2$ and $x = y^2$ are shown below. Notice again that the inverse is the reflection image of $y = x^2$ over the line with equation $y = x$.

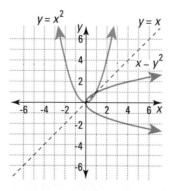

c. The inverse is not a function because both $(4, 2)$ and $(4, -2)$ are on the graph of $x = y^2$. The graph fails the Vertical-Line Test for a function.

From Examples 1 and 2, notice that you can tell by looking at the graph of the *original* function whether or not its inverse represents a function.

> **Theorem (Horizontal-Line Test for Inverses)**
> The inverse of a function *f* is itself a function if and only if no horizontal line intersects the graph of *f* in more than one point.

Example 3

Tell whether the inverse of each function is a function. Explain your reasoning.

a.

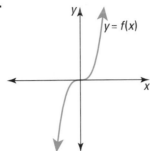

b.

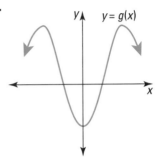

Solution

Apply the Horizontal-Line Test to each graph.

a. The graph passes the Horizontal-Line Test, so the inverse of f is a function.

b. The x-axis intersects the graph in more than one point. So the graph does not pass the Horizontal-Line Test. Thus, the inverse of g is not a function.

QUESTIONS

Covering the Reading

1. How can the inverse of a relation be found from the coordinates of the relation?

In 2 and 3, let $f = \{(4, 8), (2, 4), (3, 6), (-1, -2), (-5, -10)\}$.

2. **a.** Find the inverse of *f*.
   b. Graph the relation *f* and its inverse on the same set of axes.
   c. How are the two graphs related?
   d. Write an equation for the relation *f*.
   e. Write an equation for the inverse of *f*.

3. Give the elements of each set.
   a. the domain of *f*
   b. the range of *f*
   c. the domain of the inverse of *f*
   d. the range of the inverse of *f*.

4. Explain how the graphs of any relation f and its inverse are related.

In 5 and 6, give an equation for the inverse of the relation.

5. $y = 3x$

6. $y = 9x^2 + 12x - 6$

7. Refer to Example 2. Find two points other than those mentioned in the Example that show the inverse is not a function.

8. To tell if the inverse of a function is a function:
   a. apply the __?__ Test if you have graphed the function.
   b. apply the __?__ Test if you have graphed the inverse.

In 9–11, a graph of a function is given. Is the inverse of the graphed function a function?

9.

10.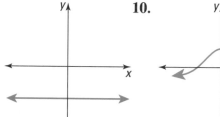

11.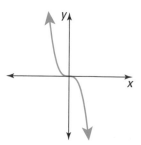

Applying the Mathematics

12. Consider the function f with equation $f(x) = 2$.
   a. Give 5 ordered pairs in f.
   b. Find 5 ordered pairs in the inverse of f.
   c. Plot graphs of f and its inverse on the same set of axes.
   d. What transformation maps the graph of f onto the graph of its inverse?

13. a. Draw the graph of $y = 4x + 9$.
   b. Find an equation for its inverse.
   c. Graph the inverse on the same set of axes.
   d. Is the inverse a function?
   e. How are the slopes of the function and its inverse related?

14. a. Graph the inverse of the absolute-value function $y = |x|$.
   b. Is the inverse a function? Explain why or why not.
   c. What rule describes the inverse of the absolute-value function?

15. In October 1994, a U.S. dollar was worth about 3.4 Mexican pesos. That means an item costing 1 U.S. dollar would cost about 3.4 Mexican pesos. Let U = cost of an item in U.S. dollars and M = cost of an item in Mexican pesos.
   a. Find an equation for U in terms of M.
   b. Find an equation for M in terms of U.

16. Let x = the cost of an item. Then $t(x) = 1.05x$ is its cost after a sales tax of 5%, and $d(x) = .8x$ is its cost after a discount of 20%.
  a. Evaluate $t(d(50))$.
  b. Explain in words what $t(d(x))$ represents.
  c. Evaluate $d(t(50))$.
  d. Explain what $d(t(x))$ represents.
  e. An item is on sale for 20% off, but you must pay a 5% sales tax. Does it matter whether the discount or tax is calculated first? Why or why not?   *(Lesson 8-1)*

17. Let $f(x) = 3x + 4$, and $g(x) = \frac{1}{3}(x - 4)$.
  a. Find $g \circ f(15)$
  b. Find $g \circ f(x)$
  c. What is another name for the function $g \circ f$?   *(Lesson 8-1)*

18. Given $f(x) = x^2 + 1$ and $g(x) = x^2 - 1$, find $f \circ g(x)$.   *(Lessons 6-1, 8-1)*

19. *True or false.*   For all real numbers $x, |x| = x$. Justify your answer. *(Lesson 6-2)*

20. How can you tell if two 2×2 matrices A and B are inverses?   *(Lesson 5-5)*

21. In the figure at the left lines ℓ and m are parallel. Find the measures of the numbered angles.   *(Previous course)*

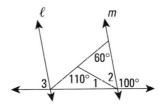

22. Another way to obtain a rule for the inverse of a function is to do the opposite operations of the rule for the function in the reverse order.
  a. In the morning, Beth put on her socks and then put on her shoes. In the evening she did the inverse. What did she do?
  b. Pick a routine that you do every day and see if you also do its inverse. Try these or one of your own.
  Is the way you get home the inverse of the way you go to school?
  Is your routine for getting ready to go to sleep the inverse of getting up in the morning?
  Is getting into a parking spot by the curb the inverse of getting out of the same spot?
  c. Pick an equation with three operations such as $y = \frac{2x - 3}{7}$. List in the correct order the steps you use to evaluate y given a value for x. Construct a rule for the inverse by listing the opposite operations in the reverse order.

"I said that the opposite of daytime is prime time, and she marked it wrong!"

8-3

Properties of Inverse Functions

The function f with equation $f(x) = 4x + 5$ is graphed at the right. This graph passes the Horizontal-Line Test, so its inverse is a function. Let's call its inverse g.

The graph of g is easy to find. Just reflect the graph of f over the line $y = x$.

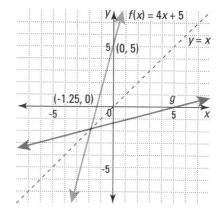

Formulas for Inverses Using $f(x)$ Notation

There are many ways to obtain a formula for $g(x)$. One way is to begin with an equation for f, but written using x and y: $\quad y = 4x + 5$.
Now switch x and y to get an equation for g: $\quad x = 4y + 5$.
In this equation, $y = g(x)$. Solve for y: $\quad x - 5 = 4y$
$$\tfrac{1}{4}(x - 5) = y$$

Now substitute $g(x)$ for y. This gives the form of the equation we want.

$$\tfrac{1}{4}(x - 5) = g(x)$$

We can check this using the graph. The formula for $g(x)$ tells us that g is a linear function with slope $\tfrac{1}{4}$ and y-intercept $-\tfrac{5}{4}$.

Composites of Inverse Functions

Another way to check that g is the inverse of f is to take their composites in both directions.

$$f \circ g(x) = f\left(\tfrac{1}{4}(x - 5)\right) = 4 \cdot \tfrac{1}{4}\left(x - 5\right) + 5 = x$$
$$g \circ f(x) = g(4x + 5) = \tfrac{1}{4} \cdot (4x + 5 - 5) = x$$

This result holds for any functions f and g that are inverses. Its converse is also true.

> **Inverse Functions Theorem**
> Two functions f and g are inverse functions, if and only if:
>   (1) For all x in the domain of f, $g \circ f(x) = x$.
> and (2) For all x in the domain of g, $f \circ g(x) = x$.

Proof
The proof has two parts. ▶

The "only if" part: Suppose f and g are inverse functions. Now let (a, b) be any ordered pair in the function f. Then $f(a) = b$. By the definition of inverse, the ordered pairs in g are the reverse of those in f, so (b, a) is an ordered pair in the function g. Thus $g(b) = a$. Now take the composites.

For any number a in the domain of f, $g \circ f(a) = g(b) = a$.
For any number b in the domain of g, $f \circ g(b) = f(a) = b$.

The "if" part: Suppose (1) and (2) in the statement of the theorem. Again let (a, b) be any point in the function f. Then $f(a) = b$ and so $g \circ f(a) = g(b)$. But using (2), $g \circ f(a) = a$, so $a = g(b)$. This means that (b, a) is in the function g. So g contains all the points obtained by reversing the coordinates of points in f.

Now suppose that (c, d) is any point in the function g. Then $g(c) = d$ and so $f \circ g(c) = f(d)$. But using (1), $f \circ g(c) = c$; so $c = f(d)$. This means that (d, c) is in the function f. So f contains all points obtained by reversing the coordinates of points in g. Thus f and g are inverse functions.

Notation for Inverse Functions

Recall that an *identity function* is a function that maps each object onto itself. Another way of stating the Inverse Functions Theorem is that the composite of two inverse functions is the identity function I with equation $I(x) = x$. When an operation on two elements of a set yields an identity element, then we call the elements *inverses*. This is the reason that we call g the "inverse" of f.

In Chapter 5, the inverse of a matrix M was designated by the symbol M^{-1}. When a function f has an inverse, we designate it by the symbol f^{-1}. For instance, when $f(x) = 4x + 5$, then $f^{-1}(x) = \frac{1}{4}(x - 5)$. The computations on page 490 show that for all x, $f \circ f^{-1}(x) = x$ and $f^{-1} \circ f(x) = x$.

Example 1

Let $g: x \rightarrow x + 500$. Find a rule for g^{-1}.

Solution

Use a process like that shown for the function f at the beginning of the lesson.

From the given information, $\quad\quad\quad g(x) = x + 500.$
Let $y = g(x)$. Then $\quad\quad\quad\quad\quad\quad y = x + 500.$
An equation for the inverse is found by switching x and y.

So an equation for the inverse g^{-1} is $x = y + 500.$
Solve this equation for y. Then $\quad\quad\quad y = x - 500.$
So $\quad\quad\quad\quad\quad\quad\quad\quad\quad\quad\quad g^{-1}(x) = x - 500.$
In mapping notation, $g^{-1}: x \rightarrow x - 500.$

The function g in Example 1 is the "adding 500" function. Its inverse is the "subtracting 500" function. Because the functions "adding h" and "subtracting h" are inverse functions, addition and subtraction are sometimes called *inverse operations*.

The inverse operations of most importance to us in this chapter are "taking the nth power" and "taking the nth root." These are the functions with equations $y = x^n$ and $y = x^{\frac{1}{n}}$. For all nonnegative numbers x, it is reasonably easy to show that they are inverse functions.

Power Function Inverse Theorem

If $f(x) = x^n$ and $g(x) = x^{\frac{1}{n}}$ and the domains of f and g are the set of nonnegative real numbers, then f and g are inverse functions.

Example 2

Prove the Power Function Inverse Theorem.

Proof

First we need to show that for all x in the domain of g, $f \circ g(x) = x$.

Just substitute. $\qquad f \circ g(x) = f\left(x^{\frac{1}{n}}\right) \qquad$ Apply g.

Since x is a nonnegative number, $x^{\frac{1}{n}}$ is always defined and we can apply f.

$$= \left(x^{\frac{1}{n}}\right)^n \qquad \text{Apply } f.$$
$$= x \qquad\qquad \text{Power of a Power Postulate}$$

Now we need to show that for all x in the domain of f, $g \circ f(x) = x$. This is left for you to do as Question 10.

An instance of this theorem for $n = 4$ is shown below at the left. That is, $f(x) = x^4$ and $f^{-1}(x) = x^{\frac{1}{4}}$ are graphed. You can see that the graphs are reflection images of each other over the line $y = x$. But notice that the domain of these functions is the set of nonnegative real numbers. The function $h(x) = x^4$ with domain the set of *all* real numbers is graphed at the right. This function does not pass the Horizontal-Line Test, so the inverse of this function is not a function.

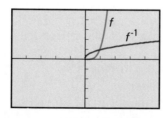

-5 ≤ x ≤ 5, x-scale = 1
-5 ≤ y ≤ 5, y-scale = 1

$f(x) = x^4$
Domain of f = set of nonnegative real numbers
$f^{-1}(x) = x^{\frac{1}{4}}$
Domain of f^{-1} = set of nonnegative real numbers

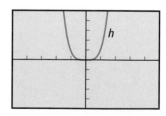

-5 ≤ x ≤ 5, x-scale = 1
-5 ≤ y ≤ 5, y-scale = 1

$h(x) = x^4$
Domain of h = set of real numbers
The inverse of h is not a function.

QUESTIONS

Covering the Reading

1. The function $h(x) = 6x - 5$ has an inverse which is a function.
   a. Find a formula for the inverse and write the rule in $f(x)$ notation.
   b. Check your answer to part **a** by finding $h \circ h^{-1}(x)$ and $h^{-1} \circ h(x)$.

2. Let $f(x) = \frac{1}{3}x - 4$ and $g(x) = 3x + 4$. Are f and g inverses of each other? Justify your answer.

3. Suppose f is a function.
   a. What does the symbol f^{-1} represent?
   b. How is the symbol f^{-1} read?

4. Consider the "adding 12" function. Call it A.
   a. What is a formula for $A(x)$?
   b. Give a formula for $A^{-1}(x)$.
   c. What is $A^{-1} \circ A(48)$?

In 5 and 6, an equation for a function f is given. The domain of f is the set of real numbers. **a.** Write an equation for the inverse. **b.** Is the inverse a function? If it is, find a rule for $f^{-1}(x)$.

5. $f(x) = 17x$ 6. $f(x) = 17$

7. For any function f that has an inverse, if x is in the range of f,
   $f \circ f^{-1}(x) = \underline{\quad?\quad}$.

8. If $f(x) = x^9$ with domain the set of nonnegative real numbers, give a formula for $f^{-1}(x)$.

9. **a.** Explain why the inverse of $g(x) = 5x^6$ is not a function when the domain of g is the set of all real numbers.
   b. Give a domain for g on which the inverse of $g(x) = 5x^6$ is a function.
   c. Find a formula for the inverse of g.

10. Complete the proof of the theorem in Example 2.

Applying the Mathematics

11. Let $g(x) = x^{\frac{2}{3}}$ and $h(x) = x^{\frac{3}{2}}$ with $x \geq 0$.
    a. Use an automatic grapher to plot g and h on the same set of axes.
    b. Describe how the graphs are related.
    c. Find $h(g(x))$ and $g(h(x))$.
    d. Are g and h inverses? Why or why not?

12. Let $m(x) = 6x$. This function could be named "multiplying by 6."
    a. Find an equation for m^{-1}.
    b. Give an appropriate name to m^{-1}.

13. If f is any function that has an inverse, what is $(f^{-1})^{-1}$? Explain your answer in your own words.

14. *Multiple choice.* Consider the absolute-value function $y = |x|$ graphed at the left. Which of the following domains gives a function whose inverse is also a function?
(a) $\{x: -2 \le x \le 5\}$ (b) $\{x: x \ge 1\}$ (c) $\{x: x \ge -4.5\}$

Review

15. Alexis makes a salary of $25,000 per year plus a commission of 40% of her ticket sales to a health and nutrition seminar. An equation for her total income $f(x)$, in terms of ticket sales x, is thus $f(x) = 25,000 + .4x$.
   a. Find a rule for the inverse of f.
   b. What does the inverse represent?   *(Lesson 8-2)*

In 16–19, suppose $f: x \to 3x - 4$ and $g: x \to x^2$.   *(Lessons 8-1, 8-2)*

16. $f \circ g: 8 \to$ _?_   **17.** $f \circ g: x \to$ _?_

18. $f(g(x)) =$ _?_   **19.** $g(f(x)) =$ _?_

20. Ricardo and Lucia and their brothers perform an acrobatic act by positioning themselves in a plane as shown below. When Ricardo jumps from the stool to the teeter-totter, Lucia is propelled up into the air. Lucia's height at time t is given by the equation $h = -\frac{1}{2}gt^2 + v_0 t + h_0$.

   a. Suppose Lucia leaves the teeter-totter at an initial velocity of 20 feet per second. Write an equation to describe her height after t seconds.
   b. Assuming Lucia has good aim and can go far enough horizontally, will she be able to land atop her two brothers? How can you tell?   *(Lessons 6-4, 6-7)*

21. What is the slope of any line that has the same nonzero number for its x- and y-intercepts?   *(Lessons 3-2, 3-4)*

Exploration

22. a. Let $f(x) = x^3$ and $g(x) = 8x - 40$. Find a formula for $(f \circ g)^{-1}(x)$.
   b. Generalize part **a.**

LESSON
8-4

Radical Notation for nth Roots

The positive nth root of a positive number x can be written as a power of x, namely, as $x^{\frac{1}{n}}$. This notation allows all the properties of powers to be used with these nth roots. But, as you know, there are nth roots that are not represented in this way. The **radical sign** $\sqrt{}$ allows all of the positive nth roots and some other numbers to be represented. The origin of the word "radical" is the Latin word "radix," which means "root."

We begin by considering only positive numbers under the radical sign. Recall that when x is positive, $\sqrt{x}$ stands for the positive square root of x. Since $x^{\frac{1}{2}}$ also stands for this square root, $\sqrt{x} = x^{\frac{1}{2}}$. Similarly the nth root of a positive number can be written in two ways.

Definition
For any nonnegative real number x and any integer $n \geq 2$, $\sqrt[n]{x} = x^{\frac{1}{n}}$.

Thus, when x is positive, $\sqrt[n]{x}$ is the positive nth root of x. When $n = 2$, we do not write $\sqrt[2]{x}$, but use the more familiar symbol $\sqrt{x}$.

Example 1

Evaluate $\sqrt[4]{81}$.

Solution
$\sqrt[4]{81} = 81^{\frac{1}{4}}$, the *positive* number whose 4th power is 81. Since $3^4 = 81$, $81^{\frac{1}{4}} = 3$. So, $\sqrt[4]{81} = 3$.

The symbol $\sqrt[n]{}$ was first used by Albert Girard around 1633. Note that $\sqrt[n]{x}$, like $x^{\frac{1}{n}}$, does not represent all nth roots of x. When x is positive and n is even, x has two real nth roots, but only the *positive* real root is denoted by $\sqrt[n]{x}$. Thus 2, -2, 2i, and -2i are fourth roots of 16, but $\sqrt[4]{16} = 2$ only. The negative fourth root can be denoted by $-16^{\frac{1}{4}}$, $-\sqrt[4]{16}$, or -2.

Activity

Most scientific and graphics calculators have either a key or a menu selection for nth roots. Check your calculator to see what to do to evaluate $\sqrt[4]{81}$.

Example 2

Estimate $\sqrt[5]{4.2}$ to the nearest hundredth.

Solution

Use the key sequence appropriate to your calculator. Our calculator displays **1.332446738** which is about **1.33**.

Check

Use your calculator to see if $(1.332446738)^5 \approx 4.2$. It is.

Radicals for Roots of Powers

Because radicals are powers, all properties of powers listed in Chapter 7 apply to radicals. In particular, because $\sqrt[n]{x} = x^{\frac{1}{n}}$ for $x \geq 0$, the mth powers of these numbers are equal. That is, $(\sqrt[n]{x})^m = \left(x^{\frac{1}{n}}\right)^m$, which equals $x^{\frac{m}{n}}$. If x is replaced by x^m in the definition, of $\sqrt[n]{x}$, the result is $\sqrt[n]{x^m} = (x^m)^{\frac{1}{n}}$, which also equals $x^{\frac{m}{n}}$. Thus, there are two radical expressions equal to $x^{\frac{m}{n}}$.

Root of a Power Theorem
For all positive integers $m > 1$ and $n \geq 2$, and all nonnegative real numbers x, $\sqrt[n]{x^m} = (\sqrt[n]{x})^m = x^{\frac{m}{n}}$.

Example 3

Suppose $x \geq 0$. Simplify $\sqrt[3]{x^{12}}$.

Solution 1

$$\sqrt[3]{x^{12}} = \left(x^{12}\right)^{\frac{1}{3}} \quad \text{definition of cube root}$$
$$= x^4 \quad\quad \text{Power of a Power Postulate}$$

Solution 2

$$\sqrt[3]{x^{12}} = x^{\frac{12}{3}} = x^4 \quad \text{Root of a Power Theorem}$$

Roots of Roots

The sequence $256, 16, 4, 2, \sqrt{2}, \ldots$, in which each number is the square root of the preceding number, can be defined recursively.

$$\begin{cases} s_1 = 256 \\ s_n = \sqrt{s_{n-1}} \text{ for integers } n \geq 2 \end{cases}$$

So, $s_2 = \sqrt{256} = 16$

$\quad s_3 = \sqrt{\sqrt{256}} = \sqrt{16} = 4$

$\quad s_4 = \sqrt{\sqrt{\sqrt{256}}} = \sqrt{\sqrt{16}} = \sqrt{4} = 2$

By rewriting the radicals as rational exponents, we have a general way of dealing with roots of roots.

Example 4

Suppose $x \geq 0$. Rewrite $\sqrt{\sqrt{\sqrt{x}}}$ using rational exponents. Is this expression an nth root of x? Justify your answer.

Solution

$\sqrt{\sqrt{\sqrt{x}}} = \left(\left(x^{\frac{1}{2}} \right)^{\frac{1}{2}} \right)^{\frac{1}{2}} = \left(x^{\frac{1}{2}} \right)^{\frac{1}{4}} = x^{\frac{1}{8}}$. So by definition of nth root, $\sqrt{\sqrt{\sqrt{x}}}$ is the positive 8th root of x.

Check

Pick a positive number for x, say 61. Compare the value your calculator gives for $\sqrt{\sqrt{\sqrt{61}}}$ and for $61^{\frac{1}{8}}$. You should get about 1.672 in both cases.

Notice that for $x \geq 0$, $x^{\frac{1}{8}} = \sqrt[8]{x}$, so that $\sqrt{\sqrt{\sqrt{x}}} = \sqrt[8]{x}$. In radical form, the preferred way to express the roots of roots is with a single radical.

QUESTIONS

Covering the Reading

1. Consider the radical expression $\sqrt[n]{d}$.
   a. For what value(s) of d and n is it defined?
   b. What power of d does it equal?

In 2–4, evaluate without a calculator.

2. $\sqrt[4]{16}$ 3. $\sqrt[3]{216}$ 4. $\sqrt[5]{10^5}$

5. Write your answer to the Activity in the Lesson.

6. a. Write a calculator key sequence to evaluate $\sqrt[4]{38.720}$.
   b. Estimate $\sqrt[4]{38.720}$ to the nearest tenth.

In 7–9, use a calculator to approximate to the nearest hundredth.

7. $\sqrt[3]{10}$ **8.** $\sqrt[4]{4}$ **9.** $\sqrt[5]{314,892}$

10. Who first used the $\sqrt[n]{}$ symbol, and in which century?

11. State the Root of a Power Theorem.

In 12–14, simplify. Assume all variables are positive.

12. $\sqrt[3]{a^{15}}$ **13.** $\sqrt[4]{c^6}$ **14.** $(\sqrt[7]{t})^{14}$

15. Rewrite $\sqrt{\sqrt{x}}$ with rational exponents, given $x \geq 0$.

In 16–18, rewrite using a single radical sign. Assume $t \geq 0$.

16. $\sqrt{\sqrt{\sqrt{256}}}$ **17.** $\sqrt{\sqrt{\sqrt{\sqrt{10^8 t^{16}}}}}$ **18.** $\sqrt{\sqrt{\sqrt{\sqrt{t^{80}}}}}$

Applying the Mathematics

Multiple choice. In 19 and 20, which of (a) to (c) is *not equal* to the others?

19. (a) $3^{\frac{5}{2}}$ (b) $\sqrt[6]{3^5}$ (c) $(\sqrt[6]{3^5})^3$ (d) All are equal.

20. (a) $\sqrt[4]{25}$ (b) $\sqrt{5}$ (c) $5^{0.5}$ (d) All are equal.

21. A cube has volume $V \, \text{cm}^3$.
   a. Express the length of an edge using radical notation.
   b. Express the length of an edge using a rational exponent.

22. Consider the formula $\frac{F}{m} = km^{\frac{-1}{3}}$ given on page 417.
   a. Solve the formula for F, and write your answer using a rational exponent.
   b. Rewrite the formula in part **a** using radical notation.

23. Simplify $\sqrt[3]{\sqrt[3]{512}}$.

In 24 and 25, for $x > 0$, write each expression in simplest radical form using no fractional exponents.

24. $\sqrt{x^{\frac{1}{2}}}$ **25.** $\dfrac{\sqrt{x}}{\sqrt{\sqrt{x}}}$

Review

26. Let $B(x) = x^2$ and $E(x) = x^{-2}$.
   a. Find $B(E(x))$.
   b. Are B and E inverses of each other? How can you tell?
   (Lessons 8-2, 8-3)

27. The time T in seconds for one complete swing of a pendulum of length L in centimeters is given by

$$T = f(L) = 2\pi\sqrt{\frac{L}{980}}.$$

Suppose the pendulum gets a little longer when it gets warmer, so that

$$L = g(C) = 100 + .003C,$$

where C is the Celsius temperature.
a. Find the length of the pendulum at 40°C.
b. Find the time it takes the pendulum to make one complete swing at 40°C.
c. Express T as a function of C.
d. Which of $f(g(C))$ or $g(f(C))$ does the answer in part **c** represent?
(Lessons 1-3, 8-1)

In 28 and 29, write without an exponent. Do not use a calculator.
(Lessons 7-3, 7-8)

28. $4^{-\frac{5}{2}}$

29. $\left(\frac{1}{7}\right)^{-3}$

30. Write the reciprocal of $2 + i$ in $a + bi$ form.   *(Lesson 6-9)*

#3 in car production.
Shown is an autobahn in Frankfort, Germany. Germany is the world's third largest producer of automobiles (behind the U.S. and Japan). In 1993, Germany produced almost 4 million automobiles.

31. If $f(x) = \dfrac{\dfrac{x+1}{x-2}}{\dfrac{x+3}{x-4}}$, what values of x are not in the domain of f?
(Lessons 1-4, 1-6)

32. In October, 1994, for every U.S. dollar you could get 1.5 DM (Deutsch Marks, the currency of Germany). If an automobile cost 60,000 DM, what was its cost in U.S. dollars?   *(Previous course)*

Exploration

33. Use an automatic grapher.
a. Graph $f(x) = \sqrt[4]{x}$, $g(x) = x^{\frac{1}{4}}$, and $h(x) = \sqrt[8]{x^2}$.
b. Explain why two of these graphs are alike and one is different.

Products
with
Radicals

Life on other planets? *Shown are Ferengi from the TV show,* Star Trek: The Next Generation. *The Ferengi are depicted as a cowardly, devious, and untrustworthy species. See Example 5.*

Recall that for all nonnegative numbers a and b,
$$\sqrt{ab} = \sqrt{a}\,\sqrt{b}.$$
This product of square roots is a special case of the Power of a Product Postulate
$$(xy)^m = x^m \cdot y^m,$$
since a square root can be considered as the $\frac{1}{2}$ power of a number. Now, if we let $m = \frac{1}{n}$, then a theorem about the product of nth roots results.

Root of a Product Theorem
For any nonnegative real numbers x and y, and any integer $n \geq 2$,
$(xy)^{\frac{1}{n}} = x^{\frac{1}{n}} \cdot y^{\frac{1}{n}}$ (power form)
$\sqrt[n]{xy} = \sqrt[n]{x} \cdot \sqrt[n]{y}$ (radical form)

Multiplying Radicals
The Root of a Product Theorem can be used to multiply nth roots.

Example 1
Find the product $\sqrt[3]{18} \cdot \sqrt[3]{12}$.

Solution
Use the Root of a Product Theorem.
$$\sqrt[3]{18} \cdot \sqrt[3]{12} = \sqrt[3]{18 \cdot 12}$$
$$= \sqrt[3]{216}$$
$$= 6$$

Example 2

Assume $x \geq 0$. Write $\sqrt[5]{5x} \cdot \sqrt[5]{3x^2}$ as a single radical.

Solution

$$\sqrt[5]{5x} \cdot \sqrt[5]{3x^2} = \sqrt[5]{5x \cdot 3x^2}$$
$$= \sqrt[5]{15x^3}$$

Simplifying Radicals

The Root of a Product Theorem can also be used to rewrite an nth root as a product. For instance, $\sqrt[3]{80}$ can be rewritten in many ways using the Root of a Product Theorem. Here are three ways.

$$(1) \quad \sqrt[3]{80} = \sqrt[3]{2 \cdot 40} \quad = \sqrt[3]{2} \cdot \sqrt[3]{40}$$
$$(2) \quad \sqrt[3]{80} = \sqrt[3]{4 \cdot 20} \quad = \sqrt[3]{4} \cdot \sqrt[3]{20}$$
$$(3) \quad \sqrt[3]{80} = \sqrt[3]{8 \cdot 10} \quad = \sqrt[3]{8} \cdot \sqrt[3]{10}$$

Notice that because 8 is a perfect cube, we can conclude from statement (3) that $\sqrt[3]{80} = 2\sqrt[3]{10}$. Some people call $2\sqrt[3]{10}$ the *simplified form* of $\sqrt[3]{80}$. In the years before calculators were widely available, people used the Root of a Product Theorem to rewrite nth roots of large numbers (such as $\sqrt[3]{80}$) as whole numbers times nth roots of smaller numbers (such as $2\sqrt[3]{10}$). Then they used tables to evaluate the nth roots of these smaller numbers. Nowadays the Root of a Product Theorem is not used much for computation, but it is used to rewrite expressions involving radicals or to recognize alternate forms of answers.

In general to **simplify an *n*th root,** factor the expression under the radical sign into perfect nth powers. Then apply the Root of a Product Theorem.

Example 3

Suppose $n \geq 0$. Simplify $\sqrt[4]{16n^{12}}$.

Solution

To simplify a 4th root, factor the expression under the radical sign into as many 4th powers as possible. When $n > 0$,

$$\sqrt[4]{16n^{12}} = \sqrt[4]{2^4 \cdot n^4 \cdot n^4 \cdot n^4}$$
$$= \sqrt[4]{2^4} \cdot \sqrt[4]{n^4} \cdot \sqrt[4]{n^4} \cdot \sqrt[4]{n^4} \qquad \text{Root of a Product Theorem}$$
$$= 2 \cdot n \cdot n \cdot n \qquad \text{Root of a Power Theorem}$$
$$= 2n^3.$$

Check

Raise the answer to the 4th power $\left(2n^3\right)^4 = 2^4 \cdot \left(n^3\right)^4 = 16n^{12}$. It checks.

Example 4

Suppose $p \geq 0$. Rewrite $\sqrt[3]{875p^7}$.

Solution

Factor the expression inside the root into as many perfect cubes as possible. Then use theorems about roots and powers.

$$\sqrt[3]{875p^7} = \sqrt[3]{125 \cdot 7 \cdot p^6 \cdot p}$$
$$= \sqrt[3]{5^3} \cdot \sqrt[3]{7} \cdot \sqrt[3]{p^6} \cdot \sqrt[3]{p} \qquad \text{Root of a Product Theorem}$$
$$= 5 \cdot \sqrt[3]{7} \cdot p^2 \cdot \sqrt[3]{p} \qquad \text{Root of a Power Theorem}$$
$$= 5p^2\sqrt[3]{7p}$$

Check 1

The inverse of taking the cube root is cubing. To check, cube the answer.

$$(5p^2\sqrt[3]{7p})^3 = 5^3(p^2)^3(\sqrt[3]{7p})^3$$
$$= 125p^6 \cdot 7p$$
$$= 875p^7$$

This is the original expression under the radical sign.

Check 2

Substitute some number for p and evaluate the given and final expressions. We let $p = 2$.

Does $\sqrt[3]{875 \cdot 2^7} = 5 \cdot 2^2\sqrt[3]{7 \cdot 2}$?

Does $\sqrt[3]{112{,}000} = 20\sqrt[3]{14}$?

Yes, both are approximately 48.20.

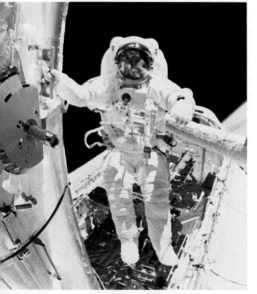

Service call in space.
Shown is F. Story Musgrave holding on to one of the handrails of the Hubble Space Telescope. Musgrave was on the first of five space walks on the HST-servicing mission in December, 1993.

The Geometric Mean

In statistics, it is common to want to describe a data set by a single number such as the mean or the median. These are called *measures of center,* or *measures of central tendency.*

Suppose a data set contains n positive numbers. If you add the numbers in the set and divide the sum by n, you will obtain the *average,* or *arithmetic mean.* If, instead, you multiply the numbers in the set and then take the nth root of the product, you will obtain the **geometric mean** of the numbers. The geometric mean may be used when numbers are quite dispersed, to keep one very large number from affecting the measure of center.

Example 5

In trying to ascertain whether there is life elsewhere in the universe, astronomers look for planets of other stars. To do this, it is helpful to have some idea of how far the planets in our solar system are from our sun. The nearest distances of the 9 planets from the sun (in millions of km) are given in the table at the top of page 503.

▶

This is part of Saturn and its ring system as photographed on August 11, 1981 by Voyager II, 8.6 million miles from Saturn.

Planet	Nearest distance to sun (perihelion)
Mercury	46.0
Venus	107.5
Earth	147.1
Mars	206.8
Jupiter	741.3
Saturn	1349.3
Uranus	2686.5
Neptune	4442.4
Pluto	4436.0

▶ What is the geometric mean of these distances?

Solution

Since there are 9 numbers, the geometric mean is the 9th root of their product. To avoid having to multiply large numbers with many digits, rewrite the numbers in scientific notation. The geometric mean is about

$$\sqrt[9]{4.60 \cdot 10^1 \cdot 1.08 \cdot 10^2 \cdot 1.47 \cdot 10^2 \cdot 2.07 \cdot 10^2 \cdot 7.41 \cdot 10^2 \cdot 1.35 \cdot 10^3 \cdot 2.69 \cdot 10^3 \cdot 4.44 \cdot 10^3 \cdot 4.44 \cdot 10^3}.$$

Then multiply the powers of 10 (by adding the exponents).

$$= \sqrt[9]{4.60 \cdot 1.08 \cdot 1.47 \cdot 2.07 \cdot 7.41 \cdot 1.35 \cdot 2.69 \cdot 4.44 \cdot 4.44 \cdot 10^{21}}$$

Now multiply the decimals.

$$\approx \sqrt[9]{8019.4 \cdot 10^{21}}$$

Now the theorems of this lesson can be applied. For the last step a calculator is used.

$$\approx \sqrt[9]{10^{18}} \cdot \sqrt[9]{8019.4 \cdot 10^3} \approx 10^2 \sqrt[9]{8{,}019{,}400} \approx 585.0$$

The geometric mean of the nearest distances of the planets from the sun is about 585 million kilometers.

QUESTIONS

Covering the Reading

1. State the Root of a Product Theorem.

2. *True or false.* $\sqrt{12} \cdot \sqrt{3} = \sqrt{2} \cdot \sqrt{18}$. Justify your answer.

In 3 and 4, multiply and simplify. Do not use a calculator.

3. $\sqrt[3]{2} \cdot \sqrt[3]{32}$

4. $\sqrt[4]{10} \cdot \sqrt[4]{1000}$

5. Write three different expressions equal to $\sqrt[3]{270}$.

In 6 and 7, find a and b.

6. $\sqrt{160} = \sqrt{a}\,\sqrt{10} = b\sqrt{10}$

7. $\sqrt[3]{56} = \sqrt[3]{a}\,\sqrt[3]{7} = b\sqrt[3]{7}$

In 8 and 9, simplify the radicals.

8. $\sqrt[3]{54}$ **9.** $\sqrt[4]{80}$

In 10–12, simplify the expression. Assume all variables are nonnegative.

10. a. $\sqrt{121x^4}$ **b.** $\sqrt{121x^5}$

11. a. $\sqrt[3]{125q^{12}}$ **b.** $\sqrt[3]{125q^{14}}$

12. a. $\sqrt[4]{48p^8}$ **b.** $\sqrt[4]{48p^{11}}$

13. How is the geometric mean of n numbers calculated?

14. The farthest distances of the 9 planets from the sun (in millions of km) are given in the table below.

Planet	Farthest distance from the sun (aphelion)
Mercury	69.8
Venus	109.0
Earth	152.2
Mars	249.4
Jupiter	815.9
Saturn	1508.8
Uranus	2992.9
Neptune	4541.1
Pluto	7324.8

What is the geometric mean of these distances?

Applying the Mathematics

In 15 and 16, simplify.

15. $\sqrt[3]{900} \cdot \sqrt[3]{30}$ **16.** $\sqrt[4]{2} \cdot \sqrt[4]{2^7}$

In 17 and 18, two expressions are given. Which is greater?

17. $\sqrt[3]{2} + \sqrt[3]{3}$ or $\sqrt[3]{5}$ **18.** $10\sqrt[3]{50}$ or $\sqrt[3]{5000}$

19. *True or false.* $\sqrt[7]{x} \cdot \sqrt[5]{y} = \sqrt[35]{xy}$. Explain your answer.

In 20 and 21, assume all variables are positive. Simplify.

20. $\sqrt[3]{750x^6y^9}$ **21.** $\sqrt[4]{405a^4b^5c^6}$

22. In 1992, there were four major earthquakes in the world. The number of deaths from these earthquakes differed considerably. On March 13 and 15 in Turkey, about 4000 people died. On June 28 in southern California, 1 person died. On October 12 in Cairo, Egypt, 450 died. And on December 12 in Flores, Indonesia, about 2500 people died. What is the geometric mean of these four numbers?

In 23 and 24, solve for n.   *(Lesson 8-4)*

23. $\sqrt[5]{10^3} = 10^n$ **24.** $\sqrt[n]{a} = a^{\frac{1}{6}}$

25. *True or false.*   The graph below could represent the inverse of an even power function. Explain your reasoning.   *(Lessons 7-1, 8-4)*

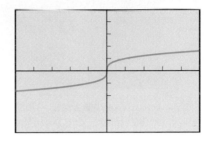

26. a. Find an equation for the inverse of the linear function
$y = mx + b$.
   b. How are the slopes of a linear function and its inverse related?
   c. When is the inverse not a function?   *(Lessons 3-2, 4-9, 8-2)*

27. In 1991, the U.S. national debt was approximately $3.683 trillion. If none of this old debt were paid off, and the interest on the debt is 5% compounded annually, what would the debt be in 2001?
(Lesson 7-4)

28. Solve and check. $\dfrac{y^{-3}}{y^{-2}} = \dfrac{1}{5}$ *(Lesson 7-3)*

29. Write $\dfrac{3 + i}{2 - 5i}$ in $a + bi$ form.   *(Lesson 6-9)*

Exploration

30. a. Write the first five terms of a geometric sequence of your own choice.
   b. Find the geometric mean of these five terms.
   c. Generalize the result of part **b,** and, if you can, prove your generalization.

Making the cut. *This sawmill is in Whiteriver, Arizona. Logging is Whiteriver's major industry, producing 100 million board feet of lumber yearly. One board foot is equal to the volume of a board one foot square and one inch thick.*

The length of either diagonal of a square is $\sqrt{2}$ times the length of a side of the square. That is, $d = s\sqrt{2}$. Consequently, the side is the length of the diagonal divided by $\sqrt{2}$. That is, $s = \dfrac{d}{\sqrt{2}}$. So the largest square piece of wood that can be cut out of a circular log with diameter d has a side of length $\dfrac{d}{\sqrt{2}}$.

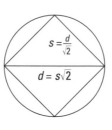

$s = \dfrac{d}{\sqrt{2}}$

$d = s\sqrt{2}$

Rationalizing the Denominator

The expression $\dfrac{d}{\sqrt{2}}$ is considered by many people to be unsimplified because there is a radical in the denominator. There is a historical reason for this. Before calculators were available, to estimate an expression like $\dfrac{3}{\sqrt{2}}$ meant having to do long division with a divisor that is the infinite decimal 1.414213562 . . . . There is no way to do this division and be certain of the accuracy of the result. So people found an equal number without a radical in its denominator. Example 1 shows how this can be done.

Example 1

Rewrite $\dfrac{3}{\sqrt{2}}$ without a radical in the denominator.

Solution

Multiply both the numerator and denominator by $\sqrt{2}$. This does not change the value of the fraction.

$$\frac{3}{\sqrt{2}} \cdot \frac{\sqrt{2}}{\sqrt{2}} = \frac{3\sqrt{2}}{2}$$

Activity

Check Example 1 by finding decimal approximations to the given and final expressions using a calculator.

The process of writing a fraction in an equivalent form without irrational numbers in the denominator is called **rationalizing the denominator.** Nowadays, we rationalize denominators to recognize equivalent expressions. To rationalize the denominator of a fraction whose denominator is $\sqrt{a}$ $(a > 0)$, multiply both the numerator and denominator by $\sqrt{a}$, because $\dfrac{\sqrt{a}}{\sqrt{a}} = 1$.

Example 2

Suppose $a > 0$. Rationalize the denominator of $\dfrac{1}{\sqrt{7a}}$.

Solution

$$\frac{1}{\sqrt{7a}} = \frac{1}{\sqrt{7a}} \cdot \frac{\sqrt{7a}}{\sqrt{7a}} = \frac{\sqrt{7a}}{7a}$$

You should be able to recognize mentally that $\dfrac{1}{\sqrt{x}} = \dfrac{\sqrt{x}}{x}$ for all positive numbers x. You will use this idea when you study trigonometry.

Example 3

Rationalize the denominator of $\dfrac{6}{\sqrt{9x^3}}$, where $x > 0$.

Solution 1

Simplify first, then rationalize.
$$\frac{6}{\sqrt{9x^3}} = \frac{6}{\sqrt{3^2 \cdot x^2 \cdot x}} = \frac{6}{\sqrt{3^2}\sqrt{x^2}\sqrt{x}} = \frac{6}{3x\sqrt{x}} = \frac{2}{x\sqrt{x}} \cdot \frac{\sqrt{x}}{\sqrt{x}} = \frac{2\sqrt{x}}{x^2}$$

Solution 2

Rationalize first, then simplify.
$$\frac{6}{\sqrt{9x^3}} \cdot \frac{\sqrt{9x^3}}{\sqrt{9x^3}} = \frac{6\sqrt{9x^3}}{9x^3}$$
$$= \frac{6\sqrt{3^2 \cdot x^2 \cdot x}}{9x^3} = \frac{6\sqrt{3^2}\sqrt{x^2}\sqrt{x}}{9x^3} = \frac{6 \cdot 3x\sqrt{x}}{9x^3} = \frac{2\sqrt{x}}{x^2}$$

Many people think that in Example 3, Solution 1 is easier than Solution 2, because smaller exponents appear earlier in the solution.

Now consider a fraction in which the radical in the denominator is added to another term, $\dfrac{n}{a + \sqrt{b}}$. To rationalize the denominator, we use a technique similar to the one used to divide complex numbers. To write $\dfrac{1}{2 + 5i}$ in $a + bi$ form, we multiplied both numerator and denominator by the complex conjugate of the denominator: $2 - 5i$. To simplify a fraction with denominator of the form $a + \sqrt{b}$, multiply both numerator and denominator by either the **conjugate** $a - \sqrt{b}$ or by the conjugate's opposite $-a + \sqrt{b}$, whichever is more convenient.

Example 4

Rationalize the denominator of $\dfrac{2}{1 + \sqrt{5}}$.

Solution

The conjugate of $1 + \sqrt{5}$ is $1 - \sqrt{5}$.

$$\frac{2}{1 + \sqrt{5}} \cdot \frac{1 - \sqrt{5}}{1 - \sqrt{5}} = \frac{2(1 - \sqrt{5})}{1 - \sqrt{5} + \sqrt{5} - 5} = \frac{2(1 - \sqrt{5})}{-4} = \frac{1 - \sqrt{5}}{-2} = \frac{\sqrt{5} - 1}{2}$$

Check

Use a calculator. $\dfrac{2}{1 + \sqrt{5}} \approx 0.61803$.

$\dfrac{\sqrt{5} - 1}{2} \approx 0.61803$. It checks.

QUESTIONS

Covering the Reading

1. What is the largest square piece that can be cut out of a circular log of diameter 30"? Answer to the nearest quarter of an inch.

2. **a.** Is $\dfrac{1}{\sqrt{3}}$ equivalent to $\dfrac{\sqrt{3}}{3}$? Justify your answer.

   b. If the expressions are equivalent, which one is written in "simplified" form?

3. What does the term *rationalize the denominator* mean?

In 4–7, rationalize the denominator. Assume $t > 0$.

4. $\dfrac{8}{\sqrt{2}}$ 5. $\dfrac{2}{\sqrt{17}}$ 6. $\dfrac{t}{\sqrt{t^5}}$ 7. $\dfrac{2t}{\sqrt{8t^5}}$

8. By rationalizing the denominator show that $\dfrac{5}{2 + \sqrt{3}} = 5(2 - \sqrt{3})$.

In 9 and 10, rationalize the denominator.

9. $\dfrac{4}{8 + \sqrt{5}}$ 10. $\dfrac{8}{\sqrt{6} + 4}$

Applying the Mathematics

In 11 and 12, a fraction is given.
a. Rationalize the denominator.
b. Check your work by finding decimal approximations.

11. $\dfrac{2 + \sqrt{3}}{2 - \sqrt{3}}$ 12. $\dfrac{6}{\sqrt{7} - \sqrt{5}}$

13. Given the triangle at the left, find the ratio $\dfrac{BC}{AB}$. Rationalize the denominator.

14. Show that the reciprocal of $\sqrt{26} - 5$ is 10 greater than $\sqrt{26} - 5$.

15. The time T that it takes a pendulum to complete one full swing is given by the formula $T = 2\pi\sqrt{\frac{L}{g}}$, where L is the length of the arm of the pendulum (in cm) and g is the acceleration due to gravity. Rationalize the denominator in this formula.

Review

In 16 and 17, v and w are nonnegative real numbers. Simplify. *(Lesson 8-5)*

16. $\sqrt[5]{64v^{11}}$

17. $\sqrt[3]{12v^2w^2}\,\sqrt{18vw^4}$

18. *True or false.* Justify your answer.

   a. $\sqrt{ab^{\frac{1}{3}}} = \sqrt[6]{a^3b}$, for $a > 0$ and $b > 0$.

   b. $\sqrt[3]{m^{\frac{1}{2}}n} = \sqrt[6]{mn^2}$, for $m > 0$ and $n > 0$. *(Lesson 8-4)*

19. A rectangular solid has edges of lengths $\sqrt{2}$, $\sqrt{3}$, and $\sqrt{5}$.

   a. Write its volume in radical form.

   b. Give the exact length of the edge of a cube whose volume is the same as the solid in part **a.** *(Lesson 8-4)*

20. Suppose $f: x \to x^3$, and $x > 0$.

   a. Find a formula for $f^{-1}(x)$.

   b. If $g: x \to 2x$, find a formula for $f \circ g(x)$. *(Lessons 8-1, 8-2, 8-3)*

21. Explain why the inverse of $f: x \to 5x^{100}$ is not a function. *(Lessons 7-1, 8-2)*

22. Let $f = \{(1, 5), (2, 7), (3, 9), (4, 7)\}$. Is the inverse of f a function? Why or why not? *(Lesson 8-2)*

23. Suppose a bond matures after 8 years, paying the investor $10,000. If the bond earned 5.5% interest compounded annually, what principal was invested? *(Lesson 7-4)*

In 24 and 25, assume p and q are positive numbers. Write each expression without negative exponents. *(Lesson 7-3)*

24. $p^{\frac{1}{3}}q^{-\frac{1}{2}}$

25. $\dfrac{p^{-\frac{1}{2}}}{q}$

In 26 and 27, write an expression to describe each situation. *(Lesson 1-1)*

26. Mae was assigned M math problems. She has done P problems. To finish her assignment, how many more must she do?

27. Mel has P pesos in his pocket. If each peso is worth D U.S. dollars, what is the value of his pesos in U.S. dollars?

Shown is a market in Manila, Philippines. Many countries, including the Philippines, Argentina, Chile, Columbia, Cuba, the Dominican Republic, and Mexico use the peso as their monetary unit. The value of the peso is different in each country.

Exploration

28. Devise a method to rationalize the denominator of $\dfrac{1}{\sqrt[3]{x}}$.

Graphs of
Radical
Functions

IN-CLASS

ACTIVITY

Work with a partner.

1 Graph $y = \sqrt{x}$. From the result, indicate the domain that your calculator seems to assume for this function.

2 Repeat Step 1 for the graphs of $y = \sqrt[3]{x}$, $y = \sqrt[4]{x}$, and $y = \sqrt[5]{x}$.

3 **a.** Use the trace function on the last graph of Step 2 to estimate $\sqrt[5]{-20}$ to the nearest tenth.
b. Use your technology to find another way to estimate $\sqrt[5]{-20}$ to the nearest tenth.

4 **a.** For what values of n does $y = \sqrt[n]{x}$ seem to have as its domain the set of all real numbers?
b. For what values of n does $y = \sqrt[n]{x}$ seem to have as its domain the set of all nonnegative real numbers?

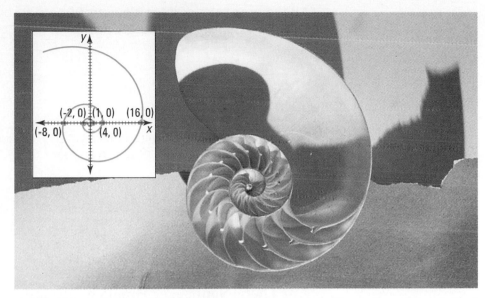

LESSON

8-7

Powers and Roots of Negative Numbers

A spiral in a shell. *This spiral (inset) has a shape similar to that of the shell of the chambered nautilus, a squid-like animal. Its x-intercepts are the powers of -2.*

You have already calculated some powers and some square roots of negative numbers. First we review the powers.

Integer Powers of Negative Numbers

You can calculate positive integer powers of negative numbers using repeated multiplication. Here are the first few positive integer powers of -8.

$$(-8)^1 = -8$$
$$(-8)^2 = (-8)(-8) = 64$$
$$(-8)^3 = (-8)(-8)(-8) = -512$$
$$(-8)^4 = (-8)(-8)(-8)(-8) = 4096$$

Notice that these integer powers alternate between positive and negative numbers. The even exponents produce positive numbers, while odd exponents result in negative numbers. The same is true if zero and negative powers of -8 are considered, since $(-8)^{-n}$ is the reciprocal of $(-8)^n$.

$$(-8)^0 = 1$$
$$(-8)^{-1} = -\frac{1}{8}$$
$$(-8)^{-2} = \frac{1}{64}$$
$$(-8)^{-3} = -\frac{1}{512}$$
$$(-8)^{-4} = \frac{1}{4096}$$

Integer powers of negative numbers satisfy the order of operations. So $-8^4 = -4096$ because the power is calculated before taking the opposite. However, as shown above, $(-8)^4 = 4096$. So $-8^4 \neq (-8)^4$.

All the properties of integer powers of positive bases studied in Chapter 7 apply also to integer powers of negative bases.

Example 1

Write as a power: $(-8)^4 \cdot (-8)^{-3}$.

Solution 1

Use the Product of Powers Property.

$(-8)^4 \cdot (-8)^{-3} = (-8)^{4+-3} = (-8)^1 = -8$

Solution 2

Use the definition of nth power and the values above.

$(-8)^4 \cdot (-8)^{-3} = 4096 \cdot \left(-\frac{1}{512}\right) = -8$

Are There Noninteger Powers of Negative Numbers?

Many times in this course, you have seen that powers and roots of negative numbers do not have the same properties that powers and roots of positive numbers do. Here are some of the properties that are different.

When x is positive, $\sqrt{x}$ is a real number, but $\sqrt{-x}$ is an imaginary number.

If both x and y are negative, $\sqrt{x} \cdot \sqrt{y} \neq \sqrt{xy}$. (The left side is the product of two imaginary numbers and is negative; the right side is positive.)

If x is negative, $x^3 \neq \sqrt{x^6}$. (Again the left side is negative; the right side is positive.)

These examples indicate that powers and roots of negative numbers have to be dealt with very carefully.

For this reason, we do not define x^m when x is negative and m is not an integer. *Noninteger powers of negative numbers are not defined in this book.* That is, an expression such as $(-3)^{\frac{1}{2}}$ is not defined. You have seen expressions that are not defined before. Years ago you learned that division by zero, as in $\frac{3}{0}$, is not defined. And you have learned in this book that 0^0 is not defined.

The Expression $\sqrt[n]{x}$ When x Is Negative and n Is Odd

When x is positive, the radical expression $\sqrt[n]{x}$ stands for the unique positive nth root of x. It would be nice to use the same expression for an nth root of a negative number. This can be done for *odd* roots of negative numbers. If a number is negative, then it has exactly one real odd root. For instance, -8 has one real cube root, namely -2. Consequently, we define the symbol as follows:

For instance, since $(-2)^3 = -8$, we write $\sqrt[3]{-8} = -2$. Since $(-10)^7 = -10{,}000{,}000$, we can write $\sqrt[7]{-10{,}000{,}000} = -10$. To evaluate nth roots of negative numbers without a calculator, you can use numerical or graphical methods.

Example 2

Evaluate $\sqrt[5]{-32}$.

Solution 1

Think: $\sqrt[5]{-32}$ represents the real 5th root of -32. That is, solve $x^5 = -32$. Since $(-2)^5 = -32$, $\sqrt[5]{-32} = -2$.

Solution 2

Think of the graph of the 5th power function $y = x^5$. The line $y = -32$ intersects this graph at only one point. The x-coordinate of this point is the real 5th root of -32. So $\sqrt[5]{-32} = -2$.

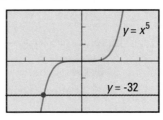

$-4 \leq x \leq 4$, x-scale $= 1$
$-48 \leq y \leq 48$, y-scale $= 16$

Solution 3

Think of the graph of the 5th root function $y = \sqrt[5]{x}$. Trace along the curve until $x = -32$. The y-coordinate of the point is the 5th root of -32.

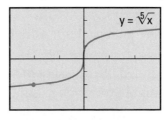

$-48 \leq x \leq 48$, x-scale $= 16$
$-4 \leq y \leq 4$, y-scale $= 1$

Check

$(-2)^5 = -32$.

Notice that the graphs of the function $y = x^5$ and its inverse $x = y^5$ or $y = \sqrt[5]{x}$ verify that every real number has exactly one real 5th root. Thus, -2 is the only real 5th root of -32.

The Expression $\sqrt[n]{x}$ When x Is Negative and n Is Even

The expression $\sqrt{x}$ is defined when x is negative; it equals $i\sqrt{-x}$. For example, $\sqrt{-3} = i\sqrt{3}$. These and all other square roots of negative numbers are not real numbers and they do not satisfy all the properties that square roots of positive numbers have. These difficulties are shared by 4th roots, 6th roots, and so on. So the nth root expression $\sqrt[n]{x}$ is not defined when x is negative and n is an even number greater than 2.

Let us summarize the use of the $\sqrt[n]{}$ symbol. When x is nonnegative, $\sqrt[n]{x}$ is defined for any integer $n > 2$. It equals the positive nth root of x. When x is negative, $\sqrt[n]{x}$ is defined and is a real number only for odd integers n. It equals the negative nth root of x. This may seem unnecessarily complicated, but there is a bonus. Expressions with radical signs can be handled in much the same way that square roots are handled.

Theorem
When $\sqrt[n]{x}$ and $\sqrt[n]{y}$ are defined and are real numbers, then $\sqrt[n]{xy}$ is also defined and $\sqrt[n]{xy} = \sqrt[n]{x} \cdot \sqrt[n]{y}$.

Example 3
Simplify $\sqrt[3]{-2000}$.

Solution
Look for perfect-cube factors of -2000. We use -8.
$$\sqrt[3]{-2000} = \sqrt[3]{-8} \cdot \sqrt[3]{250}$$
$$= -2 \cdot \sqrt[3]{250}$$
250 has the perfect-cube factor 125.
$$= -2 \cdot \sqrt[3]{125} \cdot \sqrt[3]{2}$$
$$= -2 \cdot 5 \cdot \sqrt[3]{2}$$
$$= -10\sqrt[3]{2}.$$

One final caution: Although $\sqrt[n]{x^m} = x^{\frac{m}{n}}$ when x is positive, this property does not necessarily hold when x is negative. For instance, $\sqrt[4]{-32}$ is not defined. As you should have found in the preceding In-class Activity, the domain of $y = \sqrt[n]{x}$ is the set of nonnegative real numbers.

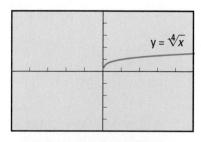

$-5 \leq x \leq 5$, x-scale $= 1$
$-5 \leq y \leq 5$, y-scale $= 1$

Covering the Reading

1. Calculate $(-6)^n$ for all integer values of n from 3 to -3.

2. Tell whether the number is positive or negative.
   a. $(-2)^3$ b. $(2)^{-3}$ c. $(-2)^{-3}$

In 3–5, evaluate.

3. $\sqrt[3]{-1000}$ 4. $\sqrt[5]{-1}$ 5. $\sqrt[7]{-128}$

In 6 and 7, write as a power.

6. $(-3)^4 \cdot (-3)^9$ 7. $((-5)^3)^6$

8. Tell whether the expression is defined or not. If defined, tell whether the number is real or nonreal. If real, tell whether the number is positive or negative.
   a. $\sqrt{-243}$ b. $\sqrt[3]{-243}$
   c. $\sqrt[4]{-243}$ d. $\sqrt[5]{-243}$

9. *True or false.* $\sqrt[3]{(-6)^6} = -6$. Explain your answer.

10. a. For what values of x is $\sqrt[5]{x}$ defined?
    b. What are the domain and the range of the function with equation $y = \sqrt[5]{x}$?

In 11–13, simplify.

11. $\sqrt[3]{-24}$ 12. $\sqrt[5]{-10^{40}}$ 13. $\sqrt[7]{-1}$

14. Explain why the expression $\sqrt[8]{-8}$ is not defined.

Applying the Mathematics

In 15–17, simplify.

15. $\sqrt[3]{-27} \cdot \sqrt[3]{-1}$ 16. $\sqrt[3]{-64} + \sqrt[3]{-8}$ 17. $\sqrt{-16} \cdot \sqrt{-1}$

18. What is the geometric mean of the numbers -8, 4, and 16?

19. a. Show that $2 + 2i$ is a 4th root of -64.
    b. Show that $-2 - 2i$ is a 4th root of -64.
    c. Show that $2 - 2i$ is a 4th root of -64.
    d. Find the one other 4th root of -64, and verify your finding.
    e. How are parts **a–d** related to the fact that $\sqrt[4]{-64}$ is not defined?

20. Explain why the graphs of $y = \sqrt[5]{x}$ and $y = \sqrt[10]{x^2}$ are not the same.

21. Let $f: x \rightarrow \sqrt[3]{x}$ and $g: x \rightarrow \sqrt[5]{x}$. What is a formula for $f \circ g(x)$?

Towering structure.
Shown is the Canadian National (CN) Tower in Toronto, Canada. The tower, used for communications and observation, is the world's tallest free-standing structure at 533 meters.

In 22 and 23 rationalize the denominator and simplify. *(Lesson 8-6)*

22. $\dfrac{3}{\sqrt{6}}$ **23.** $\dfrac{8}{5 + \sqrt{3}}$

24. a. Write a calculator key sequence to evaluate $\sqrt[3]{\sqrt{2000}}$.
   b. Use a calculator to evaluate the expression in part **a**.
   (Lessons 8-4, 8-6)

25. a. Simplify without using a calculator: $(\sqrt{17} - \sqrt{15})(\sqrt{17} + \sqrt{15})$.
   b. Check by approximating $\sqrt{17}$ and $\sqrt{15}$ with a calculator.
   (Lessons 6-2, 8-5)

26. Solve $m^{\frac{3}{2}} = 27^{-1}$. *(Lesson 7-8)*

In 27 and 28, use this information. The maximum distance d you can see from a building of height h is given by the formula $d \approx k\sqrt{h}$.

27. The CN Tower in Toronto is about 2.5 times as tall as the Los Angeles City Hall. About how many times as far can you see from the top of the CN Tower than from the top of L.A. City Hall? *(Lesson 7-1)*

28. About how many times as far can you see from the 108th floor of the World Trade Center than from the sixth floor? (Assume floors have the same height.) *(Lesson 7-1)*

29. A rectangle has vertices at $(0, 0)$, $(-1, 3)$, $(6, 2)$, and $(5, 5)$. What is its area? *(Previous course, Lesson 6-1)*

30. Use your results from part **a** to answer the other parts.
   a. Sketch the graphs of $y = \sqrt[3]{x^3}$, $y = \sqrt[4]{x^4}$, $y = \sqrt[5]{x^5}$, and $y = \sqrt[6]{x^6}$.
   b. For what values of n does $\sqrt[n]{x^n} = x$ for every real number x?
   c. For what values of n does $\sqrt[n]{x^n} = |x|$ for every real number x?
   d. Does $y = \sqrt{x^2}$ follow the pattern of other nth root of nth power functions?
   e. Simplify.
      i. $\sqrt[11]{(-4)^{11}}$ ii. $\sqrt[8]{c^8}$ iii. $\sqrt[4]{16x^{12}}$

Earthly swing. *Shown is a Foucault pendulum, named for the French scientist Jean Foucault. This pendulum verifies that the Earth rotates. The plane of the swing appears to change as the Earth goes through its daily rotation.*

Remember that to solve an equation with a single rational power, such as $x^{\frac{2}{3}} = 5$, you can take both sides to the power of the reciprocal of that exponent.

$$\left(x^{\frac{2}{3}}\right)^{\frac{3}{2}} = 5^{\frac{3}{2}}$$

and so $x = 5^{\frac{3}{2}}$, which a calculator shows to be about 11.18. This checks, for $11.18^{\frac{2}{3}} \approx 4.999898 \ldots$.

Similarly, because the radical $\sqrt[n]{}$ involves an nth root, to solve an equation containing only this single radical, you can take both sides to the nth power.

Example 1

The time t (in seconds) that it takes a pendulum to complete one full swing is given by the formula

$$t = 2\pi\sqrt{\frac{L}{g}},$$

where L is the length of the arm of the pendulum (in cm) and g is the acceleration due to gravity. Suppose a ball on a string, swinging like a pendulum, takes 2 seconds to complete one swing back and forth. If $g = 980$ cm/sec^2, find the length of the string.

Solution

Here $t = 2$ sec, $g = 980$ cm/sec^2. Substitute the given values into the formula $t = 2\pi\sqrt{\frac{L}{g}}$. Solve for L.

▶

$$2 = 2\pi\sqrt{\frac{L}{980}}$$

$$\frac{1}{\pi} = \sqrt{\frac{L}{980}} \qquad \text{Divide both sides by } 2\pi$$

$$\left(\frac{1}{\pi}\right)^2 = \left(\sqrt{\frac{L}{980}}\right)^2 \qquad \text{Square both sides of the equation.}$$

$$\frac{1}{\pi^2} = \frac{L}{980}$$

$$\frac{980}{\pi^2} = L \qquad \text{Multiply each side by 980.}$$

$$99.29 \approx L \qquad \text{Estimate with a calculator.}$$

The string is about 99 cm long, a little short of one meter.

Check

$$2\pi\sqrt{\frac{99.29}{980}} \approx 1.99995 \approx 2. \text{ The answer checks.}$$

Extraneous Solutions

There is only one difficulty that may occur when taking an nth power to solve equations with radicals. The new equation may have more solutions than the original equation does. So you must be careful to check each solution back in the original equation. If a solution to a later equation does not check in the original equation, it is called **extraneous,** and it is not a solution to the original equation.

Example 2

Solve $3 - \sqrt[4]{y} = 10$.

Solution

$$-\sqrt[4]{y} = 7 \qquad \text{Add -3 to each side.}$$

$$\left(-\sqrt[4]{y}\right)^4 = 7^4 \qquad \text{Raise both sides to 4th power}$$

$$y = 2401 \qquad \text{Simplify.}$$

To check, substitute $y = 2401$.

Does $3 - \sqrt[4]{2401} = 10$?

$$3 - 7 = 10? \text{ No.}$$

So 2401 is not a solution. It is extraneous.

The sentence $3 - \sqrt[4]{y} = 10$ has no solutions.

Notice that, in the solution of Example 2, as soon as you write the equation

$$-\sqrt[4]{y} = 7,$$

you might see that there are no solutions. The left side represents a negative number or zero, so it cannot equal the positive number 7.

Equations from the Distance Formula

The distance formula $\sqrt{(x_1 - x_2)^2 + (y_1 - y_2)^2}$ can lead to equations involving square roots. Even though the equations may look quite complicated, they can be solved in the same way as simpler equations are solved.

Example 3

Find the two points on the line $y = 5$ that are 6 units away from the point $(3, 1)$.

Solution

Draw a picture. Let a desired point be (x, y).

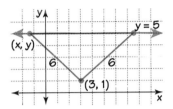

Since the distance from (x, y) to $(3, 1)$ is 6,

$$\sqrt{(x - 3)^2 + (y - 1)^2} = 6.$$

Since $y = 5$, substitute 5 for y. $\sqrt{(x - 3)^2 + 16} = 6.$

Now square both sides. $\qquad (x - 3)^2 + 16 = 36$

$$(x - 3)^2 = 20$$

This equation can be solved easily by taking the two square roots.

$$x - 3 = \sqrt{20} \qquad \text{or} \qquad x - 3 = -\sqrt{20}$$
$$x = 3 + \sqrt{20} \qquad \text{or} \qquad x = 3 - \sqrt{20}$$
$$x \approx 7.47 \qquad \text{or} \qquad x \approx -1.47$$

The two points are exactly $(3 + \sqrt{20}, 5)$ and $(3 - \sqrt{20}, 5)$, or approximately $(7.47, 5)$ and $(-1.47, 5)$.

QUESTIONS

Covering the Reading

In 1–4, find all real solutions.

1. $\sqrt[3]{w} = 4$

2. $4\sqrt[5]{x} = -2$

3. $50 - 8\sqrt[6]{m} = 34$

4. $12 + \sqrt[3]{z} = 9$

5. Refer to Example 1. If a pendulum takes 4 seconds to complete one full swing, how long is the pendulum?

6. Explain why you do not have to solve $\sqrt{m - 3} = -10$ in order to know that the equation has no solutions.

7. What is an *extraneous* solution to an equation?

In 8 and 9, find all real solutions.

8. $\sqrt{3x - 4} = 10$

9. $5 + \sqrt[4]{2x - 5} = 1$

10. Find the two points on the line $y = 8$ that are 7 units from $(3, -4)$.

Applying the Mathematics

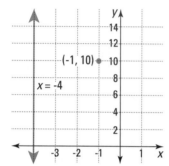

11. Find the two points on the line $x = -4$ that are 5 units from $(-1, 10)$.

12. When traveling at a fast rate, a ship's speed s (in knots) varies directly as the seventh root of the power p (in horsepower) generated by the engine. Suppose the equation $s = 6.5\sqrt[7]{p}$ describes the situation. If a ship is traveling at a speed of 25 knots, about how much horsepower is the engine generating?

13. A formula that police use for finding the speed s (in mph) that a car was going from the length L (in feet) of its skid marks can be written as $s = 2\sqrt{5L}$.
  a. In setting a world land speed record in 1964, a jet-powered car left skid marks nearly 6 miles long. According to the formula, how fast was the car going?
  b. About how far does an automobile travel if it skids from 30 mph to a stop?

14. A cube had edges of length s millimeters. Then 0.5 millimeter was shaved off each dimension. What was the original length of an edge of the cube if the shaved cube has a volume of 1000 cubic millimeters?

15. Find all real solutions to $\sqrt[4]{z} + 9 = 10\sqrt[4]{z}$.

Review

In 16 and 17, simplify each expression. Assume the variables are nonnegative. *(Lesson 8-7)*

16. $\sqrt[3]{-125x^6}$

17. $\sqrt{50a^3} \cdot \sqrt{8b^4}$

18. Give a counterexample to the statement "$\sqrt[4]{x^4} = x$ for all real numbers x." *(Lesson 8-7)*

In 19 and 20, *true or false.* Justify your answer. *(Lesson 8-6)*

19. $\dfrac{1}{\sqrt{5}} = \dfrac{\sqrt{5}}{5}$

20. $\dfrac{\sqrt{28}}{\sqrt{8}} = \dfrac{\sqrt{56}}{\sqrt{16}}$

21. *Multiple choice.* If a and k are positive, which of these is a solution to $a(x - h)^n = k$? *(Lessons 6-3, 8-6)*

  (a) $\sqrt[n]{\dfrac{k}{a - h}}$ (b) $\sqrt[n]{\dfrac{k}{a}} + h$

  (c) $\sqrt[n]{\dfrac{k - a}{n}}$ (d) $\dfrac{\sqrt[n]{k}}{a} + h$

22. Let $m(x) = \sqrt{x}$ and $n(x) = x^4$.   *(Lessons 8-1, 8-4)*
   a. Find a formula for $m(n(x))$. What is the domain of $m \circ n$?
   b. Find a formula for $n(m(x))$. What is the domain of $n \circ m$?

23. Explain why the inverse of the function $f: x \to x^2 - 5$ is not a function.   *(Lessons 8-2, 8-3)*

24. In 1980, the world was using energy from petroleum at a rate of $1.35 \cdot 10^{20}$ J/yr, where J is a unit of energy called a *joule*. At that time the world's supply of petroleum was estimated to be enough to create about 10^{22} J of energy. If the global consumption of energy has remained constant, about how long will the world's supply of petroleum last?   *(Lesson 7-2)*

25. Write a system of three inequalities whose solutions fill the shaded region graphed below.   *(Lesson 5-7)*

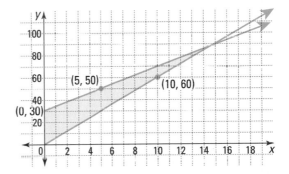

Exploration

Oil's well that ends well.
This 1859 photograph (top) is of the first oil well in the U.S. and its owner, Edwin Drake (with the tall hat). The well, built that year, was located in Titusville, PA. Shown (bottom) is an aerial view of a modern off-shore oil platform in the Gulf of Mexico near the coast of Louisiana.

26. Refer to the formula $s = 2\sqrt{5L}$ in Question 13 for the speed of a car as a function of the length of the skid mark. Explain how this formula is related to the formula $d = \dfrac{s^2}{16}$ from Lesson 2-5, which gives braking distances as a function of speed.

A project presents an opportunity for you to extend your knowledge of a topic related to the material of this chapter. You should allow more time for a project than you do for typical homework questions.

1 Composites of Types of Functions

a. Pick any two linear functions, for instance, $f(x) = 3x - 5$ and $g(x) = x + 2$. Find $f(g(x))$ and $g(f(x))$. Are $f \circ g$ and $g \circ f$ linear? If not, what type of functions are $f \circ g$ and $g \circ f$? Make a conjecture about the composite of two linear functions. Prove your conjecture.

b. Pick any two quadratic functions, for instance, $f(x) = x^2 + 4$ and $g(x) = 2x^2 + x$. Find $f \circ g$ and $g \circ f$. Make a conjecture about the composite of two quadratic functions.

c. Investigate the composite of one linear function and one quadratic function. What type of function results?

d. Choose another type of function you have studied, for instance, an absolute value function or an inverse-variation function with equation of the form $y = \frac{k}{x}$ or $y = \frac{k}{x^2}$. Find some composites of your functions with both linear and quadratic functions. What type of function results? Can you prove any of your conjectures?

e. Summarize the results of your investigations.

2 Irrationality of Some Square Roots

a. From the time of Pythagoras, mathematicians have known that $\sqrt{2}$ is irrational. That is, it cannot be written in the form $\frac{a}{b}$ where a and b are integers ($b \neq 0$) and a and b have no common factors. Give a proof that $\sqrt{2}$ is not rational.

b. Adapt this proof to show that $\sqrt[3]{2}$ is irrational.

3 Properties of Irrational Numbers

In Appendix A, the Field Properties of real numbers are listed. Investigate which of these properties apply to the set of irrational numbers. Include examples and counterexamples to illustrate your conclusions.

5 Square Roots of Imaginary Numbers

a. Find a square root of i, and hence a fourth root of -1, by solving the equation $i = (a + bi)^2$ for a and b. It will help to recall that $i = 0 + i$, and that two complex numbers are equal if and only if their real parts are equal and their imaginary parts are equal. This allows you to set up a system to solve for a and b. Write the square root of i in radical form.

b. We know that 4 has another square root besides the one denoted $\sqrt{4}$. Similarly, i has another square root. Use what you know about the other square root of 4 to hypothesize about the other square root of i. Check your hypothesis by squaring your result.

c. Find the square roots of some other nonreal numbers.

4 Radicals and Heights

a. Build a model of a regular tetrahedron. (A regular tetrahedron has four faces, all of which are equilateral triangles.)

b. On your model show where you would measure the slant height.

   i. Measure the slant height (in the same units as the sides of the triangles).

   ii. Calculate the slant height in radical form, using what you know about an equilateral triangle.

   iii. Compare parts **i** and **ii**.

c. On your model show where you would measure the altitude of your tetrahedron.

   i. Measure the height.

   ii. Calculate the height in radical form, based on right triangles that are formed.

   iii. Compare parts **i** and **ii**.

Shown is Alexander Graham Bell holding a model of his tetrahedral kite. Bell's development of this kite resulted from his interest in and his experiments with flight.

Chapter 8 *Projects* **523**

SUMMARY

Every relation has an inverse, which can be found by switching the coordinates of its ordered pairs. The graphs of any relation and its inverse are reflection images of each other over the line $y = x$.

Inverses of some functions are themselves functions. If the function has a graph in the coordinate plane, then its inverse is a function if and only if no horizontal line intersects the graph of the original function in two points. In general, two functions f and g are inverses of each other if and only if $f \circ g$ and $g \circ f$ are defined and $f(g(x)) = x$ for all values of x in the domain of g and $g(f(x)) = x$ for all values of x in the domain of f.

Consider the function f with domain the set of all real numbers and equation of the form $y = f(x) = x^n$. If n is an odd integer ≥ 3, its inverse is the nth root function with equation $x = y^n$ or $y = \sqrt[n]{x}$.

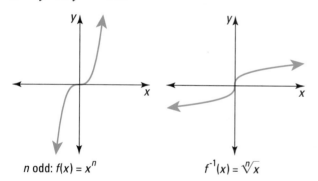

n odd: $f(x) = x^n$ $f^{-1}(x) = \sqrt[n]{x}$

If n is an even integer ≥ 2, the graph of $y = x^n$ does not pass the Horizontal-Line Test, so the inverse of $y = f(x) = x^n$ is not a function. However, if the domain of f is restricted to the set of nonnegative reals, the inverse of f is the, nth root function with equation $y = \sqrt[n]{x} = x^{\frac{1}{n}}$.

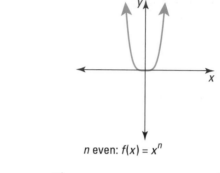

n even: $f(x) = x^n$

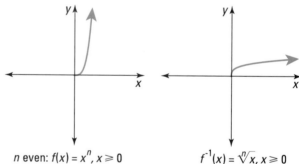

n even: $f(x) = x^n, x \geq 0$ $f^{-1}(x) = \sqrt[n]{x}, x \geq 0$

When n is odd and greater than or equal to 3, the expression $\sqrt[n]{x}$ stands for the one real nth root of x. When n is even and greater than or equal to 2 and x is nonnegative the symbol $\sqrt[n]{x}$ stands for the nonnegative real nth root of x.

All properties of powers listed in Chapter 7 apply to these radicals. They lead to the following theorems for any real numbers x and y and integers m and n for which the symbols are defined and represent real numbers.

Root of a Power Theorem: $x^{\frac{m}{n}} = \sqrt[n]{x^m} = (\sqrt[n]{x})^m$

Root of a Product Theorem: $\sqrt[n]{xy} = \sqrt[n]{x}\,\sqrt[n]{y}$

These properties are helpful in simplifying radical expressions and in solving equations with radicals. To solve such an equation, you need to raise each side to the nth power. When you do this, you may obtain extraneous solutions.

Always check every possible answer to make sure that no extraneous solutions have been included.

Radicals appear in many formulas. For example, the nth root of the product of n numbers is the geometric mean of the numbers. When a radical appears in the denominator of a fraction, multiplying both the numerator and denominator by a well-chosen number can make the new denominator rational. This is called rationalizing the denominator.

VOCABULARY

Below are the most important terms and phrases for this chapter. You should be able to give a definition for those terms marked with *. For all other terms you should be able to give a general description and a specific example.

Lesson 8-1
radical notation $\sqrt[n]{}$
composite of s and f, $s \circ f$
function composition

Lesson 8-2
inverse of a relation
Horizontal-Line Test for Inverses

Lesson 8-3
Inverse Functions Theorem
f^{-1}
identity function

Lesson 8-4
radical sign, radical
$\sqrt[n]{x}$ when $x \geq 0$
Root of a Power Theorem

Lesson 8-5
Root of a Product Theorem
simplified form
simplify an nth root
geometric mean

Lesson 8-6
rationalizing the denominator
conjugate

Lesson 8-7
$\sqrt[n]{x}$ when $x < 0$

Lesson 8-8
extraneous solution

PROGRESS SELF-TEST

Take this test as you would take a test in class. You will need a calculator. Then check your work with the solutions in the Selected Answers section in the back of the book.

In 1 and 2, let $f(x) = x^2 + 6$ and $g(x) = 2x - 3$.

1. $f \circ g\colon -1 \to \underline{\ ?\ }$.

2. Find $f(g(x))$.

3. a. Find a formula for the inverse of the function defined by $r(x) = 5x - 6$.

   b. Is the inverse a function? Justify your answer.

In 4 and 5, refer to the function graphed below.

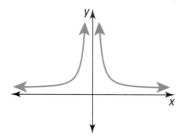

4. Graph the inverse of the function.

5. How can you restrict the domain of the original function so that the inverse is also a function?

6. *True or false.* $\sqrt[6]{64} = -2$. Justify your answer.

7. Recall the formula $T = 2\pi\sqrt{\dfrac{L}{g}}$ for the time T (in seconds) it takes a pendulum to complete one full swing, where L is length (in centimeters) and g is the acceleration due to gravity. How long to the nearest centimeter is a pendulum that takes 1 second to swing? Use 980 cm/sec² for g.

In 8–10, simplify. Assume $x > 0$ and $y > 0$.

8. $\sqrt[4]{625x^4y^8}$ **9.** $\sqrt[5]{-96x^{15}y^3}$ **10.** $\dfrac{2x^4}{\sqrt{16x}}$

11. A square with side s is inscribed in a circle with diameter 9 cm. On a test, students were asked to find s. Three students' answers are below.

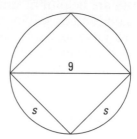

Amanda: $s = \sqrt{\dfrac{81}{2}}$

Bruce: $s = \dfrac{9}{\sqrt{2}}$

Carlos: $s = \dfrac{9\sqrt{2}}{2}$

Which answer(s) is (are) correct? Justify your response.

12. Rewrite $\sqrt[3]{\sqrt[4]{7}}$ using a simple rational exponent.

13. Rationalize the denominator and simplify: $\dfrac{6}{\sqrt{10} - \sqrt{4}}$

14. Use the formula $s = 6.5\sqrt[7]{p}$ to determine the horsepower p generated by the engine of a certain ship traveling at a speed of 20 knots.

In 15 and 16, solve.

15. $400 = 9\sqrt{3t}$ **16.** $12 + \sqrt[3]{x+5} = 9$

17. Suppose x can be any real number and n is a positive integer. For what values of n is the inverse of the function f with equation $y = f(x) = x^n$ also a function?

18. Give the domain and the range of the function with equation $y = \sqrt[8]{x}$, if x and y are real numbers.

CHAPTER REVIEW

Questions on SPUR Objectives

SPUR stands for **S**kills, **P**roperties, **U**ses, and **R**epresentations. The Chapter Review questions are grouped according to the SPUR Objectives for this chapter.

SKILLS DEAL WITH THE PROCEDURES USED TO GET ANSWERS.

Objective A: *Find values and rules for composites of functions.* *(Lesson 8-1)*

1. In the symbolism $f \circ g$, which function is applied first?

In 2–4, let $t(x) = x^2 + 1$ and $m(x) = x - 6$.

2. **a.** Find $t(m(5))$ **b.** Find $t(m(x))$.

3. **a.** Find $m(t(5))$ **b.** Find $m(t(x))$.

4. The function $t \circ m$ maps -10 onto what number?

In 5 and 6, rules for functions f and g are given. Does $f \circ g = g \circ f$? Justify your response.

5. $f: x \to -\frac{2}{7}x$; $g: x \to -\frac{7}{2}x$

6. $f(x) = \sqrt{x}$; $g(x) = \frac{4}{x}, x > 0$

Objective B: *Find the inverse of a relation.* *(Lessons 8-2, 8-3)*

7. A function has equation $y = 4x - 2$. In slope-intercept form, what is an equation for its inverse?

8. A function has equation $y = |x|$. What is an equation for its inverse?

9. If $f: x \to 2x + 7$, then $f^{-1}: x \to \underline{\ ?\ }$.

10. Show that $f: x \to 2x + 3$ and $g: x \to \frac{1}{2}x - 3$ are not inverse functions.

11. If $g(x) = -x^2$ for $x \le 0$, then $g^{-1}(x) = \underline{\ ?\ }$.

12. Suppose $f(x) = x^3$. Find an equation for the inverse of this function.

Objective C: *Evaluate radicals.* *(Lesson 8-4)*

In 13–16, write as a whole number or a simple fraction.

13. $\sqrt[4]{625}$ 14. $\sqrt[3]{-8}$ 15. $\sqrt[3]{\left(\frac{8}{125}\right)^2}$

16. $(\sqrt{19} + \sqrt{17})(\sqrt{19} - \sqrt{17})$

In 17–20, estimate to the nearest hundredth.

17. $\sqrt[4]{4}$ 18. $\sqrt[3]{27 + 64}$

19. $\sqrt[3]{-80}$ 20. $\sqrt[10]{346}$

Objective D: *Rewrite or simplify expressions with radicals.* *(Lessons 8-5, 8-6)*

In 21–28, simplify. Assume variables under the radical sign are positive.

21. $\sqrt{a^6}$ 22. $\sqrt[3]{54p^3}$

23. $\sqrt[6]{128x^8y^7}$ 24. $\sqrt[3]{-80c^9}$

25. $\sqrt[5]{-b^{14}c^{30}}$ 26. $\sqrt{7v^3} \cdot \sqrt{14v}$

27. $\sqrt{\sqrt{\sqrt{h}}}$ 28. $\frac{5f}{\sqrt{3f^2}}$

In 29–32, rationalize the denominator and simplify if possible.

29. $\frac{7}{\sqrt{7}}$ 30. $\frac{6}{\sqrt{2}}$ 31. $\frac{3}{\sqrt{5} - 1}$

32. $\frac{a}{\sqrt{a} + \sqrt{b}}$ $(a \ne b, a > 0, b > 0)$

Objective E: *Solve equations with radicals.* *(Lessons 8-7, 8-8)*

In 33–36, find all real solutions. Round irrational answers to the nearest tenth.

33. $\sqrt[3]{a} = 1.5$ 34. $13 = 5\sqrt[4]{b}$

35. $14 = \frac{1}{4}\sqrt{9 - y}$ 36. $\sqrt[3]{x + 1} - 9 = 16$

37. $4 + \sqrt[4]{3n} = 2$ 38. $\sqrt{5x} + 2\sqrt{5x} = 12$

PROPERTIES DEAL WITH THE PRINCIPLES BEHIND THE MATHEMATICS.

Objective F: *Apply properties of inverse relations and inverse functions.*   *(Lessons 8-2, 8-3)*

In 39–41,   *true or false.*

39. If two functions f and g are inverses of each other, then $f \circ g(x) = x$ for all x in the domain of g.

40. When the domain of f is the set of positive real numbers, then the inverse of $f(x) = x^4$ has equation $y = \sqrt[4]{x}$.

41. The Horizontal-Line Test fails for the function $y = x^6$.

42. Suppose the domain of a linear function is $\{x: x \geq 0\}$ and the range is $\{y: y = 6\}$. What are the domain and range of the inverse?

In 43 and 44, suppose f and g are inverses of each other.

43. If (a, k) is a point on the graph of f, what point must be on the graph of g?

44. If the domain of g is the set of all negative integers, what can you conclude about the domain or the range of f?

Objective G: *Apply properties of radicals and nth root functions.*   *(Lessons 8-4, 8-5, 8-7)*

45. If x is negative, for what values of n is $\sqrt[n]{x}$ a real number?

46. *Multiple choice.*   Which symbol is not defined?
(a) $\sqrt[3]{64}$ (b) $\sqrt[3]{-64}$
(c) $\sqrt[6]{64}$ (d) $\sqrt[6]{-64}$

47. Tell why the statement $\sqrt[5]{a} = a^{1/5}$ is not true for all real numbers a.

48. For what values of x is $\sqrt[3]{x^3} = x$?

49. Give a counterexample to the statement "For all real numbers x, $\sqrt[6]{x^6} = x$."

In 50 and 51, tell whether the statement $\sqrt[n]{a} \cdot \sqrt[n]{b} = \sqrt[n]{ab}$ is true for the stated values.

50. a and b are negative, $n = 2$.

51. a and b are negative, $n = 3$.

USES DEAL WITH APPLICATIONS OF MATHEMATICS IN REAL SITUATIONS.

Objective H: *Solve real-world problems which can be modeled by equations with radicals.* *(Lessons 8-4, 8-8)*

52. The maximum distance d you can see from a building with height h is approximated by the formula $d = k\sqrt{h}$. Apartments A and B are 3 and 6 stories high, respectively. If these two apartment buildings have the same height per floor, about how many times as far can you see from the top of apartment B than the top of apartment A?

53. The diameter of a spherical balloon varies as the cube root of its volume. If one balloon holds 10 times as much air as a second balloon, how do their diameters compare?

54. Use the formula $s = 6.5 \sqrt[7]{p}$ to determine the horsepower p generated by the engine of a certain ship traveling at a speed of 300 knots.

55. Recall that skid marks of length L feet indicates that a car was traveling at least s mph, where $s \approx 2\sqrt{5L}$. If a car skids to a stop from a speed of 60 mph, about how long would its skid marks be?

REPRESENTATIONS DEAL WITH PICTURES, GRAPHS, OR OBJECTS THAT ILLUSTRATE CONCEPTS.

Objective I: *Make and interpret graphs of inverses of relations.* *(Lessons 8-2, 8-4)*

56. Let $f = \{(-4, 5), (-3, 4), (-2, 3), (-1, 2)\}$.
   a. Graph f^{-1}.
   b. What transformation maps f to f^{-1}?

57. Graph the inverse of the relation shown at the right.

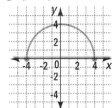

58. Graph the inverse of the relation with equation $y = |x|$.

59. a. *Multiple choice.* For which of the relations graphed below is the inverse not a function?

(a)
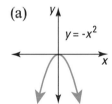
$y = -x^2$

(b)
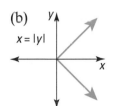
$x = |y|$

(c)
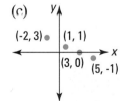
$(-2, 3)$ $(1, 1)$
$(3, 0)$ $(5, -1)$

(d)
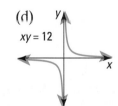
$xy = 12$

   b. How can you restrict the domain of the function in your answer to part **a** so that the inverse is a function?

60. Draw a graph of a function with domain $\{x: -1 < x < 1\}$ that has an inverse which is not a function.

61. Let $g(x) = x^3$.
   a. Graph $y = g(x)$ and $y = g^{-1}(x)$.
   b. What is the domain of g^{-1}?

62. a. Graph $h(x) = \sqrt[4]{x}$.
   b. State the domain and the range of h.

63. a. Graph $f(x) = \sqrt[7]{x}$.
   b. State the domain and the range of f.

CHAPTER

9

EXPONENTIAL AND LOGARITHMIC FUNCTIONS

The bar graph below gives the U.S. population for each census from 1790 to 1990.

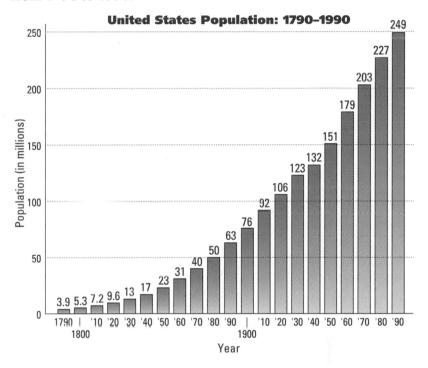

United States Population: 1790–1990

The population P (in millions) is closely approximated by the equations

$$P = \begin{cases} 13(1.03)^{x-1830}, & \text{for } 1790\text{–}1860, \\ 63(1.02)^{x-1890}, & \text{for } 1870\text{–}1910, \\ 151(1.012)^{x-1950}, & \text{for } 1920\text{–present} \end{cases}$$

where x is the year of the census. Each equation defines an *exponential function.* Three different equations are needed because the yearly *growth factor* of the population of the United States has changed from about 1.03 to about 1.012 since 1790.

In this chapter you will learn about exponential functions and their inverses, called *logarithmic functions,* and their many applications.

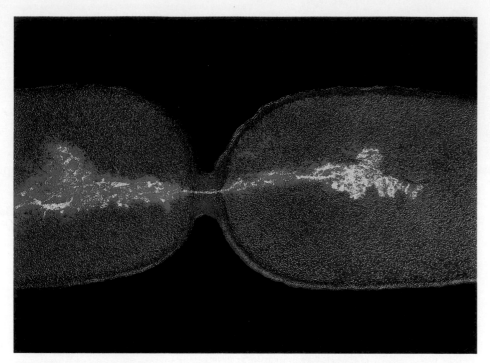

The power of growth. *Shown is the bacterium* Salmonella typhimurium *undergoing division by binary fission. This bacterium is a common cause of food poisoning in humans.*

The Meaning of Exponential Growth

Many bacteria reproduce very quickly. Suppose an experiment begins with 300 bacteria, and that the population doubles every hour. (This is a reasonable assumption over a short period of time.) Below is the population y after x hours for $x = 0, 1, 2,$ and 3.

x	0	1	2	3
y	300	600	1200	2400

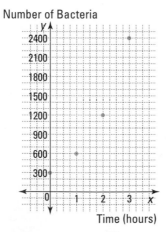

Number of Bacteria

Time (hours)

As x takes on integer values increasing by 1, the values of y form a geometric sequence with constant ratio 2. A formula for this sequence is $y = 300 \cdot 2^x$. The points (x, y) are graphed above.

Of course, the bacteria population does not double all at once. Using noninteger exponents, you can estimate the population at intermediate times. For instance, after half an hour, $y = 300 \cdot 2^{.5} \approx 424$.

Activity 1

Copy and complete the table using the formula $y = 300 \cdot 2^x$.

x	.25	.5	.8	1.4	2.6
y		424			

You can also estimate population values *before* the experiment started. For instance, an hour before the experiment $x = -1$, so $y = 300 \cdot 2^{-1} = 150$. A half hour before the start, at $x = -.5$, $y = 300 \cdot 2^{-.5} \approx 212$.

Activity 2

Use the formula $y = 300 \cdot 2^x$ to estimate the population 1.5 hours before the experiment began.

In Activities 1 and 2, you should have found that intermediate rational values of x produce intermediate values of y. As more intermediate values of x are chosen, more intermediate values of y result. Consider the graphs below

Number of Bacteria

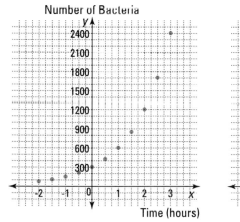

Number of Bacteria

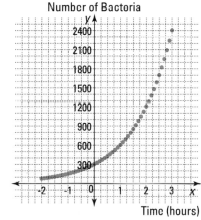

Number of Bacteria

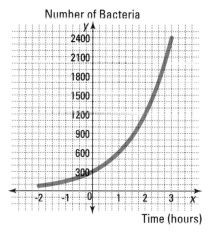

In the leftmost graph $y = 300 \cdot 2^x$ is plotted for values of x from -2 to 3, increasing by .5. The middle graph shows values for x from -2 to 3, increasing by .1. The rightmost graph shows values for x from -2 to 3, increasing by .01. This last graph looks almost like a smooth curve. And it suggests how to find $300 \cdot 2^x$ when x is irrational.

For instance, the value $x = \sqrt{5}$ yields $y = 300 \cdot 2^{\sqrt{5}}$.

We know $\quad\quad\quad\quad\quad 2.2 < \quad \sqrt{5} \quad < 2.3$.

So, $\quad\quad\quad\quad 300 \cdot 2^{2.2} < 300 \cdot 2^{\sqrt{5}} < 300 \cdot 2^{2.3}$.

Since $300 \cdot 2^{2.2} \approx 1378$ and $300 \cdot 2^{2.3} \approx 1477$,

$$1378 < 300 \cdot 2^{\sqrt{5}} < 1477.$$

To find a closer estimate to $\sqrt{5}$, we begin with

$$2.23 < \quad \sqrt{5} \quad < 2.24.$$

Then $\quad\quad\quad 300 \cdot 2^{2.23} < 300 \cdot 2^{\sqrt{5}} < 300 \cdot 2^{2.24}$.

Now, $300 \cdot 2^{2.23} \approx 1407$ and $300 \cdot 2^{2.24} \approx 1417$.

So, $\quad\quad\quad\quad\quad 1407 < 300 \cdot 2^{\sqrt{5}} < 1417$.

A calculator shows $300 \cdot 2^{\sqrt{5}} \approx 1413$.

You can extend this reasoning to calculate powers involving any irrational values of x. So we can think of the function $y = 300 \cdot 2^x$ as being defined for all real values of x. This results in the function graphed below. The shape of an exponential curve is different from the shape of a parabola, a hyperbola, or an arc of a circle.

Number of Bacteria

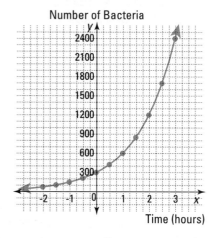

Time (hours)

Notice that the range of $y = 300 \cdot 2^x$ is the set of positive real numbers. Its graph never intersects the x-axis but gets closer and closer to it as x gets smaller and smaller; the x-axis is an asymptote to the graph. Substituting $x = 0$ into the equation yields a y-intercept of 300. This represents the number of bacteria present at the start of the experiment.

Example 1

Use the exponential curve on page 534 to estimate:
a. the number of bacteria after 1 hour and 15 minutes.
b. the time when 1500 bacteria were present.

Solution

a. 1 hour and 15 minutes = $1\frac{1}{4}$ hour. On the graph, When x = $1\frac{1}{4}$, y ≈ 700. After 1 hour and 15 minutes there were about 700 bacteria.

b. When y = 1500, x ≈ $2\frac{1}{4}$. So there were 1500 bacteria after about 2 hours and 15 minutes.

Check

Graph the function defined by $y = 300 \cdot 2^x$ using an automatic grapher.
a. Use the trace feature of your grapher to trace along the curve until x ≈ 1.25. You should find y ≈ 714. It checks.
b. Trace along the curve until y ≈ 1500. We found x ≈ 2.3. Three tenths of an hour is .3(60) = 18 minutes. Our estimate is very good.

The equation $y = 300 \cdot 2^x$ defines a function in which the independent variable x is the exponent. Such a function is called an *exponential function*. Its graph is called an *exponential curve*. This particular exponential curve models *exponential growth;* that is, as time increases, so does the population of bacteria.

> **Definition**
> A function f defined by the equation $f(x) = ab^x$
> ($a \neq 0, b > 0, b \neq 1$) is an **exponential function.**

In the equation $y = ab^x$, b is the **growth factor.** This is the amount by which y is multiplied for each unit increase in x. The situation above with bacteria involves a growth factor $b = 2$. In general, when $b > 1$, exponential growth occurs. In the next lesson we consider situations when $0 < b < 1$.

In Chapter 7 you studied two situations that lead to exponential functions. The geometric sequence with nth term $g_n = g_1 r^{n-1}$ is an exponential function. In it, n is the independent variable, g_n is the dependent variable, and r is the growth factor. When $r > 1$, exponential growth occurs. The compound interest formula $A = P(1 + r)^t$, when P and r are fixed, also defines an exponential function. In this case, t is the independent variable, $A = f(t)$ is the dependent variable, and $1 + r$ is the growth factor. Since $1 + r$ is always greater than one, compound interest always yields exponential growth.

If you know the initial amount a and growth factor b, you can write an equation for the exponential function described.

Example 2

In 1993, the population of the U.S. was about 258 million. During the early 1990s, the population of the U.S. was growing at a rate of about 1% annually. Suppose this growth rate continues. Let $P(x) =$ the population x years after 1993.
a. Find a formula for $P(x)$.
b. What would the population of the U.S. be in the year 2013?
c. In about what year will the population of the U.S. reach 400 million?

Solution

a. A constant growth rate implies an exponential function. An annual growth rate of 1% means that each year the population is 101% of the previous year's population. So in the equation $y = ab^x$, $b = 1.01$. You are given $a = 258{,}000{,}000$. So
$$P(x) = 258{,}000{,}000(1.01)^x = 258 \cdot 10^6(1.01)^x.$$
b. The year 2013 is 20 years after 1993. Find $P(20)$.
$$P(20) = 258 \cdot 10^6(1.01)^{20}$$
$$\approx 315 \cdot 10^6$$
The population in the year 2013 would be about 315 million.
c. For an estimate, graph the equation $y = 258 \cdot 10^6(1.01)^x$. Use the trace feature to find the value of x when $y = 400$ million. We entered $y = 258 \cdot (1.01)^x$ and looked for x when $y = 400$. The graph on the window $-100 \le x \le 100$ and $0 \le y \le 500$ is at the right. Using the trace feature, we find $x \approx 44$. So 44 years after 1993, or in the year 2037, the population of the U.S. will be about 400 million according to this model.

$-100 \le x \le 100, \quad x\text{-scale} = 10$
$0 \le y \le 500, \quad y\text{-scale} = 50$

Proud to be Americans.
Shown is a Naturalization Ceremony at Mount Rushmore in South Dakota. Each year about 125,000 people become naturalized U.S. citizens.

Later in this chapter you will see how to get the exact solution for questions like those in part **c** of Example 2.

QUESTIONS

Covering the Reading

In 1–4, refer to the bacteria experiment at the beginning of the lesson.

1. How many bacteria are there after 2.2 hours?

2. a. What does $x = -1.5$ represent in terms of the experiment?
   b. What was the population of bacteria 1.5 hours before the experiment began?

3. Use the exponential curve to estimate the number of bacteria present after 1.75 hours.

4. After about how many hours were 1700 bacteria present?

5. *True or false.* Because $\sqrt{7}$ is between 2.64 and 2.65, $2^{2.64} < 2^{\sqrt{7}} < 2^{2.65}$.

6. Use your calculator to evaluate each power.
  a. $2^{1.73}$ **b.** $2^{1.74}$ **c.** $2^{\sqrt{3}}$

7. Define *exponential function.*

8. *Multiple choice.* Which equation has a graph that is an exponential curve?
  (a) $y = 2x + 5$ (b) $y = x^2 + 5$ (c) $y = 2.5^x$ (d) $y = 2x^5$

9. Let $f(x) = 4^x$.
  a. Evaluate the following.
    (i) $f(2)$ (ii) $f(1.4)$ (iii) $f(0)$ (iv) $f(-2)$
  b. Graph $y = f(x)$.
  c. *True or false.* f is an exponential function.

In 10–12, refer to the exponential curve with equation $y = ab^x$, where $b > 1$.

10. The y-intercept is ___?___.

11. The constant growth factor is ___?___.

12. Which line is an asymptote to the graph?

13. Suppose $g_n = 10 \cdot 5^{n-1}$. Identify the growth factor.

14. *True or false.* $A = P(1 + r)^t$, where P and t are fixed, defines an exponential function.

In 15 and 16, use the assumptions of Example 2.

15. Estimate the population of the U.S. in the year 2076.

16. In about what year will the U.S. population first reach 500 million?

Applying the Mathematics

---

17. Refer to the graph of $y = 300 \cdot 2^x$ which is given in the Lesson.
  a. Find the rate of change between $x = 1$ and $x = 2$.
  b. Find the rate of change between $x = 2$ and $x = 3$.
  c. Find the rate of change between $x = 3$ and $x = 4$.
  d. What conclusion can you draw from your answers in parts **a–c?**

18. The *1994 Information Please Almanac* states that Australia is experiencing an .8% natural increase in population every year.
  a. Find the growth factor for the exponential function describing this situation.
  b. The 1993 population of Australia was about 17,800,000. According to the stated growth rate, what should the projected population be in the year 2000?
  c. The projected population given in the almanac for the year 2000 is 19,100,000. Does this agree with your result from part **b?** If not, what growth rate would give the projected population?

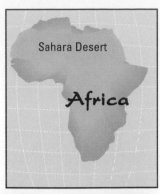

Sahara Desert

Africa

19. In 1992, the sub-Saharan region of Africa had the fastest growing population of the world. By 2020, it is expected to more than double its population. Estimate the annual growth rate.

20. **a.** Graph the equation $y_1 = 10 + 20x$ on the window $-3 \le x \le 30$, $0 \le y \le 500$.
    b. Graph the equation $y_2 = 10 \cdot 1.2^x$ on the same window.
    c. Compare and contrast the linear growth in part **a** with the exponential growth in part **b**, by describing the intervals where each growth pattern gives the larger values.

Review

21. Suppose $f(x) = 4x - 9$.
    a. Find an equation for f^{-1}.
    b. Graph $y = f(x)$ and $y = f^{-1}(x)$ on the same set of axes.
    c. *True or false.* The graphs in part **b** are reflection images of each other. *(Lessons 8-2, 8-3)*

22. Suppose you invest $2500 at 4.5% annual interest compounded monthly. What amount is the investment worth at the end of 3 years? *(Lesson 7-4)*

23. Suppose the matrix $\begin{bmatrix} -2 & -2 & 2 \\ 1 & 2 & -3 \end{bmatrix}$ represents triangle *TRI*.
    a. Give the matrix for the image of $\triangle TRI$ under the translation $T_{3,-4}$.
    b. Graph the image and preimage on the same set of axes. *(Lessons 4-1, 4-10)*

24. A teacher finds that due to an unusually difficult test, her students' grades are low and need to be rescaled. A 100 will remain a 100, but a 65 will become an 80.
    a. If the rescaling is linear, find a relationship between the new score y and the old score x. (Hint: write the scores as ordered pairs.)
    b. What will an old score of 51 become? *(Lesson 3-7)*

25. Suppose a new car costs $15,000 in 1995. Find the cost of the car one year later, if in 1996 the following is true.
    a. The car is worth 85% of its purchase price.
    b. The car depreciated 20% in value.
    c. The value of the car depreciated r%. *(Previous course)*

Shown is a market in Dakar, Senegal. Senegal is a country in the sub-Saharan region of Africa (the portion below the Sahara Desert on the map). Some of the most populous countries in the sub-Saharan region are Nigeria, Cameroon, and Kenya.

Exploration

26. China and India are the two most populous countries in the world. Find an estimate for the current population and growth rate in each country. Use these figures to estimate what their populations will be in the year 2010.

27. For what reasons might a country's population growth rate decrease? increase? Which of these factors were present in the United States between 1920 and 1980?

Some cars never die. *Most cars depreciate in value rapidly. Some car enthusiasts and collectors restore cars in order to increase their value. Old cars have other uses, like this one at a trendy New York restaurant.*

In each population situation studied in Lesson 9-1, the growth factor was greater than one, so each population increased over time. Sometimes a growth factor is between zero and one. When this is true, the value of the function decreases as the independent variable increases. This happens in situations of **exponential decay.**

Depreciation as an Example of Exponential Decay

Automobiles and other manufactured goods often decrease in value over time. This decrease is called **depreciation.** If the decrease is $r\%$ annually, then each year the item is worth $(100 - r)\%$ of its previous value.

Example 1

In 1988, a new Firebird Trans Am cost $15,798. Suppose the car depreciates 13% each year.
a. Find an equation that gives its value x years after 1988.
b. Predict the car's value in 1995.

Solution

a. Because the car is depreciating in value by a constant factor, the situation can be modeled by an exponential function with equation $V(x) = ab^x$. Here $V(x)$ is the value of the car at a time x years after 1988; a is the original value, 15,798; b is the growth factor. Since $(100 - 13)\% = 87\% = .87$, b = 0.87. Thus an equation giving the value is $V(x) = 15,798(0.87)^x$.
b. The year 1995 is 7 years after 1988, so $x = 7$.
$$V(x) = 15,798(0.87)^x$$
$$V(7) = 15,798(0.87)^7 \approx 5,960.$$
So the model predicts the car's value to be $5,960 in 1995. The actual book value of this car in 1995 was $5,985, so this model gives a fairly accurate prediction.

Radioactive Decay and Half-Life

The amount of a radioactive substance decreases over time. The **half-life** of a substance is the amount of time it takes half of the material to decay. Half-life can range from a few seconds, as in the 3.82 seconds half-life of radon, to millions of years, as in the 4.47×10^9 year half-life of uranium 238.

Wall painting. Carbon dating, which is based on half-life, could be used to find the age of paintings recently discovered in Avignon, France in the Chauvet cave (shown). The artists were the Cro-Magnon people who migrated to this region about 40,000 years ago.

Activity

Strontium 90 (^{90}Sr) has a half-life of 29 years. This means that in each 29 year period, half of the ^{90}Sr decays and half remains. Suppose you have 1000 grams of ^{90}Sr.

a. Complete the table below to find the amount of ^{90}Sr remaining after 1 to 10 half-life periods.

Number of years	Number of half-life periods	Amount of ^{90}Sr remaining
—	—	1000 g
29	1	500 g
58	2	.
.	3	.
.	.	.
.	.	.

b. If you start with 1000 g of ^{90}Sr, how much will remain after ten half-life periods?

c. How many years equal ten half-life periods of ^{90}Sr?

d. What kind of sequence occurs in the rightmost column?

If an amount a of a radioactive substance has a known half-life period, then one half-life period later the amount left is 50% of a or $.5a$. During each additional half-life period the amount decays by another factor of .5. So after x half-life periods the amount of radioactive material left is $a(.5)^x$. Thus, radioactive decay is another example of exponential decay.

Radioactive Carbon 14 (^{14}C) also decays exponentially. The amount of ^{14}C in a plant or animal fossil is often used to date the fossil.

Example 2

^{14}C has a half-life of 5730 years.

a. Determine an equation for the percent of ^{14}C remaining in the original sample after x half-life periods.

b. Graph the equation.

c. If an artifact has only 20% of the ^{14}C it had originally, about how old is it?

▶

Solution

a. Use the exponential equation $y = ab^x$, where x is the number of half-life periods and y is the percent of ^{14}C remaining. Since half remains in a fixed time period, $b = \frac{1}{2}$ or 0.5. The initial amount is 100%, so $a = 100$. The equation is
$$y = 100(0.5)^x,$$
where x is the number of 5730-year intervals.

b. Evaluate y for various values of x which correspond to the number of half-life periods. Draw the graph using an automatic grapher.

x	y
-1	200
0	100
1	50
2	25
3	12.5
4	6.25

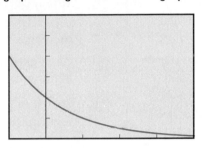

$-1 \leqslant x \leqslant 4, \quad x$-scale = 1
$0 \leqslant y \leqslant 300, \quad y$-scale = 50

c. Use the graph to find x when $y = 20$. From the table, $2 < x < 3$. Use the trace feature to get a more accurate estimate. We find $x \approx 2.3$ half-life periods or $(2.3)(5730)$ years. In about 13,200 years after the measurement, only 20% would be left. An artifact with only 20% of its original ^{14}C is about 13,200 years old.

Check

c. Evaluate $100(.5)^{2.3} \approx 20.3$. It checks.

Growth vs. Decay

The examples of exponential growth from Lesson 9-1 and the examples of exponential decay from this lesson fit a general model called the *Exponential Growth Model.*

Exponential Growth Model
If a quantity a grows by a factor b ($b > 0$, $b \neq 1$) in each unit period, then after a period of length x, there will be ab^x of the quantity.

When $b > 1$, the situation is one of exponential growth. The values of y increase as x increases.

When $0 < b < 1$, the situation is one of exponential decay. The values of y decrease as x increases.

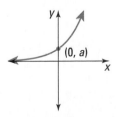

$y = ab^x, a > 0, b > 1$

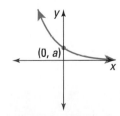

$y = ab^x, a > 0, 0 < b < 1$

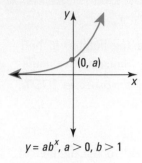

$y = ab^x, a > 0, b > 1$

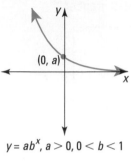

$y = ab^x, a > 0, 0 < b < 1$

The exponential functions $y = ab^x$ have the following properties.

1. The domain of each function is the set of real numbers.
2. The range of each function is the set of positive real numbers.
3. The y-intercept of each graph is a.
4. The graph never crosses the x-axis. Exponential curves have no x-intercepts.
5. In one of its two quadrants, the graph gets closer and closer to the x-axis. The x-axis is an asymptote of the graph.
6. When the constant growth factor b is greater than one, as x increases, the value of the function increases. If $0 < b < 1$, as x increases, the value of the function decreases.
7. Each exponential growth curve is the reflection image of an exponential decay curve over the y-axis.

Properties of powers can be used to prove that some functions are exponential functions.

Example 3

Let $f(x) = 3^{-x}$. Prove that f is an exponential decay function.

Solution
$f(x) = 3^{-x} = (3^{-1})^x$. Using the Negative Exponent Theorem, $(3^{-1})^x$ can be rewritten as $\left(\frac{1}{3}\right)^x$. So $f(x) = \left(\frac{1}{3}\right)^x$. This is an exponential function with growth factor $\frac{1}{3}$ between 0 and 1. So the function is an exponential decay function.

Check
Use an automatic grapher with window $-2 \leq x \leq 5, 0 \leq y \leq 10$. Entering $y = 3^{-x}$ should result in the graph below. The shape of the curve suggests exponential decay.

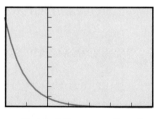

$-2 \leq x \leq 5, \quad x$-scale $= 1$
$0 \leq y \leq 10, \quad y$-scale $= 1$

Covering the Reading

1. Refer to Example 1. If you assume that the depreciation model continues to be valid, what would the Firebird be worth in 1996?

2. Suppose a boat purchased for $28,000 decreases in value by 10% each year.
   a. What is the yearly growth factor?
   b. How much is the boat worth after 2 years?
   c. How much is the boat worth after x years?

3. What does the term "half-life" mean?

4. Refer to the Activity in the lesson. Answer parts **a** to **d**.

In 5 and 6, refer to Example 2.

5. Find how much ^{14}C remains after 17,190 years.

6. How old is an artifact which now has 60% of the ^{14}C it had originally?

7. If $y = ab^x$ and $0 < b < 1$, then y __?__ as x increases.

8. *Multiple choice.* Pick the equation for an exponential decay function and explain why it is that kind of function.
   (a) $f(x) = \frac{1}{5}^x$
   (b) $f(x) = 5^{-x}$
   (c) $f(x) = 5^x$
   (d) $f(x) = \left(\frac{1}{5}\right)^{-x}$

9. **a.** Graph $y = 5^x$ for values of x between -3 and 3.
   b. Graph $y = 5^{-x}$ for values of x between -3 and 3 on the same axes.
   c. The graphs in parts **a** and **b** are related to each other. How are they related?

10. *Multiple choice.* Which graph could represent exponential decay?

(a)
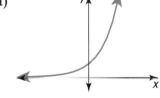

(b)
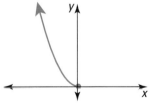

(c)
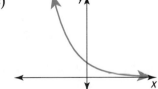

(d)
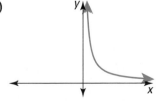

In 11 and 12, *true or false.*

11. The graph of an exponential function has both an *x*-intercept and a *y*-intercept.

12. The *y*-axis is an asymptote for the graph of an exponential function.

Applying the Mathematics

13. Suppose a car costs $10,000 and loses value every year. Let *N* be its value after *t* years.
  a. Assume the depreciation is exponential with 20% of the value lost per year. Then $N = 10{,}000(0.8)^t$. Copy and complete the table below.

t	1	2	3	4
N				

  b. Assume the depreciation is linear with $1000 lost per year. Then $N = 10{,}000 - 1000t$. Copy and complete the table below.

t	1	2	3	4
N				

  c. If this were your car and you were trading it in after 4 years, explain why you would probably prefer that the car dealer assumed the model of part **b?**
  d. Under what circumstances would you prefer that the dealer use the equation in part **a?** Justify your answer.

14. The half-life of iodine 123 (^{123}I) is about 13 hours. Suppose you begin with a sample of 20 grams.
  a. Write an equation to model the decay.
  b. Copy and complete the table below.

x = number of 13-hour periods	-2	-1	0	1	2	3
y = amount of ^{123}I						

  c. Graph the exponential function containing these points. Show these points on your graph.
  d. Use the graph to estimate the number of hours needed for 20 g of ^{123}I to decay to 4 g.

Review

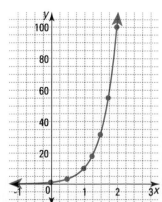

In 15–18, use the graph at the left. *(Lessons 1-2, 9-1)*

15. When $x = \underline{\ ?\ }$, $y = 100$.

16. *Multiple choice.* Which could be an equation for the graph?
  (a) $y = 10x$
  (b) $y = 10 + x$
  (c) $y = 10^x$
  (d) $y = \left(\frac{1}{10}\right)^x$

Shown are people in Chapultepec Park, Mexico City's largest park. The park includes art and history museums, a zoo, an amusement park, and Chapultepec Castle—the residence of some former Mexican presidents.

17. What is the domain of the function?

18. What is the range of the function?

19. In the 1980s and 1990s, Mexico City was one of the most rapidly growing cities in the world. In 1991, its population was 20,899,000, and the average annual growth rate was about 2.69%. Use these data to estimate the population of Mexico City in the year 2001. *(Lesson 9-1)*

20. Suppose an experiment begins with 120 bacteria and that the population of bacteria doubles every hour.
    a. About how many bacteria will there be after 3 hours?
    b. What is an equation for the number y of bacteria after x hours? *(Lesson 9-1)*

21. The inflation rate is typically reported monthly. Suppose a monthly rate of 0.5% is reported for January. Assume this rate continues for 1 year. *(Lessons 5-1, 7-2, 9-1)*
    a. What is the inflation rate for the year?
    b. The value 0.5% has been rounded. The actual value could be any number equal to or greater than 0.45%, and less than 0.55%. Write an inequality for r, the annual inflation rate, based on those two extreme values.

22. a. Give a decimal approximation to the nearest hundredth. *(Lessons 7-2, 7-7, 9-1)*
    i. 4^3 ii. $4^{\frac{7}{2}}$ iii. $4^{\sqrt{10}}$
    b. Explain why the answer to part iii should be between the answers to i and ii. *(Lesson 9-1)*

23. Write $\dfrac{10^{10.8}}{10^{8.4}}$
    a. as a power of ten.
    b. as a decimal. *(Lessons 7-2, 7-7)*

24. Simplify $gr^{n-1} \cdot r$. *(Lesson 7-2)*

Exploration

25. a. Graph $y = 10^x$ using an automatic grapher. Use the graph to estimate solutions to the following pairs of equations.
    (i) $10^a = 3$ and $10^b = \frac{1}{3}$
    (ii) $10^c = 2$ and $10^d = \frac{1}{2}$
    b. What generalization have you verified?

*The
Number e*

IN·CLASS
ACTIVITY

Work with a partner on this Activity.

Recall that the General Compound Interest Formula
$$A = P\left(1 + \frac{r}{n}\right)^{nt},$$

gives the amount A an investment is worth when the principal P is invested in an account paying an annual interest rate r with interest compounded n times per year for t years.

Suppose a bank pays 100% interest annually (don't we wish!), and you invest $P = \$1$. As you know, as the number of compounding periods within the year increases, the amount you earn increases.

1 Copy and complete the table below to show the value of A at the end of one year ($t = 1$) for successively shorter compounding periods.

type of compounding	n = number of compoundings per year	$P\left(1 + \frac{r}{n}\right)^{nt}$	A
annually	1	$1\left(1 + \frac{1}{1}\right)^1$	\$2.00
semi-annually	2	$1\left(1 + \frac{1}{2}\right)^2$	\$2.25
quarterly			\$2.44141
monthly			
daily			
hourly			
by the second			

2 Suppose you could compound interest continuously or instantaneously, that is, over smaller and smaller periods of time. The sequence of values for the total amount gets closer and closer to the number e, which is approximately equal to 2.71828. Thus, the number e is the value of \$1 after one year of being invested at 100% interest compounded continuously. Like π, the number e can be found on virtually every graphics calculator. What value does your calculator display for e? (Hint: Look for a key or menu selection labeled e^x, and evaluate e^1.)

The number e studied in the previous Activity is named after Euler, who proved that the sequence of numbers of the form $\left(1 + \frac{1}{n}\right)^n$ approaches a particular number as n increases. Like π, e is an irrational number which can be expressed as an infinite, nonrepeating decimal. Here are the first fifty places of e.

$e \approx 2.71828182845904523536028747135266249775724709369995 \ldots$

The number e arises in situations in which interest is compounded continuously.

Interest Compounded Continuously

Suppose a bank pays 5% interest on $1 for one year. Here are some values of A for different compounding periods.

compounding method	$P\left(1 + \frac{r}{n}\right)^{nt}$	A
annually	$1\left(1 + \frac{.05}{1}\right)^1$	$1.05
semi-annually	$1\left(1 + \frac{.05}{2}\right)^2$	$1.050625
quarterly	$1\left(1 + \frac{.05}{4}\right)^4$	$1.050945
daily	$1\left(1 + \frac{.05}{365}\right)^{365}$	$1.051267
hourly	$1\left(1 + \frac{.05}{8760}\right)^{8760}$	$1.051271

Notice that the total amount seems to be getting closer to $1.051271. . . . This number is very close to the value of $e^{0.05}$. In fact, $1 compounded continuously at 5% annual interest for one year will be worth exactly $\$e^{0.05}$.

Activity

Evaluate $e^{.05}$ on your calculator. If your calculator does not have an $\boxed{e^x}$ function, you can evaluate $e^{.05}$ by using the approximation given above.

At a rate of 6.5% annual interest, an amount P compounded continuously would be worth $P \cdot e^{.065}$. In t years, it would be worth $P \cdot (e^{.065})^t$. For situations where interest is *compounded continuously,* the General Compound Interest Formula can be greatly simplified.

Continuously Compounded Interest Formula
If an amount P is invested in an account paying an annual interest rate r compounded continuously, the amount A in the account after t years will be

$$A = Pe^{rt}.$$

Example 1

If $850 is invested at an annual interest rate of 6% compounded continuously, what is the amount in the account after 10 years?

Solution

Use the formula $A = Pe^{rt}$, where $P = 850$, $r = 0.06$ and $t = 10$.

$$A = 850e^{0.06(10)}$$
$$A = 850e^{0.6}$$
$$A \approx 1548.80.$$

After 10 years, the amount in the account is $1548.80.

Hot mail? *Shown is mail, arriving in West Germany in May 1986, being tested for radioactive contamination. After the Chernobyl nuclear disaster in the Ukraine in April 1986, many countries tested items coming out of that area for radioactive contamination.*

Other Uses of the Number e

The exponential function $y = e^x$ has special properties that make it particularly suitable for applications. Some of these properties are studied in calculus. For now, you only need to know that many formulas for growth and decay are written using e as the base. This is why most scientific and graphics calculators have a key to find values of e^x.

Example 2 illustrates continuous exponential decay.

Example 2

The amount L of a certain radioactive substance remaining after t years decreases according to the formula $L = Be^{-0.0001t}$. If 2000 μg (micrograms) are left after 6000 years, how many micrograms were present initially?

Solution

When $t = 0$, $L = Be^{-0.0001(0)} = B \cdot 1 = B$, so B is the initial amount present. When $t = 6000$, $L = 2000$. Substitute these values and solve for B.

$$L = Be^{-0.0001t}$$
$$2000 = Be^{-0.0001(6000)}$$
$$2000 \approx B(0.54881)$$
$$3644 \approx B$$

About 3600 μg were present initially.

Formulas such as those in Examples 1 and 2,

$$A = 850e^{0.06t}$$
$$\text{and } L = Be^{-0.0001t}$$

are instances of a general model for situations involving continuous change. The *Continuous-Change Model* is often described using function notation. Let N_0 (read "N subzero" or "N naught") be the initial amount, and let r be the growth factor by which this amount continuously grows or decays per unit time t. Then $N(t)$, the amount at time t, is given by the equation

$$N(t) = N_0e^{rt}.$$

This equation can be rewritten as $N(t) = N_0(e^r)^t$. So it is an exponential equation of the form

$$y = ab^x$$

where $a = N_0$, $x = t$, and the growth factor $b = e^r$. If r is positive, then $e^r > 1$ and there is exponential growth. If r is negative, then $0 < e^r < 1$ and there is exponential decay.

QUESTIONS

Covering the Reading

1. In whose honor is the number e named?

2. Approximate e to the nearest hundred-thousandth.

In 3 and 4, use the General Compound Interest Formula and let $P = \$1$, $r = 100\%$, and $t = 1$ year.

3. What is the value of $\$1$ invested for one year at 100% interest compounded hourly?

4. As n increases, the value of A becomes closer and closer to what number?

5. Give the value of $e^{.05}$ obtained on your calculator rounded to the nearest millionth.

6. If $\$1$ is invested at 10% interest compounded continuously, find its value at the end of one year.

7. Suppose $\$700$ is invested at an annual interest rate of 5% compounded continuously. How much is in the account after 10 years?

8. Use the formula $L = Be^{-.0001t}$ given in Example 2. If at the end of 2000 years there are 1000 micrograms of the substance remaining, how many micrograms were present initially?

9. Consider the function $N(t) = N_0e^{rt}$.
   a. What does N_0 represent?
   b. What does e^r represent?
   c. What is true about r when this function models exponential decay?

Applying the Mathematics

10. a. Graph each function for values of x between -3 and 3, inclusive.
    (i) $y_1 = e^x$ (ii) $y_2 = e^{-x}$ (iii) $y_3 = \left(\frac{1}{e}\right)^x$
    b. Explain why two of the graphs in part a coincide.
    c. Which of the functions in part a describe exponential growth?
    d. Which describe exponential decay?

This is a traffic jam in Karachi, the largest city and chief port of Pakistan, In 1994, the population of Pakistan was 121,856,000.

11. Complete with >, <, or = : π^e _?_ e^π.

12. In 1991, the population of the metropolitan area of Karachi, Pakistan, was about 8 million. At that time, the formula $N(t) = N_0 e^{0.039t}$ was being used to project the population t years later.
   a. What annual rate of population increase was assumed?
   b. Find $N(9)$ and explain what this number represents.
   c. Calculate $N_0(1.039)^t$ for $t = 9$, and explain what this number represents.

13. A machine used in an industry depreciates so that its value after t years is given by $N(t) = N_0 e^{-.25t}$.
   a. What is the annual rate r of depreciation of the machine?
   b. If after 3 years the machine is worth $12,000, what was its original value?

14. Some people have theorized that a rumor spreads like an epidemic at a rate directly proportional to the number of people who have heard the rumor (and thus perpetuate it). Under this assumption, rumor spreading can be modeled by the equation

$$H = \frac{C}{1 + \left(\frac{C - S}{S}\right)e^{-0.4t}},$$

where C is the total number of people in the community, S is the number of people who initially spread the rumor, and H is the number of people who have heard the rumor after t minutes. In a school of 1800 students, one student on Friday overhears the principal saying the following Monday there will be a surprise school holiday. About how many students will have heard the rumor after 45 minutes?

Review

15. Radium 226 (^{226}Ra) has a half-life of 1600 years. Suppose 100% of ^{226}Ra is present initially.
   a. Make a table of values showing how much ^{226}Ra will be left after 1, 2, 3, 4, and 5 half-life periods.
   b. Write a formula for the amount A of ^{226}Ra left after x half-life periods.
   c. Write a formula for the amount A of ^{226}Ra left after t years.
   (Lesson 9-2)

16. Let $f(x) = 9^x$.   *(Lessons 7-3, 7-6, 8-2, 9-1)*
   a. Evaluate $f(-2)$, $f(0)$, and $f\left(\frac{3}{2}\right)$.
   b. Identify the domain and range of f.
   c. Give an equation for the reflection image of the graph of $y = f(x)$ over the y-axis.

17. Rationalize the denominator of $\dfrac{\sqrt{3} + 1}{\sqrt{3} - 1}$.   *(Lesson 8-6)*

18. Solve $9m^3 = 16$.   *(Lessons 7-6, 8-4)*

19. Use the diagram at the right. Midpoints of a 12-by-16 rectangle have been connected to form a rhombus. Then midpoints of the rhombus are connected to form a rectangle, and so on.

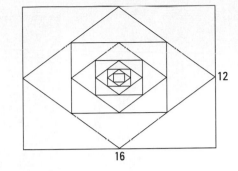

a. List the perimeters of the first 6 rectangles formed (including the largest rectangle).
b. What kind of sequence do the perimeters of the first 10 rectangles form? *(Previous course, Lesson 7-5)*

20. For what values of m does the equation $mx^2 + 12x + 9 = 0$ have exactly one solution for x? *(Lesson 6-10)*

21. Choose all that apply. What kind of number is $\sqrt{-4}$? *(Lessons 6-2, 6-8, 6-9)*
    (a) rational   (b) irrational   (c) imaginary

22. The table below gives the number of African-Americans elected to the U.S. Congress or state legislatures.

year	1970	1975	1980	1985	1990
number of elected officials	179	299	326	407	440

a. Let x = the year after 1970 and y = the number of elected officials. Find an equation for the line of best fit for this data set.
b. Use the equation in part **a** to predict the number of African-Americans that will be in Congress or a state legislature in the year 2005. *(Lesson 3-6)*

Exploration

23. Another way to get an approximate value of e is to evaluate the infinite sum
$$1 + \frac{1}{1!} + \frac{1}{2!} + \frac{1}{3!} + \ldots.$$
(Recall that $n!$ is the product of all integers from 1 to n inclusive).
a. Use your calculator to calculate each of the following.
$$1 + \frac{1}{1!} + \frac{1}{2!} + \frac{1}{3!}$$
$$1 + \frac{1}{1!} + \frac{1}{2!} + \frac{1}{3!} + \frac{1}{4!}$$
$$1 + \frac{1}{1!} + \frac{1}{2!} + \frac{1}{3!} + \frac{1}{4!} + \frac{1}{5!}$$
b. How many terms must you add to approximate $e = 2.71828\ldots$ to the nearest thousandth? What is the last term you need to add to do this?

LESSON

$9\text{-}4$

*Fitting
Exponential
Models to
Data*

***Kelp*, I'm trapped below the surface!** *Kelp fields often partially block out sunlight and form underwater forests. Shown are harvested giant kelp fields at Nugget Point, New Zealand. When harvested, the plants are brought to shore.*

In previous chapters, you have learned how to fit linear or quadratic models to data. In this lesson, we study how to fit exponential models to data.

Finding Exponential Equations for Data Known to be Exponential

As you know, exponential functions have the form $y = ab^x$, where a is the value of y when $x = 0$ and b is the growth factor during each unit period. If data are known to be exponential and if a and b are known, you can write an exponential model by substituting the values of a and b into the equation $y = ab^x$. If a and b are not known you can find them using the techniques illustrated in Example 1.

Example 1

The amount of sunlight that reaches the plant life underwater determines the amount of photosynthesis that takes place. The table and graph below give the percentage of sunlight that is present at various depths in a part of an ocean. The percentage of sunlight is known to depend exponentially on depth. Find an exponential equation to model the data.

depth (in meters)	percentage of light
0	100
10	10.7
20	1.15
30	.12
40	.01
50	.001

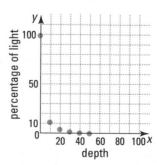

Solution 1

We need to find a and b in the equation $y = ab^x$. Using the point $(0, 100)$, $100 = ab^0$. Thus $a = 100$ and the equation is of the form $y = 100b^x$. Now we choose another point and substitute. We choose $(20, 1.15)$.

$1.15 = 100b^{20}$ Substitute.

$0.0115 = b^{20}$ Divide each side by 100.

                Take the 20th root of each side

$0.8 \approx b$ $\left(\text{or raise to the } \frac{1}{20} \text{ power}\right)$.

So an equation for the data is $y \approx 100 \cdot (.8)^x$.

Solution 2

The point $(0, 100)$, which tells us the y-intercept, was very convenient to use in Solution 1 because $b^0 = 1$, and the value of a could be easily found. If you do not know the y-intercept, the following method can be used to find a and b in $y = ab^x$. Choose two data points and substitute into $y = ab^x$. The result is a system of two equations in a and b.

Substitute $(10, 10.7)$ into $y = ab^x$. $10.7 = ab^{10}$ (1)

Substitute $(40, .010)$ into $y = ab^x$. $.010 = ab^{40}$ (2)

Divide (2) by (1). $\dfrac{.010}{10.7} = \dfrac{ab^{40}}{ab^{10}}$

Simplify. $9.3458 \cdot 10^{-4} \approx b^{30}$

Take the 30th root of each side. $.7925 \approx b$

Substitute $b \approx .7925$ into one of the first two equations. We use equation (1). $10.7 = a(.7925)^{10}$.

Solve for a. $109.5 \approx a$

An equation for the data is $y = 109.5(.7925)^x$.

Notice that the equations in Solutions 1 and 2 are not identical. This indicates that all the data points do not fit exactly on the same exponential curve. Let $y_1 = 100(.8)^x$ and $y_2 = 109.5(.7925)^x$. To decide which is better, you could make a table of values comparing the values predicted by these models to the actual percentage y_A of sunlight.

Each equation predicts values close to the actual value, y_A. Alternately, you could graph each equation and see how close the actual y-values are to the curves.

x	y_A	y_1	y_2
0	100	100	109.5
10	10.7	10.74	10.70
20	1.15	1.153	1.046
30	.12	.124	.1022
40	.01	.013	.01
50	.001	.0014	.00098

Many automatic graphers have the ability to find an exponential equation that models a set of data. You should check to see if your grapher has such a feature, and if so, learn how to use it. Our grapher gives

$$y = 107.9(.794)^x$$

as the best exponential model for the data in Example 1.

a. Let $y_3 = 107.9(.794)^x$. Evaluate y_3 for the values of x in the table on the previous page.

b. Which model, y_1, y_2, or y_3 is more accurate at a depth of 50 meters?

Deciding Whether an Exponential Model Is Appropriate

Sometimes you may not be sure that an exponential model is appropriate. If that is the case, consider the growth factor between various data points. If it is constant, an exponential model is appropriate; if not, another model must be found. Consider again the population of the United States from 1790 to 1990 given on page 531. Below we give data from only the first eight and last three censuses.

Year	Population	Decade Growth Factor
1790	3,930,000	
1800	5,300,000	1.349
1810	7,240,000	1.366
1820	9,640,000	1.331
1830	12,870,000	1.335
1840	17,070,000	1.326
1850	23,190,000	1.359
1860	31,440,000	1.356
1970	203,300,000	1.134
1980	226,540,000	1.114
1990	248,710,000	1.098

The **decade growth factor** in the table is the ratio of the population in a specific year to the population 10 years earlier.

For example, the decade growth factor for 1820 is $\frac{9,640,000}{7,240,000}$ or approximately 1.331, indicating a 33.1% population increase from 1810 to 1820. The **yearly** or **annual growth factor** for a given decade is the positive number b such that b^{10} gives the decade growth factor.

Example 2

Calculate the annual growth factor between 1810 and 1820.

Solution

The decade growth factor is 1.331. Solve $b^{10} = 1.331$.
Take the 10th root of each side or raise each side to the $\frac{1}{10}$ power. Only the positive root is appropriate here. $\sqrt[10]{1.331} \approx 1.029$.
This indicates that the annual growth factor is about 2.9% from 1810 to 1820. In other words, the population in 1811 was about 1.029 times the population in 1810.

Notice that during the years 1790 to 1860 the decade growth factors are almost constant. Over that seventy-year period, the annual growth factor was about 3%. However, in recent years the decade growth factors are lower, indicating a lower annual growth rate. This means that a single exponential model does not fit the complete set of data. But different exponential models can be used for smaller time intervals.

Shown is an 1860 painting by Charles Hargems depicting the first Pony Express Ride.

Example 3

Find an exponential model for the U.S. population between 1790 and 1860.

Solution

We know the annual growth rate from the previous discussion, so $b \approx 1.03$ in the equation $y = ab^x$. The growth factor needs to be raised to a power that gives the number of years it has been applied. We choose 1830, a starting point close to the middle of this time period. If x is the year, x – 1830 gives the number of years before or after 1830. The exponent is 0 where $x = 1830$, so the initial value is about 13 million. An equation for the population in millions is

$$y = 13(1.03)^{x - 1830}.$$

The equation given in the solution to Example 3 is not the only exponential equation that can be used to model the data. If you want greater accuracy you can use

$$y = 12.87(1.03)^{x - 1830}.$$

If you choose a different starting point, say 1840, you will get a different equation. You are asked to experiment with other equations in the Questions. But as was the case in the two Solutions to Example 1, each will be a reasonably good model for the data set.

QUESTIONS

Covering the Reading

In 1–3, refer to Example 1.

1. **a.** Use the point (30, .12) and the method of Solution 1 to find an equation to model the data.
   b. Use this equation to predict y when $x = 30$.

2. **a.** Use the points (20, 1.15) and (50, .001) and the method of Solution 2 to find an equation to model the data.
   b. Use this equation to predict y when $x = 30$.

3. Write your answers to the Activity in the Lesson.

In 4 and 5, refer to the U.S. population data given in the lesson.

4. Find the annual growth factor from 1830 to 1840.

5. **a.** Find a model for the population between 1790 and 1860 for which the starting point is 1840.
   b. Use the equation in part **a** to predict the U.S. population in 1860.
   c. Use the equation in Example 3 to predict the population in 1860.
   d. Which equation gives a better prediction for the population in 1860?

6. **a.** Use the equation in Example 3 to estimate the U.S. population in the year 1990.
   b. Why is the answer to part **a** such a bad estimate?

age	life expectancy
0	70.85
10	71.48
20	71.87
40	73.43
60	77.08

Applying the Mathematics

7. Refer to the graph and data on page 531.
   a. Show how the equation for 1870–1910 was obtained.
   b. Explain why this model is not a good fit for the data after 1910.

8. The table at the left gives the life expectancy for males in Finland as a function of their age.
   a. Make a scatterplot of these data.
   b. Fit an exponential model to this data.
   c. Fit a linear model to this data.
   d. Use each of the models to predict the life expectancy of a 30-year-old Finnish male.
   e. Over this time period, compare the differences in the two models. For what ages do the two models begin to give quite different values?

Different expectations.
When Daisy Fuentes (top) was born in 1966, the life expectancy was 69 yr. When Jaleel White (middle) was born in 1977, the life expectancy was 66 yr. When George Burns (bottom) was born in 1896, the life expectancy was 51 yr.

Review

9. a. Graph $y = e^{3x}$ for $-2 < x < 2$. Label the coordinates of three points on the graph.
   b. State the domain and range of this function.   *(Lesson 9-3)*

10. The amount A of radioactivity from a nuclear explosion is given by $A = A_0 e^{-2t}$, where t is measured in days after the explosion. What percent of the original radioactivity is present 5 days after the explosion?   *(Lesson 9-3)*

11. Tina invests $850 in an account with a 2.75% annual interest rate. What will be her balance if she leaves the money untouched for 4 years compounded as follows?   *(Lessons 7-4, 9-3)*
    a. annually   b. daily   c. continuously

12. Consider the sequence defined by $\begin{cases} g_1 = .375 \\ g_n = .8g_{n-1}, \text{ for integers } n \geq 2. \end{cases}$
    (Lessons 3-1, 7-5, 9-1, 9-2)
    a. List the first five terms of this sequence.
    b. Which phrase applies to this sequence: exponential growth, exponential decay, constant increase, constant decrease?

In 13 and 14, simplify without using a calculator.   *(Lessons 7-3, 7-6, 7-7)*

13. $\left(\frac{1}{3}\right)^{-4}$ 14. $64^{\frac{7}{6}}$

15. Simplify $(a + 2b)^2 - (2a + b)^2$.   *(Lesson 6-1)*

16. Suppose m varies inversely with n, and $m = 18$ when $n = 18$. Find m when $n = 3$.   *(Lesson 2-2)*

Exploration

17. Find the life expectancy for either U. S. males or U. S. females of various ages. Try to model the data with an exponential function.

LESSON
9-5

Common Logarithms

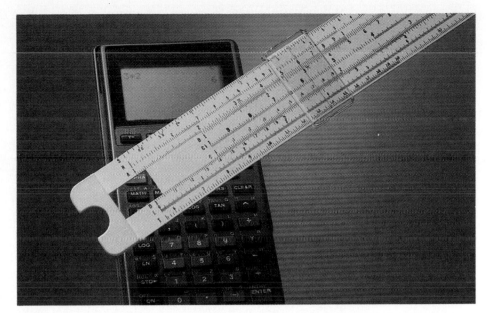

Slide rules. *Before calculators became readily available, the slide rule was widely used to do many mathematical operations, such as multiplication, division, square roots, and logarithms.*

The Inverse of $y = 10^x$

In the last four lessons you studied exponential functions. In this lesson, you will investigate the inverse of the exponential function defined by $y = 10^x$ and see how it is related to the title of this lesson.

Activity 1

a. Using a calculator, make a table of values for points on the graph of $y = 10^x$ from $x = -2$ to $x = 2$. Round values to the nearest hundredth.

x	-2	-1.5	-1	-.5	0	.5	1	1.5	2
$y = 10^x$	.01			.32					

b. On a sheet of graph paper, label the x- and y-axes and number each from -2 to 10.

c. Plot the points from part **a** and connect them with a smooth curve. Label the curve with its equation $y = 10^x$.

d. Make a new table of values by switching the x- and y-coordinates in part **a**.

x	.01			.32					
y	-2	-1.5	-1	-.5	0	.5	1	1.5	2

This is a table of values for the inverse of the function with equation $y = 10^x$.

e. Plot the points from part **d** and connect them with a smooth curve.

f. What equation relates the points in part **e**?

g. Draw the line $y = x$ on the same set of axes.

Your graphs from the Activity should look like those below.

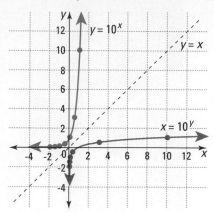

The graph of $y = 10^x$ passes the Horizontal-Line Test, so its inverse describes a function. An equation for this inverse function is $x = 10^y$.

To solve $x = 10^y$ for y, we define a new function.

Definition
y is the **logarithm of x to the base 10**, or **log of x to the base 10**, written

$$y = \log_{10}x,$$

if and only if $10^y = x$.

Thus, the inverse of $f(x) = 10^x$ can be written in the following ways.

$$x = 10^y$$
$$y = \log_{10}x$$
$$f^{-1}(x) = \log_{10}x$$

This last sentence can be read as "f inverse of x equals the logarithm of x with base 10," or ". . . log base 10 of x." The curve defined by these equations is called a *logarithmic curve*.

Evaluating Common Logarithms

Logarithms to the base 10 are called **common logarithms** or **common logs**. We often write common logs without indicating the base. That is, $\log_{10}x = \log x$. A calculator can be used to evaluate $\log x$ for any positive real number x. On most graphics calculators, to evaluate $\log_{10}100$, press [log] first, then 100. Use your calculator to do Example 1.

Example 1

Evaluate the following.
a. $\log_{10}100$ b. $\log_{10}5$

Solution

Use your calculator. You should get the following.
a. $\log_{10}100 = 2$ b. $\log_{10}5 = .6989700043 \ldots$

▶

► **Check**

Use the definition of logarithm to the base 10: $\log_{10} x = y$ if and only if $10^y = x$.

a. Does $10^2 = 100$? Yes. It checks.

b. Does $10^{.6989700043} \approx 5$? Use the powering key on your calculator. One calculator shows $10^{.6989700043} = 5$. It checks.

Common logarithms of powers of 10 can be found without using a calculator.

Example 2

Evaluate without a calculator.

a. $\log_{10} 1,000,000$

b. $\log .1$

c. $\log \sqrt[3]{10}$.

Solution

First write each number as a power of ten. Then apply the definition of logarithm.

a. $1,000,000 = 10^6$. Because $1,000,000$ is the 6th power of 10, $\log_{10} 1,000,000 = 6$.

b. You need to find n such that $10^n = .1$. Because $.1 = 10^{-1}$, $\log .1 = -1$.

c. $\sqrt[3]{10} = 10^{\frac{1}{3}}$. So, by definition, $\log \sqrt[3]{10} = \frac{1}{3}$.

Check

Use your calculator to check.

You can estimate the common logarithm of a number by putting it in scientific notation.

Example 3

Between what two consecutive integers is log 7598?

Solution 1

Because 7598 is between $10^3 = 1000$ and $10^4 = 10,000$, log 7598 is between 3 and 4.

Solution 2

$7598 = 7.598 \cdot 10^3$. This indicates that log 7598 will be between 3 and 4.

Check

A calculator shows that log 7598 = 3.88 . . . .

The Common Logarithm Function

The function that maps x onto $\log_{10} x$ for all positive numbers x is called the **logarithm function to the base 10** or the **common logarithm function.**

Activity 2

Use an automatic grapher to graph $y = 10^x$ and $y = \log x$ on the same set of axes. Use the window $-2 \le x \le 12$ and $-2 \le y \le 12$. You should see the same curves you drew by hand in Activity 1. Trace along the curve $y = \log x$.
a. When $x = 10$, what is y?
b. When $y = .5$, what is x?

Properties of $y = 10^x$ and $y = \log x$

By looking at the graphs of the functions defined by $y = 10^x$ and $y = \log x$, you may observe several other properties.

1. The domain of the exponential function is the set of real numbers; its range is the set of positive real numbers. Consequently, the domain of the logarithm function is the set of positive real numbers; its range is the set of real numbers.

2. The graph of $y = 10^x$ never touches the x-axis; the x-axis is an asymptote of the graph. Consequently, the graph of $y = \log x$ never touches the y-axis; the y-axis is an asymptote of the graph.

3. The y-intercept of $y = 10^x$ is 1. Consequently, the x-intercept of $y = \log x$ is 1.

Because they are inverses, each property of the exponential function $y = 10^x$ corresponds to a property of its inverse, the common logarithm function $y = \log x$.

Solving Logarithmic Equations

By using the definition of common logarithms you can solve *logarithmic equations.* In this chapter you may use either radical form or a decimal approximation for answers unless told otherwise.

Example 4

Solve for x: $\log x = 1.5$.

Solution

$\log x = 1.5$ if and only if $10^{1.5} = x$.
So $x = 10^{1.5} = 31.6227766 \ldots$.

Check

Evaluate $\log 31.6227766$ with your calculator.
$\log 31.6227766 = 1.5$
It checks.

QUESTIONS

Covering the Reading

1. Show your tables and graph for Activity 1.

2. Consider the graph of $y = \log_{10} x$.
   a. Name its x- and y-intercepts, if they exist.
   b. Name three points on its graph.
   c. Name the three corresponding points on the graph of $y = 10^x$.

3. What are the domain and range of the common logarithm function?

4. The functions f and g with equations $f(x) = \log_{10}x$ and $g(x) = \underline{\ ?\ }$ are inverses of each other.

5. Give the answers to the questions in Activity 2.

6. a. How is the expression $\log_{10}6$ read?
   b. Evaluate $\log_{10}6$ with a calculator.

In 7–9, evaluate to the nearest thousandth with a calculator.

7. $\log 2$　　　　　8. $\log 0.00046$　　　　　9. $\log 4{,}600{,}000$

10. If $m = \log_{10}n$, what other relationship exists between m, 10 and n?

In 11–16, evaluate without using a calculator.

11. $\log 1{,}000{,}000$　　12. $\log 10^8$　　　　13. $\log .00001$

14. $\log \sqrt{10}$　　　　15. $\log 1$　　　　　16. $\log 10$

17. *Multiple choice.* Without using a calculator, what is the approximate value of the common logarithm of 1,000,529?
   (a) 3　　　　(b) 6　　　　(c) 7　　　　(d) 10

In 18 and 19, solve for x.

18. $\log x = 4$　　　　19. $\log x = 2.5$

Applying the Mathematics

20. If a number is between 100 and 1000, its common logarithm is between what two consecutive integers?

21. Without using a calculator, explain how to determine the closest integer to $\log .0012$.

22. The common logarithm of a number is -2. What is the number?

23. If $f(x) = 10^x$ and $g(x) = \log_{10}x$, what must $f \circ g\,(x)$ equal? Why?

24. If $4\log x = 6$, what is the value of x?

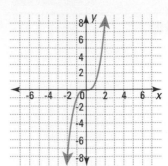

25. Refer to the graph at the left of the cubing function f defined by $f(x) = x^3$. *(Lessons 7-1, 7-2, 7-6, 8-4)*
    a. What are the domain and range of this function?
    b. Graph the inverse.
    c. Is the inverse a function? Why or why not?
    d. What name is usually given to the inverse function?

In 26 and 27, simplify without a calculator. *(Lessons 7-2, 7-3, 7-6)*

26. $(8^2 \cdot 8^{-4})^5$

27. $\sqrt[3]{10^{12}}$

28. Find an equation of the form $y = ab^x$ that passes through the points (1, 6) and (3, 20). *(Lesson 9-4)*

29. The power output P (in watts) of a satellite is given by the equation $P = 50e^{\frac{-t}{250}}$, where t is the time in days. How much power will be available at the end of two years? *(Lesson 9-3)*

30. Between 1983 and 1994, geneticists used high-speed computers and specialized software for what is known as "computational biology" to identify about 200,000 DNA sequences. In September, 1994, GenBank, a database at the National Center for Biotechnology Information (part of the National Institutes of Health) that stores this information, was doubling in size every 21 months. Assume this trend continues.
    a. Estimate the month and year when GenBank will have 400,000 DNA sequences identified.
    b. About how many DNA sequences will be identified by September, 2001?
    c. Write an equation to predict the number n of DNA sequences that GenBank will hold m months after September 1994.
    (Lesson 9-1)

DNA fingerprinting.
Shown are scientists analyzing the genetic code of Pseudorabies virus. DNA fingerprinting divides the DNA into fragments that form a pattern of dozens of parallel bands that reflect the composition of the DNA. It is virtually impossible that the complete DNA pattern of one person could match that of another.

Exploration

31. Refer to the U.S. census bar graph on page 531.
    a. Determine log P for each population P.
    b. Graph x on the horizontal axis, and log P on the vertical axis.
    c. What do you notice about the points you plotted?

x	P (millions)	log P
1790	3.9	
1800	5.3	
1810	7.2	
.		
.		
.		
1990	249	

LESSON 9-6

Logarithmic Scales

A sound experience. *Shown is a rock group performing at the Woodstock 25th-Anniversary Concert. The concert, held in Saugerties, New York in 1994, was plagued with much rain—just as the original concert in 1969 was.*

Logarithmic scales are useful when all numbers are positive and cover a range of values from small to very large. Two commonly used logarithmic scales are the *decibel scale,* which measures sound intensity, and the *pH scale,* which measures the concentration of hydrogen ions in a substance.

The Decibel Scale

The quietest sound that a human can hear has an intensity of about 10^{-12} watts per square meter (w/m^2). The human ear can also hear sounds with an intensity as large as 10^2 w/m^2. Because the range from 10^{-12} to 10^2 is so large, it is convenient to use another unit, the *decibel* (dB), to measure sound intensity. A *decibel* is $\frac{1}{10}$ of a *bel,* a unit named after Alexander Graham Bell (1847–1922), the inventor of the telephone.

The chart at the right gives the sound intensity in w/m^2 and the corresponding decibel values for some common sounds.

Sound Intensity (watts/square meter)		Relative Intensity (decibels)
10^2	jet plane (30 m away)	140
10^1	pain level	130
10^0	amplified rock music	120
10^{-1}		110
10^{-2}	noisy kitchen	100
10^{-3}	heavy traffic	90
10^{-4}		80
10^{-5}		70
10^{-6}	normal conversation	60
10^{-7}	average home	50
10^{-8}		40
10^{-9}	soft whisper	30
10^{-10}		20
10^{-11}		10
10^{-12}	barely audible	0

A formula which relates the sound intensity N in w/m^2 to its relative intensity D in decibels is

$$D = 10 \log \left(\frac{N}{10^{-12}} \right).$$

With this formula, if you know the intensity of a sound, you can find its relative intensity in decibels.

Example 1

Find the relative intensity in decibels of a conversation in which the sound intensity is 3.16×10^{-6} w/m^2.

Solution

Substitute $N = 3.16 \times 10^{-6}$ into the formula given on page 563.

$$D = 10 \log \left(\frac{3.16 \times 10^{-6}}{10^{-12}} \right) = 10 \log 3{,}160{,}000 \approx 65$$

The relative intensity of the conversation is about 65 dB.

Notice that $10^{-6} < 3.16 \times 10^{-6} < 10^{-5}$, and the relative intensity of 65 decibels found in Example 1 falls midway between 60 and 70. This agrees with the table.

To convert from the decibel scale to the w/m^2 scale you can solve a logarithmic equation.

Example 2

A very loud rock band played at a relative intensity of 125 dB. What was the intensity of the sound in w/m^2?

Solution

Substitute $D = 125$ into the formula:

$$125 = 10 \log \left(\frac{N}{10^{-12}} \right)$$

Divide each side by 10.

$$12.5 = \log \left(\frac{N}{10^{-12}} \right)$$

Use the definition of common logarithm.

$$10^{12.5} = \frac{N}{10^{-12}}$$

Multiply each side by 10^{-12}.

$$10^{12.5} \cdot 10^{-12} = N$$

$$10^{.5} = N$$

Music of 125 dB has a sound intensity of about 3.16 w/m^2.

Check

Refer to the chart on the previous page. Notice that 125 dB is between 120 dB and 130 dB on the right-hand scale. This should correspond to a sound intensity between 10^0 and 10^1 w/m^2. Since $1 < 3.16 < 10$, it checks.

Refer again to the chart on page 563. Notice that as the decibel values in the right column increase by 10, the corresponding intensities in the left column are multiplied by 10. Thus, if the number of decibels is increased by 20, the sound intensity is multiplied by $100 = 10^2$. If you increase the sound intensity by 40 dB, you multiply the watts per square meter by $10{,}000 = 10^4$. *In general, an increase of n dB multiplies the sound intensity by $10^{\frac{n}{10}}$.*

Example 3

Rock music played at 125 dB is how many times as intense as music played at 105 dB?

Solution

Use the generalization on the previous page. The difference in the decibel level is 125 – 105 = 20 dB. So the 125 dB music is $10^{\frac{20}{10}} = 10^2 = 100$ times as intense as the 105 dB music.

Check

Use the formula $D = 10\left(\frac{N}{10^{-12}}\right)$ to solve for the intensity N when $D = 125$ and $D = 105$. Then find the ratio of the two values. When $D = 125$ dB, we found in Example 2 that $N = 10^5$ w/m^2. By a similar process, when $D = 105$,

$$105 = 10 \log\left(\frac{N}{10^{-12}}\right).$$

To solve for N, rewrite the equation.

$$10^{10.5} = \frac{N}{10^{-12}}$$

so $$10^{-1.5} \text{ w/m}^2 = N.$$

Dividing the values, $\frac{10^5}{10^{-1.5}} = 10^{5-(-1.5)} = 10^2 = 100$. This checks.

Linear Scales versus Logarithmic Scales

Refer again to the chart on page 563. The decibel scale on the right is a *linear scale.* On a **linear scale,** the units are spaced so that the *difference* between successive units is the same.

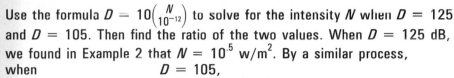

In the w/m^2 scale, an example of a **logarithmic scale,** the units are spaced so that the *ratio* between successive units is the same.

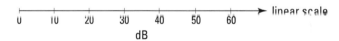

The pH Scale

The *acidity* or *alkalinity* of a substance is measured on another logarithmic scale, called the *pH scale.*

$$\mathbf{pH} = -\log(\mathbf{H}^+),$$

where H^+ is the concentration of hydrogen ions (in moles/liter) of the substance.

The pH of a substance can range from 0 to 14. If the pH of a substance is less than 7, it is called *acidic.* If the pH equals 7, it is called *neutral.* If the pH is greater than 7, the substance is called *alkaline.*

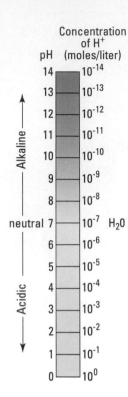

Concentration
of H⁺
pH (moles/liter)

pH	Concentration
14	10^{-14}
13	10^{-13}
12	10^{-12}
11	10^{-11}
10	10^{-10}
9	10^{-9}
8	10^{-8}
neutral 7	10^{-7} H_2O
6	10^{-6}
5	10^{-5}
4	10^{-4}
3	10^{-3}
2	10^{-2}
1	10^{-1}
0	10^{0}

Alkaline

Acidic

The figure at the left shows the corresponding concentration of hydrogen ions. Read the scales from top to bottom. Notice that each decrease of one unit of pH increases the concentration of hydrogen ions by a factor of 10. A decrease of three units of pH, say from pH = 7 to pH = 4, increases the concentration of hydrogen ions by $10^3 = 1000$. This idea generalizes. *A decrease of x units of the pH scale increases the concentration of hydrogen ions by a factor of 10^x.*

Example 4

A family has a garden plot. A test shows that the soil is slightly acidic, with pH = 5.8. Cabbage prefers slightly alkaline soil with pH = 7.5.
a. Which soil, that of the garden plot or that preferred by cabbage, has a higher concentration of hydrogen ions?
b. How many times as great is it?

Solution

a. Refer to the above scales. A higher concentration of hydrogen ions corresponds to a lower pH. So The soil with pH = 5.8 (the garden plot) has a higher concentration of hydrogen ions.
b. The difference in pH between the soil types is 7.5 − 5.8 = 1.7. So by the generalization above, The garden plot has a concentration of hydrogen ions that is $10^{1.7} \approx 50$ times as great as that preferred by cabbage.

Few plants will survive in soils more acidic than pH = 4 or more alkaline than pH = 8. Except for parts of the digestive tract and a few other isolated areas, most cells in an animal function best when conditions are nearly neutral. This is why many people are concerned about acid rain and other changes to the pH of substances in our environment.

Some other examples of logarithmic scales include the scales used on radio dials and the scale for measuring the magnitude (brightness) of stars. Earthquake intensity is often reported using a logarithmic scale called the *Richter Scale*. However, recent scientific evidence suggests that the Richter model is not accurate for very powerful earthquakes, so it is being replaced by other models.

QUESTIONS

Covering the Reading

In 1 and 2, use the formula $D = 10 \log \left(\frac{N}{10^{-12}}\right)$ for intensity of sound.

1. Find the relative intensity in decibels of an explosion which has a sound intensity of 1.65×10^{-2} w/m².

2. Find the intensity, in w/m², of a sound which has a relative intensity of 45 dB.

In 3–5, refer to the chart of sound intensity levels or to the formula for D in terms of N.

3. What is the intensity of a sound which is barely audible to human beings?

4. Give an example of a sound that is 100 times more intense than a noisy kitchen.

5. How many times more intense is normal conversation than a soft whisper?

6. The intensity level of a jet plane at a distance of 600 m is 20 dB more than that of a pneumatic drill 15 m away. How many times more intense is the sound of the jet?

7. The relative intensity of one engine is 35 dB higher than that of a second engine. How many times more intense is the first than the second?

8. Describe the major difference between a linear scale and a logarithmic scale.

9. Why is a logarithmic scale better than a linear scale for illustrating the data below?
5×10^{-34} kg; 1.6726×10^{-27} kg; 10^{-21} kg; 3.15 kg; 1.38×10^{5} kg

10. The gastric juice in your digestive system has a pH of 1.7 and many soft drinks have a pH of 3.0.
   a. Which is more acidic, gastric juice or soft drinks?
   b. What is the concentration of hydrogen ions in the more acidic solution?

11. Seawater has a pH of 8.5.
   a. Is seawater acidic or alkaline?
   b. What is the concentration of hydrogen ions in seawater?
   c. Rewrite your answer to part b in scientific notation.

12. A solution consists only of acid and water. It is changed in acidity from pH 5 to pH 1.
   a. Which has been added, acid or water?
   b. The concentration of hydrogen ions in the solution with pH 1 is how many times the H^{+} concentration in the solution with pH 5.

13. a. Which name, linear scale or logarithmic scale, best describes the number line below? Justify your answer.

   b. Plot the points $A = 5$ and $B = 3.2 \cdot 10^{-2}$ on this line. Explain why you placed the numbers where you did.

Applying the Mathematics

In 14–17, use the following information. The Richter scale is a logarithmic scale in which the value x on the Richter scale corresponds to a measured amplitude of $K \cdot 10^x$, where the constant K depends on the units being used to measure the quake. The table below gives a brief description of the effects of earthquakes of different magnitudes.

Richter magnitude	Description
1	cannot be felt except by instruments
2	cannot be felt except by instruments
3	cannot be felt except by instruments
4	like vibrations from a passing train
5	strong enough to wake sleepers
6	very strong; walls crack, people injured
7	ruinous; ground cracks, houses collapse
8	very disastrous; few buildings survive, landslides

Earth shattering. *Shown is the Mitsubishi Bank after the January 17, 1995 earthquake that hit Kobe, Japan. The earthquake had a Richter magnitude of 7.2. It destroyed over 20,000 homes and killed over 5000 people.*

14. On January 17, 1994, an earthquake measuring 6.6 on the Richter scale struck southern California. Was this a strong quake?

15. An increase of one unit on the Richter scale corresponds to multiplying the measured amplitude of a quake by what number?

16. To what factor does an increase of two units on the Richter scale correspond?

17. A series of earthquakes in New Madrid, Missouri, in 1811–1812 had a Richter magnitude estimated at 8.7. The famous San Francisco earthquake of 1906 is estimated to have had a Richter magnitude of 8.3. Find the ratio of the measured amplitudes of these two earthquakes.

Review

18. The "Haugh unit" is a measure of egg quality that was introduced in 1937 in the *U.S. Egg and Poultry Magazine.* The number U of Haugh units of an egg is given by the formula

$$U = 100 \log [H - \tfrac{1}{100} \sqrt{32.2} \, (30W^{.37} - 100) + 1.9],$$

where W is the weight of the egg in grams, and H is the height of the albumen in millimeters when the egg is broken on a flat surface. Find the number of Haugh units of an egg that weighs 58.8 g and for which $H = 6.3$ mm. *(Lesson 9-5)*

In 19 and 20, explain how to evaluate without using a calculator. *(Lesson 9-5)*

19. $\log_{10}100{,}000$ **20.** $\log_{10}10^{-7}$

21. Solve. *(Lesson 9-5)*
a. $\log x = 5$ **b.** $\log 5 = x$

22. Give two equations for the inverse of the function with equation $y = 10^x$. *(Lesson 9-5)*

23. Give an equation for a function of exponential decay and sketch its graph. *(Lesson 9-2)*

24. *Multiple choice.* A culture of 8000 bacteria triples every 40 minutes. Let $P =$ the population and $t =$ the number of minutes after the start. Which equation models the population size? *(Lesson 9-1)*
(a) $P = 8000 + 40t$ (b) $P = 40t^2 + 8000$
(c) $P = 8000 \cdot 3^{\frac{t}{40}}$ (d) $P = 3 \cdot 8000^{\frac{t}{40}}$
(e) $P = 8000 + 3 \cdot 40t$

25. Suppose that the inverse of function f is also a function. If the domain of f is the set of all real numbers and the range of f is the set of positive real numbers, find the domain and range of f^{-1}. *(Lesson 8-2)*

In 26 and 27, assume all variables represent positive numbers. Simplify. *(Lessons 7-2, 7-6)*

26. $\dfrac{a^3b^2}{(bc)^3}$ **27.** $\left(q^3\right)^{\frac{1}{3}}\left(r^6\right)^{\frac{2}{3}}$

Exploration

28. Acid rain is a serious environmental issue in many parts of the world. Find the pH levels of acid rain and non-acid rain.

Logarithms to Bases Other Than 10

John Napier

Defining Logarithms with Other Bases

Logarithms were invented by the English mathematician John Napier (1550–1617) in the early 1600s. Henry Briggs, also from England, first used common logarithms about 1620. In England even today logs to the base 10 are sometimes called Briggsian logarithms. In the 18th century Euler was the first person to realize that *any* real number could be an exponent. He was also the first to relate logarithms to exponents. Today most people study real exponents before logarithms, but that is not the order in which they developed historically.

Any positive number except 1 can be the base of a logarithm.

Definition
Let $b > 0$ and $b \neq 1$. Then n is the **logarithm of m to the base b,** written $n = \log_b m$, if and only if
$$b^n = m.$$

For example, since $2^5 = 32$, we can say that "5 is the logarithm of 32 with base 2" or "5 is log 32 to the base 2". We write $\log_2 32 = 5$.

Here are some other powers of 2 and the related logs to the base 2.

Exponential Form			means	Logarithmic Form		
2^4	=	16	means	$\log_2 16$	=	4
2^3	=	8	means	$\log_2 8$	=	3
2^2	=	4	means	$\log_2 4$	=	2
2^1	=	2	means	$\log_2 2$	=	1
2^0	=	1	means	$\log_2 1$	=	0
2^{-1}	=	$\frac{1}{2}$	means	$\log_2 \left(\frac{1}{2}\right)$	=	-1
2^{-2}	=	$\frac{1}{4}$	means	$\log_2 \left(\frac{1}{4}\right)$	=	-2
2^{-3}	=	$\frac{1}{8}$	means	$\log_2 \left(\frac{1}{8}\right)$	=	-3
2^y	=	x	means	$\log_2 x$	=	y

Graphs of Logarithm Functions

Both the exponential form $2^y = x$ and the logarithmic form $y = \log_2 x$ in the above columns describe the *inverse* of the function $y = 2^x$. The graphs of these functions are shown below. In general, the **logarithm function with base *b*, *g*(x) = $\log_b x$**, is the inverse of the exponential function with base b, $f(x) = b^x$.

Recall that the domain of the exponential function $y = 2^x$ is the set of all real numbers. Consequently, logarithms to the base 2 can be negative. However, the range of $y = 2^x$ is the set of positive real numbers. This means you cannot have the logarithm of a nonpositive number.

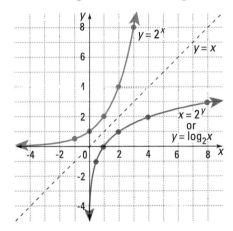

Every function with an equation of the form $y = \log_b x$, where $b > 0$ and $b \neq 1$, has the following properties.

1. The domain is the set of positive real numbers.
2. The range is the set of all real numbers.
3. The x-intercept is 1; there is no y-intercept.
4. The y-axis ($x = 0$) is an asymptote to the curve.

You may wonder, in the definition of the logarithm of m with base b, why b cannot equal 1. We do not consider $b = 1$ a suitable base for a logarithm function because the inverse of $y = 1^x$ is not a function.

Evaluating Logarithms in Bases Other Than 10

The methods of evaluating logarithms and solving logarithmic equations of logarithms with bases other than 10 are very similar to the methods you used in Lesson 9-5 with common logarithms. When x is a power of the base b, $\log_b x$ can be found without a calculator.

Example 1

Evaluate the following.
a. $\log_2 16$
b. $\log_6 \sqrt{6}$
c. $\log_5 \frac{1}{5}$

Solution

a. Let $\qquad\qquad\qquad\qquad\qquad\qquad\log_2 16 = x.$

Apply the definition of $\log_b m$ to rewrite $\qquad 2^x = 16$
the equation in exponential form.

Rewrite 16 as a power of 2. $\qquad\qquad\qquad 2^x = 2^4$

So, $\qquad\qquad\qquad\qquad\qquad\qquad\qquad\quad x = 4.$

Therefore, $\qquad\qquad\qquad\qquad\qquad\log_2 16 = 4.$

b. Let $\qquad\qquad\qquad\qquad\qquad\qquad\log_6 \sqrt{6} = x.$

Use the definition of $\log_b m$. $\qquad\qquad 6^x = \sqrt{6}$

Write $\sqrt{6}$ as a power of 6. $\qquad\qquad 6^x = 6^{\frac{1}{2}}$

So, $\qquad\qquad\qquad\qquad\qquad\qquad\qquad\quad x = \frac{1}{2}.$

Thus, $\qquad\qquad\qquad\qquad\qquad\quad \log_6 \sqrt{6} = \frac{1}{2}.$

c. Let $\qquad\qquad\qquad\qquad\qquad\qquad \log_5 \frac{1}{5} = x.$

Rewrite in exponential form. $\qquad\qquad 5^x = \frac{1}{5}$

Rewrite $\frac{1}{5}$ as a power of 5. $\qquad\qquad 5^x = 5^{-1}$

So, $\qquad\qquad\qquad\qquad\qquad\qquad\qquad\quad x = -1.$

Then $\qquad\qquad\qquad\qquad\qquad\qquad \log_5 \frac{1}{5} = -1.$

Solving Logarithmic Equations

To solve a logarithmic equation with base b, just as with base 10 logarithms, it often helps to use the definition of logarithm to rewrite the equation in exponential form.

Example 2

Solve for c: $\log_{81} c = \frac{5}{4}$.

Solution

Use the definition of logarithm. $\qquad\qquad 81^{\frac{5}{4}} = c$

Simplify the left side. $\qquad\qquad\qquad\quad 243 = c$

To solve for the base in a logarithmic equation, apply the techniques you learned in Chapters 7 and 8 for solving equations with nth powers.

Example 3

Find z if $\log_z 8 = \frac{3}{4}$.

Solution

Rewrite the equation in exponential form. $z^{\frac{3}{4}} = 8$

Take the $\frac{4}{3}$ power of each side. $\left(z^{\frac{3}{4}}\right)^{\frac{4}{3}} = 8^{\frac{4}{3}}$

Simplify each side of the equation. $z = 16$

Check

Does $\log_{16} 8 = \frac{3}{4}$? It will if $16^{\frac{3}{4}} = 8$, which is the case.

QUESTIONS

Covering the Reading

1. **a.** $\log_6 216$ is the logarithm of __?__ with base __?__.
   b. $\log_6 216 = $ __?__ because __?__ to the __?__ power equals 216.

2. Suppose $b > 0$ and $b \neq 1$. When $b^n = m$, $n = $ __?__.

3. Write the equivalent logarithmic form for $8^7 = 2,097,152$.

In 4 and 5, write the equivalent exponential form for the sentence.

4. $\log_2 0.5 = -1$ 5. $\log_b a = c$

6. **a.** Calculate $3^{-2}, 3^{-1}, 3^0, 3^1, 3^2$, and 3^3.
   b. Write the six logarithmic equations that are suggested by the calculations.

7. Write two equations for the inverse of the function $y = 2^x$.

8. State the domain and the range of the function $y = \log_2 x$.

9. Name one point that is on the graph of every logarithm function with equation $y = \log_n x$, if $n > 0$ and $n \neq 1$.

In 10–12, evaluate.

10. $\log_{1000} 100$ 11. $\log_3\left(\frac{1}{27}\right)$ 12. $\log_5 \sqrt{5}$

In 13–18, solve.

13. $\log_a 3 = \frac{1}{2}$ 14. $\log_b 32 = 5$ 15. $\log_{100} c = -1.5$

16. $\log_6 d = 3$ 17. $\log_{17} x = 0$ 18. $\log_t\left(\frac{1}{243}\right) = -\frac{5}{6}$

19. a. Graph $y = 3^x$ and $y = \log_3 x$ on the same set of axes.
(Hint: Use the values you found in Question 6.)
   b. *True or false.* The domain of $y = 3^x$ is the range of $y = \log_3 x$.
   c. Does the graph of $y = \log_3 x$ have any asymptotes? If yes, write the equation(s) for the asymptotes.

20. The population P (in billions) of the Earth in the year Y, can be approximated by the equation
$$P = 2^{\left(\frac{Y - 1975}{35} + 2\right)}.$$

   a. Write this equation in logarithmic form. (Hint: What is the base of the logarithmic equation?)
   b. When $P = 8$, what is Y?
   c. What does your answer in part **b** mean?

21. a. Evaluate $\log_5 125$ and $\log_{125} 5$ without a calculator.
   b. Evaluate $\log_4 16$ and $\log_{16} 4$ without a calculator.
   c. Generalize the results of parts **a** and **b**.

Review

22. Suppose one sound has an intensity of 80 decibels and a second sound has an intensity of 40 decibels. How many times more intense is the first sound? (The answer is *not* 2.) *(Lesson 9-6)*

In 23 and 24, use this information. In astronomy, the *magnitude* (brightness) *m of a star* is measured not by the energy I meeting the eye, but by its logarithm. In this scale, if one star has radiation energy I_1 and magnitude m_1, and another star has energy I_2 and magnitude m_2, then

$$m_1 - m_2 = -2.5 \log\left(\frac{I_1}{I_2}\right). \quad \textit{(Lesson 9-6)}$$

23. The star Rigel in the constellation Orion radiates about 45,000 times as much energy as the sun. The sun has magnitude 4.8. Using $\frac{I_1}{I_2} = 45,000$ and $m_2 = 4.8$, find the magnitude of Rigel.

24. Suppose the difference $m_1 - m_2$ in absolute magnitudes of two stars is 5. Find $\frac{I_1}{I_2}$, the ratio of the energies they radiate.

In 25 and 26, evaluate without a calculator. *(Lesson 9-5)*

25. $\log 10^5$ **26.** $\log .00001$

27. Radon 222 (^{222}Rn) has a half-life of 3.82 days.
  a. Write an equation modeling the decay.
  b. Sketch a graph of the equation.
  c. Use the graph to estimate the number of days needed for only 10% of the original radon to be present.   *(Lesson 9-2)*

28. *Skill sequence.*   Solve for x.   *(Lessons 1-5, 6-2, 8-7)*
  a. $3x + 6(x - 7) = 93$ **b.** $3x^2 + 6(x - 7)^2 = 93$
  c. $3 + 6(x - 7)^{1.25} = 93$

29. Simplify each expression.   *(Lessons 7-2, 8-4)*
  a. $x^{10} \cdot x^2$ **b.** $\dfrac{x^2}{x^{10}}$
  c. $(x^{10})^2$ **d.** $\sqrt{x^{10}}$

Radon detector. *Radon, a radioactive element, leaks into houses through the cracks in basement floors and walls. Highly concentrated radon can cause lung cancer if inhaled in large quantities. Radon detectors, like the one shown, are usually placed in basements.*

Exploration

---

30. a. Find all values of x such that $\log_3 x = \log_5 x$.
  b. Generalize the idea of part **a.**

Properties of Logarithms

Shanghai bikers. *Shown are pancho-clad bicyclists on a busy street on a rainy day in Shanghai, China. See Question 20.*

As you have seen, decimal or fraction values of some logarithms can be found without a calculator using the definition of logarithm. In this lesson, you will learn other properties of logarithms which can be used to rewrite expressions or solve problems. Because every logarithm in base b is an exponent of b, it should not surprise you that the properties of logarithms are derived from the properties of powers.

Basic Properties of Logarithms

Recall that the base b can be any positive number other than 1. You know that $b^0 = 1$ for any nonzero b. When we rewrite this equation in logarithmic form it becomes $\log_b 1 = 0$. This proves the following theorem.

> **Theorem (Logarithm of 1)**
> For every base b, $\log_b 1 = 0$.

You also know that the common log of 10^7 is 7. That is, $\log_{10} 10^7 = 7$. This is an instance of the following theorem.

> **Theorem (Log$_b$ of b^n)**
> For every base b and any real number n, $\log_b b^n = n$.

Proof
Let $$\log_b b^n = x.$$
By the definition of logarithm, $$b^x = b^n.$$
If two positive powers with the same base are equal, the exponents are equal. So, $$x = n.$$
Thus $$\log_b b^n = n.$$

In words, if a number can be written as the power of the base, the exponent of the number is its logarithm.

Example 1

Evaluate $\log_4 4^{15}$.

Solution

Use the Log_b of b^n Theorem.
$$\log_4 4^{15} = 15$$

Notice that it is not necessary to calculate 4^{15} in order to find its logarithm in base 4.

The Logarithm of a Product

For any base b ($b > 0$, $b \neq 1$) and any real numbers m and n, if
$$x = b^m, \quad y = b^n, \text{ and } \quad z = b^{m+n},$$
then
$$xy = z$$
by the Product of Powers Postulate.

By the definition of logarithm,
$$\log_b x = m, \log_b y = n, \text{ and } \log_b z = m + n.$$

Then by substitution, we can find $\log_b (xy)$ as follows.
$$\log_b (xy) = \log_b z$$
$$= m + n$$
$$= \log_b x + \log_b y$$

We have proved the following theorem.

Theorem (Product Property of Logarithms)
For any base b and for any positive real numbers x and y,
$$\log_b (xy) = \log_b x + \log_b y.$$

In words, the logarithm of a product equals the sum of the logarithms of the factors.

Example 2

Find $\log_6 2 + \log_6 108$.

Solution

By the Product Property of Logarithms,
$$\log_6 2 + \log_6 108 = \log_6 (2 \cdot 108)$$
$$= \log_6 216.$$
But since $216 = 6^3$, $\log_6 216 = 3$.
So $\log_6 2 + \log_6 108 = 3$.

The Logarithm of a Quotient

A Quotient Property of Logarithms follows from the related Quotient of Powers Postulate,

$$b^m \div b^n = b^{m-n}.$$

The proof of the Quotient Property of Logarithms is very similar to that of the Product Property of Logarithms.

> **Theorem (Quotient Property of Logarithms)**
> For any base b and for any positive real numbers x and y,
> $$\log_b\left(\frac{x}{y}\right) = \log_b x - \log_b y.$$

You are asked to complete the proof in the Questions.

Example 3

Recall that the formula $D = 10 \log\left(\frac{N}{10^{-12}}\right)$ is used to compute the number of decibels D from the sound intensity N measured in watts/m^2. Use the Quotient Property of Logarithms to simplify the original formula.

Solution

By the Quotient Property of Logarithms,

$$\log\left(\frac{N}{10^{-12}}\right) = \log N - \log(10^{-12}).$$

Thus, $\qquad\qquad\qquad D = 10 \log\left(\frac{N}{10^{-12}}\right)$ is equivalent to

$$D = 10(\log N - \log(10^{-12})).$$

But $\log(10^{-12}) = -12.$ So

$$D = 10(\log N - (-12))$$
$$= 10(\log N + 12).$$

The Logarithm of a Power

Recall the Power of a Power Property

$$(b^m)^n = b^{mn}.$$

This leads to the following theorem about logarithms.

> **Theorem (Power Property of Logarithms)**
> For any base b and for any positive real number x,
> $$\log_b(x^n) = n \log_b x.$$

You are asked to complete the proof of this theorem in the Questions.

Example 4

Suppose $x > 0$ and $x \neq 1$. Rewrite $\log x^7$ in terms of $\log x$.

Solution

By the Power Property of Logarithms, $\log x^7 = 7 \cdot \log x$ or $7 \log x$.

Check

Let $x = 2$. Does $\log 2^7 = 7 \cdot \log 2$? $2^7 = 128$, so $\log 2^7 = \log 128 \approx 2.10721$. $\log 2 \approx 0.30103$, so $7 \cdot \log 2 \approx 2.10721$. It checks.

Nowadays the properties of logarithms are used mainly to rewrite expressions or equations so they can be simplified or solved.

Example 5

Solve for p: $\log p = 3\log 5 + \log 2$.

Solution

Use the theorems of this lesson to rewrite the right side.

$$\log p = 3\log 5 + \log 2$$
$$= \log 5^3 + \log 2 \qquad \text{Power Property of Logarithms}$$
$$= \log (5^3 \cdot 2) \qquad \text{Product Property of Logarithms}$$
$$\log p = \log 250 \qquad \text{Arithmetic}$$

So $\qquad p = 250$.

Check

Does $\log 250 = 3\log 5 + \log 2$? Use a calculator to evaluate each side. Our calculator shows each side to equal approximately 2.39794. It checks.

QUESTIONS

Covering the Reading

In 1–3, simplify.

1. $\log_7 7^{26.8}$

2. $\log_m(m^n)$

3. $\log_\pi 1$

4. What property of powers is used to justify the theorem stating that $\log_b 1 = 0$?

In 5–8, an expression is given. **a.** Simplify it. **b.** Name the property or properties you used.

5. $\log_2 25 + \log_2 7$

6. $\log_{12} 3 + \log_{12} 4$

7. $\log_5 40 - \log_5 8$

8. $2\log_6 6\sqrt{6}$

In 9 and 10, rewrite the expression as the logarithm of a single number.

9. $\log 85 - \log 17 + \frac{1}{2} \log 25$.

10. $\log_b x + \log_b y - \log_b z$

In 11–15, *true or false*. If false, give a counterexample, and then correct the statement to make it true.

11. $\log (M + N) = \log M \cdot \log N$

12. $\dfrac{\log p}{\log q} = \log \left(\dfrac{p}{q}\right)$

13. $\log x^{10} = 10 \log x$

14. $\log_b(3x) = 3 \log_b x$

15. $\log(x - y) = \log x - \log y$

In 16 and 17, solve.

16. $\log x = 4 \log 2 + \log 3$

17. $\log y = \frac{1}{2} \log 25 - \log 2$

Applying the Mathematics

Shown are commuters in China's capital city of Beijing. Cars are too expensive for the average citizen to purchase, so people use bicycles for transportation.

18. Fill in the blanks in this proof of the Quotient Property of Logarithms.
Let $x = b^m$, $y = b^n$, and $z = b^{m-n}$.   Assume $b > 0$, $b \neq 1$.
  a. Since $x = b^m$, ___.   definition of logarithm
  b. Since $y = b^n$, ___.   definition of logarithm
  c. $\dfrac{x}{y} = $ ___   Quotient of Powers Postulate
  d. $\log_b\left(\dfrac{x}{y}\right) = $ ___   definition of logarithm
  $\log_b\left(\dfrac{x}{y}\right) = \log_b x - \log_b y$ substitution

19. The Henderson-Hasselbach formula $\text{pH} = 6.1 + \log \left(\dfrac{B}{C}\right)$ gives the pH of a patient's blood as a function of the bicarbonate concentration B and the carbonic acid concentration C. The normal pH is about 7.4.
  a. Rewrite this equation using the Quotient Property of Logarithms.
  b. A sick patient has a bicarbonate concentration of 24 and a pH reading of 7.2. Find the concentration of carbonic acid. (Hint: first solve the equation in part **a** for $\log C$.)

In 20 and 21, use this information. The quantity $\log\left(\dfrac{x}{y}\right)$, when rounded to the nearest whole number, is called the "number of orders of magnitude difference between x and y." (Each order of magnitude roughly means a substantial difference between the things being measured.) Find the number of orders of magnitude difference between the two given numbers in two ways: **a.** by calculating $\log \left(\dfrac{x}{y}\right)$ directly, and **b.** by calculating $\log x - \log y$.

20. 1.2 billion, the population of China, and 1.2 million, the population of Maine.

21. 40,000,000,000,000 km, the distance from the Earth to Proxima Centauri, the second nearest star, and 149,600,000 km, the distance from the Earth to the Sun, the nearest star.

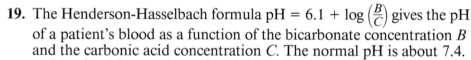

22. Justify each step in the proof of the Power Property of Logarithms given here. Let $\log_b x = m$.

a. Then $\quad x = b^m$.

$x^n = (b^m)^n \qquad\qquad$ _____ substitution

b. $\qquad\qquad x^n = b^{mn} \qquad\qquad$ _____

c. $\qquad\qquad x^n = b^{nm} \qquad\qquad$ _____

d. $\qquad \log_b x^n = nm \qquad\qquad$ _____

e. $\qquad \log_b x^n = n \log_b x \qquad$ _____

Review

23. Simplify these logs. *(Lesson 9-7)*

a. $\log_{64} 64$ **b.** $\log_{64} 8$ **c.** $\log_{64} 2$

d. $\log_{64} 1$ **e.** $\log_{64} \frac{1}{64}$ **f.** $\log_{64} \frac{1}{8}$

In 24 and 25, solve. *(Lessons 9-5, 9-7)*

24. $\log_x 81 = 4$

25. a. $\log y = -1$ **b.** $\log(-1) = y$

26. A foundation decided to hold the following contest. It made a test with ten very hard questions. It offered 2¢ to anyone correctly answering one question. It would give ten times that for two questions correctly answered; ten times that for 3 questions and so on. Construct a logarithmic scale showing the number of questions and the amount of money offered. *(Lesson 9-6)*

27. At the right is a chart showing the foreign exchange rate from 1970 to 1994 for the currency of India (the rupee). *(Lesson 9-4)*

Year	# rupees = $1.00
1970	7.576
1980	7.887
1981	8.681
1982	9.485
1983	10.104
1984	11.348
1985	12.332
1986	12.597
1987	12.943
1988	13.899
1989	16.213
1990	17.492
1991	22.712
1992	28.043
1993	31.291
1994	31.374

Face value. *The rupee is a monetary unit of India. On the front of the 100-rupee bill, the fourteen official languages of India are printed.*

a. Plot the data on the window $1970 \le x \le 1995$, $0 \le y \le 33$.

b. Why does this data suggest an exponential model?

c. One equation which models this data is $y = 5.48(1.06)^{x-1970}$. Graph this equation.

d. State a possible domain where the model is most accurate

e. Another equation which models this data is $y = 6.12(1.096)^{x-1970}$. Which equation would you use to predict the exchange rate in 2000? Why?

28. A student invested $800 in an account compounded continuously at an annual rate of 4.7% for five years. What was the final balance? *(Lesson 9-3)*

29. *Multiple choice.* The charge to park a car at a city lot is 90¢ for the first hour and 75¢ for each additional hour or fraction thereof. Let $t = $ the number of hours parked. Which equation models this situation? *(Lesson 3-9)*
(a) $f(t) = 90 + 75 \lfloor t + 1 \rfloor$ (b) $f(t) = 90 - 75 \lfloor 1 - t \rfloor$
(c) $f(t) = 75 + 90t$ (d) $f(t) = 90 + 75(t \cdot 60)$

30. a. Find the rate of change between $x = -4$ and $x = -3$ for the equation $y = \frac{1}{2}x^2$.
b. Draw a picture and use it to explain what you calculated in part **a.** *(Lesson 2-5)*

Exploration

31. When asked why he memorized that log 2 is about 0.301 and log 3 is about 0.477, a student answered, "Of course I know log 1 and log 10. Using log 2 and log 3, I can get the logs of all but one of the other integers from 1 to 10." Which logs between 4 and 9 can be found from the logs of 2 or 3 using the properties of this lesson, and what are they?

Log this one in. *Shown is the Space Shuttle* Discovery *landing on February 11, 1995 after an eight-day mission. Piloting the* Discovery *was Captain Eileen M. Collins, the first female to pilot a mission. See the Example.*

What Are Natural Logarithms?

Any positive number except 1 can be the base of a logarithm. One number which is often used as the base for logarithms in real-world applications is the number *e*, which you studied in Lesson 9-3.

Logarithms to the base *e* are called **natural logarithms.** Sometimes they are called *Napierian logarithms* after Napier. Just as log *x* (without any base named) is a shorthand for $\log_{10} x$, we abbreviate $\log_e x$ as **ln *x.***

> **Definition**
> $n = \ln m$, the **natural logarithm of *m*,** if and only if $m = e^n$.

The symbol "ln *x*" is often read "the natural log of *x*".

Natural logarithms of powers of *e* can be determined mentally from the definition.

$$\ln 1 = \log_e 1 = 0 \text{ because } 1 = e^0.$$
$$\ln e = \log_e e = 1 \text{ because } e = e^1.$$
$$\ln e^3 = \log_e e^3 = 3 \text{ because } e^3 = e^3.$$

Notice that $\ln(e^x) = x$ is a special case of the $\log_b$ of b^n Theorem: $\log_b (b^x) = x$.

Evaluating Natural Logarithms

To determine natural logarithms of numbers not in e^x form, you need a calculator, computer, or table of values. The natural logarithm key on scientific calculators is usually labeled In or LN.

Activity 1

Use a calculator to evaluate each of the following.
a. ln 1   **b.** ln 10   **c.** ln 2   **d.** ln .5

You should find that the sum of the answers to Activity 1 is 2.302585 . . .

All of the properties of logarithms proved in Lesson 9-8 apply to natural logarithms. In particular, for all $x > 0$ and $y > 0$,

$$\ln (xy) = \ln x + \ln y,$$
$$\ln \left(\frac{x}{y}\right) = \ln x - \ln y,$$
$$\text{and } \ln (x^n) = n \ln x.$$

Activity 2

Verify with a calculator that $\ln(2^6) = 6 \ln 2$.

Computers can also be used for finding natural logarithms. But beware! In BASIC and some other computer languages, the natural logarithm function is denoted LOG. This can be confusing because log in other places usually refers to base 10.

The Graph of the Function $y = \ln x$

Just as the function with equation $y = \log x$ is the inverse function of $y = 10^x$, and $y = \log_2 x$ is the inverse of $y = 2^x$, similarly, $y = \ln x$ is the inverse of $y = e^x$. This can be seen in the graph below. The graph of each function is the reflection image of the other over the line with equation $y = x$.

$y = e^x$

x	y
-1	0.37
0	1.00
1	$e \approx 2.72$
1.6	4.95

$y = \ln x$

x	y
0.37	-1
1.00	0
$e \approx 2.72$	1
4.95	1.6

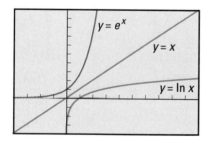

$-4 \le x \le 10, \ x\text{-scale} = 1$
$-4 \le y \le 10, \ y\text{-scale} = 1$

The function $f(x) = \ln x$ has all the properties of other logarithm functions. In particular, the domain of the natural-logarithm function is the set of positive reals, and its range is the set of all real numbers.

Some Uses of Natural Logarithms

Natural logarithms are frequently used in formulas.

Captain Eileen M. Collins

Example

Ignoring the force of gravity, the maximum velocity v of a rocket is given by the formula $v = c \cdot \ln R$, where c is the velocity of the exhaust and R is the ratio of the mass of the rocket with fuel to its mass without fuel. To achieve a stable orbit 160 km above Earth, a spacecraft must attain a velocity of about 7.8 km/s.

a. With a small payload, a solid-propellant rocket could have a mass ratio of about 19. A typical exhaust velocity for such a rocket might be about 2.4 kilometers per second. Could a spacecraft propelled by this rocket achieve a stable orbit?

b. Find R for a rocket if $c = 1963$ m/sec and $v = 2021$ m/sec.

Solution

a. For the solid-propellant rocket $R \approx 19$ and $c \approx 2.4$ km/s. Find v using the formula above.

$$v \approx 2.4 \cdot \ln 19$$
$$\approx 2.4\,(2.944)$$
$$\approx 7.1$$

Notice that the maximum velocity of this rocket is less than the velocity needed for orbit. So this rocket could not propel a spacecraft into orbit.

b. Substitute the given values for c and v, and solve for R.

$$v = c \cdot \ln R$$
$$2021 = 1963 \cdot \ln R$$

To solve for R you must first solve for $\ln R$. Divide both sides by 1963.

$$1.0295 \approx \ln R$$

Now, use the definition of natural logarithm to solve for R.

$$e^{1.0295} \approx R$$
$$R \approx 2.7997$$

The mass of the rocket with fuel is about 2.8 times the mass of the rocket without fuel. (Thus the mass of the fuel is 1.8 times the rocket's mass.)

QUESTIONS

Covering the Reading

1. What is the base of natural logarithms?

2. *Multiple choice.* Which of the following is equivalent to $y = \ln x$?
   (a) $x = \log_e y$
   (b) $x = \log_y e$
   (c) $y = \log_e x$
   (d) $y = \log_x e$

In 3 and 4, write the equivalent exponential form.

3. $\ln 1 = 0$

4. $\ln 300 \approx 5.70$

In 5 and 6, write in logarithmic form.

5. $e^2 \approx 7.39$ **6.** $e^{0.06} \approx 1.06$

7. Approximate ln 400 to the nearest thousandth.

8. What values did you get in Activity 1 of this lesson?

9. *Skill sequence.* Evaluate without a calculator.
    a. $\ln e$ **b.** $\ln e^2$ **c.** $\ln e^{100}$

10. Give your response to Activity 2 in this lesson.

11. The graph of what function is the reflection image of the graph of $y = \ln x$ over the line $y = x$?

In 12–14, use the formula $v = c \cdot \ln R$ from the Example.

12. The space shuttle has an R value of about 3.5. Its main engines can produce an exhaust velocity of about 4.6 km/s. Can the Space Shuttle achieve a stable orbit with its main engines only?

13. One of the Viking rockets used in the 1950s had an exhaust velocity of about 1.22 km/s and traveled without fuel at a maximum rate of about 1.79 km/s. Find its mass ratio R.

14. If the maximum velocity of a rocket is 2195 m/sec and the mass ratio is 2.5, what is the velocity of the exhaust?

Applying the Mathematics

15. At what point does the line with equation $x = \frac{1}{2}$ intersect the graph of $y = \ln x$? Explain how you got your answer.

In 16 and 17, suppose $\ln x = 8$ and $\ln y = 4$. Evaluate.

16. $\ln(3xy)$ **17.** $\ln\left(\sqrt[4]{\dfrac{x}{y}}\right)$

In 18 and 19, use this information. Most of today's languages are thought to be descended from a few common ancestral languages. The longer the time since the languages split from the ancestral language, the fewer common words there are in the descendent languages. Let c = the number of centuries since two languages split from an ancestral language. Let w = the percentage of words from the ancestral language that are common to the two descendent languages. In linguistics, the equation below (in which $r = .86$ is the index of retention) has been used to relate c and w.

$$\frac{10}{c} = \frac{2 \log r}{\log w}$$

18. If about 15% of the words in an ancestral language are common to two different languages, about how many centuries ago did they split from the ancestral language?

19. If it is known that two languages split from an ancestral language about 1500 years ago, about what percentage of the words in the ancestral language are common to the two languages?

20. a. What happens when you try to find ln (-2) on your calculator?
    b. Explain why the calculator displayed what you saw in part a.

21. a. Does the graph of $y = \ln x$ have an x-intercept?
    b. If the intercept exists identify it. If the intercept does not exist, explain why it does not.

Review

22. If $y = 3 \cdot 5^x$ write an expression equal to log y.   *(Lesson 9-8)*

23. *True or false.* $\log(1.7 \times 10^3) = (\log(1.7)(\log 10^3))$. Justify your answer.   *(Lesson 9-7)*

24. Solve for x: $\log_x 7 = 2$.   *(Lesson 9-7)*

In 25 and 26, determine the pH of each of the following substances with the given H^+ concentration.   *(Lesson 9-6)*

25. black coffee: 1×10^{-5} 26. baking soda: 1×10^{-9}

27. In 1982, it was projected that t years later the population of the Philippines would be given by $P(t) = 50e^{0.02t}$ million. In 1992, the population was 67,114,000. How good was the projection? *(Lesson 9-3)*

28. Consider the function with equation $y = -300(2)^x$.
    a. Sketch a graph using the domain $-4 \leq x \leq 4$.
    b. Is this an example of exponential decay? Why or why not?   *(Lesson 9-2)*

29. Two solid models of statues, A' and A, are similar. The ratio of similitude is 5, with A' larger than A. If the volume of A is 400 cm^3, what is the volume of A'?   *(Previous course, Lesson 2-4)*

With a population of about 1.6 million, Manila is the capital and largest city of the Philippines

---

Exploration

30. Natural logarithms can be calculated using the series

$$\ln (1 + x) = x - \frac{x^2}{2} + \frac{x^3}{3} - \frac{x^4}{4} + \frac{x^5}{5} - \cdots .$$

    a. Substitute 0.5 for x to estimate ln 1.5.
    b. How close is this value to the one determined by using the In key on your calculator?
    c. How many more terms of the series are needed to get a value that is within 0.001 of the actual value?

9-10

Using Logarithms to Solve Exponential Equations

Solving Equations of the Form $b^x = a$

In previous lessons, you have learned to solve equations of the form $b^x = a$ by trial-and-error or by using graphs. In this lesson, we focus on the use of logarithms to obtain exact solutions to equations of this form. The process involves three steps.

1. Take logarithms (with the same base) of each side.
2. Apply the Power Property of Logarithms.
3. Solve the resulting linear equation.

Example 1

Solve $5^x = 20$
a. by taking common logarithms of each side.
b. by taking natural logarithms of each side.

Solution

Solutions **a** and **b** are written side-by-side. First read Solution **a** (the columns in the left and center). Then read Solution **b** (the center and right columns). Then reread both solutions by reading across each line.

a. $5^x = 20$

$\log 5^x = \log 20$ Take the logarithm of each side.

$x \log 5 = \log 20$ Power Property of Logarithms

$x = \dfrac{\log 20}{\log 5}$ Divide both sides by the coefficient of x.

$x \approx \dfrac{1.3010}{.6990}$ Evaluate logarithms with a calculator.

$x \approx 1.861$

b. $5^x = 20$

$\ln 5^x = \ln 20$

$x \ln 5 = \ln 20$

$x = \dfrac{\ln 20}{\ln 5}$

$x \approx \dfrac{2.9957}{1.6094}$

$x \approx 1.861$

Check

Note that $5^1 = 5$ and $5^2 = 25$, so you should expect $1 < x < 2$. Use the powering key on your calculator to verify that $5^{1.861} \approx 20$.

In the solutions above, logarithms to bases 10 and e were taken. In fact, any base for the logarithms could have been used. Because the same results are *always* obtained by using either common or natural logarithms, you may choose either one for a given situation. In some cases, one is more efficient or easier to use than the other. When the base of the exponential equation is 10, it is easier to use common logarithms. When the base is e, it is easier to use natural logarithms.

Example 2

At what rate of interest, compounded continuously, would you have to invest your money so that it would triple in 10 years?

Solution

Use the Continuous Compound Interest Model.
$$A = Pe^{rt}$$
Because A, the total amount desired, is three times the starting amount P,
$$A = 3P. \text{ Here } t = 10.$$

Substitute. $\qquad\qquad 3P = Pe^{10r}$

Simplify by dividing by P. $\qquad 3 = e^{10r}$

Because the base in the equation is e, take the natural logarithm of each side. (This gives the same result as applying the definition of the natural logarithm.)
$$\ln 3 = \ln (e^{10r})$$

$\log_b$ of b^n Theorem $\qquad \ln 3 = 10r$

Divide both sides by 10. $\qquad r = \dfrac{\ln 3}{10} \approx \dfrac{1.0986}{10} = 0.10986$

It takes an interest rate of about 11%, compounded continuously, to triple your money in 10 years.

Check

Suppose you began with $1000. At 10.986% interest compounded continuously for 10 years, it will grow to
$$1000e^{10(.10986)} \text{ dollars.}$$
A calculator shows $1000e^{10(.10986)} \approx 2999.96$. This is almost exactly three times the starting amount.

Decay or depreciation problems modeled by exponential equations with negative exponents can also be solved by the above techniques.

Example 3

The intensity L_t of light transmitted through ordinary glass of thickness t (in centimeters) is modeled by the exponential equation
$$L_t = L_0 \, 10^{-0.0022t}$$
where L_0 is the intensity before entering the glass. How thick must the glass be to block out 10% of the light?

Solution

Blocking out 10% of the light means allowing 90% of the light in.

$$L_t = .90L_0$$

$.90L_0 = L_0 10^{-0.0022t}$	Substitute into formula.
$.90 = 10^{-0.0022t}$	Divide by L_0.
$\log .90 = \log 10^{-0.0022t}$	Take logarithm of both sides.
$\log .90 = -0.0022t$	$\log_b$ of b^n Theorem
$t = \dfrac{\log .90}{-0.0022} \approx 20.79886$	

So to block out 10% of the light ordinary glass must be about 20.8 cm thick.

In addition to glass thickness, another factor affecting light intensity is color. Stained-glass windows block out more of the sun's intensity than do ordinary windows of the same thickness.

Changing the Base of a Logarithm

Most calculators have keys for only two types of logarithms—common logarithms $\boxed{\log}$ and natural logarithms $\boxed{\ln}$. So you cannot immediately evaluate expressions such as $\log_7 25$. However, with the following theorem, you can convert logarithms with any base b to a ratio of either common logarithms or natural logarithms.

Theorem
(Change of Base Property)
For all positive real numbers a, b, and t, $b \neq 1$ and $t \neq 1$,
$$\log_b a = \frac{\log_t a}{\log_t b}.$$

Proof:

Suppose	$\log_b a = x.$
By the definition of logarithm,	$b^x = a.$
Take the logarithm with base t of each side.	$\log_t b^x = \log_t a$
Apply the Power Property of Logarithms.	$x\log_t b = \log_t a$
Divide both sides by $\log_t b$	$x = \frac{\log_t a}{\log_t b}$
Use the Transitive Property of Equality.	$\log_b a = \frac{\log_t a}{\log_t b}$

The Change of Base Property enables you to find the logarithm of any number with any base.

Example 4

Approximate $\log_7 25$ to the nearest thousandth.

Solution 1

Use the Change of Base Property with common logarithms.
$$\log_7 25 = \frac{\log 25}{\log 7} \approx \frac{1.39794}{0.84510} \approx 1.654$$

Solution 2

Use the Change of Base Property with natural logarithms.
$$\log_7 25 = \frac{\ln 25}{\ln 7} \approx \frac{3.2189}{1.9459} \approx 1.654$$

Check

By definition $\log_7 25 \approx 1.654$ is equivalent to $7^{1.654} \approx 25$. Our calculator shows that $7^{1.654} \approx 25$. It checks.

QUESTIONS

Covering the Reading

1. Solve $7^x = 15$ to three decimal places
   a. by taking common logarithms.
   b. by taking natural logarithms.

2. Solve $2^r = 3$
   a. to the nearest tenth by making a graph.
   b. to the nearest thousandth by using logarithms.

In 3 and 4, solve and check.

3. $\log_3 12 = y$

4. $25.6^z = 2.89$

In 5 and 6, refer to Example 2.

5. a. Solve this problem using common logarithms.
   b. Why is this problem more efficiently solved with natural logarithms than with common logarithms?

6. What interest rate would it take to double your money in 7 years?

7. Refer to Example 3. Find the thickness of normal glass needed to block out 4% of the light.

In 8 and 9, approximate the logarithm to the nearest thousandth and check your answer.

8. $\log_2 50$

9. $\log_3 50$

Applying the Mathematics

10. Suppose you invest $200 in a savings account paying 4.25% interest compounded continuously. How long would it take for your account to grow to $300, assuming that no other deposits or withdrawals are made?

11. Suppose a colony of bacteria grows according to $N = N_0 e^{2t}$, where N_0 is the initial number of bacteria and t is the time in hours. How long does it take the colony to quadruple in size?

In 12 and 13, use the following information. In 1992 the population of Japan was about 124.5 million, and was growing at an annual rate of about 0.3%. At the same time the population of Mexico was 92.4 million and growing at a rate of 2.4% annually. Assume that these growth rates hold indefinitely, and that population grows continuously.

12. In about what year will the population of Mexico reach 100 million?

13. a. In about what year will the populations of Mexico and Japan be equal?
    b. What will their populations be in that year?
    c. Which country will have the higher population after that year?

14. The amount A of radioactivity from a nuclear explosion is estimated to decrease exponentially by $A = A_0e^{-2t}$, where t is measured in days. How long will it take for the radioactivity to reach $\frac{1}{1000}$ of its original intensity?

In 15 and 16, solve.

15. $5^{2y} = 1993$

16. $\log_8 256 = -4x$

Review

17. For a small 3-stage rocket, the formula $V = c_1 \cdot \ln R_1 + c_2 \cdot \ln R_2 + c_3 \cdot \ln R_3$ is used to find the velocity of the rocket at the final burnout. If $R_1 = 1.46$, $R_2 = 1.28$, $R_3 = 1.41$, $c_1 = 2255$ m/sec, $c_2 = c_3 = 2470$ m/sec, find V. *(Lesson 9-9)*

In 18 and 19, solve and check. *(Lesson 9-8)*

18. $\log z = \frac{2}{3} \log 8 + \log 3$.

19. $\log 64 = x \log 2$

20. Find the error in the following "proof" that $5 < 2$. *(Lessons 5-1, 9-5, 9-8)*

Proof:

    a. $\frac{1}{32} < \frac{1}{4}$

    b. $\log\left(\frac{1}{32}\right) < \log\left(\frac{1}{4}\right)$

    c. $\log\left[\left(\frac{1}{2}\right)^5\right] < \log\left[\left(\frac{1}{2}\right)^2\right]$

    d. $5 \log\left(\frac{1}{2}\right) < 2 \log\left(\frac{1}{2}\right)$

    e. $5 < 2$

21. Write $\log(pq^2)$ in terms of $\log p$ and $\log q$. *(Lesson 9-7)*

22. Lemons have a pH of 2.3 and milk of magnesia has a pH of 10.5. Which of these has a higher concentration of H^+? *(Lesson 9-6)*

In 23 and 24, suppose $f(x) = \log x$. Then find: *(Lessons 8-3, 9-5)*

23. $f(0.1)$

24. $f^{-1}(3)$

25. Sketch the graphs of $f(x) = 6^x$ and $g(x) = 6^{-x}$ on the same axes. Describe how the graphs are related. *(Lessons 1-2, 8-3, 8-4)*

26. a. Graph g when $g: x \rightarrow 3x + 2$.
    b. Give a formula for g^{-1} and graph $y = g^{-1}(x)$ on the same axes.
    c. Find $g \circ g^{-1}$. *(Lessons 3-2, 7-1, 7-3)*

Exploration

27. Suppose $a^x = b$ and $b^y = a$. How are x and y related? (Hint: If you cannot figure this out in general, start by letting a and b have certain values and solving for x and y.)

A project presents an opportunity for you to extend your knowledge of a topic related to the material of this chapter. You should allow more time for a project than you do for typical homework questions.

PROJECTS
9
CHAPTER NINE

1 Car Loans

a. Consult a car dealer, insurance agent, books, or magazines about how automobile depreciation is typically calculated. Is depreciation typically described by a linear, exponential, or some other model? Explain.

b. Gather data on the "book value" of two or three cars that interest you. Include only cars whose values you can determine for at least the past five years. Which of these data sets, if any, seem to illustrate exponential decay? Find equations that model the value of these cars over time. Use the equations to predict the value of each car five years from now. How reasonable do you think these estimates are? What are some limitations of your mathematical model?

2 Modeling the Growth of HIV

Communicable diseases often spread in a population according to an exponential model, until treatments or methods of slowing the spread of the disease are found. The *Historical Statistical Abstract of the U.S.* contains data about the number of people in the U.S. that have tested positive for HIV, the virus that most scientists believe causes AIDS. These data exhibit what appears to be an exponential increase during a certain period, but after a point, the number of HIV-positive people in the U.S. seems to grow in a linear fashion.

a. Find these data in the *Historical Statistical Abstract of the U.S.* (or some other source).

b. Make a scatterplot showing the number of HIV-positive people on the *y*-axis with the corresponding years on the *x*-axis.

c. Assume the data can be split into two pieces such that prior to some year the data grow exponentially, but following that year the data grow linearly. Pick a year that you feel best splits the data in this fashion. Then find an equation of a function which fits the data in the first piece to an exponential model, and similarly, find the equation of a line which fits the data in the second piece. Combine these equations into a piecewise definition of a function that models all of the data. Graph the function over the scatterplot.

d. Examine your plot and determine whether or not you agree that the data split into an "exponential piece" and a "linear piece." If not, explain why this model is not appropriate for the data.

e. Give some possible reasons for changes in the growth of the spread of HIV in the U.S.

▶

3 Predicting Cooling Times

You can predict how long it will take a soft drink to cool to a desired temperature on a particular shelf in your refrigerator using either exponential or logarithmic functions. You will need a thermometer that can measure a wide range of temperatures.

a. Measure the temperature T_R in your refrigerator. (Let's say it is 42°F.) T_R is constant. Fill a cup with very hot tap water. Measure the temperature T_M of the water. (Say it is 135°F.) Place the cup in your refrigerator. Make a table and record your first measurement.

t (min) time since cooling began	T_m measured temp. of water (°F)	$T_m - T_R$ diff. between measured temp. and refrig. temp. (°F)
0	135	$135 - 42 = 93$

b. Measure the temperature of the water periodically. At first you will want to take a measurement every 5 minutes. Later you might wait longer between measurements. Record the temperatures in the table. Continue taking measurements for at least 4 hours. Be sure to take measurements the same way each time.

c. Plot the ordered pairs $(t, T_m - T_R)$. Describe the shape of the graph. Use the modeling features of a graphics calculator or statistics package to fit a reasonable curve to your data. Explain why you chose the model you did.

d. Use technology which will fit a logarithmic model to your data. This time, let $(T_m - T_R)$ be your independent variable and t be your dependent variable. The statistics package might give you a model like $t = a + b \cdot \ln(T_m - T_R)$, where a and b are parameters for the model. Is this model equivalent to the one you found in part **c**? Why or why not?

e. Use your model from parts **c** and **d** to predict how long it takes for a can of soft drink in your refrigerator to cool from 70°F (about room temperature) to 48°F (an acceptable drinking temperature). Comment on the answer(s) you get.

4 Logarithms for Calculation

Before calculators became widely available in the 1970s, difficult multiplications, divisions, and powerings were usually done using properties of logarithms and tables of common logarithms.

a. Find a table of logarithms. Explain how to use the table to perform a multiplication, a division, and the power of a number.

b. Show how to use tables and properties of logarithms to estimate $\dfrac{86400\sqrt[3]{365.25}}{.079^{11}}$.

c. Show how to use tables and properties of logarithms to determine the number of digits in $2^{859,433} - 1$, the largest prime known as of 1994. (This prime number was found by David Slowinski and Paul Gage.)

SUMMARY

A function with an equation of the form $y = ab^x$, where $b > 0$ and $b \neq 1$, is an exponential function. In the formula $A = P\left(1 + \frac{r}{n}\right)^{nt}$, when P, r, and n are given, A is an exponential function of t. By continuously compounding an initial amount of \$1.00 at 100% interest, the value of the investment after one year approaches $e \approx 2.71828$. In general, the formula $A = Pe^{rt}$ can be used to calculate the value of an investment of P at $r\%$ interest compounded continuously for t years. All geometric sequences are also exponential functions.

Some exponential functions represent exponential growth or decay situations. In an exponential growth situation, the growth factor is greater than one. In an exponential decay situation, the growth factor is between 0 and 1. Over short periods of time, many populations grow exponentially, and the value of many items depreciates exponentially. Real data from these and other contexts can be modeled using exponential growth or decay functions.

The function $f\colon x \to b^x$ is the exponential function with base b. Its inverse $f^{-1}\colon b^x \to x$ is the logarithm function with base b. Thus $b^x = a$ if and only if $x = \log_b a$. Because exponential and logarithm functions are inverses, their graphs are reflection images of each other. The properties of the logarithm functions are derived from their being inverses of exponential functions.

Logarithm functions are used to scale data having a wide range (e.g., pH or decibel levels) and to solve equations of the form $b^x = a$, where b and a are positive. To solve equations of this type, take the logarithm of both sides; the solution is $x = \frac{\log a}{\log b}$. The base of a logarithm function can be any positive real number not equal to 1, but the most often used bases are 10 and e. When the base is 10, the values of the log function are called common logarithms. When the base is $e \approx 2.71828$, the values of the log function are called natural logarithms.

Exponential Growth Functions
$$y = b^x, b > 1$$
for example $y = 2^x$

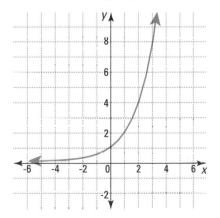

Logarithmic Functions
$$y = \log_b x, b > 1$$
for example, $y = \log_2 x$

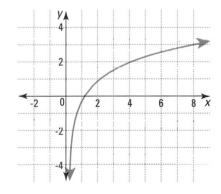

set of all real numbers	Domain	set of all positive real numbers
set of all positive real numbers	Range	set of all real numbers
y-intercept is 1	Intercepts	x-intercept is 1
no x-intercept		no y-intercept
the x-axis ($y = 0$)	Asymptotes	the y-axis ($x = 0$)

SUMMARY

The basic properties of logarithms correspond to properties of powers. Here is the correspondence. Let $x = b^m$ and $y = b^n$, and take the logarithms of both sides of each power property. The result is a logarithm property.

Power property	Logarithm property
$b^0 = 1$	$\log_b 1 = 0$
$b^m \cdot b^n = b^{m+n}$	$\log_b (xy) = \log_b x + \log_b y$
$\dfrac{b^m}{b^n} = b^{m-n}$	$\log_b \left(\dfrac{x}{y}\right) = \log_b x - \log_b y$
$(b^m)^a = b^{am}$	$\log_b (x^a) = a \log_b x$

VOCABULARY

Below are the most important terms and phrases for this chapter. You should be able to give a definition or statement for those terms marked with an *. For all other terms you should be able to give a general description and a specific example of each.

Lesson 9-1
exponential function
exponential curve
exponential growth

Lesson 9-2
exponential decay
depreciation
half-life
Exponential Growth Model

Lesson 9-3
continuous compounding
instantaneous compounding
e
Continuously Compounding Interest Formula
Continuous Change Model

Lesson 9-4
decade growth factor
yearly (annual) growth factor

Lesson 9-5
logarithm of x with base 10, log x
logarithmic curve
*common logarithm
common logarithm function
logarithmic equations

Lesson 9-6
logarithmic scale
Richter scale
bel, decibel
pH
base, acid
acidic
alkaline
linear scale

Lesson 9-7
*logarithm of m with base b, $\log_b m$

Lesson 9-8
Logarithm of 1 Property
$\log_b$ of b^n Property
Product Property of Logarithms
Quotient Property of Logarithms
Power Property of Logarithms

Lesson 9-9
*natural logarithm of x, ln x

Lesson 9-10
Change of Base Property

PROGRESS SELF-TEST

Take this test as you would take a test in class. Then check your work with the solutions in the Selected Answers section in the back of the book.

1. A classic car is increasing in value at an annual rate of 17%. If it is valued at $43,500 now, what will it be worth (to the nearest hundred dollars) in three years?

2. Suppose each of a certain type of bacterium splits into two every half hour. If there are 5 bacteria initially, about how many will there be after 24 hours?

3. Consider $f(x) = 4^x$.
   a. Draw a graph when $-2 \le x \le 2$.
   b. Evaluate $f(\pi)$ to the nearest tenth.
   c. Does the graph have any asymptotes? If yes, state the equations for all asymptotes. If no, explain why not.

4. a. Use a calculator to find $\ln(42.7)$ to the nearest hundredth.
   b. Write a statement about powers that could be used to check your work.

In 5–8, evaluate each expression exactly without using a calculator. Explain how you got your answer.

5. $\log(1,000,000)$ 6. $\log_4\left(\frac{1}{16}\right)$

7. $\ln e^{-6}$ 8. $\log_2 1$

9. Radioactive ^{14}C has a half-life of 5730 years. Suppose a substance starts with 40 milligrams of ^{14}C.
   a. How many mg are left after 3 half-life periods?
   b. Find a formula for the number of mg left after t half-life periods.

In 10 and 11, solve. If necessary, round to the nearest hundredth.

10. $\log_x 8 = \frac{3}{4}$

11. $6^y = 32$

In 12 and 13, *true or false*. Justify your reasoning.

12. $\log\left(\frac{M}{N^2}\right) = \log M - 2 \log N$

13. $\log_3 x \cdot \log_3 y = \log_3(x \cdot y)$

14. If $1500 is invested at 4% compounded continuously and no additional deposits or withdrawals are made, after how many years will the investment grow to $3000?

15. Recall that the difference in the magnitudes (brightness) of two stars varies as the logarithm of the ratio of their radiation energy, or $m_1 - m_2 = -2.5 \log\left(\frac{I_1}{I_2}\right)$, where I_1 is the radiation energy for the magnitude m_1 and I_2 that for m_2. The magnitude of the sun as seen from the Earth is -26. The magnitude of the full moon is -13. How many times brighter is the sun than the full moon?

16. Suppose one sound measures 95 decibels while a second measures 125 decibels. How many times more intense is the second than the first?

In 17 and 18, assume that a bacteria population decays according to the model $y = A(.92)^x$ where A cells of bacteria become y cells x hours later.

17. If you start with 12,000 bacteria, how many will remain after 8 hours?

18. *Multiple choice.* Which graph could represent this situation?

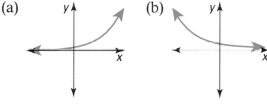

(a)   (b)

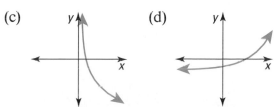

(c)   (d)

PROGRESS SELF-TEST

19. The 1994 *Information Please Almanac* gives the 1993 population of Peru as 22,900,000. The projected population for the years 2000 and 2010 are given below. In the table, t represents the number of years after 1993.

t (years after 1993)	0	7	17
p (population in millions)	22.9	26.4	30.1

a. Fit an exponential model to these data. Explain how you get your equation.

b. According to the model in part **a,** what will the population of Peru be in the year 2025? (Round your answer to the nearest hundred thousand.)

c. The 1994 *Information Please Almanac* gives the projected population for Peru for 2025 as 35,600,000. What are some factors that might explain why your answer in part **b** is different than 35.6 million?

20. Consider the function defined by $y = \log_3 x$.

a. State the coordinates of three points on the graph.

b. State the domain and range of the function.

c. Graph the function.

d. State an equation for its inverse.

e. Graph the inverse on the same axes you used in part **c.**

CHAPTER REVIEW

Questions on SPUR Objectives

SPUR stands for **S**kills, **P**roperties, **U**ses, and **R**epresentations. The Chapter Review questions are grouped according to the SPUR Objectives for this chapter.

SKILLS DEAL WITH THE PROCEDURES USED TO GET ANSWERS.

Objective A: *Determine values of logarithms.*
(Lessons 9-5, 9-7, 9-9)

In 1–8, write each number as a decimal. Do not use a calculator.

1. $\log 1000$
2. $\log (.000001)$
3. $\ln e^9$
4. $\log_3 243$
5. $\log_{11} (11^{15})$
6. $\ln 1$
7. $\log_{\frac{1}{2}} (8)$
8. $\log_5 \sqrt[3]{5}$

In 9–14, use a calculator to find each logarithm to the nearest hundredth.

9. $\log 97{,}234$
10. $\ln (100.95)$
11. $\ln 87$
12. $\log (.0003)$
13. $\ln (-4.1)$
14. $\ln 10$

Objective B: *Use logarithms to solve exponential equations.* *(Lessons 9-6, 9-10)*

In 15–22, solve. If necessary, round to the nearest hundredth.

15. $7^x = 343$
16. $9^y = 27$
17. $1000(1.05)^n = 2000$
18. $3 \cdot 2^x = 1$
19. $e^z = 22$
20. $(0.4)^w = e$
21. $12^{a+1} = 1000$
22. $3^{-2b} = 51$

Objective C: *Solve logarithmic equations.*
(Lessons 9-5, 9-6, 9-7, 9-8)

In 23–30, solve. If necessary, round to the nearest hundredth.

23. $\log_x 37 = \log_{11} 37$
24. $\ln (4y) = \ln 9 + \ln 12$
25. $\log z = 4$
26. $\log x = 2.91$
27. $2 \ln 15 = \ln x$
28. $\log_8 x = \frac{3}{4}$
29. $\log_x 64 = 3$
30. $\log_x 5 = 10$

PROPERTIES DEAL WITH THE PRINCIPLES BEHIND THE MATHEMATICS.

Objective D: *Recognize properties of exponential functions.* *(Lessons 9-1, 9-2, 9-3)*

31. What is the domain and range of the function defined by $f(x) = e^x$?
32. What is the domain and range of the function defined by $f(x) = 2^x$?
33. When does the function $f(x) = a^x$ describe exponential growth?
34. What must be true about the value of b in the equation $y = ab^x$, if the equation models exponential decay?

35. Which lines are asymptotes of the graph of $y = 30(1.03)^x$?
36. *Multiple choice.* Which situation does the function $y = e^x$ describe?
    (a) constant increase
    (b) constant decrease
    (c) exponential growth
    (d) exponential decay

Objective E: *Identify or apply properties of logarithms.* *(Lessons 9-5, 9-7, 9-8, 9-9)*

37. What is the inverse of f, when $f(x) = e^x$?

38. State the inverse of the function with equation $y = \log_2 x$.

In 39 and 40, *true or false.*

39. The domain of the log function with base 5 is the range of the exponential function with base 5.

40. Negative numbers are not included in the domain of $f(x) = \log_b x$.

In 41–44, write in exponential form.

41. $\log_6\left(\frac{1}{216}\right) = -3$ 42. $\ln(6.28) \approx 1.8$

43. $\log a = b$ 44. $\log_b m = n$

In 45–48, write in logarithmic form.

45. $10^{-1.2} \approx 0.0631$ 46. $e^4 \approx 54.5982$

47. $x^y = z$, $x > 0$, $x \neq 1$

48. $3^n = 12$

In 49–54, state the general property used in simplifying the expression.

49. $\ln 3 + \ln 4 = \ln 12$

50. $\log 40 - \log 4 = \log 10$

51. $\log_{16}(13^{-2}) = -2\log_{16} 13$

52. $\ln e = 1$ 53. $\log_{92} 92^{81} = 81$

54. $\log_{2.1} 1 = 0$

In 55 and 56, rewrite as the logarithm of a single quantity.

55. $\log x - 3 \log y$

56. $\log a + \log b + .5 \log c$

USES DEAL WITH APPLICATIONS OF MATHEMATICS IN REAL SITUATIONS.

Objective F: *Apply exponential growth and decay models.* *(Lessons 9-1, 9-2, 9-3, 9-10)*

In 57–60, use the following information. In 1991 the population of the Tokyo-Yokohama region in Japan was about 27.245 million. The average annual growth rate was 0.86%. In 1991 the population of Sao Paulo, Brazil, was 18.701 million. Sao Paulo was growing at an annual rate of 2.83%. Suppose these rates continue indefinitely.

57. Find the population of the Tokyo-Yokohama area in 2001.

58. Find the population of the Sao Paulo area in 2001.

59. In what year will the population of Sao Paulo reach 25 million?

60. Estimate the year in which Sao Paulo's metropolitan population will first exceed Tokyo-Yokohama's population.

61. The population of a certain strain of bacteria grows according to $N = N_0 3^{0.827t}$ where t is the time in hours. How long will it take for 30 bacteria to increase to 500 bacteria?

62. A new car costing $13,000 is predicted to depreciate at a rate of 15% per year. About how much will the car be worth in six years?

63. Strontium 90 (^{90}Sr) has a half-life of 29 years.

   a. How much will be left of 5 grams of ^{90}Sr after 116 years?

   b. How much will be left of 5 grams after t years?

64. The power output P (in watts) of a satellite is given by the equation $P = 50e^{-\frac{t}{250}}$, where t is the time in days. If the equipment aboard a satellite requires 15 watts of power, how long will the satellite continue to operate?

Objective G: *Fit an exponential model to data.* *(Lesson 9-4)*

65. Find an equation for the exponential curve passing through (2, 21) and (5, 64).

66. A bacteria population was counted every hour for 7 hours with the following results.

Hour (h)	1	2	3	4	5	6	7
Population (p) (in hundreds)	6	12	23	45	93	190	390

   a. Construct a scatterplot of these data.

   b. Fit an exponential model to these data.

   c. Use your model to estimate the population on the 10th hour.

67. A new substance is known to decay at the following rate.

Days (d)	1	2	3	4	5	6	7	8	9	10
Amount present (grams)	800	640	512	410	327	260	209	167	134	107

   a. Construct a scatterplot of these data.

   b. From the data, what is the approximate half-life of this new substance?

   c. Fit an exponential model to these data.

   d. In 20 days, how much of the substance will be present?

Objective H: *Apply logarithmic scales (pH, decibel), models, and formulas.* *(Lessons 9-6, 9-9)*

In 68–70, use the formula $D = 10 \log \left(\frac{I}{10^{-12}} \right)$ that relates sound intensity I in w/m^2 and relative intensity D in decibels.

68. Find D when $I = 2.48 \times 10^9$.

69. What sound intensity corresponds to a relative intensity of 90 decibels?

70. How many times more intense is a 90 dB sound than a 70 dB sound?

71. Under certain conditions, the height h in feet above sea level can be approximated by knowing the atmospheric pressure P in pounds per square inch (psi) using the model

$$\frac{\ln P - \ln 14.7}{-0.000039} = h.$$

Human blood at body temperature will boil at 0.9 psi. At what height would your blood boil in an unpressurized cabin?

72. Sea water has a pH value of 8 while pure water has a pH value of 7. How many times more acidic is pure water than sea water?

REPRESENTATIONS DEAL WITH PICTURES, GRAPHS, OR OBJECTS THAT ILLUSTRATE CONCEPTS.

Objective I: *Graph exponential functions.*
(Lessons 9-1, 9-2, 9-3)

73. Graph $y = 5^x$ using at least 5 points.

74. Graph $y = \left(\frac{1}{5} \right)^x$ using at least 5 points.

75. At the right are the graphs of the equations $y = 2^x$ and $y = 3^x$. Which equation corresponds to the graph

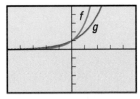

   a. of f?

   b. of g?

   -5 ≤ x ≤ 5, x-scale = 1
   -5 ≤ y ≤ 5, y-scale = 1

   c. Describe how the graph of $y = e^x$ is related to the graphs of f and g.

76. Graph $g(x) = \left(\frac{1}{2} \right)^x$ and $h(x) = \left(\frac{1}{2} \right)^{2x}$ on the same set of axes. Which function has greater values when

   a. $x > 0$?

   b. $x < 0$?

Objective J: *Graph logarithmic curves.*
(Lessons 9-5, 9-7, 9-9)

77. Use the results of Question 73 to plot at least 5 points on the graph of $y = \log_5 x$.

78. a. Plot $y = 10^x$ and $y = \log_{10} x$ on the same set of axes.

   b. Identify all intercepts of these curves.

79. a. Graph 5 points on $y = \ln x$.

   b. Name its inverse function.

80. The graph below has the equation $y = \log_Q x$. Find Q.

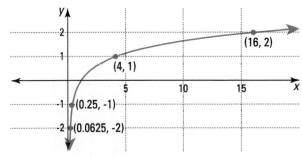

81. What is the x-intercept of the graph of $y = \log_b x$, where $b > 1$?

TRIGONOMETRY

The word *trigonometry* is derived from Greek words meaning "triangle measure." As early as 1500 B.C., the Egyptians had sun clocks. Using their ideas, the ancient Greeks created sundials by erecting a gnomon, or staff, in the ground. The shadows and the height of the gnomon created triangles that could be used to measure the angle of the sun. Using these measurements, the Greeks could determine the duration of a year.

By measuring shadows and the angle of the sun, ancient peoples were also able to measure heights of objects. "Shadow reckoning" was used by the Greeks to measure heights of the Egyptian pyramids. Today the shadows cast by the sun are employed to find the depths of craters on the moon and the heights of dust tornadoes on Mars.

Trigonometry is also used to describe wave-like patterns. For instance, when a spacecraft is launched from Cape Canaveral, its position with respect to the equator has a graph like the curve below. This curve can be described with trigonometric functions.

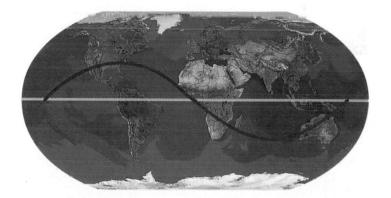

This chapter introduces the three fundamental trigonometric ratios: sine, cosine, and tangent. It proceeds as the history of trigonometry did, starting with right-triangle relationships, moving to the study of all triangles, and then considering the trigonometric functions whose graphs include the above curve.

Me and my shadow. *If you knew the latitude at which this picture was taken and the date, the shadows of these penguins could be used to tell the possible times of the day when the picture was taken. Trigonometry would be needed.*

Suppose a flagpole casts a 12-ft shadow when the sun is at an angle of 64° with the ground. What is the height of the pole? Problems like this one led to the development of *trigonometry.*

Three Trigonometric Ratios

Consider the two right triangles ABC and $A'B'C'$, with $\angle A \cong \angle A'$.

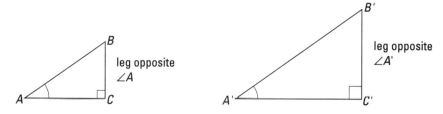

By the AA Similarity Theorem, these triangles are similar, so ratios of the lengths of corresponding sides are equal. In particular,

$$\frac{B'C'}{BC} = \frac{A'B'}{AB}.$$

Exchanging the means gives the equivalent proportion,

$$\frac{B'C'}{A'B'} = \frac{BC}{AB}.$$

Look more closely at these ratios:

$$\frac{B'C'}{A'B'} = \frac{\text{length of leg opposite } \angle A'}{\text{length of hypotenuse of } \triangle A'B'C'}$$

and

$$\frac{BC}{AB} = \frac{\text{length of leg opposite } \angle A}{\text{length of hypotenuse of } \triangle ABC}.$$

Thus, in every right triangle with an angle congruent to ∠A, the ratio of the length of the leg opposite that angle to the length of the hypotenuse of the triangle is the same.

In similar right triangles, any other ratio of corresponding sides is also constant. These ratios are called *trigonometric ratios.* There are six possible trigonometric ratios. All six have special names, but three of them are more important. They are defined here. The Greek letter θ (theta) is customarily used to refer to an angle or its measure.

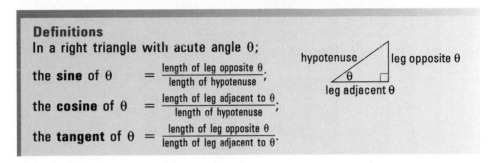

Definitions
In a right triangle with acute angle θ:

the **sine** of θ $= \dfrac{\text{length of leg opposite θ}}{\text{length of hypotenuse}}$,

the **cosine** of θ $= \dfrac{\text{length of leg adjacent to θ}}{\text{length of hypotenuse}}$,

the **tangent** of θ $= \dfrac{\text{length of leg opposite θ}}{\text{length of leg adjacent to θ}}$.

To follow a practice begun by Euler, we use the abbreviations **sin** θ, **cos** θ, and **tan** θ to stand for these ratios, and abbreviate them as below.

$$\sin θ = \frac{\text{opposite}}{\text{hypotenuse}} = \frac{\text{opp.}}{\text{hyp.}}$$

$$\cos θ = \frac{\text{adjacent}}{\text{hypotenuse}} = \frac{\text{adj.}}{\text{hyp.}}$$

$$\tan θ = \frac{\text{opposite}}{\text{adjacent}} = \frac{\text{opp.}}{\text{adj.}}$$

Sine, Cosine, and Tangent Functions

The correspondences

$$θ \rightarrow \sin θ$$
$$θ \rightarrow \cos θ$$
$$θ \rightarrow \tan θ$$

between an angle measure θ in a right triangle and the three right triangle ratios define three functions called the sine, cosine, and tangent functions. The domain of each of these functions is a set of angle measures.

In the early history of trigonometry, mathematicians calculated values of these functions and published the values in tables. Today, calculators give the values of these functions, using formulas derived from calculus. Many calculators allow you to enter angle measures in three different units: degrees, radians, or grads. You will learn about radians in Lesson 10-10. We do not use grads in this book.

For now, you should make sure that any calculator you use is set to degrees. To evaluate sin n° on most graphics calculators, you key in ⟨sin⟩ *n;* on most scientific calculators you key in *n* ⟨sin⟩. Values of the cosine and tangent functions are found in the same way. Use your calculator to verify the answers in Example 1.

Example 1

Find tan 64°, sin 64°, and cos 64° to the nearest thousandth.

Solution

One calculator gives tan 64° ≈ 2.0503038. Rounded to the nearest thousandth, tan 64° ≈ 2.050. Similarly, we find sin 64° ≈ 0.899 and cos 64° ≈ 0.438.

Check

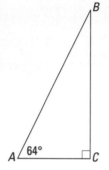

Use the definition, and draw a right triangle with a 64° angle. In right triangle *ABC* at the left, we measured the sides and found *AB* ≈ 44 mm, *BC* ≈ 39 mm, and *AC* ≈ 19 mm.

$$\tan 64° = \frac{\text{leg opposite } \angle A}{\text{leg adjacent to } \angle A} = \frac{BC}{AC} \approx \frac{39}{19} \approx 2.053$$

$$\sin 64° = \frac{\text{leg opposite } \angle A}{\text{hypotenuse}} \approx \frac{39}{44} \approx 0.886$$

$$\cos 64° = \frac{\text{leg adjacent to } \angle A}{\text{hypotenuse}} \approx \frac{19}{44} \approx 0.432.$$

These are close enough, given the accuracy of the drawing.

In this book, we usually give values of the trigonometric functions to the nearest thousandth. But when a trigonometric value appears as part of a longer calculation we do not round the calculator's values until the end of the calculation.

Using Trigonometry to Find Sides of Right Triangles

Example 2

Find the height of the flagpole mentioned in the first paragraph of this lesson.

Solution

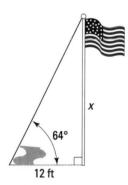

With respect to the 64° angle, the adjacent leg is known and the opposite leg is needed. Consequently, use the tangent ratio to set up an equation.

$$\tan 64° = \frac{opposite}{adjacent}$$

$$\tan 64° = \frac{x}{12}$$

Solve for *x*. x = 12 · tan 64°
From Example 1, we know tan 64° ≈ 2.050.
Substitute. x ≈ 12(2.050) = 24.6
The flagpole is about 24.6 ft high.

Check

Recall from geometry that within a triangle, longer sides are opposite larger angles. We have found that the side opposite the 64° angle is about 24.6 feet long. The angle opposite the 12-foot side is 26°, which is less than 64°. So the answer makes sense.

The tangent ratio was used in Example 2 because one leg was known and another leg was desired. In Example 3, the hypotenuse of the triangle is known. Either the cosine or the sine can be used to determine a leg.

Example 3

A 5″–54 caliber projectile with range of 13 miles is fired at sea on a bearing of 30°. (A **bearing** is the angle measured clockwise from due north.)

a. How far north of its original position will the artillery land?

b. How far east of its original position will the artillery land?

Solution

a. Call the original position Q and the landing position L. Let N be the point due north of Q and due west of L. Draw right triangle QNL. The hypotenuse QL is known, and the leg adjacent to $\angle Q$, QN, is needed. Use the cosine ratio.

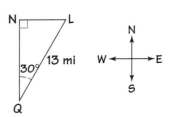

$$\cos Q = \frac{\text{adj.}}{\text{hyp.}} = \frac{QN}{QL}$$
$$\cos 30° = \frac{QN}{13}$$
$$QN = 13 \cdot \cos 30°$$

Use the calculator to compute $13 \cdot \cos 30°$.
$$QN \approx 11.258$$

The artillery will land about 11.3 mi. north of its original position.

b. The leg opposite $\angle Q$, NL, is needed. The hypotenuse QL is known. Use the sine ratio.

$$\sin Q = \frac{\text{opp.}}{\text{hyp.}} = \frac{NL}{QL}$$
$$\sin 30° = \frac{NL}{13}$$
$$NL = 13 \cdot \sin 30°$$
$$NL = 6.5$$

The artillery will land about 6.5 mi. east of its original position.

Check 1

$\triangle QNL$ is a 30-60-90 right triangle. The leg opposite the 30° angle should be half the hypotenuse, which it is.

Check 2

The sides should agree with the Pythagorean Theorem.
Does $(11.258)^2 + (6.5)^2 = (13)^2$?
Does $168.992564 \approx 169$?
Yes. Slight differences are due to rounding.

This is the USS Normandy, a guided missile cruiser commissioned by the Navy in 1989.

Drawing Auxiliary Lines to Create Right Triangles

When you wish to find the length of a segment and no right triangle is given, you can sometimes draw an auxiliary line to create a right triangle.

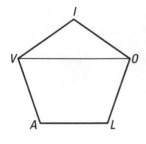

Example 4

Each side of regular pentagon *VIOLA* is 7.8 cm long. Find the length of diagonal $\overline{VO}$.

Solution

Create a right triangle by drawing the perpendicular to $\overline{VO}$ from *I*. Call the intersection point *P*, as in the drawing at the right. Recall that each angle in a regular pentagon has measure

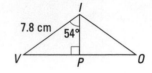

$$\frac{180(5 - 2)}{5} = 108°. \text{ So } m\angle VIO = 108°.$$

$\overline{IP}$ bisects $\angle VIO$. So $m\angle VIP = 54°$. Now, since the hypotenuse of $\triangle VIP$ is known and the opposite leg, *VP*, is needed, use the sine ratio.

$$\sin 54° = \frac{opposite}{hypotenuse}$$

$$\sin 54° = \frac{VP}{7.8}$$

$$VP = 7.8 \cdot \sin 54°$$

$$VP \approx 6.3$$

The perpendicular $\overline{IP}$ bisects $\overline{VO}$, so the diagonal is about 2(6.3), or 12.6 cm long.

QUESTIONS

Covering the Reading

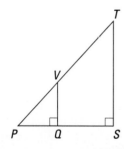

1. What is the origin of the word "trigonometry"?

2. Name one current application where "shadow reckoning" is used.

3. State the AA Similarity Theorem.

4. Consider triangles $\triangle PQV$ and $\triangle PST$ at the left. Put a <, =, or > sign to make the statement true.

$$\frac{ST}{PT} \underline{\quad ? \quad} \frac{QV}{PV}$$

5. Refer to the triangle below. Fill in each blank with the correct ratio.
   a. $\sin \theta = \underline{\quad ? \quad}$
   b. $\cos \theta = \underline{\quad ? \quad}$
   c. $\tan \theta = \underline{\quad ? \quad}$

6. What is the domain of the function $\theta \rightarrow \sin \theta$ in Question **5a**?

7. Refer to △*ABC* at the right.
  a. $\overline{BC}$ is the __?__ of the triangle.
  b. Which is the leg opposite ∠*B?*
  c. Which is the leg adjacent to ∠*B?*
  d. $\overline{AB}$ is the leg opposite which angle?
  e. $\frac{AC}{AB} = $ __?__ B
  f. $\frac{AC}{BC} = $ __?__ B
  g. $\frac{AB}{BC} = $ __?__ B

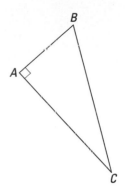

In 8–10, use your calculator to estimate the trigonometric value to the nearest millionth and to the nearest thousandth.

  8. sin 64°   **9.** tan 45°   **10.** cos 87.5°

11. a. Measure the length of each side of △*DEF* at the left to the nearest millimeter, and then estimate sin *D*, cos *D*, and tan *D* using these lengths.
  b. m∠*D* ≈ 25°. Check your answers from part **a** by finding sin *D*, cos *D*, and tan *D* on a calculator.

12. Refer to Example 2. If the shadow were 20 feet long, what would be the height of the flag pole? (Assume the angle of the sun is still 64°.)

13. A ship sails 80 kilometers on a bearing of 75°.
  a. Sketch a diagram to represent this situation.
  b. How far north of its original position is the ship?
  c. How far east of its original position is the ship?

14. In Example 4, find *IP* to the nearest tenth of a centimeter.

Applying the Mathematics

15. A 20-ft ladder is placed against a wall at an angle of 72° with the ground.
  a. Draw and label a diagram to represent this situation.
  b. How far from the base of the wall is the bottom of the ladder?

16. The building code in one city specifies that ramps must form an angle with the horizontal with a measure α (alpha) no greater than 5°. A porch is 6 ft high.

  a. What is the length of the shortest ramp that will meet the code?
  b. How far from the base of the porch will the ramp extend?

17. Suppose each side in the regular octagon at the right has a length of 4 cm. Find the length of $\overline{AC}$.

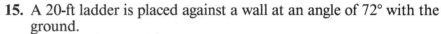

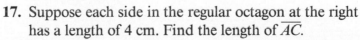

Towing the line. *Shown is the crew of* America[3] *who sailed in the America's Cup in 1995. The crew of* America[3] *included twenty-three women.*

18. The half-life of carbon-14 (^{14}C) is about 5730 years. How many years does it take 1 kg of ^{14}C to decay to 250 g? *(Lessons 9-2, 9-10)*

19. Solve for p: $7\sqrt[3]{p-4} = 21$. *(Lesson 8-8)*

20. Assume $x \neq 0$. Write two different expressions equal to $\dfrac{x}{x\sqrt{3}}$. *(Lesson 8-6)*

21. Suppose g is a function and g^{-1} is its inverse. Simplify the following. *(Lesson 8-3)*

    a. $g(g^{-1}(5))$ **b.** $g(g^{-1}(x))$

22. A singing group buys sunglasses and caps for one of their acts. Four pairs of sunglasses and 10 caps cost $103.70. Five pairs of sunglasses and 5 caps cost $69.70. If all sunglasses are the same price, and all caps are the same price, what is the cost of one cap? *(Lesson 5-4)*

In 23 and 24, *true or false.* Refer to the figure at the left where $j \parallel k$. *(Previous course)*

23. $\angle 3 \cong \angle 6$ **24.** $\angle 2 \cong \angle 5$.

25. In the figure at the right m$\angle HTC = 15°$ and $\overrightarrow{TH} \parallel \overline{BC}$. Find

    a. m$\angle BTC$ **b.** m$\angle C$ *(Previous course)*

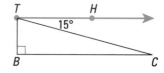

26. Three trigonometric ratios not mentioned in this book are the secant, cosecant, and cotangent. Find the definitions of these ratios and the abbreviations commonly used for them. Give values to the nearest thousandth of the secant of 40°, the cosecant of 40°, and the cotangent of 40°.

10-2

More Right-Triangle Trigonometry

A towering castle. *Shown is the Neuschwanstein Castle of Ludwig II, King of Bavaria, located atop the Bavarian Alps in Germany. Construction of the castle began in 1869 and took 17 years to complete. See Example 2.*

In the last lesson, you learned how to find lengths of sides in a right triangle using trigonometric ratios. In this lesson, you will learn to use trigonometric ratios to find angle measures in right triangles.

Finding an Angle Measure Using a Trigonometric Ratio

Like other functions, trigonometric functions have inverses. On restricted domains, these inverses are functions denoted by $\sin^{-1}$, $\cos^{-1}$, and $\tan^{-1}$. To find angle measures using trigonometric ratios, you need to use these functions. For instance, if you know that

$$\cos \theta = 0.899,$$

then you can take the inverse cosine of each side.
$$\cos^{-1}(\cos \theta) = \cos^{-1}(0.899).$$

Since, in general, $f^{-1}(f(x)) = x$,
$$\theta = \cos^{-1}(0.899).$$

The $\cos^{-1}$ function on most calculators is performed by pressing INV or 2nd and the cos key. Our calculator shows 25.97306856 . The inverse sine and inverse tangent functions work in the same way.

Activity

Use a calculator to find the angle θ. Round to the nearest degree.
a. $\tan \theta = 0.25$ **b.** $\sin \theta = 0.61$ **c.** $\cos \theta = 0.80$

Example 1

Find the measures of the acute angles of a 3-4-5 right triangle.

Solution

Draw a diagram such as the one at the right. You must find m∠A and m∠B. Because you know the lengths of all three sides, you can use any of the trigonometric ratios. For instance,

$\cos B = \frac{4}{5} = .8$. Use the INV or 2nd key with cos. m∠B ≈ 36.870° ≈ 37°.

To find ∠A use another ratio or use the Triangle-Sum Theorem. m∠A ≈ 53°.

Check

You are asked to check this Example in the Questions by finding m∠A and m∠B in another way.

Finding Angles of Elevation and Depression

The **angle of elevation** of the sun is the angle between the horizontal and the observer's *line of sight* to the sun. From ancient times to the present, people have used the angle of elevation to determine the time of day. Using other stars, you can also tell time at night.

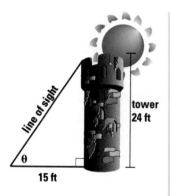

Example 2

A 24-foot high tower casts a 15-foot shadow. What is the angle of elevation of the sun?

Solution

Let θ be the angle of elevation. Given are the values of the sides opposite and adjacent to θ, so use the tangent ratio.

$$\tan \theta = \frac{opposite}{adjacent} = \frac{24}{15} = 1.6$$

$$\tan^{-1}(\tan \theta) = \tan^{-1}(1.6)$$

So
$$\theta = \tan^{-1}(1.6) \approx 57.994617.$$

The angle of elevation is about 58°.

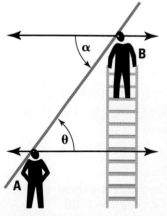

Related to the angle of elevation is another angle. In the figure at the left, if *A* looks up at *B*, then θ represents the angle of elevation. If *B* looks down at *A*, the angle between *B*'s line of sight and the horizontal is called the **angle of depression.** The angle of depression is labeled α (the Greek letter alpha). The line of sight between *A* and *B* is a transversal for the parallel horizontal lines. Thus θ and α are alternate interior angles and must be congruent. *So the angle of elevation is equal to the angle of depression.*

Example 3

A person on top of a building finds that there is a 28° angle of depression to the head of a 6-foot-tall assistant. If the assistant is 40 ft from the building, how tall is the building?

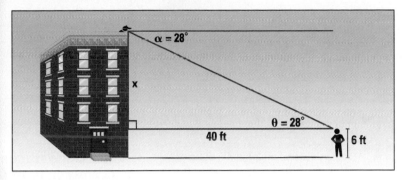

Solution

The angle of depression α is not inside a right triangle, so you cannot use it directly to set up a trigonometric ratio. But the angle of depression is congruent to the angle of elevation, which is θ in the drawn triangle. To find x, use the tangent ratio because the adjacent side is known and the opposite side is needed. Then find the height of the building by adding x to the height of the assistant.

$$\tan \theta = \tan 28° = \frac{opposite}{adjacent} = \frac{x}{40}$$

$$x = (40)(\tan 28°) \approx 21$$

The height of the building is about $21 + 6 = 27$ ft.

QUESTIONS

Covering the Reading

1. Write a key sequence for a calculator you use to find θ if $\cos \theta = .866$. (Identify the calculator.)

2. What values did you find for θ in the Activity in this lesson?

In 3 and 4, evaluate the following to three decimal places.

3. $\cos^{-1}(.766)$

4. $\sin^{-1} .5$

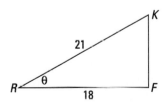

5. Refer to $\triangle RFK$ at the left. Find θ to the nearest degree.

6. Refer to Example 1.
   a. Find $m\angle B$ using the tangent ratio.
   b. Find $m\angle A$ using the sine ratio.

7. Find the measures of the acute angles of a 5-12-13 right triangle.

8. The angle of elevation is the angle made between the line of sight of an observer and the __?__.

9. If a tower 37 meters high casts a shadow 6.2 meters long, what is the angle of elevation of the sun?

10. In the picture below, a person is standing on a cliff looking down at a boat. Which angle, θ or α, is the angle of depression?

11. *True or false.* The angle of elevation from a point *A* to a point *B* equals the angle of depression from *B* to *A*.

12. Refer to Example 3. Suppose the same assistant stands 50 ft from another building, and the angle of depression is 65°. How tall is this new building?

Applying the Mathematics

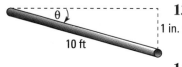

13. To ensure that water and waste are not trapped in a drain pipe, drain pipes are installed so that for every 10 feet of pipe, there is a drop of 1″. What angle does the pipe make with the horizontal?

14. To avoid a steep descent, a plane flying at 35,000 ft starts its descent 150 miles from the airport. At what constant angle of descent θ should the plane descend?

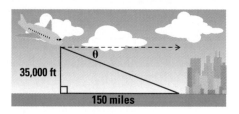

15. A certain ski slope is 580 meters long with a vertical drop of 150 m. At what angle does the skier descend?

16. After a major storm, a forester noted that a large tree had been blown 13° from the vertical. When the forester stands directly under the top of the tree, he is about 10 feet from the base of the tree.
a. How far above the ground was the top of the tree before the storm?
b. How far above the ground was the top of the tree after the storm?

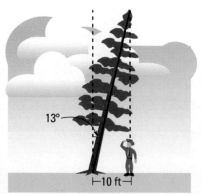

17. To estimate the distance across a river, Sir Vayer marks point A near one bank, sights a tree T growing on the opposite bank, and measures off a distance AB of 100 ft. At B he sights T again. If $m\angle A = 90°$ and $m\angle B = 76°$, how wide is the river? *(Lesson 10-1)*

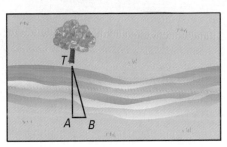

18. a. Use triangle ABC at the left and a calculator to complete the chart.

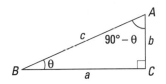

θ	10°	20°	30°	40°	50°	60°	70°	80°
$\sin\theta$								
$\cos\theta$								

b. Make a conjecture. For all θ between 0° and 90°, $\sin\theta = \underline{\quad?\quad}$.
c. Prove your conjecture. (You may wish to use the triangle at the left.) *(Lesson 10-1)*

In 19–21, use triangle SKY at the right.
(Previous course, Lessons 4-5, 10-1)

19. Find the coordinates of the vertices of the image triangle under the transformation R_{270}.

20. Find SK.

21. Find $m\angle SKY$.

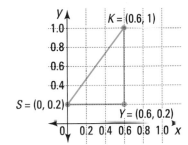

22. Simplify $\dfrac{12}{\sqrt{3}}$. *(Lesson 8-5)*

23. If an isosceles right triangle has a leg of length x, how long is its hypotenuse? *(Previous course, Lesson 8-5)*

24. Solve for x: $mx^2 + px + t = 0$. *(Lesson 6-7)*

Master of the sea. *Bark sailboats, like the* Gloria *shown here, have three or more masts. This ship, built in Columbia, is 255 ft in length.*

Exploration

25. A *British nautical mile* is defined as the length of a minute of arc of a meridian. A minute is $\frac{1}{60}$ of a degree. In feet, it is approximated by
$$6{,}077 - 31\cos(2\theta)$$
where θ is the latitude in degrees.
a. Find the length of a British nautical mile where you live. (You need first to find your latitude.)
b. The *U.S. nautical mile* is defined to be 6080.2 feet. At what north latitude do the two definitions agree?

10-3

Properties of Trigonometric Ratios

Complementary colors on color wheels. *Two colors that appear opposite each other on the color wheel are called complementary. When two complementary-colored pigments are combined, the resulting color is gray.*

In this lesson, you will see four important theorems relating sines, cosines, and tangents, and use these theorems to compute exact values for some trigonometric functions.

The Complements Theorem

Activity 1

1. Use your calculator to find these values.
   a. sin 17°   **b.** cos 73°   **c.** sin 65°   **d.** cos 25°

Activity 2

2. Find another pair of angle measures *x* and *y* that illustrates the pattern cos *x* = sin *y*.

The pairs of numbers in the Activity are instances of the following theorem.

> **Complements Theorem**
> For all θ between 0° and 90°,
> sin θ = cos (90° − θ) and cos θ = sin (90° − θ).

Proof
Consider △*ABC* with right angle *C*. Then
m∠*A* + m∠*B* = 90°.
So if m∠*A* = θ, then
m∠*B* = 90° − θ
Notice that sin θ = $\frac{a}{c}$ and also
$$\cos (90° − θ) = \frac{a}{c}.$$
So sin θ = cos (90° − θ).
Similarly, both cos θ and sin (90°− θ) equal $\frac{b}{c}$. So cos θ = sin (90°− θ).

In words, if two angles are complementary, the sine of one angle equals the cosine of the other. This is how the name "cosine" arose; cosine is short for complement's sine. For instance, cos 23° = sin(90° − 23°) = sin 67°. You should check with your calculator that both cos 23° and sin 67° are approximately 0.921.

The Pythagorean Identity

Consider $\triangle ABC$ on the previous page again.

Because $\sin \theta = \frac{a}{c}$ and $\cos \theta = \frac{b}{c}$,

$$(\sin \theta)^2 + (\cos \theta)^2 = \left(\frac{a}{c}\right)^2 + \left(\frac{b}{c}\right)^2$$

$$= \frac{a^2}{c^2} + \frac{b^2}{c^2} \qquad \text{Power of a Quotient Property}$$

$$= \frac{a^2 + b^2}{c^2}$$

$$= \frac{c^2}{c^2} \qquad \text{Pythagorean Theorem}$$

$$= 1$$

This proves the theorem called the **Pythagorean Identity.**

> **Theorem (Pythagorean Identity)**
> For all θ between 0° and 90°, $(\cos \theta)^2 + (\sin \theta)^2 = 1$.

The Pythagorean Identity can be used to find the value of sin θ if only cos θ is known, or vice versa.

Example 1

Suppose θ is an acute angle in a right triangle, and sin θ = 0.6. Find cos θ.

Solution 1

From the Pythagorean Identity, you know that $(\cos \theta)^2 + (\sin \theta)^2 = 1$. Substitute 0.6 for sin θ and solve for cos θ.

$$(\cos \theta)^2 + 0.60 = 1$$
$$(\cos \theta)^2 + 0.36 = 1$$
$$(\cos \theta)^2 = 0.64$$
$$\cos \theta = \pm 0.8$$

For acute angles, cos θ is always positive because it represents the ratio of lengths. So cos θ = 0.8.

Solution 2

First find θ with a calculator. Since sin θ = 0.6, $\theta = \sin^{-1}(0.6) \approx 36.87°$. Now find cos θ. cos 36.87° ≈ .8.

The Tangent Theorem

An important theorem relates the values of three trigonometric functions.

Tangent Theorem

For all θ between $0°$ and $90°$, $\tan \theta = \frac{\sin \theta}{\cos \theta}$.

Proof

Consider any right triangle, say $\triangle ABC$ at the right. By the definition of tangent,

$$\tan \theta = \frac{a}{b}.$$

Dividing the numerator and denominator each by the hypotenuse c, we get

$$\tan \theta = \frac{\frac{a}{c}}{\frac{b}{c}}$$

$$\text{So } \tan \theta = \frac{\sin \theta}{\cos \theta}.$$

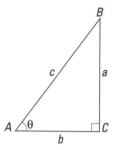

Example 2

Find the tangent of angle θ in Example 1.

Solution

You are given $\sin \theta = 0.6$ and found $\cos \theta = 0.8$.

$$\tan \theta = \frac{\sin \theta}{\cos \theta} = \frac{0.6}{0.8} = 0.75$$

Exact Values of Some Trigonometric Ratios

Calculators must give approximations for most values of sines and cosines. For some angles, however, the sine and cosine have simple exact values. You should know these.

Exact-Value Theorem

a. $\sin 30° = \cos 60° = \frac{1}{2}$

b. $\sin 45° = \cos 45° = \frac{\sqrt{2}}{2}$

c. $\sin 60° = \cos 30° = \frac{\sqrt{3}}{2}$

Proof

a. Examine a $30°$-$60°$-$90°$ triangle. Recall that the length of the side opposite a $30°$ angle is half the length of the hypotenuse.

So, $\sin 30° = \cos 60° = \frac{BC}{AB} = \frac{\frac{1}{2}h}{h} = \frac{1}{2}$.

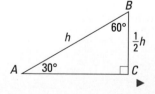

b. Recall also from geometry that the hypotenuse of a 45°-45°-90° triangle is $\sqrt{2}$ times the length of either leg. Therefore:

$$\sin 45° = \cos 45° = \frac{EF}{DE}$$

$$= \frac{x}{x\sqrt{2}} = \frac{1}{\sqrt{2}} = \frac{1}{\sqrt{2}} \cdot \frac{\sqrt{2}}{\sqrt{2}} = \frac{\sqrt{2}}{2}.$$

c. This part is left for you to do in the Questions.

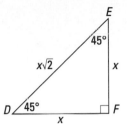

If you know the exact values of $\sin\theta$ and $\cos\theta$ for $\theta = 30°$, $45°$, and $60°$ you can calculate the exact values of the tangent of these angles.

Example 3

Find the exact value of tan 60°.

Solution

$$\tan 60° = \frac{\sin 60°}{\cos 60°} \qquad \text{Tangent Theorem}$$

$$= \frac{\frac{\sqrt{3}}{2}}{\frac{1}{2}} \qquad \text{Exact Value Theorem}$$

$$= \frac{\sqrt{3}}{2} \cdot \frac{2}{1} \qquad \text{Definition of Division}$$

$$= \sqrt{3} \qquad \text{Arithmetic}$$

The exact value of tan 60° is $\sqrt{3}$.

Check

Use a calculator to get a decimal approximation.

$$\tan 60° \approx 1.73205$$
$$\sqrt{3} \approx 1.73205$$

It checks.

In this course you need to know the exact values of sine, cosine, and tangent of 30°, 45°, and 60°. The decimal approximations 0.707, 0.866, and 1.732 also appear quite often. You should recognize that $0.707 \approx \frac{\sqrt{2}}{2}$, $0.866 \approx \frac{\sqrt{3}}{2}$, and $1.732 \approx \sqrt{3}$.

QUESTIONS

Covering the Reading

1. *Multiple choice.* Which is the measure of the complement of an angle with measure θ?
   (a) $\theta - 90°$ (b) $180° - \theta$ (c) $90° - \theta$ (d) $90° + \theta$

2. What did you answer for Question 2 in the Activity on page 616?

In 3 and 4, copy and complete with the measure of an acute angle.

3. $\cos 40° = \sin$ __?__   **4.** $\sin 72° = \cos$ __?__

5. State the Pythagorean Identity.

6. a. Without using a calculator, give the value of $(\cos 15°)^2 + (\sin 15°)^2$.
   b. Check by using a calculator.

In 7 and 8, assume that θ is an acute angle in a right triangle. Use the Pythagorean Identity to find $\cos \theta$ when $\sin \theta$ has the given value.

7. $\sin \theta = .28$ **8.** $\sin \theta = \frac{\sqrt{3}}{2}$

9. *True or false.* For all θ between 0° and 90°, $\frac{\sin \theta}{\cos \theta} = \tan \theta$.

10. Complete this chart with exact values.

θ	$\cos \theta$	$\sin \theta$	$\tan \theta$
30°			
45°			
60°			

Applying the Mathematics

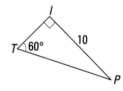

11. Use the triangle at the left to complete the proof that $\sin 60° = \cos 30° = \frac{\sqrt{3}}{2}$.

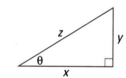

12. In the triangle at the right, find $\frac{\sin \theta}{\cos \theta}$.

13. In $\triangle TIP$ at the left below, $IP = 10$. Find exact values of:
   a. *IT.*   **b.** *PT.*

Pipe down! *Shown is a water pipeline in southern California. Pipelines transport such things as water, wheat, sawdust, or petroleum.*

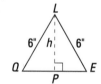

14. $\triangle QLE$ on the right above is an equilateral triangle with sides 6" long.
   a. Find the exact height h.
   b. Find the exact area of $\triangle QLE$.

15. Find the exact value of $(\tan 30°)^2$.

16. *Multiple choice.* The angle of elevation of a pipeline up the side of a mountain is 45°. If the pipe is 20 meters long, which of these is not the vertical rise of the pipe?
   (a) $20 \cdot \sin 45°$ (b) $\frac{20}{\sqrt{2}}$
   (c) $10\sqrt{2}$ (d) $20\sqrt{2}$

17. A private plane flying at an altitude of 1 mile begins its descent to an airport when it is a ground distance of 6 miles away. At what constant angle of depression would it need to descend? *(Lesson 10-2)*

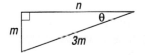

18. In the triangle at the left find each value. *(Lesson 10-1)*
    a. $\sin \theta$ b. $\cos \theta$ c. $\tan \theta$

19. a. Solve the system $\begin{cases} y = x^2 - 2x - 1 \\ y = 5 \end{cases}$ by graphing.

    b. Check your work by using some other method. *(Lessons 5-2, 5-3, 6-7)*

20. Find the coordinates of each image. *(Lessons 4-7, 4-8)*
    a. $R_{90}(1, 0)$ b. $R_{180}(1, 0)$
    c. $R_{270}(1, 0)$ d. $R_{-90}(1, 0)$

21. Ms. Driver bought a new set of tires for her car after 59,000 miles of driving. The tire store suggested that the tires be rotated after every 6000 miles. Assume Ms. Driver takes this advice.
    a. What will the odometer show when the tires should be first rotated?
    b. What will the odometer show when the tires should be rotated for the nth time?
    c. How many times will the tires have been rotated when the odometer reaches 100,000 miles? *(Lessons 3-1, 3-7, 3-8, 3-9)*

22. State the quadrants (I, II, III, or IV) in which a point (x, y) may be found if
    a. x is negative and y is positive.
    b. x is positive and y is negative.
    c. $x = y$ and $xy \neq 0$. *(Previous course)*

23. a. Verify that
       (i) $\sin 60° = 2 \cdot \sin 30° \cdot \cos 30°$.
       (ii) $\sin 84° = 2 \cdot \sin 42° \cdot \cos 42°$.
    b. Generalize the result of part a and verify your generalization with some other values. (The result is called the Double Angle Formula for the sine.)

Rotations, Sines, and Cosines

IN-CLASS
ACTIVITY

Materials: Each pair should have a good quality compass, a protractor, graph paper, and a calculator. Work on this Activity with a partner.

1 Draw a set of coordinate axes on the graph paper. Let each side of a square on your coordinate grid represent 0.1 unit. With the origin as the center, use a compass to draw a circle with radius 1. Label the positive x-intercept of the circle as A_0. Your circle should look like the one at the right.

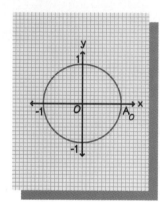

2 a. With a protractor, locate the image of $A_0 =$ (1, 0) under R_{20}. Label this point A_{20}.
b. Use the grid to estimate the x-coordinate and y-coordinate of A_{20}.
c. Use a calculator to find cos 20° and sin 20°.

3 a. With a protractor, locate $R_{40}(1, 0)$. Label it A_{40}.
b. Use the grid to estimate the x- and y-coordinates of A_{40}.
c. Use a calculator to find cos 40° and sin 40°.

4 a. Locate $R_{55}(1, 0)$. Label it A_{55}.
b. Estimate the x- and y-coordinates of this point.
c. Use a calculator to evaluate cos 55° and sin 55°.

5 a. Look back at your work for Questions 2 to 4. What relation do you see between the x- and y-coordinates of $R_\theta(1, 0)$, cos θ, and sin θ?
b. Use the relation to estimate the values of cos 73° and sin 73° from your figure without a calculator.
c. How close are your predictions to the actual values?

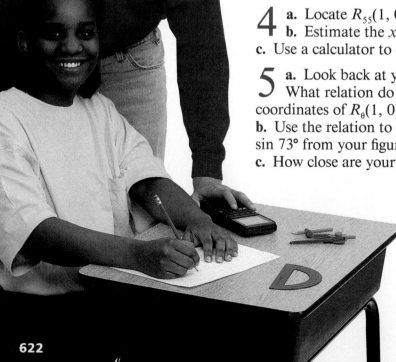

Trigonometry and the Unit Circle

Converting circular motion. *Shown is the* Delta Queen *on the Mississippi River. This steamboat was built in 1926 and uses a paddle wheel as a means of propulsion.*

The sine, cosine, and tangent functions can be defined for all real numbers. However, in a right triangle, the two angles other than the right angle measure between 0° and 90°. So the definitions of cosine and sine given in Lesson 10-1 only apply to measures between 0° and 90°. To define cosines and sines for all real numbers, we use rotations with center (0, 0).

The Unit Circle, Sines, and Cosines

The circle you drew in the In-class Activity on page 622 is a *unit circle*. The **unit circle** is the circle with center at the origin and radius 1 unit. If the point (1, 0) on the circle is rotated around the origin with a magnitude θ, then the image point (x, y) is also on the circle. The coordinates of the image point can be found using sines and cosines, and verify what you should have found in the Activity.

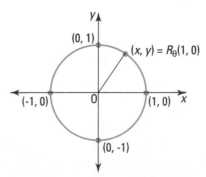

Example 1

What are the coordinates of the image of (1, 0) under R_{70}?

Solution

Let $A = (x, y) = R_{70}(1, 0)$. In the figure at
right, $OA = 1$. Since the radius of the unit
circle is 1. Draw the segment from A to the
point $B = (x, 0)$. $\triangle ABO$ is a right triangle
with legs of length x and y, and hypotenuse
of length 1. Now use the definitions of sine
and cosine.

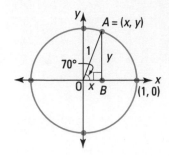

$$\cos 70° = \frac{adj}{hyp} = \frac{x}{1} = x$$

$$\sin 70° = \frac{opp}{hyp} = \frac{y}{1} = y$$

The first coordinate is cos 70° and the second coordinate is
sin 70°. Thus (x, y) = (cos 70°, sin 70°) ≈ (.342, .940). That is,
the image of (1, 0) under R_{70} is (cos 70°, sin 70°), or about
(.342, .940).

Check

Use the Pythagorean Identity with cos 70° and sin 70°.
Is $(.342)^2 + (.940)^2 \approx 1^2$? .117 + .884 = 1.001 ≈ 1, so it checks.

The idea of Example 1 is generalized to define the sine and cosine for
any magnitude θ. Since any real number can be the magnitude of a
rotation, this definition extends these trigonometric functions to have
the set of all real numbers as their domain.

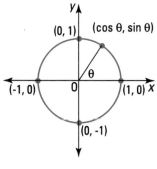

> **Definition**
> Let R_θ be the rotation with center (0, 0) and magnitude θ. Then, for
> any θ, the point **(cos θ, sin θ)** is the image of (1, 0) under R_θ.

Stated another way, cos θ is the x-coordinate of the image of $R_\theta(1, 0)$;
sin θ is the y-coordinate of $R_\theta(1, 0)$.

Rotations of Multiples of 90°

Sines and cosines of angles whose measures are multiples of 90° can be
found from the definition without using a calculator.

Example 2

Explain how to use the unit circle to find
a. sin 90°.   **b.** cos (-180°).

Solution

a. Think: sin 90° is the y-coordinate of $R_{90}(1, 0)$ and
$R_{90}(1, 0) = (0, 1)$. So sin 90° = 1.

▶

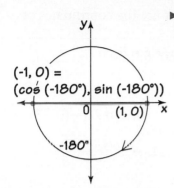

b. $R_{-180}(1, 0) = (-1, 0)$. Since $\cos(-180°)$ is the x-coordinate of this point, $\cos(-180°) = -1$.

Check

Check these values on your calculator.

Rotations of More than 360°

Recall that rotations of magnitude greater than 360° refer to more than one complete revolution.

Example 3

a. Find sin 630°.   **b.** Find sin 385°.

Solution

For each value, first find the magnitude between 0° and 360° corresponding to the given magnitude.

a. $630° - 360° = 270°$. So R_{630} equals one complete revolution R_{360} around the circle, followed by R_{270}.
$R_{630}(1, 0) = R_{270}(1, 0) = (0, -1)$. So $\sin 630° = -1$.

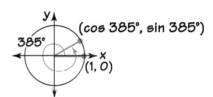

b. $385° - 360° = 25°$. R_{385} equals one complete revolution followed by a rotation of 25°. Thus, $\cos 385° = \cos 25° \approx .906$.

Money revolution. *On the TV game show* Wheel of Fortune, *the contestant must spin the wheel more than 360°.*

QUESTIONS

Covering the Reading

1. If (1, 0) is rotated θ around the origin,
   a. cos θ is the __?__-coordinate of its image.
   b. sin θ is the __?__-coordinate of its image.

2. *True or false.*   The image of (1, 0) under R_{23} is (sin 23°, cos 23°).

3. $R_0(1, 0) = $ __?__, so cos 0° = __?__ and sin 0° = __?__.

4. Explain how to use the unit circle to find sin 180°.

In 5–7, use the unit circle to find the value.

5. cos 90°   6. sin (-90°)   7. cos 270°

8. If (1, 0) is rotated -42° about the origin, what are the coordinates of its image, to the nearest thousandth?

9. a. A rotation of 540° equals a rotation of 360° followed by ___?___.
 b. The image of (1, 0) under R_{540} is ___?___.
 c. Evaluate sin 540°.

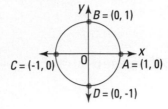

In 10–12, suppose $A = (1, 0)$, $B = (0, 1)$, $C = (-1, 0)$, and $D = (0, -1)$. Which point is the image of (1, 0) under the rotation?

10. R_{450} **11.** R_{540} **12.** R_{-720}

In 13 and 14, evaluate without using a calculator.

13. cos 450° and sin 450°   **14.** cos(-720°) and sin(-720°)

In 15 and 16, use a calculator. Approximate to the nearest thousandth.

15. cos 392°   **16.** sin 440°

Applying the Mathematics

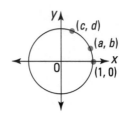

In 17–22, which letter on the figure at the left could stand for the indicated value of the trigonometric function?

17. cos 80°   **18.** sin 80°   **19.** cos (-280°)

20. sin 800°   **21.** cos 380°   **22.** sin (-340°)

23. Suppose $0° \leq \theta \leq 360°$.
 a. What is the largest possible value of cos θ?
 b. What is the smallest possible value of sin θ?

In 24 and 25, verify by substitution that the statement holds for the given value of θ.

24. $(\cos \theta)^2 + (\sin \theta)^2 = 1$; $\theta = 630°$.

25. $\sin \theta = \cos (90° - \theta)$; $\theta = -90°$.

26. Explain how to find the exact value of cos 420° without using a calculator.

Review

In 27 and 28, solve for θ, if $0° < \theta < 90°$. *(Lesson 10-3)*

27. $\sin 14° = \cos \theta$ **28.** $(\sin \theta)^2 + (\cos 74°)^2 = 1$

29. Find the exact value. *(Lesson 10-3)*
 a. cos 30°
 b. sin 30°
 c. tan 30°

30. While stationed 65 feet up in a lookout tower, a ranger sighted a fire. The angle of depression to the fire measured 4°. How far from the base of the tower was the fire? *(Lesson 10-2)*

65 ft

This is a view of a forest fire in Yellowstone National Park as might be seen from a lookout tower.

31. Refer to the diagram below. A roof has a pitch (slope) of $\frac{1}{12}$. What angle θ does the roof make with the horizontal? *(Lesson 10-2)*

θ

12 □ 1

32. Without graphing, explain how to determine the number of x-intercepts of the graph of $y = 5x^2 - 7x + 4$. *(Lessons 6-4, 6-10)*

33. Use the formula $d = \sqrt{(x_2 - x_1)^2 + (y_2 - y_1)^2}$ to find the distance between (-3, 5) and (1, -9). *(Previous course)*

Exploration

34. Fill in this spreadsheet.
Find at least four patterns in the numbers in the table.

θ	cos θ	sin θ	$(\cos θ)^2 + (\sin θ)^2$	$(\cos θ)^2 - (\sin θ)^2$
0°				
20°				
40°				
60°				
80°				
100°				
120°				
140°				
160°				
180°				

10-5

Cosines and Sines in Quadrants II–IV

The Sign of cos θ or sin θ

With a calculator, it is easy to determine the values of cos θ and sin θ for any value of θ. If $0° < θ < 90°$, you can also find these values by drawing a right triangle. You can find the other values using the unit circle.

Where is the image of (1, 0) under $R_θ$? Every value of cos θ or sin θ is a coordinate of a point on the unit circle, because (cos θ, sin θ) is the image of (1, 0) under $R_θ$. Unless θ is a multiple of 90°, the image is in one of the four quadrants. As the figure below on the left shows, each quadrant is associated with a range of values of θ.

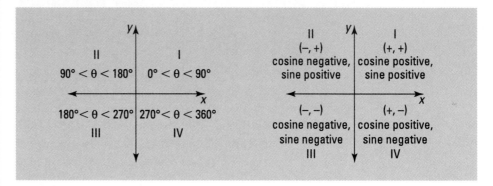

What are the signs of cos θ and sin θ? The quadrants enable you to determine quickly whether cos θ and sin θ are positive or negative. Refer to the figure above on the right. Because cos θ is the first coordinate, or *x*-coordinate, of the image, it is positive in Quadrants I and IV and negative in Quadrants II and III. Sin θ, which is the *y*-coordinate of the image, is positive in Quadrants I and II and negative in Quadrants III and IV.

You do *not* need to memorize these signs. You can always rely on the definition of cos θ and sin θ and visualize a rotation of θ on the unit circle.

cos θ and sin θ when $90° < θ < 180°$

Example 1

Use the unit-circle definition of sin θ to tell whether sin 172° is positive or negative.

▶

Once you know the *sign* of the cosine or sine of an angle, you can find its value by referring to points in the first quadrant. In the following examples we show how knowledge of transformations can help you to find values of sin θ and cos θ when $R_\theta(1, 0)$ is in Quadrant II, III, or IV.

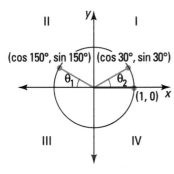

When θ is between 90° and 180°, $R_\theta(1, 0)$ is in Quadrant II. Every point on the unit circle in Quadrant II is the image of a point on the circle in Quadrant I under a reflection over the y-axis. For instance, the reflection image of the point (cos 150°, sin 150°) over the y-axis is (cos 30°, sin 30°), which is in the first quadrant. Notice that the acute angles formed by these two points with the x-axis are congruent. Recall that under $r_{y\text{-axis}}$ the image of (x, y) is (-x, y). So, the first coordinates of these points are opposites.

Thus cos 150° = -cos 30° = $-\frac{\sqrt{3}}{2}$.

The second coordinates are equal, so sin 150° = sin 30° = $\frac{1}{2}$.

cos θ and sin θ when 180° < θ < 270°

When a point is in Quadrant I, rotating it 180° gives a corresponding point in Quadrant III. Thus, to find sin θ or cos θ when 180° < θ < 270°, think of the angle with measure θ − 180°.

Example 2

Show why sin 235° = -sin 55°.

Solution

Make a sketch. P' = (cos 235°, sin 235°) is the image of P = (cos 55°, sin 55°) under R_{180}. Because the image of (x, y) under a rotation of 180° is (-x, -y), P' also has coordinates (-cos 55°, -sin 55°). Thus sin 235° = -sin 55°.

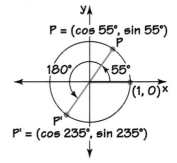

Check

A calculator shows that sin 235° ≈ -0.819 and -sin 55° ≈ -0.819, so it checks.

cos θ and sin θ when 270° < θ < 360°

Points in Quadrant IV are reflection images over the *x*-axis of points in the first quadrant.

Example 3

Find an exact value for cos 315°.

Solution

Cos 315° is the first coordinate of a point in Quadrant IV, so the cosine is positive. Reflect (*cos 315°, sin 315°*) over the x-axis. Since 360° – 315° = 45°, the image point is (*cos 45°, sin 45°*). Since the first coordinates of these points are equal,

$$\cos 315° = \cos 45° = \frac{\sqrt{2}}{2}.$$

Check

A calculator shows cos 315° ≈ 0.707. Recall that 0.707 is an approximation to $\frac{\sqrt{2}}{2}$, so it checks.

If you add or subtract multiples of 360° to a magnitude of a rotation, you will get the same rotation. For instance, $R_{79} = R_{439} = R_{-281} = R_{-641}$. For this reason, cos 79° = cos 439° = cos (-281°) = cos (-641°) and sin 79° = sin 439° = sin (-281°) = sin (-641°). The fact that values of sines and cosines repeat every 360° is a very important property. You will learn more about this in Lesson 10-8.

QUESTIONS

Covering the Reading

In 1 and 2, *multiple choice.* Select from the following choices.
(a) is always positive
(b) is always negative
(c) may be positive or negative

1. If $R_\theta(1, 0)$ is in Quadrant II, then cos θ __?__.

2. When 180° < θ < 360°, sin θ __?__.

3. When cos θ is positive, in which quadrant(s) is $R_\theta (1, 0)$?

In 4 and 5, a trigonometric value is given. **a.** Draw the corresponding point on the unit circle. **b.** Without using a calculator, state whether the value is positive or negative.

4. sin 343°   **5.** cos 217°

6. a. Evaluate cos 118° and sin 118° with a calculator.
   b. Explain why cos 118° = -cos 62° and sin 118° = sin 62°.

In 7 and 8, the given statement is true. **a.** Use the unit circle and transformations to explain why the statement is true. **b.** Verify the statement with a calculator.

7. $\sin 182° = -\sin 2°$ **8.** $\cos 295° = \cos 65°$

In 9–12, find the exact value.

9. $\sin 315°$ **10.** $\cos 240°$

11. $\cos (-150°)$ **12.** $\sin 135°$

Applying the Mathematics

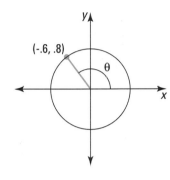

13. Copy and complete with "positive" or "negative." If $\angle B$ is obtuse, then $\cos B$ is __?__ and $\sin B$ is __?__.

14. Refer to the unit circle at the left. Find θ to the nearest degree.

15. Find C such that $0° < C < 180°$ and $\cos C = -0.251$.

In 16–19, use this information. The **tangent of θ, tan θ,** is defined for all values of θ as $\frac{\sin \theta}{\cos \theta}$. This agrees with its right triangle definition when $0° < \theta < 90°$ and extends the relationships between $\tan \theta$, $\sin \theta$, and $\cos \theta$ to all real values.

16. a. Use your calculator to evaluate $\sin 100°$, $\cos 100°$, and $\tan 100°$.
  b. Verify that $\frac{\sin 100°}{\cos 100°} = \tan 100°$.

17. What is the sign of $\tan \theta$ when $270° < \theta < 360°$? Justify your answer using the definition of $\tan \theta$.

18. Without a calculator, evaluate $\tan 180°$.

19. Without a calculator, give the exact value of $\tan 120°$.

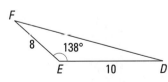

20. Refer to $\triangle ABC$ and $\triangle DEF$ at the left.
  a. Give an area formula for $\triangle ABC$.
  b. Give a formula for $\sin B$ as a ratio.
  c. By substituting your answer to part **b** for $\sin B$, prove that Area of $\triangle ABC = \frac{1}{2}ac \sin B$.
  d. Use the result of part **c** to find the area of $\triangle DEF$.

Review

21. Explain how to evaluate $\sin(-630°)$ without a calculator.   *(Lesson 10-4)*

22. If $0° < \theta < 90°$ and $\sin 83.5° = \cos \theta$, find θ.   *(Lesson 10-3)*

23. On her 5th birthday, Jamie received $500. The money is to be invested so that on her 18th birthday she will have $1800.
  a. What rate of interest compounded annually will allow this?
  b. If interest is compounded continuously, what rate will Jamie's parents need?   *(Lessons 8-4, 9-3, 9-10)*

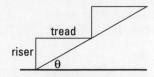

tread

riser

θ

24. In many cities in the United States, building codes specify that stairs in homes must be built with an $8\frac{1}{4}$-inch riser and 9-inch tread. Currently, falls are the leading cause of non-fatal injuries in the U.S. To reduce the number of falls, some architects are proposing changing building codes to require a 7-inch riser and 11-inch tread.

a. By how many degrees would this proposal change the angle θ that the stairs make with the horizontal?

b. Why do you think the architects feel the new stairs would be safer? *(Lesson 10-2)*

25. Solve $3(x - 4)^6 = 2187$ for x. *(Lesson 7-6)*

26. Expand and simplify: $(xa - b)^2 + (xa)^2$. *(Lesson 6-1)*

27. Find the distance between (a, b) and $(-6, 5)$. *(Previous course)*

28. Determine whether or not the triangles in each pair are congruent. *(Previous course)*

a.

b.

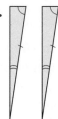

c.

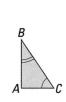

d.

Rising to the occasion.
An important reason why there is a code for building stairs is to help reduce accidents.

Exploration

29. a. Copy and complete the following chart using a calculator or spreadsheet.

b. You should find that $\sin \theta - \tan \theta$ gets closer and closer to 0. Explain why this happens.

θ	$\sin \theta$	$\tan \theta$	$\sin \theta - \tan \theta$
10°			
5°			
2°			
1°			
0.5°			
0.1°			

10-6

The Law of Cosines

One of the main applications of trigonometry is in finding lengths or distances that are difficult or impossible to measure directly. In previous lessons, you have learned to use trigonometric ratios to find unknown sides or angles of *right* triangles. In this lesson and the next, you will learn to determine some unknown sides or angles in *any* triangle, provided you are given enough information.

If you know the measures of two sides and the included angle (the SAS condition) or the measures of three sides of a triangle (the SSS condition), you can use the *Law of Cosines* to find other measures in the triangle.

Theorem (Law of Cosines)
In any triangle ABC,
$c^2 = a^2 + b^2 - 2ab \cdot \cos C$.

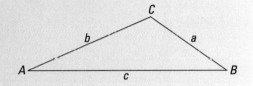

Proof
Set up $\triangle ABC$ on a coordinate plane so that $C = (0, 0)$ and $A = (b, 0)$. To find the coordinates of B, notice that B is a times as far from the origin as is the intersection of the unit circle and $\overline{CB}$. Since that intersection has coordinates $(\cos C, \sin C)$, $B = (a \cos C, a \sin C)$. All that remains is to use the distance formula to find c and square the result.

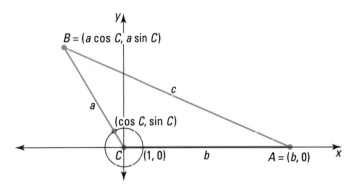

$$c = \sqrt{(a \cos C - b)^2 + (a \sin C - 0)^2} \qquad \text{Distance formula}$$
$$c^2 = (a \cos C - b)^2 + (a \sin C - 0)^2 \qquad \text{Square both sides.}$$
$$c^2 = a^2 (\cos C)^2 - 2ab \cdot \cos C + b^2 + a^2 (\sin C)^2 \qquad \text{Expand.}$$
$$c^2 = a^2 (\cos C)^2 + a^2 (\sin C)^2 + b^2 - 2ab \cdot \cos C \qquad \text{Commutative Property of Addition}$$
$$c^2 = a^2 ((\cos C)^2 + (\sin C)^2) + b^2 - 2ab \cdot \cos C \qquad \text{Distributive Property}$$
$$c^2 = a^2 + b^2 - 2ab \cdot \cos C \qquad \text{Pythagorean Identity}$$

The Law of Cosines applies to *any* two sides and their included angle. Thus it is also true that, in $\triangle ABC$,

$$a^2 = b^2 + c^2 - 2bc \cdot \cos A \quad \text{and} \quad b^2 = a^2 + c^2 - 2ac \cdot \cos B.$$

In words, the Law of Cosines says that in any triangle, the sum of the squares of two sides minus twice the product of these sides and the cosine of the included angle equals the square of the third side. You should memorize the Law of Cosines.

Using the Law of Cosines to Find a Length

Here is how to use the Law of Cosines to find the third side of a triangle when two sides and the included angle are known (the SAS condition).

Example 1

Find the distance between ships A and B below.

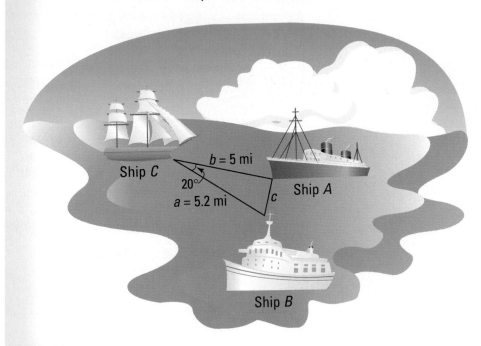

Solution

The unknown side is c and the two known sides are $a = 5.2$ miles and $b = 5$ miles. The included angle is $20°$. Use the Law of Cosines.

$c^2 = a^2 + b^2 - 2ab \cdot \cos C$

$c^2 = (5.2)^2 + 5^2 - 2(5.2)(5) \cos 20°$ Substitute.

$c^2 \approx 27.04 + 25 - 52(.940)$ (You will not see this step if you use a calculator.)

$c^2 \approx 3.16$ Evaluate

$c \approx \pm \sqrt{3.16}$ Take square roots of each side.

$c \approx \pm 1.78.$

Because c is a distance, only the positive solution is acceptable. **The two ships are about 1.8 miles apart.**

Using the Law of Cosines to Find an Angle Measure

If the lengths of all three sides of a triangle are known (the SSS condition), the Law of Cosines can be used to find the measure of any angle of the triangle.

Example 2

A triangle has sides of length 4, 5 and 8.5. To the nearest degree, what is the measure of its largest angle?

Solution

Draw a figure whose sides measure 4, 5, and 8.5. The largest angle is opposite the longest side. We name that angle A. Then a = 8.5 is the longest side. Let b = 5 and c = 4.

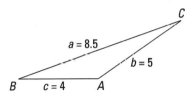

By the Law of Cosines,

Substitute.

Solve for cos A.

$$a^2 = b^2 + c^2 - 2bc \cdot \cos A$$
$$(8.5)^2 = 5^2 + 4^2 - 2(5)(4) \cos A$$
$$72.25 = 25 + 16 - 40 \cos A$$
$$31.25 = -40 \cos A$$
$$-.78125 = \cos A$$
$$m\angle A = \cos^{-1}(-.78125)$$

Use your calculator to find m∠A. Your display should read 141.375 . . . . Thus m∠A ≈ 141°.

Check

You can draw a triangle with the given sides using ruler and compass.

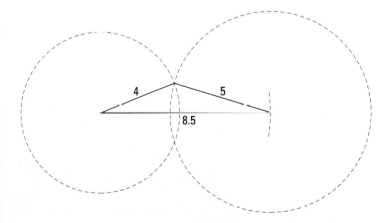

Measure the largest angle with a protractor. It seems to be about 141°.

Covering the Reading

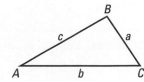

In 1–3, according to the Law of Cosines, in $\triangle ABC$, what expression is equal to the following?

1. $c^2 + b^2 - 2bc \cdot \cos A$

2. $a^2 + c^2 - 2ac \cdot \cos B$

3. c^2

4. *Multiple choice.* Which of the following verbally describes the Law of Cosines?
   (a) The third side of a triangle equals the sum of the squares of the other two sides minus the product of the two sides and the included angle.
   (b) The square of the third side of a triangle equals the sum of the squares of the other two sides minus the product of the two sides and the cosine of the included angle.
   (c) The square of the third side of a triangle equals the sum of the squares of the other two sides minus twice the product of the two sides and the cosine of the included angle.
   (d) none of these

5. If you know two sides and the included angle of a triangle, which other measure(s) can you find using the Law of Cosines?

6. In the proof of the Law of Cosines, why does
$$a^2 ((\cos C)^2 + (\sin C)^2) = a^2?$$

7. Refer to Example 1. Suppose later in the day that ship C is 1.1 miles from ship A and 2.4 miles from ship B and the angle between the two sightings is 135°. How far apart are ships A and B?

8. Refer to Example 2.
   a. Use the Law of Cosines to find the measure of $\angle B$.
   b. Find the measure of $\angle C$ using the Law of Cosines.
   c. How can part **b** be used to check part **a**?

Applying the Mathematics

9. The water molecule H_2O can be modeled by the diagram at the right. The angle between the two oxygen-hydrogen (O—H) bonds is 105°. If the average distance between the oxygen and hydrogen nuclei is p units, how far apart, on average, are the two hydrogen nuclei?

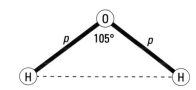

10. Refer to the diagram at the right. If two planes leave Dallas, one flying toward Bismarck and the other flying toward Chicago, by approximately what angle θ do their headings differ?

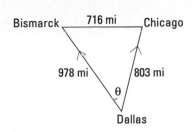

11. Sirius and Alpha Centauri are the two brightest stars (other than our Sun) near the Earth. The distance from Earth to Sirius is about 8.8 light-years. The distance from Earth to Alpha Centauri is about 4.3 light-years. The angle between these stars, with the Earth as vertex, is about 44°.
   a. What is the approximate distance between Sirius and Alpha Centauri?
   b. Would Sirius appear as bright to an observer near Alpha Centauri as it does as seen from Earth? Justify your answer.

12. Use the Law of Cosines to get a formula for cos C in terms of a, b, and c.

13. Refer to the triangle at the right.
   a. Find the value of b.
   b. Use your answer from part **a** to find the measure of θ.

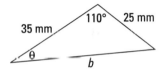

14. At a criminal trial, a witness gave the following testimony: "The defendant was 20 ft from the victim. I was 50 ft from the defendant and 75 ft from the victim when the shooting occurred. I saw the whole thing."
   a. Use the Law of Cosines to show that the testimony has errors.
   b. How else could you know that the testimony has errors?

15. Suppose △ABC has a = 7, b = 24, and C = 90°.
   a. Use the Law of Cosines to find c.
   b. How else could you have found c?

Shown is a scene from the TV show Matlock. *The show stars Andy Griffith (at left) as a defense attorney.*

Review

16. If sin 160° = sin θ and 0° < θ < 90°, what is θ?   *(Lesson 10-5)*

In 17–20, give an exact value. Do not use a calculator.
(Lessons 10-3, 10-4, 10-5)

17. cos 30°

18. sin 420°

19. sin 150°

20. tan (-45°)

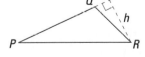

21. *Multiple choice.*   In △PQR at the left, with h the altitude to $\overline{PQ}$, h =
   (a) PR sin P.
   (b) QR tan P.
   (c) PQ cos P.
   (d) none of these
   (Previous course, Lessons 1-6, 10-1)

22. The equation $2 \cdot 3^x = 17$ can be solved in at least three ways: (1) by trial and error, (2) by graphing, and (3) by using logarithms. Solve using your favorite method. *(Previous course, Lessons 9-1, 9-10)*

23. If the population of a state is now 5,000,000 and the population grows by 5% each year, in how many years will the population reach 7,000,000? *(Lessons 9-1, 9-10)*

24. If $\log x = 3.5$, find x to the nearest tenth. *(Lesson 9-5)*

25. Suppose $\frac{p}{x} = \frac{q}{y}$, $x \neq 0$, and $y \neq 0$. *(Previous course, Lesson 1-6)*
 a. Solve for p.   **b.** Solve for x.

Exploration

26. The Law of Cosines has been described as "the Pythagorean Theorem with a correction term." Explain why this is an appropriate description.

Fire triangle. *The main goal of suppressing a forest fire is to break the "fire triangle" of fuel, temperature, and oxygen. This is often done through the application of dirt, water, or chemicals and by partial removal or separation of fuels.*

You may have seen smoke in a certain direction and wondered, "Where is the fire?" Sometimes the location is not known and needs to be determined. Consider the following situation. Two forest rangers are in their stations S and T, 25 miles apart. On a certain day, the ranger at S sees a fire F at an angle of 40° with the line connecting the stations. The ranger at T sees the fire at an angle of 60° with $\overline{ST}$. How far is the fire from each ranger's station?

The given information here is an instance of the ASA condition, because the known side is the side included between two known angles. However, since the Law of Cosines involves the three sides and only one angle of a triangle, it is not useful when only one side is known. If you would try to use the Law of Cosines to solve this problem, you would find that there are two unknowns in a single equation.

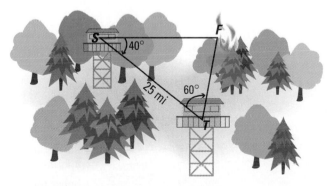

However, there is an extraordinarily beautiful, simple theorem that you can use to find the missing distance. It is called the *Law of Sines.*

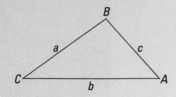

Theorem (Law of Sines)

In any triangle ABC,

$$\frac{\sin A}{a} = \frac{\sin B}{b} = \frac{\sin C}{c}.$$

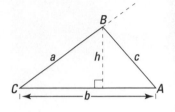

Proof

The proof involves finding the area of $\triangle ABC$ in three ways. Draw the altitude h to side $\overline{AC}$. Note that $\sin C = \frac{h}{a}$, so $h = a \sin C$.

$$\text{Area } (\triangle ABC) = \frac{1}{2}bh = \frac{1}{2}b(a \sin C).$$

$$= \frac{1}{2}ab \sin C$$

Similarly, calculate $\sin A$. $\sin A = \frac{h}{c}$, so $h = c \sin A$.

$$\text{Area } (\triangle ABC) = \frac{1}{2}bh = \frac{1}{2}b(c \sin A) = \frac{1}{2}bc \sin A$$

Drawing the altitude to $\overline{AB}$ and computing the area of $\triangle ABC$ yet another way yields

$$\text{Area } (\triangle ABC) = \frac{1}{2}ac \sin B.$$

So $\frac{1}{2}ab \sin C = \frac{1}{2}ac \sin B = \frac{1}{2}bc \sin A$.

$ab \sin C$	$= ac \sin B$	$= bc \sin A$	Multiply all sides by 2.
$\frac{ab \sin C}{abc}$	$= \frac{ac \sin B}{abc}$	$= \frac{bc \sin A}{abc}$	Divide all sides by abc.
$\frac{\sin C}{c}$	$= \frac{\sin B}{b}$	$= \frac{\sin A}{a}$	Simplify the fractions.

In words, the Law of Sines states that in any triangle, the ratios of the sines of its angles to the lengths of their opposite sides are equal.

Using the Law of Sines to Find a Side

The Law of Sines can be applied directly in an AAS situation, that is, in a triangle in which two angles and a non-included side are given.

Example 1

In $\triangle XYZ$, $m\angle X = 25°$, $m\angle Y = 75°$, and $x = 4$. Find y.

Solution

Sketch the figure.

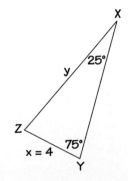

$$\frac{\sin X}{x} = \frac{\sin Y}{y} \qquad \text{Law of Sines}$$

$$\frac{\sin 25°}{4} = \frac{\sin 75°}{y} \qquad \text{Substitute the given information.}$$

$$y = \frac{4 \sin 75°}{\sin 25°} \qquad \text{Solve for } y.$$

$$y \approx \frac{4(.966)}{.423} \approx 9.1$$

Example 2 illustrates how to use the Law of Sines when you are given two angles and the included side (the ASA condition) and want to find the length of another side.

Example 2

In the situation described at the beginning of this lesson, find the distance from the ranger at station T to the fire.

Solution

Let s be the desired length. The angle opposite s is $\angle S$, with measure $40°$. To use the Law of Sines, you need the values of another angle and its opposite side. Because the sum of the measures of the angles in a triangle is $180°$, $\angle F$ has measure $80°$. Now you know $m\angle S = 40°$, $m\angle F = 80°$, and $f = 25$ miles.

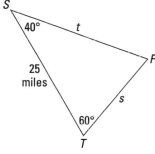

$$\frac{\sin S}{s} = \frac{\sin F}{f} \qquad \text{Law of Sines}$$

$$\frac{\sin 40°}{s} = \frac{\sin 80°}{25} \qquad \text{Substitute the given information.}$$

$$s = \frac{25 \sin 40°}{\sin 80°} \qquad \text{Solve for } s.$$

$$s \approx \frac{25(.643)}{.985} \approx 16.3$$

The fire is about 16 miles from the ranger at T.

In general, if you need to find a side or an angle of a triangle, try methods involving simpler computations before trying trigonometry. But if that does not work:

1. Use right-triangle trigonometric ratios when the missing side or angle is part of a right triangle.
2. Use the Law of Sines on any triangle when you have the ASA, AAS, or SSA condition.
3. Use the Law of Cosines when you have the SAS or SSS condition.

The Law of Sines was known to Ptolemy in the 2nd century A.D. A theorem equivalent to the Law of Cosines is in Euclid's *Elements* written four centuries earlier. The Greeks used these theorems as the forest ranger used them in Example 1, to locate landmarks. This practice led eventually to **triangulation,** the process of dividing a region into triangular pieces, making a few accurate measurements, and using trigonometry to determine most of the distances. Triangulation made it possible for reasonably accurate maps of parts of the Earth to be drawn well before the days of artificial satellites.

QUESTIONS

Covering the Reading

1. *Multiple choice.* Which of the following verbally describes the Law of Sines?
   (a) In a triangle, the ratios of the measures of its angles to the lengths of their opposite sides are equal.
   (b) In a triangle, the ratios of the sines of its angles to the lengths of their adjacent sides are equal.
   (c) In a triangle, the ratios of the sines of its angles to the lengths of their opposite sides are equal.
   (d) None of the above describes the Law of Sines.

In 2 and 3, refer to the proof of the Law of Sines.

2. What does the expression $\frac{1}{2} ab \sin C$ represent for triangle ABC?

3. Draw a $\triangle ABC$ and the altitude to $\overline{AB}$. Explain why the area of $\triangle ABC$ equals $\frac{1}{2} ac \sin B$.

4. Refer to the forest-fire situation at the beginning of the lesson. Find the distance from the fire to the ranger at station S.

In 5 and 6, find y.

5.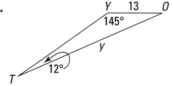

6.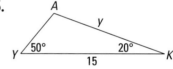

7. With information satisfying the given condition, which theorem—the Law of Cosines or the Law of Sines—is more useful for finding other parts of a triangle?
   a. SAS   **b.** ASA   **c.** AAS   **d.** SSS

8. In $\triangle ABC$, suppose $m\angle A = 45°$, $m\angle B = 60°$, and $a = 24$. Find the exact value of b.

Applying the Mathematics

9. Refer to $\triangle PQR$ at the left.
   a. Find RQ.
   b. Use your answer to part **a** and the Law of Sines to find $m\angle R$.
   c. Use the Law of Cosines and your answer to part **a** to find $m\angle R$.
   d. *True or false.* In an SAS situation, once you find the third side, you can use either the Law of Sines or the Law of Cosines to find a second angle. Explain your answer.

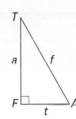

10. In $\triangle AFT$ at the left, $m\angle F = 90°$.
    a. By the Law of Sines, $\frac{f}{\sin F} = \frac{a}{\sin A}$. Solve this expression for $\sin A$ and simplify it by calculating $\sin 90°$.
    b. Solve $\frac{f}{\sin F} = \frac{t}{\sin T}$ for $\sin T$, and simplify.
    c. How do your answers to parts **a** and **b** compare to the trigonometric ratios?

11. When a beam of light in air strikes the surface of water, it is **refracted**, or bent, as shown below.

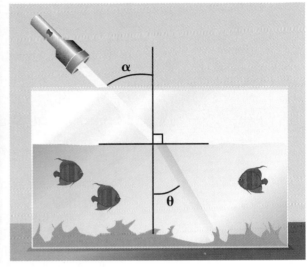

Refraction. *Shown is a beam of light passing from air through water. The different density of the substances is the cause for the light "bending."*

The relationship between α and θ is known as *Snell's Law,* $\frac{\sin \alpha}{\text{speed of light in air}} = \frac{\sin \theta}{\text{speed of light in water}}$. The speed of light in air is about 3.00×10^8 m/sec. If $\alpha = 45°$ and $\theta = 32°$, estimate the speed of light in water.

12. While using the Law of Sines, a student came up with $\sin A = 1.234$. What can you tell the student about his or her solution?

13. Because surveyors cannot get to the inside center of a mountain, its height must be measured in a more indirect way. Refer to the diagram below. All labeled points lie in a single plane.

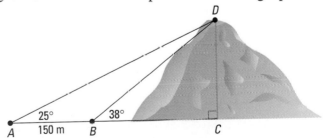

a. Find the measures of $\angle ABD$ and $\angle ADB$. (You need to use only your basic knowledge from geometry.)
b. Find BD.
c. Find DC, the height of the mountain.

14. *Multiple choice.*   Which of the following is the Law of Cosines for △*ABC*?   *(Lesson 10-6)*
    (a) $a^2 = b^2 + c^2 + 2bc \cdot \cos A$
    (b) $a^2 = b^2 + c^2 - bc \cdot \cos A$
    (c) $a^2 = b^2 + c^2 - 2 \cdot \cos A$
    (d) $a^2 = b^2 + c^2 - 2bc \cdot \cos A$

15. Suppose $\sin \theta = \frac{4}{5}$.
    a. Find the two possible values of $\cos \theta$.
    b. Graph the two points $(\cos \theta, \sin \theta)$.   *(Lessons 10-2, 10-5)*

In 16 and 17, give the exact value without using a calculator.
(Lessons 10-4, 10-5)

16. $\cos 180°$ 17. $\sin 225°$

18. Give the coordinates of $R_{60}(1, 0)$
    a. exactly.
    b. to the nearest thousandth.   *(Lesson 10-4)*

19. A rock is thrown upward with an initial velocity of 30 $\frac{\text{ft}}{\text{sec}}$ from a height of 12 ft.
    a. Write an equation to describe the height h of the rock (in feet) with respect to time t (in seconds).
    b. Graph the equation.
    c. What is the maximum height of the rock?
    d. When does the rock hit the ground?   *(Lessons 6-4, 6-5, 6-7)*

In 20 and 21, consider $A = \begin{bmatrix} -100 & 5 \\ -80 & 4 \end{bmatrix}$.

20. **a.** Find det A.
    b. Does A^{-1} exist? If so, find it. If not, explain why it does not exist. *(Lesson 5-5)*

21. **a.** Find an equation for the line through the two points in matrix A.
    b. What kind of variation is described by the line in part **a**?
    (Lessons 2-1, 3-5, 4-1)

22. What is the measure of the angle formed by the hands of a clock at 7:00 P.M.?   *(Previous course)*

23. Refer to the triangle shown below.
    a. Measure the sides of this triangle in centimeters and the angles in degrees.
    b. Find the sines of the angles.
    c. Substitute the values you get into the Law of Sines.
    d. How nearly equal are the fractions?

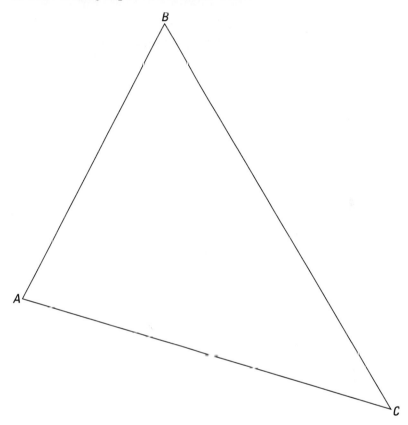

*Graphing
$y = \cos \theta$*

IN·CLASS

ACTIVITY

Work with a partner on this activity.

1 Set your calculator to degree mode. Make a table of values of the
form $(\theta, \cos \theta)$ for $0° \leq \theta \leq 360°$ in increments of 15°. Round
cosines to the nearest hundredth. The first few pairs in the table are
given below.

θ (in degrees)	0	15	30	45	60	75	. . .	345	360
cos θ	0	.26	.50	.71			. . .		0

2 Make a graph using the values in part **a.** Plot θ on the horizontal
axis and cos θ on the vertical axis.

3 Describe some patterns you notice in the table or graph.

4 Does the graph in part **b** represent a function? Why or why not?

5 What is the maximum value cos θ can have? Explain why cos θ
can never be larger than this number.

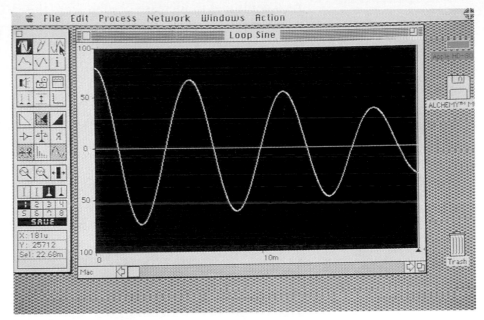

Musical sine wave. *Shown is an oscillographic display of a sine wave as displayed by a musical computer program. The sine wave, which is decreasing in amplitude, represents a pure musical tone.*

Recall that when (1, 0) is rotated θ degrees around the origin, its image is the point (cos θ, sin θ). We can set up a correspondence θ → cos θ. This correspondence associates the magnitude of the rotation with the first coordinate of the image of (1, 0). This correspondence is a function, because for each θ there is only one value for cos θ. Similarly, the correspondence θ → sin θ is a function which associates θ with the second coordinate of the image of (1, 0) under R_θ.

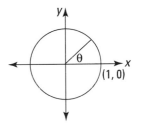

> f: θ → cos θ is called the **cosine function.**
> g: θ → sin θ is called the **sine function.**

The Graph of the Sine Function

Some ordered pairs of the function $g(\theta) = \sin \theta$ are given in the table below and graphed on the following page. For instance, since sin 30° = 0.5, the point (30°, 0.5) is graphed.

θ	0	15	30	45	60	75	90	105	120	135	150	165	180
sin θ*	0	.26	.50	.71	.87	.97	1	.97	.87	.71	.50	.26	0

θ		195	210	225	240	255	270	285	300	315	330	345	360
sin θ*		−.26	−.5	−.71	−.87	−.97	−1	−.97	−.87	−.71	−.5	−.26	0

*decimal approximation

As θ continues to increase beyond 360°, the rotation images of (1, 0) coincide with previous ones. For instance, the rotation of magnitude 390° is the same as a rotation of magnitude 30°. So sin 390° = sin 30°. In general, the values of sin θ repeat every 360°. As a result, the values of the function $g(\theta) = \sin \theta$ repeat every 360°. A more complete graph is shown below.

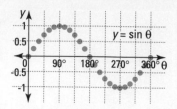

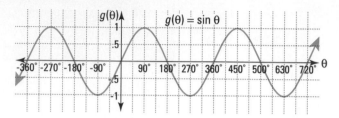

The Graph of the Cosine Function

The graph of the cosine function can be constructed by a similar process, using the first coordinate of the rotation image of (1, 0). This is what you were asked to do in the Activity preceding this lesson. For instance, because $\cos 30° = \dfrac{\sqrt{3}}{2} \approx .87$, the point (30°, .87) is on the graph of $f(\theta) = \cos \theta$. Below is a graph of $f(\theta) = \cos(\theta)$. The graph you constructed in the Activity should have points from the part of this graph where $0° \leq \theta \leq 360°$.

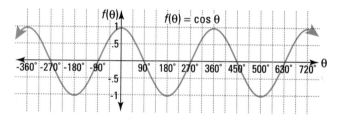

Properties of the Sine and Cosine Functions when θ Is Measured in Degrees

The functions $f(\theta) = \cos \theta$ and $g(\theta) = \sin \theta$ have the following properties.

1. a. Since θ, the magnitude of rotation, may be any real number, the domain of both the sine and cosine functions is the set of real numbers.

    b. Because the numbers cos θ and sin θ are coordinates of points on the unit circle, the range of these functions is the set of real numbers from -1 to 1.

2. a. The sine graph has y-intercept 0 and x-intercepts . . . , -180°, 0°, 180°, 360°, 540°, . . . , that is, the even multiples of 90°.

    b. The cosine graph has y-intercept 1 and x-intercepts . . . , -270°, -90°, 90°, 270°, 450°, . . . , that is, the odd multiples of 90°.

▶

▶ 3. Both the sine and cosine functions are *periodic*. A function is a **periodic function** if there is a smallest positive number *p* such that its graph can be mapped to itself under a horizontal translation with magnitude *p*. This *p* is the **period** of the function. For each of $f(\theta) = \cos \theta$ and $g(\theta) = \sin \theta$, the period is 360°. That is, both functions repeat values every 360°. This means that under a horizontal translation of magnitude 360, the graph of $g(\theta) = \sin \theta$ coincides with itself. Similarly, under this translation the graph of $f(\theta) = \cos \theta$ coincides with itself.

4. The graph of *f* can be mapped onto the graph of *g* by a horizontal translation of 90°. So the graphs of $f(\theta) = \cos \theta$ and $g(\theta) = \sin \theta$ are congruent.

The graph of $g(\theta) = \sin \theta$ is called a *sine wave*.

> **Definition**
> A **sine wave** is a graph which can be mapped onto the graph of $g(\theta) = \sin \theta$ by any composite of translations, scale changes, or reflections.

Because the graph of $f(\theta) = \cos \theta$ is a translation image of the graph of $g(\theta) = \sin \theta$, its graph is a sine wave. Situations that lead to sine waves are said to be **sinusoidal.**

Sine waves have many applications. Pure sound tones travel in sine waves; these can be pictured on an oscilloscope. The location of a satellite relative to the equator as it travels around the earth is an approximate sine wave, and the time of sunrise for a given location over the year also shows sinusoidal behavior.

QUESTIONS

Covering the Reading

1. The function $f: \theta \rightarrow \cos \theta$ maps θ onto the __?__ coordinate of the image of (1, 0) under R_θ.

2. The function $g: \theta \rightarrow \sin \theta$ maps θ onto the __?__ coordinate of the image of (1, 0) under R_θ.

3. Refer to a graph of $g(\theta) = \sin \theta$.
   a. As θ increases from 0° to 90°, sin θ increases from __?__ to __?__.
   b. As θ increases from 90° to 180°, sin θ decreases from __?__ to __?__.
   c. As θ increases from 180° to 270°, does the value of $g(\theta) = \sin \theta$ increase or decrease?
   d. As θ increases from 270° to 360°, how do the values of sin θ change?

In 4 and 5, for each function assume that θ is in degrees.
a. Name two points on the function. b. Give its domain. c. Give its range.
4. $f(\theta) = \cos \theta$ 5. $g(\theta) = \sin \theta$

In 6 and 7, assume θ is in degrees.

6. Name four *x*-intercepts—two positive and two negative—of the curve $g(\theta) = \sin \theta$.

7. Name four negative *x*-intercepts of the curve $f(\theta) = \cos \theta$.

8. Define *periodic function*.

In 9–11, *true or false*.

9. The function defined by $f(\theta) = \cos \theta$ is periodic.

10. The graphs of $f(\theta) = \cos \theta$ and $g(\theta) = \sin \theta$ are congruent.

11. The graph of $f(\theta) = \cos \theta$ is a sine wave.

Applying the Mathematics

12. a. On the same set of axes, graph $f(\theta) = \cos \theta$ and $g(\theta) = \sin \theta$ over the interval $-360° \le \theta \le 360°$. (Use an automatic grapher if you wish.)
   b. Find all values of θ between $-360°$ and $360°$ such that $\cos \theta = \sin \theta$.

13. *Multiple choice.* Which choice completes a symbolic definition of "periodic function"?
f is periodic if and only if there is a smallest positive number *p* such that for all *x*:
(a) $f(x + p) = f(x)$. (b) $p \cdot f(x) = f(px)$.
(c) $f(x) + f(p) = f(x + p)$. (d) $f(x) + p = f(x)$.

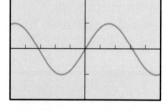

14. Output from an automatic grapher is shown at the left. What equation related to the ideas in the lesson may be graphed?

In 15–18, part of a function is graphed. **a.** Does the function seem to be periodic? **b.** If so, what is the period?

$-450° \le x \le 450°$, *x*-scale = 90°
$-2 \le y \le 2$, *y*-scale = 1

15.

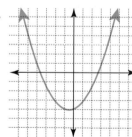

16.

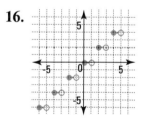

17.

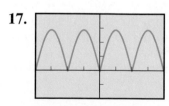

$-8 \le x \le 8$, *x*-scale = 2
$-3 \le y \le 4$, *y*-scale = 1

18.

$-8 \le x \le 8$, *x*-scale = 2
$-3 \le y \le 4$, *y*-scale = 1

19. In $\triangle HJK$, $m\angle K = 81°$, $k = 21$, and $j = 20$. Find $m\angle J$.   *(Lesson 10-7)*

20. An observer in a lighthouse on the shore sees a ship in distress. The ship is 15 miles away at an angle of 20° with the shoreline. A Coast Guard station is on the shoreline 30 miles away from the lighthouse.

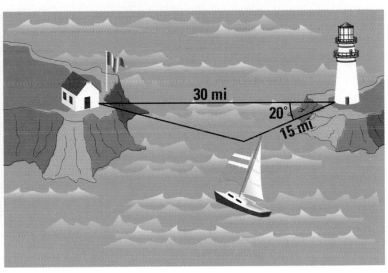

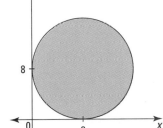

Helping you see at sea.
Pictured is the Portland Head Light in Cape Elizabeth, Maine. The lighthouse is operated by computer from South Portland by the Coast Guard.

a. How far will a Coast Guard rescue ship have to travel from the station to reach the ship?
b. The path of the rescue vessel should be at what angle to the shoreline? (Use your answer to part **a.**)   *(Lessons 10-7, 10-6)*

In 21 and 22, complete each statement with a trigonometric expression to make the equation true.   *(Lesson 10-3)*

21. $(\sin \theta)^2 + (\underline{\ ?\ })^2 = 1$

22. $\sin (90° - \theta) = \underline{\ ?\ }$

23. Estimate each to the nearest thousandth.   *(Lessons 9-5, 9-7, 9-9)*
    a. $\log 64$ b. $\ln 64$ c. $\log_2 64$

24. The circle at the left is tangent to the axes at $(8, 0)$ and $(0, 8)$. Find its area and circumference.   *(Previous course)*

25. a. Write a table of values for the tangent function $y = \tan \theta$ from 0° to 360° in increments of 15°.
    b. Graph the function $y = \tan \theta$ for $-720° < \theta < 720°$.
    c. What is the domain of the tangent function?
    d. What is the range of the tangent function?
    e. Describe some features of the graph of the tangent function.

*The Law
of Sines
When SSA
Is Given*

IN·CLASS
ACTIVITY

Materials: Each person will need a ruler, a compass, a protractor, and a calculator. Work on this Activity with a partner.

The *SSA Condition* is a situation where two sides and a *nonincluded* angle of a triangle are given.

1 **a.** Draw $\triangle ABC$ with $a = 7$ cm, $c = 8$ cm, and m$\angle A = 41°$.
   b. Use a protractor to estimate m$\angle C$.

2 **a.** Draw $\triangle DEF$ with $d = 7$ cm, $f = 8$ cm, and m$\angle D = 41°$ that is *not* congruent to $\triangle ABC$. (Recall from geometry that the SSA condition does not always yield congruent triangles.)
   b. Use a protractor to find m$\angle F$ in this triangle.
   c. How is m$\angle F$ in Question 2 related to m$\angle C$ in Question 1?

3 Use the Law of Sines with the values $a = 7$, $c = 8$, and m$\angle A = 41°$. Does the value it predicts for m$\angle C$ agree with your work in Questions 1 and 2? Why or why not?

***Sine* of the times.** *Shown is a crowd at the XXV Olympiad in Barcelona, Spain in 1992 doing "the wave." There are similarities between these waves and sine waves.*

To use the Law of Sines or the Law of Cosines to find the measure of an angle of a triangle, you have to solve an equation of the form $\cos \theta = k$ or $\sin \theta = k$. We now examine these equations in more detail.

The Solutions to cos θ = *k*

When θ is an angle in a triangle, it has a measure between 0° and 180°. For $0° < \theta < 180°$, the equation $\cos \theta = k$ has a unique solution. To see why, consider the graph of $y = \cos \theta$ on this interval. For any value of k from -1 to 1, the graph of $y = k$ intersects $y = \cos \theta$ at a single point. The θ-coordinate of this point is the solution to $\cos \theta = k$.

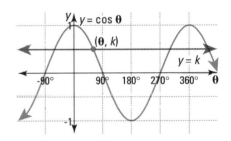

The Solutions to sin θ = *k*

The situation is different for the equation $\sin \theta = k$. On the interval $0° < \theta < 180°$, for any value of k between 0 and 1, the graph of $y = k$ intersects $y = \sin \theta$ in two points. In the graph at the right, we call these points (θ_1, k) and (θ_2, k). The numbers θ_1 and θ_2 are the solutions to $\sin \theta = k$.

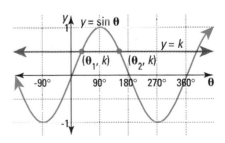

The points (θ_1, k) and (θ_2, k) are reflection images of each other over the vertical line $\theta = 90°$. This is because when $0 < k < 1$, the two solutions to $\sin \theta = k$ between 0° and 180° are supplementary angles.

The Supplements Theorem can be proved from the definition of sin θ. You are asked to do this in Question 14.

Example 1

If sin θ = .624, find the two values of θ between 0° and 180°.

Solution

$\theta_1 = \sin^{-1} .624 \approx 38.6°$. The second value
$\theta_2 = 180° - \theta_1 \approx 180° - 38.6° = 141.4°$ is the supplement of θ_1.
So if sin θ = .624, θ ≈ 38.6° or 141.4°.

Check 1

Does sin 38.6° = sin 141.4° ≈ .624? Yes, they check.

Check 2

Graph y = sin x and y = .624, as shown at the left. The graphs intersect twice. Tracing shows the intersections to be near 39° and 141°.

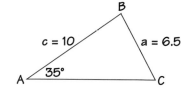

-90° ≤ x ≤ 270°, x-scale = 90°
-1.5 ≤ y ≤ 1.5,   y-scale = .5

The Supplements Theorem and the Law of Sines

The Supplements Theorem is critical when using the Law of Sines to find measures of angles.

Example 2

In △ABC, a = 6.5, c = 10, and m∠A = 35°. Find m∠C.

Solution

Sketch a picture. It is natural to use the Law of Sines.

$$\frac{\sin A}{a} = \frac{\sin C}{c}$$

$$\frac{\sin 35°}{6.5} = \frac{\sin C}{10}$$

$$\frac{10 \sin 35°}{6.5} = \sin C$$

$$.882 \approx \sin C$$

There are two solutions to this equation. We call them C_1 and C_2.
A calculator shows $m\angle C_1 \approx 61.9°$.

Then △ABC can look like the first triangle at the left.
By the Supplements Theorem,
$$m\angle C_2 = 180° - m\angle C_1$$
$$\approx 180° - 61.9°$$
$$= 118.1°.$$
The bottom triangle at the left shows a second solution to the problem.

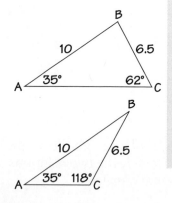

654

The SSA Condition does not always determine a unique triangle. Sometimes there are two noncongruent triangles which have two pairs of sides and a pair of corresponding nonincluded angles congruent. Example 2 and the In-class Activity on page 652 provide examples of this.

However, sometimes there is a unique triangle determined by the SSA Condition. *If two triangles have two pairs of sides and a pair of corresponding angles congruent, and the congruent sides opposite the congruent angles are longer than the other congruent sides, then the triangles are congruent.* This is the SsA Triangle Congruence Theorem, and we call its condition the SsA condition. (The small *s* indicates that the side of the angle is smaller than the side opposite the angle.) Example 3 deals with the SsA condition.

Example 3

In $\triangle SPX$, $m\angle S = 75°$, $s = 11$, and $x = 9$. Find $m\angle X$.

Solution

Use the Law of Sines.

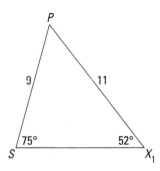

$$\frac{\sin S}{s} = \frac{\sin X}{x}$$

$$\frac{\sin 75°}{11} = \frac{\sin X}{9}$$

$$\frac{9 \sin 75°}{11} = \sin X$$

$$.790 \approx \sin X$$

$m\angle X_1 \approx \sin^{-1} .790 \approx 52°$. The triangle is pictured above.
A second angle X_2 with $\sin X_2 \approx .790$ is the supplement of 52°, which is $180° - 52° = 128°$. However, $75° + 128° > 180°$, so a 75° angle and a 128° angle cannot be in the same triangle, and $x \approx 128°$ is not a solution. The only solution is $m\angle X \approx 52°$.

Using trigonometry to find all the missing measures of sides and angles of a triangle is called **solving the triangle.** By using the Law of Cosines and the Law of Sines, you can solve any triangle.

QUESTIONS

Covering the Reading

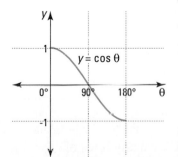

1. Copy the graph at the left.
   a. Draw a horizontal line to estimate the solution(s) to $\cos \theta = .34$ between 0° and 180°.
   b. How many solutions are there to $\cos \theta = .34$ between 0° and 180°?
   c. Find all solutions to $\cos \theta = .34$ between 0° and 180°.

2. Draw a graph of $y = \sin \theta$ between 0° and 180°. Repeat the three parts of Question 1 for the equation $\sin \theta = .34$.

3. If A is the measure of an angle in degrees, give another expression equal to $\sin(180° - A)$.

4. Solve $\sin \theta = \frac{1}{5}$ for θ, where $0° < \theta < 180°$.

5. In $\triangle ABC$, $m\angle B = 42°$, $c = 13$, and $b = 18$.
   a. Use the Law of Sines to find the measure of $\angle C$.
   b. Explain why there is only one solution to part **a.**

6. In $\triangle ABC$, $a = 5.5$, $b = 7$ and $m\angle A = 22°$.
   a. Find two possible values for $m\angle B$.
   b. Sketch two triangles to illustrate the possible solutions to part **a.**

Applying the Mathematics

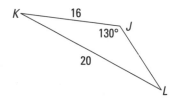

7. Solve $\triangle JKL$ at the left. Approximate each value to the nearest tenth.

8. Consider $\triangle RST$, where $\angle R = 52°$, $r = 20.1$, and $s = 23.1$.
   a. Find all possible values for the measure of $\angle S$.
   b. For each solution in part **a,** find the length of the third side.

In 9–12, suppose you know the three measures of the indicated condition in a triangle. Tell whether the Law of Sines or the Law of Cosines is more useful for finding the fourth measure indicated.

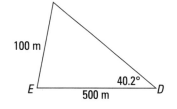

9. ASA, find second side

10. SAS, find third side

11. SSA, find second angle

12. SSS, find any angle

13. A surveyor marks off points D, E, and F and records that $m\angle D = 40.2°$, $d = 100$ m and $f = 500$ m. Explain why there is a problem with the surveyor's measurements.

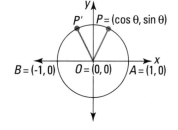

14. Use the diagram of the unit circle at the left. Let $P = (\cos \theta, \sin \theta)$, and let P' be the reflection image of P over the y-axis.
   a. Explain why the coordinates of P' are $(-\cos \theta, \sin \theta)$.
   b. Explain why $P' = (\cos(180° - \theta), \sin(180° - \theta))$.
   c. Use parts **a** and **b** to prove the Supplements Theorem.

15. Prove the SsA Triangle Congruence Theorem: If, in two triangles, two sides and the angle opposite the larger side of one are congruent respectively to two sides and the angle opposite the larger side of the other, then the triangles are congruent.

   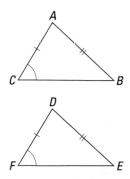

   Take as given: $AB = DE$, $AC = DF$, $\angle C \cong \angle F$, and $AB > AC$. Use the Law of Sines to prove that $\angle B \cong \angle E$ and thus that the triangles are congruent.

Review

16. The period of the sine function is __?__ degrees. *(Lesson 10-8)*

17. Does the graph of the function $f(x) = \cos x$ have any lines of symmetry? If so, give an equation for one such line. *(Lesson 10-8)*

18. a. Is the relation graphed below a function? Why or why not?
  b. Is the relation periodic? If so, what is its period?
  c. Is the inverse of this relation a function? Why or why not?
  (Lessons 1-2, 8-2, 10-8)

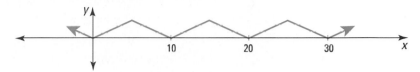

19. To determine the distance to a cabin *C* across a lake, some campers marked off points *A* and *B* 300 feet apart on one shore. They then found m∠*CAB* = 87.5° and m∠*ABC* = 91°. About how far is it from *B* to *C*? *(Lessons 10-6, 10-7)*

Scouting out a campsite.
This picture is from the National Boy Scout Jamboree in Pennsylvania. Over 27,000 scouts participate in the Jamboree each time it is held.

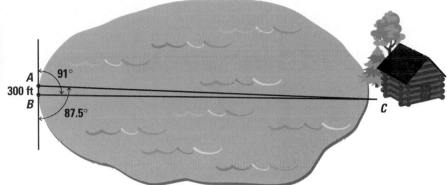

In 20–22, give exact values without using a calculator.
(Lessons 10-1, 10-3, 10-4, 10-5)

20. sin (-90°) **21.** cos (-60°) **22.** tan 30°

23. What is the circumference of the unit circle?
  (Previous course, Lesson 10-4)

In 24 and 25, refer to circle *O* at the left. *(Previous course)*

24. If θ = 80°, what fraction of the circle's area is the area of the shaded sector?

25. If the length of $\overset{\frown}{AB}$ is $\frac{7}{24}$ of the circumference of the circle, find m∠θ.

Exploration

---

26. Draw a circle and a triangle *PQR* with vertices on that circle.
  a. Measure ∠*P* and side *p*. Verify that the ratio $\frac{p}{\sin P}$ equals the diameter of the circle.
  b. What does $\frac{q}{\sin Q}$ equal?

Radiating plans. *Some architects use radian measurement in finding the grading of land. The slope, or grade, of land is important to ensure that water will drain correctly.*

So far in this chapter you have learned to evaluate sin *x*, cos *x*, and tan *x* when *x* has been given in degrees. Angles, magnitudes, and rotations may also be measured in *radians*. In fact, in some areas of mathematics, radians are used more than degrees.

What Is a Radian?

Since the radius of a unit circle is the number 1, the circumference of the unit circle is 2π units. Thus, on a unit circle, a 360° arc has a length of 2π. Similarly, a 180° arc has a length of $\frac{1}{2}(2\pi) = \pi$, and a 90° arc has the length $\frac{1}{4}(2\pi) = \frac{\pi}{2}$.

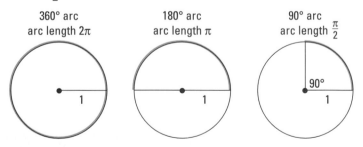

360° arc
arc length 2π

180° arc
arc length π

90° arc
arc length $\frac{\pi}{2}$

The radian is a measure created so that the arc *measure* and the arc *length* use the same number.

Definition
The radian is a measure of an angle, arc, or rotation such that
π radians = 180 degrees.

Notice that a 180° arc has measure π radians and length π. A 90° angle has measure $\frac{\pi}{2}$ radians, and its arc has length $\frac{\pi}{2}$. *The measure of an angle, arc, or rotation in radians equals the length of its arc on the unit circle.*

Conversion Factors for Degrees and Radians

The definition of radian can be used to give conversion factors for changing degrees into radians and vice versa. Begin with the equation

$$\pi \text{ radians} = 180°.$$

Dividing each side by π radians, gives

$$\frac{\pi \text{ radians}}{\pi \text{ radians}} = \frac{180°}{\pi \text{ radians}}$$

So,
$$1 = \frac{180 \text{ degrees}}{\pi \text{ radians}}.$$

Similarly, dividing each side by 180 degrees, gives

$$\frac{\pi \text{ radians}}{180°} = \frac{180°}{180°}$$

$$\frac{\pi \text{ radians}}{180 \text{ degrees}} = 1.$$

We summarize these results below.

Conversion Factors for Degrees and Radians

To convert *radians* to degrees, multiply by $\frac{180 \text{ degrees}}{\pi \text{ radians}}$.

To convert *degrees* to radians, multiply by $\frac{\pi \text{ radians}}{180 \text{ degrees}}$.

Converting from Radians to Degrees

You may wonder "How big is a radian?" Example 1 gives an answer.

Example 1

Convert 1 radian to degrees.

Solution

Because radians are given, multiply by the conversion factor with radians in the denominator.

$$1 \text{ radian} = 1 \text{ radian} \cdot \frac{180°}{\pi \text{ radians}}.$$

$$= \frac{180°}{\pi}$$

$$\approx 57.3°$$

Notice that one radian is much larger than one degree.

Check

Recall that the measure of an angle in radians equals the length of its arc on the unit circle. So an angle of 1 radian determines an arc of length 1 on a unit circle. A unit circle has circumference $2\pi \approx 6.28$, so an arc length of 1 is a little less than $\frac{1}{6}$ of the circle's circumference. So the angle $\theta = 1$ radian measures a little less than $\frac{1}{6}$ of 360° or 60°. It checks.

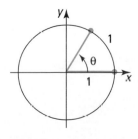

Some calculators can convert from radians to degrees without having to use a conversion factor. Find out if yours is able to do this. You should see 1 radian ≈ 57.295° . . . .

Converting from Degrees to Radians

Example 2

a. Convert 45° to its exact radian equivalent.
b. Estimate 45° to the nearest hundredth of a radian.

Solution

a. Multiply 45° by one of the conversion factors. Because you want radians, choose the ratio with radians in the numerator.

$$45° \cdot \frac{\pi \text{ radians}}{180°} = \frac{45°}{180°}\pi \text{ radians}$$

$$= \frac{\pi}{4} \text{ radian}$$

$45° = \frac{\pi}{4}$ radians, exactly.

b. Use the answer to part **a.** A calculator gives $\frac{\pi}{4} ≈ 0.785398$. So 45° ≈ .79 radian.

Check

a. Since 180° = π radians, and 45° is $\frac{1}{4}$ of 180°, 45° is also $\frac{1}{4}$ of π radians. So $45° = \frac{1}{2} \cdot 90° = \frac{1}{2} \cdot \frac{\pi}{2} = \frac{\pi}{4}$ radian.

b. In Example 1 we found that 57.3° ≈ 1 radian. So 45° ≈ .8 radian seems reasonable.

Radian expressions are often left as multiples of π because this form gives an exact value. Usually in mathematics, the word *radian* or the abbreviation *rad* is omitted. In trigonometry, when no degree symbol or other unit is specified, we assume that the measure of the angle, arc, or rotation is radians.

θ = 2° means "the angle (or the arc or the rotation) θ has measure 2 degrees."
θ = 2 means "the angle (or the arc or the rotation) θ has measure 2 radians."

Refer back to Example 2. From the fact that $\frac{\pi}{4} = 45°$, we can conclude that $\frac{3\pi}{4} = 3 \cdot \frac{\pi}{4} = 3 \cdot 45° = 135°$. Similarly, $\frac{5\pi}{4} = 5 \cdot 45° = 225°$. The diagram at right shows other common equivalences of degrees and radians.

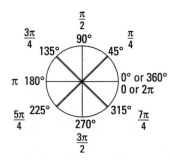

Trigonometric Values in Radians

Every scientific calculator can evaluate sin θ, cos θ, and tan θ, where θ is in radians. Learn how to set your calculator to radian mode.

Example 3

Evaluate cos 2.

Solution

In radian mode, enter cos 2. You should find $\cos 2 \approx -0.416$.

Check

From Example 1, you know that 1 radian ≈ 57.3°, so 2 radians ≈ 2 · (57.3°) = 114.6°. A calculator gives cos 114.6° ≈ -0.416.

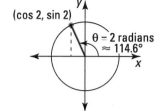

The multiples of π and the simplest fractional parts of π $\left(\frac{\pi}{2}, \frac{\pi}{3}, \frac{\pi}{4}, \frac{\pi}{6}\right)$, and their multiples correspond to those angle measures which give exact values of sines, cosines, and tangents.

Example 4

Evaluate $\sin\left(\frac{\pi}{4}\right)$ on your calculator.

Solution 1

Since there is no degree symbol, we assume $\frac{\pi}{4}$ means $\frac{\pi}{4}$ radian. Put your calculator in radian mode. So $\sin \frac{\pi}{4} \approx 0.707$.

Solution 2

From Example 2, $\frac{\pi}{4}$ = 45°. You know $\sin 45° = \frac{\sqrt{2}}{2}$, so $\sin \frac{\pi}{4} = \frac{\sqrt{2}}{2}$ which is about .707.

Graphs of the Sine and Cosine Functions Using Radians

The cosine and sine functions are graphed below with the rotation x in radians rather than in degrees. Notice that each function is still periodic, but that each period is 2π rather than 360°.

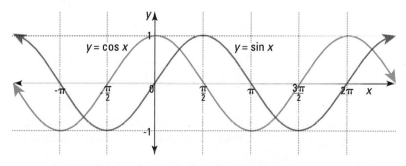

Covering the Reading

1. A circle has a radius of 1 unit.
   Give the length of an arc with the given measure.
   a. 360°   **b.** 180°   **c.** 90°

2. Fill in the blanks.
   a. π radians   = __?__°.
   b. 1 radian   = __?__°.
   c. __?__ radian(s) =   1°.

In 3 and 4, convert the radian measure to degrees.

3. $\frac{\pi}{6}$ 4. $\frac{-5\pi}{4}$

In 5–8, convert to radians. Give your answer as a rational number times π.

5. 90°   6. 60°   7. 225°   8. 330°

9. **a.** Explain the difference between the meaning of sin 4 and sin 4°.
   b. Evaluate sin 4.
   c. Check your work in part **b.**

In 10 and 11, **a.** Evaluate directly using radian mode. **b.** Check your answer using degrees.

10. $\sin \frac{3\pi}{2}$ 11. $\tan \frac{\pi}{6}$

In 12 and 13, suppose x is in radians.

12. What is the period of the function $y = \cos x$?

13. Name three x-intercepts of the function $y = \sin x$.

Applying the Mathematics

14. *Multiple choice.* Suppose x is in radians. Which transformation maps the graph of $y = \sin x$ onto itself?
    (a) reflection over the x-axis
    (b) reflection over the y-axis
    (c) translation of π to the right
    (d) translation of 2π to the right

15. What angle measure is formed by the hands of a clock at 2:00
    a. in degrees?   **b.** in radians?

16. Six equally-spaced diameters are drawn on a unit circle as shown at the left. Copy and complete this figure, giving equivalent degrees and radians at the end of each radius.

In 17 and 18, find the exact values.

17. $\sin \left(\frac{7\pi}{6}\right)$ 18. $\cos \left(\frac{15\pi}{4}\right)$

length of arc = rx

In 19–21, use this relationship between radian measure and arc length. In a circle of radius r, a central angle of x radians has an arc of length rx.

19. a. How long is the arc of a $\frac{\pi}{4}$ radian angle in a circle of radius 20?

  b. How long is a 45° arc in a circle of radius 20?

20. On a circle of radius 1 meter, find the length of a 45° arc.

21. How long is the arc of a $\frac{2\pi}{3}$ radian angle in a circle of radius 6 feet?

Review

22. Suppose $0° < \theta < 180°$. Solve $\sin \theta = .76$.   *(Lesson 10-9)*

23. In $\triangle ABC$, $m\angle B = 16°$, $b = 10$, and $c = 24$. Explain why there are two possible measures for $\angle C$, and find both of them.   *(Lesson 10-9)*

24. In $\triangle MAP$, $m = 22$, $m\angle M = 149°$, and $m\angle P = 23°$. Find the lengths of p and a.   *(Lessons 10-7, 10-8)*

In 25 and 26, evaluate the expression without using a calculator. *(Lessons 10-3, 10-4, 10-5)*

25. $(\sin 450°)^2 + (\cos 450°)^2$ **26.** $\tan 135°$

27. The newspaper article at the left is from the *Detroit Free Press*, January 29, 1985. Use the drawings below to explain why the construction technique leads to an angle of 26.5°.   *(Lesson 10-2)*

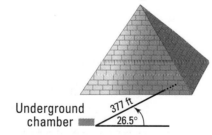

Underground chamber   377 ft   26.5°

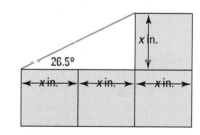

28. One of Murphy's Laws is that the time a committee spends debating a budget item is inversely proportional to the number of dollars involved. Suppose the function is $t = \frac{15}{d}$, where t is measured in hours and d is in dollars.
  a. How much time is spent on a $300 item?
  b. How much time is spent on a $3000 item?   *(Lesson 2-2)*

Exploration

29. When x is in radians, $\sin x$ can be estimated by the formula
$\sin x = x - \frac{x^3}{6} + \frac{x^5}{120} - \frac{x^7}{5040}$.
  a. How close is the value of this expression to $\sin x$ when $x = \frac{\pi}{4}$?
  b. To get greater accuracy, you can add $\frac{x^9}{362,880}$ to the value you got in part **a.** Where does the 362,880 come from?
  c. How close is this to $\sin \frac{\pi}{4}$?

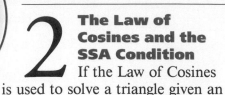

PROJECTS

10

CHAPTER TEN

A project presents an opportunity for you to extend your knowledge of a topic related to the material of this chapter. You should allow more time for a project than you do for typical homework questions.

1 Triangulation and Surveying

As you know, triangulation is the technique of dividing part of a plane into triangles in which almost all angles and a few sides are measured. From these measurements, all other distances are determined. Find out more about how surveyors use triangulation and other aspects of trigonometry in their work. You may want to concentrate on current applications such as how a new subdivision will be surveyed, or you might want to do some historical research on how a community near you was originally surveyed.

2 The Law of Cosines and the SSA Condition

If the Law of Cosines is used to solve a triangle given an SSA Condition, a quadratic equation results. Show that this quadratic equation will always give all of the correct answers. Include some specific examples and, if possible, a general proof.

3 Spherical Trigonometry

When three great circles intersect, they determine regions called *spherical triangles.* Consult reference books to report on how the Law of Sines and Law of Cosines are modified to apply to spherical triangles.

4 Benjamin Banneker

Benjamin Banneker was one of the first African American mathematicians. Among other things, he was responsible for surveying Washington, D.C., when it was first being developed. Find out more about this man and how he used mathematics, particularly trigonometry, in his work.

5 Sunrise and Sunset Times

The following table lists the time of sunset each Sunday of 1994 for Denver, Colorado. (Daylight savings time has been ignored.)

1/2	4:47	4/3	6:26	7/3	7:32	10/2	5:41
1/9	4:54	4/10	6:33	7/10	7:29	10/9	5:30
1/16	5:01	4/17	6:40	7/17	7:26	10/16	5:19
1/23	5:09	4/24	6:47	7/24	7:21	10/23	5:09
1/30	5:17	5/1	6:54	7/31	7:14	10/30	5:00
2/6	5:26	5/8	7:01	8/7	7:06	11/6	4:53
2/13	5:34	5/15	7:08	8/14	6:58	11/13	4:46
2/20	5:42	5/22	7:14	8/21	6:48	11/20	4:41
2/27	5:50	5/29	7:20	8/28	6:37	11/27	4:37
3/6	5:57	6/5	7:25	9/5	6:27	12/4	4:35
3/13	6:05	6/12	7:28	9/12	6:15	12/11	4:36
3/20	6:12	6/19	7:31	9/19	6:04	12/18	4:37
3/27	6:19	6/26	7:32	9/26	5:52	12/25	4:41

a. Graph these data and describe what you get.

b. Find some data for the time of sunrise and the time of sunset in a recent year for some city that interests you. Make sure you have data for at least one day every week.

c. Make an accurate graph of each set of your data on the same set of axes. Draw a smooth curve through each set of points.

d. Use the data from part **b** to determine the number of hours of daylight at various times during the year in the city you chose. Make a graph showing how the number of hours of daylight varies during the year.

e. Describe what the graph tells you about the daylight in the city you chose.

f. From your data, estimate what the graph for sunrise in Denver should look like.

g. Think about some other places in the world. Which places should have graphs just like the city you chose? Why? Which places should have graphs quite different from the city you chose? Why?

6 Area Under the Graph of a Sine Curve

a. Carefully draw a graph of the equation $y = \sin x$ from 0 to π radians on graph paper. Let one unit equal 0.1 on each axis.

b. Using the scale of your graph, what is the area of each square?

c. How many whole squares are between the sine curve and the x-axis? Estimate the number of whole squares you can make from the remaining partial squares, as best you can. Add these to estimate the total number of squares between the sine curve and the x-axis.

d. Calculate $\frac{\text{area}}{\text{square}} \times$ (number of squares) to estimate the total area under the graph of $y = \sin x$. The final answer should be surprisingly simple!

e. Predict the area under the curve $y = \cos x$ from $x = 0$ to $x = \frac{\pi}{2}$. Devise a method to test your prediction, and carry it out.

f. Summarize what you found.

SUMMARY

Trigonometry is the study of relations among sides and angles in triangles. In a right triangle, three important trigonometric ratios are the sine, cosine, and tangent of an acute angle θ, defined as follows:

$$\sin θ = \frac{\text{leg opposite } θ}{\text{hypotenuse}}$$

$$\cos θ = \frac{\text{leg adjacent to } θ}{\text{hypotenuse}}$$

$$\tan θ = \frac{\text{leg opposite } θ}{\text{leg adjacent to } θ}.$$

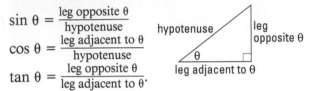

The sine, cosine, and tangent ratios are frequently used to find lengths or angle measures in situations involving right triangles.

Trigonometric ratios for angles that are multiples of 30° or 45° can be calculated exactly.

$$\sin 30° = \cos 60° = \tfrac{1}{2} \qquad \tan 30° = \frac{\sqrt{3}}{3}$$

$$\sin 45° = \cos 45° = \frac{\sqrt{2}}{2} \qquad \tan 45° = 1$$

$$\sin 60° = \cos 30° = \frac{\sqrt{3}}{2} \qquad \tan 60° = \sqrt{3}$$

Others can be approximated by decimals or found using the following theorems. For all real numbers θ,

$$(\cos θ)^2 + (\sin θ)^2 = 1 \quad \text{Pythagorean Identity}$$

$\sin θ = \cos (90° - θ)$ Complements Theorem
$\cos θ = \sin (90° - θ)$ Complements Theorem
$\sin θ = \sin (180° - θ)$ Supplements Theorem

Lengths of sides and measures of angles in any triangle are related. In any triangle ABC,

$$c^2 = a^2 + b^2 - 2ab \cos C.$$
(Law of Cosines)
$$\frac{\sin A}{a} = \frac{\sin B}{b} = \frac{\sin C}{c}.$$
(Law of Sines)

These theorems can be used to find unknown sides and angle measures in triangles. The Law of Cosines is most useful when an SAS or SSS condition is given; the Law of Sines is used in all other situations that determine triangles. When the Law of Sines is used to find an angle in an SSA condition, one or two solutions may occur.

Trigonometric values can be defined for any real number. For any real number θ:

$\cos θ = $ the x-coordinate of $R_θ(1, 0)$
$\sin θ = $ the y-coordinate of $R_θ(1, 0)$
$\tan θ = \frac{\sin θ}{\cos θ}$, provided $\cos θ \neq 0$.

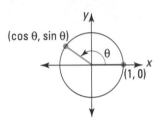

The correspondences
$$f\colon θ \to \cos θ \text{ and } g\colon θ \to \sin θ$$
are functions whose domains are the set of real numbers and whose ranges are $\{y\colon -1 \leq y \leq 1\}$. When θ is in degrees the graphs of $f(θ) = \cos θ$ and $g(θ) = \sin θ$ are sine waves with period equal to 360°. When θ is in radians, the period is $2π$, because radians are defined so that $π$ radians = 180°.

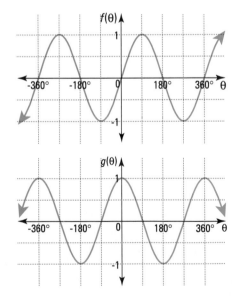

VOCABULARY

Below are the most important terms and phrases for this chapter. You should be able to give a definition for those terms marked with an *. For all other terms you should be able to give a general description and a specific example of each.

Lesson 10-1
trigonometry
trigonometric ratios
* sine of θ, sin θ, sin
* cosine of θ, cos θ, cos
* tangent of θ, tan θ, tan
bearing

Lesson 10-2
inverse trigonometric functions, $\sin^{-1}$, $\cos^{-1}$, $\tan^{-1}$
angle of elevation
line of sight
angle of depression

Lesson 10-3
Complements Theorem
Pythagorean Identity
Tangent Theorem
Exact-Value Theorem

Lesson 10-4
unit circle

Lesson 10-5
signs of sine and cosine in quadrants II-IV

Lesson 10-6
Law of Cosines

Lesson 10-7
Law of Sines
triangulation
refracted
Snell's law

Lesson 10-8
* cosine function
* sine function
periodic function, period
sine wave
sinusoidal

Lesson 10-9
Supplements Theorem
solving a triangle

Lesson 10-10
radian, rad
Conversion Factors for Degrees and Radians

PROGRESS SELF-TEST

Take this test as you would take a test in class. You will need a calculator. Then check your work with the solutions in the Selected Answers section in the back of the book.

In 1 and 2, use the triangle at the right. Evaluate the following to the nearest thousandth.

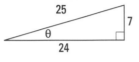

1. $\cos \theta$

2. $\sin \theta$

3. The roof of a gazebo has dimensions as in the figure below. A carpenter cut an angle θ in the rafter so the rafter fits snugly on the beam. At what angle θ was the cut made?

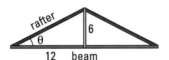

In 4 and 5, refer to the unit circle at the right. Name the letter equal to the value of the trigonometric function.

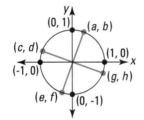

4. $\cos 70°$

5. $\sin 160°$

6. For what value of x such that $0° < x \leq 180°$ and $x \neq 57°$ does $\sin 57° = \sin x$?

7. Find an exact value for $\sin 210°$.

8. Runner A is 110 meters from an observer and runner B is 85 meters away. The angle between the sightings is 40°. How far apart are the runners from each other?

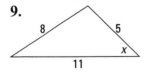

In 9 and 10, find x.

9.

10.

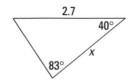

11. In $\triangle SLR$, $m\angle L = 31°$, $s = 525$, and $l = 421$. Find $m\angle S$.

12. A 2-foot-tall eagle was perched on an 8-foot-high sign. It then flew directly to its nest. The angle of elevation of the flight path (from the eagle's beak) was about 70°. Suppose the eagle flew for 5 seconds at a speed of about 26 ft/sec. So it traveled 130 ft. About how high off the ground is the nest?

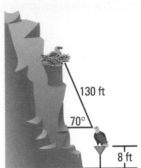

13. *Multiple choice.* Which of the following statements is *not* true for all values of θ? Justify your answer by giving a counterexample.
    (a) $\sin (90° - \theta) = \cos \theta$
    (b) $\sin (\theta + 360°) = \sin \theta$
    (c) $(\sin \theta)^2 + (\cos \theta)^2 = 1$
    (d) $\cos (\theta + 180°) = \cos \theta$

14. Show how to convert 24° to radians without using a calculator.

15. Find the exact value of $\cos \left(\frac{5\pi}{6}\right)$.

In 16 and 17, use the graph below.

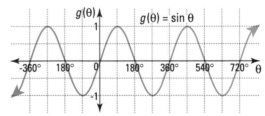

16. What is the period of this function?

17. As θ increases from 90° to 180°, the value of $\sin \theta$ decreases from _?_ to _?_.

In 18 and 19, consider the function $y = \cos x$.

18. Sketch a graph when $\frac{-\pi}{2} \leq x \leq \frac{3\pi}{2}$.

19. State the range of this function.

20. Solve $\sin \theta = .848$ for $0° < \theta < 180°$.

CHAPTER REVIEW

Questions on SPUR Objectives

SPUR stands for **S**kills, **P**roperties, **U**ses, and **R**epresentations. The Chapter Review questions are grouped according to the SPUR Objectives for this chapter.

SKILLS DEAL WITH THE PROCEDURES USED TO GET ANSWERS.

Objective A: *Approximate values of trigonometric functions using a calculator.*
(Lessons 10-1, 10-5, 10-10)

In 1–6, evaluate to the nearest thousandth.

1. $\sin 17°$ **2.** $\cos 143°$ **3.** $\sin(-50°)$

4. $\cos \frac{\pi}{3}$ **5.** $\sin \frac{11\pi}{6}$ **6.** $\tan 211°$

In 7–9, use the triangle below. Approximate each trigonometric value to the nearest thousandth.

7. $\sin \theta$ **8.** $\cos \theta$ **9.** $\tan \theta$

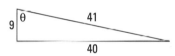

Objective B: *Find exact values of trigonometric functions of multiples of 30° or 45° or their radian equivalents.* *(Lessons 10-3, 10-4, 10-5, 10-10)*

In 10–15, give exact values.

10. $\cos 45°$ **11.** $\sin 405°$ **12.** $\tan 30°$

13. $\cos \frac{-\pi}{6}$ **14.** $\sin \frac{3\pi}{2}$ **15.** $\sin \frac{\pi}{4}$

Objective C: *Determine the measure of an angle given its sine, cosine, or tangent.*
(Lessons 10-2, 10-3, 10-9)

In 16–19, find all θ between 0° and 180° with the given trigonometric value.

16. $\cos \theta = .5$ **17.** $\sin \theta = \frac{\sqrt{2}}{2}$

18. $\sin \theta = 1$ **19.** $\tan \theta \approx -1.732$

In 20 and 21, solve for all θ between 0 and $\frac{\pi}{2}$ radians.

20. $\cos \theta \approx .309$ **21.** $\sin \theta = \frac{2}{3}$

22. Find two values of $\cos \theta$ if $\sin \theta = .36$.

23. If $\sin \theta = .8$ and θ is obtuse, find $\cos \theta$.

Objective D: *Convert angle measures from radians to degrees or from degrees to radians.*
(Lesson 10-10)

In 24–27, convert to radians.

24. 30° **25.** 105° **26.** 360° **27.** 540°

In 28–31, convert to degrees.

28. π **29.** $\frac{9\pi}{4}$ **30.** $\frac{5\pi}{3}$ **31.** $-\frac{\pi}{8}$

PROPERTIES DEAL WITH THE PRINCIPLES BEHIND THE MATHEMATICS.

Objective E: *Identify and use definitions and theorems relating sines, cosines, and tangents.*
(Lessons 10-3, 10-8)

In 32–35, *true or false.*   If false, change the statement so that it is true.

32. $(\sin \theta)^2 + (\cos \theta)^2 = 1$

33. $\cos(90° - \theta) = \sin \theta$

34. $\sin(180° - \theta) = \cos \theta$

35. $\tan \theta = \frac{\cos \theta}{\sin \theta}$

In 36 and 37, copy and complete with the measure of an acute angle.

36. $\sin 73° = \cos \underline{\ ?\ }$

37. $\cos(90° - \underline{\ ?\ }) = \sin 41°$

USES DEAL WITH APPLICATIONS OF MATHEMATICS IN REAL SITUATIONS.

Objective F: *Solve real-world problems using the trigonometry of right triangles.*
(Lessons 10-1, 10-2)

38. How tall is the building pictured below if a person 6 feet tall sights the top of the building at 49° while standing 53 feet away?

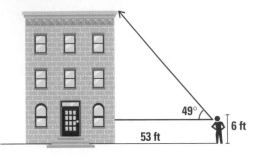

39. A ship sails 695 kilometers on a bearing of 75°. How far east of its original position is the ship?

40. A wheel-chair ramp is to be built with a slope of $\frac{1}{12}$. What angle will the ramp make with the horizontal?

41. An airplane begins a smooth final descent to the runway from an altitude of 5,000 feet when it is 30,000 horizontal feet away. At what angle of depression will the plane descend?

42. The ancient Greeks carved amphitheaters out of the sides of hills. Suppose one amphitheater went down 200 feet vertically while extending 300 feet horizontally. At what angle of depression did they dig?

Objective G: *Solve real-world problems using the Law of Sines or Law of Cosines.*
(Lessons 10-6, 10-7, 10-9)

43. Observers in two ranger stations 8 miles apart spot a fire ahead. The observer in station *A* spots the fire at an angle of 44° with the line between the two stations, while the observer in station *B* spots the fire at an angle of 105° with the same line. Which station is nearer the fire, and how near is it?

44. Two observers are in lighthouses 25 miles apart, as shown below. The observer in the northern lighthouse spots a ship in distress at an angle of 15° with the line between the lighthouses. The other observer spots the ship at an angle of 35°. How far is the ship from each lighthouse?

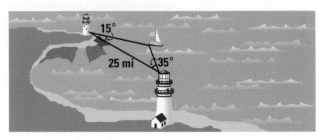

45. Pictured below is a chandelier with 12 spokes equally spaced around a center. If the spokes are each 2′ long, find the perimeter of the chandelier.

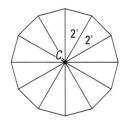

46. Nancy and Byron plan to sail from the marina to a picnic site on an island. To avoid some smaller islands, they plan to sail 10 miles on a bearing of 50°, and then change to a bearing of 135° for 7 miles. How far is the picnic site from the marina as the crow flies?

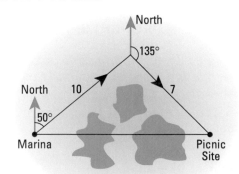

REPRESENTATIONS DEAL WITH PICTURES, GRAPHS, OR OBJECTS THAT ILLUSTRATE CONCEPTS.

Objective H: *Find missing parts of a triangle using the Law of Sines or the Law of Cosines.*
(Lessons 10-6, 10-7, 10-9)

47. Find *BC*.

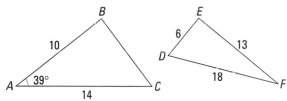

48. Find m∠*E*.

49. Find *GH*.

50. Find *JK*.

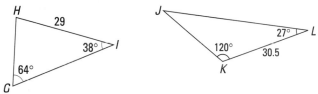

51. Suppose that in △*WET*, m∠*W* = 112°, *w* = 9, and *e* = 7. Solve the triangle.

52. In △*JHS*, *j* = 2, *s* = 3, and m∠*J* = 25°.
   a. Find all possible measures of m∠*S*.
   b. Draw a sketch to illustrate each possibility.

Objective I: *Use the properties of a unit circle to find trigonometric values.* *(Lessons 10-4, 10-5, 10-10)*

In 53 and 54, use the sketch below.

53. What is the value of sin θ?

54. Find θ to the nearest degree.

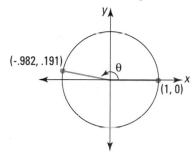

In 55–57, use the unit circle below. Which letter stands for the given number?

55. sin 25°   **56.** sin 225°   **57.** cos $\frac{2\pi}{3}$

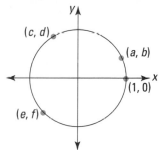

Objective J: *Identify properties of the sine, cosine, and tangent functions using their graphs.*
(Lesson 10-8)

58. **a.** Graph *f*: θ → sin θ, for 0° < θ < 360°.
   b. State the domain and the range of the sine function.

59. **a.** Graph *g*: θ → cos θ with the *x*-axis given in radians, for -2π < *x* < 2π.
   b. What is the period of the cosine function?
   c. At what points do the graph of *y* = cos *x* intersect the *x*-axis?

60. As θ increases from 0° to 180°, cos θ decreases from ⸏?⸏ to ⸏?⸏.

61. The graph of *y* = cos *x* is an image of the graph of *y* = sin *x* under what translation?

62. What is the period of the function graphed below? (Assume the graph continues in both directions.)

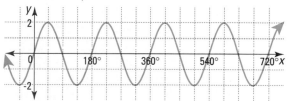

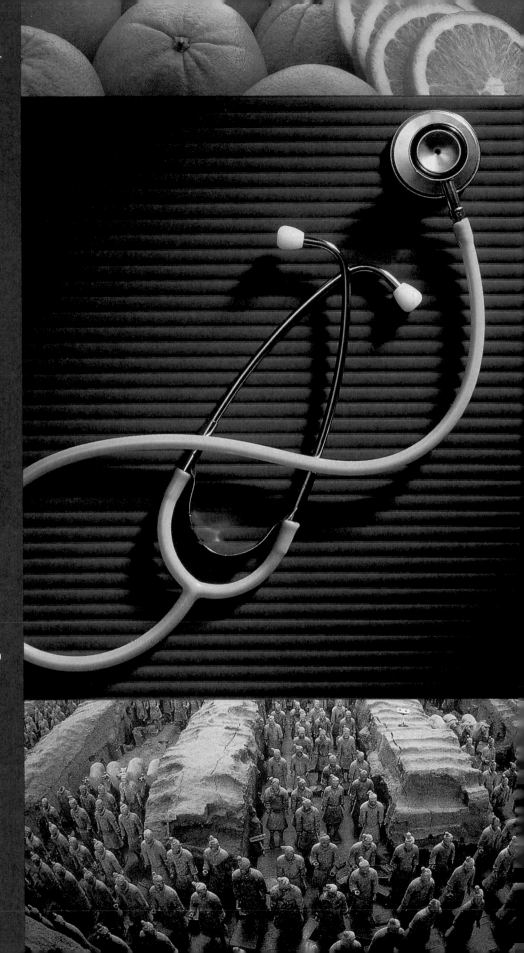

CHAPTER 11

POLYNOMIALS

The population of Manhattan Island, a part of New York City, has gone up and down over the past 100 years. Below are a table and a graph showing the population every 20 years from 1890 to 1990.

Year	Population
1890	1,441,000
1910	2,332,000
1930	1,867,000
1950	1,960,000
1970	1,539,000
1990	1,488,000

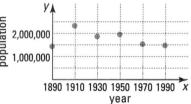

None of the kinds of functions you have studied fit these points very well. However, using techniques similar to those you used to fit quadratic models to data, an equation can be found for the population as a function of year. If $x =$ the number of 20-year periods since 1890, then the population $P(x)$ of Manhattan (in millions) at these times is given by the equation

$$P(x) = \frac{2471}{60,000}x^5 - .53625x^4 + \frac{30,083}{12,000}x^3 - 5.06275x^2 + 3.9419x + 1.441.$$

This equation is a *polynomial equation,* and the function P is a *polynomial function.* Although the formula for $P(x)$ is quite complicated, mathematicians would not be surprised by it. For any finite set of points, no two of which are on the same vertical line, there is a polynomial function whose graph contains those points.

A graph of $y = P(x)$ is shown below. On this graph, $x = 0$ corresponds to the year 1890 and $y = 1.441$ corresponds to the population 1,441,000; $x = 1$ corresponds to the year 1910 and $y = 2.332$ corresponds to the population 2,332,000; and so on.

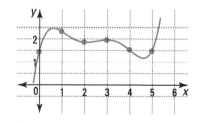

In this chapter, you will study situations that lead to polynomial functions. You will see how to graph and analyze such functions, learn when data can be described by a *polynomial model,* and learn how to find the model for specific data points. Along the way, you will encounter various properties of polynomials, some of which you have studied before.

Doctors in residence. *Shown are some of the doctors from the TV show* ER *looking at a patient. Usually three years of residency after medical school must be completed before a doctor is licensed. See Example 3.*

What Is a Polynomial?

The expression

$$\frac{2471}{60,000} x^5 - .53625x^4 + \frac{30,083}{12,000} x^3 - 5.06275x^2 + 3.9419x + 1.441$$

from page 673 is a *polynomial in the variable x*. When the polynomial contains only one variable, the largest exponent of the variable is the **degree** of the polynomial. This polynomial has degree 5. The expressions $\frac{2471}{60,000} x^5$, $-.53625 x^4$, $\frac{30,083}{12,000} x^3$, $-5.06275x^2$, $3.9419x$, and 1.441 are the **terms** of the polynomial.

Definition

A **polynomial in x** is an expression of the form
$$a_n x^n + a_{n-1}x^{n-1} + a_{n-2}x^{n-2} + \ldots + a_1 x^1 + a_0,$$
where n is a positive integer and $a_n \neq 0$.

Vocabulary Used with Polynomials

The **standard form** of the general polynomial is the one displayed in the definition, that is, with the terms written in descending order by degree. The number n is the degree of the polynomial, and we sometimes call this an *n*th degree polynomial. The numbers $a_n, a_{n-1}, a_{n-2}, \ldots, a_0$ are its **coefficients.** The number a_n is called the **leading coefficient** of the polynomial. For instance, when $n = 4$,

$$a_n x^n + a_{n-1}x^{n-1} + a_{n-2}x^{n-2} + \ldots + a_1 x^1 + a_0$$
$$\text{becomes}$$
$$a_4 x^4 + a_3 x^3 + a_2 x^2 + a_1 x^1 + a_0,$$

which is the standard form of the polynomial with degree 4. It has coefficients $a_4, a_3, a_2, a_1,$ and a_0. Its leading coefficient is a_4.

Writing a product of polynomials or a power of a polynomial as a polynomial is called **expanding the polynomial.**

Example 1

a. Expand $(3x^2 + 4)^2$ and put the result in standard form.
b. What is its degree?
c. What is the leading coefficient?

Solution

a. $(3x^2 + 4)^2 = (3x^2)^2 + 2 \cdot 3x^2 \cdot 4 + 4^2$ Expand.
                $= 9x^4 + 24x^2 + 16$ Put in standard form.
b. The highest exponent is 4. The polynomial has *degree 4.*
c. The coefficient of x^4 is 9. So *The leading coefficient is 9.*

A **symbol manipulator** is computer software or a calculator preprogrammed to perform operations on variables. Most symbol manipulators can do all operations with polynomials. If you have access to such technology, you should learn how to use it to expand polynomials.

Polynomials of the first degree, such as $mx + b$, are called **linear polynomials.** Those of the second degree, such as $ax^2 + bx + c$, are called **quadratic polynomials,** and those of the third degree, such as $ax^3 + bx^2 + cx + d$, are **cubic polynomials.** Polynomials of the fourth degree, such as that in Example 1, are sometimes called **quartic** polynomials.

Nonzero constants can be considered as polynomials of degree zero. This is because the constant k can be written as $k \cdot x^0$, which is a polynomial of degree zero. However, the constant 0 is not considered to be a polynomial because a polynomial's leading coefficient must be nonzero.

Polynomial Functions

A **polynomial function** is a function of the form $x \rightarrow P(x)$, where $P(x)$ is a polynomial. You can evaluate or graph polynomial functions in the same way that you evaluate or graph other functions.

Example 2

Consider the polynomial function defined by
$P(x) = 6x^5 - 3x^4 + 4x^2 - 2x - 70.$
a. What is $P(2)$?
b. Graph this function on the window $-5 \le x \le 5$, $-120 \le y \le 120$.

Solution

a. The value of the function when x = 2 is
    $P(2) = 6(2)^5 - 3(2)^4 + 4(2)^2 - 2(2) - 70 = 86.$
b. The graph is shown at the left. This curve is related to the graph of $y = x^5$. However, it is not a translation image of that graph.

Check

a. Use the trace function on an automatic grapher. It should show that (2, 86) is on the graph.

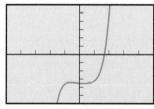

$-5 \le x \le 5$,   x-scale = 1
$-120 \le y \le 120$,   y-scale = 20

Savings and Polynomials

The calculation of compound interest can lead to polynomial functions of any degree. Recall that the compound interest formula $A = P(1 + r)^t$ gives the value to which P dollars grow if invested at an annual yield r for t years. If amounts of money are deposited for different lengths of time, then this formula must be applied to each amount separately. Example 3 illustrates such a situation.

Example 3

Starting with the summer after her senior year in high school, Yolanda Fish worked to save money for medical school. At the end of each summer, she put her money in a savings account with an annual yield of 6%. Yolanda is planning to go to medical school in the fall following her 4th year in college. How much will be in her account when she goes to medical school, if no other money is added or withdrawn?

summer	saved
after senior year	$1500
after 1st year of college	2200
after 2nd year of college	2100
after 3rd year of college	3000
after 4th year of college	3300

Solution

The money put in the bank after her senior year in high school earns interest for 4 years. It is worth $1500(1.06)^4$ when Yolanda goes to medical school. Similarly, the amount saved at the end of her first year of college is worth $2200(1.06)^3$. Adding the values from each summer gives the total amount that will be in Yolanda's account.

$$1500(1.06)^4 + 2200(1.06)^3 + 2100(1.06)^2 + 3000(1.06)^1 + 3300$$

| after senior year | after 1st year | after 2nd year | after 3rd year | after 4th year |

Evaluating this expression shows that Yolanda will have about $13,354 in her account when she goes to medical school.

Higher learning. *Some of these graduates will go on to medical school after they finish college. Approximately 15,000 students graduate from medical school each year.*

In Example 3, if you don't know the interest rate Yolanda can receive, you could replace 1.06 with x. Then when Yolanda goes to medical school she will have (in dollars)

$$1500x^4 + 2200x^3 + 2100x^2 + 3000x + 3300.$$

This expression gives the amount in the account for any annual yield. If the annual yield is r, just substitute $1 + r$ for x and find the new total. If the first deposit has earned interest for n years, the result is a polynomial of degree n. Caution: When constructing an expression in a situation like this, you must be careful to check whether or not the last amount saved earns interest.

Covering the Reading

In 1–3, tell whether or not the expression is a polynomial. If it is, state its degree and its leading coefficient. If it is not a polynomial, explain why not.

1. $4x + 7$ **2.** $7x^4 - 12x^5 + 100$ **3.** $x^{-2} + x^{-1} + 1$

4. Refer to the definition of an nth degree polynomial and the polynomial $5x^7 + 4x^6 - 8x^3 + 1.3x^2 - x$.
State the values of each of the following.
  a. n **b.** a_n **c.** a_{n-1} **d.** a_0
  e. a_1 **f.** a_2 **g.** a_5

In 5–7, write the standard form of the polynomial.

5. a cubic polynomial in the variable x

6. a fifth degree polynomial in the variable y

7. an nth degree polynomial in the variable z

8. *True or false.* The number 17 is a polynomial. Justify your answer.

In 9 and 10, an expression is given.
a. Expand the expression and put the result in standard form.
b. State its degree.
c. State its leading coefficient.

  9. $(x + 2)(x^2 - 3)$ **10.** $(6 + 5a^3)^2$

11. Refer to the population function P for Manhattan Island given on page 673.
  a. Evaluate $P(3)$.
  b. How close is $P(3)$ to the 1950 population?

12. Let $f(x) = x^3 - 6x^2 + 3x + 10$.
  a. Evaluate $f(-1)$, $f(2)$, and $f(5)$.
  b. What do all the values in part **a** have in common?
  c. Sketch a graph of $y = f(x)$ on the window $-2 \leq x \leq 6$, $-30 \leq y \leq 30$.

13. Refer to Example 3. Suppose that in successive summers beginning after eighth grade, Ethan put $200, $500, $1475, $1600, and $1300 into a bank account.
  a. Assume Ethan goes to college in the fall immediately after finishing high school. Let $x = 1 + r$. If the annual yield is r and no other money is added or withdrawn, how much is in his account when he goes to college?
  b. Evaluate your answer to part **a** when $x = 1.04$.
  c. What is the degree of the polynomial you found in part **b**?

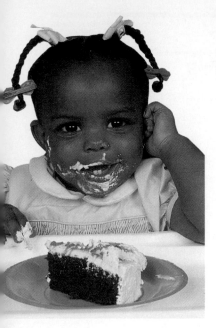

Applying the Mathematics

14. On her first birthday, Jennifer got $25. On each successive birthday she got twice as much money as she had gotten for her preceding birthday. All money was put into an account with annual yield r. No additional money was added or withdrawn.
    a. Write a polynomial expression to give the total amount in Jennifer's account on her sixth birthday. (The money from her sixth birthday earns no interest.) Let $x = 1 + r$.
    b. Evaluate your answer to part **a** when $x = 1.045$.

In 15 and 16, recall the formula for the height h of an object thrown upward: $h = -\frac{1}{2}gt^2 + v_0t + h_0$, where t is the number of seconds after being thrown, h_0 is the initial height, v_0 is the initial velocity, and g is the acceleration due to gravity (32 ft/sec^2). This formula describes a polynomial function in t.

15. What is the degree of this polynomial?

16. Suppose a ball is thrown upward from the ground with initial velocity 45 ft/sec. Find its height after .9 second.

17. Consider $f(x) = 10^x$ and $g(x) = x^{10}$.
    a. Which function, f or g, is a polynomial function?
    b. Which function, f or g, is an exponential function?
    c. Explain how to tell the difference between an exponential function and a polynomial function.

Review

18. Of the sequences A and B below, one is arithmetic, the other geometric.   *(Lessons 3-7, 3-8, 7-5)*

    $A:$ 16, 4, 1, $\frac{1}{4}$ . . . $B:$ 53, 41, 29, 17 . . .

    a. Write the next two terms of each sequence.
    b. Write an explicit formula for the geometric sequence.
    c. Write a recursive formula for the arithmetic sequence.
    d. Which sequence might approximate the sequence of consecutive highest points a bouncing ball could reach?

19. Solve this system of equations.   *(Lessons 5-3, 5-6)*
    $$\begin{cases} x = 3z \\ y = x - 4 \\ z = x + 2y \end{cases}$$

20. A cheetah trots along at 5 mph for a minute, spies a small deer and speeds up to 60 mph in just 6 seconds. After chasing the deer at this speed for 30 seconds, the cheetah gives up and, over the next 20 seconds, slows to a stop. Graph this situation, plotting time on the horizontal axis and speed on the vertical axis. *(Lesson 3-1)*

In 21 and 22, refer to the population of Manhattan given on page 673. *(Lesson 2-4)*

21. Find the rate of change in population per year for the period 1890 to 1910.

22. What was the average change in population per year between 1950 and 1970?

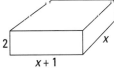

23. Find the volume of the rectangular solid at the left. *(Previous course)*

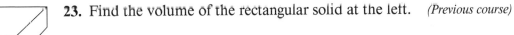

Exploration

24. What did the keeper say to the parrot who needed to go on a diet?

25. a. If you have access to a symbol manipulator, use it to expand $(a - 2b^2 + 3c^3)^2$.
   b. Check the result by hand by letting $a = 2$, $b = 3$, and $c = 10$.

11-2

Polynomials and Geometry

Next-door neighbors. *Shown are town houses in Jersey City, New Jersey. Town houses have at least two stories and share a common wall with the next house. See Example 1.*

Classifying Polynomials by the Number of Terms

Sometimes polynomials are classified by degree; sometimes polynomials are classified according to the number of terms they have after combining like terms. A **monomial** is a polynomial with one term; a **binomial** is a polynomial with two terms, and a **trinomial** is a polynomial with three terms. Below are some examples.

$$\text{monomials:} \quad -7, x^2, 3y^4, x + 8x$$
$$\text{binomials:} \quad x^2 - 11, 3y^4 + y, 12a^5 + 4a^3$$
$$\text{trinomials:} \quad x^2 - 5x + 6, 10y^6 - 9y^5 + 17y^2$$

Notice that monomials, binomials, and trinomials can be of any degree. No special name is given to polynomials with more than three terms.

When a polynomial in one variable is added to or multiplied by a polynomial in another variable, the result is a polynomial in several variables. The **degree of a polynomial in several variables** is the largest sum of the exponents of the variables in any term. For instance, $x^3 + 8x^2y^3 + y^4$ is a trinomial in x and y of degree 5.

Polynomials and Area

Some applications of polynomials arise from formulas for area.

Example 1

The widths of the town houses pictured below are x, y, and z. Each has height $f + s$. Find a polynomial for A, the total surface area of the fronts of the three town houses.

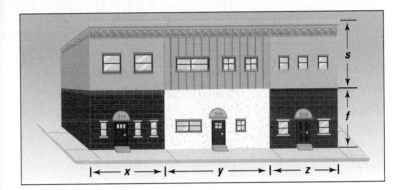

Solution 1

Think of the surface area of the front of the town houses as the sum of the areas of the six smaller rectangles (three first floors, three second floors).

$$A = fx + fy + fz + sx + sy + sz$$

Solution 2

Think of the surface area of the front of the town houses as the area of one large rectangle with base $x + y + z$ and height $f + s$. Thus,

$$A = (x + y + z)(f + s)$$

To expand this expression, use the Distributive Property, considering $x + y + z$ as a chunk.

$$A = (x + y + z)f + (x + y + z)s$$

Now apply the Distributive Property again.

$$A = xf + yf + zf + xs + ys + zs$$

Check

The two solutions check each other.

Because of the multiple use of the Distributive Property we say that Solution 2 of Example 1 illustrates the *Extended Distributive Property*.

The Extended Distributive Property
To multiply two polynomials, multiply each term in the first polynomial by each term in the second.

If one polynomial has m terms and the second n terms, there will be mn terms in their product before combining like terms.

Example 2

Expand $(5x^2 - 4x + 3)(x - 7)$.

Solution

Multiply each term in the first polynomial by each in the second. There will be six terms in the product.

$$(5x^2 - 4x + 3)(x - 7)$$

$$= 5x^2 \cdot x + 5x^2 \cdot (\text{-}7) + (\text{-}4x) \cdot x + (\text{-}4x) \cdot (\text{-}7) + 3 \cdot x + 3 \cdot (\text{-}7)$$

$$= 5x^3 - 35x^2 - 4x^2 + 28x + 3x - 21$$

Now simplify by combining like terms.

$$= 5x^3 - 39x^2 + 31x - 21$$

Polynomials and Volume

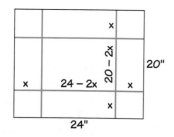

Other applications of polynomials arise from volume. Consider a piece of metal 24″ by 20″ which is to be folded into an open box after a square is cut from each corner. What will be the volume of the box?

The volume depends on the length x of the side of each square cut out. Suppose $x = 2″$. Then the box will be 2″ high and its other dimensions will be 20″ and 16″. Its volume is then $2″ \cdot 20″ \cdot 16″$, or 640 cubic inches.

Activity

Suppose a square with sides of length $x = 3″$ is cut from each corner of a 24″-by-20″ rectangle and the figure is folded to form a box.
a. What are the dimensions of the box?
b. What is the volume of the box?

Which value of x gives the largest possible volume? To answer that question, we need the volume in terms of x.

Example 3

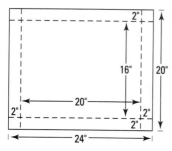

A piece of metal 24 inches by 20 inches is made into a box by cutting out squares of side x from each corner. Let $V(x)$ be the volume of the box. Find a polynomial formula in standard form for $V(x)$.

Solution

The volume of a rectangular box is the product of the dimensions, $V = \ell wh$. When the metal is folded up, the dimensions of the box in inches will be $(24 - 2x)$ long by $(20 - 2x)$ wide by x high.

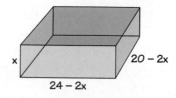

▶

682

▶ So $V(x) = (24 - 2x)(20 - 2x)x$
$V(x) = (24 - 2x)(20x - 2x^2)$ Distribute the *x*.
$V(x) = 480x - 48x^2 - 40x^2 + 4x^3$ Distributive Property
$V(x) = 4x^3 - 88x^2 + 480x.$ Combine like terms; write in standard form.

Check

Choose a particular value of *x*, and calculate the volume using the formula.
When $x = 2''$, $V(x) = 4 \cdot 2^3 - 88 \cdot 2^2 + 480 \cdot 2$
$= 32 - 352 + 960 = 640$ cubic inches.
When $x = 2''$, the box has dimensions 2″, 20″ and 16″. Its volume is 640 in³.

Boxes like this Russian lacquer box are used for decorative purposes or to store keepsakes.

Because volume is 3-dimensional, you should expect a volume formula to involve a third power of length. When a formula for $V(x)$ is known, the function *V* can be graphed. The graph below shows that $x = 2$ does not give the largest volume. A slightly larger volume occurs when $x = 3$, and the largest volume occurs when *x* is a little less than 4. You are asked to find a better estimate to this value in Question 13.

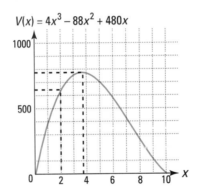

$V(x) = 4x^3 - 88x^2 + 480x$

QUESTIONS

Covering the Reading

In 1–6, a polynomial is given. **a.** State whether the polynomial is a monomial, a binomial, a trinomial, or none of these. **b.** Give its degree.

1. $x^9 - x$ **2.** $3x^5 + x^2 + x + 1$ **3.** $a^3 - b^3$

4. $5x + x$ **5.** $x^2 + 7xy - 8$ **6.** $173x^2y^3z$

7. Give an example of a fourth degree binomial.

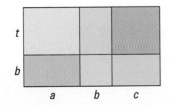

8. Find the total area *A* of the rectangles at the left by
    a. summing the areas of the six small rectangles.
    b. using the formula $A = \ell w$ directly and applying the Extended Distributive Property.

In 9 and 10, expand and put in standard form.

9. $(3x^2 + x - 5)(x + 2)$ **10.** $(4a^2 + 2a + 1)(2a - 1)$

11. What answer did you get for the Activity in the Lesson?

In 12 and 13, refer to the box of Example 3.

12. a. Use the equation $V(x) = 4x^3 - 88x^2 + 480x$ to complete the table below for $x = 1$ to 5.

x	1	2	3	4	5	6	7	8	9	10	11
$V(x)$		640							108		

    b. $V(x) = (24 - 2x)(20 - 2x)x$ is called the *factored form* of the polynomial of Example 3. Use the factored form to check the value of the polynomial for $x = 9$.

    c. Use the factored form to calculate the value of the polynomial for the other integer values of x from $x = 6$ to 11.

    d. For what integer value of x between 1 and 11 is $V(x)$ greatest? least?

    e. Interpret your results for $x = 10$ in terms of the box.

    f. What is a reasonable domain for the function V?

13. a. Use an automatic grapher to estimate to the nearest tenth the value of x at which V achieves its maximum value.

    b. Use this value to find the dimensions of the box with the largest possible volume.

In a zone. *Residential areas, like this one in Houston, Texas, are regulated by local governments. An important reason for zoning laws is to restrict the building of commercial or industrial buildings in residential areas.*

Applying the Mathematics

14. Suppose a rectangular piece of metal is 18 inches by 10 inches and squares of side x are cut out of the corners. An open box is formed from the remaining metal.

    a. Find a polynomial formula for the volume $V(x)$ of the box.

    b. Calculate $V(2)$.

    c. Find a value of x such that $V(x) > V(2)$.

15. A town's zoning ordinance shows a figure like the one at the right. Distances a, b, and c are the minimum setbacks allowed to the street, the side lot lines, and the rear lot lines, respectively. If a rectangular lot has 75′ frontage and is 150′ deep, what is the maximum ground area possible for a one-story house?

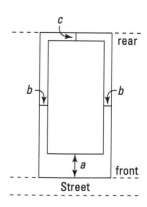

In 16 and 17, multiply and put in standard form.

16. $(x^2 - 2x + 2)(x^2 + 2x + 2)$ **17.** $(a + b - c)(a - b + c)$

18. For 6 years, after each birthday, Devin invested his money in an account which compounded interest annually at rate r. He saved
$$56x^5 + 32x^4 + 40x^3 + 47x^2 + 61x + 59 \text{ dollars,}$$
where $x = 1 + r$.
   a. What is the degree of this polynomial?
   b. How much money would Devin have if the money were invested at a rate of 7.25%? *(Lesson 11-1)*

number of deer

N(t)

100,000

50,000

0 5 10 15 20 25 t
number of years
after 1905

19. During the early part of the twentieth century, the deer population of the Kaibab Plateau in Arizona grew rapidly, because hunters had reduced the number of natural predators. Later, the increase in deer population depleted the food supply and the number of deer declined quickly. The number $N(t)$ of deer from 1905 to 1930 is approximated by
$$N(t) = -.125t^5 + 3.125t^4 + 4000,$$
where t is the time in years after 1905. This function is graphed at the left.
   a. What is the degree of this polynomial function?
   b. To the nearest thousand, what was the deer population in 1905?
   c. To the nearest thousand, what was the deer population in 1930?
   d. Over what time period (between 1905 and 1930) was the deer population increasing?
   e. Approximately when did the deer population start to decline? *(Lessons 1-4, 3-1, 11-1)*

20. *Skill sequence.* *(Lesson 6-1)*
   a. Expand $(x + 10)^2$.
   b. Rewrite $x^2 + 14x + 49$ as the square of a binomial.
   c. Factor $x^2 - 10x + 25$.

In 21 and 22, consider that for diamonds of similar quality, the price tends to vary directly as the square of the weight. Suppose a half-carat diamond costs $2200. *(Lessons 2-1, 2-3)*

21. Estimate the cost of a $1\frac{1}{2}$ carat diamond of similar quality.

22. How many times as much will a 2-carat diamond cost than a half-carat diamond of similar quality?

23. In Question 16, the product of the two trinomials simplifies to be a binomial. Find another pair of trinomials whose product is a binomial.

24. Some automatic graphers have the ability to find the maximum value(s) of functions. Find out whether yours has this feature. If it does, use it to answer Question 13.

Polynomials are, by definition, a sum.

$$P(x) = a_n x^n + a_{n-1} x^{n-1} + a_{n-2} x^{n-2} + \ldots + a_2 x^2 + a_1 x^1 + a_0.$$

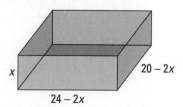

However, sometimes a polynomial may be rewritten as a product. For instance, recall that the polynomial $4x^3 - 88x^2 + 480x$ from Example 3 of Lesson 11-2 is the volume of a box with sides of length x, $24 - 2x$, and $20 - 2x$. Thus the polynomial can be rewritten in the *factored form* $(24 - 2x)(20 - 2x)x$. By factoring out a 2 from each of the binomials, we see that all of the following are equivalent expressions.

$$4x^3 - 88x^2 + 480x = (24 - 2x)(20 - 2x)x$$
$$= 4x(12 - x)(10 - x)$$

In Lesson 11-2, you multiplied the polynomials on the right side above to obtain the polynomial. In this lesson we present a number of ways to undo this process, that is, to rewrite a polynomial as a product of two or more factors. This rewriting is called **factoring** the polynomial.

There are four common ways of factoring.

   1. Factor the greatest common monomial factor.
   2. Use special formulas.
   3. Use trial and error.
   4. Use the Factor Theorem.

You may have seen these in earlier courses. We review the first three types in this lesson and introduce the Factor Theorem in Lesson 11-5.

Common Monomial Factoring

Common monomial factoring is an application of the Distributive Property.

Example 1

Factor $6x^3 - 18x^2$.

Solution

First look for the greatest common monomial factor of the terms. The greatest common factor of 6 and -18 is 6; x^2 is the highest power of x that divides each term. So $6x^2$ **is the greatest common monomial factor of** $6x^3$ **and** $-18x^2$. Now apply the Distributive Property.

$$6x^3 - 18x^2 = 6x^2(\underline{\,?\,})$$
$$= 6x^2(x - 3)$$

$6x^2(x - 3)$ cannot be factored further.   ▶

▶ **Check 1**

Test a special case. Let $x = 2$. Do the given expression and the factored answer have the same values?
When $x = 2$, $6x^3 - 18x^2 = 6 \cdot 8 - 18 \cdot 4 = -24$.
When $x = 2$, $6x^2(x - 3) = 6 \cdot 4(-1) = -24$. Yes.

Check 2

Graph $f(x) = 6x^3 - 18x^2$ and $g(x) = 6x^2(x - 3)$. Check that the graphs coincide. On the window at the left, the graphs seem to be identical.

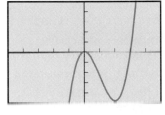

$-5 \le x \le 5$, x-scale = 1
$-25 \le y \le 25$, y-scale = 5

Special Factoring Relationships

There are three factoring relationships you should memorize. The first comes from the Binomial-Square Theorem.

Binomial-Square Factoring
For all a and b,
$$a^2 + 2ab + b^2 = (a + b)^2$$
and $a^2 - 2ab + b^2 = (a - b)^2$

The sum of two squares, $a^2 + b^2$, can only be factored using complex numbers. However, the difference of two squares, $a^2 - b^2$, has a well-known factorization.

Difference-of-Squares Factoring
For all a and b,
$$a^2 - b^2 = (a + b)(a - b).$$

Example 2 shows the use of the Difference-of-Squares Factoring Theorem.

Example 2

Factor $9m^2n^2 - 49$.

Solution

Each term of the polynomial can be written as a perfect square.
$$9m^2n^2 = (3mn)^2$$
$$49 = 7^2$$
Thus the Difference-of-Squares Factoring Theorem can be applied using $a = 3mn$ and $b = 7$.
$$9m^2n^2 - 49 = (3mn)^2 - 7^2$$
$$= (3mn + 7)(3mn - 7)$$

Check
$$(3mn + 7)(3mn - 7) = 9m^2n^2 - 21mn + 21mn - 49$$
$$= 9m^2n^2 - 49$$

Sometimes more than one factoring technique may be applied to rewrite a polynomial.

Example 3

Factor $81x^5 - 16x$ completely.

Solution

$$81x^5 - 16x = x(81x^4 - 16)$$

Factor out the common monomial factor.

$$= x((9x^2)^2 - 4^2)$$

Rewrite perfect squares.

$$= x(9x^2 - 4)(9x^2 + 4)$$

Difference-of-Squares Theorem

$$= x((3x)^2 - 2^2)(9x^2 + 4)$$

Rewrite perfect squares.

$$= x(3x - 2)(3x + 2)(9x^2 + 4)$$

Difference-of-Squares Theorem

Check

The check is left to you.

Quadratic Trinomial Factoring

By the Quadratic Formula, if $ax^2 + bx + c = 0$ and $a \neq 0$, then

$$x = \frac{-b \pm \sqrt{b^2 - 4ac}}{2a}.$$

Recall from Lesson 6-10 that if a, b, and c are real numbers, then the discriminant $D = b^2 - 4ac$ indicates whether or not there are real roots. In particular, there is at least one real root if $D \geq 0$.

The discriminant and the Quadratic Formula can also help you factor quadratic polynomials. Consider the polynomial $ax^2 + bx + c$, where a, b, and c are integers. To have linear factors with integer coefficients, the quadratic equation $ax^2 + bx + c = 0$ must have rational roots. However, the roots will not be rational unless $\sqrt{b^2 - 4ac}$ is rational; that is, unless the value of $b^2 - 4ac$ is a perfect square. This proves the following theorem.

> **Discriminant Theorem for Factoring Quadratics**
> Suppose a, b, and c are integers with $a \neq 0$, and let $D = b^2 - 4ac$. Then the polynomial $ax^2 + bx + c$ can be factored into first degree polynomials with integer coefficients if and only if D is a perfect square.

Example 4

Can $3y^2 - 2y - 5$ be factored into first degree polynomials with integer coefficients? If so, factor it.

Solution

Here $a = 3$, $b = -2$, and $c = -5$. $b^2 - 4ac = 64$, which is a perfect square. So the polynomial is factorable. First write the form of the linear factors.

$$3y^2 - 2y - 5 = (\underline{\ ?\ }y + \underline{\ ?\ } \times \underline{\ ?\ }y + \underline{\ ?\ })$$

▶

▶ The coefficients of y must multiply to 3. Thus they are 3 and 1. The constant terms must multiply to -5. So they are either 1 and -5, or -1 and 5. Here are all the possibilities.

$$(3y + 1)(y - 5)$$
$$(3y - 1)(y + 5)$$
$$(3y - 5)(y + 1)$$
$$(3y + 5)(y - 1)$$

At most, you need to do these four multiplications. If one of them gives $3y^2 - 2y - 5$, then that is the correct factoring. We show all four products. You can see that the desired one is third.

$$(3y + 1)(y - 5) = 3y^2 - 14y - 5$$
$$(3y - 1)(y + 5) = 3y^2 + 14y - 5$$
$$(3y - 5)(y + 1) = 3y^2 - 2y - 5$$
$$(3y + 5)(y - 1) = 3y^2 + 2y - 5$$

So, $3y^2 - 2y - 5 = (3y - 5)(y + 1)$.

Prime Polynomials

If none of the multiplications in the solution to Example 4 resulted in $3y^2 - 2y - 5$, then that polynomial would be *prime,* or *irreducible,* over the rational numbers. A polynomial is **prime,** or **irreducible,** over the set of rational numbers if it cannot be factored into polynomials of lower degree whose coefficients are rational numbers. A similar definition applies to primeness or irreducibility over the set of real numbers.

Taking the *byte* out of factoring. *Today's technology can be used to factor polynomials of various degrees.*

Example 5

a. Is $x^2 - 6$ prime over the set of rational numbers?
b. Is $x^2 - 6$ prime over the set of real numbers?

Solution

a. Test using the Discriminant Theorem for Factoring.
$$x^2 - 6 = x^2 + 0x + -6$$
So $b^2 - 4ac = 0^2 - 4 \cdot 1 \cdot (-6) = 24$. This is not a perfect square. So $x^2 - 6$ cannot be factored using rational coefficients. Thus $x^2 - 6$ is prime over the set of rational numbers.

b. Use the Difference-of-Squares Factoring Theorem.
$$x^2 - 6 = x^2 - (\sqrt{6})^2 = (x - \sqrt{6})(x + \sqrt{6})$$
So $x^2 - 6$ is not prime over the set of real numbers.

Factoring Using Technology

Most symbol manipulators can do all operations with polynomials and can also factor polynomials over the sets of rational, real, or complex numbers. Normally, to factor a polynomial, you must enter the polynomial and indicate the set over which you wish it to be factored. If you have access to such technology, you should try it out on the examples of this lesson.

QUESTIONS

Covering the Reading

1. Copy and complete: $9d^2 + 3ed - 6d^3 = 3d(\underline{\ ?\ } + \underline{\ ?\ } + \underline{\ ?\ })$.

In 2 and 3, factor.

2. $21x^3 - 28x$

3. $-62x^5y^2 + 124x^4y^3$

In 4–9, a polynomial is given. **a.** Tell whether the polynomial is a perfect square, a difference of squares, or a sum of squares. **b.** Factor, if possible.

4. $x^2 - y^2$

5. $a^2 - 2ab + b^2$

6. $x^2 - 256$

7. $25a^2 - 36b^2$

8. $49a^2 - 42ab + 9b^2$

9. $x^2 + 25$

10. **a.** Factor $x^3 - 16x$ into linear factors. **b.** Check by multiplying.

11. **a.** Factor $9x^3 - 25x$ completely. **b.** Check your answer.

12. Check the result of Example 3.

13. If $a, b,$ and c are integers, when is $ax^2 + bx + c$ factorable into linear factors with integer coefficients?

In 14–17, a trinomial is given. **a.** Determine whether the trinomial is factorable into linear factors with integer coefficients. **b.** If so, factor.

14. $5x^2 + 8x - 4$

15. $y^2 + 3y + 4$

16. $3x^2 - 9x - 10$

17. $7z^2 - z - 8$

18. *True or false.* $x^2 + 3x + 2$ is a prime polynomial over the set of rational numbers.

19. **a.** Is $x^2 - 8$ prime over the set of real numbers? If not, factor it.
    b. Is $x^2 - 8$ prime over the set of rational numbers? If not, factor it.

Applying the Mathematics

20. One factor of $6x^2 + 7x - 10$ is $(x + 2)$. Find the other factor.

21. *Multiple choice.* Which of the following is a perfect square trinomial? Justify your answer.
    (a) $9x^2 + 60x + 25$
    (b) $a^4 + 24a^3 + 144$
    (c) $y^2 + 9$
    (d) $4q^2 + r^2 - 4qr$

22. **a.** Write $x^4 - 81$ as the product of two binomials.
    b. Write $x^4 - 81$ as the product of three binomials.

In 23 and 24, a polynomial is given. **a.** First factor out the greatest common monomial factor. Then complete the factorization. **b.** Check by graphing or by multiplying.

23. $4x^3 - 88x^2 + 480x$

24. $1000x^3 - 90xy^2$

25. *Multiple choice.* Which is a factorization of $x^2 + y^2$ over the complex numbers?
(a) $(x + y)(x + y)$
(b) $(x + iy)(x + iy)$
(c) $(x - iy)(x - iy)$
(d) $(x + iy)(x - iy)$

Review

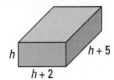

In 26 and 27, consider a closed rectangular box with dimensions h, $h + 2$, and $h + 5$. Write a polynomial in standard form for

26. $S(h)$, the surface area of the box. *(Lesson 11-2)*

27. $V(h)$, the volume of the box. *(Lesson 11-2)*

28. A graphic designer works with sheets of paper 11 in. by 17 in. Suppose the designer lays out a rectangular design in the center of the sheet with a border of x in. on each side. *(Lesson 11-2)*
a. Find the area of the design if $x = 3$.
b. Write an expression for $A(x)$, the area of the design.

29. a. Let $f(x) = x^3 - x^2 - 12x$. Construct a table of values for $f(x)$ with x an integer from -5 to 5.
b. Plot the eleven points in part **a**. Estimate what the graph of $y = f(x)$ looks like by drawing a smooth curve through the points.
c. Check your work in part **b** by using an automatic grapher.
(Lesson 11-1)

30. Express as the logarithm of a single number: $\log_5 100 - \log_5 25$. *(Lesson 9-8)*

31. Evaluate $\log_3 \frac{1}{3}$. *(Lesson 9-5)*

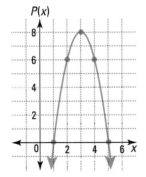

32. A function P is graphed at the left. *(Lessons 6-3, 6-4, 6-7, 6-10)*
a. Which word best describes the function: constant, linear, quadratic, or exponential? Explain how you know.
b. For what values of x does $P(x) = 0$?
c. How many solutions does the equation $P(x) = 9$ have?

Exploration

33. a. Prove the Difference-of-Cubes Factoring Theorem: For all numbers a and b, $(a - b)(a^2 + ab + b^2) = a^3 - b^3$.
b. Use the result to factor $x^3 - 8$.

34. a. Multiply $(a + b)(a^2 - ab + b^2)$.
b. Use the result of part **a** to factor $x^3 + 27$.
c. Give a name to the result of part **a**.

LESSON

11-4

*Estimating
Solutions to
Polynomial
Equations*

Summer job. *Shown is a college student doing reforestation after a fire in the Boise National Forest. Many high school and college students work summer jobs to help finance their college education.*

Linear and quadratic equations can be solved exactly by hand, with paper and pencil. There are also formulas for finding exact solutions to third and fourth degree polynomial equations, but they are quite complicated. These and higher degree polynomial equations are now seldom solved by hand. Instead, a variety of calculator and computer technology is used, including graphing, tables and spreadsheets, and special keys for solving equations.

Solving Polynomial Equations Using Graphs

Recall Yolanda Fish's savings situation (Example 3 of Lesson 11-1). In five consecutive summers Yolanda saves $1500, $2200, $2100, $3000, and $3300. With a 6% annual yield at the end of the 5th summer she accumulates

$$1500(1.06)^4 + 2200(1.06)^3 + 2100(1.06)^2 + 3000(1.06) + 3300 = \$13,354$$

for medical school. Suppose Yolanda wonders: What rate of interest is needed to accumulate $20,000? To answer this question, she would have to solve

$$1500x^4 + 2200x^3 + 2100x^2 + 3000x + 3300 = 20,000,$$

where $x = 1 + r$, and r is the annual yield.

Example 1

---

a. Use graphs to solve the equation above for *x*.
b. Use the solution(s) to find the yield *r* which Yolanda would have to obtain in order to have $20,000 at the end of the 5th summer. ▶

Solution

a. $1500x^4 + 2200x^3 + 2100x^2 + 3000x + 3300 = 20,000$

Divide each side by 100 to deal with smaller numbers.

$$15x^4 + 22x^3 + 21x^2 + 30x + 33 = 200$$

To find the solutions graph the equation

$$f(x) = 15x^4 + 22x^3 + 21x^2 + 30x + 33$$

(the function determined by the left side), and $g(x) = 200$ (the function determined by the right side) on the same set of axes. These graphs are shown at the left. Notice that there are two points of intersection, point S near $x = -2$, and point T near $x = 1$. You can use an automatic grapher to estimate the x-coordinates of these points more accurately. You should find that, to the nearest hundredth, $x \approx -2.24$ or $x \approx 1.31$.

b. You were given that $x = 1 + r$, so solve

$$\begin{array}{ccc} 1 + r \approx -2.24 & \text{or} & 1 + r \approx 1.31. \\ r \approx -3.24 & \text{or} & r \approx 0.31. \end{array}$$

Only $r \approx 0.31 = 31\%$ could be an interest rate. Yolanda would need a yield of almost 31% compounded annually to have $20,000 at the end of the 5th summer. This is an unrealistic expectation.

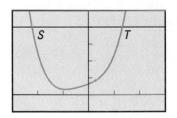

$-3 \leqslant x \leqslant 3, \quad x\text{-scale} = 1$
$-50 \leqslant y \leqslant 250, \quad y\text{-scale} = 50$

Solving Polynomial Equations Using Tables

You could also use tables of values to solve the equation

$$1500x^4 + 2200x^3 + 2100x^2 + 3000x + 3300 = 20,000.$$

Some automatic graphers can create tables of values. If your automatic grapher does not make tables, you can write a program for your calculator or computer to do so, or you could use a spreadsheet. Here is a BASIC program that lists values of any function $y = f(x)$ for $x = A$ to $x = B$ in increments of C.

```
10 REM PROGRAM TO PRINT TABLE OF FUNCTIONAL VALUES
20 INPUT "ENDPOINTS A AND B OF DOMAIN"; A, B
30 INPUT "STEP SIZE"; C
40 PRINT "X", "Y"
50 FOR X = A TO B STEP C
60 REM TYPE IN YOUR OWN FUNCTION AT LINE 70
70 Y =
80 PRINT X, Y
90 NEXT X
100 END
```

To use this program, first type the formula for y as a function of x in line 70. Then each time you run the program, enter values for A, B, and C when prompted.

Example 2

From Example 1, you know that the polynomial function
$$P(x) = 15x^4 + 22x^3 + 21x^2 + 30x + 33$$
equals 200 for some value of x between $x = 1$ and $x = 2$. Use a table generator to estimate this solution to the nearest hundredth.

Solution

If you do not have software to generate tables, use the program similar to the one above, and for line 70 type
$$Y = 15*X^4 + 22*X^3 + 21*X^2 + 30*X + 33.$$
Run the program using $A = 1$ and $B = 2$. A step size of 0.1 will locate the solution to the tenths place. The table of values follows.

X	Y
1	121
1.1	142.65
1.2	168.36
1.3	**198.67**
1.4	**234.15**
1.5	275.44
1.6	323.18
1.7	378.06
1.8	440.81
1.9	512.19
2.0	593

The table shows that the equation has a solution between 1.3 (where the value of $y < 200$) and 1.4 (where $y > 200$). It appears to be closer to 1.3 than to 1.4. To achieve more accuracy, enter 1.3 for A, 1.4 for B, and .01 for C. The new table is below.

X	Y
1.3	**198.67**
1.31	**201.97**
1.32	205.33
1.33	208.74
1.34	212.2
1.35	215.72
1.36	219.3
1.37	222.93
1.38	226.61
1.39	230.35
1.4	234.15

This shows there is a solution between 1.30 and 1.31.
Thus, rounded up to the nearest hundredth, the solution to $P(x) = 15x^4 + 22x^3 + 21x^2 + 30x + 33$, between 1 and 2, is 1.31.

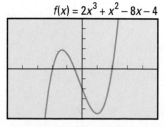

$f(x) = 2x^3 + x^2 - 8x - 4$

$-5 \leqslant x \leqslant 5,$ x-scale = 1
$-10 \leqslant y \leqslant 10,$ y-scale = 2

The techniques illustrated in Examples 1 and 2 apply to any polynomial equation, and can be used together. For instance, to find the x-intercepts of the function $f(x) = 2x^3 + x^2 - 8x - 4$, you might graph the function as shown at the left. From the graph you can see that this polynomial function has at least three x-intercepts. They appear to be at -2, 2, and somewhere between -1 and 0. Thus, solutions to $2x^3 + x^2 - 8x - 4 = 0$ lie in these intervals, and you could generate tables to estimate them more closely.

Using a Solve Key with Polynomial Equations

Many graphics calculators and computer programs have a way to find real solutions to polynomial equations of the form $P(x) = 0$. To use these automatic solvers, you need to enter the polynomial $P(x)$ and an interval in which the solution lies. For instance, for the function f graphed on page 694, you would enter $2x^3 + x^2 - 8x - 4$ and the interval from -1 to 0 to find the solution to $2x^3 + x^2 - 8x - 4 = 0$ in that interval. Then you press a **SOLVE** instruction. Some calculators and symbol manipulators are even more powerful. If you enter an equation, they can give all real and complex solutions. You should examine the technology available to you to determine what capabilities it has. If you have the technology, you should learn to use it, for it can help you solve equations that would be difficult to solve otherwise. You can also use this technology to check answers to equations solved by other methods.

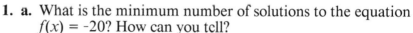

QUESTIONS

Covering the Reading

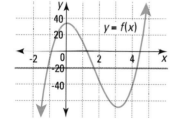

In 1 and 2, refer to the graph of the polynomial function f at the left.

1. **a.** What is the minimum number of solutions to the equation $f(x) = -20$? How can you tell?
   b. Between which pairs of consecutive integers do these solutions occur?

2. **a.** What is the minimum number of x-intercepts f has?
   b. Between which pairs of consecutive integers do they occur?

3. Refer to the polynomial equation from Examples 1 and 2:
   $P(x) = 15x^4 + 22x^3 + 21x^2 + 30x + 33 = 200$.
   a. How many solutions does the equation have?
   b. Estimate each solution to the nearest hundredth.
   c. Which solution yields an answer to Yolanda Fish's question?

4. Using either a graph or a table of values, determine r, the yield Yolanda Fish would need in order to accumulate $15,000 by the end of the fifth summer. Give your answer to the nearest whole percent.

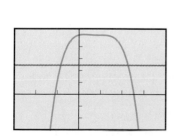

-3 $\leqslant x \leqslant$ 4, x-scale = 1
-6 $\leqslant y \leqslant$ 11, y-scale = 2

5. As shown at the left there are two values of x at which the graphs of $y = 5$ and $y = -x^4 + 3x^3 - 3x^2 + x + 10$ intersect.
   a. Use an automatic grapher to estimate these values to the nearest tenth.
   b. Solve the equation $-x^4 + 3x^3 - 3x^2 + x + 10 = 5$ to the nearest tenth.

6. Estimate the solutions to the equation $-2x^3 + 4x^2 = 1$ by using either a graph or a table.

x	y
-10	3102
-8	1728
-6	826
-4	300
-2	54
0	-8
2	18
4	36
6	-50
8	-336
10	-918

7. Vernon used a table generator to obtain values of the function $f(x) = -2x^3 + 11x^2 - x - 8$ from $x = -10$ to 10 and got the output at the left.
   a. Between which pairs of consecutive even integers must the x-intercepts of f occur?
   b. Round the largest x-intercept up to the nearest tenth.
   c. Round the smallest x-intercept down to the nearest tenth.

Applying the Mathematics

8. The data below are from a study of alcohol tolerance conducted at San Diego State University. They show the average blood alcohol level (BAL, as a percent of the maximum) of the people studied at t minutes after drinking a given amount of alcohol.

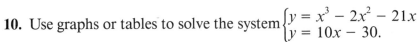

time (minutes)	0	30	42	54	66	78	90	102	114	126	138	150	162
BAL (% of max)	0	79.7	91.9	96.8	98.7	99.0	100.0	99.0	94.3	91.3	88.7	85.5	82.1

These data can be described by the polynomial function A with equation
$$A(x) = -.00000077x^4 + .000343x^3 - .057x^2 + 4.01x + .429$$
where $A(x)$ is the BAL in an average person's bloodstream x minutes after drinking a fixed amount of alcohol.
   a. According to the table, how many minutes after drinking alcohol is the BAL the highest?
   b. Graph $y = A(x)$ over an appropriate domain.
   c. Using the graph, estimate the time at which the BAL is the greatest.
   d. Estimate the length of time the BAL is at least 50% of maximum.

9. If $P(x)$ is a polynomial in x, describe how to use a graph to solve the equation $P(x) = k$.

10. Use graphs or tables to solve the system $\begin{cases} y = x^3 - 2x^2 - 21x \\ y = 10x - 30. \end{cases}$

   Round noninteger solutions to the nearest integer.

SADD. *The organization Students Against Drunk Driving (SADD) was founded in 1981 and consists of 26,000 local groups. Its purpose is to save lives by educating people not to drink and drive.*

Review

In 11 and 12, a polynomial is given. **a.** Tell whether the polynomial is a difference of squares or the square of a difference. **b.** Factor the polynomial.   *(Lesson 11-3)*

11. $49x^2 - 25y^2$

12. $9p^2 - 12p + 4$

13. Factor $x^4 + x^3 - 20x^2$ completely, and check your work.   *(Lesson 11-3)*

In 14 and 15, expand and simplify.   *(Lesson 11-2)*

14. $(x^2 - y^2)(x^2 + y^2)$

15. $(a + b + c)(a - b)$

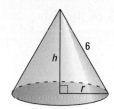

16. Suppose that the lateral height of a cone is 6 cm and its height is h.
(Lessons 1-6, 11-2)
  a. Express the volume in terms of r and h.
  b. Express the radius r of the cone in terms of h.
  c. Substitute the result of part **b** into the formula in part **a**.
  d. *True or false.* The volume of this cone is a polynomial function of r.

17. Give an example of a cubic binomial. *(Lesson 11-1)*

18. a. Graph $f(x) = \sin x$, $0° \le x \le 360°$.
  b. For what values of x in this domain does $f(x) = 0$? *(Lesson 10-8)*

19. Consider square *MATH* shown below. Use slope to prove that the diagonals of *MATH* are perpendicular to each other. *(Lesson 4-9)*

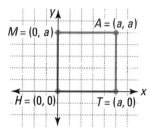

Exploration

20. Find technology that enables you to use a `solve` key to solve the equation of Example 1, and tell what answer the technology gives.

*Factors
and Graphs*

IN - CLASS

ACTIVITY

Work on this activity in small groups.

1 **a.** Graph the function $f(x) = x^3 + x^2 - 12x$ on the window $-5 \le x \le 5, -35 \le y \le 30$.
b. Identify the x-intercepts of the graph.
c. Check that these are correct by substituting into the given equation.
d. Factor the expression $x^3 + x^2 - 12x$.
e. How are the x-intercepts of the graph of $y = f(x)$ related to the factors of $x^3 + x^2 - 12x$?

2 **a.** Graph $g(x) = (x - 1)(x + 1)(x - 3)(x + 4)$ on the window $-5 \le x \le 5, -60 \le y \le 30$.
b. Rewrite $(x - 1)(x + 1)(x - 3)(x + 4)$ as a polynomial in standard form.
c. Check your multiplication in part **b** by graphing.
d. Estimate the x-intercepts of the graph. Check your work by substitution.
e. How are the x-intercepts of the graph of $y = g(x)$ related to the factors of $(x - 1)(x + 1)(x - 3)(x + 4)$?

3 Look back at your work in Questions 1 and 2. Make a conjecture about the factors of a polynomial and the x-intercepts of the graph of the function it determines. Make up a problem to test your conjecture.

LESSON

11-5

The Factor Theorem

Giftwrapped polynomials. *Many department stores provide unfolded boxes when you buy their merchandise. If the unfolded part is 12 by 20, and if squares of size x are folded up, then the volume* V(x) *is as given in Example 1.*

In Lesson 11-3, you learned how to factor many polynomials. In Lesson 11-4, you learned how to solve polynomial equations using graphs. In this lesson, we connect these two ideas. To develop this connection, recall that a product of numbers equals 0 if and only if one of the factors equals 0. This result is called the *Zero-Product Theorem.*

> **Zero-Product Theorem**
> For all a and b, $ab = 0$ if and only if $a = 0$ or $b = 0$.

This theorem is true when a and b are expressions, so it holds for polynomials. For example, it applies to the polynomial equation first used as Example 3 of Lesson 11-2.

Example 1

Let $V(x) = 4x^3 - 88x^2 + 480x$. Solve $V(x) = 0$ for x.

Solution 1

Make a graph. Graph the function with equation $y = V(x)$, as shown at the right. $V(x) = 0$ at the x-intercepts, that is, when the graph intersects the x-axis. A graph indicates that these intersections are at $x = 0$, 10, and 12. So the solution set is $\{0, 10, 12\}$.

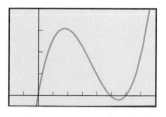

$-4 \le x \le 16$, x-scale = 2
$-125 \le y \le 1000$, y-scale = 250

▶

Solution 2

Factor the polynomial, and apply the Zero-Product Theorem.
$$4x^3 - 88x^2 + 480x = 4x(x^2 - 22x + 120) = 4x(x - 12)(x - 10) = 0$$
So, at least one of these conditions must be true:

$4x = 0$	or	$x - 12 = 0$	or	$x - 10 = 0$	
Thus $x = 0$	or	$x = 12$	or	$x = 10$	

Solution 3

Recall the situation that led to this polynomial. The equation $V(x) = 4x^3 - 88x^2 + 480x$ gives the volume of a box with sides of dimensions x, $24 - 2x$, and $20 - 2x$. *The volume of the box will be 0 exactly when any side has length 0,* leading to the same values as in the other solutions.

Check

Substitute each solution into the given equation.
$$V(0) = 4 \cdot 0^3 - 88 \cdot 0^2 + 480 \cdot 0 = 0$$
$$V(10) = 4 \cdot 10^3 - 88 \cdot 10^2 + 480 \cdot 10 = 4000 - 8800 + 4800 = 0$$
$$V(12) = 4 \cdot 12^3 - 88 \cdot 12^2 + 480 \cdot 12 = 6912 - 12{,}672 + 5760 = 0$$
They all check.

The *x*-intercepts of the graph of a function are called the **zeros** of the function. Notice that in Example 1 the function V has zeros at 0, 10, and 12. The polynomial $V(x)$ has *factors* $x - 0$, $x - 10$, and $x - 12$. The general relationship, which holds for any polynomial function, is simple and elegant. You may have found this relationship in the In-class Activity on page 698.

> **Factor Theorem**
> $x - r$ is a factor of a polynomial $P(x)$ if and only if $P(r) = 0$.

> **Proof**
> The "only if" direction: If $x - r$ is a factor of $P(x)$, then for all x, $P(x) = (x - r)Q(x)$, where $Q(x)$ is some polynomial. So $P(r) = (r - r)Q(r) = 0 \cdot Q(r) = 0$.
>
> The "if" direction requires more work. First consider the case when $r = 0$ and $P(r) = 0$. This means we begin with $P(0) = 0$. Then the graph of $y = P(x)$ contains $(0, 0)$. Also, because $P(0)$ always equals the constant term of the polynomial $P(x)$, the constant term of $P(x)$ is 0. This means that x is a factor of each term of $P(x)$, so x is a factor of $P(x)$.
>
> Now consider the case when $r \neq 0$ and $P(r) = 0$. Then the graph of $y = P(x)$ contains $(r, 0)$. So we can consider the graph of $y = P(x)$ to be a translation image r units to the right of a polynomial function $y = G(x)$ that contains $(0, 0)$. By the Graph-Translation Theorem, $P(x)$ can be formed by replacing x in $G(x)$ by $x - r$. Since x is a factor of $G(x)$, $x - r$ is a factor of $P(x)$.

Finding Zeros by Factoring

The words "if and only if" in the Factor Theorem mean that the theorem can be split into two parts. One part is: if $x - r$ is a factor of $P(x)$, then $P(r) = 0$. Thus, if you can factor a polynomial $P(x)$, you can easily obtain the zeros of the polynomial function P.

Example 2

Find the zeros of $P(x) = x^4 - x^3 - 20x^2$ by factoring.

Solution

First, factor out the greatest common monomial factor, x^2. Then factor the remaining quadratic polynomial.

$P(x) = x^2(x^2 - x - 20) = x^2(x - 5)(x + 4)$

The zeros of $P(x)$ occur when $x^2(x - 5)(x + 4) = 0$.
By the Zero-Product Theorem, $x = 0$, $x = 0$, $x - 5 = 0$, or $x + 4 = 0$.
So $x = 0$ or $x = 0$ or $x = 5$ or $x = -4$. The zeros are 0, 5, and -4.

Check

Using an automatic grapher shows zeros at approximately -4, 0, and 5.

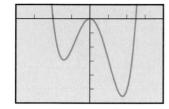

$-8 \leqslant x \leqslant 8$, x-scale = 2
$-150 \leqslant y \leqslant 25$, y-scale = 25

Factoring by Finding Zeros

The other part of the Factor Theorem is that if you know the zeros of a polynomial function, then you can determine the factors of the polynomial.

Example 3

Factor $x^4 - 14x^3 - 87x^2 + 1080x$ by graphing.

Solution

Graph $f(x) = x^4 - 14x^3 - 87x^2 + 1080x$. From the graph, we see that The zeros appear to be -9, 0, 8, and 15. By the Factor Theorem, The factors are $x - (-9)$, $x - 0$, $x - 8$, and $x - 15$. Thus,
$x^4 - 14x^3 - 87x^2 + 1080x =$
$x(x + 9)(x - 8)(x - 15)$.

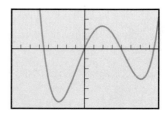

$-16 \leqslant x \leqslant 16$, x-scale = 2
$-6000 \leqslant y \leqslant 4000$, y-scale = 1000

Check 1

Expand.
$x(x + 9)(x - 8)(x - 15) = (x^2 + 9x)(x^2 - 23x + 120)$
$= x^4 - 23x^3 + 120x^2 + 9x^3 - 207x^2 + 1080x$
$= x^4 - 14x^3 - 87x^2 + 1080x$

Check 2

Graph the function $g(x) - x(x + 9)(x - 8)(x - 15)$ on the same set of axes as the function f above. The two graphs coincide.

Finding Equations from Zeros

Different polynomial functions can have the same zeros.

Activity

Consider the equations below.

$$y_1 = x(x - 3)(x + 2)$$
$$y_2 = 2x(x - 3)(x + 2)$$
$$y_3 = 5x^2(x - 3)(x + 2)^2$$

a. What do the equations have in common?
b. Without making a graph, predict the x-intercepts of the graphs of these equations.
c. Check by graphing.

You should find that every function of the form

$$P(x) = kx^a(x - 3)^b(x + 2)^c,$$

where k is a real number, and a, b, and c are positive integers, has the same zeros.

Example 4 shows how to find an equation for a polynomial function with given zeros.

Example 4

Find the general form of an equation for a polynomial function with zeros at -4, $\frac{7}{2}$, and $\frac{5}{3}$.

Solution

Call the polynomial p(x). Since it is given that the zeros of p are -4, $\frac{7}{2}$, and $\frac{5}{3}$, $p(-4) = 0$, $p\left(\frac{7}{2}\right) = 0$, and $p\left(\frac{5}{3}\right) = 0$. By the Factor Theorem, $(x - -4)$, $\left(x - \frac{7}{2}\right)$, and $\left(x - \frac{5}{3}\right)$ must be factors of p(x).
Thus

$$p(x) = k(x + 4)\left(x - \frac{7}{2}\right)\left(x - \frac{5}{3}\right)$$

where k is any constant or polynomial in x.

Check

Substitute -4 for x. Is $p(-4) = 0$?

$$p(-4) = k(-4 + 4)\left(-4 - \frac{7}{2}\right)\left(-4 - \frac{5}{3}\right)$$
$$= k(0)\left(-\frac{15}{2}\right)\left(-\frac{17}{3}\right)$$

So, $\quad p(-4) = 0$.

Similarly, $p\left(\frac{7}{2}\right) = 0$ and $p\left(\frac{5}{3}\right) = 0$.

Notice that the degree of $p(x)$ in Example 4 is at least 3. However, from the given information we cannot determine the value of k, nor even whether k is a constant or an expression. Thus, we cannot be sure of the degree of $p(x)$. Many polynomials go through the points $(-4, 0)$, $\left(\frac{7}{2}, 0\right)$, and $\left(\frac{5}{3}, 0\right)$. Three examples are

$$f(x) = 6(x + 4)\left(x - \frac{7}{2}\right)\left(x - \frac{5}{3}\right) = 6x^3 - 7x^2 - 89x + 140,$$

$$g(x) = (x + 4)\left(x - \frac{7}{2}\right)\left(x - \frac{5}{3}\right) = x^3 - \frac{7}{6}x^2 - \frac{89}{6}x + \frac{70}{3},$$

and $h(x) = x^2(x + 4)\left(x - \frac{7}{2}\right)\left(x - \frac{5}{3}\right) = x^5 - \frac{7}{6}x^4 - \frac{89}{6}x^3 + \frac{70}{3}x^2.$

Graphs of these three functions are shown below.

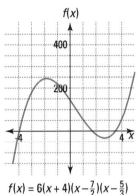

$f(x) = 6(x + 4)(x - \frac{7}{2})(x - \frac{5}{3})$

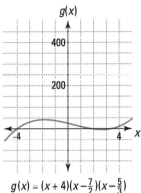

$g(x) = (x + 4)(x - \frac{7}{2})(x - \frac{5}{3})$

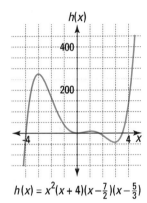

$h(x) = x^2(x + 4)(x - \frac{7}{2})(x - \frac{5}{3})$

QUESTIONS

Covering the Reading

1. State the Zero-Product Theorem.

In 2 and 3, solve.

2. $(x + 9)(3x + 4) = 0$

3. $\left(-\frac{5}{7}k + 2\right)(k - 2)(k - .9) = 0$

4. If $f(x) = x(x + 7)(x - 3)$, solve $f(x) = 0$.

5. Suppose that $P(x)$ is a polynomial and $x - 4$ is a factor of $P(x)$. According to the Factor Theorem, what can you conclude?

In 6–9, an equation for a polynomial function is given.
a. Factor the polynomial. **b.** Find the zeros of the function.

6. $j(x) = x^2 - 10x - 24$

7. $k(a) = 2a^3 - 17a^2 + 8a$

8. $g(t) = 2t^3 + 11t^2 - 63t$

9. $f(r) = 3r^4 - 108r^2$

10. *True or false.* If the graph of a polynomial function crosses the x-axis at $(3, 0)$ and $(-4, 0)$, then $(x + 3)$ and $(x - 4)$ are factors of the polynomial.

In 11 and 12, an equation for a polynomial function is given.
a. Find the zeros of the function by graphing.
b. Use this information to factor the polynomial.
c. Check your answer.

11. $y = x^3 - 5x^2 - 28x + 32$ **12.** $P(n) = 6n^3 + 5n^2 - 24n - 20$

13. Answer the questions in the Activity within the lesson.

14. A polynomial function f has zeros of 9, 10, and -4.
   a. Find the general form of an equation for this function.
   b. What is the smallest possible degree of f?
   c. Name three different polynomial functions with these zeros.

15. Find the general form of an equation for a polynomial function whose zeros are 8, 0, -10, and 2.4.

Applying the Mathematics

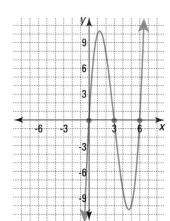

16. At the left is the graph of a third-degree polynomial function with zeros at 0, 3, and 6 and leading coefficient 1. What is an equation for this function?

17. Suppose $f(x) = (x - a)(x - b)(x + c)(x + d)$. What are the zeros of the function f?

18. A horizontal beam has its left end built into a wall, and its right end resting on a support, as shown in the figure below. The beam is loaded with weight uniformly distributed along its length. As a result, the beam sags downward according to the equation
$$y = -x^4 + 24x^3 - 135x^2,$$
where x is the distance (in meters) from the wall to a point on the beam, and y is the distance (in hundredths of a millimeter) of the sag from the x-axis to the beam.

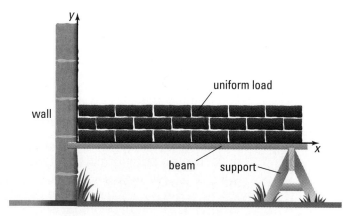

   a. What is an appropriate domain for x if the beam is 9 meters long?
   b. Find the zeros of this function.
   c. Tell what the zeros represent in this situation.

19. Solve the system $\begin{cases} y = 2 \\ y = 25x^4 - 4x^2. \end{cases}$ *(Lessons 5-2, 11-4)*

In 20–22, factor if possible.   *(Lesson 11-3)*

20. $27x^3 - 18x^2y$ **21.** $a^2 + 14ab + 49b^2$ **22.** $3x^2 - 3y^2$

In 23–25, complete the expression to form a perfect square.
(Lessons 6-5, 11-3)

23. $x^2 + \underline{\ ?\ } + 100$ **24.** $n^2 - 18n + \underline{\ ?\ }$ **25.** $y^2 + 5y + \underline{\ ?\ }$

26. Rewrite in standard form: $(x + 2)(x + 3)(x + 4)$.   *(Lessons 11-1, 11-3)*

27. Clark has a piece of construction paper 9 in. by 12 in. Suppose he cuts squares with side x inches from each corner, and folds the paper as in Lesson 11-2 to make an open box. Let $V(x) = $ the volume of the box and $S(x) = $ the surface area of the box. Find a polynomial formula for
 a. $V(x)$.
 b. $S(x)$.   *(Lesson 11-2)*

28. Graphs of cubic functions may have any one of the four types of shapes below.

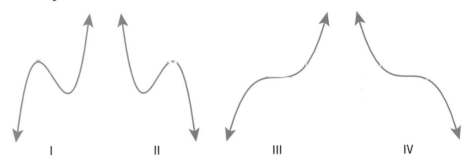

I   II   III   IV

 a. Using an automatic grapher and the Factor Theorem, find an equation for a cubic function other than those given in this chapter
   i. with three x-intercepts whose graph looks like I.
   ii. with three x-intercepts whose graph looks like II.
   iii. with one x-intercept whose graph looks like III.
   iv. with one x-intercept whose graph looks like IV.
 b. Can the graph of a cubic polynomial function ever have two x-intercepts? If so, give an equation for such a function. If not, explain why not.

11-6

Factoring Quadratic Trinomials and Related Polynomials

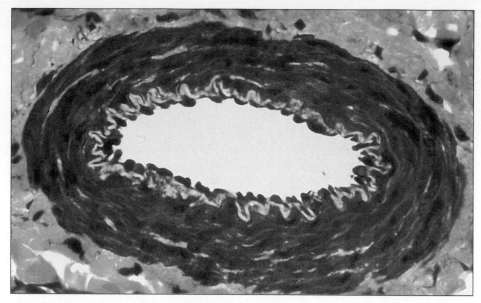

Inner-tubes. *Shown is a magnification of a cross-section of an artery. Arteries help deliver blood to the various parts of the body. See Question 20.*

The Factor Theorem enables you to factor a polynomial $P(x)$ if you know the solutions to $P(x) = 0$. If $P(x)$ is a quadratic polynomial, you can always solve $P(x) = 0$ by using the Quadratic Formula. This enables you to factor a quadratic polynomial even if its coefficients are very large.

Example 1

Factor $x^2 + x - 2162$.

Solution

Let $P(x) = x^2 + x - 2162$, and solve $P(x) = 0$.
$$x^2 + x - 2162 = 0$$
Use the Quadratic Formula. Here $a = 1$, $b = 1$, and $c = -2162$.
$$x = \frac{-1 \pm \sqrt{1 - 4 \cdot 1 \cdot (-2162)}}{2}$$
$$= \frac{-1 \pm \sqrt{8649}}{2}$$
$$= \frac{-1 \pm 93}{2}$$
So $x = \frac{-1 + 93}{2} = 46$ or $x = \frac{-1 - 93}{2} = -47$.
Since 46 and -47 are the zeros of $P(x) = x^2 - x - 2162$,
$x - 46$ and $x + 47$ are factors of $P(x)$.
$$x^2 + x - 2162 = (x - 46)(x + 47)$$

Check

Multiply. $(x - 46)(x + 47) = x^2 - 46x + 47x - 46 \cdot 47 = x^2 + x - 2162$.

The Factor Theorem also can be used to factor a quadratic whose leading coefficient is not 1, but then a few extra steps are needed.

Example 2

Factor $6x^2 - x - 12$ using the Factor Theorem.

Solution

Use the Quadratic Formula to find the zeros of $P(x) = 6x^2 - x - 12$.

$$x = \frac{+1 \pm \sqrt{1^2 - 4(6)(-12)}}{2(6)} = \frac{+1 \pm \sqrt{289}}{12} = \frac{+1 \pm 17}{12}$$

So $x = \frac{1 + 17}{12} = \frac{3}{2}$ or $x = \frac{1 - 17}{12} = \frac{-4}{3}$.

So the zeros of P are $\frac{3}{2}$ and $\frac{-4}{3}$. By the Factor Theorem,

$\left(x - \frac{3}{2}\right)$ and $\left(x - \frac{-4}{3}\right)$ are factors of $P(x)$.

However, $\left(x - \frac{3}{2}\right)\left(x - \frac{-4}{3}\right) = x^2 - \frac{1}{6}x - 2 \neq P(x)$.

Notice that the leading coefficient of $P(x)$ is 6, and the coefficients of $x^2 - \frac{1}{6}x - 2$ are $\frac{1}{6}$ the coefficients of $P(x)$. Thus, $P(x)$ is 6 times the product of the factors obtained. Thus,

$$6x^2 - x - 12 = 6\left(x - \frac{3}{2}\right)\left(x + \frac{4}{3}\right)$$
$$= 2\left(x - \frac{3}{2}\right) \cdot 3\left(x + \frac{4}{3}\right)$$
$$= (2x - 3)(3x + 4).$$

Check 1

Multiply. $(2x - 3)(3x + 4) = 6x^2 + 8x - 9x - 12 = 6x^2 - x - 12$

Check 2

Graph $y = 6x^2 - x - 12$ and $y = (2x - 3)(3x + 4)$ on the same axes. As shown at the left, the graphs appear to coincide.

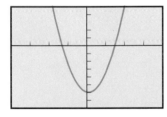

$-4 \leq x \leq 4,$ x-scale = 1
$-16 \leq y \leq 10,$ y-scale = 2

From the graph it is difficult to see that the zeros of P are $-\frac{4}{3}$ and $\frac{3}{2}$. That is why we didn't graph $P(x) = 6x^2 - x - 12$ to find the zeros and then the factors. In the next lesson you will see a theorem which helps in factoring this type of polynomial.

The Factor Theorem enables polynomials to be factored even when the factors have irrational or nonreal coefficients.

Example 3

Factor $z^2 - 5$ using the Factor Theorem.

Solution

Let $P(z) = z^2 - 5$. Solve $P(z) = 0$.

$$z^2 - 5 = 0$$
$$z^2 = 5$$
$$z = \sqrt{5} \quad \text{or} \quad z = -\sqrt{5}$$

By the Factor Theorem, $(z - \sqrt{5})$ and $(z + \sqrt{5})$ are factors of $z^2 - 5$.

$$z^2 - 5 = (z - \sqrt{5})(z + \sqrt{5})$$

Any quadratic polynomial can be factored using the Quadratic Formula.

Example 4

Explain how to factor $x^2 + x + 1$.

Solution

Let $f(x) = x^2 + x + 1$. Solve $f(x) = 0$ using the Quadratic Formula.

$$x^2 + x + 1 = 0$$

implies
$$x = \frac{-1 \pm \sqrt{1 - 4 \cdot 1 \cdot 1}}{2}.$$

So $x = \frac{-1 + \sqrt{-3}}{2}$ or $x = \frac{-1 - \sqrt{-3}}{2}$.

That is, $x = -\frac{1}{2} + \frac{i\sqrt{3}}{2}$ or $x = -\frac{1}{2} - \frac{i\sqrt{3}}{2}$.

By the Factor Theorem, if $f(r) = 0$, then $x - r$ is a factor of $f(r)$.

So $\left(x + \frac{1}{2} - \frac{i\sqrt{3}}{2}\right)$ and $\left(x + \frac{1}{2} + \frac{i\sqrt{3}}{2}\right)$ are factors of $x^2 + x + 1$.

So $x^2 + x + 1 = \left(x + \frac{1}{2} - \frac{i\sqrt{3}}{2}\right)\left(x + \frac{1}{2} + \frac{i\sqrt{3}}{2}\right)$.

Factoring Polynomials of Degree $n \geq 3$

The techniques used to factor quadratic trinomials can sometimes be used to factor polynomials of higher degree and to solve polynomial equations.

Example 5

Solve $3x^3 + 5x^2 - 28x = 0$.

Solution 1

x is a common monomial factor, so factor it.
$$x(3x^2 + 5x - 28) = 0$$
Now use the Zero Product Theorem.
$$x = 0 \text{ or } 3x^2 + 5x - 28 = 0$$
Use the Quadratic Formula to solve the quadratic equation.

$$x = 0 \text{ or } x = \frac{-5 \pm \sqrt{25 - 4 \cdot 3 \cdot (-28)}}{2 \cdot 3}$$

$$x = 0 \text{ or } x = \frac{-5 \pm 19}{6}$$

$$x = 0 \text{ or } x = -4 \text{ or } x = \frac{7}{3}$$

Solution 2

Use an automatic grapher to plot $y = 3x^3 + 5x^2 - 28x$. Use the zoom and trace or other features of the grapher to estimate the zeros. To the nearest tenth, $x \approx -4.0$, $x \approx 0.0$, and $x \approx 2.3$.

Check

Substitute and evaluate the expression.

When $x = 0$, $3 \cdot 0^3 + 5 \cdot 0^2 - 28 \cdot 0 = 0$.

When $x = \frac{7}{3}$, $3 \cdot \left(\frac{7}{3}\right)^3 + 5 \cdot \left(\frac{7}{3}\right)^2 - 28 \cdot \left(\frac{7}{3}\right) = 0$.

When $x = -4$, $3 \cdot (-4)^3 + 5 \cdot (-4)^2 - 28 \cdot (-4) = 0$.

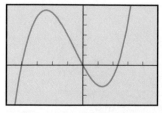

$-5 \leq x \leq 5$, x-scale = 1
$-40 \leq y \leq 60$, y-scale = 10

Covering the Reading

1. **a.** Factor $x^2 - 9x + 14$ using trial and error.
    b. Solve $x^2 - 9x + 14 = 0$ using the Quadratic Formula.
    c. Use the Factor Theorem and the result of part **b** to factor $x^2 - 9x + 14$.

2. **a.** Solve $t^2 - 2t - 15 = 0$ by using the Quadratic Formula.
    b. Use the result of part **a** to factor $t^2 - 2t - 15$.
    c. Check your work by graphing $y = t^2 - 2t - 15$.

3. Refer to Example 2. Which check do you prefer, and why?

4. **a.** Graph $y = x^2 - 5$, and estimate its x-intercepts to the nearest tenth.
    b. How is your answer to part **a** related to the solution to Example 3?

5. Refer to Example 5. Which solution do you prefer, and why?

6. Consider the polynomial $P(x) = 8x^2 - 26x + 15$.
    a. Use the Quadratic Formula to find the zeros of the function P.
    b. Use the Factor Theorem and your answer to part **a** to factor $P(x)$.
    c. Check your work.

In 7–10, factor the polynomial using any method you prefer.

7. $2a^2 - 13a - 24$

8. $9n^2 + 25n - 6$

9. $12c^2 - 28c + 15$

10. $6r^3 + 13r^2 + 6r$

In 11 and 12, solve the equation.

11. $2t^3 - 3t^2 - 20t = 0$

12. $0 = 9v^3 - 42v^2 + 49v$

In 13 and 14, factor into linear factors.

13. $x^2 - 2$

14. $x^2 + 5x + 8$

Applying the Mathematics

15. Let $f(x) = x^4 - 18x^2 + 81$.
    a. Find the zeros of f by graphing.
    b. Factor $x^4 - 18x^2 + 81$ completely.

16. **a.** Expand and simplify $(x + 5)(x^2 - 5x + 25)$.
    b. Factor $x^3 + 125$.
    c. Solve $x^3 + 125 = 0$. (You should find three solutions.)
    d. Graph $f(x) = x^3 + 125$.
    e. Which of the solutions to $x^3 + 125 = 0$ cannot appear on a graph of $f(x) = x^3 + 125$ no matter what window is used?

In 17 and 18, let $P(x) = (x + 3)(2x - 9)(x - 7)$. *(Lessons 11-1, 11-2, 11-5)*

17. What are the zeros of P?

18. a. Rewrite $P(x)$ in the standard form of a polynomial.
b. What is the degree of $P(x)$?

19. a. Find equations for two distinct functions whose graphs pass through the points $(-3, 0)$, $(5, 0)$ and $(8, 0)$.
b. Check your work by graphing.
c. Write the general form of a polynomial function with zeros at -3, 5, and 8. *(Lesson 11-5)*

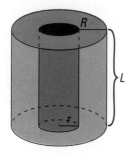

20. An artery can be modeled by a cylindrical solid of outer radius R from which another cylindrical solid of inner radius r has been removed. The figure at the left shows part of an artery.
a. Suppose the piece of artery has length L. Find a formula for the volume V of the artery's wall in terms of r, R, and L.
b. Suppose $R = 1$ cm and $L = 5$ cm. Express V as a function of r.
c. What is the degree of $V(r)$ in part **b**? *(Previous course, Lessons 11-1, 11-2)*

21. Find x to the nearest tenth of a meter. *(Lesson 10-1)*

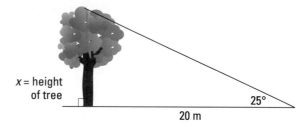

$x =$ height of tree

20 m

25°

In 22 and 23, a sequence is described. **a.** Identify the sequence as a geometric or an arithmetic sequence. **b.** Give an explicit and a recursive definition for the sequence. *(Lessons 3-4, 7-2, 7-5)*

22. The first twenty multiples of 2, starting with 2

23. The first twenty powers of 2, starting with $2^1 = 2$

24. Solve the following system. *(Lessons 5-3, 5-6)*
$$\begin{cases} x + y + z = 3 \\ y + z = -5 \\ z = 4 \end{cases}$$

Exploration

25. Because $1^n = 1$ for all n, $x^n = 1$ has the solution 1 for all n. Thus, for all n, by the Factor Theorem, $x^n - 1$ has the factor $x - 1$.
a. $x^2 - 1$ is the product of $x - 1$ and __?__.
b. $x^3 - 1$ is the product of $x - 1$ and __?__.
c. $x^4 - 1$ is the product of $x - 1$ and __?__.
d. Generalize parts **a–c**.

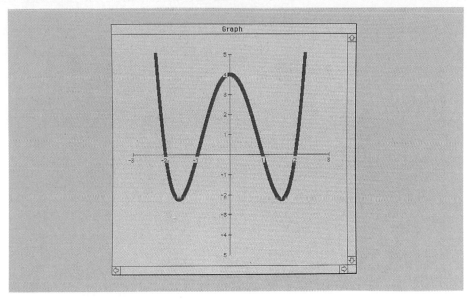

Zeroing in. *The computer image shows the four zeros (1, -1, 2, -2) of the function f defined by $f(x) = x^4 - 5x^2 + 4$. Integer zeros can be found easily by using a zoom feature.*

You have seen that it is possible to use an automatic grapher to estimate zeros of polynomial functions to a high degree of accuracy. However in many cases it is difficult to find exact values. But, when the zeros are rational numbers, it is possible to find them exactly using the *Rational-Zero Theorem*. Here is the idea.

Consider the polynomial function P with equation in standard form,

$$P(x) = 6x^4 - 7x^3 - 43x^2 + 23x + 21.$$

In this form, it is difficult to tell what the zeros are. Even if you graph the function and zoom repeatedly, it may be hard to find the *exact* rational zeros. For instance, the graph below on the left shows the graph of $y = P(x)$ on the window $-5 \leq x \leq 5$ and $-50 \leq y \leq 50$. From it, you can see that P has a zero between -3 and -2. The graph below on the right shows the result of scaling an automatic grapher to find this zero. (Notice that the *x*-axis appears on this graph but the *y*-axis does not.)

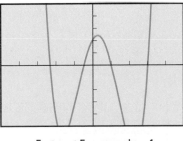

$-5 \leq x \leq 5$, *x*-scale = 1
$50 \leq y \leq 50$, *y*-scale = 10

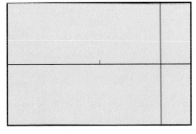

$-2.4 \leq x \leq -2.3$, *x*-scale = .05
$-5 \leq y \leq .5$, *y*-scale = .5

Tracing on the graph at the right above indicates that P has a zero near $x = -2.3336$. But what is the *exact* value?

In factored form, $P(x) = (3x + 7)(2x + 1)(x - 1)(x - 3)$. With the factored form, it is possible to find the zeros of P quickly; they are $-\frac{7}{3}$, $-\frac{1}{2}$, 1, and 3. Thus, the zero between -2.4 and -2.3 is exactly $-\frac{7}{3}$, or $-2\frac{1}{3}$.

Now we ask: How could we know the factors of $P(x)$? For the answer, look again at the factored form of $P(x)$.

$$(3x + 7)(2x + 1)(x - 1)(x - 3)$$

Now consider how the leading coefficient and constant term of a polynomial in standard form are determined. The leading coefficient is the product of the coefficients of x in each factor. The constant term is the product of the constants of the factors.

$$P(x) = (3x + 7)(2x + 1)(x - 1)(x - 3) = 3 \cdot 2 \cdot 1 \cdot 1 \cdot x^4 + \ldots + (7)(1)(-1)(-3)$$

leading coefficient

constant term

$$= 6x^4 + \ldots + 21$$

From the factored form, you know that the zeros of this function are $-\frac{7}{3}$, $-\frac{1}{2}$, 1, and 3.

Notice that the denominators of the zeros, 3, 2, 1, and 1, are factors of the leading coefficient 6. The numerators of the zeros, -7, -1, 1, and 3, are factors of the constant term 21. The generalization of this pattern is the Rational-Zero Theorem.

In words, the Rational-Zero Theorem states that if a simple fraction in lowest terms (a rational number) is a zero of a polynomial, then the numerator of the rational zero is a factor of the constant term of the polynomial and the denominator of the rational zero is a factor of the leading coefficient of the polynomial.

Rational-Zero Theorem
Suppose that all the coefficients of the polynomial function defined by

$$f(x) = a_n x^n + a_{n-1} x^{n-1} + \ldots + a_2 x^2 + a_1 x + a_0,$$

are integers with $a_n \neq 0$ and $a_0 \neq 0$. Let $\frac{p}{q}$ be a rational number in lowest terms. If $\frac{p}{q}$ is a zero of f, then p is a factor of a_0 and q is a factor of a_n.

Proof
1. Suppose that $\frac{p}{q}$ is a rational number in lowest terms that is a zero of f. Then by the definition of a zero,

$$a_n\left(\frac{p}{q}\right)^n + a_{n-1}\left(\frac{p}{q}\right)^{n-1} + \ldots + a_2\left(\frac{p}{q}\right)^2 + a_1\left(\frac{p}{q}\right) + a_0 = 0.$$

2. To clear the fractions multiply both sides of the equation by q^n.

$$a_n p^n + a_{n-1} p^{n-1} q + \ldots + a_2 p^2 q^{n-2} + a_1 p q^{n-1} + a_0 q^n = 0$$

▶

3. Solve for $a_n p^n$. This requires two steps: subtracting all terms but $a_n p^n$ from each side and factoring out q.

$$a_n p^n = -a_{n-1} p^{n-1} q - \ldots - a_2 p^2 q^{n-2} - a_1 p q^{n-1} - a_0 q^n$$
$$= q\left(-a_{n-1} p^{n-1} - \ldots - a_2 p^2 q^{n-3} - a_1 p q^{n-2} - a_0 q^{n-1}\right)$$

4. This equation shows that q is a factor of $a_n p^n$. But q and p have no common factors because $\frac{p}{q}$ is in lowest terms. So q must be a factor of a_n.

5. To complete the proof, solve the equation in step 2 for $a_0 q^n$. Again it takes two steps.

$$a_0 q^n = p\left(-a_n p^{n-1} - a_{n-1} p^{n-2} q - \ldots - a_2 p q^{n-2} - a_1 q^{n-1}\right)$$

6. So, p is a factor of $a_0 q^n$. As before, p and q cannot have any common factors. So p must be a factor of a_0.

The following example shows how the Rational-Zero Theorem can help you find the exact rational zeros of a polynomial.

Example 1

Find all the rational zeros of $f(x) = 3x^4 - 10x^2 - 8x + 15$.

Solution

Apply the Rational-Zero Theorem to identify possible rational zeros.

Let $\frac{p}{q}$ in lowest terms be a rational zero of f.

Then p is a factor of 15 and q is a factor of 3.

So p equals ± 1, ± 3, ± 5, or ± 15, and q equals ± 1 or ± 3.

Thus the possible rational zeros are ± 15, ± 5, ± 3, ± 1, $\pm\frac{5}{3}$, and $\pm\frac{1}{3}$.

You could check these 12 possibilities algebraically, but it is faster to use an automatic grapher. Note that the smallest of the possible zeros is -15, the largest is 15. So draw a graph of the function for $-15 \le x \le 15$.

The graph at the left shows that there are two real zeros on this interval. We see that $x = 1$ is a possible rational zero because the graph of f appears to go through the point (1, 0). We calculate $f(1)$ to equal 0, so

$$x = 1 \text{ is one rational zero.}$$

The other real zero looks to be between 1 and 2. But the only rational zero between 1 and 2 is $\frac{5}{3}$.

$$\text{Since } f\left(\frac{5}{3}\right) \ne 0, \frac{5}{3} \text{ is not a rational zero.}$$

Note that the graph of f does not intersect the x-axis at any other point.

$$\text{So, 1 is the only rational zero.}$$

Any other real zeros must be irrational numbers.

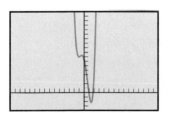

$-15 \le x \le 15$, x-scale = 1
$-10 \le y \le 50$, y-scale = 3

In Example 1, the number 1 is a zero of *f*. From the Factor Theorem, we can conclude that $x - 1$ is a factor of $f(x)$. In this way, the Rational-Zero Theorem can also be useful in factoring polynomials. Example 2 shows how.

Example 2

Factor $g(x) = 4x^4 - 45x^2 + 20x + 21$ over the set of polynomials with integer coefficients.

Solution

The factors of 21 are ±1, ±3, ±7, and ±21. The factors of 4 are ±1, ±2, and ±4. According to the Rational-Zero Theorem,

possible zeros are ±21, ±7, ±3, ±1, $\pm\frac{21}{4}$, $\pm\frac{7}{4}$, $\pm\frac{3}{4}$, $\pm\frac{1}{4}$, $\pm\frac{21}{2}$, $\pm\frac{7}{2}$, $\pm\frac{3}{2}$, $\pm\frac{1}{2}$.

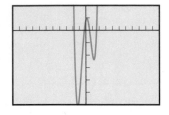

-22 ≤ *x* ≤ 22, *x*-scale = 2
-150 ≤ *y* ≤ 50, *y*-scale = 25

Graph the function on an interval that includes all possible rational zeros. The graph shows four real zeros. There appear to be integer zeros at 1 and 3. So calculate $g(1)$ and $g(3)$. $g(1) = 0$ and $g(3) = 0$. Thus, 1 and 3 are zeros and $(x - 1)$ and $(x - 3)$ are factors of $g(x)$. From the graph we also see there is another zero between -1 and 0.

Tracing indicates that this zero is close to -.5. So we Try $-\frac{1}{2}$.

$$g\left(-\frac{1}{2}\right) = 0.$$

Thus, $-\frac{1}{2}$ is a zero, and $\left(x - -\frac{1}{2}\right) = \left(x + \frac{1}{2}\right)$ is a factor.

Similarly, we notice from the graph that there is a zero between -4 and -3. So we try $-\frac{7}{2}$, the only possible rational zero in this interval.

Because $g\left(-\frac{7}{2}\right) = 0$, $-\frac{7}{2}$ is a zero and $\left(x - -\frac{7}{2}\right) = \left(x + \frac{7}{2}\right)$ is a factor. Thus $g(x)$ has at least four factors: $(x - 1)(x - 3)$, $\left(x + \frac{1}{2}\right)$, and $\left(x + \frac{7}{2}\right)$.

So we can write $g(x) = k\left(x + \frac{7}{2}\right)\left(x + \frac{1}{2}\right)(x - 1)(x - 3)$.

Because the original polynomial has degree 4, *k* is a constant. But the leading coefficient of $P(x)$ is 4.

So, $g(x) = 4\left(x + \frac{7}{2}\right)\left(x + \frac{1}{2}\right)(x - 1)(x - 3)$.

It is customary to rewrite the right side of this equation so that there are no fractions inside the parentheses. This can be done as follows:

$$g(x) = 2\left(x + \frac{7}{2}\right)2\left(x + \frac{1}{2}\right)(x - 1)(x - 3)$$

So, $g(x) = (2x + 7)(2x + 1)(x - 1)(x - 3)$.

Check

Use the Extended Distributive Property.
$$
\begin{aligned}
(2x + 7)(2x + 1)(x - 1)(x - 3) &= (4x^2 + 16x + 7)(x^2 - 4x + 3) \\
&= (4x^2 + 16x + 7)x^2 + (4x^2 + 16x + 7) \\
&\quad (-4x) + (4x^2 + 16x + 7)3 \\
&= 4x^4 + 16x^3 + 7x^2 - 16x^3 - 64x^2 - \\
&\quad 28x + 12x^2 + 48x + 21 \\
&= 4x^4 - 45x^2 + 20x + 21 \text{ It checks.}
\end{aligned}
$$

Covering the Reading

1. Suppose $\frac{p}{q}$ is a rational number in lowest terms and $\frac{p}{q}$ is a zero of
$f(x) = 10x^4 + ax^3 + bx^2 + cx + 7$.
   a. What must be true about p?
   b. What must be true about q?
   c. List the possible rational zeros of f.

2. Refer to Example 1. Why is it unnecessary to check whether $\frac{1}{3}$ is a rational root?

3. Consider the polynomial function $P(x) = 2x^3 + 3x^2 + 3x + 1$.
   a. Use the Rational-Zero Theorem to determine the possible rational zeros.
   b. Graph $y = P(x)$.
   c. How many real zeros does P have?
   d. Identify any rational zeros of $P(x)$.

In 4 and 5, a polynomial equation is given. **a.** Use the Rational-Zero Theorem to list the possible rational zeros. **b.** Find all of the rational zeros.

4. $3x^3 + 2x^2 + 3x + 2 = 0$ 5. $6x^3 + 19x^2 - 24x - 16 = 0$

6. Check Example 2, by graphing $y = (2x + 7)(2x + 1)(x - 1)(x - 3)$ and $y = 4x^4 - 45x^2 + 20x + 21$. What happens?

7. Suppose $P(x) = 3x^4 + 9x^3 - 24x^2 - 36x + 48$.
   a. List all possible rational zeros of P.
   b. Graph $y = P(x)$.
   c. Verify that -2 is a zero of P.
   d. Factor $3x^4 + 9x^3 - 24x^2 - 36x + 48$.

8. Find all rational zeros of $g(x) = 3x^3 - 2x^2 - 7x - 4$.

Applying the Mathematics

9. Consider $g(x) = x^3 - 4x$.
   a. Explain what happens if you try to use the Rational-Zero Theorem to find the rational zeros of g.
   b. Use some other techniques to find the rational zeros of g.

10. Consider $f(n) = 10n^5 - 3n^2 + n - 6$.
    a. List all possible rational zeros of this function.
    b. Use a graph to explain why f has exactly one real zero, and that it is an irrational number.

11. Find all rational roots of $7m^3 + 24m = m^4 + 13m^2 + 80$.

In 12 and 13, factor. *(Lessons 11-3, 11-6)*

12. $3n^4 - 30n^3 + 75n^2$ **13.** $4x^2 - 100$

14. Solve $3x^3 + x^2 - 14x = 0$. *(Lessons 11-3, 11-4, 11-6)*

15. Suppose three zeros of a polynomial function $V(x)$ occur at -3, 4, and -2. Find an equation for $V(x)$. *(Lesson 11-5)*

x	$f(x)$
-2	-17
-1.5	-3.625
-1	3
-.5	5.125
0	5
.5	4.875
1	7
1.5	13.625
2	27

16. At the left is a table of values for a function $f(x) = 3x^3 - x + 5$. Between which two consecutive x-values in the table must a real zero lie? Explain how you got your answer. *(Lesson 11-4)*

17. Rodney invests money in a savings account paying interest annually. He claims the following equation describes his situation. Four years ago he deposited $95; he just deposited $300.

$$F(x) = 95(1.025)^4 + 100(1.025)^3 + 250(1.025) + 300$$

a. What is the annual interest rate of the savings account?
b. How much did Rodney invest 2 years ago?
c. What is the total amount in Rodney's account at this time?
(Lesson 11-1)

18. Explain how you can determine the number of real solutions to the equation $a^2 - 6a = -9$ without actually finding them. *(Lessons 6-6, 6-10)*

19. Let $u = 2 - 5i$ and $v = 4 + 10i$. Find $\frac{uv}{2}$. *(Lesson 6-9)*

20. a. Let $f(x) = a_3x^3 + a_2x^2 + a_1x + a_0$ and
$g(x) = 5a_3x^3 + 5a_2x^2 + 5a_1x + 5a_0$, where a_3, a_2, a_1, and a_0 are integers with $a_3 \neq 0$ and $a_0 \neq 0$. How are the zeros of these functions related? Explain your reasoning.
b. Let $f(x)$ be defined as in part **a** and $h(x) = k f(x)$, where k is a nonzero constant. How are the roots of f and h related?

A poet and a scholar. *Omar Khayyam (1048–1122), Persian poet, mathematician, and astronomer, is well known not only for his work in algebra, but also for his* quatrains—*verses of four lines.*

What Types of Numbers Are Needed to Solve Polynomial Equations?

You know how to find an exact solution to any linear equation with real coefficients. That solution is always a real number. You also know how to solve any quadratic equation exactly: use the Quadratic Formula. But solutions to quadratic equations with real coefficients sometimes are not real. It is natural to wonder whether all polynomial equations can be solved *exactly* and whether any new types of numbers beyond the complex numbers are needed to solve them.

These questions occupied mathematicians even before today's notation for polynomials was invented. Much early work on polynomials was done in Turkey, Italy, and China. By the 12th century, Omar Khayyam had shown how to solve many cubic (3rd degree) equations. In the 13th century, Chinese mathematicians developed ways of estimating roots to polynomial equations of higher degree, but these methods did not reach Europe. In the 16th century, Scipione del Ferro (1465–1526) discovered how to solve some types of cubic equations exactly. (His method is too complicated to be discussed in this book.) Independently, Niccolo Tartaglia (1500–1557) in 1535 discovered a method for solving all cubic equations. A little later, Girolamo Cardano's secretary, Ludovico Ferrari (1522–1565), discovered how to solve any *quartic* (fourth degree polynomial) equation. Girolamo Cardano, whom we have mentioned in Lessons 6-8 and 6-10, published del Ferro's method in 1545 in his book *Ars magna.* The amazing thing was that no numbers beyond complex numbers were needed to solve cubic or quartic equations. Then, for over 250 years mathematicians tried unsuccessfully to find a formula for solving any *quintic* (fifth-degree polynomial) equation. Perhaps new numbers were needed.

However, new numbers are not needed. In 1797, at the age of 18, the great German mathematician Karl Gauss proved the following theorem, whose name indicates its significance. (Remember that complex numbers include the real numbers.)

> **The Fundamental Theorem of Algebra**
> Every polynomial equation $P(x) = 0$ of any degree with complex number coefficients has at least one complex number solution.

Gauss published five proofs of this theorem during his life. All require college-level mathematics. From the Fundamental Theorem of Algebra and the Factor Theorem, it is possible to prove that *every* solution to a polynomial equation is a complex number. Thus, no new type of number is needed to solve higher degree polynomials. So, for instance, the solutions to $x^5 + 3x^3 - ix^2 + 4 - 3i = 0$ are complex numbers.

How Many Complex Solutions Does a Given Polynomial Have?

Recall that the linear equation $ax + b = 0$ has one solution, or root: $x = -\frac{b}{a}$. The quadratic equation $ax^2 + bx + c = 0$ generally has two roots: $x = \frac{-b \pm \sqrt{b^2 - 4ac}}{2a}$. However, when the discriminant is 0, the two roots are equal. When this happens this root is considered to be a **double root.** For instance, when $x^2 - 8x + 16 = 0$, then $x = \frac{-b \pm \sqrt{(-8)^2 - 4(1)(16)}}{2(1)} = \frac{8 \pm \sqrt{0}}{2} = 4$. So $x = 4$ is the only root of $x^2 - 8x + 16 = 0$, and the number 4 is said to be a double root.

Notice that $x^2 - 8x + 16 = (x - 4)^2$. That is, $x - 4$ appears twice as a factor. We say that 4 is a root with **multiplicity** 2. In general, the **multiplicity of a root** r is the highest power of $x - r$ that appears as a factor of the polynomial. For instance, the equation $(x - 3)^{10}(x + 1)^2 = 0$ is an equation with only two roots: 3 has a multiplicity of 10 and -1 has a multiplicity of 2.

In general, the Factor Theorem states that if r is a root of a polynomial equation $P(x) = 0$, then $(x - r)$ is a factor of $P(x)$. This means that if r is a root of the polynomial equation $P(x) = 0$, there is some polynomial $Q(x)$ such that $P(x) = (x - r) \cdot Q(x)$ and the degree of $Q(x)$ is one less than the degree of $P(x)$. For instance, when $P(x)$ is cubic, then $Q(x)$ is quadratic. Thus when $P(x)$ is cubic, $P(x) = 0$ has three roots: one from the linear factor $(x - r)$, and two from the quadratic factor $Q(x)$. Of course, one of these might be a multiple root.

Similarly, any 4th-degree polynomial can be rewritten as the product of a linear and a cubic polynomial, or of two quadratic polynomials. Thus, 4th-degree polynomial equations have 4 complex roots. By extending this pattern we know we can express any higher degree polynomial equation as a product of lower degree polynomials.

These observations are summarized in the following theorem.

Example 1

How many roots does each equation have?
a. $x^5 - 7x^3 + 15x^2 + 3 = 10$
b. $-2ix^4 - ex^2 + \pi x - 12 = 0$

Solution

a. The degree is 5, so the equation has 5 roots.
b. The degree is 4, so the equation has 4 roots.

The question of whether a formula exists for solving all quintic equations was essentially settled in 1799 by an Italian mathematician, Paolo Ruffini (1765–1822). He gave almost all the details of a proof that the general quintic equation cannot be solved by formulas. A Norwegian mathematician, Niels Henrik Abel (1802–1829), gave a complete proof in 1824. A few years later, a young French mathematician, Évariste Galois (1811–1832), described a method for determining exactly which polynomial equations of degree five or higher can be solved using formulas.

Finding the Real Solutions to Polynomial Equations

The two theorems in this lesson tell you how many roots a polynomial equation $P(x) = k$ has, and that all roots can be expressed as complex numbers. They do not tell you how to find the roots, nor do they tell you how many of the roots are real. To answer these questions, you can apply the methods studied in this chapter for finding and analyzing zeros of polynomial functions.

Example 2

Consider the equation from Example 1a, $x^5 - 7x^3 + 15x^2 + 3 = 10$.
a. How many of its solutions are real?
b. Classify the solutions as rational or irrational.

Solution

a. Set one side of the equation equal to zero by subtracting 10 from each side.
$$x^5 - 7x^3 + 15x^2 - 7 = 0$$
By the theorems in this lesson, this equation has five roots if multiple roots are counted separately. The solutions to this equation are the zeros of the function $f(x) = x^5 - 7x^3 + 15x^2 - 7$. So there will be at most five real zeros. Use approximation methods from Lesson 11-4. First, look at the behavior of the function over a large domain.

▶

x	y
-50	-311587507
-40	-101928007
-30	-24097507
-20	-3138007
-10	-91507
0	-7
10	94493
20	3149993
30	24124493
40	101975993
50	311662493

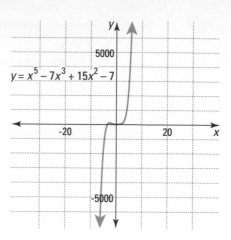

The sign changes in y-values in the table and the graph show that a zero or zeros occur between $x = -10$ and $x = 10$. Also, when $x < -10$ or $x > 10$, the value of x^5 dominates the value of f, so there can be no zeros outside $-10 < x < 10$. A further search with different windows shows that *Zeros occur only on the intervals $-4 \leq x \leq -3$, $-1 \leq x \leq 0$, and $0 \leq x \leq 1$.*

x	y
-5	-1882
-4	-343
-3	74
-2	77
-1	14
0	-7
1	2
2	29
3	182

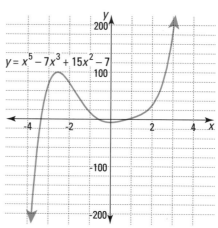

Thus, there are three real roots. The other two roots are nonreal and are not indicated on this graph.

b. By the Rational-Zero Theorem, if $f(x) = x^5 - 7x^3 + 15x^2 - 7$ has rational zeros, they must be ±1 or ±7. In the table above, you see that $f(1) = 2$ and $f(-1) = 14$. So neither 1 nor -1 is a zero. By substitution, $f(-7) \neq 0$ and $f(7) \neq 0$. So none of the zeros are rational. All three zeros are irrational.

Further use of tables or graphs allows you to approximate these irrational zeros by rational numbers. To the nearest tenth, $f(x) = x^5 - 7x^3 + 15x^2 - 7$ has zeros at -3.4, -0.6, and 0.8.

Covering the Reading

1. Name three 16th-century mathematicians who worked on solving cubic or quartic equations.

2. State the Fundamental Theorem of Algebra.

3. *True or false.* The equation $\sqrt{3}\,v^4 - .5v + \frac{1}{9} = 0$ has at least one complex number solution.

4. Who first proved the Fundamental Theorem of Algebra?

In 5 and 6, $a \neq 0$. Solve for x.

5. $ax + b = 0$

6. $ax^2 + bx + c = 0$

In 7–9, an equation is given. **a.** Solve. **b.** Identify any multiple roots.

7. $x^2 - 10x + 25 = 0$

8. $y^3 - 25y = 0$

9. $(x - 1)^2(x + 5)^3(2x - 1) = 0$

10. Every polynomial equation of degree n has exactly __?__ roots, provided that __?__.

In 11 and 12, state the number of roots each equation has. Do not solve.

11. $x^5 + x^3 + x = 0$

12. $17y^2 + \pi y^7 + iy^3 = 12$

13. State one result about polynomials discovered by Galois.

14. Consider the equation $2x^5 - 3x^3 - x = 1$.
    a. At most, how many solutions does it have?
    b. How many of these solutions are real?
    c. How many solutions are rational?
    d. Approximate the real solutions to the nearest tenth by using an automatic grapher or a table generator.

Applying the Mathematics

In 15 and 16, solve.

15. $-3x + 7i = 0$

16. $2ix^2 + 8x + 5i = 0$

17. Find all the roots of $z^4 - 1 = 0$ by factoring and solving the resulting quadratic equations.

Review

18. A polynomial function of degree 3 with leading coefficient 4 has roots equal to $-\frac{1}{2}$, 4, and $\frac{5}{2}$.
    a. Find an equation for the polynomial.
    b. Check your work. *(Lessons 11-5, 11-7)*

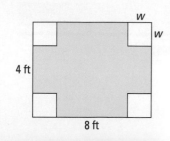

19. Find all zeros of $f(t) = t^4 - 6t^3 - 7t^2$. *(Lessons 11-3, 11-5)*

20. A $4' \times 8'$ sheet of metal is made into a box by cutting out squares of sides w from each corner. Let $V(w)$ represent the volume of the box.
  a. Find a polynomial formula for $V(w)$.
  b. Can this process yield a box with volume of one cubic yard? Why or why not?
  c. Determine the value of w that will yield the box with the largest volume. *(Lessons 11-2, 11-4)*

21. The sum of the cube and the square of a number is 1. To the nearest hundredth, what is the number? *(Lessons 11-1, 11-4)*

22. A person sights a pier directly across a river, then walks 100 meters along the shore and sights the pier at an angle of 70°. How wide is the river at the pier? *(Lesson 10-1)*

In 23 and 24, solve. *(Lessons 9-4, 10-2, 10-9)*

23. $\log x = 5$

24. $\sin x = 5$

25. Brianna, a traffic engineer, wanted to know how much force F would be needed to keep a car of weight w traveling at S mph from skidding on a curve of radius r. She knew that the force varied jointly as the weight and the square of the speed. But she still needed to find the relationship between the force and the radius.
  a. With a 2000 lb car traveling at 30 mph, she obtained the following data.

radius of curve (ft)	125	250	375	500	625
force (lb)	963	481	321	241	193

  Graph these data points.
  b. How does F vary with r?
  c. Write an equation relating F, r, S, and w. Do not find the constant of variation k. *(Lessons 2-8, 2-9)*

26. Graph and state the equation of the image of $y = 3x^2$ under $T_{-3,4}$. *(Lessons 2-5, 4-10, 6-3)*

Exploration

27. The Fundamental Theorem of Algebra is discussed in this lesson. There is also a Fundamental Theorem of Arithmetic. Look in a mathematics encyclopedia or other reference book to find out what theorem this is. (You may have known this theorem but didn't know it has this name.)

*Examining
Difference
Patterns for
Polynomial
Functions*

IN·CLASS

ACTIVITY

Work in small groups.

1 Consider the linear polynomial function $y = 4x + 5$. A table of
values, when x is an integer from 1 to 5, is shown below.

x	1	2	3	4	5	. . .
$y = 4x + 5$	9	13	17	21	25	. . .

Notice that the differences of the consecutive y-values (right minus
left), as shown below, are all equal.

$$9 \underbrace{\quad}_{4} 13 \underbrace{\quad}_{4} 17 \underbrace{\quad}_{4} 21 \underbrace{\quad}_{4} 25 \ldots$$

a. Each person in your group should pick a linear function of the
form $y = mx + b$, and make a table of x- and y-values, using 1, 2,
3, 4, 5, . . . for x.
b. Take the differences of the consecutive values (right minus left) as
shown above.
c. Compare your work to the work of others in your group. What
pattern(s) do you notice in the differences?

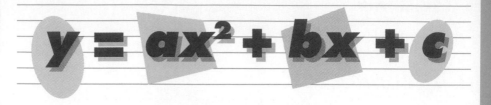

$$y = ax^2 + bx + c$$

2 Consider the quadratic polynomial function $y = 5x^2$. For x, use the consecutive integer values 1, 2, 3, 4, 5, 6, ... As before, find the values of the polynomial. Then find the differences of consecutive terms.

x	1	2	3	4	5	6	. . .
$y = 5x^2$	5	20	45	80	125	180	. . .

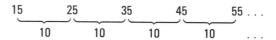

1st differences   15   25   35   45   55 . . .

The differences are not all equal, but notice what happens if differences are taken a second time. The second differences are equal.

15   25   35   45   55 . . .

    10   10   10   10   . . .

a. Pick any quadratic function of the form $y = ax^2 + bx + c$, and make a table of x- and y-values. Let x take the integer values from 1 to 6.
b. Take differences of the consecutive values, again subtracting right minus left.
c. Compare your data to the data of others. What pattern(s) do you observe?

3 **a.** Make a conjecture about difference patterns in a table of values for the function h with equation $h(x) = x^3 - 5x^2 + 10x + 50$.
b. Test your conjecture.

4 Look back at the work you did in Questions 1–3. Can you make a conjecture about difference patterns that applies to any polynomial of degree n? If so, discuss it with your group, and write the conjecture in your own words.

LESSON

11-9

Finite Differences

You have seen many situations in which an equation describes data. In some cases, graphs can be used to find such an equation.

For instance, in Chapter 2, you used this idea to find equations for functions of variation. Consider the data points below and the graph at the right.

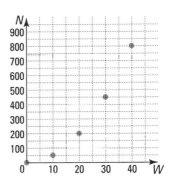

W	0	10	20	30	40
N	0	50	200	450	800

Because the graph looks like a parabola through the origin, you might think that N varies directly as the square of W. This would give the equation $N = kW^2$, which is a quadratic polynomial function.

However, as you have seen in this book, other functions have graphs which resemble the one above. Is it possible to determine in a conclusive way which function fits? The answer is *yes* when the function is a polynomial function.

To see why this is true, it helps to look at some specific polynomial functions, and to examine the way the function values increase. Here are the linear, quadratic, and cubic functions from the In-class Activity.

linear:

x	1	2	3	4	5 ...
$y = 4x + 5$	9	13	17	21	25 ...

The 1st differences are equal.   4   4   4   4 ...

quadratic:

x	1	2	3	4	5	6 ...
$y = 5x^2$	5	20	45	80	125	180 ...

15   25   35   45   55...

The 2nd differences are equal.   10   10   10   10...

cubic:

x	1	2	3	4	5	6 ...
$y = x^3$	1	8	27	64	125	216 ...

7   19   37   61   91...

12   18   24   30...

The 3rd differences are equal.   6   6   6...

You should also have found that the 3rd differences were equal for the function h in the In-class Activity. In general, if you evaluate a polynomial of degree n for consecutive integer values of x, and take differences between consecutive y-values after n sets of differences, you will get a constant.

In fact, we can generalize more than this. Consider again the linear function $y = 4x + 5$. For x-values, instead of using consecutive integers, use the arithmetic sequence -2, 1, 4, 7, 10, ...

x	-2	1	4	7	10	...
$y = 4x + 5$	-3	9	21	33	45	...

1st differences   12   12   12   12...

Again the 1st differences are all equal.

The Polynomial-Difference Theorem

Each of the polynomial functions considered at the beginning of this lesson is an instance of the following theorem. Its proof requires ideas from calculus beyond the scope of this book, and so it is omitted.

> **Polynomial-Difference Theorem**
> $y = f(x)$ is a polynomial function of degree n if and only if, for any set of x-values that form an arithmetic sequence, the nth differences of corresponding y-values are equal and the $(n - 1)$st differences are not equal.

The Polynomial-Difference Theorem provides a technique to determine whether a set of points is part of a polynomial function of a particular degree. The technique suggested by this theorem is called **the method of finite differences.** That is, from a table of y-values corresponding to an arithmetic sequence of x-values, take differences of consecutive y-values. Only if those differences are eventually constant is the function polynomial, and the number of the differences indicates the polynomial's degree.

Example 1

Consider the data points at the beginning of the lesson. Use the method of finite differences to show that N is a polynomial function of W with degree 2.

Solution

Notice that the values of the independent variable W form an arithmetic sequence, so the Polynomial-Difference Theorem applies.

W	0	10	20	30	40
N	0	50	200	450	800

1st differences   50   150   250   350

2nd differences   100   100   100

N is a polynomial function of W because the differences eventually are all equal. Because the 2nd differences are equal, the degree of the polynomial is 2.

Be careful not to overgeneralize the Polynomial-Difference Theorem.

Consider the exponential function $y = 2^x$. Make a table of x- and y-values, using integer values for x between 0 and 10. Analyze the y-values, using the method of finite differences. Describe the patterns in the differences.

Calculating differences can also be used to test whether a sequence can be described with an explicit polynomial formula.

Example 2

The recursive formula
$$\begin{cases} a_1 = 4 \\ a_n = 2a_{n-1} - 1, \text{ for integers } n \geq 2 \end{cases}$$
generates the sequence
$$4, 7, 13, 25, 49, 97, 193, \ldots.$$
Is there an explicit polynomial formula for this sequence?

Solution

Take differences between consecutive terms.

a_n	4		7		13		25		49		97		193...
1st differences		3		6		12		24		48		96...		
2nd differences			3		6		12		24		48...			

The pattern will continue to repeat and will never yield constant differences. So there is no polynomial formula for this sequence.

In the next lesson you will see how to find the particular equation for a polynomial function if y-values corresponding to consecutive x-values are known.

QUESTIONS

Covering the Reading

In 1–3, refer to the Polynomial-Difference Theorem.

1. If the y-values are all equal for the 10th set of differences of consecutive x-values and not equal for the 9th set of differences, what is the degree of the polynomial?

2. *True or false.*   The technique of finite differences takes the differences of consecutive x-values.

3. State a reason why this theorem is important.

In 4 and 5, refer to Example 1.

4. How do we know that N is a polynomial function of W?

5. Why must the degree of the polynomial be 2?

6. Refer to the Activity in the lesson.
   a. Write the values of the 1st differences.
   b. What pattern(s) did you notice in the sets of differences?

In 7–9, use the data points listed in each table below.
a. Determine if y seems to be a polynomial function of x.
b. If the function is a polynomial function, find the degree of the polynomial.

7.

x	1	2	3	4	5	6	7	8	9
y	3	11	31	69	131	223	351	521	739

8.

x	0	1	2	3	4	5	6	7	8	9
y	1	1	3	7	15	31	63	127	255	511

9.

x	1	4	9	16	25	36	49	64	81	100
y	1	2	3	4	5	6	7	8	9	10

10. a. According to the Polynomial-Difference Theorem, what differences will be equal for the polynomial function $y = x^4 + x^2$?
    b. Construct a table of x- and y-values for the function in part **a** using integer values of x between -3 and 4.
    c. Use the technique of finite differences to justify your response to part **a**.

In 11 and 12, a sequence is given. **a.** Generate its first seven terms. **b.** Tell whether the sequence can be described explicitly by a polynomial function of degree less than 6. **c.** If it can, state its degree.

11. $\begin{cases} a_1 = 7 \\ a_n = 5a_{n-1} + 3, \text{ for integers} \\ \qquad n \geq 2 \end{cases}$

12. $\begin{cases} a_1 = 4 \\ a_n = (2a_{n-1})^2 - 10, \text{ for integers} \\ \qquad n \geq 2 \end{cases}$

Applying the Mathematics

x	y
0	5
1	11
2	17
3	23
4	29

13. a. Find the values of the first differences of the function represented by the data points on the left.
    b. Find the degree of the function.
    c. Plot the data points.
    d. Find an equation of the line passing through these points.
    e. Make a generalization about what first differences represent on this graph. Does your generalization apply to other linear functions?

14. a. Let $f(x) = ax^2 + bx + c$. Find $f(1), f(2), f(3), f(4)$, and $f(5)$.
    b. Prove that the 2nd differences of these values are constant.

15. Consider the following pattern:
    $f(1) = 1^2 = 1$
    $f(2) = 1^2 + 2^2 = 5$
    $f(3) = 1^2 + 2^2 + 3^2 = 14$
    $f(4) = 1^2 + 2^2 + 3^2 + 4^2 = 30$
    a. Find $f(5)$ and $f(6)$.
    b. Using the Polynomial-Difference Theorem, what is the degree of the polynomial $f(n)$?

16. Consider the polynomial sequence
4, 15, 38, 79, 144, 239, ...
By using finite differences, predict the next term.

Review

In 17–20, consider the function $y = x^3 + x^2 - 144x - 144$.
(Lessons 11-4, 11-7, 11-8)

x	y
-10	396
-8	560
-6	540
-4	384
-2	140
0	144
2	-420
4	-640
6	-756
8	-720

17. According to the Fundamental Theorem of Algebra, how many zeros does the function have? Justify your answer.

18. According to the Rational-Zero Theorem, what are the possible rational zeros of this function?

19. A table of some values for this function is at the left. According to the table, between which two even numbers must a zero occur?

20. a. Sketch a graph of this function.
   b. Find all zeros of this function. Write all rational roots as fractions. Approximate all irrational roots to the nearest tenth.

21. Does $x^2 - 3\sqrt{2}\,x + 4 = 0$ have any nonreal roots? Why or why not?
(Lesson 11-8)

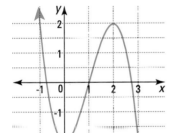

22. At the left is the graph of the polynomial function
$P(x) = (x - 1)(-x^2 + 2x + 2)$.
   a. Rewrite $P(x)$ in standard form.
   b. How many x-intercepts does P have? Find the exact value of the largest of these. *(Lessons 6-6, 11-1, 11-6)*

23. *Multiple choice.* Which of the following equations could describe the relationships graphed below where k is a constant? *(Lesson 2-9)*
   (a) $y = \dfrac{kwz}{x^2}$ (b) $y = kwzx$ (c) $y = \dfrac{kwz^2}{x}$ (d) $y = \dfrac{kwx^2}{z}$

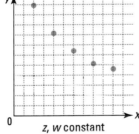
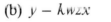
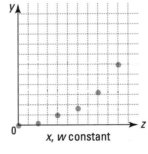
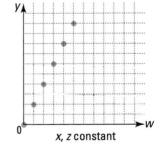

z, w constant   x, w constant   x, z constant

24. Refer to the figure at the left. A boat sails 20 mi from A to B, then turns as indicated and sails 15 mi to point C. How far is A from C directly? *(Lessons 10-6, 10-7)*

25. Solve the system. $\begin{cases} x + 4y - 3z = 6 \\ 2y + z = 9 \\ z = 8 \end{cases}$ *(Lessons 5-3, 5-6)*

Exploration

26. Find a sequence in which the 3rd differences are all 30.

LESSON

11-10

Modeling Data with Polynomials

Pizza pieces. *Why would anyone slice a pizza like this? Read on to find out.*

The employees at Primo's Pizzeria liked to cut pizza into oddly shaped pieces. In so doing, they noticed that there is a maximum number of pieces that can be formed from a given number of cuts.

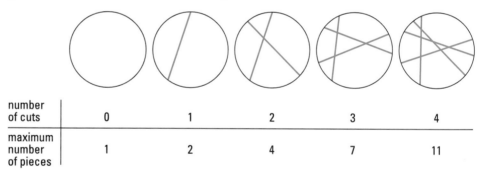

number of cuts	0	1	2	3	4
maximum number of pieces	1	2	4	7	11

Primo's employees wondered if there was a formula relating p, the maximum number of pieces that could be obtained from x, a given number of cuts. As in the last lesson, they found differences between consecutive terms.

x	0	1	2	3	4
p	1	2	4	7	11
1st differences		1	2	3	4
2nd differences			1	1	1

Because they had to take differences two times to get equal differences, they knew that a quadratic polynomial could be used to model these points. That is, they knew that

$$p = ax^2 + bx + c,$$

but they did not know the values of the coefficients a, b, or c.

Finding Quadratic Models

In Lesson 6-6, we showed that it is possible to find the values of a, b, and c by solving a system of equations. We review that method here. We substitute three known ordered pairs (x, p) into the equation. It is usually easiest to use three small values of x in an arithmetic sequence. Here we use 0, 1, and 2.

In general, $p = ax^2 + bx + c$.

When $\quad x = 0, p = 1$, so $1 = a(0)^2 + b(0) + c = c$.

$\quad\quad\quad x = 1, p = 2$, so $2 = a(1)^2 + b(1) + c = a + b + c$.

$\quad\quad\quad x = 2, p = 4$, so $4 = a(2)^2 + b(2) + c = 4a + 2b + c$.

The three pairs (x, p) have produced a system of three equations in three unknowns. To solve this system, reorder the equations so that the largest coefficients are on the top line. Then subtract each equation from the one immediately above it.

$$\begin{cases} 4a + 2b + c = 4 \\ a + b + c = 2 \\ c = 1 \end{cases} \Rightarrow \begin{cases} 3a + b = 2 \\ a + b = 1 \end{cases} \Rightarrow 2a = 1$$

Because $2a = 1$, $a = \frac{1}{2}$. To find b, substitute $a = \frac{1}{2}$ into $a + b = 1$, so $b = \frac{1}{2}$. From the first system, $c = 1$. Now substitute the values of a, b, and c into the general form of the quadratic equation. Thus

$$p = \tfrac{1}{2}x^2 + \tfrac{1}{2}x + 1$$

models the data from Primo's Pizzeria.

Example 1

a. Show that the formula $p = \frac{1}{2}x^2 + \frac{1}{2}x + 1$ correctly describes the relation between number of cuts and maximum number of pieces for $x = 3$.

b. Predict the maximum number of pieces that can result from 5 cuts. Check your answer with a drawing.

Solution

a. When $x = 3$,

$$p = \tfrac{1}{2}(3)^2 + \tfrac{1}{2}(3) + 1 = \tfrac{9}{2} + \tfrac{3}{2} + 1 = \tfrac{12}{2} + 1 = 7$$

This agrees with the data that three cuts can produce 7 pieces of pizza.

b. When $x = 5$,

$$p = \tfrac{1}{2}(5)^2 + \tfrac{1}{2}(5) + 1 = \tfrac{25}{2} + \tfrac{5}{2} + 1 = 16$$

You can get 16 pieces of pizza with 5 cuts.

Check

b. Draw a picture. The figure at the left shows one way to get 16 pieces with 5 cuts.

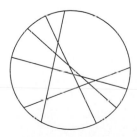

Finding Higher-Degree Polynomials

The following example shows how to use finite differences and systems of equations to find a polynomial function of degree greater than 2.

An "orangement" of oranges.

Example 2

A display of oranges can be stacked in a square pyramid in the following way: 1 orange is in the top level, 4 oranges are in the second level, 9 are in the third level, 16 are in the fourth level, and so on. How many oranges are needed for a display with n rows?

Solution

First, list some values showing how the total number of oranges depends on the number of rows. The total number in the display is as follows:

top row	1
top two rows	$1 + 4 = 5$
top three rows	$1 + 4 + 9 = 14$
top four rows	$1 + 4 + 9 + 16 = 30$
top five rows	$1 + 4 + 9 + 16 + 25 = 55$

Second, use the method of finite differences to determine whether a polynomial model fits the data points.

Number of Rows	1	2	3	4	5	6 ...
Number of Oranges	1	5	14	30	55	91 ...
1st Differences		4	9	16	25	36...
2nd Differences			5	7	9	11...
3rd Differences				2	2	2...

The 3rd differences are constant. Thus the data can be represented by a polynomial function of degree three.

Third, use a system of equations to find a polynomial model. Let n be the number of rows and $f(n)$ the total number of oranges in n rows. We know the polynomial is of the form

$$f(n) = an^3 + bn^2 + cn + d.$$

Substitute $n = 4, 3, 2,$ and 1 into the equation and solve the system as before. That is, subtract pairs of equations to eliminate coefficients $d, c,$ and b in order.

$$\begin{cases} f(4) = 64a + 16b + 4c + d = 30 \\ f(3) = 27a + 9b + 3c + d = 14 \\ f(2) = 8a + 4b + 2c + d = 5 \\ f(1) = a + b + c + d = 1 \end{cases} \Rightarrow \begin{cases} 37a + 7b + c = 16 \\ 19a + 5b + c = 9 \\ 7a + 3b + c = 4 \end{cases} \Rightarrow$$

$$\begin{cases} 18a + 2b = 7 \\ 12a + 2b = 5 \end{cases} \Rightarrow 6a = 2$$

From the equation $6a = 2$, we know $a = \frac{1}{3}$. By substitution into $12a + 2b = 5$, $b = \frac{1}{2}$. Another substitution into $7a + 3b + c = 4$ gives $c = \frac{1}{6}$. Finally, using $a + b + c + d = 1$, we find $d = 0$. Thus, $f(n) = \frac{1}{3}n^3 + \frac{1}{2}n^2 + \frac{1}{6}n$ gives the number of oranges needed for a display with n rows.

▶

Check

You should check that this equation fits the data points. For instance, if $n = 5$, then $f(n) = \frac{1}{3}(5)^3 + \frac{1}{2}(5)^2 + \frac{1}{6}(5) = \frac{125}{3} + \frac{25}{2} + \frac{5}{6} = 55$, which checks.

Limitations of Polynomial Modeling

When using finite differences, you must have a sufficient number of data points to check the formula you get. For instance, suppose you are given only the data below.

x	1	2	3
y	1	2	4

The 1st differences are 1 and 2 and there is only one 2nd difference. So you cannot tell whether the second differences are constant. If the y-values for $x = 4$ and $x = 5$ are 7 and 11 respectively, then the second differences are equal.

x	1	2	3	4	5
y	1	2	4	7	11
	1	2	3	4	
		1	1	1	

A polynomial model for these data is

$$y = \frac{x^2 - x + 2}{2}.$$

However, if 8 and 15 are the next y-values, a polynomial equation modeling the data could be

$$y = \frac{x^3 - 3x^2 + 8x}{6}.$$

These are only two of many polynomial models fitting the data points (1, 1), (2, 2), (3, 4).

Modeling Manhattan's Population

Recall page 673, where the population of Manhattan in 20-year intervals was given. Now we can describe how we found the formula $P(x)$ on that page. The data we had are shown at the left.

We know that a fifth-degree polynomial equation fits these data, because there is sufficient data and after five differences there is only one number left, so all differences are the same! Thus we seek a formula of the form

$$P(x) = ax^5 + bx^4 + cx^3 + dx^2 + ex + f$$

where x is the number of 20-year periods since 1890 and $P(x)$ is the population (in millions). For instance, for the year 1930, $x = 2$ and so

$$P(2) = 1.867 = a \cdot 2^5 + b \cdot 2^4 + c \cdot 2^3 + d \cdot 2^2 + e \cdot 2 + f.$$

In a similar fashion, we found the other five equations needed. We then solved the system.

Year	Population
1890	1,441,000
1910	2,332,000
1930	1,867,000
1950	1,960,000
1970	1,539,000
1990	1,488,000

Covering the Reading

In 1–3, refer to Primo's data at the beginning of this lesson.

1. What general polynomial equation models these data?

2. Primo had three coefficients to find, so he needed to solve a system of __?__ equations.

3. Show that the formula $p = \frac{1}{2}x^2 + \frac{1}{2}x + 1$ is correct for $x = 4$.

In 4–6, refer to Example 2.

4. In which equation(s) could you substitute $a = \frac{1}{3}$ and $b = \frac{1}{2}$ to find $c = \frac{1}{6}$?

5. How many oranges are needed for 6 rows of oranges?

6. Predict the number of oranges in a display with 15 rows.

7. Consider the table below.

x	1	2	3	4	5	6
y	3	16	39	72	115	168

   a. Determine the degree of a polynomial function that models these data.
   b. Find a formula for the function.

8. Suppose that the data in the table below have a formula of the form $y = ax^3 + bx^2 + cx + d$. What four equations are satisfied by a, b, c, and d?

x	2	4	6	8	$\ldots$
y	0	40	168	432	$\ldots$

Shown is a colorized photo of 5th Avenue at 51st Street in Manhattan in 1900. The two buildings in the foreground are part of the Vanderbilt mansion.

In 9 and 10, solve each system.

9. $\begin{cases} x + y + z = -2 \\ 4x + 2y + z = 7 \\ 9x + 3y + z = 5 \end{cases}$

10. $\begin{cases} p + q + r + s = 4 \\ 8p + 4q + 2r + s = 15 \\ 27p + 9q + 3r + s = 40 \\ 64p + 16q + 4r + s = 85 \end{cases}$

11. In finding the 5th-degree equation on page 673 that models Manhattan's population, what linear equation was determined by the 1990 population?

12. Using the technique of finite differences, Norma determined $y = n^2 - n + 2$ to be a formula for the data below.

n	1	2	3	. . .
y	2	4	8	. . .

   a. Check that these data satisfy Norma's equation.
   b. Can Norma be assured that her equation is the correct one? If so, why? If not, find another formula which Norma's data also satisfy.

13. Recall the sequence of triangular numbers, whose first five terms are illustrated below.

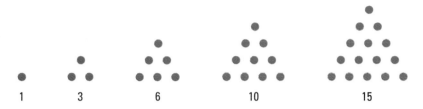

1   3   6   10   15

   a. Show how to use the method of finite differences to find a formula which will generate any triangular number t_n in terms of n, its position in the sequence.
   b. Compare your result in part **a** to the formula given in Lesson 1-9. Are they equivalent? Why or why not?
   c. Describe in words what this formula will tell you when $n = 8$.

14. One cross-section of a honeycomb is a tessellation of regular hexagons, with three hexagons meeting at each vertex. The tessellation is formed by starting with one hexagon, surrounding it with six more hexagons, and then surrounding these with another "circle" of 12 hexagons. If this pattern were to continue, find

   a. the number of hexagons in the 4th circle.
   b. the total number of hexagons in the first four circles.
   c. a polynomial equation which expresses the total number of hexagons h as a function of the number of circles n.
   d. the total number of hexagons in a honeycomb with 10 circles.

15. Consider the data below.

m	1	2	3	4	5	6	7	. . .
n	10	5	2	1	2	5	10	. . .

   a. Can the data be modeled by a polynomial function? How can you tell?
   b. If so, what is the degree of the polynomial? *(Lesson 11-9)*

In 16 and 17, factor completely. *(Lessons 11-3, 11-5, 11-6, 11-7)*

16. $16z^4 - 1$ **17.** $10x^4 + 7x^3 - 45x^2$

18. Consider the system $\begin{cases} y = 5 - 5x^2 \\ y = x^3 - 4x. \end{cases}$

    a. Sketch a graph of the system for $-3 \le x \le 3$.
    b. Solve the system. Approximate your solutions to the nearest tenth. *(Lessons 5-2, 11-3, 11-4)*

19. How much larger is the volume of a cube with dimensions $(x + 3)$ than the volume of a cube with dimensions $(x + 2)$? *(Lesson 11-2)*

20. Describe three properties of the graph of $f(x) = \cos x$. *(Lesson 10-8)*

21. Sergei earns 0.5% interest a month on every dollar he saves from working. He receives interest on the interest, paid at the end of each month. His deposits, made on the first of each month, were $77 in March, $51 in April, $37 in May, $86 in June, $39 in July, and $35 in August. On August 1st, how much will he have altogether, including the interest? *(Lesson 11-1)*

22. Express as the logarithm of a single number. *(Lessons 9-7, 9-8)*

    a. $\log_3 40 - \log_3 10$ **b.** $3\log 64$ **c.** $\frac{1}{3} \log 64$

23. Ben Spender recently had his credit limit reduced by 25%. What percent increase would he need to return to his original amount? *(Previous course)*

Exploration

24. In Example 1b, explore how to cut the pizza so that the 16 pieces are more nearly the same size.

25. Some automatic graphers and statistical packages can generate a best-fit equation for specific data points. Use the five data points from Example 2 and compare the best-fit equation to the one calculated in the lesson.

A project presents an opportunity for you to extend your knowledge of a topic related to the material of this chapter. You should allow more time for a project than you do for typical homework questions.

1 Modeling Manhattan's Population

On page 673, an equation modeling Manhattan's population is given. The process for obtaining that equation is described in Lesson 11-10. Carry out that process—it requires solving a system of 5 linear equations in 5 unknowns. How close do you come to the equation that is displayed on page 673? If your equation is different, why do you think it is different?

2 Proving that Certain *n*th Roots are Irrational

According to the Rational Zero Theorem, the only rational roots possible for the equation $x^2 - 5 = 0$ are ± 1 and ± 5. None of these work, so the solutions to this equation must be irrational. This proves that $\sqrt{5}$ and $-\sqrt{5}$ are irrational. Use this idea to prove the following theorems.

a. $\sqrt{37}$ is irrational.

b. $\sqrt[3]{7}$ is irrational.

c. $\sqrt[5]{2}$ is irrational.

d. Prove that some other number, of your own choosing, is irrational.

e. Explain why this process does not work to prove that $\sqrt{49}$ is irrational.

3 Factoring Using Trial and Error

The polynomial $6x^2 - x - 12$ was factored in Lesson 11-6 using the Quadratic Formula and the Factor Theorem. Suppose you tried to factor this polynomial directly, by writing
$6x^2 - x - 12 = (ax + b)(cx + d)$.

a. What are the possible pairs of values of *a* and *c*?

b. What are the possible pairs of values of *b* and *d*?

c. Multiply all the possible combinations of linear factors to show that only one gives this quadratic, and thus there is only one way to factor this quadratic (except for switching the order of the factors).

d. Compare this trial-and-error method with the method given in Lesson 11-6. Which do you prefer, and why?

4 Synthetic Substitution

The polynomial $ax^4 + bx^3 + cx^2 + dx + e$ can be written as $(((ax + b)x + c)x + d)x + e$. This latter expression can be easier to evaluate, particularly by computers, since it does not involve exponents and it can be described by an iterative, or repeating, algorithm. For any particular value for x, multiply the first coefficient by x, then add the next coefficient. Multiply that sum by x, then add the next coefficient. And so on, until the last coefficient has been added. The process can be done without a computer, and is then called *synthetic substitution*. First write down the coefficients a, b, c, d, and e. Then follow the arrows.

$$
\begin{array}{ccccc}
a & b & c & d & e \\
\downarrow \searrow ax & \searrow (ax+b)x & \searrow (ax^2+bx+c)x & \searrow (ax^3+bx^2+cx+d)x \\
a & ax+b & ax^2+bx+c & ax^3+bx^2+cx+d & ax^4+bx^3+cx^2+dx+e
\end{array}
$$

For instance, to find $P(5)$ for $P(x) = 2x^4 - 9x^3 + 4x - 7$, you would write the following:

$$
\begin{array}{ccccc}
2 & -9 & 0 & 4 & -7 \\
\downarrow \nearrow 10 & \nearrow 5 & \nearrow 25 & \nearrow 145 \\
2 & 1 & 5 & 29 & 138
\end{array}
$$
and find that $P(5) = 138$.

a. Verify that $P(5) = 138$ by substituting in the formula for $P(x)$.
b. Use synthetic substitution to evaluate $Q(7)$ when $Q(x) = 3x^4 + 2x^3 - 20x^2 - 3x + 12$.
c. Make up three other examples with polynomials of degree 3 or more to show synthetic substitution. Verify each result by direct substitution.

5 Volumes of Boxes

Begin with at least 4 rectangular pieces of paper of the same reasonable size.
a. Cut off squares of size x from each corner of one rectangle and fold up the remaining parts to form an open box. Calculate the volume of this box.

b. Do the same, with different values of x, for the other three pieces of paper.
c. Find a polynomial for the volume $V(x)$ of your boxes in terms of x.
d. Graph the function V, and from the graph determine the value of x that gives the maximum value of V within the range of allowable values for x.
e. Do any of the boxes come close to having the maximum possible volume? How far off is the closest box?

SUMMARY

A polynomial in x is an expression which can be written in the form
$a_n x^n + a_{n-1} x^{n-1} + \ldots + a_2 x^2 + a_1 x + a_0$, where a_n is the leading coefficient, and n is the degree of the polynomial. Polynomial functions include the linear and quadratic functions and the direct variation and power functions you studied in previous chapters. Polynomials arise directly from compound interest situations and questions of surface area and volume. They can model many other real-world situations when only a finite set of data points is given. Then the degree of the polynomial can be found by the method of finite differences, and the polynomial itself can be found using a system of linear equations.

When a polynomial of degree n is set equal to zero, the resulting equation has n roots. The roots of a polynomial equation $P(x) = 0$ are the zeros of the function P. The Fundamental Theorem of Algebra guarantees that for all polynomials, $P(x) = 0$ has at least one complex root. Sometimes they can be found exactly by factoring, by trial and error, or by the Factor Theorem. The Rational-Zero Theorem provides a technique to identify all the rational roots of a polynomial with integer coefficients. Real roots can be approximated by making a table or drawing a graph. The most efficient tables and graphs are created with the help of a calculator or computer.

VOCABULARY

Below are the most important terms and phrases for this chapter. You should be able to give a definition for those terms marked with *. For all other terms you should be able to give a general description and a specific example of each.

Lesson 11-1
*polynomial in x
*degree of a polynomial term
*standard form of a polynomial
*coefficients of a polynomial
leading coefficient
expanding a polynomial
*linear, quadratic, cubic, quartic polynomials
polynomial equation, polynomial function
symbol manipulator

Lesson 11-2
monomial
binomial
trinomial
degree of a polynomial in several variables
Extended Distributive Property

Lesson 11-3
factored form
factoring
Binomial Square Factoring Theorem
Difference-of-Squares Factoring Theorem
Discriminant Theorem for Factoring Quadratics
prime polynomial, irreducible polynomial

Lesson 11-4
solve key

Lesson 11-5
Zero-Product Theorem
*zero of a function
factor
Factor Theorem

Lesson 11-7
Rational-Zero Theorem

Lesson 11-8
quartic, quintic equations
Fundamental Theorem of Algebra
double root, multiplicity of a root
Number of Roots of a Polynomial Equation Theorem

Lesson 11-9
Polynomial-Difference Theorem

PROGRESS SELF-TEST

Take this test as you would take a test in class. Then check your work with the solutions in the Selected Answers section in the back of the book.

In 1 and 2, use these facts. When Beth turned 16, she began saving money from her summer jobs. After the first summer, she saved $750. After the second summer, she saved $600. After the third, she saved $925, and the following two summers she saved $1075 and $800, respectively. Beth invested all this money at an annual yield of r, compounded annually, and did not add or withdraw any other money.

1. If $x = 1 + r$, write a polynomial in terms of x which gives the final amount of money in her account the summer after her 21st birthday.

2. How much money would she have the summer after her 21st birthday if she had been able to invest all the money at an annual interest rate of 4%?

3. Pedro has a rectangular piece of cardboard with dimensions 40 in. by 60 in. He forms a box by cutting out squares with sides of length x from each corner and folding up the sides. Find a polynomial formula for the volume $V(x)$ of the box.

4. Expand and write in standard form: $(a^2 + 3a - 7)(5a + 2)$.

In 5 and 6, factor completely.
5. $10s^7t^2 + 15s^3t^4$ 6. $25y^2 + 60y + 36$
7. Find all solutions to $z^3 - 216z = 0$.

In 8–10, consider the polynomial function P where $P(x) = x^4 + 9x^2 - 3 - 8x^5$.

8. **a.** What is the degree of the polynomial?
   b. Is $P(x)$ a monomial, binomial, trinomial, or none of these?

9. State all possible rational roots of $P(x)$.

10. Sketch a graph of $P(x)$.

11. Find the zeros of the polynomial function h with equation
$$h(x) = 4x^3(5x - 11)(x + \sqrt{7}).$$

12. Use $f(x) = 3x^4 - 12x^3 + 9x^2$.
    a. Factor $f(x)$.   **b.** Find the zeros of f.

In 13 and 14, consider the polynomial function with equation $y = x^3 - 3x^2 - 3x + 9$. The table at the right gives some values of this function.

x	y
-3	-36
-2	-5
-1	8
0	9
1	4
2	-1
3	0
4	13

13. How many real zeros does this polynomial have and how do you know?

14. **a.** According to the table, between what pairs of consecutive integers must the zeros of the polynomial be located?
    b. Find the smallest noninteger zero, rounded up to the nearest tenth.

In 15 and 16, *multiple choice.*
15. Which polynomial equations of degree 11 have 12 complex roots?
    (a) all   (b) some   (c) none

16. When $f(x) = x^4 + 3x - 22$, $f(2) = 0$. Which is a factor of $x^4 + 3x - 22$?
    (a) 0   (b) 2   (c) $x + 2$ (d) $x - 2$

17. Write a possible formula $P(x) = \ldots$ for the 4th degree polynomial function P with integer zeros graphed at the right.

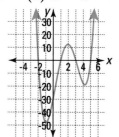

18. Refer to the table below.

n	1	2	3	4	5	6	7	8
t	2	5	9	14	20	27	35	44

    a. Can the above data points be modeled by a polynomial function of degree ≤ 5?
    b. If so, what is the smallest possible degree of the polynomial? If not, why not?

19. Find an equation for a polynomial function which is described by the data points below.

x	-2	-1	0	1	2	3	4
z	12	4	0	0	4	12	24

20. State the Fundamental Theorem of Algebra and identify who first proved it.

CHAPTER REVIEW

Questions on SPUR Objectives

SPUR stands for **S**kills, **P**roperties, **U**ses, and **R**epresentations. The Chapter Review questions are grouped according to the SPUR Objectives for this chapter.

SKILLS DEAL WITH THE PROCEDURES USED TO GET ANSWERS.

Objective A: *Use the Extended Distributive Property to multiply polynomials.*   *(Lesson 11-2)*

In 1–4, expand and write in the standard form of a polynomial.

1. $(x^2 + x + 3)(x - 1)$ **2.** $(a + 6)(a + 7)(a + 8)$
3. $(2y + 5)^3$ **4.** $(2x^2 - x + 4)(3x - 10)$

In 5 and 6, multiply and simplify.

5. $(2x^2 - y)(3x + y)$ **6.** $(p + q + r)(p + q - r)$

Objective B: *Factor polynomials.*
(Lessons 11-3, 11-6)

7. Fill in the blank.
$7a^5b^2 - 63a^2b^4 = 7a^2b^2 (\underline{\ ?\ } + \underline{\ ?\ })$.

8. Fill in the blank with the value(s) which will make a perfect square trinomial:
$w^2 + \underline{\ ?\ } + 25$.

In 9–18, factor completely over the set of polynomials with integer coefficients.

9. $x^2 - 14x + 49$ **10.** $a^2 - b^2$
11. $r^4s^4 - 81$ **12.** $16m^2 - 88m + 121$
13. $x^2 - 9x + 14$ **14.** $40 + 3n - n^2$
15. $4p^2 + 4p - 15$ **16.** $6x^2 + 26x + 8$
17. $x^4 - 32x^2 + 256$ **18.** $8a^2 + 24a + 16$

In 19 and 20, factor into linear factors.

19. $z^2 + 27$ **20.** $x^2 + x + 1$

Objective C: *Find zeros of polynomial functions by factoring.*   *(Lessons 11-3, 11-5, 11-6)*

In 21 and 22, find the exact zeros of the polynomial function.

21. $f(x) = x^2(x - .5)(3x + 1)$
22. $P(x) = x^3 - 36x$

In 23–26, **a.** solve. **b.** Identify any multiple roots.

23. $0 = 5x(x + 4)(9x + 7)$
24. $0 = (y - 1)^3(y - 2)^2$
25. $n^3 + 16n^2 + 64n = 0$
26. $m^4 - 81 = 0$

Objective D: *Determine an equation for a polynomial function from data points.*
(Lessons 11-5, 11-9, 11-10)

In 27–29, is the function defined a polynomial function? If so, find an equation for the polynomial. If not, explain why not.

27.

x	1	2	3	4	5	6
y	7	12	18	25	33	42

28.

x	1	2	3	4	5	6
y	1	3	7	15	31	63

29. the function (n, a_n) where $a_1 = 5$ and $a_n = a_{n-1} - 6$, for integers $n \geq 2$

30. Find an equation for a quadratic function whose graph crosses the *x*-axis at (-69, 0) and (-4.5, 0).

31. Find equations for two different polynomial functions whose zeros are -12, 0, $\frac{1}{4}$ and $\frac{1}{6}$.

32. Consider the polynomial function of smallest degree described by the data points below.

x	1	2	3	4	5	6
y	5	19	43	77	121	175

  a. What is the degree?

  b. *Multiple choice.* Which system of equations could be solved to find the coefficients of the polynomial?

  (i) $\begin{cases} 9a + 3b + c = 43 \\ 4a + 2b + c = 19 \\ a + b + c = 5 \end{cases}$ (ii) $\begin{cases} 3x^2 + 3x + 3 = 43 \\ 2x^2 + 2x + 2 = 19 \\ x^2 + x + 1 = 5 \end{cases}$

  (iii) $\begin{cases} 2a + b = 19 \\ a + b = 5 \end{cases}$ (iv) none of these

  c. Determine an equation for the polynomial function.

PROPERTIES DEAL WITH THE PRINCIPLES BEHIND THE MATHEMATICS.

Objective E: *Use technical vocabulary to describe polynomials.* (Lessons 11-1, 11-2)

In 33 and 34, state: **a.** the degree and **b.** the leading coefficient of the polynomial.

33. $7c^5 + 3c^2 - 15$ **34.** $1 + d - 12d^2 - 8d^9$

In 35–38, *multiple choice.* State whether the polynomial is (a) a monomial, (b) a binomial, (c) a trinomial, or (d) none of (a)–(c).

35. $e^5 - 6$ **36.** $32f^2g^3$

37. $\frac{6}{h^2}$ **38.** $b^2 + b + 7$

39. Give an example of a trinomial with degree 6.

40. Give an example of a binomial of degree 4.

Objective F: *Apply the Zero-Product Theorem, Factor Theorem, and Fundamental Theorem of Algebra.* (Lessons 11-5, 11-8)

In 41 and 42, explain why the Zero-Product Theorem cannot be used directly on the given equation.

41. $(x - 8)(x + 11) = 10$

42. $2(t + 2) - \left(t + \frac{2}{3}\right) = 0$

43. Every polynomial equation of degree n has exactly __?__ roots, provided that __?__ roots are counted separately.

44. Suppose $f(m) = (m - 3)^2(m - 4)^3$. Then 3 is a zero with multiplicity __?__ and 4 is a zero with multiplicity __?__.

In 45 and 46, *true or false.*

45. If $(x - 7)$ is a factor of some polynomial function P, then $P(7) = 0$.

46. If a polynomial has a double root, then it has a zero with multiplicity 2.

In 47–49, *multiple choice.*

47. If $xyz = 0$, then which of the following is true?
  (a) $x = 0$
  (b) $x = 0$ or $y = 0$ or $z = 0$
  (c) $x = 0$ and $y = 0$ and $z = 0$
  (d) none of these

48. Suppose $p(x)$ is a polynomial, $p(r) = 0$, $p(s) = 0$, and $p(t) = 7$. Which of the following is *not* true?
  (a) $p(r) \cdot p(s) = 0$
  (b) $k(x - r)(x - s)(x - t) = p(x)$
  (c) r and s are x-intercepts of the graph of $p(x)$.
  (d) r and s are roots of the equation $p(x) = 0$.

49. Suppose $x - r$ and $x - s$ are factors of a quadratic polynomial $p(x)$. Which of the following is *not* true for all x?
  (a) $p(r) = 0$ (b) $k(x - r)(x - s) = p(x)$
  (c) $p(s) = 0$ (d) $(x - r)(x - s) = 0$

Objective G: *Apply the Rational-Zero Theorem.* (Lesson 11-7)

50. *True or false.* $P(x) = 3x^2 - 5x + 2$ could have a rational zero at $-\frac{3}{2}$.

51. **a.** List all possible rational zeros of $R(x) = 3x^2 - 5x - 2$.
  b. Find the rational zeros of $R(x)$.

52. **a.** $U(x) = x^3 - 2x^2 - x + 2$ has one rational zero at -1. Name the other two rational zeros.
  b. Factor $U(x) = x^3 - 2x^2 - x + 2$.

53. Use the Rational-Zero Theorem to factor $M(z) = z^6 - 14z^4 + 49z^2 - 36$.

USES DEAL WITH APPLICATIONS OF MATHEMATICS IN REAL SITUATIONS.

Objective H: *Use polynomials to model real-world situations.*   *(Lessons 11-1, 11-9, 11-10)*

54. Consider $100x^3 + 200x^2 + 300x + 400$.

   a. Make up a question involving money that could be answered by using this expression.

   b. Answer your question.

55. Each birthday from age 9 on, Charles decided to save $150 of his gifts. He puts the money in a savings account at an interest rate of r, compounded annually, without withdrawing or adding any other money.

   a. Write a polynomial in x, where $x = 1 + r$, that represents the amount of money he would have after his 16th birthday.

   b. If the account pays 6% interest annually, calculate how much money Charles would have after his 16th birthday.

   c. At about what rate of interest would Charles have to invest in order to have $2000 on his 16th birthday?

In 56 and 57, suppose that a manufacturer determines that n employees on a certain production line will produce $f(n)$ units per month, where $f(n) = 80n^2 - 0.1n^4$.

56. How many units will be produced monthly by

   a. 3 employees?   **b.** 10 employees?

57. Sketch a graph of f, and determine a reasonable domain for f in this model.

58. Recall that when a beam of light in air strikes the surface of water it is refracted or bent (see page 643). At the right are the earliest known data on the relation between i, the angle of incidence in degrees, and r, the angle of refraction in degrees. The data are recorded in the Optics of Ptolemy, a Greek scientist who lived in the 2nd century A.D.

i	r
10	8
20	15.5
30	22.5
40	29
50	35
60	40.5
70	45.5
80	50

   a. Can these data be modeled by a polynomial function?

   b. If so, what is the degree of the function? If not, explain why a polynomial function is not a good model.

59. The total number of games G needed for n chess players to play each other twice (once with black pieces, once with white pieces) is given by the following table.

n	2	3	4	5	6	. . .
G	2	6	12	20	30	. . .

   a. Find a polynomial formula relating n and G.

   b. Use the formula to find G when $n = 20$.

Objective I: *Use polynomials to describe geometric situations.*   *(Lesson 11-2)*

60. Refer to the rectangle below.

   a. What are the dimensions of the rectangle?

   b. What is the area of the rectangle?

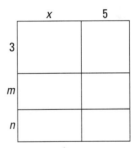

In 61–63, from a sheet of paper measuring 11×17, squares of side x are removed from each corner, and an open box is formed.

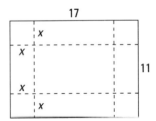

61. Write an expression for the area of the bottom of the box.

62. Write a volume formula $V(x)$ for the box.

63. Write a formula $S(x)$ for the surface area of the box.

In 64 and 65, consider that a worker cuts a square out of each corner of a piece of sheet metal which measures 1 m × 1.5 m. If the length of the side of the square is x meters long, find a polynomial for each quantity.

64. the volume, $V(x)$, of the box when folded

65. $S(x)$, the surface area of the open box

66. A right circular cone has slant height $s = 17$.

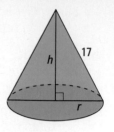

  a. Express its radius r in terms of its altitude h.

  b. Use the result of part **a** to express the volume $V(h)$ of this cone as a polynomial function in h.

REPRESENTATIONS DEAL WITH PICTURES, GRAPHS, OR OBJECTS THAT ILLUSTRATE CONCEPTS.

Objective J: *Graph polynomial functions.*
(Lessons 11-1, 11-5)

In 67 and 68, **a.** graph the function; **b.** use the graph to factor the polynomial.

67. $f(x) = 2x^3 - x^2 - 18x + 9$

68. $g(x) = 3x^3 - 7x^2 - 20x$

69. A polynomial function f of degree 3 has zeros at -5, 3, and -2.

  a. Find an equation for one function satisfying these conditions.

  b. Graph the function in part **a**.

  c. Write the general form of an equation for f.

  d. What do the graphs of all functions in part **c** have in common with the graph in part **b**?

70. A polynomial function g with degree 3 has zeros at $-\frac{10}{3}$, 0, and $\frac{13}{4}$.

  a. Express an equation for g in factored form.

  b. Suppose the leading coefficient of $g(x)$ is 12. Find an equation for g.

  c. Graph the equation in part **b**.

Objective K: *Estimate zeros of functions of polynomials using tables or graphs.*
(Lessons 11-4, 11-5, 11-9)

In 71 and 72, estimate the real zeros of the function with the given equation to the nearest tenth.

71. $f(x) = -9x^3 + 5x^2 - 7$ **72.** $y = x^4 + 3x^3 - 20$

In 73 and 74, the graph of $y = P(x)$ at the right contains (-3, 0), (0, 24), (2, 0), and (3, 0).

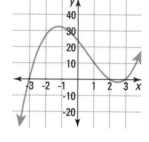

73. *True or false.*
  $P(2) = P(-2)$.

74. Suppose
  $P(x) = a_3x^3 + a_2x^2 + a_1x + a_0$.

  a. What is the value of a_0?

  b. Find the values of a_1, a_2, and a_3.

  c. Give the value of $P(5)$.

75. Use the table at the right for values of the function J with $J(x) = -x^3 - x^2 + 7x + 18$.

x	$J(x)$
-10	848
-8	410
-6	156
-4	38
-2	8
0	18
2	20
4	-34
6	-192
8	-502
10	-1012

  a. According to the table, what is the least number of x-intercepts this polynomial function has?

  b. According to the Fundamental Theorem of Algebra, what is the maximum number of real zeros this function has?

  c. Use technology to estimate each x-intercept to the nearest tenth.

  d. Which, if any, of the x-intercepts are rational? How can you tell?

76. The polynomial function with equation $y = f(x)$ graphed below has integer zeros.

   a. What are the zeros?

   b. What does your answer to part **a** imply about the degree of f?

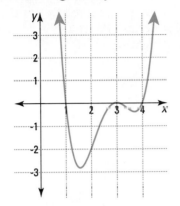

77. Refer to the graph of the function below. Name two pairs of consecutive integers between which a zero of f must occur.

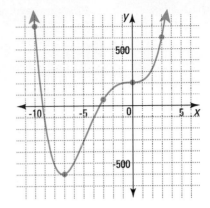

CULTURE DEALS WITH THE PEOPLES AND THE HISTORY RELATED TO THE DEVELOPMENT OF MATHEMATICAL IDEAS.

Objective L: *Be familiar with the history of the solving of polynomial equations.* *(Lesson 11-8)*

78. Identify three countries in which there were mathematicians who developed methods for solving polynomial equations of degree 3 or higher.

79. It is now known that there does not exist a formula for the solving of all polynomial equations. In what century was this discovered?
   (a) 12th   (b) 16th   (c) 18th   (d) 19th

80. Why is the Fundamental Theorem of Algebra so significant?

CHAPTER

12

QUADRATIC RELATIONS

A quadratic equation in two variables x and y is an equation of the form

$$Ax^2 + Bxy + Cy^2 + Dx + Ey + F = 0,$$

where A, B, C, D, E, and F are real numbers, and at least one of A, B, or C is not zero. The set of ordered pairs that satisfy a sentence in the above form, with the equal sign ($=$) or one of the inequality symbols ($>$, $<$, $\geq$, $\leq$) is called a **quadratic relation in two variables.**

Quadratic relations have connections with a wide variety of ideas you already know. They include the parabolas you studied in Chapters 2 and 6 and the hyperbolas you saw in Chapter 2. Quadratic relations also describe circles; the orbits of comets, satellites, and planets; and the shapes of communication receivers and mirrors used in car headlights.

Quadratic relations may also be defined geometrically, as the intersection of a plane and a *double cone.* Such cross-sections of a double cone are usually called *conic sections,* or simply *conics.* They include hyperbolas, parabolas, and ellipses. In this description, circles are special cases of ellipses.

plane intersecting both cones

hyperbola

plane // to edge of cone

parabola

plane intersecting one cone not // to edge

ellipse

In this chapter you will study quadratic relations both algebraically and geometrically, that is, as equations or inequalities and as figures with certain properties. You will also learn how to solve systems of quadratic equations.

12-1

Parabolas

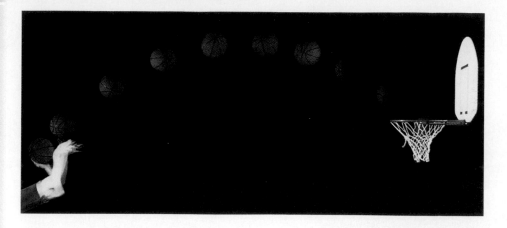

What Is a Parabola?

In Chapter 6, you saw that the path of a tossed object follows a path called a *parabola*. For instance, the path of a basketball shot is part of a parabola from the time it leaves the shooter's hands until it hits some other object.

In order to determine whether a curve is or is not a parabola, a definition of parabola is necessary. Parabolas can be defined geometrically.

> **Definition**
> Let ℓ be a line and F be a point not on ℓ. A **parabola** is the set of every point in the plane of ℓ and F whose distance from F equals its distance from ℓ.

F is the **focus** and ℓ is the **directrix** of the parabola. Thus a parabola is the set of points in a plane equidistant from its focus and its directrix.

Neither the focus nor the directrix is on the parabola. At the right is a sketch of a parabola. The four points V, P_1, P_2, and P_3 on the parabola are identified. Note that each is equidistant from the focus F and the directrix ℓ.

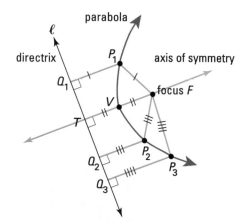

To understand the definition, you must remember that the distance from a point P to a line ℓ is the length of the perpendicular from P to ℓ. In the sketch, $P_1Q_1 \perp \ell$ and $P_1Q_1 = P_1F$. Also, $P_2Q_2 \perp \ell$ and $P_2Q_2 = P_2F$, and so on. The line through the focus perpendicular to the directrix is called the **axis of symmetry**. The point V on the axis of symmetry is the **vertex** of the parabola.

Drawing a Parabola

Example 1

Find five points on the parabola with focus *F* and directrix *m*.

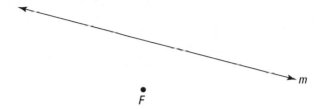

Solution

Use a ruler and compass to find five points *V*, P_1, P_2, P_3, and P_4 as shown.

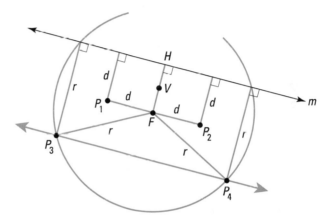

1. First find the vertex. The vertex *V* is the midpoint of the perpendicular segment $\overline{FH}$ from *F* to *m*.
2. Now find points P_1 and P_2 on the line through *F* parallel to *m*. P_1 and P_2 are vertices of squares with $\overline{FH}$, as the common side. So they are the same distance *d* from *F* as from *m*.
3. To find two other points, first draw a circle with center *F* and any radius $r > \frac{d}{2}$. Then draw a line that is parallel to *m* and a distance *r* from *m*. The intersection of that line with the circle gives two points, P_3 and P_4, on the parabola. To find other points, use a circle with a different radius $r > \frac{d}{2}$.

Activity

Trace line *m* and point *F* above. Find four other points on the parabola with focus *F* and directrix *m*. Connect those points and the points P_3, P_1, *V*, P_2, and P_4 to sketch part of a parabola.

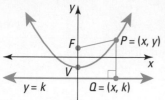

$y = k$ $Q = (x, k)$

Equations for Parabolas

Suppose you know the coordinates of the focus and an equation for the directrix of a parabola. You can find an equation for the parabola by using the definition of parabola and the distance formula.

Examine the diagram at the left in which the horizontal line $y = k$ is the directrix and point F on the y-axis is the focus. Let $P = (x, y)$ be a point on the parabola and $Q = (x, k)$. Then $\overline{PQ}$ is a vertical segment, so it is perpendicular to the directrix $y = k$. So, by the definition of a parabola, $PF = PQ$. When this equation is expressed using the distance formula, an equation in x and y results. This is shown in Example 2.

Example 2

Find an equation for the parabola with focus (0, 5) and directrix $y = -5$.

Solution

Sketch the given information. Let $P = (x, y)$ be any point on the parabola. If $Q = (x, -5)$, then we must have $PF = PQ$.

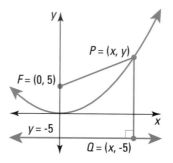

$PF = PQ$	Definition of parabola
$\sqrt{(x - 0)^2 + (y - 5)^2} = \sqrt{(x - x)^2 + (y - -5)^2}$	Distance Formula
$x^2 + (y - 5)^2 = (y + 5)^2$	Square both sides.
$x^2 + y^2 - 10y + 25 = y^2 + 10y + 25$	Expand.
$x^2 - 10y = 10y$	Add $-y^2 - 25$ to both sides.
$x^2 = 20y$	Add $10y$ to both sides.
$y = \frac{1}{20}x^2$	Solve for y.

An equation for the parabola is $y = \frac{1}{20}x^2$.

Check

Pick a point on $y = \frac{1}{20}x^2$. We use the point $A = (30, 45)$. Now show that A is equidistant from (0, 5) and $y = -5$.

$$AF = \sqrt{(30 - 0)^2 + (45 - 5)^2} = \sqrt{30^2 + 40^2} = \sqrt{2500} = 50.$$

The distance from A to $y = -5$ is the distance from (30, 45) to (30, -5), which is also 50. So A is on the parabola with focus (0, 5) and directrix $y = -5$.

In Example 2, if you were to replace $(0, 5)$ by $\left(0, \frac{1}{4}\right)$ and $y = -5$ by $y = -\frac{1}{4}$, the equation for the parabola would be $y = x^2$. If $(0, 5)$ is replaced by $\left(0, \frac{1}{4a}\right)$ and $y = -5$ is replaced by $y = -\frac{1}{4a}$, then the parabola has equation $y = ax^2$. The derivation of both of these equations uses the same steps as Example 2 and proves the following theorem.

> **Theorem**
> The graph of $y = ax^2$ is the parabola with focus $\left(0, \frac{1}{4a}\right)$ and directrix $y = -\frac{1}{4a}$.

Because the image of the graph of $y = ax^2$ under the translation $(x, y) \rightarrow (x + h, y + k)$ is the graph with equation $y - k = a(x - h)^2$, the graph of any quadratic equation of the form $y = a(x - h)^2 + k$ or $y = ax^2 + bx + c$ is also a parabola.

When $a < 0$, you have learned that the parabola opens down. In this case, the directrix is above the x-axis, and the focus is below the x-axis.

Paraboloids

If a parabola is rotated in space around its line of symmetry, the three-dimensional figure it creates is called a **paraboloid.** The focus of a paraboloid is the focus of the rotated parabola. Two examples are the paths of the jets of a fountain and a satellite receiving dish.

This fountain consists of many whirling sprinkler heads.

These parabolic dish antennas receive and transmit signals to and from satellites.

Covering the Reading

1. Define *quadratic equation in the two variables x and y*.

2. A *conic section* is the intersection of a __?__ and a __?__.

3. Name three types of conic sections.

4. Define *parabola*.

5. Show the results you obtained from the Activity in this lesson.

6. Trace the figure at the left and draw five points on the parabola with focus F and directrix ℓ.

In 7–9, refer to the parabola at the left with focus F and directrix ℓ. P_1, P_2, P_3, and P_4 are points on the parabola.

7. *True or false.*
   a. $P_1F = FG_2$ **b.** $FP_4 = G_4P_4$

8. Name the vertex of the parabola.

9. Is the focus on the parabola?

10. **a.** Graph the parabola with equation $y = \frac{1}{20}x^2$ in Example 2.
    b. Name its focus, vertex, and directrix.
    c. Verify that the point (2, 0.2) is equidistant from the focus and directrix, and, therefore, a point on the parabola $y = \frac{1}{20}x^2$.
    d. Find another point on the graph of $y = \frac{1}{20}x^2$. Show that it is equidistant from the focus and directrix.

11. Verify that the graph of $y = x^2$ is a parabola with focus at $\left(0, \frac{1}{4}\right)$ and directrix $y = -\frac{1}{4}$ by choosing a point on the graph and showing that two appropriate distances are equal.

12. Given $F = (0, 2)$ and line ℓ with equation $y = -2$, what is an equation for the set of points equidistant from F and ℓ?

13. **a.** What is a *paraboloid*?   **b.** Give an example of a paraboloid.

Applying the Mathematics

In 14 and 15, use the information given in Question 11.

14. What are the focus and the directrix of $y - 4 = (x - 5)^2$?

15. Give the focus and the directrix of the parabola with equation $y = -x^2$.

16. Prove the theorem of this lesson.

In 17 and 18, an equation for a parabola is given. **a.** Tell whether the parabola opens up or down. **b.** Give the focus of the parabola.

17. $y = -4x^2$ **18.** $y = \frac{1}{4}x^2$

19. The drawing below at the left shows a geometric construction described around the year 1000 A.D. by Arab mathematician Ibrahim ibn Sina. With this method, only a straightedge and compass are used to construct points on a parabola. The drawing below at the right shows a parabola constructed using ibn Sina's method. Which equation describes this parabola?

(a) $y = 2x^2$ (b) $y = \frac{1}{2}x^2$ (c) $y = 4x^2$ (d) $y = \frac{1}{4}x^2$

Shown is Ibrahim ibn Sina (980–1037), also known as Avicenna. In addition to being a mathematician, he was a physician, philosopher, astronomer, and poet.

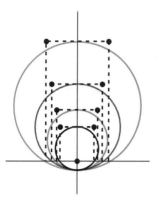

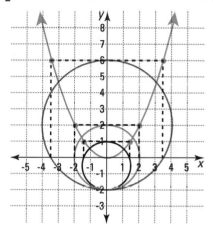

Review

20. An isotope has a half-life of 94 seconds. How long will it take 50 mg of this isotope to decay to 5 mg? *(Lesson 9-2)*

21. *Skill sequence.* Solve for y. *(Lessons 6-2, 6-6, 6-10, 8-8)*
     a. $y^2 = 7$ **b.** $(y + 2)^2 = 7$
     c. $\sqrt{y} = 7$ **d.** $\sqrt{y + 2} = 7$

22. Determine whether the set of pairs (x, y) satisfying the given equation is a function. Justify your response.
(Lessons 1-2, 1-4, 2-5, 6-4, 8-2)
     a. $y = x^2$ **b.** $x = y^2$

23. Simplify. *(Lessons 7-7, 7-8)*
     a. $\left(\frac{9}{25}\right)^{\frac{1}{2}}$ **b.** $(.001)^{-\frac{2}{3}}$

24. Name 4 points with a distance of 30 from $(10, 25)$. *(Previous course)*

Exploration

25. Consider the method of ibn Sina for drawing parabolas, as shown in Question 19 above. Write an explanation of the method and use it to draw a different parabola.

Circles occur in many situations. You have probably noticed that when you throw a pebble into a calm body of water, *concentric* circles soon form around the point where the pebble hit the water. Similarly, when an earthquake occurs, various kinds of *seismic waves* radiate in concentric circles from the *epicenter,* the point on the earth's surface above the point where the earthquake began. These waves are recorded by an instrument called a *seismograph.*

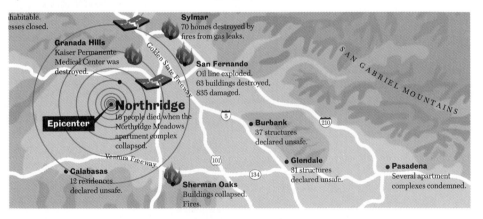

Seismic waves travel at the same speed in all directions. The fastest seismic waves, called *compressional waves,* travel at a speed of about 8 kilometers (5 miles) a second. So the points reached at any given instant of time by compressional waves from a particular earthquake will lie on a circle whose center is the epicenter of the quake. This follows from the definition of a circle.

Definition
A **circle** is the set of all points in a plane at a given distance (its **radius**) from a fixed point (its **center**).

Equations for Circles

From its definition and the Distance Formula, you can find an equation for any circle.

Example 1

Let the unit of a graph be in kilometers. Place the epicenter of a quake at the origin. Find an equation for the set of points that will be reached by compressional waves in 10 seconds.

Solution

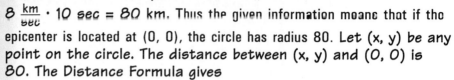

In 10 seconds, a compressional wave travels about

$8 \frac{km}{sec} \cdot 10 \ sec = 80 \ km$. Thus the given information means that if the epicenter is located at (0, 0), the circle has radius 80. Let (x, y) be any point on the circle. The distance between (x, y) and (0, 0) is 80. The Distance Formula gives

$$\sqrt{(x - 0)^2 + (y - 0)^2} = 80.$$

Square both sides.

$$(x - 0)^2 + (y - 0)^2 = 80^2$$

That is,

$$x^2 + y^2 = 6400.$$

Notice that the equation determined in Example 1 is a quadratic equation in x and y. This is also true of circles not centered at the origin.

Example 2

Find an equation for the circle with radius 130 and center (150, 100).

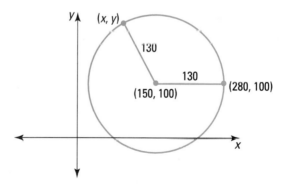

Solution

Let (x, y) be any point on the circle. By the Distance Formula,

$$\sqrt{(x - 150)^2 + (y - 100)^2} = 130.$$

Square both sides.

$$(x - 150)^2 + (y - 100)^2 = 16{,}900$$

Check

The point (280, 100) is 130 units from (150, 100) and its coordinates should satisfy the equation.

Does $\quad (280 - 150)^2 + (100 - 100)^2 = 16{,}900$?

Does $\qquad\qquad\qquad 130^2 + 0^2 = 16{,}900$? Yes

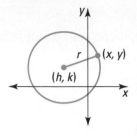

Example 2 can be generalized to determine an equation for *any* circle. Let (h, k) be the center of a circle with radius r, and let (x, y) be any point on the circle. Then by the definition of a circle, the distance between (x, y) and (h, k) equals r.

By the Distance Formula,

$$\sqrt{(x - h)^2 + (y - k)^2} = r.$$

Squaring gives

$$(x - h)^2 + (y - k)^2 = r^2.$$

This argument proves the following theorem.

> **Theorem (Center-Radius Equation for a Circle)**
> The circle with center (h, k) and radius r is the set of points (x, y) that satisfy
> $$(x - h)^2 + (y - k)^2 = r^2.$$

When the center of a circle is the origin, $(h, k) = (0, 0)$. The equation

$$(x - h)^2 + (y - k)^2 = r^2$$

becomes

$$(x - 0)^2 + (y - 0)^2 = r^2$$

or

$$x^2 + y^2 = r^2.$$

This proves the following special case of the Center-Radius Equation for a Circle.

> **Corollary**
> The circle with center at the origin and radius r is the set of points (x, y) that satisfy the equation $x^2 + y^2 = r^2$.

Example 3

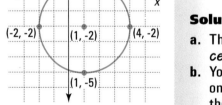

a. Find the center and the radius of the circle with equation
$(x - 1)^2 + (y + 2)^2 = 9$.
b. Graph this circle.

Solution

a. The equation is in the center-radius form for a circle, with
center $= (h, k) = (1, -2)$ and radius $r = 3$.
b. You can sketch this circle by locating the center and then four points on the circle whose distance from the center is 3, as illustrated at the left.

If you know an equation for a circle and one coordinate of a point on the circle, you can determine the other coordinate of that point.

Example 4

Refer to the circle in Example 3. There are two points at which $x = 3$. Find the y-coordinate of each point.

▶

Solution

Substitute $x = 3$ into the equation for the circle, and solve for y.

$$(3 - 1)^2 + (y + 2)^2 = 9$$
$$4 + (y + 2)^2 = 9$$
$$(y + 2)^2 = 5$$
$$y + 2 = \pm\sqrt{5}$$
$$y = -2 \pm \sqrt{5}$$

So $y \approx 0.236$ or $y \approx -4.236$.

Check

Refer to the graph on the previous page. (3, 0.236) and (3, -4.236) seem to be on the circle.

QUESTIONS

Covering the Reading

1. What is the *epicenter* of an earthquake?

In 2 and 3, consider the circle with equation $x^2 + y^2 = 60^2$.

2. What is the radius of this circle?

3. Tell whether the point is on the circle.
   a. (0, 0)   **b.** (0, 60)   **c.** (-60, 0)   **d.** (30, 30)

4. The circle with equation $(x - h)^2 + (y - k)^2 = r^2$ has center ___?___ and radius ___?___.

In 5 and 6, an equation for a circle is given. **a.** Give its center. **b.** Give its radius. **c.** Sketch it.

5. $x^2 + y^2 = 100$ 6. $(x - 5)^2 + (y - 1)^2 = 36$

7. Consider the circle of Question 5. Find the y-coordinates of all points where $x = -6$.

8. Find an equation for the circle with center (0, 0) and radius 9.

9. Find an equation for the circle with center (-3, 2) and radius 8.

Shop till you drop.
Shown is an architect's rendering of the Grand Court Area in the Woodfield Shopping Center in Schaumburg, Illinois. In 1995, Woodfield surpassed The Mall of America as the largest U.S. mall in retail square feet.

Applying the Mathematics

In 10 and 11, consider a map in which your school is at the origin, and the unit of the map is miles. Suppose Bill lives 2 miles west and 1 mile north of the school.

10. If Annie lives $\frac{1}{2}$ mile from Bill, give an equation for the circle on which her residence must lie.

11. Suppose a shopping mall is known to be 3 miles from your school, and 5 miles away from Bill's house. Is this enough information to locate the mall? Why or why not?

12. A circle has center at the origin and radius $\sqrt{10}$.
  a. Find an equation for this circle.
  b. Identify eight points with integer coordinates on the circle.
  c. Graph the circle.

13. a. Expand the binomials in the equation for the circle of Example 2. Then simplify to get an equation of the form
  $Ax^2 + Bxy + Cy^2 + Dx + Ey + F = 0$.
  b. What are the values of A, B, C, D, E, and F?

Review

14. A parabola has focus $(0, 1)$ and directrix $y = -1$.
  a. What is its vertex?
  b. Tell whether the vertex is a maximum or a minimum.
  c. Give an equation for the parabola.
  d. Give an equation for its axis of symmetry.   *(Lessons 2-5, 6-4, 12-1)*

15. Consider the parabola with directrix $y = 2x + 1$ and vertex $(3, 0)$.
  a. Is this parabola the graph of a relation? Why or why not?
  b. Is this parabola the graph of a function? Why or why not?
  (Lessons 1-4, 12-1)

16. Give the focus and directrix of the parabola with equation $y = 3x^2$.
  (Lesson 12-1)

17. What are the zeros of the polynomial function graphed at the left?
  (Lesson 11-4)

18. *Skill sequence.* Expand and simplify.   *(Lessons 6-1, 11-2)*
  a. $(x + 3)^2 + y^2$ **b.** $(x + y)^2 + 3^2$ **c.** $(x + y + 3)^2$

19. In 1985 an investor deposited $6000 in a retirement account paying interest compounded quarterly. No additional deposits or withdrawals were made. In 1995 the account was valued at $10,900. What was the annual rate of interest during this period?
  (Lessons 7-4, 7-6, 7-7)

20. *Skill sequence.*   Solve for y.   *(Lessons 1-6, 6-2)*
  a. $y^2 = 100$ **b.** $25 + y^2 = 100$ **c.** $x^2 + y^2 = 100$

21. *Multiple choice.*   If x is a real number, then $\sqrt{x^2} = \underline{\ ?\ }$.
  (a) x (b) $-x$ (c) $|x|$ (d) none of these   *(Lesson 6-2)*

22. Find an equation for the line containing the origin and $(5, -4)$.
  (Lesson 3-5)

Exploration

23. A **lattice point** is a point with integer coordinates. If possible, find an equation for a circle that passes through
  a. no lattice points.   **b.** exactly one lattice point.
  c. exactly two lattice points.   **d.** exactly three lattice points.
  e. more than ten lattice points.

LESSON

12-3

Semicircles, Interiors, and Exteriors of Circles

Roman influence. *Arch bridges, like this one in Europe, were common during the Roman Empire. The Romans built semicircular forms, wedged stones in place from each end, and worked toward the top. See Example 2.*

Semicircles

Many vertical lines intersect a circle in two points. Thus, the Vertical-line Test shows that a circle is a relation but not a function. For this reason, many automatic graphers cannot graph a circle directly. With these graphers, you need to think of the circle as the union of two *semicircles,* each of which is the graph of a function.

Example 1

Graph the circle with equation $(x - 1)^2 + (y + 2)^2 = 9$ using an automatic grapher.

Solution

Solve the equation for y. $(y + 2)^2 = 9 - (x - 1)^2$

Take the square root of each side. One way to write this is as
$$|y + 2| = \sqrt{9 - (x - 1)^2}.$$

By the definition of absolute value,
$$y + 2 = \pm \sqrt{9 - (x - 1)^2}.$$

So either   (1) $y = -2 + \sqrt{9 - (x - 1)^2}$

or   (2) $y = -2 - \sqrt{9 - (x - 1)^2}.$

Since (1) and (2) are equations of functions, they can be graphed. Graph them on the same window. (You may need to call the first equation $y_1 = \ldots$ and the second $y_2 = \ldots$) Make certain that the window allows you to see the entire circle.

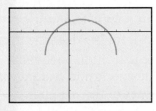

-5 ≤ x ≤ 7, x-scale = 1
-6 ≤ y ≤ 2, y-scale = 1

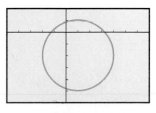

-5 ≤ x ≤ 7, x-scale = 1
-6 ≤ y ≤ 2, y-scale = 1

-5 ≤ x ≤ 7, x-scale = 1
-6 ≤ y ≤ 2, y-scale = 1

Lesson 12-3 *Semicircles, Interiors, and Exteriors of Circles*   **759**

Semicircles occur often in architecture.

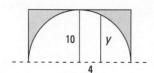

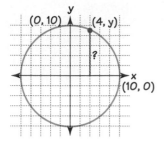

Example 2

A semicircular arch over a street has radius 10 feet. How high is the arch at a point whose ground distance is 4 feet from the center?

Solution

Superimpose a coordinate system. Suppose the x-axis represents the street with the origin as the center of the arch. Then the circle determined by the arch has center (0, 0) and radius 10, so its equation is $x^2 + y^2 = 10^2$. The height of the circle 4 feet from the center equals the y-coordinate of the point on the graph where x = 4. Now substitute 4 for x in the equation and solve for y.

$$16 + y^2 = 100$$
$$y^2 = 84$$
$$y = \sqrt{84} \text{ or } -\sqrt{84}.$$

We must reject $-\sqrt{84}$ because y, the height above the ground, cannot be negative. Thus,

$$y = \sqrt{84} \approx 9.17.$$

The bridge is about 9.17 ft (about 9 ft 2 in.) high at a point whose ground distance is 4 feet from its center.

Check 1

Examine the graph. This value looks about right.

Check 2

Substitute (4, 9.17) into $x^2 + y^2 = 100$. Is $4^2 + (9.17)^2 \approx 100$? Yes.

Interiors and Exteriors of Circles

Every circle separates the plane into three regions. The region inside the circle is called the **interior** of the circle. The region outside the circle is called the **exterior** of the circle. The circle itself is the **boundary** between these two regions.

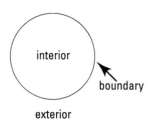

Regions bounded by concentric circles are often used in target practice. Consider the target shown at the left.

To describe the colored regions mathematically, we place the target on a coordinate system with the center at (0, 0). Note that the ▓▓ region worth 50 points is the interior of the circle with radius 3. All points in this region are less than 3 units from the origin. Thus if (x, y) is a point in the ▓▓ region, then from the distance formula you can conclude that $\sqrt{x^2 + y^2} < 3$. Notice that in the inequality, the expressions on both sides are positive. Whenever a and b are positive and a < b, then $a^2 < b^2$. Thus, when both sides of the inequality $\sqrt{x^2 + y^2} < 3$ are squared, the sentence becomes $x^2 + y^2 < 9$. So $x^2 + y^2 < 9$ also describes the region worth 50 points.

Similarly, the region worth less than 50 points is the exterior of the circle with radius 3. All (x, y) in this region satisfy the sentence $\sqrt{x^2 + y^2} > 3$, or $x^2 + y^2 > 9$.

The two instances above are generalized in the following theorem.

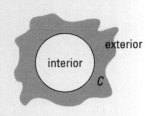

Theorem (Interior and Exterior of a Circle)
Let C be the circle with center (h, k) and radius r.
Then the interior of C is described by
$$(x - h)^2 + (y - k)^2 < r^2.$$
The exterior of C is described by
$$(x - h)^2 + (y - k)^2 > r^2.$$

If $\geq$ or $\leq$ is used in the theorem above, the circle is included.

Example 3

Graph the points satisfying
$(x - 3)^2 + (y + 5)^2 \geq 16.$

Solution

The sentence represents the union of a circle with center at $(3, -5)$ and radius 4 and its exterior. The shaded region and the circle at the left make up the graph.

Example 4

Consider the target on the previous page. Write a sentence to describe all points in the region worth 30 points, the ▮▮▮ region.

Solution

The ▮▮▮ region is the intersection of the interior of the circle with radius 9 and the exterior of the circle with radius 6.
The interior of the circle with radius 9 is the set $\{(x, y): x^2 + y^2 < 81\}$, and the exterior of the circle with radius 6 is the set $\{(x, y): x^2 + y^2 > 36\}$. So $\{(x, y): 36 < x^2 + y^2 < 81\}$ describes the 30-point region.

QUESTIONS

Covering the Reading

1. Sketch by hand on separate axes.
   a. $x^2 + y^2 = 49$ **b.** $y = \sqrt{49 - x^2}$ **c.** $y = -\sqrt{49 - x^2}$

2. Refer to Example 2. How high is the arch at a point whose ground distance is 2 feet from the center?

3. What is the region outside a circle called?

4. *Multiple choice.* On the target shown in the lesson, all (x, y) in the region worth 50 points lie
(a) in the interior of the circle with radius 3.
(b) on the circle with radius 3.
(c) in the exterior of the circle with radius 3.

5. Write a sentence to describe the set of points (x, y) in the 40-point region of the target.

6. Graph all points (x, y) satisfying $(x - 3)^2 + (y + 5)^2 < 16$.

In 7 and 8, *multiple choice.* Given a circle with center (h, k) and radius *r,* state which of the following the given sentence describes.
(a) the interior of the circle   (b) the exterior of the circle
(c) the union of the circle   (d) the union of the circle
     and its interior   and its exterior

7. $(x - h)^2 + (y - k)^2 \geq r^2$ **8.** $(x - h)^2 + (y - k)^2 < r^2$

Applying the Mathematics

In 9 and 10, use an automatic grapher to graph.

9. $x^2 + y^2 = 8$ **10.** $(x + 3)^2 + (y + 4)^2 = 25$

11. The BASIC program at the left printed the output at the right.

```
10 INPUT "RADIUS"; R
20 PRINT "X", "Y1", "Y2"
30 FOR X = -R TO R STEP 0.5
40 Y1 = SQR(R^2 - X^2)
50 Y2 = -1*SQR(R^2 - X^2)
60 PRINT X, Y1, Y2
70 NEXT X
80 END
```

X	Y1	Y2
-2	0	0
-1.5	1.322876	-1.322876
-1	1.732051	-1.732051
-0.5	1.936492	-1.936492
0	2	-2
0.5	1.936492	-1.936492
1	1.732051	-1.732051
1.5	1.322876	-1.322876
2	0	0

a. Plot the points (x, y_1) and (x, y_2) given in the output.
b. Find the value of R input by the user.
c. Find an equation for the circle containing the points graphed in part **a.**
d. What lines in the program generate the points on the semicircle above the x-axis?

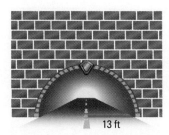

13 ft

12. A moving van 6 ft wide and 12 ft high is approaching a semicircular tunnel with radius 13 ft.
a. Explain why the truck cannot pass through the tunnel if it stays in its lane.
b. Can the truck fit through the tunnel if it is allowed to drive anywhere on the roadway? Justify your answer.

13. In sumo wrestling, the participants wrestle in the interior of a circle. A wrestler wins by pushing his opponent out of the circle. Suppose the circle has radius r and center $(0, 0)$.
a. What sentence describes positions that are "in bounds"?
b. What sentence describes losing positions?

In 14 and 15, define the term. *(Lessons 12-1, 12-2)*

14. circle

15. parabola

16. a. Find an equation for the circle with center at (0, 0) and radius 1.
(Lesson 12-2)
   b. What is this circle called? *(Lesson 10-4)*

17. A circle with center at the origin passes through the point (3, -4).
   a. Find the radius of the circle.
   b. Find an equation for the circle. *(Lesson 12-2)*

18. Identify the focus and the directrix of the parabola $y - \frac{1}{1000}x^2$.
(Lesson 12-1)

19. Expand and simplify $(x - \sqrt{x^2 + y^2})^2$. *(Lessons 6-1, 8-5)*

In 20 and 21, recall that in football a touchdown is worth 6 points; a field goal, 3 points; a safety, 2 points; and a kicked point-after-touchdown, 1 point.

20. If a team gets T touchdowns, F field goals, S safeties, and P points-after-touchdowns by kicking, how many points does it have in all?
(Lesson 3-3)

21. A team had no safeties and no points-after-touchdowns and scored at most 27 points. Also, the team did not run for any 2 points-after-touchdown. Graph the set of possible ways this could have happened. *(Lesson 3-9)*

22. Can you draw a circle with a ruler? "Of course not," you may think. "A circle is round and a ruler is straight."
   a. Try this. Mark a point P on a sheet of plain paper. Take a ruler and put one edge so that it goes through the point. Then draw a line along the other edge.
   b. Repeat this twice using the same point and the same ruler. Your drawing may look like the center diagram at the left.
   c. Draw more lines in the same way. You will begin to see something like the picture at the bottommost left.
   d. The lines are tangent to the same circle and are said to form an *envelope* for the circle. No line drawn as suggested above will intersect the interior of this circle. Where is the center of the circle you've formed? What is its radius?
   e. If you were to repeat this process using a ruler of a different width, how would the outcome be affected?

IN·CLASS
ACTIVITY

Graph paper consisting of two intersecting sets of concentric circles makes it easy to draw some conic sections. Such graph paper is sometimes called *conic graph paper*. In the conic graph paper below, the centers of the two sets of circles are 12 units apart. These centers become the *foci* F_1 and F_2 of an ellipse.

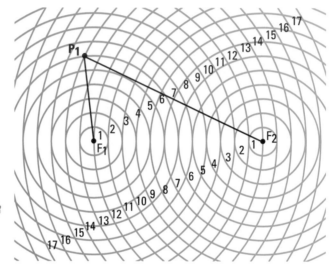

1 On a sheet of conic graph paper, draw F_1, F_2, P_1, $\overline{F_1P_1}$, and $\overline{F_2P_1}$, as shown above.

2 **a.** What is the distance from P_1 to F_1?
b. What is the distance from P_1 to F_2?
c. What is the distance $P_1F_1 + P_1F_2$?
d. Find and mark another point P_2 such that $P_1F_1 = P_2F_1$ and $P_1F_2 = P_2F_2$.

In 3 and 4, mark all points P satisfying the given condition.

3 $PF_1 = 15$ and $PF_2 = 5$.

4 $PF_2 = 16$ and $PF_1 = 4$.

5 Plot 10 other points such that $PF_1 + PF_2 = 20$.

6 **a.** Connect the dots to form an ellipse.
b. Draw the symmetry lines for the ellipse.

LESSON

12-4

Ellipses

The shape of things around the capital. *Shown is an aerial view of Washington, D.C. An elliptical park, known as the Ellipse, lies between the White House and the Washington Monument.*

What Is an Ellipse?

Recall that a parabola is determined by a point (its focus) and a line (its directrix). An *ellipse* is determined by two points, its foci (pronounced "foe sigh," plural of focus), and a number, the *focal constant*. On page 764, the points F_1 and F_2 are the foci. The *focal constant* is the constant sum of the distances from any point P to the foci. The focal constant of the ellipse in the activity is 20.

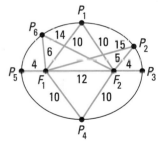

The set of all points P such that $PF_1 + PF_2 = 20$ is an ellipse. The drawing at the right shows the ellipse and six points on it, P_1, P_2, P_3, P_4, P_5, and P_6. You should verify that $P_nF_1 + P_nF_2 = 20$ for each n.

> **Definition**
> Let F_1 and F_2 be any two points in a plane and d be a constant with $d > F_1F_2$. Then the **ellipse with foci F_1 and F_2 and focal constant d** is the set of points P in the plane which satisfy $PF_1 + PF_2 = d$.

For any point P on the ellipse, $PF_1 + PF_2$ has to be greater than F_1F_2 because of the Triangle Inequality. This is why $d > F_1F_2$.

Equations for Some Ellipses

To find an equation for the ellipse in the In-class Activity, consider a coordinate system with $\overleftrightarrow{F_1F_2}$ as the *x*-axis and with the origin midway between the foci on an axis. Then the foci are $F_1 = (-6, 0)$ and $F_2 = (6, 0)$. This is the *standard position* for the ellipse.

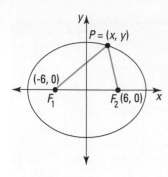

If $P = (x, y)$ is on the ellipse, then because the focal constant is 20,
$$PF_1 + PF_2 = 20.$$

So by the Distance Formula,
$$\sqrt{(x + 6)^2 + (y - 0)^2} + \sqrt{(x - 6)^2 + (y - 0)^2} = 20$$
or
$$\sqrt{(x + 6)^2 + y^2} + \sqrt{(x - 6)^2 + y^2} = 20.$$

This equation for the ellipse is quite complicated. Surprisingly, to find an equation for this ellipse, and all others with their foci on an axis, it is easier to consider a more general case. The resulting equation is well worth the effort it takes to derive it.

Theorem (Equation for an Ellipse)
The ellipse with foci $(c, 0)$ and $(-c, 0)$ and focal constant $2a$ has equation $\dfrac{x^2}{a^2} + \dfrac{y^2}{b^2} = 1$, where $b^2 = a^2 - c^2$.

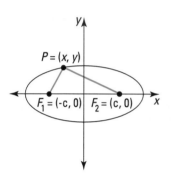

Proof
Let $F_1 = (-c, 0)$, $F_2 = (c, 0)$, and $P = (x, y)$. We number the steps for reference.
1. By the definition of an ellipse,
$$PF_1 + PF_2 = 2a.$$
By the Distance Formula, this becomes
$$\sqrt{(x + c)^2 + y^2} + \sqrt{(x - c)^2 + y^2} = 2a.$$
2. Subtract one of the square roots from both sides.
$$\sqrt{(x - c)^2 + y^2} = 2a - \sqrt{(x + c)^2 + y^2}$$
3. Square both sides (the right side is a binomial).
$$(x - c)^2 + y^2 = 4a^2 - 4a\sqrt{(x + c)^2 + y^2} + (x + c)^2 + y^2$$
4. Expand the binomials and do appropriate subtractions.
$$-2cx = 4a^2 - 4a\sqrt{(x + c)^2 + y^2} + 2cx$$
5. Use the Addition Property of Equality and rearrange terms.
$$4a\sqrt{(x + c)^2 + y^2} = 4a^2 + 4cx$$
6. Multiply both sides by $\frac{1}{4}$.
$$a\sqrt{(x + c)^2 + y^2} = a^2 + cx.$$
7. Square a second time.
$$a^2[(x + c)^2 + y^2] = a^4 + 2a^2cx + c^2x^2$$
8. Expand the binomial and subtract $2a^2cx$ from both sides.
$$a^2x^2 + a^2c^2 + a^2y^2 = a^4 + c^2x^2$$
9. Subtract a^2c^2 and c^2x^2 from both sides. Then factor.
$$(a^2 - c^2)x^2 + a^2y^2 = a^2(a^2 - c^2)$$
10. Since $c > 0$, $F_1F_2 = 2c$, and $2a > F_1F_2$, we have $2a > 2c > 0$. So $a > c > 0$. Thus $a^2 > c^2$, and $a^2 - c^2$ is positive. So $a^2 - c^2$ can be considered as the square of some real number, say b. Now let $a^2 - c^2 = b^2$ and substitute.
$$b^2x^2 + a^2y^2 = a^2b^2$$
11. Divide both sides by a^2b^2.
$$\frac{x^2}{a^2} + \frac{y^2}{b^2} = 1$$

The Standard Form for an Ellipse in Standard Position

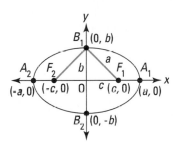

The equation $\frac{x^2}{a^2} + \frac{y^2}{b^2} = 1$ is in the **standard form** for an equation of this ellipse. By substitution, it is easy to check that $(a, 0)$, $(-a, 0)$, $(0, b)$, and $(0, -b)$ are on this ellipse. These points help graph it.

The segments $\overline{A_1A_2}$ and $\overline{B_1B_2}$ are, respectively, the **major axis** and **minor axis** of the ellipse. (The major axis contains the foci and is always longer than the minor axis.) The axes lie on the symmetry lines and intersect at the **center** O of the ellipse. The diagram at the left illustrates the following theorem. It applies to all ellipses centered at the origin with foci on one of the coordinate axes.

> **Theorem**
>
> In the ellipse with equation $\frac{x^2}{a^2} + \frac{y^2}{b^2} = 1$, $2a$ is the length of the horizontal axis, and $2b$ is the length of the vertical axis.

The length of the major axis is the focal constant. If $a > b$, then the major axis is horizontal and $(c, 0)$ and $(-c, 0)$ are the foci. Then the focal constant is $2a$, and by the Pythagorean Theorem $b^2 = a^2 - c^2$ as in the ellipse above. If $b > a$, then the major axis is vertical. So the foci are $(0, c)$ and $(0, -c)$, the focal constant is $2b$, and $a^2 = b^2 - c^2$.

The endpoints of the major and minor axes are called the **vertices** of the ellipse. For the ellipse with equation $\frac{x^2}{a^2} + \frac{y^2}{b^2} = 1$, the vertices are $(a, 0)$, $(-a, 0)$, $(0, b)$, and $(0, -b)$.

Graphing an Ellipse in Standard Form

> **Example 1**
>
> Consider the ellipse with equation $\frac{x^2}{4} + \frac{y^2}{9} = 1$.
> **a.** Identify the vertices and the lengths of the major and minor axes.
> **b.** Graph the ellipse.
>
> **Solution**
>
>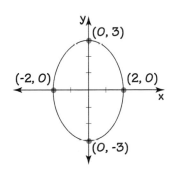
>
> **a.** $a^2 = 4$ and $b^2 = 9$. So $a = 2$ and $b = 3$. Since $b > a$, the foci of the ellipse are on the y-axis. The vertices are $(2, 0)$, $(-2, 0)$, $(0, 3)$, and $(0, -3)$. The length of the major axis is 6; the minor axis has length 4.
> **b.** A graph is at the left.
>
> Example 2 results in the ellipse drawn on page 766.

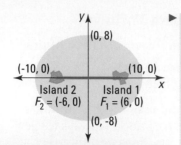

Shown is one of the ferries that link the San Juan Islands in Puget Sound to the mainland in Washington state.

Example 2

A tour boat operates between two small islands 12 miles apart. Because of fuel restrictions, the boat cannot travel more than 20 miles in going from one island to the other. Suppose a coordinate system is drawn as shown at the left. Find a sentence to describe the set of points that the boat can travel.

Solution

The boat can travel in the interior of the ellipse with foci (6, 0) and (-6, 0) and focal constant 20. This ellipse is in standard position, so it has an equation of the form

$$\frac{x^2}{a^2} + \frac{y^2}{b^2} = 1.$$

Only the values of a^2 and b^2 are needed. From the given information, $c = 6$ and $2a = 20$. So $a = 10$, and thus $a^2 = 100$. Now $b^2 = a^2 - c^2 = 100 - 6^2 = 64$. Thus an equation is

$$\frac{x^2}{100} + \frac{y^2}{64} = 1.$$

This is a simpler equation than the one given on page 766 for the same ellipse. The boat can go anywhere on or in the interior of this ellipse, described by the inequality

$$\frac{x^2}{100} + \frac{y^2}{64} \leq 1.$$

In general, if the equal sign in the equation for an ellipse is replaced by $<$ or $>$, the resulting inequality represents the interior or the exterior of the ellipse, respectively.

QUESTIONS

Covering the Reading

In 1–4, refer to the ellipse from the In-class Activity.

1. The distance between the foci is __?__.

2. The focal constant is __?__.

3. $F_1 P_1 + F_2 P_2 = $ __?__.

4. **a.** An equation for the ellipse with foci (6, 0) and (-6, 0) and focal constant 20 is $\sqrt{(x - 6)^2 + y^2} + $ __?__ $ = $ __?__.
   b. The equation of part **a** can be simplified to what equation in standard form?
   c. Verify that part **a** is equivalent to part **b**.

In 5–8, use the ellipse at left. The foci are *F* and *G*. Name its

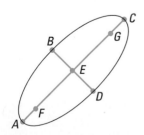

5. major axis.

6. minor axis.

7. center.

8. vertices.

In 9–11, consider the ellipse with equation $\frac{x^2}{a^2} + \frac{y^2}{b^2} = 1$.
Identify

9. its center.

10. the length of the major and minor axes.

11. the endpoints of the major and minor axes.

In 12 and 13, graph the ellipse with the given equation.

12. $\frac{x^2}{4} + \frac{y^2}{25} = 1$

13. $\frac{x^2}{9} + y^2 = 1$

14. Find an equation in standard form for the ellipse with focal constant 25 and foci (10, 0) and (-10, 0).

In 15 and 16, refer to the boat in Example 2.

15. What inequality describes the region in which the boat *cannot* travel?

16. Suppose the boat has enough fuel to travel 30 miles in going from one island to another. Write a sentence in standard form for the possible positions that the boat can travel.

Applying the Mathematics

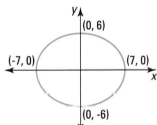

17. Refer to the ellipse graphed at the left.
    a. Find an equation for the ellipse.
    b. What sentence describes the interior of this ellipse?

18. *Multiple choice.* Which of the following describes the set of points P whose distances from (7, 2) and (3, 4) sum to 12?
    (a) $(x - 7)^2 + (y - 2)^2 + (x - 3)^2 + (y - 4)^2 = 12$
    (b) $(x + 7)^2 + (y + 2)^2 + (x + 3)^2 + (y + 4)^2 = 12$
    (c) $\sqrt{(x - 7)^2 + (y - 2)^2} + \sqrt{(x - 3)^2 + (y - 4)^2} = 12$
    (d) $\sqrt{(x - 7)^2 + (y + 2)^2} + \sqrt{(x + 3)^2 + (y + 4)^2} = 12$

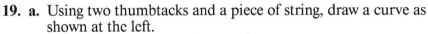

19. a. Using two thumbtacks and a piece of string, draw a curve as shown at the left.
    b. Explain why the curve is an ellipse.
    c. What part of your equipment represents the focal constant of the ellipse?

20. The orbits of the planets are nearly elliptical with the sun at one focus. The orbit of Mars can be approximated by the equation

$$\frac{x^2}{20{,}093} + \frac{y^2}{19{,}917} = 1,$$

where x and y are in millions of miles.
    a. What is the farthest Mars gets from the sun?
    b. What is the closest Mars gets to the sun?

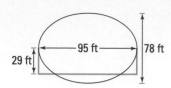

21. In the United States Capitol is a gallery called Statuary Hall. It is an elliptical chamber in which a person whispering while standing at one focus can be easily heard by another person standing at the other focus. The gallery is 78 ft wide and about 95 ft long. A diagram of the gallery's floor is at the left.

a. Find an equation which could describe the ellipse of the gallery.

b. A politician noted this feature of the chamber because the desk of the opposing party's floor leader was at one focus. How far from the floor leader's desk could the politician stand and overhear the floor leader's whispered conversations?

c. How far from the closest end of the gallery would the politician be?

Review

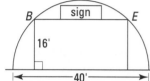

22. The figure at the left shows a cross-section of a semicircular tunnel with diameter 40 feet. A sign [Entering Tunnel—Do Not Pass] must be placed 16 feet above the roadway. Find the length BE of the beam that will support the bottom of the sign. *(Lesson 12-3)*

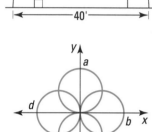

In 23–26, each circle drawn at the left has radius 4 and its center on either the x- or y-axis. *(Previous course, Lessons 12-2, 12-3)*

23. Write an equation for circle a.

24. Find an inequality describing the interior of circle b.

25. Find the circumference of each circle.

26. Find the area of each circle.

27. Use the triangle at the left. Find the length of $\overline{QR}$ to the nearest tenth. *(Lesson 10-6)*

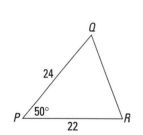

28. a. Draw the triangle ABC with vertices $A = (-2, 3)$, $B = (4, 3)$, and $C = (4, -1)$.

b. Draw its image $\triangle A'B'C'$ under the scale change $S_{2,3}$.

c. Is $\triangle ABC \sim \triangle A'B'C'$? Why or why not?

d. Find the area of $\triangle ABC$.

e. Find the area of $\triangle A'B'C'$.

f. The area of $\triangle A'B'C'$ is how many times as large as the area of $\triangle ABC$? *(Previous course, Lesson 4-5)*

Exploration

29. a. Use the Graph-Translation Theorem to predict what the graph of $\frac{(x-2)^2}{9} + \frac{(y+6)^2}{25} = 1$ will look like.

b. Check your conjecture with an automatic grapher.

c. Graph some other equations of the form $\frac{(x-h)^2}{a^2} + \frac{(y-k)^2}{b^2} = 1$.

d. Write a paragraph summarizing your work.

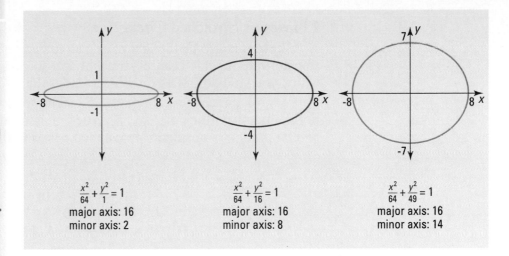

$\frac{x^2}{64} + \frac{y^2}{1} = 1$
major axis: 16
minor axis: 2

$\frac{x^2}{64} + \frac{y^2}{16} = 1$
major axis: 16
minor axis: 8

$\frac{x^2}{64} + \frac{y^2}{49} = 1$
major axis: 16
minor axis: 14

Circles as Special Ellipses

In some ellipses, the major axis is much longer than the minor axis. In others, the two axes are almost equal. Consider the three cases above. An ellipse whose major and minor axes are equal is a special kind of ellipse.

At the right is an ellipse with major axis 16 and minor axis 16. It has equation $\frac{x^2}{64} + \frac{y^2}{64} = 1$ which can be rewritten as $x^2 + y^2 = 64$. Thus, this ellipse is a circle.

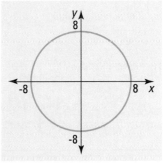

This can be generalized. Consider the standard form of an equation for an ellipse, $\frac{x^2}{a^2} + \frac{y^2}{b^2} = 1$. If the major and minor axes each have length $2r$, then $2a = 2r$ and $2b = 2r$, so we may substitute r for a and r for b. The equation becomes $\frac{x^2}{r^2} + \frac{y^2}{r^2} = 1$. Multiplying both sides of the equation by r^2 yields $x^2 + y^2 = r^2$. This is an equation for the circle with center at the origin and radius r. So, a circle is a special kind of ellipse whose major and minor axes are equal.

Ellipses as Circles in Perspective

Circles are related to noncircular ellipses. If you look at a circle on an angle, then it appears to be a noncircular ellipse. Notice how the circular hoop in the photo at the right appears to be taller than it is wide. Artists who want to draw circles in perspective must actually draw noncircular ellipses.

Ellipses as Stretched Circles

An ellipse can also be thought of as a stretched circle. The basic transformation which causes stretches and shrinks is the scale change, which you studied in Lesson 4-5. Consider the circle with equation $x^2 + y^2 = 1$ under the scale change $S_{2,3}$. The scale change $S_{2,3}$ has a horizontal magnitude of 2 and a vertical magnitude of 3. The images of several points on the circle are graphed below at the left.

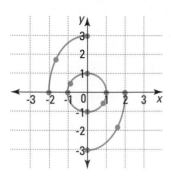

$$
\begin{array}{ll}
(1, 0) & \rightarrow (2, 0) \\
(0, 1) & \rightarrow (0, 3) \\
(-1, 0) & \rightarrow (-2, 0) \\
(0, -1) & \rightarrow (0, -3) \\
(-0.8, 0.6) & \rightarrow (-1.6, 1.8) \\
(0.6, -0.8) & \rightarrow (1.2, -2.4)
\end{array}
$$

From these six points you can see that the image of the unit circle under this scale change is not a circle. It appears to be an ellipse with foci on the y-axis.

Example 1

Find an equation for the image of the circle $x^2 + y^2 = 1$ under $S_{2,3}$.

Solution

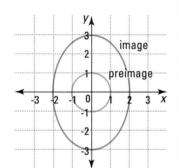

To find an equation of the image of the circle, let (x', y') be the image of (x, y). Then $(x', y') = (2x, 3y)$.

So, $x' = 2x$ and $y' = 3y$.

Thus, $x = \frac{x'}{2}$ and $y = \frac{y'}{3}$.

We know that $x^2 + y^2 = 1$. Substituting for x and y in that equation, an equation for the image is

$$\left(\frac{x'}{2}\right)^2 + \left(\frac{y'}{3}\right)^2 = 1.$$

Since (x', y') represents a point on the image, we can rewrite the equation of the image as

$$\frac{x^2}{4} + \frac{y^2}{9} = 1.$$

This is an equation for an ellipse with a minor axis of length 4 and a major axis of length 6.

Check

Substitute some points known to be on the image. Do their coordinates satisfy this equation?

Try $(2, 0)$. $\frac{2^2}{4} + \frac{0^2}{9} = 1 + 0 = 1$ It checks.

Try $(-1.6, 1.8)$. $\frac{(-1.6)^2}{4} + \frac{(1.8)^2}{9} = \frac{2.56}{4} + \frac{3.24}{9}$
$= 0.64 + 0.36 = 1$ It checks.

The argument in Example 1 can be repeated using a in place of 2 and b in place of 3. It shows that any ellipse in standard form can be thought of as a stretched circle.

The previous theorem is a special case of the Graph Scale-Change Theorem, which is analogous to the Graph-Translation Theorem you studied in Chapter 6.

A Formula for the Area of an Ellipse

Because the ellipse is related in so many ways to the circle it should not surprise you that the area of an ellipse is related to the area of a circle. In general, the scale change $S_{a,b}$ multiplies the area of the preimage by ab. Since the area of a unit circle, which has radius 1, is $\pi(1)^2 = \pi$, the area of the ellipse that is its image under $S_{a,b}$ has area $\pi \cdot (ab) = \pi ab$.

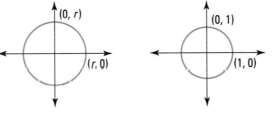

area: πr^2 square units   area: π square units

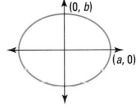

area: πab square units

Example 2

Find the area of the ellipse in Example 1.

Solution

The length of the major axis is 6 and the length of the minor axis is 4. So $a = 2$ and $b = 3$.

$$A = \pi ab = \pi \cdot \left(\frac{1}{2} \cdot 6\right) \cdot \left(\frac{1}{2} \cdot 4\right) = \pi \cdot 3 \cdot 2 = 6\pi$$

The area is 6π square units.

Covering the Reading

1. An ellipse in which the major and minor axes are equal in length is called a __?__.

In 2 and 3, *true or false.*

2. Every circle is an ellipse.

3. All ellipses are circles.

In 4 and 5, consider the circle $x^2 + y^2 = 1$ and the scale change $S_{4,3}$.

4. a. Find the image of $(1, 0)$.
b. Find the image of $(0, 1)$.
c. What is an equation for the image of the circle under $S_{4,3}$?

5. a. What is the area of the circle?
b. What is the area of its image?

In 6 and 7, consider the ellipse drawn at the left.

6. a. What scale change maps the unit circle to this ellipse?
b. Find an equation for this ellipse.

7. Find its area.

Applying the Mathematics

8. a. *True or false.* Under a scale change, a figure is similar to its image.
b. Justify your answer to part **a** by using an example from this lesson.

9. Prove that the image of the unit circle under the scale change $S_{a,b}$ is the ellipse $\frac{x^2}{a^2} + \frac{y^2}{b^2} = 1$. (Hint: Follow the idea of Example 1.)

10. a. Sketch a circle that has area 16π square units.
b. Sketch three noncongruent ellipses whose areas are also 16π.

11. In Australia, a type of football is played on elliptical regions called Aussie Rules fields. One such field has a major axis of length 185 m and minor axis of length 155 m. A track 1 meter wide surrounding the field is to be covered with turf. Surrounding it is an elliptical fence with major axis of length 187 m and minor axis of length 157 m. Find the area of the track.

In 12–14, use this definition. The **eccentricity** of an ellipse is the ratio of the distance between its foci to its focal constant.

12. What is the eccentricity of the ellipse in Example 1?

13. Sketch an ellipse with eccentricity $\frac{1}{2}$.

14. Why must the eccentricity of an ellipse be a number greater than or equal to 0 but less than 1?

Bon moat. *Not all moats are circular. Shown is Tanlay Chateau with its square moat in Burgundy, France.*

15. Use conic graph paper with foci F_1 and F_2 and $F_1F_2 = 12$.
    a. Draw the set of points P with $PF_1 + PF_2 = 16$.
    b. Give an equation for this ellipse if F_1 and F_2 are on the x-axis and the midpoint of $\overline{F_1F_2}$ is the origin. *(Lesson 12-4)*

In 16–20, match each equation with the best description. A letter may be used more than once. Do not graph. *(Lessons 12-1, 12-2, 12-4)*

16. $x^2 + y^2 = 25$

17. $\frac{x^2}{25} + \frac{y^2}{81} = 1$

18. $4x^2 + y^2 = 100$

19. $x^2 + y^2 < 25$

20. $\frac{x^2}{81} + \frac{y^2}{25} > 1$

(a) circle
(b) ellipse
(c) interior of circle
(d) interior of ellipse
(e) exterior of circle
(f) exterior of ellipse

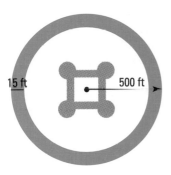

21. At the left is an overhead view of a castle surrounded by a circular moat 15 feet wide. The distance from the center of the castle to the outside of the moat is 500 feet. If the center of the castle is considered the origin, write a system of inequalities to describe the set of points on the surface of the moat. *(Lesson 12-2)*

22. A vacuum pump is designed so that each stroke leaves only 97% of the gas in the chamber. *(Lessons 7-5, 9-10)*
    a. What percent of the gas remains after 2 strokes?
    b. Write an equation that gives the percent P of the gas left after s strokes.
    c. How many strokes are necessary so that only 5% of the gas remains?

In 23 and 24, consider the line ℓ with equation $y = -\frac{1}{2}x + 4$ and the point $P = (3, -1)$.

23. Find an equation for the line through P perpendicular to ℓ. *(Lesson 4-9)*

24. Find an equation for the line through P parallel to ℓ. *(Lesson 3-5)*

Exploration

25. a. Below are equations of the ellipses shown at the start of this lesson. Locate their foci using the relationship $a^2 - c^2 = b^2$.
    (i) $\frac{x^2}{64} + \frac{y^2}{1} = 1$ (ii) $\frac{x^2}{64} + \frac{y^2}{16} = 1$ (iii) $\frac{x^2}{64} + \frac{y^2}{49} = 1$
    b. As the distance between the foci decreases, what happens to the shape of an ellipse?
    c. Find the distance between the foci for the circle $\frac{x^2}{64} + \frac{y^2}{64} = 1$.
    d. Are your answers to parts **b** and **c** consistent? Explain.

IN-CLASS

A C T I V I T Y

*Drawing a
Hyperbola*

Materials: Conic graph paper

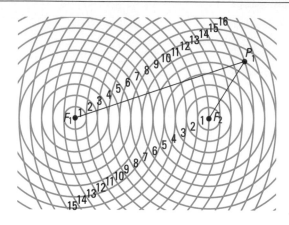

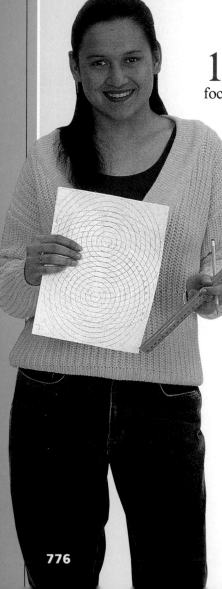

1 On a sheet of conic graph paper, draw F_1, F_2, P_1, $\overline{F_1P_1}$, and $\overline{F_2P_1}$ as shown above. F_1 represents one focus and F_2 represents another focus. Note that $F_1F_2 = 12$.

2
a. What is P_1F_1?
b. What is P_1F_2?
c. What is $|P_1F_2 - P_1F_1|$?
d. Mark another point P_2 such that $P_1F_2 = P_2F_2$ and $P_1F_1 = P_2F_1$.

In 3 and 4, find and mark all points satisfying the given conditions.

3 $PF_1 = 13$ and $PF_2 = 3$.

4 $PF_2 = 11$ and $PF_1 = 1$.

5 Plot 16 other points such that $|PF_1 - PF_2| = 10$.

6
a. Connect the dots to form two branches of a hyperbola.
b. Draw the line $\overleftrightarrow{F_1F_2}$. Label the points on the hyperbola which intersect this line as V_1 and V_2. These are the *vertices* of the hyperbola.

Making hyperbolas. *When a hexagonally-shaped pencil is sharpened, one branch of a hyperbola is formed because slices are made that are not parallel to an edge.*

What Is a Hyperbola?

Like an ellipse, a hyperbola is determined by two foci and a focal constant. The set of all points P such that $|PF_1 - PF_2| = 10$ is a hyperbola. Notice that the hyperbola you drew in the In-class Activity has two branches. One branch comes from $PF_1 - PF_2 = 10$, the other from $PF_1 - PF_2 = -10$. The notion of absolute value enables both branches to be described with one equation.

> **Definition**
> Let F_1 and F_2 be any two points and d be a constant with $0 < d < F_1F_2$. Then the **hyperbola with foci F_1 and F_2 and focal constant d** is the set of points P in a plane which satisfy $|PF_1 - PF_2| = d$.

The **vertices** of the hyperbola are the points of intersection of $\overleftrightarrow{F_1F_2}$ and the hyperbola.

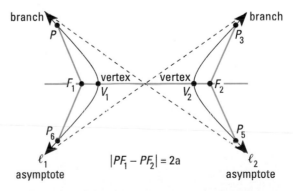

At first glance, it may look as if each branch of the hyperbola is a parabola. However, this is not true because each branch of a hyperbola has *asymptotes*. In the figure on page 777, ℓ_1 and ℓ_2 are asymptotes; that is, the points on the curve farther from the vertex are nearer (but do not meet) one of the asymptotes. In contrast, parabolas do not have asymptotes.

The Standard Form for an Equation of a Hyperbola

If the foci of a hyperbola are taken to be $(c, 0)$ and $(-c, 0)$ and the general focal constant is $2a$, as for the ellipse, an equation arises which resembles the standard form of the equation of an ellipse.

Theorem (Equation for a Hyperbola)
The hyperbola with foci $(c, 0)$ and $(-c, 0)$ and focal constant $2a$ has equation $\dfrac{x^2}{a^2} - \dfrac{y^2}{b^2} = 1$, where $b^2 = c^2 - a^2$.

Proof
The proof is almost identical to the proof of the Equation for an Ellipse Theorem in Lesson 12-4. Let $P = (x, y)$ be any point on the hyperbola. By the definition of a hyperbola, we begin with
$$\left| PF_1 - PF_2 \right| = 2a.$$
By the definition of absolute value, we know that this equation is equivalent to $PF_1 - PF_2 = \pm 2a$.

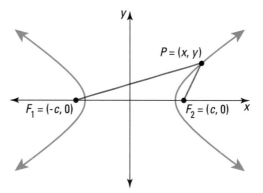

Now substitute $P = (x, y)$, $F_1 = (-c, 0)$, and $F_2 = (c, 0)$, to get
$$\sqrt{(x + c)^2 + (y - 0)^2} - \sqrt{(x - c)^2 + (y - 0)^2} = \pm 2a.$$
Doing algebraic manipulations similar to those in steps 2–8 of the proof in Lesson 12-4, the same equation in step 8 results.
$$(a^2 - c^2)x^2 + a^2 y^2 = a^2(a^2 - c^2)$$
For hyperbolas, $c > a > 0$, so $c^2 > a^2$. Thus, $c^2 - a^2$ is positive and we can let $b^2 = c^2 - a^2$. So $-b^2 = a^2 - c^2$. This accounts for the minus sign in the equation.
$$\frac{x^2}{a^2} - \frac{y^2}{b^2} = 1$$

Example 1

Consider the hyperbola with $F_1F_2 = 12$ and $|PF_1 - PF_2| = 10$. (This is the hyperbola drawn with conic graph paper in the In-class Activity on page 776.) Suppose a rectangular coordinate system is placed so that the x-axis coincides with $\overline{F_1F_2}$, and such that the y-axis bisects $\overline{F_1F_2}$. Find an equation for this hyperbola.

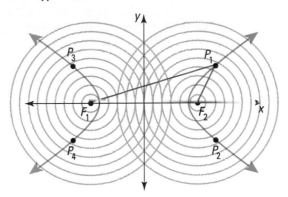

Solution

You are given $F_1F_2 = 12$; so $2c = 12$, and $c = 6$. The focal constant is 10, so $2a = 10$, so $a = 5$ and $a^2 = 25$. Now, $b^2 = 6^2 - 5^2 = 11$. Thus, An equation for this hyperbola is

$$\frac{x^2}{25} - \frac{y^2}{11} = 1.$$

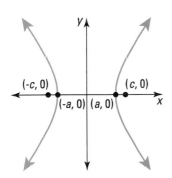

We say that the equation $\frac{x^2}{a^2} - \frac{y^2}{b^2} = 1$ is the **standard form** for the equation of a hyperbola. To graph $\frac{x^2}{a^2} - \frac{y^2}{b^2} = 1$, notice that $(a, 0)$ and $(-a, 0)$ satisfy the equation. These are the vertices of the hyperbola. Because the foci are on the x-axis, the hyperbola is symmetric about that axis. When $x = 0$, no real value of y satisfies the equation, so the hyperbola does not intersect the y-axis. A sketch using this information is given at the left. To make an accurate graph, more points or the equations of the asymptotes are needed.

The Hyperbola $x^2 - y^2 = 1$

To find equations for the asymptotes of $\frac{x^2}{a^2} - \frac{y^2}{b^2} = 1$, it helps to examine the simplest hyperbola of this kind, that is, the hyperbola with equation $x^2 - y^2 = 1$. Some points on the graph of $x^2 - y^2 = 1$ are given below at the left and graphed at the right.

$$A = (1, 0)$$
$$B = (2, \sqrt{3}) \approx (2, 1.73)$$
$$C = (3, \sqrt{8}) \approx (3, 2.83)$$
$$D = (4, \sqrt{15}) \approx (4, 3.87)$$
$$E = (5, \sqrt{24}) \approx (5, 4.90)$$

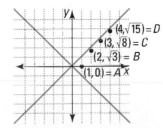

Notice that as x gets larger, the points on $x^2 - y^2 = 1$ get closer to the line with equation $y = x$. For instance, the point $(100, \sqrt{9999}) \approx (100, 99.995)$ is on the hyperbola. Thus, the line with equation $y = x$ appears to be an asymptote.

Think about what happens in quadrants II, III, and IV. Because of the hyperbola's symmetry, each point in the first quadrant has reflection images on the hyperbola in other quadrants. Below we show the images of A, B, C, and D after reflection over the x-axis and y-axis. The line with equation $y = -x$ also appears to be an asymptote.

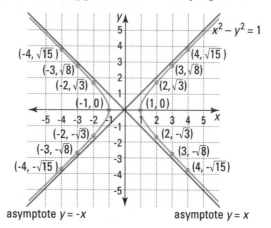

asymptote $y = -x$ asymptote $y = x$

The asymptotes can be verified algebraically. When $x^2 - y^2 = 1$,
$$y^2 = x^2 - 1.$$
So
$$y = \pm\sqrt{x^2 - 1}.$$

As x gets larger, $\sqrt{x^2 - 1}$ gets closer to $\sqrt{x^2}$, which is $|x|$. However, since $\sqrt{x^2 - 1} \neq \sqrt{x^2}$, the curve never intersects the line $y = x$. So y gets closer to x or $-x$ but never reaches it. Thus $y = x$ and $y = -x$ are asymptotes of $x^2 - y^2 = 1$.

In general, using the Graph Scale-Change Theorem, the scale change $S_{a,b}$ maps $x^2 - y^2 = 1$ onto $\frac{x^2}{a^2} - \frac{y^2}{b^2} = 1$. The asymptotes $y = \pm x$ of $x^2 - y^2 = 1$ are mapped onto the lines $\frac{y}{b} = \pm\frac{x}{a}$. These lines are the asymptotes of $\frac{x^2}{a^2} - \frac{y^2}{b^2} = 1$.

Theorem
The asymptotes of the hyperbola with equation
$\frac{x^2}{a^2} - \frac{y^2}{b^2} = 1$ are $\frac{y}{b} = \pm\frac{x}{a}$ $\left(\text{or } y = \pm\frac{b}{a}x\right)$.

Graphing a Hyperbola in Standard Form

Example 2

Graph $\frac{x^2}{9} - \frac{y^2}{16} = 1$.

Solution

The equation is in the form $\frac{x^2}{a^2} - \frac{y^2}{b^2} = 1$. Thus, $a^2 = 9$, so $a = 3$. Thus the vertices are $(3, 0)$ and $(-3, 0)$. From the theorem, we find that the asymptotes are $\frac{y}{4} = \pm\frac{x}{3}$ or $y = \pm\frac{4}{3}x$. Carefully graph the vertices and asymptotes. Then sketch the hyperbola.

Check

From the equation, find coordinates of a point other than the vertex on the hyperbola. For instance, if $x = 5$, then $\frac{25}{9} - \frac{y^2}{16} = 1$, from which $y = \pm\frac{16}{3}$. The points $\left(5, \frac{16}{3}\right)$ and $\left(5, -\frac{16}{3}\right)$ do seem to be on the hyperbola. It checks.

To graph a hyperbola in standard form with an automatic grapher, you may need to solve its equation for y first. As with the ellipse there are two parts to graph. But they are not necessarily the two branches! You are asked to explore the possibilities in the Questions.

QUESTIONS

Covering the Reading

1. A hyperbola with foci $(c, 0)$ and $(-c, 0)$ has an equation of the form ___?___.

2. Why is the hyperbola with equation $x^2 - y^2 = 1$ so useful?

In 3 and 4, an equation for a hyperbola is given. **a.** Name its vertices. **b.** Identify its asymptotes.

3. $1 = x^2 - y^2$

4. $\frac{x^2}{a^2} - \frac{y^2}{b^2} = 1$

In 5 and 6, *true or false.*

5. The focal constant of a hyperbola equals the distance between the foci.

6. If F_1 and F_2 are the foci of a hyperbola, then $\overleftrightarrow{F_1F_2}$ is a line of symmetry for the curve.

7. Consider the hyperbola with equation $\frac{x^2}{25} - \frac{y^2}{11} = 1$ from Example 1.
   a. Name its vertices.   **b.** State equations for its asymptotes.

In 8 and 9, consider the hyperbola with equation $\frac{x^2}{9} - \frac{y^2}{16} = 1$.

8. Graph this hyperbola.

9. Find two points on the curve with x-coordinate equal to 6.

10. Finish the proof of the Equation for a Hyperbola Theorem by showing that $\sqrt{(x + c)^2 + (y - 0)^2} - \sqrt{(x - c)^2 + (y - 0)^2} = \pm 2a$ is equivalent to $(a^2 - c^2)x^2 + a^2y^2 = a^2(a^2 - c^2)$.

Applying the Mathematics

11. Solve $x^2 - y^2 = 1$ for y. Use your solution to graph $x^2 - y^2 = 1$ using an automatic grapher.

12. *True or false.* A single equation for *both* asymptotes of $x^2 - y^2 = 1$ is $y = |x|$.

13. a. Graph the set of points satisfying $y^2 - x^2 = 1$.
b. Is the graph a hyperbola? Why or why not?

14. The point $(-7, 4)$ is on a hyperbola with foci $(5, 0)$ and $(-5, 0)$.
a. Find the focal constant of the hyperbola.
b. Give an equation for this hyperbola in standard form. (Hint: Find b using $b^2 = c^2 - a^2$.)
c. Graph this hyperbola.

15. Consider the graphs of $\frac{x^2}{25} - \frac{y^2}{9} = 1$ and $\frac{x^2}{9} - \frac{y^2}{25} = 1$.
a. What do these hyperbolas have in common?
b. How do these hyperbolas differ?

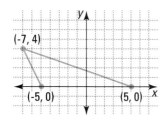

Review

16. An ellipse has foci F_1 and F_2 on the x-axis, center $(0, 0)$, and $F_1F_2 = 4$. Also $PF_1 + PF_2 = 7$. Find
a. the length of the major axis.
b. an equation in standard form for the ellipse.
c. the area of the ellipse. *(Lessons 12-4, 12-5)*

17. An exhibition tent is in the form of half a cylindrical surface with each cross-section a semiellipse (half an ellipse) having base 20 ft and height 8 ft. How close to either side of the tent can a person 5 ft tall stand straight up? *(Lesson 12-4)*

In 18 and 19, sketch a graph. *(Lessons 12-2, 12-4)*

18. $\{(x, y): x^2 + y^2 = 144\}$ **19.** $\{(x, y): 9x^2 + y^2 = 144\}$

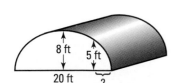

20. Solve $x^5 - 81x = 0$. *(Lessons 6-5, 11-3)*

21. What number can be put into the blank to make the expression $y^2 - 13y + \underline{\ ?\ }$ a perfect square? *(Lessons 6-5, 11-3)*

22. Consider the system $\begin{cases} y = 4x \\ 2x - 3y = -15. \end{cases}$
a. Name three methods you can use to solve this system.
b. Solve the system using any method. *(Lessons 5-2, 5-3, 5-4, 5-6)*

23. A water-storage tank has a slow leak. Suppose the water level starts at 100 inches and falls $\frac{1}{2}$ inch per day.   *(Lesson 3-1)*

    a. Write an equation relating the number N of days the tank has been leaking and the water level L.

    b. After how many days will the tank be empty?

24. Refer to the graph below which shows the number of calories of heat needed to raise the temperature of 1 g of ice so that ice will turn to water and eventually to steam.

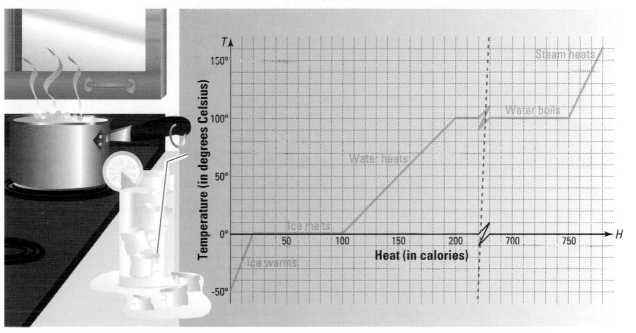

    a. At what temperature in °C does ice melt?

    b. *True or false.*   Water boils at a constant temperature.

    c. Find the slope of the line between the points (100, 0) and (200, 100).

    d. Write a sentence explaining your answer to part **c**.   *(Lessons 2-4, 3-1)*

Exploration

25. The words *ellipsis* and *hyperbole* have meanings in grammar. What are these meanings?

Conic sections. *A plane intersecting the base forms either a parabola (left) or a branch of a hyperbola (right). Ellipses (center) are formed by planes not parallel to the base.*

Generating Hyperbolas from Inverse Variation

In Chapter 2, you studied hyperbolas that arose from situations modeled by inverse variation of the form $y = \frac{k}{x}$. For instance, a person traveling on a highway with mileage markers can use the formula $r = \frac{d}{t}$ to check a speedometer by driving at a constant speed for one mile and timing how long it takes.

$$\text{rate in } \frac{\text{miles}}{\text{hour}} = \frac{\text{distance in miles}}{\text{time in hours}}$$

If the distance is 1 mile, the above equation becomes

$$\text{rate in } \frac{\text{miles}}{\text{hour}} = \frac{1}{\text{time in hours}}.$$

There are 3600 seconds in an hour, so when time is measured in seconds this relationship is equivalent to

$$\text{rate in } \frac{\text{miles}}{\text{hour}} = 3600 \cdot \frac{1}{\text{time in seconds}} = \frac{3600}{\text{time in seconds}}.$$

That is,
$$r = \frac{3600}{t}.$$

For instance, if it takes 90 seconds to travel a mile, then the average rate is $\frac{3600}{t} = 40 \frac{\text{miles}}{\text{hour}}$.

At the right are some pairs of numbers satisfying the equation $r = \frac{3600}{t}$.

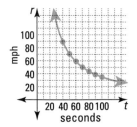

time t (sec)	40	50	60	70	80	90	100
rate r $\left(\frac{\text{mi}}{\text{hr}}\right)$	90	72	60	51.4	45	40	36

To show that this graph is one branch of a hyperbola, we need to show that any point on the graph of $y = \frac{k}{x}$ satisfies the geometric definition of hyperbola given in the last lesson. Because this branch is reflection-symmetric to the line $y = x$, the foci must be on the line $y = x$. Because the entire hyperbola is rotation-symmetric to the origin, the foci must also be rotation-symmetric to the origin. Example 1 shows how an equation of the form $y = \frac{k}{x}$ arises when the foci meet these criteria.

Example 1

Find an equation of the form $y = \frac{k}{x}$ for the hyperbola with foci $F_1 = (6, 6)$ and $F_2 = (-6, -6)$ and focal constant 12.

Solution

Let $P = (x, y)$ be a point on the hyperbola. Then, by the definition of hyperbola, one branch of the curve is the set of points P such that
$$PF_1 - PF_2 = d.$$
Use the Distance Formula with $F_1 = (6, 6)$, $F_2 = (-6, -6)$, and $d = 12$.
$$\sqrt{(x - 6)^2 + (y - 6)^2} - \sqrt{(x + 6)^2 + (y + 6)^2} = 12$$
Add one of the square roots to both sides.
$$\sqrt{(x - 6)^2 + (y - 6)^2} = 12 + \sqrt{(x + 6)^2 + (y + 6)^2}$$
Square both sides. Notice that the right side is like a binomial.
$$(x - 6)^2 + (y - 6)^2 = 144 + 24\sqrt{(x + 6)^2 + (y + 6)^2} + (x + 6)^2 + (y + 6)^2$$
Expand the binomials, combine like terms, and simplify.
$$-24x - 24y - 144 = 24\sqrt{(x + 6)^2 + (y + 6)^2}$$
Divide by -24, and then square both sides.
$$(x + y + 6)^2 = (x + 6)^2 + (y + 6)^2$$
$$x^2 + y^2 + 36 + 2xy + 12x + 12y = x^2 + 12x + 36 + y^2 + 12y + 36$$
Combine like terms and simplify.
$$2xy = 36$$
Solve for y.
$$y = \frac{18}{x}$$
The other branch also satisfies this equation.

Check

Pick a point that satisfies $y = \frac{18}{x}$, say $P = (2, 9)$. Show that this point satisfies the conditions in the geometric definition of a hyperbola.

Does $|PF_1 - PF_2| = d$?
$$\left| \sqrt{(2 - 6)^2 + (9 - 6)^2} - \sqrt{(2 - -6)^2 + (9 - -6)^2} \right| = 12?$$
$$\left| \sqrt{(-4)^2 + 3^2} - \sqrt{8^2 + 15^2} \right| = |5 - 17| = 12? \text{ Yes}$$

The following theorem below can be proved in a similar way.

Theorem:

The graph of $y = \frac{k}{x}$ or $xy = k$ is a hyperbola. When $k > 0$, this is the hyperbola with foci $(\sqrt{2k}, \sqrt{2k})$ and $(-\sqrt{2k}, -\sqrt{2k})$ and focal constant $2\sqrt{2k}$.

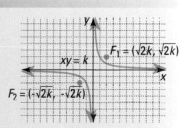

By reversing the process used in Example 2, you can conclude that the graph of $r = \frac{3600}{t}$ is one branch of the hyperbola with foci at $(\sqrt{7200}, \sqrt{7200})$ and $(-\sqrt{7200}, -\sqrt{7200})$ and focal constant $2\sqrt{7200}$.

Recall that the x- and y-axes are asymptotes of all equations of the form $y = \frac{k}{x}$, where $k \neq 0$. A hyperbola with perpendicular asymptotes is called a **rectangular hyperbola.** Thus, graphs of equations of the form $y = \frac{k}{x}$ are rectangular hyperbolas.

The Conic Sections

On page 747, we mentioned that parabolas, hyperbolas, and ellipses can all be formed by intersecting a plane with a double cone. Let k be the measure of the acute angle between the axis of the double cone and an edge. Let θ be the measure of the smallest angle between the axis and the intersecting plane. The three possible relations between θ and k determine the three types of conic sections.

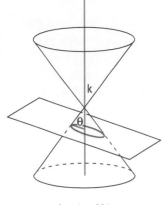

$k < \theta \leq 90°$
ellipse

The Standard Form for a Quadratic Relation

You have now seen equations for all the different types of conics. Here are some of these equations in standard form.

$$y = ax^2 + bx + c \qquad \text{parabola}$$
$$(x - h)^2 + (y - k)^2 = r^2 \qquad \text{circle}$$
$$\frac{x^2}{a^2} + \frac{y^2}{b^2} = 1 \qquad \text{ellipse}$$
$$\frac{x^2}{a^2} - \frac{y^2}{b^2} = 1 \text{ or } xy = k \qquad \text{hyperbola}$$

Although the equations for the hyperbola and ellipse look similar, the others look different. However, all these equations contain only terms with x^2, xy, y^2, x, or y. Thus all the conic sections are special types of quadratic relations. That is, their equations can be rewritten in the **standard form for a quadratic relation**

$$Ax^2 + Bxy + Cy^2 + Dx + Ey + F = 0,$$

where A, B, C, D, E, and F are real numbers, and at least one of A, B, or C is nonzero.

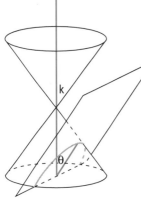

$\theta = k$
parabola

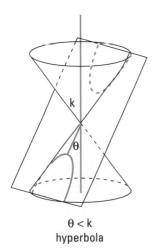

$\theta < k$
hyperbola

Example 2

Show that the circle with equation $(x - 3)^2 + y^2 = 14$ is a quadratic relation.

Solution

To do this, the equation for this circle must be put into the general form of a quadratic relation. So first expand the binomial.
$$x^2 - 6x + 9 + y^2 = 14$$
Now add -14 to both sides. Then use the Commutative Property of Addition to reorder the terms so that they are in the order x^2, xy, y^2, x, y, and constants.
$$x^2 + 0xy + y^2 - 6x + 0y - 5 = 0$$
This is in the desired form with A = 1, B = 0, C = 1, D = -6, E = 0, and F = -5. Since at least one of A, B, or C is nonzero, this is a quadratic relation.

QUESTIONS

Covering the Reading

1. If it takes 75 seconds to drive a mile, what is the average speed in miles per hour?

2. If it takes t seconds to drive a mile, what is the average speed in miles per hour?

3. At the left is a hyperbola with foci A and B. What must be true about $|Q_1A - Q_1B|$ and $|Q_2A - Q_2B|$?

In 4–6, consider the hyperbola with equation $xy = k$ where $k > 0$. Name its

4. foci.   5. asymptotes.   6. focal constant.

7. Graph the hyperbola with equation $xy = 18$. On your graph, identify the foci, asymptotes, and focal constant.

8. **a.** Find an equation for the hyperbola with foci at (10, 10) and (-10, -10) and focal constant 20.
   b. Verify that the point (-2, -25) is on this hyperbola.

9. What is a rectangular hyperbola?

10. Which hyperbola mentioned in this lesson is a rectangular hyperbola?

In 11–14, an equation is given.
a. Is it an equation for a quadratic relation?
b. If so, put the equation in standard form for a quadratic relation. If not, tell why not.

11. $x^2 + 4xy^2 = 6$ 12. $\frac{1}{2}y - 13x^2 = \sqrt{5}\,x$

13. $xy - 8 = 2xy$ 14. $x^2 + 2xy + 3y^2 + 4x + 5y = 6$

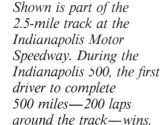

Shown is part of the 2.5-mile track at the Indianapolis Motor Speedway. During the Indianapolis 500, the first driver to complete 500 miles—200 laps around the track—wins.

Applying the Mathematics

In 15 and 16, show that the equation describes a quadratic relation by rewriting it in standard form for a quadratic relation. Give the values of A, B, C, D, E, and F.

15. $\frac{x^2}{4} - \frac{y^2}{9} = 1$ 16. $y = 3(x + 1)^2 - 8$

17. The graph of $rt = 3600$ is a hyperbola. What are its vertices?

18. A car travels the 2.5 miles around the Indianapolis Speedway in t seconds at an average rate of r mph. Racing fans with stopwatches can calculate how fast a car is traveling if they know the value of the constant rt. What is that value?

19. Sketch a graph.
   a. $xy > 8$ **b.** $xy \leq 8$

20. Consider the hyperbola with equation $\frac{x^2}{25} - \frac{y^2}{36} = 1$.
   a. What are its foci?
   b. Name its vertices.
   c. State equations for its asymptotes.   *(Lesson 12-6)*

21. Use conic graph paper. Draw a hyperbola with foci F_1 and F_2, where $F_1F_2 = 10$, and $|PF_1 - PF_2| = 4$.   *(Lesson 12-6)*

In 22–24, choose the best response from the following.
(a) circle   (b) ellipse
(c) parabola   (d) hyperbola   *(Lessons 12-1, 12-2, 12-4, 12-6)*

22. Which is the set of points satisfying the equation
$$\left| \sqrt{(x-3)^2 + (y-3)^2} - \sqrt{(x+3)^2 + (y+3)^2} \right| = 6?$$

23. Which is the set of points equidistant from a given focus and directrix?

24. Which is the set of points satisfying the equation $4x^2 + 5y^2 = 100$?

In 25 and 26, give the singular form of each word.   *(Lesson 12-4)*

25. foci   26. vertices

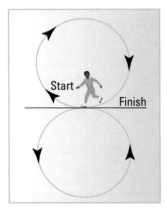

In 27 and 28, consider that in the past, figure skaters had to perform compulsory figures in competitions. The compulsory circles made by a figure skater needed to have radius 1.5 times the height of the skater. Assume a coordinate system shown in the figure at the left. Write a sentence to describe the following points for a skater 160 cm tall.
(Lessons 12-2, 12-3)

27. the points on the upper circle

28. the points in the interior of the lower circle

Exploration

29. **a.** Refer to the diagram at the left. To do the following you will need 3 tacks, a ruler, a piece of string of length shorter than that of the ruler, a pencil, and a small board.
   (i) Tack the ruler to the board so that the ruler is pivoted at point A.
   (ii) Tack one end of the piece of string to the other end of the ruler (point B).
   (iii) Take the other end of the string and tack it to the board at point C. The distance between tacks A and C must be greater than the difference between the length of the ruler and the length of the string.
   (iv) Holding the string taut against the ruler with a pencil, rotate the ruler about point A.
   b. Explain why the resulting curve must be a part of a hyperbola.

No more jet lag? *Shown is the Concorde, the first supersonic airliner. The Concorde flies at speeds of up to Mach 2.2—which is 2.2 times as fast as the speed of sound—or about 2340 km per hour. See Question 15.*

What Is a Quadratic System?

A **quadratic system** is a system that involves at least one quadratic sentence. A quadratic system with at least one linear sentence is called a **quadratic-linear system.** As with linear systems, you may solve quadratic systems by

   (1) graphing,

   (2) substitution,

or   (3) linear combinations.

No new properties are needed to solve quadratic-linear systems. Geometrically, the task is to find the intersection of a conic section and a line. You solved some systems like this in Lesson 5-2.

Solving Quadratic-Linear Systems by Substitution

Example 1

Find exact solutions to the system $\begin{cases} y - 3x = 1 \\ \quad xy = 10. \end{cases}$

Solution
Solve the first sentence for y.

$$y = 3x + 1$$

Substitute the expression $3x + 1$ for y in the second sentence.

$$x(3x + 1) = 10 \qquad \blacktriangleright$$

This is a quadratic equation that you can solve by the Quadratic Formula or by factoring.

$$3x^2 + x = 10$$
$$3x^2 + x - 10 = 0$$
$$x = \frac{-1 + \sqrt{1 - 4 \cdot 3(-10)}}{2 \cdot 3} = \frac{-1 \pm \sqrt{121}}{6}$$
$$x = \frac{-1 - 11}{6} \text{ or } x = \frac{-1 + 11}{6}$$
$$x = -2 \text{ or } x = \frac{5}{3}$$

Now remember that $y = 3x + 1$.

When $x = -2$, $y = 3(-2) + 1 = -5$. And when $x = \frac{5}{3}$, $y = 3\left(\frac{5}{3}\right) + 1 = 6$.

So the solutions are $(-2, -5)$ and $\left(\frac{5}{3}, 6\right)$.

Check

Substitute the coordinates of each point into each equation.

(1) Does $-5 - 3(-2) = 1$? Yes   (2) Does $6 - 3\left(\frac{5}{3}\right) = 1$? Yes

Does $(-2)(-5) = 10$? Yes   Does $\left(\frac{5}{3}\right)(6) = 10$? Yes

So $(-2, -5)$ checks.   So $\left(\frac{5}{3}, 6\right)$ checks.

In Example 1, the substitution of the linear quantity into the quadratic relation resulted in a quadratic equation in one variable. Because a quadratic equation may have 2, 1, or 0 solutions, a quadratic-linear system may also have 2, 1, or 0 solutions.

Example 2

At the left are graphs of $6x^2 + y^2 = 100$ and $y = -12x + 50$.

It appears they intersect in only one point. Is this so? Justify your answer.

Solution

Solve the system $\begin{cases} 6x^2 + y^2 = 100 \\ y = -12x + 50. \end{cases}$

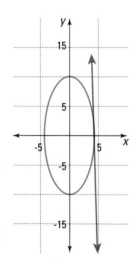

The second sentence is already solved for y. Substitute for y in the first sentence.
$$6x^2 + (-12x + 50)^2 = 100$$

Expand and rewrite in the general form of a quadratic equation.
$$6x^2 + 144x^2 - 1200x + 2500 = 100$$
$$150x^2 - 1200x + 2400 = 0$$

Divide each side by 150 to simplify.
$$x^2 - 8x + 16 = 0$$

The left side is a perfect square.
$$(x - 4)^2 = 0$$

So $x = 4$ is the only solution. When $x = 4$ in the first sentence,
$$6(4)^2 + y^2 = 100$$
$$y^2 = 4$$
$$y = \pm 2.$$

Thus, there are two possible solutions: (4, 2) and (4, -2). The point (4, 2) satisfies the equation $y = -12x + 50$, but the point (4, -2) does not. (It is an extraneous solution.) Therefore, there is only one solution to this system. So the ellipse and line intersect at exactly one point.

The line $y = -12x + 50$ is tangent to the ellipse $6x^2 + y^2 = 100$.

Inconsistent Quadratic Systems

Like linear systems, quadratic systems can be inconsistent. The signal for inconsistency is that the solutions to the quadratic system are not real.

Example 3

Find the points of intersection of the line $y = x - 1$ and the parabola $y = x^2$.

Solution 1

Graphs of the line and parabola, shown at the right, show there is no solution.

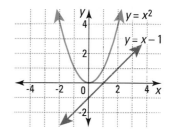

Solution 2

Solve the system $\begin{cases} y = x - 1 \\ y = x^2. \end{cases}$

Substitute x^2 for y in the first sentence.
$$x^2 = x - 1$$
Thus $x^2 - x + 1 = 0.$

By the Quadratic Formula,
$x = \dfrac{1 \pm \sqrt{1 - 4 \cdot 1 \cdot 1}}{2} = \dfrac{1 \pm \sqrt{-3}}{2}$. The nonreal solutions indicate there are no points of intersection in the real coordinate plane.

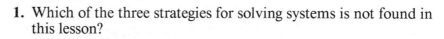

QUESTIONS

Covering the Reading

1. Which of the three strategies for solving systems is not found in this lesson?

2. How many solutions may a system of one linear and one quadratic equation have?

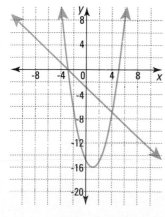

3. A graph of the system $\begin{cases} y = x^2 - 2x - 15 \\ x + y = -3 \end{cases}$ is shown at the left.

   a. How many solutions are there?
   b. Approximate the solutions.
   c. Check your answers.

In 4 and 5, a system is given. **a.** Estimate the solutions by graphing.
b. Find exact solutions by substitution.

4. $\begin{cases} xy = 18 \\ \quad y = 3x + 12 \end{cases}$ 5. $\begin{cases} x^2 + y^2 = 25 \\ \quad y = \frac{3}{4}x \end{cases}$

6. Find the point(s) of intersection of the line $y = x + 2$ and the parabola $y = x^2$.

7. Find the point(s) of intersection of the line $y = x - 1$ and the parabola $y = 2x^2$.

8. **a.** What name is given to a system which has no solutions?
   b. Give an example of such a system.

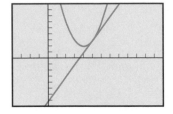

In 9 and 10, consider the figure at the left which suggests that the parabola $y = x^2 - 8x + 18$ and the line $y = 2x - 7$ intersect near the point (5, 3).

9. Check by substitution that this point is on both curves.

10. Solve the system algebraically to verify that this is the only solution.

$-4 \leq x \leq 13,$ x-scale = 1
$-8 \leq y \leq 9,$ y-scale = 1

Applying the Mathematics

11. Phillip has 150 m of fencing material and wants to form a rectangle whose area is 1300 square meters.
    a. Let x = the width of the field and y = its length. Write a system of equations that models this situation.
    b. Use graphing to estimate the dimensions of this region.
    c. Solve this system using substitution.

12. Someone claims that the sum of two real numbers is 10 and their product is 30. Use equations and graphs to explain why this is impossible.

In 13 and 14, **a.** Solve the system algebraically. **b.** Check your work.

13. $\begin{cases} x^2 + y^2 = 9 \\ 2x + y = 2 \end{cases}$ 14. $\begin{cases} \dfrac{x^2}{25} + \dfrac{y^2}{9} = 1 \\ \quad y = \frac{1}{4}x \end{cases}$

Review

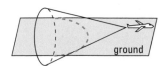

15. A supersonic jet traveling parallel to the ground generates a shock wave in the shape of a cone. The sonic boom is felt simultaneously on all the points located on the intersection of the cone and the ground. These points all lie on what type of conic section? *(Lesson 12-7)*

16. Find an equation for a hyperbola with foci at (2, 2) and (-2, -2) and focal constant 4. *(Lesson 12-7)*

17. Consider the hyperbola with equation $\frac{x^2}{36} - \frac{y^2}{25} = 1$.
   a. Identify its foci.
   b. Identify its vertices.
   c. State equations for its asymptotes.   *(Lesson 12-6)*

18. Give an equation for a hyperbola that
   a. is the graph of a function.   **b.** is not the graph of a function.
   (Lessons 1-2, 2-6, 12-6, 12-7)

19. Halley's comet has an elliptical orbit with the sun at one focus. Its closest distance to the sun is about $9 \cdot 10^7$ km, while its farthest distance is about $5.34 \cdot 10^9$ km.

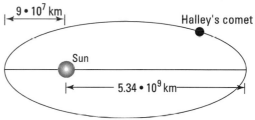

9 • 10⁷ km — Halley's comet — Sun — 5.34 • 10⁹ km

   a. Find the length of the major axis of Halley's comet's orbit.
   b. What is the length of its minor axis?   *(Lesson 12-4)*

20. Factor completely: $x^4 - 4x^2y^2$.   *(Lesson 11-3)*

21. *Skill sequence.*   Simplify.   *(Lessons 8-5, 8-6)*
   a. $\sqrt{6}\,\sqrt{24}$ **b.** $\sqrt{2400}$ **c.** $\dfrac{8 \pm \sqrt{2400}}{6}$

22. A lemonade stand reports the following monthly profit P (in hundreds of dollars) in relation to the average monthly temperature T (in degrees Celsius).

T	-10	0	10	20
P	-125	-50	25	100

LEMONADE
35¢ SMALL
50¢ LARGE

   a. Does a linear function fit these data?
   b. If so, write P as a function of T, and tell what quantity the slope represents. If not, tell why not.   *(Lessons 2-4, 3-1, 3-6, 11-7)*

Exploration

23. Draw an example of a system with exactly one solution that involves a hyperbola and an oblique line.

12-9

Quadratic-Quadratic Systems

Quadratic-quadratic systems involve the intersection of curves represented by quadratic relations: circles, ellipses, hyperbolas, and parabolas. They are a bit more complicated to solve algebraically than linear systems or quadratic-linear systems are. A quadratic-quadratic system may have 0, 1, 2, 3, 4, or infinitely many solutions. The first two examples illustrate systems with 4 solutions each.

To find exact solutions, the first goal is always the same: *work to get an equation in one variable.* In Example 1 we solve by substitution.

Example 1

At the right is pictured the system
$$\begin{cases} x^2 + y^2 = 25 \\ y = x^2 - 13. \end{cases}$$
Find the four solutions shown in the graph.

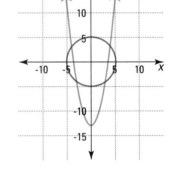

Solution

Although we could substitute $x^2 - 13$ for y in the first equation, this results in a 4th degree polynomial to solve. There is an easier approach. Since both equations include a term with x^2, solve the second equation for x^2.
$$x^2 = y + 13$$
Now substitute into the first equation.
$$(y + 13) + y^2 = 25$$
$$y^2 + y - 12 = 0$$
$$y = \frac{-1 \pm \sqrt{1 - 4(1)(-12)}}{2}$$
$$= \frac{-1 \pm 7}{2}.$$
So $y = -4$ or $y = 3$
When $y = -4$, $x^2 = -4 + 13 = 9$.
So $x = \pm 3$. Hence $(-3, -4)$ and $(3, -4)$ are solutions.
When $y = 3$, $x^2 = 3 + 13 = 16$.
So $x = \pm 4$. Hence $(4, 3)$ and $(-4, 3)$ are solutions.

Check

These four points seem to be near the points of intersection shown on the graph. You should check that they satisfy both equations.

In the next example, we use the linear-combination method to solve the system. Because the graphs of both relations in the system are symmetric to the *x*- and *y*-axes, so is the graph of the set of solutions.

Example 2

Find all points of intersection of the ellipse $\frac{x^2}{16} + \frac{y^2}{9} = 1$ and the hyperbola $x^2 - y^2 = 7$.

Solution

A sketch shows that there are four points. We call them A, B, C, and D. To find their coordinates, multiply the first equation by $16 \cdot 9 = 144$ to remove fractions. The system becomes

$$\begin{cases} 9x^2 + 16y^2 = 144 \\ x^2 - y^2 = 7. \end{cases}$$

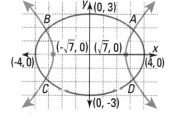

Use the linear-combination method. Multiply the second equation by -9 and add the equations.

$$\begin{array}{r} 9x^2 + 16y^2 = 144 \\ -9x^2 + 9y^2 = -63 \\ \hline 25y^2 = 81 \\ y^2 = \frac{81}{25} \end{array}$$

Solve for y.
$$y = \pm\sqrt{\frac{81}{25}} = \pm\frac{9}{5}$$

Now substitute these y-values in $x^2 - y^2 = 7$ to find x. Since $x^2 - \frac{81}{25} = 7$, $x^2 = \frac{256}{25}$. Thus for each value of y, $x = \pm\sqrt{\frac{256}{25}} = \pm\frac{16}{5}$.

So points of intersection are $A = \left(\frac{16}{5}, \frac{9}{5}\right) = (3.2, 1.8)$, $B = \left(\frac{-16}{5}, \frac{9}{5}\right) = (-3.2, 1.8)$, $C = \left(\frac{-16}{5}, \frac{-9}{5}\right) = (-3.2, -1.8)$, and $D = \left(\frac{16}{5}, \frac{-9}{5}\right) = (3.2, -1.8)$.

Check

The graph shows these coordinates to be quite reasonable.

To solve systems of quadratic equations involving hyperbolas, multiple substitutions are sometimes needed. Example 3 involves two rectangular hyperbolas.

Example 3

One month, Wanda's Western Wear took in $12,000 from boot sales. The next month, although Wanda sold 40 fewer pairs of boots, the store took in $12,800 from boot sales because the price had been raised by $20 per pair. Find the price of a pair of boots in each month.

Solution

Let n = the number of pairs of boots sold in the first month.
   p = the price of a pair of boots in the first month.
The equations for total sales in the first and second months respectively, are:
(1) $np = 12,000$
(2) $(n - 40)(p + 20) = 12,800$
From (1), $p = \frac{12,000}{n}$.
From (2), $np + 20n - 40p - 13,600 = 0$.   ▶

The two forms of equation (1) allow us to make two substitutions into equation (2), namely 12,000 for np and $\frac{12{,}000}{n}$ for p, to get:

$$12{,}000 + 20n - 40\left(\frac{12{,}000}{n}\right) - 13{,}600 = 0.$$

Simplify. $20n - 1600 - \frac{480{,}000}{n} = 0$

Multiply by n. $20n^2 - 1600n - 480{,}000 = 0$

Divide by 20. $n^2 - 80n - 24{,}000 = 0$

Factor. $(n - 200)(n + 120) = 0$

Use the Zero Product Property. $n = 200$ or $n = -120$

The number of pairs of boots can only be positive, so we use the positive answer and substitute in equation (1) to find the price.

$$200p = 12{,}000$$
$$p = 60$$

The boots were priced at $60 the first month, and $p + 20 = \$80$ the second month.

Check

First month: Does $(200)(60) = 12{,}000$? Yes, it checks.
Second month: Does $(200 - 40)(60 + 20) = (160)(80)$
    $= 12{,}800$? Yes, it checks.

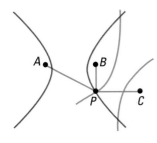

Intersections of hyperbolas are the basis for the LORAN system. In this system, two LOng RAnge Navigational stations A and B simultaneously send electronic signals to a ship at sea. The ship receives these signals at slightly different times. By measuring the time differential and by taking into account the speed of the radio waves, the ship P can be located on a hyperbola with foci at A and B. A similar process locates the ship on a hyperbola with foci at stations B and C. The intersection of the two hyperbolas gives the ship's location.

QUESTIONS

Covering the Reading

1. **a.** How many solutions may a system of two quadratic equations in x and y have?
   b. Which of these possibilities is illustrated in Example 1?

In 2 and 3, refer to the relations $x^2 + y^2 = 9$ and $x^2 + 4y^2 = 16$ graphed at the left.

2. Estimate the solutions from the graph.

3. Find the exact solutions algebraically.

In 4 and 5, refer to Example 2.

4. What in the original equations signifies that the solutions will be symmetric to the y-axes?

5. The four solutions are vertices of what figure?

6. Find all points of intersection of the circle $x^2 + y^2 = 9$ and the hyperbola $\frac{x^2}{4} - y^2 = 1$.

In 7 and 8, refer to Example 3.

7. What were the two substitutions that transformed the second equation into an equation with only one variable?

8. If, in the second month, Wanda had instead raised prices by $30 per pair and earned $10,800 from 80 fewer sales than the previous month, what would have been the price of boots in each month?

This is a photograph of Nat Love, a cowboy who lived in the late 1800s.

Applying the Mathematics

9. Solve the system
$$\begin{cases} y = x^2 - 4x + 3 \\ y = x^2 - 9. \end{cases}$$

10. Consider this situation. The product of two numbers is 1073. If one number is increased by 3 and the other is decreased by 7, the new product is 960.
  a. *Multiple choice.* Which of the following systems represents this situation?
  (i) $\begin{cases} xy = 960 \\ (x + 3)(y - 7) = 1073 \end{cases}$
  (ii) $\begin{cases} xy = 1073 \\ (x - 3)(y + 7) = 960 \end{cases}$
  (iii) $\begin{cases} xy = 1073 \\ (x + 3)(y - 7) = 960 \end{cases}$
  (iv) $\begin{cases} xy = 960 \\ (x - 3)(y - 7) = 960 \end{cases}$
  b. Find the numbers.

11. Without doing any calculations, solve the system
$$\begin{cases} (x - 3)^2 + y^2 = 4 \\ (x + 3)^2 + y^2 = 4. \end{cases}$$
Explain how you found your answer.

12. A circle and a parabola can intersect in at most how many points?

13. Two different circles can intersect in at most how many points?

14. One earthquake monitoring station determines that the center of a quake is 30 miles away. A second station 40 miles east and 10 miles north of the first finds that it is 20 miles from the quake's center.
  a. Suppose the first monitoring station is at the origin of a coordinate system. Write a system of equations to describe this situation.
  b. Solve the system and find the coordinates of all possible locations of the quake's center.

15. By graphing, estimate all solutions to the system $\begin{cases} xy = 10 \\ y = 4x + 5. \end{cases}$
(Lesson 12-8)

16. Give an example of an inconsistent quadratic-linear system.
(Lesson 12-8)

In 17–19, the shape of the light beam from a flashlight is a cone. When that cone of light hits a flat surface, the outline is a conic section. Tell which conic section is formed when the flashlight is held in the following manner. *(Lesson 12-7)*

17. perpendicular to the wall

18. at an angle of about 75° to the wall

19. touching the wall, with the axis of the flashlight parallel to the wall.

20. Determine equations for all asymptotes of the hyperbola $x^2 - y^2 = 7$ from Example 2 in this lesson. *(Lesson 12-7)*

In 21 and 22, consider the ellipse pictured at the left.

21. Find its area.

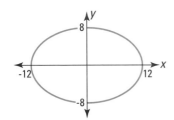

22. Give an equation for it in standard form. *(Lessons 12-4, 12-5)*

23. In early Roman construction the gate to a building often consisted of a semicircular arch built over a square opening. Suppose such a gate is 3.5 m wide.
  a. How high is it?
  b. Can a truck 2 m wide and 4 m high fit through the gate? *(Lesson 12-3)*

24. A skydiver jumping from a plane falls about 16 ft the first second, 48 ft the next second, and 80 ft the third second. The sequence of distances is arithmetic if air resistance is ignored. How many feet will the diver fall in the thirtieth second? *(Lessons 3-7, 3-8)*

Taking a dive. *Shown are skydivers in a free fall. Skydivers usually jump at altitudes of up to 15,000 ft and fall at speeds of more than 100 mph. They open their parachutes at altitudes of 2000–3000 ft and glide to the ground at about 10 mph.*

25. Give an equation for a quadratic relation that intersects the unit circle $x^2 + y^2 = 1$
  a. in no points.
  b. in exactly one point.
  c. in exactly two points.
  d. in exactly three points.
  e. in exactly four points.

PROJECTS
12
CHAPTER TWELVE

A project presents an opportunity for you to extend your knowledge of a topic related to the material of this chapter. You should allow more time for a project than you do for typical homework questions.

1 Arches

Semicircular arches were popular with the early Romans. Other kinds of arches have been popular at other times and places. Prepare a brief report on the various kinds of arches. Include arches from the Roman and Renaissance eras and modern arches such as the Gateway Arch in St. Louis. If possible, give equations for curves that resemble these arches.

2 Whispering Galleries

A whispering gallery is a large room in which a person whispering while standing in one special spot can easily be heard by another person standing in another special spot. There are whispering galleries in the United States Capitol; in the church of St. John Lateran in Rome, Italy; in the Louvre museum in Paris, France; and in many other public buildings and places of worship around the world. Prepare a report or a poster about these whispering galleries, describing their sizes, why they were built, and their relation to the conic sections.

3 Sections of Cones

On page 747, an ellipse, a hyperbola, and a parabola are shown as the intersections of a plane and a double cone. Make four models of a double cone out of paper that is thin enough to be bent yet thick enough to be sturdy. Intersect these models with planes to form each type of conic and also a circle.

4 Reflection Properties of the Conics

Parabolas, ellipses, and hyperbolas are used in telescopes, whispering galleries, navigation, satellite dishes, and headlights because they have reflection properties. Describe the reflection property for each of these conic sections, and illustrate the property with accurate drawings.

Above: Very Large Array (VLA), National Radio Astronomy in Socorro, New Mexico.
Lower Right: Mauna Kea Observatory in Hawaii.

5 Orbits of the Planets

The orbits of planets around the sun are ellipses with the sun at one focus.

a. Describe these ellipses, giving the major and minor axes for the orbit of each planet and indicating the nearest and farthest distances of each planet to the sun (these are called the planet's *perihelion* and *aphelion,* respectively).

b. Draw two accurate pictures of these orbits, one with the inner four planets, the other with the outer five planets.

c. The closeness of an ellipse to a circle is measured by the *eccentricity* of the ellipse. Give the eccentricity of each orbit and tell how it can be calculated.

6 Graphing Conic Sections with an Automatic Grapher

a. Pick at least one circle, one noncircular ellipse, and one hyperbola from this chapter and graph them using an automatic grapher. Show the instructions needed and the graphs.

b. Repeat part a for a quadratic-linear system and a quadratic-quadratic system.

c. Write a short essay comparing the automatic grapher procedures with paper-and-pencil methods for graphing and solving.

7 Asymptotes

How fast does a hyperbola get close to its asymptotes? Consider the hyperbola $x^2 - y^2 = 1$.

a. How large does x have to be in order for the point (x, y) to be within a vertical distance of 0.1 to the hyperbola?

b. Answer part a for a vertical distance of 0.01, 0.001, and 0.0001.

c. Extend the pattern you find in part b to indicate how one could estimate how large x has to be in order for the point (x, y) to be within 10^{-n} of the hyperbola, for large values of n.

SUMMARY

In this chapter you studied quadratic relations in two variables, their graphs, and geometric properties of these figures. A quadratic equation in two variables is of the form $Ax^2 + Bxy + Cy^2 + Dx + Ey + F = 0$, where A, B, and C are not all zero.

Conic section	Circle	Ellipse	Parabola	Hyperbola
Geometric definition:	given C, set of points P such that $PC = r$	given F_1 and F_2, set of points P such that $PF_1 + PF_2 = 2a$	given F and ℓ, set of points P equidistant from F and ℓ	given F_1 and F_2, set of points P such that $\lvert PF_1 - PF_2 \rvert = 2a$
Equation in standard form:	$(x - h)^2 + (y - k)^2 = r^2$	$\dfrac{x^2}{a^2} + \dfrac{y^2}{b^2} = 1$	$y - k = a(x - h)^2$	$\dfrac{x^2}{a^2} - \dfrac{y^2}{b^2} = 1$ or $xy = k$

Graph:

Circle:

center: (h, k)
radius: r

Ellipse:
If $a > b$:

foci: $(-c, 0)$, $(c, 0)$
length of major axis (focal constant): $2a$
length of minor axis: $2b$
$b^2 = a^2 - c^2$

If $b > a$:

foci: $(0, -c)$, $(0, c)$
length of major axis (focal constant): $2b$
length of minor axis: $2a$
$a^2 = b^2 - c^2$

Parabola:
$y - k = a(x - h)^2$

axis of symmetry: $x = h$
vertex: (h, k)

$y - ax^2$

axis of symmetry: $x = 0$
vertex: $(0, 0)$
focus: $\left(0, \dfrac{1}{4a}\right)$
directrix: $y = -\dfrac{1}{4a}$

Hyperbola:
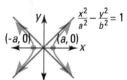 $\dfrac{x^2}{a^2} - \dfrac{y^2}{b^2} = 1$
foci: $(-c, 0)$, $(c, 0)$, $c^2 = a^2 + b^2$
focal constant: $2a$
asymptotes: $\dfrac{y}{b} = \pm\dfrac{x}{a}$

 $xy = k$
foci: $\left(\sqrt{2k}, \sqrt{2k}\right)$, $\left(-\sqrt{2k}, -\sqrt{2k}\right)$
focal constant: $2\sqrt{2k}$
asymptotes: $x = 0$, $y = 0$

Conic section:

SUMMARY

Conic sections appear naturally as orbits of planets and comets, in paths of objects thrown into the air, as energy waves radiating from the epicenter of an earthquake, and in many manufactured objects such as tunnels, windows, and satellite receiver dishes.

Systems of equations with quadratic sentences are solved in much the same way as linear systems, that is, by graphing, by substitution, or by using linear combinations. A system of one linear equation and one quadratic equation may have 0, 1, or 2 solutions; a system of two quadratic equations may have 0, 1, 2, 3, 4, or infinitely many solutions.

VOCABULARY

Below are the most important terms and phrases for this chapter. You should be able to give a definition for those terms marked with *. For all other terms you should be able to give a general description and a specific example.

Lesson 12-1
*quadratic equation in two variables
*quadratic relation in two variables
double cone
conic section, conic
*parabola
focus, directrix
axis of symmetry
vertex
paraboloid

Lesson 12-2
*circle, radius, center
concentric circles
*Center-Radius Equation for a Circle Theorem

Lesson 12-3
*interior, exterior, of a circle
*boundary
*Interior and Exterior of a Circle Theorem

Lesson 12-4
conic graph paper
*ellipse
foci, focal constant
standard position for an ellipse
Equation for an Ellipse Theorem
standard form of equation for an ellipse
*major axis, minor axis, center of an ellipse
vertex, vertices of an ellipse

Lesson 12-5
Graph Scale-Change Theorem
area of an ellipse
eccentricity of an ellipse

Lesson 12-6
*hyperbola
foci, focal constant
vertices of a hyperbola
asymptotes of a hyperbola
Equation for a Hyperbola Theorem
*standard form for an equation of a hyperbola

Lesson 12-7
*rectangular hyperbola
*standard form for a quadratic relation

Lesson 12-8
quadratic system
quadratic-linear system

Lesson 12-9
quadratic-quadratic system

PROGRESS SELF-TEST

Take this test as you would take a test in class. You will need regular graph paper, conic graph paper, and a ruler. Then check your work with the solutions in the Selected Answer section in the back of the book.

1. Determine an equation for the circle with center (-3, 13) and radius 10.

2. Graph the set of points (x, y) such that $x^2 + y^2 < 25$.

In 3 and 4, consider the image of $x^2 + y^2 = 1$ under the scale change $S_{3,4}$.

3. Write an equation for the image.

4. Graph the preimage and the image.

In 5 and 6, refer to the ellipse drawn below.

5. Determine an equation for this ellipse.

6. Find its area.

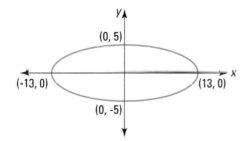

In 7 and 8, consider the system
$$\begin{cases} y = x - 2 \\ y = 4x - x^2. \end{cases}$$

7. Estimate the solutions by graphing the system.

8. Find the exact solutions.

9. Pluto has an elliptical orbit with the Sun at one focus. Its closest distance to the Sun is about 2.8 billion miles, while its farthest distance is about 4.6 billion miles. What is the length of the major axis of Pluto's orbit?

10. Use conic graph paper with centers 12 units apart to sketch the set of points P such that $|PF_1 - PF_2| = 5$.

11. a. Rewrite the equation $y = 2(x + 4)^2 - 9$ in the standard form for a quadratic relation.

b. Identify the conic section represented by this equation.

12. A forest fire is sighted 400 meters from one reporting station. A second reporting station 400 meters due east of the first station sights the same fire at a distance of 500 meters.

a. Set up a system of equations that represents this situation.

b. Solve the system to find the possible locations of the forest fire in relation to your system.

13. The graph of $xy = 2$ is a hyperbola. Find equations for the asymptotes of this hyperbola.

14. Graph the set of points (x, y) satisfying
$$\frac{x^2}{9} - \frac{y^2}{4} = 1.$$

15. The vertex of the parabola below is (3, 5). The directrix is the line $y = 6$. What are the coordinates of the focus?

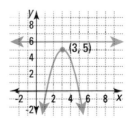

CHAPTER REVIEW

Questions on SPUR Objectives

SPUR stands for **S**kills, **P**roperties, **U**ses, and **R**epresentations. The Chapter Review questions are grouped according to the SPUR Objectives for this chapter.

SKILLS DEAL WITH THE PROCEDURES USED TO GET ANSWERS.

Objective A: *Rewrite an equation for a conic section in the general form of a quadratic equation in two variables.* *(Lesson 12-7)*

In 1–6, rewrite in the form
$Ax^2 + Bxy + Cy^2 + Dx + Ey + F = 0.$

1. $(x - 3)^2 + (y + 7)^2 = 100$

2. $y = 4(x - 2)^2 + 6$

3. $\dfrac{x^2}{25} - \dfrac{y^2}{9} = 1$ 4. $\dfrac{x^2}{6} + \dfrac{y^2}{5} = 1$

5. $y = \pm\sqrt{25 - x^2}$ 6. $y = \dfrac{20}{x}$

Objective B: *Write equations or inequalities for quadratic relations given sufficient conditions.*
(Lessons 12-1, 12-2, 12-4, 12-6, 12-7)

In 7 and 8, find an equation for the circle satisfying the conditions.

7. center at origin, radius 6

8. center is (-7, 5), radius 12

9. Give an equation for the upper semicircle of the circle $x^2 + y^2 = 20$.

10. What inequality describes the interior of the ellipse with equation $x^2 + 3y^2 = 75$?

11. What sentence describes the exterior of the ellipse with equation $x^2 + 3y^2 = 75$?

In 12 and 13, write an equation for the ellipse satisfying the given conditions.

12. foci: (0, 5) and (0, -5); focal constant: 26

13. The endpoints of the major and minor axes are (3, 0), (-3, 0), (0, 6), and (0, -6).

In 14 and 15, find an equation for the hyperbola satisfying the given conditions.

14. vertices: (1, 1) and (-1, -1)

15. foci: (7, 0) and (-7, 0); focal constant: 8

Objective C: *Find the area of an ellipse.*
(Lesson 12-5)

In 16 and 17, find the area of the ellipse satisfying the given conditions.

16. Its equation is $\dfrac{x^2}{121} + \dfrac{y^2}{9} = 1$.

17. The endpoints of its major and minor axes are (0, 10), (0, -10), (5, 0), and (-5, 0).

18. Which has a larger area: a circle of radius 5 or an ellipse with major and minor axes of lengths 12 and 8, respectively? Justify your answer.

19. Find the area of the shaded region below, which is between an ellipse with major axis of length 10 and minor axis of length 8, and a circle with diameter 8.

Objective D: *Solve systems of one linear and one quadratic equation or two quadratic equations by substitution or linear combination.*
(Lessons 12-8, 12-9)

In 20–27, solve.

20. $\begin{cases} y = x^2 + 5 \\ y = -x^2 + 5x + 8 \end{cases}$

21. $\begin{cases} 2x + y = 23 \\ y = 2x^2 - 7x + 5 \end{cases}$

22. $\begin{cases} y = x^2 + 3x - 4 \\ y = 2x^2 + 5x - 3 \end{cases}$

23. $\begin{cases} x^2 + y^2 = 1 \\ x^2 + y^2 = 9 \end{cases}$

24. $\begin{cases} (x - 3)^2 + y^2 = 25 \\ x^2 + (y - 1)^2 = 25 \end{cases}$

25. $\begin{cases} x^2 - y^2 = 9 \\ \dfrac{x^2}{50} + \dfrac{y^2}{32} = 1 \end{cases}$

26. $\begin{cases} xy = 12 \\ y = 3x - 1 \end{cases}$

27. $\begin{cases} y = 2x^2 \\ x + 2y = 5 \end{cases}$

PROPERTIES DEAL WITH THE PRINCIPLES BEHIND THE MATHEMATICS.

Objective E: *Find points on a conic section using the definition of the conic.* *(Lessons 12-1, 12-4, 12-6)*

28. Graph the set of points equidistant from the point $(3, 2)$ and line $y = -2$.

$F \bullet$

29. Copy the figure at the right. Find five points on the parabola with focus F and directrix d, including the vertex of the parabola.

d

In 30 and 31, use conic graph paper with centers 10 units apart to draw the set of points P satisfying the given condition.

30. $PF_1 + PF_2 = 18$ **31.** $|PF_1 - PF_2| = 8$

Objective F: *Identify characteristics of parabolas, circles, ellipses and hyperbolas.* *(Lessons 12-1, 12-2, 12-4, 12-6, 12-7)*

In 32 and 33, identify the center and radius of the circle with the given equation.

32. $(x + 8)^2 + y^2 = 196$ **33.** $x^2 + y^2 = 5$

34. Consider the ellipse with equation $\dfrac{x^2}{169} + \dfrac{y^2}{400} = 1.$

    a. Name its vertices.

    b. State the length of its minor axis.

35. Consider the parabola with equation $y = \frac{1}{2}x^2$. Name the

    a. focus.   **b.** vertex.   **c.** directrix.

36. Consider the ellipse with equation $\dfrac{x^2}{100} + \dfrac{y^2}{36} = 1.$

    a. Find the foci F_1 and F_2.

    b. Suppose P is on this ellipse. Find the value of $PF_1 + PF_2$.

37. Consider the hyperbola with equation $\dfrac{x^2}{16} - \dfrac{y^2}{4} = 1.$

    a. Name its vertices.

    b. State equations for its asymptotes.

38. Identify the asymptotes of the hyperbola $xy = 5$.

Objective G: *Classify curves as circles, ellipses, parabolas, or hyperbolas using algebraic or geometric properties.* *(Lessons 12-1, 12-3, 12-4, 12-5, 12-6, 12-7)*

In 39 and 40, consider two fixed points F_1 and F_2 and a focal constant d. Identify the set of points P satisfying the given conditions.

39. $F_1P + F_2P = d$, where $d > F_1F_2$

40. $|F_1P - F_2P| = d$, where $d < F_1F_2$

41. The figure at the right shows a double cone intersected by four planes A, B, C, and D. Plane B is parallel to an edge of the cone; plane D is perpendicular to an axis of the cone. Identify the curve produced by each intersection.

In 42–45, *true or false.*

42. Every circle is an ellipse.

43. The image of the unit circle under a scale change is an ellipse.

44. A hyperbola can be considered as the union of two parabolas.

45. All types of quadratic relations in two variables can be determined from the intersection of a plane and a double cone.

46. a. What equation describes the image of the circle $x^2 + y^2 = 1$ under the scale change $S: (x, y) \rightarrow (6x, 9y)$?

  b. What kind of curve is the image in part **a**?

USES DEAL WITH APPLICATIONS OF MATHEMATICS IN REAL SITUATIONS.

Objective H: *Use circles, ellipses, and hyperbolas to solve real-world problems.*
(Lessons 12-2, 12-3, 12-4, 12-5, 12-6, 12-7)

47. A truck 10 ft high and 5 ft wide approaches a semicircular tunnel with a radius of 12 ft. Will the truck fit through the tunnel if it travels to one side of the center line? Justify your answer.

48. The elliptically shaped pool below is to be surrounded by tile so that the outer boundary of the tile is also an ellipse. The tiler needs to know the area of the shaded region to determine how much tile to buy. The major axis of the pool is 15 m, and the minor axis of the pool is 8 m. The major axis AB is 18 m, and the minor axis DC is 11 m. What is the area of the shaded region?

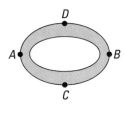

49. The orbit of the Earth around the Sun is elliptical with the Sun as one focus. The closest and farthest distances of the Earth from the Sun are 91.4 and 94.5 million miles, respectively.

  a. How far is F_2, the second focus, from the Sun?

  b. What is the length of the minor axis of the Earth's orbit?

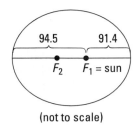

(not to scale)

In 50 and 51, refer to the figure at the right. A computer programmer needs to write instructions to draw such a figure with concentric circles with radii of 10, 30, and 50 pixels. The center of the circles is at the point (200, 100).

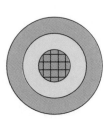

50. What sentence does the checkerboard region satisfy?

51. What sentence does the light-shaded ring satisfy?

52. A person bought a certain number n of pencils at the same cost c for each pencil and spent \$10. Which conic section contains all possible points (n, c)?

Objective I: *Use systems of quadratic equations to solve real-world problems.* *(Lessons 12-8, 12-9)*

53. A rectangular Oriental rug has an area of 200 square feet and a perimeter of 60 feet. Find the dimensions of the rug.

In 54 and 55, suppose the epicenter of an earthquake is about 50 miles away from monitoring station 1. The quake is also 50 miles away from another monitoring station, which is 60 miles east and 40 miles south of station 1.

54. Find the possible locations for the epicenter.

55. The same quake is about 106 miles away from station 3, which is 70 miles west and 20 miles north of station 1. Where is the actual epicenter of the earthquake?

56. The demand function for Peewee's Sports Company is $xp = 250$, where x is the number of baseballs in hundreds, and p is the unit price of a baseball. The supply function for the Giant Baseball Manufacturer is $p = 2x^2$. Find the equilibrium point, that is, the point where supply and demand intersect.

57. Eileen's Eye Extravaganza took in $5600 in sales of sunglasses for last year. This year Eileen lowered the price by two dollars, sold seventy more pairs of sunglasses, and took in $5880.

  a. How much is she selling her sunglasses for now?

  b. How many pairs did she sell this year?

REPRESENTATIONS DEAL WITH PICTURES, GRAPHS, OR OBJECTS THAT ILLUSTRATE CONCEPTS.

Objective J: *Graph quadratic relations given sentences for them, and vice versa.*
(Lessons 12-1, 12-2, 12-3, 12-4, 12-6, 12-7)

In 58–61, sketch a graph.

58. $\frac{x^2}{16} + \frac{y^2}{81} = 1$ **59.** $\frac{x^2}{16} - \frac{y^2}{81} = 1$

60. $xy = 12$ **61.** $x^2 + y^2 \geq 9$

In 62 and 63, state an equation for the curve.

62. the circle tangent to the coordinate axes as shown at the right

63. the ellipse drawn below

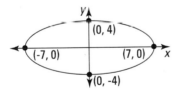
(0, 4)
(-7, 0)   (7, 0) *x*
(0, -4)

In 64 and 65, *multiple choice.* Select an equation that best describes each graph.

  (a) $\frac{x^2}{a^2} + \frac{y^2}{b^2} = 1$ (b) $\frac{x^2}{a^2} - \frac{y^2}{b^2} = 1$

  (c) $y = ax^2$ (d) $xy = a$

64.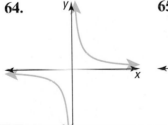

65.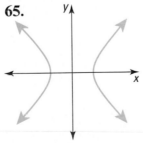

Objective K: *Solve systems of quadratic equations graphically.* *(Lessons 12-8, 12-9)*

In 66 and 67, **a.** Solve the system by graphing. **b.** Check your work.

66. $\begin{cases} y = x^2 - 10 \\ y = 11 - x \end{cases}$ **67.** $\begin{cases} x^2 + y^2 = 81 \\ x^2 + (y + 18)^2 = 81 \end{cases}$

In 68 and 69, draw an example showing how the situation can occur.

68. a circle and a hyperbola that intersect in 4 points

69. two parabolas that do not intersect

70. Refer to the graphs below of the curves $\frac{x^2}{40} + \frac{y^2}{10} = 1$ and $x + y = 1$.

  a. Estimate the points of intersection from this sketch.

  b. Use an automatic grapher to estimate the solutions to the nearest tenth.

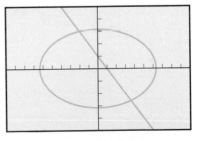

$-10 \leq x \leq 10,\ \ x\text{-scale} = 1$
$-5 \leq y \leq 5,\ \ y\text{-scale} = 1$

SERIES AND COMBINATIONS

Addition and multiplication are as fundamental in advanced mathematics as in arithmetic. There are formulas for sums of terms of various sequences, and special notations for sums and for certain products.

Many of these applications are related to the triangular array known as Pascal's triangle.

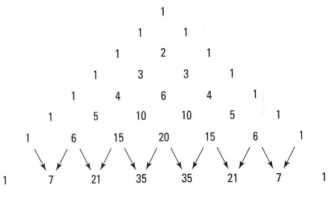

The arrows in the array show how the sum of two elements in one row equals an element in the next row. Yet, an explicit formula for each element involves products.

The entries in Pascal's triangle have many applications. They help to expand the power of any binomial $(x + y)^n$. They answer counting questions involving combinations. They can be used to determine the probability that a person will win a lottery. They give guidance for determining how many people should be questioned in an election poll or for television ratings. In this chapter, you will learn about these and other applications of sums and products. Along the way, you will have the opportunity to revisit many ideas you have seen earlier in this book.

Sum mathematician! *Shown is Carl Friedrich Gauss on the terrace of the Gottingen Observatory where he was a professor.*

Sums of Consecutive Integers

There is a story often told about the famous mathematician Gauss, whose third grade class misbehaved. The teacher gave the following problem as punishment.

"Add the whole numbers from 1 to 100."

The story is that Gauss solved the problem in almost no time at all. He wrote only the number 5050 on his slate. The teacher recognized that Gauss was extraordinary and gave him some advanced books to read.

Gauss's method was something like the following. Let S be the desired sum.

$$S = 1 + 2 + 3 + \ldots + 98 + 99 + 100$$

Using the Commutative and Associative Properties, you can rewrite the sum in reverse order.

$$S = 100 + 99 + 98 + \ldots + 3 + 2 + 1$$

Now add corresponding terms in the equations above.

So $\quad 2S = \underbrace{101 + 101 + 101 + \ldots + 101 + 101 + 101}_{100 \text{ terms}}.$

Thus $\quad 2S = 100 \cdot 101$
$\qquad\ \ S = 5050$

Gauss's method of solution is the basis for the proof of the next theorem.

Theorem

The sum of the integers from 1 to n is $\frac{1}{2}n(n + 1)$.

Proof:

Let
$$S = 1 + 2 + \ldots + (n - 1) + n.$$
Reorder terms to get
$$S = n + (n - 1) + \ldots + 2 + 1.$$
Add equations to get
$$2S = (1 + n) + (2 + n - 1) + \ldots + (n - 1 + 2) + (n + 1).$$
Simplify.
$$2S = \underbrace{(n + 1) + (n + 1) + \ldots + (n + 1) + (n + 1)}_{n \text{ terms}}$$
$$2S = n(n + 1)$$
$$S = \tfrac{1}{2}n(n + 1).$$

By the theorem, the sum of the integers from 1 to 100 is $\frac{1}{2}(100)(101) = 5050$, the answer Gauss gave.

Example 1

Part of a popular Christmas carol is "On the 12th day of Christmas my true love gave to me

12 drummers drumming	6 geese-a-laying
11 pipers piping	5 golden rings
10 lords-a-leaping	4 calling birds
9 ladies dancing	3 French hens
8 maids-a-milking	2 turtle doves, and
7 swans-a-swimming	a partridge in a pear tree."

How many gifts did the true love give on the 12th day of Christmas?

Solution

You must find the sum $S = 1 + 2 + 3 + \ldots + 12.$
Use the theorem for the sum of integers 1 to n, with $n = 12$.

$$S = \frac{1}{2} \cdot 12 \cdot 13$$
$$S = 78$$

The true love gave 78 gifts on the 12th day of Christmas.

Check

Add the whole numbers from 1 to 12 in your head.

Notice that the sum $\frac{1}{2}n(n + 1)$ of the integers from 1 to n is a quadratic expression in the variable n. Hence, if you are given this sum, to find n, you can solve a quadratic equation.

Example 2

How many consecutive integers beginning with 1 would you have to add in order to get 2080?

Solution

Let n be the last integer you would add. Then

$$1 + 2 + 3 + \ldots + n = 2080.$$ Apply the theorem on page 811.

$$\frac{1}{2}n(n + 1) = 2080$$ Multiply both sides by 2.

$$n(n + 1) = 4160$$ Distributive Property

$$n^2 + n = 4160$$ Add −4160 to both sides.

$$n^2 + n - 4160 = 0$$

$$n = \frac{-1 \pm \sqrt{1^2 - 4 \cdot 1 \cdot -4160}}{2 \cdot 1}$$ Quadratic Formula

$$n = \frac{-1 + 129}{2} = 64 \text{ or } n = \frac{-1 - 129}{2} = -65$$

Since $n > 1$, you would have to add 64 consecutive integers beginning with 1 to get 2080.

Check

Calculate $1 + 2 + 3 + \ldots + 64$ using the theorem.

Does $\frac{1}{2}(64)(65) = 2080$? Yes, it checks.

Sums of Terms in an Arithmetic Sequence

Recall that an arithmetic, or linear, sequence is a sequence in which the difference between consecutive terms is constant and has the form

$$a_1, a_1 + d, a_1 + 2d, \ldots, a_1 + (n - 1)d, \ldots.$$

The integers from 1 to 100 form a finite arithmetic sequence with $a_1 = 1$, $n = 100$, and $d = 1$. Reasoning similar to that of Gauss can be used to find the sum of consecutive terms of any finite arithmetic sequence.

Example 3

Find the sum of the first 30 terms of the arithmetic sequence 4, 11, 18, 25, ...

Solution

First, calculate the 30th term. The first term a_1 is 4, the number of terms n is 30, and the common difference d is 7.

In general $a_n = a_1 + d(n - 1).$

So $a_{30} = 4 + 7(30 - 1) = 207.$

The 30th term is 207.

Thus, $S = 4 + 11 + \ldots + 200 + 207.$

Now do what Gauss did. Reverse the order of the terms being added.

Add the equations. $S = 207 + 200 + \ldots + 11 + 4$

So, $2S = \underbrace{211 + 211 + \ldots + 211 + 211.}_{30 \text{ terms}}$

$$2S = 30 \cdot 211$$

$$S = \frac{1}{2}(30)(211) = 3165$$

The sum of the first 30 terms is 3165.

What Is an Arithmetic Series?

A series is an indicated sum of terms of a sequence. If the terms form an arithmetic sequence with first term a_1 and common difference d, the indicated sum of the terms is called an arithmetic series. The sum of the first n terms, represented as S_n, is

$$S_n = a_1 + a_2 + a_3 + \ldots + a_{n-2} + a_{n-1} + a_n.$$

We find a formula for S_n by writing the arithmetic series in two ways:

(1) Start with the first term a_1 and successively *add* the common difference d.

$$S_n = a_1 + (a_1 + d) + (a_1 + 2d) + \ldots + [a_1 + (n - 1)d]$$

(2) Start with the last term a_n and successively *subtract* the common difference d.

$$S_n = a_n + (a_n - d) + (a_n - 2d) + \ldots + [a_n - (n - 1)d]$$

Now add corresponding pairs of terms of these two formulas, as Gauss did. Then each of the n pairs has the same sum, $a_1 + a_n$.

$$S_n + S_n = \underbrace{(a_1 + a_n) + (a_1 + a_n) + (a_1 + a_n) + \ldots + (a_1 + a_n)}_{n \text{ terms}}$$

So $\qquad 2S_n = n(a_1 + a_n).$

Thus, $\qquad S_n = \frac{n}{2}(a_1 + a_n).$

> **Theorem**
>
> Let $S_n = a_1 + a_2 + \ldots + a_n$ be an arithmetic series. Then
>
> $$S_n = \frac{n}{2}(a_1 + a_n)$$

The formula in the theorem is convenient if the first and nth terms are known. If the nth term is not known, another formula can be used. Start with the formula for the nth term of an arithmetic sequence.

$$a_n = a_1 + (n - 1)d$$

Substitute this expression for a_n in the right side of the formula for S_n.

$$S_n = \frac{n}{2}[a_1 + (a_1 + (n - 1)d)]$$

That is, $\qquad S_n = \frac{n}{2}[2a_1 + (n - 1)d].$

This argument proves the following corollary.

> **Corollary**
>
> Let $S_n = a_1 + a_2 + \ldots + a_n$ be an arithmetic series with constant difference d. Then:
>
> $$S_n = \frac{n}{2}[2a_1 + (n - 1)d].$$

Shown is the auditorium in Hong Kong's Cultural Center. The auditorium can seat 2200.

Example 4

An auditorium has 15 rows, with 20 seats in the front row and 2 more seats in each row thereafter. How many seats are there in all?

Solution

Since the first term, the constant difference, and the number of terms are given, use the formula

$$S_n = \frac{n}{2}(2a_1 + (n - 1)d).$$

Then

$$S_{15} = \frac{15}{2}(2 \cdot 20 + (15 - 1) \cdot 2)$$

$$= \frac{15}{2}(40 + 28) = 510.$$

Check

Use the formula $\quad S_n = \frac{n}{2}(a_1 + a_n).$

In this case, $\quad a_n = a_{15} = 20 + 14 \cdot 2 = 48.$

$$S_{15} = \frac{15}{2}(20 + 48) = \frac{15}{2} \cdot 68 = 510$$

So there are 510 seats in the auditorium.

QUESTIONS

Covering the Reading

1. What problem was Gauss given in 3rd grade, and what is its answer?

2. Find the sum of the integers from 1 to 1000.

3. If the sum of n consecutive integers beginning with 1 is 1540, how many integers are being added?

4. Consider the expressions $20 + 18 + 16 + 14$ and 20, 18, 16, 14.
   a. Which is an arithmetic sequence?
   b. Which is an arithmetic series?

5. Consider the arithmetic sequence with first term a_1 and constant difference d.
   a. Write a formula for the nth term.
   b. Write a formula for the sum of the first n terms.

6. The Jewish holiday Chanukah is celebrated by lighting candles in a menorah for eight days. On the first night, two candles are lit; on the second night three candles are lit; and on each successive night an additional candle is lit. Each night new candles are lit. How many candles are needed for all eight nights?

Holiday candles. *Shown is a menorah lit on the eighth night of Chanukah. On each night a special candle (here in the center) is lit and then used to light the other candles for that day of Chanukah.*

7. **a.** Write all the terms in the arithmetic series $5 + 9 + 13 + \ldots + 37$.
   b. How many terms are there?
   c. What is the sum of all the terms?

8. **a.** Find the sum of the first 60 terms of the sequence in Example 3.
   b. *True or false.* $S_{60} = 2S_{30}$.

9. Consider the arithmetic sequence $1, 3, 5, 7, \ldots$.
   a. Find a_{50}.   **b.** Find S_{50}.

10. Suppose a theater has 26 seats in the first row and that each row has 4 more seats than the previous row. If there are 30 rows in the theater, how many seats are there in all?

In 11 and 12, suppose cans are stacked for a display in a store as shown at the left. There is one can at the top, and in each successive row there is one more can than in the preceding row.

11. If there are 20 rows in the display, how many cans are used?

12. The store manager wants to display exactly 500 cans in this way. Can this be done? If it can, determine how many rows are needed. If it cannot, justify your answer, and describe how else the manager might display 500 cans.

Applying the Mathematics

13. The following BASIC program generates recursively the terms of an arithmetic sequence and the sum of the terms of that sequence.

```
10 REM PROGRAM TO PRINT TERMS OF ARITHMETIC SEQUENCE AND SUM OF
   SERIES
15 LET N = 1
20 LET TERM = 10
25 LET SUM = 0
30 LET SUM = SUM + TERM
35 PRINT "N", "TERM", "SUM"
40 PRINT N, TERM, SUM
45 FOR N = 2 TO 15
50 TERM = TERM + 3
55 SUM = SUM + TERM
60 PRINT N, TERM, SUM
65 NEXT N
70 END
```

   a. Run this program or a similar one for your technology and list the last line of output.
   b. What explicit formulas could have been used to calculate the last term and sum directly?
   c. Modify the program so it generates the sequence and series determined by Questions 17 and 18.

14. Finish this sentence: The sum of the n terms of an arithmetic sequence equals the average of the first and last term multiplied by __?__.

15. An organization has new officers for the year and is ordering new stationery. In January, a mailing is sent to the 325 current members. If the membership increases by 5 members each month, how many envelopes will be needed for monthly mailings for the entire year?

16. a. How many odd integers are there from 25 to 75?
b. Find the sum of the odd integers from 25 to 75.

In 17 and 18, let S_n be the sum of the first n terms of the sequence defined by $a_n = 11n - 3$.

17. Find the indicated sum.
a. S_2 **b.** S_3 **c.** S_{25}

18. Find the smallest value of n such that $S_n \geq 5000$.

Review

19. Consider the geometric sequence 8, -12, 18, -27, ... Determine
a. the common ratio.
b. the next term.
c. an explicit formula for the nth term.   *(Lesson 7-5)*

20. Suppose an account pays 5.25% annual interest compounded monthly.   *(Lesson 7-4)*
a. Find the annual yield of the account.
b. Find the value of a $1500 deposit after 4 years.

21. Find an equation for the parabola that contains the points (0, 10), (1, 16), and (4, 10).   *(Lesson 6-6)*

22. The function with equation $y = 3(x - 4)^2 + k$ contains the point (2, -1).
a. What is the value of k?
b. Describe the graph of this function.   *(Lessons 6-3, 6-4)*

23. a. Find the inverse of $\begin{bmatrix} 3 & 2 \\ 1 & n \end{bmatrix}$.
b. For what value(s) of n does the inverse not exist?   *(Lesson 5-5)*

. . . and a partridge in a pear tree. *Shown is the chukar partridge. It is native to Europe and Asia.*

Exploration

24. The number 9 can be written as the sum of an arithmetic sequence $9 = 1 + 3 + 5$. What other numbers from 1 to 100 can be the sum of an arithmetic sequence whose terms are positive integers? (Assume the sequence must have at least three distinct terms.)

25. How many gifts would the true love have given in all if the true love did exactly what the song *The Twelve Days of Christmas* says? (Count two turtle doves as two gifts.)

A reward that went against the grain. *Shown is a chess piece for the king—and a very small portion of the wheat grain that would be needed to reward the game's inventor as discussed in the legend below.*

A Famous Legend

Legend has it, that when he first learned to play chess, the King of Persia was so impressed he summoned the game's inventor to offer a reward. The inventor pointed to the chessboard, and said that for a reward he would like one grain of wheat on the first square, two on the second, four on the third, eight on the fourth, and so on, for all sixty-four squares. The king protested that this was not enough reward, but the inventor insisted. What do you think? Is this a large or small reward?

Notice that the situation is one of exponential growth. In fact, the terms in this situation form a geometric sequence with first term 1 and growth factor, or constant ratio, 2. The nth term of this sequence is 2^{n-1}. The total number of grains of wheat on the chessboard is the sum of the first 64 terms of this sequence. Call this sum S_{64}.

$$S_{64} = 1 + 2 + 4 + 8 + \ldots + 2^{62} + 2^{63}$$

An indicated sum of successive terms of a geometric sequence, like this formula for S_{64}, is called a **geometric series.**

To evaluate S_{64}, you can use a method similar to that used in the previous lesson for an arithmetic series. Notice that if each term of S_{64} is doubled, many values are identical to those in the original formula.

$$2S_{64} = 2 + 4 + 8 + \ldots + 2^{62} + 2^{63} + 2^{64}.$$

Subtracting the first equation from the second gives

$$2S_{64} - S_{64} = 2^{64} + (2^{63} - 2^{63}) + (2^{62} - 2^{62}) + \ldots + (8 - 8) + (4 - 4) + (2 - 2) - 1.$$

That is, $S_{64} = 2^{64} - 1.$

Shown is a grain elevator. Some large grain elevators can store up to 1 million bushels of grain. Electrically operated buckets raise the grain to the top of the elevator where it is weighed and cleaned. It is then moved to the storage bins below.

Activity 1

a. Evaluate $2^{64} - 1$ with a calculator, and express the number of grains of wheat in scientific notation.

b. Assume that each grain weighs 0.008 gram. What is the total weight of wheat requested by the inventor?

c. It was estimated that about $5.5 \cdot 10^8$ metric tons of wheat were produced in the world in 1991. One metric ton = 10^3 kilograms. How do you think the king should have responded to the inventor's request?

A Formula for Any Finite Geometric Series

The above procedure can be generalized to find the value S_n of any finite geometric series. Let S_n be the geometric series with first term g, constant ratio $r \neq 1$, and number of terms n.

$$S_n = g + gr + gr^2 + \ldots + gr^{n-1}$$

Multiply by r.

$$rS_n = gr + gr^2 + \ldots + gr^{n-1} + gr^n$$

Subtract the 2nd equation from the first.

$$S_n - rS_n = g - gr^n$$

Use the Distributive Property.

$$(1 - r)S_n = g(1 - r^n)$$

Divide each side by $1 - r$.

$$S_n = \frac{g(1 - r^n)}{1 - r}$$

The constant ratio r cannot be 1 in this formula. (Do you see why?) But that is not a problem. If $r = 1$, the series is $g + g + g + \ldots + g$, with n terms, and its sum is ng. This argument proves the following theorem.

Theorem:

Let S_n be the sum of the first n terms of the geometric sequence with first term g_1 and constant ratio $r \neq 1$. Then

$$S_n = \frac{g_1(1 - r^n)}{1 - r}.$$

Example 1

Evaluate $18 + 6 + 2 + \frac{2}{3} + \frac{2}{9} + \frac{2}{27} + \frac{2}{81}$.

Solution

This is a geometric series with first term $g_1 = 18$, constant ratio $r = \frac{6}{18} = \frac{1}{3}$, and the number of terms $n = 7$.

$$S_n = \frac{g_1(1 - r^n)}{1 - r}$$

So,

$$S_7 = \frac{18\left(1 - \left(\frac{1}{3}\right)^7\right)}{1 - \frac{1}{3}}$$

$$= \frac{18\left(1 - \left(\frac{1}{2187}\right)\right)}{\frac{2}{3}}$$

$$= \frac{18 \cdot \frac{2186}{2187}}{\frac{2}{3}} = \frac{2186}{81} = 26\frac{80}{81}.$$

▶

Check 1

Here is a rough check. Because each of the fractions $\frac{2}{3}, \frac{2}{9}, \frac{2}{27}$, and $\frac{2}{81}$ is less than 1, the sum is between $18 + 6 + 2 = 26$ and $18 + 6 + 2 + 1 + 1 + 1 + 1 = 30$.

Check 2

Use a calculator. Add the decimal approximations for the fractions.

The formula for a geometric series works even when the constant ratio is negative.

Example 2

Find the sum of the first 100 terms of the geometric series whose first five terms are $5 - 10 + 20 - 40 + 80$.

Solution

$S_n = \frac{g_1(1 - r^n)}{1 - r}$. In this case $g_1 = 5$, $r = -\frac{10}{5} = -2$, and $n = 100$.

$$S_{100} = \frac{5(1 - (-2)^{100})}{1 - (-2)}$$

$$= \frac{5 - 5 \cdot 2^{100}}{3}$$

$$S_{100} \approx -2.11 \times 10^{30}$$

An Equivalent Formula

When the constant ratio $r > 1$, it is often more convenient to use a different formula for a geometric series.

Corollary

Let S_n be the sum of the first n terms of the geometric sequence with the first term g_1 and a constant ratio $r \neq 1$. Then

$$S_n = \frac{g_1(r^n - 1)}{r - 1}.$$

Activity 2

Explain how the corollary follows from the theorem in this lesson.

Geometric series that arise from compound-interest situations can often be evaluated more quickly using the formula in the Corollary than by adding each deposit's yield.

Example 3

Suppose $100 is deposited on January 1st of the years 1990, 1991, 1992, and so on, to 1999, with an annual yield of 7%. How much will there be on January 1, 2000?

Solution

The growth factor in this situation is 1.07. Ten deposits are made, one in each of the years 1990, 1991, . . . , 1999.

The deposit made in 1990 will have earned interest for 10 years. So on January 1, 2000, it will be worth $100(1.07)^{10}$.

The deposit made in 1991 will have earned interest for 9 years. On January 1, 2000, it will be worth $100(1.07)^{9}$.

$$\vdots$$

The deposit made in 1999 will have earned interest for 1 year. On January 1, 2000, it will be worth $100(1.07)^{1}$.

So on January 1st, 2000, there will be
$$100(1.07)^{10} + 100(1.07)^{9} + . . . + 100(1.07)^{2} + 100(1.07).$$

This is a geometric series with the first term $a = 100(1.07)$ and constant ratio $r = 1.07$. There are 10 terms, so $n = 10$. Use the Corollary.

$$S_{10} = \frac{g_1(r^{10} - 1)}{r - 1} = \frac{100(1.07)(1.07^{10} - 1)}{1.07 - 1} = \frac{107(1.07^{10} - 1)}{0.07} \approx 1478.36$$

On January 1, 2000, there will be $1478.36.

QUESTIONS

Covering the Reading

1. According to the story about the King of Persia, how many grains of wheat were on the first two rows of the chess board?

2. Answer the questions in Activity 1 in the lesson.

3. **a.** State a formula for the sum of the first n terms of a geometric series with first term g and constant ratio r.
   b. In the formula for the value of a geometric series, what value can r not have?
   c. Why can it not have this value?

In 4–7, a geometric series is given. **a.** How many terms does the series have? **b.** Give the value of the series.

4. $3 + 12 + 48 + . . . + 3 \cdot 4^{9}$

5. $50 + 10 + 2 + \frac{2}{5} + \frac{2}{25} + \frac{2}{125}$

6. $50 - 10 + 2 - \frac{2}{5} + \frac{2}{25} - \frac{2}{125}$

7. $1 + b + b^{2} + b^{3} + . . . + b^{16}$

8. Consider the geometric series in Example 2.
 a. Calculate the following sums.
   i. $S_2 = 5 - 10$
   ii. $S_3 = 5 - 10 + 20$
   iii. $S_4 = 5 - 10 + 20 - 40$
   iv. $S_5 = 5 - 10 + 20 - 40 + 80$
   v. S_{99}
 b. How can you tell whether S_n is positive or negative?

9. Find the sum of the first 20 terms of the geometric series with first term 12 and common ratio 1.

10. Give your answer for Activity 2 in the lesson.

11. Suppose $200 is deposited on January 1st for five consecutive years and earns an annual yield of 4%.
 a. Write a geometric series that represents the value of this investment on January 1st of the 6th year.
 b. Evaluate the series of part **a.**

Applying the Mathematics

12. a. Write the first 8 terms of the geometric series $g_1 + g_2 + \ldots$ if
$$\begin{cases} g_1 = 6 \\ g_n = \tfrac{2}{3}g_{n-1}, \text{ for integers } n \geq 2. \end{cases}$$
 b. Find the sum of these terms.

A snail's pace. *Shown is a tree snail. There are more than 80,000 kinds of snails. Some grow up to 2 feet in length.*

13. A snail is climbing straight up a wall. The first hour it climbs 16 inches; the second hour it climbs 12 inches; each succeeding hour it climbs $\frac{3}{4}$ the distance it climbed the previous hour. Assume this pattern holds indefinitely.
 a. How far does the snail climb during the 7th hour?
 b. What is the total distance climbed in 7 hours?

14. Currently on the first day of each month Mollie pays $100 on a car loan. Suppose she had no loan, but invested the $100 each month in an account that earns $\frac{1}{2}$% interest per month. How much would she have at the end of 12 months?

15. A ball is dropped from a height of 2 meters and bounces up to 90% of its height on each bounce.
 a. Draw a sketch of the path of the ball during its first three bounces.
 b. When it hits the ground for the eighth time, how far has it traveled?

Review

16. If the inventor of chess mentioned at the start of this lesson had wanted 1 grain on the first square, 2 on the second, 3 on the third, and so on, in arithmetic sequence, how many grains would have been his reward? *(Lesson 13-1)*

17. **a.** Find the sum of the odd integers from 1 to 999.
    b. Find the sum of the even integers from 2 to 1000.
    c. Verify that the sum of the answers in parts **a** and **b** equals the sum of the integers from 1 to 1000. *(Lesson 13-1)*

18. Suppose $t_n = -3n + 4$. Find $t_1 + t_2 + \ldots + t_{22}$. *(Lessons 3-8, 13-1)*

In 19–21, let $f(x) = 8^x$. *(Lessons 7-1, 7-2, 7-3, 7-7, 9-7)*

19. Evaluate $f(-2)$, $f(0)$, and $f\left(\frac{2}{3}\right)$.

20. Identify the domain and the range of f.

21. Give an equation for the reflection image of the graph of $y = f(x)$ over the line $y = x$.

22. Give an equation for the line parallel to $3x + 2y = 10$ and containing (8, 4). *(Lessons 3-2, 3-5)*

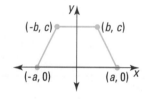

23. **a.** Identify the type of quadrilateral graphed at the left.
    b. Prove or disprove: The diagonals of this quadrilateral have the same length. *(Previous course)*

24. List all possible orders a judge might rank three cats, Buffy, Fluffy, and Muffy, in a show. *(Previous course)*

Exploration

25. Your ancestors consist of 2 parents, 4 grandparents, 8 great-grandparents, 16 great-great-grandparents, and so on. Pick some estimate for the number of years in a generation.
    a. Use that estimate to help calculate the total number of ancestors you have had in the past 2000 years.
    b. Must there have been some duplicates (people from whom you descended in two different ways)? Explain your answer.

LESSON

13-3

The Σ and ! Symbols

Getting organized. *Spreadsheets help store, organize, compile, and analyze data. They can perform a variety of complex arithmetic tasks.*

Sums and Sigma Notation

In the spreadsheet below the sum of the numbers in cells C1 through C6 is to be put in cell C7. One way to do this is to write

$$- C1 + C2 + C3 + C4 + C5 + C6$$

in cell C7. But writing all the entries and plus signs is inefficient when there are many numbers to be added. A notation used on some spreadsheets is

$$= SUM(C1:C6).$$

It is understood that all the cells from C1 to C6 are included in the sum.

	A	B	C
1	Expenses	Jan	313.29
2		Feb	86.71
3		Mar	212.43
4		Apr	65.00
5		May	111.35
6		Jun	81.92
7		Total	

The sum of $c_1, c_2, c_3, c_4, c_5,$ and c_6 is usually written

$$c_1 + c_2 + c_3 + c_4 + c_5 + c_6.$$

But, when there are many numbers to be added, this notation also is too cumbersome. You can shorten this by writing

$$c_1 + c_2 + \ldots + c_6.$$

It is understood that the terms $c_3, c_4,$ and c_5 are included. You may also use the Greek letter Σ (sigma) to denote a sum. In Σ-notation, called **sigma notation,** or **summation notation,** the above sum is

$$\sum_{i=1}^{6} c_i$$

The expression is read "the sum of the values of c sub i, from i equals 1 to i equals 6." The variable i under the Σ sign is called the **index variable** or **index.** It is common to use the letters i, j, k or n as index variables. In this book, index variables have only integer values. (In summation notation, i is not the complex number $\sqrt{-1}$ unless it is so specified.) As you can see, summation notation and spreadsheet notation are quite similar. They each denote a sum by describing all terms as a sequence and indicating the first and last terms.

Example 1

a. Write the meaning of $\displaystyle\sum_{i=5}^{11} 2^i$.

b. Evaluate $S = \displaystyle\sum_{i=5}^{11} 2^i$.

Solution

a. $\displaystyle\sum_{i=5}^{11} 2^i$ means the sum of the numbers of the form 2^i for integer values of i from 5 to 11.

b. $S = \displaystyle\sum_{i=5}^{11} 2^i = 2^5 + 2^6 + 2^7 + 2^8 + 2^9 + 2^{10} + 2^{11}$
$ = 32 + 64 + 128 + 256 + 512 + 1024 + 2048$
$ = 4064$

Check

b. The sum is a geometric series with first term $g_1 = 2^5$, common ratio $r = 2$, and number of terms $n = 7$. So,

$$S_7 = \frac{2^5(2^7 - 1)}{2 - 1} = 32 \cdot 127 = 4064.$$

Writing Formulas Using Σ-Notation

Each of the theorems studied in the last two lessons can be restated using Σ-notation. Notice that i is used as the index variable to avoid confusion with the variable n. Compare these restatements with the original wording in the theorems.

Sum of integers from 1 to n

$$\sum_{i=1}^{n} i = \frac{1}{2}n(n + 1)$$

In an arithmetic sequence $a_1, a_2, a_3, \ldots, a_n$ with constant difference d,

$$\sum_{i=1}^{n} a_i = \frac{1}{2} n(a_1 + a_n) = \frac{n}{2}[2a_1 + (n - 1)d].$$

In a geometric sequence $g_1, g_2, g_3, \ldots, g_n$ with constant ratio r,

$$\sum_{i=1}^{n} g_i = \frac{g_1(1 - r^n)}{1 - r}.$$

Example 2

Evaluate $\sum_{i=1}^{1000} (3i + 5)$.

Solution 1

This is the arithmetic series $8 + 11 + 14 + \ldots + 3005$. Its first term is $3 \cdot 1 + 5$, or 8, and the constant difference is 3. There are 1000 terms. Use the formula

$$\sum_{i=1}^{n} a_i = \frac{n}{2}(2a_1 + (n - 1)d).$$

$$\sum_{i=1}^{1000} (3i + 5) = \frac{1000}{2} (2 \cdot 8 + (1000 - 1)3)$$
$$= 500(16 + 2997)$$
$$= 1,506,500$$

Solution 2

The first term is $3 \cdot 1 + 5 = 8$ and the last term, its 1000th term, is $3 \cdot 1000 + 5 = 3005$. Use the formula

$$\sum_{i=1}^{n} a_i = \frac{1}{2}n (a_1 + a_n).$$

$$\sum_{i=1}^{1000} (3i + 5) = \frac{1}{2} \cdot 1000 (8 + 3005)$$
$$= 500 \cdot 3013$$
$$= 1,506,500$$

The Factorial Symbol

Certain products can also be written with a special symbol, called the factorial symbol. The factorial symbol is an exclamation point. The symbol $n!$ is read "n factorial."

> **Definition**
> $n!$ = product of the integers from n to 1.

The factorial function is the function with equation $f(n) = n!$. For now, we take the domain of the factorial function to be the set of positive integers. In Lesson 13-5, the domain is extended to include 0. Small values of the factorial function can be calculated by hand or in your head.

$f(1) = 1! = 1$ $f(4) = 4! = 4 \cdot 3 \cdot 2 \cdot 1 = 24$
$f(2) = 2! = 2 \cdot 1 = 2$ $f(5) = 5! = 5 \cdot 4 \cdot 3 \cdot 2 \cdot 1 = 120$
$f(3) = 3! = 3 \cdot 2 \cdot 1 = 6$ $f(6) = 6! = 6 \cdot 5 \cdot 4 \cdot 3 \cdot 2 \cdot 1 = 720$

Larger values require a calculator. Some calculators have a separate **factorial key** $\boxed{x!}$. Others have the factorial symbol listed under a menu of special functions.

> **Activity**
>
> Use your calculator to evaluate the following. Record your key sequences as well as your answers.
> **a.** 7!   **b.** 17!

An Application of the Factorial Symbol

Shown are runners at a high school state track meet in San Antonio, Texas.

An arrangement of objects in a row is called a **permutation.** With three objects A, B, and C, there are six possible permutations: *ABC, ACB, BAC, BCA, CAB,* and *CBA.* Example 3 considers permutations of four objects.

> **Example 3**
>
> Find the number of possible orders in which four runners, *A, B, C,* and *D,* might finish in a race.
>
> **Solution**
>
> The number of possible orders is the number of permutations of the four runners. List the possible orders.
>
> | | | | |
> |---|---|---|---|
> | ABCD | BACD | CABD | DABC |
> | ABDC | BADC | CADB | DACB |
> | ACBD | BCAD | CBAD | DBAC |
> | ACDB | BCDA | CBDA | DBCA |
> | ADBC | BDAC | CDAB | DCAB |
> | ADCB | BDCA | CDBA | DCBA |
>
> There are 24 permutations.

Notice the number of permutations of 3 objects is 3! and of 4 objects is 4!.

To list the possible ways in which five people could finish a race, you could begin with the list in Example 3. Call the fifth racer E. In each permutation in the list, you can insert the E at the beginning, in one of the three middle spots, or at the end. For instance, inserting E into *ABCD* yields *EABCD, AEBCD, ABECD, ABCED,* or *ABDCE.* This means that the number of permutations of 5 objects is five times the number of permutations of 4 objects. So the number of permutations of 5 objects is $5 \cdot 4!$, which equals 5!. Similarly, the number of permutations of 6 objects is $6 \cdot 5!$, which equals 6!. Extending this argument proves the following theorem.

Theorem
There are $n!$ permutations of n distinct objects.

Numbers of permutations grow quickly. With 20 objects, there are $20! \approx 2.4329 \cdot 10^{18}$ permutations.

QUESTIONS

Covering the Reading

1. Write the sum D1 + D2 + D3 + D4 + D5
   a. using spreadsheet shorthand.
   b. using sigma notation.

2. Repeat Question 1 for the sum
   A7 + A8 + A9 + A10 + A11 + A12 + A13 + A14 + A15.

3. The symbol Σ is the Greek letter __?__.

4. In Σ-notation, the variable under the Σ sign is the __?__ variable.

In 5–8, *multiple choice.*

5. $\sum\limits_{i=1}^{3} i^2 =$
   (a) 3^2 (b) $1 + 4 + 9$ (c) $1 + 2 + \ldots + 9$ (d) none of these

6. $\sum\limits_{k=1}^{4} 3^k =$
   (a) 16   (b) 30   (c) 82   (d) 120

7. $\sum\limits_{n=1}^{5} (2n + 1) =$
   (a) $3 + 5 + 7 + 9 + 11$ (b) $3 + 11$
   (c) $1 + 5 + 11$ (d) $2 + 4 + 6 + 8 + 10 + 1$

8. $3 + 6 + 9 + 12 + 15 + 18 + 21 =$

(a) $\displaystyle\sum_{i=3}^{21} i$ (b) $\displaystyle\sum_{i=1}^{7} (3i)$ (c) $\displaystyle\sum_{i=3}^{21} (3i)$ (d) none of these

9. In $\displaystyle\sum_{i=100}^{200} (4i)$ how many terms are added?

In 10–12, give the value of the sum.

10. $\displaystyle\sum_{i=1}^{36} i$ **11.** $\displaystyle\sum_{i=1}^{100} (2i - 1)$ **12.** $\displaystyle\sum_{n=1}^{3} (4 \cdot 10^n)$

13. How is the symbol $n!$ read?

In 14 and 15, refer to the Activity in the lesson.

14. a. Evaluate $7!$.
    b. Verify that $7! = 7 \cdot 6!$.

15. Evaluate $17!$.

16. a. Write out all permutations of the 4 letters *P, E, R, M*.
    b. How many permutations are there?

17. Five friends decide to have their picture taken together. In how many different orders, from left to right, can they be pictured?

Applying the Mathematics

In 18–20, rewrite using Σ-notation.

18. $2 + 4 + 6 + 8 + 10 + 12 + 14$

19. $9 + 18 + 36 + 72 + 144 + 288 + 576 + 1152$

20. the sum of the squares of the integers from 1 to 100

21. a. Translate this statement into an algebraic formula using Σ-notation: The sum of the cubes of the integers from 1 to n is the square of the sum of the integers from 1 to n.
    b. Verify part **a** when $n = 4$.

22. Write the arithmetic mean of the n numbers $a_1, a_2, a_3, \ldots, a_n$ using Σ-notation.

23. Consider the sequence
$a_1 = 1$
$a_n = n \cdot a_{n-1}$, for integers $n \geq 2$.
    a. Give the first seven terms of the sequence.
    b. What is an appropriate name for this sequence?

24. Simplify each expression.

    a. $\dfrac{4!}{3!}$ **b.** $\dfrac{15!}{14!}$ **c.** $\dfrac{100!}{99!}$ **d.** $\dfrac{(n + 1)!}{n!}$

25. Consider the following investment. Ima Saver deposits $50 on the first day of every month and earns 6% compounded monthly.
  a. How much interest will the first $50 deposited earn in 6 months?
  b. How much will there be in Ima's account just after she makes the 7th deposit? (Assume the account starts with $0, and that there are no withdrawals.)   *(Lessons 7-4, 11-1, 13-2)*

26. Let $a_n = 2n + 8$ and let $S_n = a_1 + a_2 + \ldots + a_n$.
  a. Find a_n and S_n when $n = 50$.
  b. Suppose $S_n = 7120$. Find n and a_n.   *(Lessons 3-7, 13-1)*

27. Graph $\{(x, y): x^2 + y^2 = 1\}$ and $\{(x, y): x^2 - y^2 = 1\}$ on the same axes.   *(Lessons 12-2, 12-6, 12-9)*

28. How many times as loud is a sound of 100 decibels than one of 80 decibels?   *(Lesson 9-6)*

29. If 3 blobs and 4 globs weigh 170 kg and 7 blobs and 6 globs weigh 330 kg, what will 4 blobs and 2 globs weigh?   *(Lesson 5-4)*

A glob of a blob. *Shown is a scene from the 1988 film,* The Blob. *In this re-make of the 1958 film of the same name, a small town is invaded by an amorphous space creature that devours anyone in its path.*

30. Consider the series of reciprocals of consecutive integers.
$$\sum_{i=1}^{n} \frac{1}{i} = 1 + \frac{1}{2} + \frac{1}{3} + \frac{1}{4} + \ldots + \frac{1}{n}$$
  a. How many terms of the series are needed before the sum exceeds 2?
  b. How many terms of the series are needed before the sum exceeds 3?
  c. How many terms of the series are needed before the sum exceeds 10?
  d. Do you think the sum ever gets larger than 100? Why or why not?

Describing data. *Shown is a student using statistical measures to describe the heights of the students in the classroom.*

The data set at the right shows the math scores on a college entrance exam taken by ten students in a school. Recall that in a data set, an element may be listed more than once.

Measures of Center

To describe these scores quickly, you may use a single number which in some way describes the entire set of scores. Three numbers used commonly for this purpose are the *mean,* the *median,* and the *mode.* The mean, median, and mode are examples of *statistical measures.* A **statistical measure** is a single number which is used to describe an entire set of numbers. The formula for the mean can be written using Σ-notation.

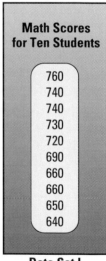

Math Scores for Ten Students
760
740
740
730
720
690
660
660
650
640

Data Set I

Definitions

Let S be a data set of n numbers $x_1, x_2, x_3, \ldots, x_n$.

mean of S = the *average* of all terms of $S = \dfrac{\displaystyle\sum_{i=1}^{n} x_i}{n}$.

median of S = the *middle* term of S when the terms are placed in increasing order.

mode of S = the number which occurs most often in the set.

Example 1

For the scores in the data set on the previous page, find
a. the mean. **b.** the median. **c.** the mode.

Solution

a. For the data set on the previous page, $n = 10$ and $\Sigma \, x_i = 6990$.
So, the mean score is $\frac{6990}{10} = 699$.

b. The median is the mean of the two middle scores, 690 and 720.
So, the median is $\frac{690 + 720}{2} = 705$.

c. The mode is the most common score. There are two modes, 660 and 740.

The mean and median are called **measures of center** or **measures of central tendency,** because they give a number which in some sense is at the "center" of the set. Because the mode may be an extreme value, we do not consider it a measure of the center of the data set.

The *mean* is most often used when the terms of the sequence are fairly closely grouped, as in finding an individual's bowling average. The *median* is used when there are a few low or high terms which could greatly affect the mean, as with personal incomes. The *mode* is particularly useful when many of the terms are the results of rounding, as often occurs when recording the ages of people.

A Measure of Spread: The Standard Deviation

At the left is a second set of math scores of ten students in a different school.

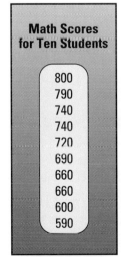

Math Scores for Ten Students

800
790
740
740
720
690
660
660
600
590

Data Set II

Activity

Find the mean, median, and mode for Data Set II.

You should find that the mean, median, and mode of Data Set II are identical to those in Data Set I. But the scores in Data Set II are more widely dispersed, or spread out, than the scores in Data Set I.

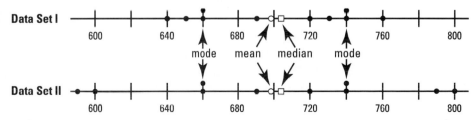

One statistical measure of spread is the *standard deviation.*

Definition:
Let S be a data set of n numbers $\{x_1, x_2, \ldots, x_n\}$. Let m be the mean of S. Then the **standard deviation**, or **s.d.**, of S is given by

$$\text{s.d.} = \sqrt{\frac{\sum_{i=1}^{n} (x_i - m)^2}{n}}.$$

Example 2

Calculate the standard deviation of the math scores of Data Set II.

Solution

The formula for standard deviation requires knowing the mean. The mean m was found above to be 699. The number of scores n is 10.

$$\text{So s.d.} = \sqrt{\frac{\sum_{i=1}^{n} (x_i - m)^2}{n}} = \sqrt{\frac{\sum_{i=1}^{10} (x_i - 699)^2}{10}}.$$

To calculate the sum under the radical, you may wish to organize your work as shown below.

i	x_i	$x_i - 699$	$(x_i - 699)^2$
1	800	101	10201
2	790	91	8281
3	740	41	1681
4	740	41	1681
5	720	21	441
6	690	-9	81
7	660	-39	1521
8	660	-39	1521
9	600	-99	9801
10	590	-109	11881

Sum = 47090

$$\text{Thus s.d.} = \sqrt{\frac{\sum_{i=1}^{10} (x_i - 699)^2}{10}} = \sqrt{\frac{47090}{10}} = \sqrt{4709} \approx 68.62.$$

The steps in finding the standard deviation of a data set are as follows:

Step 1   Calculate the mean of S.
Step 2   Subtract the mean from each term of S.
Step 3   Square these differences.
Step 4   Add the squares.
Step 5   Divide the sum by n (the number of elements).
Step 6   Find the square root of this quotient.

In the Questions you are asked to verify that for Data Set I, s.d. $= \sqrt{\frac{17490}{10}} = \sqrt{1749} \approx 41.82$. So the standard deviation for Data Set II is larger than the standard deviation for Data Set I. In general, the larger the standard deviation, the more widely dispersed are the scores. Although standard deviations are hard to calculate by hand, they are easily calculated by computers and calculators. Some calculators even have special keys or menu options to calculate standard deviations.

The measures in this lesson are not the only statistical measures in common use. You will encounter many others if you study more statistics.

QUESTIONS

Covering the Reading

1. How does a data set differ from ordinary sets?

2. Find the mean, median, and mode of the set: 1, 2, 2, 3, 3, 3, 4, 4, 4, 4.

3. Here is a set of low temperatures in degrees Fahrenheit for an Alaskan city for a week in January: -14, -14, -9, 2, 3, -4, 0.
   a. Find the mean, median, and mode of the data.
   b. Which of these numbers seems most representative of the set? Explain your answer.

4. Name a statistic which is not a measure of central tendency.

5. A person bowls games of 182, 127, 161, and 155.
   a. Which measure of center is usually used to describe bowling scores?
   b. Give that measure for this data set.

6. a. Why is *median income* often considered a better indicator of the wealth of a community than *mean income?*
   b. Why is the mode income not used at all?

7. In the lesson, the standard deviation of Data Set I is reported to be about 41.82. Do the calculations to verify this value, organizing your work as in Example 2.

In 8 and 9, calculate the mean and the standard deviation of the data set.

8. 10, 20, 30, 40, 50

9. 88, 90, 90, 90, 92

10. The larger the standard deviation of a data set, the ? the numbers in the set are.

Fish dry. *Shown is a woman drying salmon strips in Graveyard Point in the Bristol Bay Region in southwest Alaska.*

Applying the Mathematics

11. A store has two managers and nine employees. Each manager earns $35,000 a year; six employees earn $20,000 a year; and three employees earn $15,000 a year. Give the mean, median, and mode of the salaries.

12. A student has test scores of 83, 85, 93, and 88. What must the student score on the next test to have
  a. a mean of 88 for the five tests?
  b. a median of 88 for the five tests?
  c. a mode of 88 for the five tests?

13. The mean of two scores is x and of three scores is y. Find the mean of all five scores.

14. Give an example, different from the one in the lesson, of two different data sets that have the same mean but different standard deviations.

15. At the left are the heights in inches of the members of two basketball teams.
  a. Describe some ways in which the two teams are alike.
  b. Describe some ways in which the two teams are different.

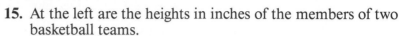

Rocketeers	Sunbursts
68	72
70	72
72	73
72	73
73	73
73	73
74	74
75	74
78	75
80	76

Review

In 16–18 evaluate. *(Lesson 13-3)*

16. a. $\displaystyle\sum_{i=1}^{4} (i^2 - 3)$ **b.** $\displaystyle\sum_{i=1}^{4} (i)^2 - 3$ **c.** $\displaystyle\left(\sum_{i=1}^{4} i\right)^2 - 3$

17. a. $3! \cdot 4!$ **b.** $3! + 4!$

18. a. $\dfrac{173!}{171!}$ **b.** $\dfrac{173!}{171! \, 2!}$

In 19 and 20, suppose a tennis ball is released from a height of 1 m above a floor. Each time it hits the floor it bounces to 90% of its previous height.

19. Suppose the ball has hit the floor four times. How high will it get on the next bounce?

20. If the ball hits the floor ten times, find the total vertical distance (up and down) it will have traveled. *(Lessons 7-5, 13-2)*

21. a. In how many ways can 10 books be arranged on a shelf?
  b. If your favorite novel and favorite biography are 2 of the 10 books in part **a**, how many of these ways have your favorite novel at the right and your favorite biography at the left of the arrangement? *(Lesson 13-1)*

22. a. For what values of x and y does $(x + y)^2 = x^2 + 2xy + y^2$?
  b. For what values of x and y does $(x + y)^2 = x^2 + y^2$? *(Previous course, Lessons 6-1, 11-3)*

23. Suppose y varies as x^4. If x is multiplied by 3, what is the effect on y? *(Lesson 2-3)*

Exploration

24. Give an example of a set of 5 scores with a mean of 50 and a standard deviation of 10.

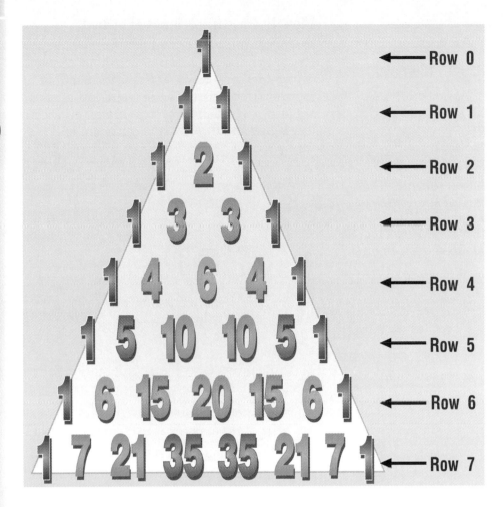

Very often a single idea has applications to many parts of mathematics. The triangular array above is such an idea. This array seems to have first appeared in the works of Abu Bakr al-Karaji, an Islamic mathematician, and Jia Xian, a Chinese mathematician, in the 11th century. The works of both of these men have been lost, but 12th-century writers refer to them. Versions of the array were discovered independently by the Europeans Peter Apianus in 1527 and Michael Stifel in 1544. But in most of the western world the array is known as **Pascal's triangle,** after Blaise Pascal (1623-1662), the French mathematician and philosopher who discovered many properties relating the elements (numbers) in the array. Pascal himself called it the *triangle arithmetique,* literally the "arithmetical triangle."

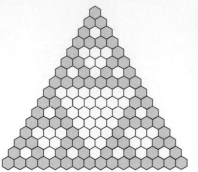

Pascal patterns. *The shaded hexagons in this rendition of Pascal's triangle represent the odd numbers in the triangle.*

How Is Pascal's Triangle Formed?

Pascal's triangle is formed in a very simple way. You can think of Pascal's triangle as a two-dimensional sequence. Each element is determined by a row and its position in that row. The only element in the top row (row 0) is 1. The first and last elements of every succeeding row are also 1. If x and y are located next to each other on a row, the element just below and directly between them is $x + y$, as illustrated below.

For instance, from row 4 you can get row 5 as follows.

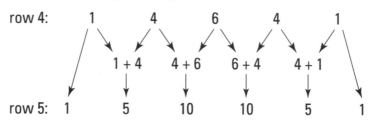

This rule provides a recursive definition for Pascal's triangle. With it, you can obtain any other row in the array.

A few different symbols for elements in Pascal's triangle are in common use. One of these symbols uses large parentheses.

Definition
The $(r + 1)$st element in row n of Pascal's triangle is denoted by $\binom{n}{r}$.

For reasons that are made clear in Lesson 13-7, we read $\binom{n}{r}$ "*n* choose *r*." For instance, the 1st element in the 7th row of Pascal's triangle is $\binom{7}{0}$. The entire 7th row consists of the following elements.

$$\binom{7}{0} \quad \binom{7}{1} \quad \binom{7}{2} \quad \binom{7}{3} \quad \binom{7}{4} \quad \binom{7}{5} \quad \binom{7}{6} \quad \binom{7}{7}$$

That is, $\binom{7}{0} = 1$, $\binom{7}{1} = 7$, $\binom{7}{2} = 21$, and so forth. We found these values by referring to row 7 as it was given in Pascal's triangle. The top row is called the 0th row and has one element, $\binom{0}{0} = 1$.

The method used above to construct Pascal's triangle applies a recursive definition of the triangle. Because the triangle is a sequence in two directions—down and across—the recursive rule involves two variables. On the next page, the rule is written using the $\binom{n}{r}$ symbol.

Definition

Pascal's triangle is the sequence satisfying

(1) $\binom{n}{0} = \binom{n}{n} = 1$ for all integers $n \geq 0$,

and (2) $\binom{n + 1}{r + 1} = \binom{n}{r} + \binom{n}{r + 1}$ for all integers r and n satisfying $0 \leq r < n$.

Part (1) of the definition gives the "sides" of the triangle. Part (2) is a symbolic way of stating that adding two adjacent elements in one row gives an element in the next row.

An Explicit Formula for the Entries in Pascal's Triangle

In order to determine the elements in the 14th row of the triangle using the recursive definition above, you would have to construct the first 13 rows of Pascal's triangle. The next theorem was known to Pascal and shows how to calculate $\binom{n}{r}$ explicity, that is, without constructing the triangle. The theorem is a surprising application of factorials.

Pascal's Triangle Explicit-Formula Theorem

If n and r are integers with $0 \leq r \leq n$, then
$$\binom{n}{r} = \frac{n!}{r!(n - r)!}.$$

Before we discuss the proof of this theorem, here are some instances of it.

Example 1

Calculate $\binom{5}{3}$.

Solution

Here $n = 5$ and $r = 3$. So $n - r = 2$.
$$\binom{5}{3} = \frac{5!}{3!(5 - 3)!} = \frac{5 \cdot 4 \cdot 3 \cdot 2 \cdot 1}{3 \cdot 2 \cdot 1(2 \cdot 1)} = 10$$

Check

This agrees with $\binom{5}{3}$ being the 4th element in the 5th row of Pascal's triangle.

Example 2

Calculate $\binom{11}{3}$.

Solution
$$\binom{11}{3} = \frac{11!}{3!(11 - 3)!} = \frac{11!}{3! \, 8!} = \frac{11 \cdot 10 \cdot 9 \cdot \cancel{8} \cdot \cancel{7} \cdot \cancel{6} \cdot \cancel{5} \cdot \cancel{4} \cdot \cancel{3} \cdot \cancel{2} \cdot \cancel{1}}{3 \cdot 2 \cdot 1 \cdot \cancel{8} \cdot \cancel{7} \cdot \cancel{6} \cdot \cancel{5} \cdot \cancel{4} \cdot \cancel{3} \cdot \cancel{2} \cdot \cancel{1}} = 165$$
Notice how easily the fraction simplifies because of the common factors.

▶

Some calculators give values of $\binom{n}{r}$ automatically. You should check to see if yours does. Notice that to evaluate $\binom{n}{0}$ using the previous theorem, you must calculate $0!$. For instance,

$$\binom{7}{0} = \frac{7!}{0!(7-0)!} = \frac{7!}{0! \cdot 7!} = \frac{1}{0!}.$$

However, $\binom{7}{0}$ is the 1st element in the 7th row, and from the triangle on page 835 we can see that $\binom{7}{0} = 1$. Thus we define $0!$ to be equal to 1.

A Proof of the Explicit Formula

For a proof of the theorem, we need to show that $\binom{n}{r} = \frac{n!}{r!(n-r)!}$, where n and r are integers and $0 \le r \le n$. It is enough to show that the factorial formula $\frac{n!}{r!(n-r)!}$ satisfies the relationships involving $\binom{n}{r}$ in the recursive definition of Pascal's triangle.

(1) For $n \ge 0$, does the formula for $\binom{n}{0}$ equal the formula for $\binom{n}{n}$ and equal 1?

$$\binom{n}{0} = 1 \text{ and } \frac{n!}{0!(n-0)!} = \frac{n!}{0!n!} = \frac{n!}{1 \cdot n!} = 1, \text{ so } \binom{n}{0} = \frac{n!}{0!(n-0)!}.$$

$$\binom{n}{n} = 1 \text{ and } \frac{n!}{n!(n-n)!} = \frac{n!}{n!0!} = \frac{n!}{n!(1)} = 1, \text{ so } \binom{n}{n} = \frac{n!}{n!(n-n)!}.$$

Thus the formula works for the "sides" of Pascal's triangle.

(2) To prove that the formula for $\binom{n+1}{r+1}$ is the sum of the formulas for $\binom{n}{r}$ and $\binom{n}{r+1}$ requires substantial algebraic manipulation.

It is omitted here. In the Questions you are asked to prove a special case.

QUESTIONS

Covering the Reading

1. In what century did the array known as Pascal's triangle first appear?

2. In what century did Pascal live?

3. Write rows 0 through 10 of Pascal's triangle. (It is a good idea to keep rows 0 through 10 handy for reference.)

In 4 and 5, consider the symbol $\binom{n}{r}$.

4. What element in which row of Pascal's triangle does this symbol represent?

5. How is this symbol read?

6. What are the rules by which Pascal's triangle is defined?

7. In terms of factorials, $\binom{n}{r} = \underline{\quad?\quad}$.

In 8–15, calculate.

8. $\binom{8}{2}$ **9.** $\binom{3}{2}$ **10.** $\binom{10}{5}$ **11.** $0!$

12. $\binom{6}{6}$ **13.** $\binom{15}{0}$ **14.** $\binom{15}{14}$ **15.** $\binom{20}{2}$

Applying the Mathematics

In 16 and 17, *true or false*. Explain your reasoning.

16. $\binom{99}{17}$ is an integer.

17. $\dfrac{n!}{(n-2)!}$ is always an integer when $n \geq 2$.

In 18 and 19, find a solution to the equation.

18. $\binom{10}{5} + \binom{10}{6} = \binom{x}{y}$

19. $\binom{9}{2} + \binom{a}{b} = \binom{10}{2}$

In 20–22, tell where in Pascal's triangle the following sequence can be found.

20. the sequence of positive integers

21. the sequence of triangular numbers: 1, 3, 6, 10, 15, . . .

22. the sequence of partial sums of triangular numbers: 1, 1 + 3, 1 + 3 + 6, 1 + 3 + 6 + 10, . . .

23. Prove that for $n \geq 5$, $\binom{n}{4} + \binom{n}{5} = \binom{n+1}{5}$. (This is a special case of the Pascal's Triangle Explicit-Formula Theorem.

Review

24. Give a set of integers whose mean is 10, whose mode is 12, and whose median is 11. *(Lesson 13-4)*

25. Find the standard deviation of the data set 1, 7, 21, 35, 35, 21, 7, 1. *(Lesson 13-4)*

26. If two sets of scores have the same mean but the standard deviation of the first set is much smaller than that for the second set, what can you conclude? *(Lesson 13-4)*

27. a. If $10 \cdot 9! = x!$, then $x = \underline{\ ?\ }$.
   b. If $(n - r)(n - r - 1)! = y!$, then $y = \underline{\ ?\ }$.   *(Lesson 13-3)*

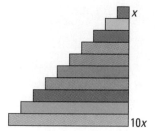

28. School children often use rods whose lengths form an arithmetic sequence $x, 2x, 3x, \ldots, 10x$.
   a. What is the sum of the lengths of these rods?
   b. If you wanted to make such rods for a child out of a piece of wood 6 feet long, how long could you make the shortest rod?
   (Lessons 3-7, 13-1)

29. *Skill sequence.* Expand and simplify.   *(Lessons 6-1, 11-2)*
   a. $(a + b)^2$ **b.** $(a + b)^3$ **c.** $(a + b)^4$

30. Multiply the complex numbers $2i$ and $2 - 2i$.   *(Lesson 6-8)*

Exploration

31. There are six elements surrounding each element not on a side of Pascal's triangle. For instance, around 15 in row 6 are the elements 5, 10, 20, 35, 21, and 6. In 1969, an amazing property about the product of these elements was discovered by Verner Hoggatt and W. Hansell of San Jose State University.
   a. Consider the number 2 in row 2. Find the product of all the elements surrounding it.
   b. Consider the number 3 in row 3. Find the product of all the elements surrounding it.
   c. Consider the numbers 4 and 6 in row 4. For each of them find the product of all the elements surrounding it.
   d. Find this product for each element not on a side of the 5th row of the triangle.
   e. What is true of these products?
   f. Verify your answer to part **e** by calculating the product for other elements in the triangle.

You have seen powers of binomials in many places within this book. Two examples are $A = (x + y)^2$, the area of a square with sides $x + y$ (*Lesson 6-1*), and $A = P(1 + r)^t$, the Compound-Interest Formula (*Lesson 7-4*).

The geometric sequence with first term 1 and the binomial $(a + b)$ as its constant ratio generates all the positive integer powers of binomials.

$$1, a + b, (a + b)^2, (a + b)^3, (a + b)^4, \ldots$$

You know the binomial $(a + b)^2 = a^2 + 2ab + b^2$. It is natural to try to write the other powers of binomials as polynomials. The results are known as **binomial expansions.**

$$(a + b)^0 = 1$$
$$(a + b)^1 = a + b$$
$$(a + b)^2 = a^2 + 2ab + b^2$$
$$(a + b)^3 = a^3 + 3a^2b + 3ab^2 + b^3$$
$$(a + b)^4 = a^4 + 4a^3b + 6a^2b^2 + 4ab^3 + b^4$$
$$\vdots \qquad\qquad \vdots$$

It may seem that there is no pattern in the expansion. But if the as and bs are ignored, and only the coefficients are written, then Pascal's triangle appears!

Powers of $(a + b)$	Coefficients of the Expansion	Row of Pascal's Triangle
$(a + b)^0$	1	0
$(a + b)^1$	1 1	1
$(a + b)^2$	1 2 1	2
$(a + b)^3$	1 3 3 1	3
$(a + b)^4$	1 4 6 4 1	4
$\vdots$	$\vdots$	$\vdots$

As a consequence, the binomial expansions can be written using the $\binom{n}{r}$ symbols.

$$(a + b)^0 = \binom{0}{0}$$
$$(a + b)^1 = \binom{1}{0} a + \binom{1}{1} b$$
$$(a + b)^2 = \binom{2}{0} a^2 + \binom{2}{1} ab + \binom{2}{2} b^2$$
$$(a + b)^3 = \binom{3}{0} a^3 + \binom{3}{1} a^2b + \binom{3}{2} ab^2 + \binom{3}{3} b^3$$
$$(a + b)^4 = \binom{4}{0} a^4 + \binom{4}{1} a^3b + \binom{4}{2} a^2b^2 + \binom{4}{3} ab^3 + \binom{4}{4} b^4$$
$$\vdots \qquad\qquad \vdots$$

Notice how easy Pascal's triangle makes the expansion of $(a + b)^n$.
(1) All the powers of a from a^n to a^0 occur in order.
(2) In each term, the exponents of a and b add to n.
(3) If the power of b is r, then the coefficient of the term is $\binom{n}{r}$.

The information on page 841 is summarized in a famous theorem, which was known to Omar Khayyam, the Persian poet, mathematician, and astronomer who died c.1123. He did not have our current notation.

Binomial Theorem
For all complex numbers a and b, and for all integers n and r with $0 \le r \le n$,

$$(a + b)^n = \sum_{r=0}^{n} \binom{n}{r} a^{n-r}b^r.$$

We do not show a formal proof of the Binomial Theorem. The proof requires mathematical induction, a powerful proof technique beyond the scope of this book and first used by Pascal when he discussed the array.

Example 1

Expand $(a + b)^7$.

Solution

First, fill in powers of a and b.
$$(a + b)^7 = \underline{}a^7 + \underline{}a^6b + \underline{}a^5b^2 + \underline{}a^4b^3 + \underline{}a^3b^4 +$$
$$\underline{}a^2b^5 + \underline{}ab^6 + \underline{}b^7$$
Second, put in the coefficients.
$$(a + b)^7 = \binom{7}{0}a^7 + \binom{7}{1}a^6b + \binom{7}{2}a^5b^2 + \binom{7}{3}a^4b^3 + \binom{7}{4}a^3b^4 +$$
$$\binom{7}{5}a^2b^5 + \binom{7}{6}ab^6 + \binom{7}{7}b^7$$
Finally, evaluate the coefficients, either by referring to row 7 of Pascal's triangle or by using the formula $\binom{n}{r} = \dfrac{n!}{r!(n-r)!}$.
$$(a + b)^7 = a^7 + 7a^6b + 21a^5b^2 + 35a^4b^3 + 35a^3b^4 + 21a^2b^5 + 7ab^6 + b^7$$

Automatic symbol manipulators have an ⟨EXPAND⟩ instruction that lets a user expand powers of polynomials. The Binomial Theorem makes it possible to expand powers of any binomial without using technology.

Example 2

Expand $(5x - 2y)^3$.

Solution
$$(a + b)^3 = \binom{3}{0}a^3 + \binom{3}{1}a^2b + \binom{3}{2}ab^2 + \binom{3}{3}b^3$$
Substituting $a = 5x$ and $b = -2y$, yields
$$(5x - 2y)^3 = 1(5x)^3 + 3(5x)^2(-2y) + 3(5x)(-2y)^2 + 1(-2y)^3$$
$$= 125x^3 - 150x^2y + 60xy^2 - 8y^3.$$

Check

Substitute specific values for x and y. We let $x = 2$ and $y = 3$. Then $(5x - 2y)^3 = (10 - 6)^3 = 64$ and $125x^3 - 150x^2y + 60xy^2 - 8y^3 = 125 \cdot 8 - 150 \cdot 4 \cdot 3 + 60 \cdot 2 \cdot 9 - 8 \cdot 27 = 64$. It checks.

Example 3

Expand $(x^2 + 1)^4$.

Solution

Think of x^2 as a, and 1 as b, and follow the form of $(a + b)^4$.

$$(x^2 + 1)^4 = \binom{4}{0}(x^2)^4 + \binom{4}{1}(x^2)^3 \cdot 1 + \binom{4}{2}(x^2)^2 \cdot 1^2 + \binom{4}{3}(x^2)^1 \cdot 1^3 + \binom{4}{4} \cdot 1^4$$
$$= x^8 + 4x^6 + 6x^4 + 4x^2 + 1$$

Check

Let $x = 2$. Then $(x^2 + 1)^4 = (4 + 1)^4 = 625$. Verify that the value of the polynomial when $x = 2$ is also 625. Also note as a check that the exponents of x in the consecutive terms form an arithmetic sequence.

Due to their use in the Binomial Theorem, the numbers in Pascal's triangle are sometimes called **binomial coefficients.** The Binomial Theorem has a surprising number of applications—estimation, counting, probability, and statistics which are studied in this chapter.

QUESTIONS

Covering the Reading

In 1–4, expand the binomial power.

1. $(a + b)^2$ **2.** $(a + b)^3$

3. $(a + b)^4$ **4.** $(a + b)^5$

5. State the Binomial Theorem.

In 6–9, expand the binomial power.

6. $(x + 1)^5$ **7.** $(a - b)^3$

8. $(2 - m)^4$ **9.** $(8x + y)^3$

Applying the Mathematics

In 10 and 11, convert to an expression in the form $(a + b)^n$.

10. $\displaystyle\sum_{r=0}^{n} \binom{n}{r} x^{n-r} 3^r$

11. $\displaystyle\sum_{i=0}^{n} \binom{n}{i} y^{n-i} (2a)^i$

12. Multiply $(a + b)^4$ by $a + b$ to check the expansion for $(a + b)^5$.

13. a. Multiply and simplify. $(a^2 + 2ab + b^2)(a^2 + 2ab + b^2)$
     b. Your answer to part **a** should be a power of $a + b$. Which one? Why?

In 14 and 15, use the Binomial Theorem to approximate some powers quickly without a calculator. Here is an example.

$$(1.002)^3 = (1 + .002)^3$$
$$= 1^3 + 3 \cdot 1^2 \cdot (.002) + 3 \cdot 1 \cdot (.002)^2 + (.002)^3$$
$$= 1 + .006 + .000012 + .000000008$$
$$= 1.006012008$$

Since the last two terms in the expansion are so small, they might be ignored in an estimate. $(1.002)^3 \approx 1.006$ to the nearest thousandth.

14. Estimate $(1.004)^3$ to the nearest thousandth. Check your answer with a calculator.

15. Estimate $(1.001)^{10}$ to fifteen decimal places.

Review

16. Write row 9 of Pascal's triangle.   *(Lesson 13-5)*

17. *Multiple choice.*   Which polynomial equals $\dfrac{(n + 2)!}{n!}$?   *(Lesson 13-3)*
   (a) $n + 1$ (b) $n + 2$ (c) $(n + 1)(n + 2)$ (d) $n(n + 1)(n + 2)$

18. a. How many permutations of the letters of the word MOUSE are possible?
   b. If all the permutations are listed in alphabetical order, which comes first and which comes last?   *(Lesson 13-3)*

19. Consider the geometric sequence whose first four terms are 64, 48, 36, and 27.   *Multiple choice.* Which represents the sum of the first six terms of the sequence?   *(Lessons 7-5, 13-2, 13-3)*
   (a) $\displaystyle\sum_{i=1}^{6} 64 \cdot \left(\frac{3}{4}\right)^i$ (b) $\displaystyle\sum_{i=0}^{6} \left(\frac{3}{4}\right)^{i-1}$
   (c) $\displaystyle\sum_{i=1}^{6} \left(\frac{3}{4}\right)^{i-1}$ (d) $\displaystyle\sum_{i=1}^{6} 64\left(\frac{3}{4}\right)^{i-1}$

20. Give the first 8 terms of the sequence $a_n = \sin\left(\frac{n}{2}\pi\right)$.
   (Lessons 1-7, 10-4, 10-10)

21. Suppose $\log x = 5$ and $\log y = \frac{1}{2}$.   *(Lessons 9-5, 9-8)*
   a. Find x.   **b.** Find y.   **c.** Find $\log(x^2 y)$.

22. If the measurements of a single brick are $3\frac{1}{2}'' \times 7\frac{3}{4}'' \times 2\frac{1}{4}''$, what is its volume, to the nearest cubic inch?   *(Previous course)*

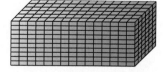

23. A pile of bricks is 10 bricks high, 6 bricks deep and 15 bricks wide. If a single brick weighs between 4 and $4\frac{1}{4}$ lb, what is the largest possible weight of the pile?   *(Previous course)*

Exploration

24. a. Calculate 11^n for $n = 0, 1, 2, 3, 4, 5,$ and 6.
   b. Explain how you can obtain *all* these powers from the binomial coefficients.

Subsets

A subcommittee of 3 people is to be chosen from the 10-person committee pictured above. In how many ways can this be done?
To answer this question, think of the committee as the set {*A, B, C, D, E, F, G, H, I, J*}. Each set of three people from this committee is a **subset** of this set. For instance, two possible subcommittees are: Alan, Barbara, and Carlos; or Frank, Delphine, and Joe. These subcommittees can be represented as the subsets {*A, B, C*} and {*F, D, J*}. Thus the question can be viewed as a problem in counting subsets. How many subsets of 3 elements are possible from a set of 10 elements?

Counting Subsets

Suppose the subsets are formed one element at a time. There are 10 possibilities for the first element. Once the first element has been selected, there are 9 possibilities for the second element. Once the first two elements have been chosen, there are 8 possibilities for the third element.

So it seems there are $10 \cdot 9 \cdot 8$ possibilities. This assumes that the order in which the elements are chosen makes a difference. But the order of the elements in a set makes no difference. {*F, D, J*} and {*D, F, J*} are the same subset, and in fact there are 3!, or 6, different orders which give rise to the same subset as {*F, D, J*}. This is true of all subsets, so the answer $10 \cdot 9 \cdot 8$ is 3! times what we need. The number of subsets with 3 elements is thus

$$\frac{10 \cdot 9 \cdot 8}{3!}.$$

▶

▶Now multiply the fraction by $\frac{7!}{7!}$. This multiplier equals 1, so it does not change the fraction's value. But it does change the way the fraction looks.

$$\frac{10 \cdot 9 \cdot 8}{3!} = \frac{10 \cdot 9 \cdot 8 \cdot 7!}{3! \cdot 7!}$$
$$= \frac{10!}{3! \cdot 7!}$$

Notice that the answer is a binomial coefficient. Evaluate this expression directly, or look for $\binom{10}{3}$ as the 4th element in the 10th row of Pascal's triangle. You will find that there are 120 possible ways to choose a subcommittee of 3 people from a group of 10 people.

The following theorem and its proof generalize the above argument.

Theorem

The number of subsets of r elements which can be formed from a set of n elements is $\frac{n!}{r!(n-r)!}$, the binomial coefficient $\binom{n}{r}$.

Proof

There are n choices for the first element in a subset. Once that element has been picked, there are $n - 1$ choices for the second element, and $n - 2$ choices for the third element. This continues until all r elements have been picked. There are $(n - r + 1)$ choices for the rth element. So, if all the possible orders are considered different, there are

$$\underbrace{n(n - 1)(n - 2) \ldots (n - r + 1)}_{r \text{ factors}}$$

ways to pick them. But each subset is repeated $r!$ times with the same elements. So the number of different subsets is

$$\frac{n(n - 1)(n - 2) \ldots (n - r + 1)}{r!}.$$

Multiplying both numerator and denominator by $(n - r)!$, the theorem results.

The above theorem has applications to a variety of counting problems.

Example 1

Five points are labeled in a plane, with no three collinear. How many triangles have these points as vertices?

Solution 1

Draw a picture. Label the points *A, B, C, D,* and *E.* Form triangles with the points as vertices. The possible triangles are (in alphabetical order) ABC, ABD, ABE, ACD, ACE, ADE, BCD, BCE, BDE, and CDE. So 10 triangles can be formed.

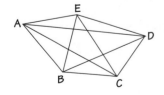

Solution 2

Because no three points are collinear, any choice of 3 points from the 5 points determines a triangle. Use the theorem with n = 5 and r = 3. The number of possible triangles is

$$\binom{5}{3} = \frac{5!}{3!\,(5-3)!} = \frac{5!}{3!\,2!} = \frac{5 \cdot 4 \cdot 3 \cdot 2 \cdot 1}{3 \cdot 2 \cdot 1 \cdot 2 \cdot 1} = 10.$$

Solution 3

Use the idea of the proof of the theorem. The first vertex of the triangle can be chosen in 5 ways. The second vertex can then be chosen in 4 ways. And the third vertex then can be chosen in 3 ways. So, if order made a difference, there would be 5 · 4 · 3, or 60, different triangles. But order doesn't make a difference and each triangle is determined 3!, or 6, times. So divide 60 by 6, yielding 10.

What Are Combinations?

Any choice of r objects from n objects is called a **combination.** The theorem of this lesson can be restated: The number of combinations of r objects from n objects is $\binom{n}{r}$. This relation between $\binom{n}{r}$ and combinations is the reason we read $\binom{n}{r}$ as "n choose r."

Example 1 shows that the number of combinations of 3 objects from 5 objects is 10; in some books, you may see the symbol $_nC_r$. Like $\binom{n}{r}$, it stands for the number of combinations of r objects from n objects. $_5C_3 = \binom{5}{3} = 10$. The subcommittee situation at the beginning of this lesson shows that $_{10}C_3 = \binom{10}{3} = 120$. Some graphics calculators have a key or menu choice for $_nC_r$.

Shown is a Vietnamese restaurant in Garden Grove, California.

Example 2

There are 30 items on a menu in a Vietnamese restaurant. A group of friends is ordering 5 different items. In how many ways can this be done?

Solution

Choosing 5 items from 30 items is equivalent to choosing a subset of 5 elements from a set of 30 elements. This can be done in $\binom{30}{5}$ ways.

$$\binom{30}{5} = \frac{30!}{5!\,25!} = \frac{\overset{6}{\cancel{30}} \cdot 29 \cdot \overset{7}{\cancel{28}} \cdot \overset{9}{\cancel{27}} \cdot \overset{13}{\cancel{26}} \cdot 25!}{5 \cdot 4 \cdot 3 \cdot 2 \cdot 1 \cdot 25!} = 142{,}506$$

There are 142,506 ways to select 5 items from a menu of 30 items.

Check 1

Use a calculator to evaluate $\frac{30!}{5!\ 25!}$.

Check 2

If your calculator can evaluate combinations, evaluate $_{30}C_5$.

Combinations enable you to determine the number of subsets of a particular size. You may wonder how many subsets there are in all. The answer is simple and surprising. For instance, consider the subsets of $\{A, B, C, D\}$.

$\{\ \}$	1 subset has 0 elements.
$\{A\}, \{B\}, \{C\}, \{D\}$	4 subsets have 1 element.
$\{A, B\}, \{A, C\}, \{A, D\}, \{B, C\}, \{B, D\}, \{C, D\}$	6 subsets have 2 elements.
$\{A, B, C\}, \{A, B, D\}, \{A, C, D\}, \{B, C, D\}$	4 subsets have 3 elements.
$\{A, B, C, D\}$	1 subset has 4 elements.

The numbers from Pascal's triangle appear again! The total number of subsets is $1 + 4 + 6 + 4 + 1$, or 16. In general, the total number of subsets of a set with n elements is the sum of the elements in the nth row of Pascal's triangle.

$$\binom{n}{0} + \binom{n}{1} + \binom{n}{2} + \ldots + \binom{n}{n} = \sum_{i=0}^{n} \binom{n}{i}$$

If we replace $\binom{n}{i}$ by $\binom{n}{i} 1^i 1^{n-i}$, the right side looks like one side of the Binomial Theorem. It equals the expansion of $(1 + 1)^n$. Consequently,

$$\binom{n}{0} + \binom{n}{1} + \binom{n}{2} + \ldots + \binom{n}{n} = \sum_{i=0}^{n} \binom{n}{i} 1^i 1^{n-i}$$
$$= (1 + 1)^n$$
$$= 2^n.$$

We have proved the following theorem.

Theorem

A set with n elements has 2^n subsets.

When $n = 4$, $2^n = 16$. This agrees with the number of subsets found above for a set with 4 elements.

QUESTIONS

Covering the Reading

1. *Multiple choice.* Which is not a subset of $\{T, E, A, M\}$?
   (a) $\{M, E, A, T\}$ (b) $\{\ \}$ (c) $\{A, M\}$ (d) $\{T, E, A, M, S\}$

2. **a.** How many subsets of $\{T, E, A, M\}$ have 3 elements?
   b. List the subsets.

In 3–5, consider the set $\{p, q, r\}$.

3. List all the subsets of $\{p, q, r\}$.

4. How many subsets are there with the indicated number of elements?
   a. 0   **b.** 1   **c.** 2   **d.** 3

5. How many subsets does $\{p, q, r\}$ have?

6. Any choice of r objects from n objects is called a(n) __?__.

7. The symbol $_nC_r$ is another way of writing __?__.

8. Copy and complete this pattern.

       1   n
       2       $n - 1$
       3       $n - 2$
   a. 4   __?__
        ⋮   ⋮
   b. r __?__

9. Simplify: $n \cdot (n - 1) \cdot \ldots \cdot (n - r + 1) \cdot ((n - r)!)$.

10. How many combinations of r objects are there from n different objects?

11. How many subcommittees of 4 people can be formed from a committee of 10?

12. Ten points are in a plane, with no three collinear. How many triangles have these points as vertices?

13. In how many ways can 6 different types of sushi be chosen from a display with 25 options?

14. a. What is the sum of the entries in row 7 of Pascal's triangle?
   b. What does that have to do with this lesson?

15. A set with 8 elements has how many subsets?

Applying the Mathematics

16. a. Suppose you are given three noncollinear points. How many triangles have these points as vertices?
   b. Does this agree with the formula for $\binom{n}{r}$? Justify your answer.

17. Simplify:
   $$_9C_0 + {}_9C_1 + {}_9C_2 + {}_9C_3 + {}_9C_4 + {}_9C_5 + {}_9C_6 + {}_9C_7 + {}_9C_8 + {}_9C_9.$$

In 18–20, recall that the U.S. Congress consists of 100 senators and 435 representatives.

18. How many five-person Senatorial Committees are possible?

19. In how many ways can a "committee of the whole" be chosen in the House of Representatives?

Choices, choices. *In many sushi restaurants, like this one in Yokosuka, Japan, the patrons can see the different types of sushi before they order.*

Pick a card . . . any card.

20. What is the total number of committees of any size (except the empty set) which can be formed in the Congress? (Assume that senators and representatives cannot be on a committee together.)

21. **a.** Suppose you pick one card from a 52-card playing deck. In how many ways can this be done?
    b. Suppose you pick two cards from a 52-card playing deck. In how many ways can this be done?
    c. If you pick 4 cards from the deck what are the chances of their being the 4 aces?

Review

In 22 and 23, expand.   *(Lessons 11-3, 13-6)*

22. $(a + b)^8$ **23.** $(2x - 3y)^3$

24. The mean of three numbers in a geometric sequence is 28. The first number is 12. What might the other numbers be?
(Lessons 6-7, 7-5, 11-6, 13-4)

25. a. A runner runs 10 miles one day and then one more mile each day until 20 miles are run in a day. What is the total number of miles run during this period?   *(Lessons 3-7, 13-1)*
   b. *Multiple choice.*   Which symbol does not stand for the sum in part **a**?   *(Lesson 13-1)*

     (i) $\displaystyle\sum_{n=1}^{10} (n + 9)$ (ii) $\displaystyle\sum_{n=10}^{20} n$

    (iii) $\displaystyle\sum_{n=1}^{11} (n + 9)$ (iv) $\displaystyle\sum_{n=0}^{10} (10 + n)$

26. a. Graph the ellipse with equation $\dfrac{x^2}{9} + \dfrac{y^2}{4} = 1$.
   b. Where are its foci?   *(Lesson 12-4)*

27. *Multiple choice.*   Which does not equal the others?   *(Lessons 9-5, 9-7)*
   (a) $\log 100$ (b) $\log_2 4$ (c) $\log_3 6$ (d) $\log_4 16$

28. Evaluate $3p^2q$ when $p = .7$ and $q = 1 - p$.   *(Previous course, Lesson 1-1)*

Exploration

29. Find a formula for one or more of these expressions, in terms of n.
    a. for $n \geq 0$, the sum of the squares of all the elements in the nth row of Pascal's triangle
    b. for $n \geq 1$, the sum of the elements in the nth row of Pascal's triangle when alternate minus and plus signs are put between the elements of the row
    c. for $n \geq 2$, the third element from the end of the nth row of Pascal's triangle.

You have seen that Pascal's triangle can be viewed as a triangle of combinations or as a triangle of binomial coefficients. In this lesson, you will see how probability helps to connect these ideas. We begin with a simple idea: the tossing of a coin.

Independent Events

Suppose a coin is thought to be weighted so that the probability of heads is 0.6. Then we know that the probability of the only other choice, tails, is $1 - 0.6$, or 0.4.

If the coin is tossed twice, then because the coin has no memory, the results of one toss have no effect on the other toss. So tails on the second toss will occur with probability 0.4 regardless of what occurs on the first toss. So the probability of the event "heads followed by tails", which we abbreviate as HT, is $0.6 \cdot 0.4$. We write
$$P(HT) = P(H) \cdot P(T) = 0.6 \cdot 0.4 = 0.24.$$

These two tosses are *independent events.* When two events are **independent events,** the probability that they both occur is found by multiplying the probabilities of the events. Here are the probabilities for the possible outcomes when the coin we have been considering is tossed twice.

Probability of 2 heads: $P(HH) = 0.6 \cdot 0.6 = 0.36$
Probability of heads followed by tails: $P(HT) = 0.6 \cdot 0.4 = 0.24$
Probability of tails followed by heads: $P(TH) = 0.4 \cdot 0.6 = 0.24$
Probability of 2 tails: $P(HH) = 0.4 \cdot 0.4 = 0.16$

The arithmetic can be checked. Since the four events cover all possibilities when a coin is tossed twice, the sum of the four probabilities should be 1.

These four events are *mutually exclusive.* **Mutually exclusive events** cannot happen at the same time, and the probability that one *or* the other will happen is found by adding their probabilities. So the probability of one head and one tail in either order can be found as follows:
$$P(HT \text{ or } TH) = P(HT) + P(TH) = 0.24 + 0.24, \text{ or } 0.48.$$

In general, if p is the probability of heads, and q is the probability of tails, then $q = 1 - p$. If the coin is tossed twice, then:
$$\begin{aligned} P(2 \text{ heads}) &= p \cdot p &&= p^2 \\ P(1 \text{ head}) &= pq + qp &&= 2pq \\ P(0 \text{ heads}) &= q \cdot q &&= q^2. \end{aligned}$$

It may come as a surprise: the three probabilities are the terms in the binomial expansion of $(p + q)^2$.

Example 1 considers this same situation, but with the coin tossed 3 times.

Example 1

Suppose a coin with $P(H) = .6$ is tossed three times. Calculate each probability.
a. $P(3 \text{ heads})$ b. $P(\text{exactly 2 heads})$
c. $P(\text{exactly 1 head})$ d. $P(0 \text{ heads})$

Solution

a. P(3 heads) = P(HHH)
            = .6 · .6 · .6 = .216
b. P(2 heads) = P(HHT or HTH or THH)
            = .6 · .6 · .4 + .6 · .4 · .6 + .4 · .6 · .6 = .432
c. P(1 head) = P(HTT or THT or TTH)
            = .6 · .4 · .4 + .4 · .6 · .4 + .4 · .4 · .6 = .288
d. P(0 heads) = P(TTT)
            = .4 · .4 · .4 = .064

Check

Does $P(3 \text{ heads}) + P(2 \text{ heads}) + P(1 \text{ head}) + P(0 \text{ heads}) = 1$? The check is left to you.

These are the terms in the binomial expansion of $(p + q)^3$. If $p = .6$, then the four probabilities in Example 1 are p^3, $3p^2q$, $3pq^2$, and p^3.

The Binomial Probability Theorem

The situations described above are called *binomial experiments*. A **binomial experiment** has the following features.

1. A situation, called a **trial,** is repeated n times, where $n \geq 2$.

2. The trials are independent.

3. For each trial there are only two outcomes, often called "success" and "failure."

4. Each trial has the same probability of success.

In general, suppose you toss a coin n times. Each toss is a trial. Let p represent the probability of tossing a head, and q represent the probability of *not* tossing a head, that is, of tossing a tail. So $q = 1 - p$. Then, the probability of *each particular* combination of r heads (and so $n - r$ tails) in the n tosses is $p^r q^{n-r}$. There are $\binom{n}{r}$ such combinations. So the probability of getting any of the combinations of r heads in n tosses is $\binom{n}{r} p^r q^{n-r}$. This argument is valid for any binomial experiment.

> **Theorem (Binomial Probability)**
> Suppose a binomial experiment has n trials. If the probability of a success is p, and the probability of failure is $q = 1 - p$, then the probability that there are r successes in n trials is $\binom{n}{r} p^r q^{n-r}$.

In the situation of Example 1, when the coin is tossed 3 times and the probability of heads is .6, the theorem gives the following probabilities:

$$\text{Probability of 3 heads} = \binom{3}{3} p^3 q^{3-3} = p^3 q^0 = (.6)^3(.4)^0 = .216$$

$$\text{Probability of 2 heads} = \binom{3}{2} p^2 q^{3-2} = 3p^2 q^1 = 3(.6)^2(.4) = .432$$

$$\text{Probability of 1 head} = \binom{3}{1} p^1 q^{3-1} = 3p^1 q^2 = 3(.6)(.4)^2 = .288$$

$$\text{Probability of 0 heads} = \binom{3}{0} p^0 q^{3-0} = p^0 q^3 = (.6)^0(.4)^3 = .064$$

These are exactly the results found in Example 1.

Notice that the probabilities are also the terms in the expansion of $(p + q)^n$. Since $p + q = 1$, $(p + q)^n = 1^n = 1$. This proves that you should get 1 as the sum.

Example 2

You take a multiple-choice test and guess on 5 questions. You estimate that you have a probability of $\frac{2}{3}$ of getting each question correct. What is the probability that you get exactly 4 of 5 questions correct?

Solution
Here the experiment that is being repeated is your guessing the correct answer. It is being repeated 5 times, so $n = 5$. The probability of success on a trial is $\frac{2}{3}$, so $p = \frac{2}{3}$. This means that the probability of failure is $\frac{1}{3}$. So $q = \frac{1}{3}$. You want to know the probability of getting 4 successes out of 5 questions, so $r = 4$. Thus,

$$P(4 \text{ successes in 5 trials}) = \binom{5}{4}\left(\frac{2}{3}\right)^4\left(\frac{1}{3}\right)^1 = 5 \cdot \frac{16}{81} \cdot \frac{1}{3} = \frac{80}{243} \approx$$

.329. There is about a 33% probability that you will get exactly 4 of 5 answers correct.

An important case of the Binomial Probability Theorem occurs when the probability of success is $\frac{1}{2}$, as is the case when considering a fair coin. Then the probability of failure is also $\frac{1}{2}$, so $p = q = \frac{1}{2}$. Then $p^r q^{n-r} = p^r p^{n-r} = p^n = \left(\frac{1}{2}\right)^n = \frac{1}{2^n}$. These values can be substituted into the theorem.

Example 3

Suppose a fair coin is tossed 4 times. Find the following.
a. the probability of getting exactly 2 heads
b. the probability of getting either 2 heads or 3 heads

Solution 1

a. Use the Corollary. The probability of getting 2 heads in 4 tosses of a fair coin is $\dfrac{\binom{4}{2}}{2^4} = \dfrac{6}{16} = \dfrac{3}{8} = .375$.

b. From part **a**, $P(2\text{ heads}) = .375$. Using the same Corollary,

$$P(3\text{ heads}) = \dfrac{\binom{4}{3}}{2^4} = \dfrac{4}{16} = \dfrac{1}{4} = .25.$$

$P(2\text{ heads or 3 heads}) = P(2\text{ heads}) + P(3\text{ heads})$
$$= .375 + .25 = .625$$

The possibility of getting 2 or 3 heads is .625.

Solution 2

a. If you forget the Binomial Probability Theorem or its corollary, you can find the answer by counting. When a fair coin is tossed 4 times, there are 16 possible ways of getting H(heads) and T(tails).

HHHH	HTHH	THHH	TTHH
HHHT	HTHT	THHT	TTHT
HHTH	HTTH	THTH	TTTH
HHTT	HTTT	THTT	TTTT

Of these, $\binom{4}{2}$, or 6, have 2 Hs. They are circled. Since the coin is fair, each outcome has equal probability. So $\frac{6}{16} = \frac{3}{8}$ is the probability of 2 heads in 4 tosses of a fair coin.

b. The $\binom{4}{3}$ or 4 ways to have 3 Hs are boxed above. Thus, there are $\binom{4}{2} + \binom{4}{3} = 6 + 4 = 10$ ways to get 2 or 3 heads. So $P(2\text{ or 3 heads}) = \frac{10}{16} = \frac{5}{8} = .625$.

Many people think the answer to Example 3a is $\frac{1}{2}$. Until the beginnings of the theory of probability were put forth by Pascal and Fermat in the 17th century, these questions were quite difficult to answer or verify.

Covering the Reading

In 1–3, tell whether events A and B are independent, mutually exclusive, or neither of these.

1. A = tossing a coin once and heads occurs
B = tossing a coin once and tails occurs

2. A = tossing a coin once and heads occurs
B = tossing a coin a second time and tails occurs

3. A = tossing a coin once and heads occurs
B = tossing a coin a second time and heads occurs

In 4–7, suppose there is a 70% probability that a coin will show tails when tossed. Find the probability of each event.

4. The coin shows heads when tossed.

5. When tossed twice, the coin shows heads both times.

6. When tossed twice, the coin shows heads the first time and tails the second time.

7. When tossed three times, the coin shows tails each time.

8. Give an example of a binomial experiment with 4 trials.

9. Use the situation of Example 2. What is the probability of getting exactly 3 questions correct?

In 10 and 11, a fair coin is tossed 6 times. Give the probability of each event.

10. exactly 3 heads   **11.** exactly 2 tails

12. A fair coin is tossed n times. State
  a. the number of ways r tails may occur.
  b. the total number of ways the coins may fall.
  c. the probability of r tails.

Applying the Mathematics

13. Give the probabilities of getting 0, 1, 2, . . . , 8 heads in 8 tosses of a fair coin.

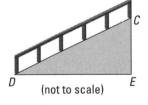

QUIZ

① In 1950, which city had the larger population, Boston or Washington, D.C.?

② Was Joan of Arc born before or after 1400?

③ Which city is farther north: Havana, Cuba, or Mexico City, Mexico?

④ Which mathematician died first, Pascal or Fermat?

In 14 and 15, a student is given the quiz at the left.

14. List all possible ways the quiz might be corrected. (Assume that all questions are answered. Use *R* for right and *W* for wrong.)

15. Assuming that the student guesses on each item, and that the probability of getting the right answer by guessing is $\frac{1}{2}$, calculate each probability.
   a. The student gets all 4 correct.
   b. The student gets exactly 3 correct.
   c. The student gets exactly 2 correct.
   d. The student gets exactly 1 correct.
   e. The student gets none correct.
   f. The student gets at least 2 questions correct.

In 16 and 17, consider a 10-question true-false test. Find the probability you will get 8 or more questions correct under the given assumption.

16. You guess on all 10 questions.

17. You correctly answer 4 questions and guess on the other 6 questions.

Review

18. How many different bridge foursomes can be formed from eight people?   *(Lesson 13-7)*

19. Expand $(p + q)^7$.   *(Lesson 13-6)*

20. a. Evaluate $\dfrac{\sum\limits_{i=1}^{5} a_i}{5}$ if $a_1 = 3$, $a_2 = 8$, $a_3 = 9$, $a_4 = -7$, and $a_5 = 2$.
   b. What does the answer to part **a** tell you about the data set $\{a_1, a_2, a_3, a_4, a_5\}$?   *(Lessons 13-3, 13-4)*

21. Solve $_nC_2 = 28$.   *(Lessons 6-7, 13-3, 13-7)*

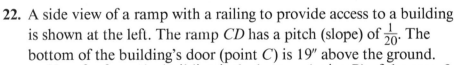

22. A side view of a ramp with a railing to provide access to a building is shown at the left. The ramp *CD* has a pitch (slope) of $\frac{1}{20}$. The bottom of the building's door (point *C*) is 19″ above the ground.
   a. How far from the building is the bottom (point *D*) of the ramp?
   b. Find the measure of angle *CDE*.
   c. Find the length of the railing.   *(Lessons 2-4, 10-1, 10-4)*

D (not to scale) *E*

Exploration

23. a. Toss 4 coins and record how many heads appear. Repeat this at least 50 times.
   b. Calculate the percent of times each of the following occurs: 0 heads, exactly 1 head, exactly 2 heads, exactly 3 heads, 4 heads.
   c. How closely do your results agree with what would be predicted by the corollary of this lesson if the coins are fair? Do you think your coins are fair?

It's all in the numbers. *Shown are the two winners who split a $90 million Powerball lottery in 1993, Charles Gill and Percy Pridgen. They will receive 20 annual payments of $2.25 million each.*

A **lottery** is a game or a procedure in which prizes are distributed among people by pure chance. The simplest lotteries are raffles in which you buy tickets, the tickets are put in a bin, and the winning tickets are picked from that bin. In recent years, however, more complicated lotteries have been designed. These lotteries pay out large amounts of money to a few individuals in order to attract bettors. Today, countries in Asia, Africa, Europe, and South America run lotteries themselves or allow private lotteries. In the United States, over 80% of the population live in states that run lotteries.

Lotteries are designed to make money, so they always pay out far less than they take in. This means that many more people will lose money in a lottery than will win. Still, the possibility—however remote—of winning a huge amount of money attracts bettors.

Because lotteries are pure chance, the probabilities of winning various prizes can be calculated. Consider a typical lottery. To participate in the Super Lotto game in the state of Ohio, a person (who must be 18 years of age or older) picks six numbers out of the set of consecutive integers {1, 2, 3, . . . , 47}, marks these on a card, and pays $1. Each Wednesday and Saturday, six balls are picked at random from balls numbered 1 through 47. These are the winning Lotto numbers. For instance, the winning numbers in the Ohio Super Lotto game on January 12, 1994, were 8, 10, 11, 20, 32, and 36. The people who pick all six winning numbers split the grand prize, which is always at least $4 million.

Example 1

What is the probability of picking the six winning numbers in the Ohio Super Lotto game?

Solution

The number of possible combinations that could be chosen is

$$\binom{47}{6} = \frac{47!}{6!41!} = \frac{47 \cdot 46 \cdot 45 \cdot 44 \cdot 43 \cdot 42}{6 \cdot 5 \cdot 4 \cdot 3 \cdot 2 \cdot 1} = 10{,}737{,}573.$$

Since each combination has the same probability, the probability that a particular combination will appear is $\frac{1}{10,737,573}$. *So the probability of picking the winning numbers is* $\frac{1}{10,737,573}$.

The answer to Example 1 explains why there is often no winner in a lottery. In the long run, only about one entry of every 11,000,000 is a winner. The answer also shows you how much money the state of Ohio takes in for every grand prize. In the long run, Ohio pays out about $4,000,000 in grand prizes for every $11,000,000 it has taken in.

Because the probability of winning the grand prize is so low, most states give smaller prizes to someone who picks almost all of the numbers. For instance, in Ohio there is a prize (but *much* smaller than the grand prize) for picking five of the six winning numbers. There is an even smaller prize given for picking four of the six winning numbers.

Example 2

What is the probability of picking exactly four of the six winning numbers in the Ohio Super Lotto game?

Solution

Of the six numbers, 4 must be picked from the six winning numbers and the other 2 must be picked from the 41 that remain. These two events can be done in $\binom{6}{4}\binom{41}{2}$ ways. The winners are chosen from the $\binom{47}{6}$ possible combinations found in Example 1. So the probability of picking four of the six winning numbers is

$$\frac{\binom{6}{4}\binom{41}{2}}{\binom{47}{6}} = \frac{12{,}300}{10{,}737{,}573} = .001146 \approx \frac{1}{873}.$$

In Example 2, the probability has been approximated by a fraction with 1 in its numerator. We did this because it is easier to understand the fraction $\frac{1}{873}$ than the decimal 0.001146 . . . . The probability given as a fraction can be read as "about 1 chance in 873." You could also say, "The odds against winning are about 872 to 1." If you first obtain a probability as a decimal, then you can use the reciprocal 1/x or x⁻¹ key to obtain the fraction.

Because probabilities of winning are so low in these lottery games, people can get discouraged from entering the game repeatedly. So states have games in which more people win, using lotteries with fewer numbers to match. Some of these games require that 3 or 4 digits (from 0 to 9) be matched exactly, in order. Because the order matters, the probability of winning involves permutations, not combinations.

Example 3

To play Minnesota's Daily 3 game, a player picks three digits each from 0 to 9 and must match a 3-digit number. What is the probability of matching the 3 winning digits?

Solution 1

There is a probability of $\frac{1}{10}$ that the first digit will be matched. There is a probability of $\frac{1}{10}$ that the second digit will be matched. And there is a probability of $\frac{1}{10}$ that the third digit will be matched. Since these events are independent, The probability of matching all three numbers is $\frac{1}{10} \cdot \frac{1}{10} \cdot \frac{1}{10}$, or $\frac{1}{1000}$.

Solution 2

Think of the three digits as forming one number. There are one thousand numbers from 000 to 999. The probability of matching one of these is $\frac{1}{1000}$.

We should note that many people allow computers to pick the numbers, but that does not change the odds of winning. Some people study the past winning numbers, but that does not change the odds. And some people use their lucky numbers or their birthdays. That doesn't change the odds, but it may make bettors feel better. There is no system for winning lotteries like these.

QUESTIONS

Covering the Reading

1. What is a lottery?

In 2–4, consider Ohio's Super Lotto game.

2. What must a person do in order to win the grand prize in this game?

3. What is the probability that someone who participates in Ohio's Super Lotto game wins the grand prize?

4. What is the probability of picking 5 of 6 winning numbers in this game?

5. Montana Ca$h gives players a chance to win a minimum jackpot of $20,000 by matching five numbers (in any order) from a field of 37 numbers.
   a. What is the probability of winning the jackpot in Montana Ca$h?
   b. What are the odds against winning the jackpot in Montana Ca$h?

6. Minnesota's Daily 3 game has a Front Pair option. To win this game, a player must match the first two of three numbers (each of which can be any digit from 0 through 9) in exact order. What is the probability of winning Front Pair?

Applying the Mathematics

7. To play the Match 5 game in Maryland a person picks 5 of 39 numbers. A person can play one Match 5 game board for $1 or three Match 5 game boards for $2.
   a. What is the probability of winning Match 5 if you play one game board?
   b. A publication of the Maryland State Lottery Headquarters claims that the probability of winning Match 5 on a $2 ticket is $\frac{1}{191,919}$. What must be true about the numbers the player picks on the three game boards in order for this to be correct?

8. POWERBALL™ is a lottery played in 15 states. To play, an individual picks six numbers from 1 to 45, one of which is designated as the "powerball" number. To win the jackpot, which is always at least $2 million, you must match all 5 numbers (in any order) plus the Powerball.
   a. What is the probability of winning the POWERBALL™ jackpot?
   b. *True or false.* The probability of winning the jackpot in POWERBALL™ is equal to the probability of winning a lottery in which a player chooses six numbers from 1 to 45. Justify your answer.

In 9 and 10, in the Quinto game in the State of Washington a player picks five playing cards of a standard 52-card deck. A sample ticket is shown at the left.

9. If your ticket's symbols match the five drawn (order doesn't matter) you win a grand prize of at least $100,000. What is the probability of winning the Quinto grand prize with one $1 ticket?

10. Washington pays $1000 for a ticket matching four of five cards, and $20 for a ticket matching three of five. Does this mean that it is exactly $\frac{1000}{20} = 50$ times as likely to match three of five winning cards than four of five winning cards? Explain why or why not.

11. Most lotteries do not pay out winning amounts all at one time. Many pay $\frac{1}{20}$ of the winnings each year. For instance, if a million dollars is won, then $50,000 is paid the winner each year for 20 years. What are some advantages of this to the winner? What are some disadvantages? Do you think this is fair? Explain why or why not.

Review

12. A family has 6 children. What is the probability that 3 are girls and 3 are boys?   *(Lesson 13-8)*

13. From previous research, scientists estimate that the probability that a new experimental treatment will cure a disease is 0.75. Find the probability that at least 8 of 10 people receiving this treatment will be cured.   *(Lesson 13-8)*

14. A teacher is asked to select three students from a class of 25 to be interviewed by a local television station. You and your two best friends would like to be chosen. If the selection is made at random, what is the probability that you and your two best friends will be chosen?   *(Lessons 13-7, 13-8)*

15. Rewrite $\binom{n}{3}$ in the standard form of a polynomial.   *(Lessons 11-1, 13-7)*

16. Expand.
    a. $(a + b)^5$ **b.** $\left(6 - \frac{x}{2}\right)^5$ *(Lesson 13-6)*

17. Evaluate and write as a decimal: $\sum\limits_{i=-3}^{2} 6 \cdot 10^i$.   *(Lesson 13-3)*

18. The five members of a jazz quintet are having their picture taken for a new album cover.
    a. How many ways are there for them to line up in a straight line?
    b. How many ways are there for them to line up with the drummer at either end?   *(Lesson 13-3)*

19. **a.** According to the Rational Zero Theorem, what are all possible rational zeros of $g(x) = 3x^2 - 5x + 2$?
    b. Find all the rational zeros of g.   *(Lessons 11-6, 11-7)*

20. **a.** Factor $13x^3 - 36x - x^5$.
    b. Find the zeros of the function defined by $f(x) = 13x^3 - 36x - x^5$.
    (Lessons 11-4, 11-4, 11-6)

Exploration

21. If your state has lottery games, find out what they are and determine the probability of winning each game. If your state does not have lottery games, use one from a neighboring state.

Normal heights. *The distribution of adult heights of any species is typically a normal distribution. These zebras live in Africa.*

A Binomial Distribution with Six Points

Suppose you toss a fair coin 5 times. Let $P(n)$ = the probability of n heads in 5 tosses of a fair coin. Then the domain of P is $\{0, 1, 2, 3, 4, 5\}$. From the Corollary to the Binomial Probability Theorem in Lesson 13-8,

$$P(n) = \frac{\binom{5}{n}}{2^5} = \frac{1}{32}\binom{5}{n}.$$

Below is a table of values for P.

n = number of heads	0	1	2	3	4	5
$P(n)$ = probability of n heads	$\frac{1}{32}$	$\frac{5}{32}$	$\frac{10}{32}$	$\frac{10}{32}$	$\frac{5}{32}$	$\frac{1}{32}$

We call P a *probability function*. A **probability function** or **probability distribution** is a function which maps a set of events onto their probabilities. Below is the graph of P.

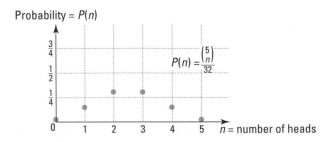

Because this function results from calculations of binomial probabilities, it is called a **binomial probability distribution,** or simply a **binomial distribution.**

A Binomial Distribution with Eleven Points

If a fair coin is tossed 10 times, the possible numbers of heads are 0, 1, 2, . . . , 10, so there are 11 points in the graph of the corresponding probability function. So, using the Corollary to the Binomial Probability Theorem, $P(x)$, the probability of x heads, is given by

$$P(x) = \frac{\binom{10}{x}}{2^{10}} = \frac{\binom{10}{x}}{1024}.$$

The 11 probabilities are easy to calculate because the numerators in the fractions are the numbers in the 10th row of Pascal's triangle. That is, they are binomial coefficients.

x = Number of heads	0	1	2	3	4	5	6	7	8	9	10
$P(x)$ = Probability of x heads	$\frac{1}{1024}$	$\frac{10}{1024}$	$\frac{45}{1024}$	$\frac{120}{1024}$	$\frac{210}{1024}$	$\frac{252}{1024}$	$\frac{210}{1024}$	$\frac{120}{1024}$	$\frac{45}{1024}$	$\frac{10}{1024}$	$\frac{1}{1024}$

This binomial distribution is graphed below.

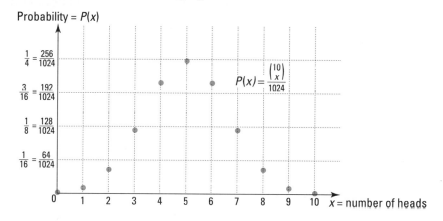

Examine this 11-point graph of $P(x) = \frac{\binom{10}{x}}{1024}$ closely. The individual probabilities are all less than $\frac{1}{4}$. Notice how unlikely it is to get no heads or 10 heads in a row $\left(\text{the probability for each is less than } \frac{1}{1000}\right)$. Even for 9 heads in 10 tosses the probability is less than $\frac{1}{100}$. Like the graph of the 6-point probability function $P(n) = \frac{\binom{5}{n}}{32}$ on the previous page, this 11-point graph has a vertical line of symmetry.

Normal Distributions

As the number of tosses of a fair coin is increased, the points on the graph more closely outline a curve shaped like a bell. Below is this bell-shaped curve in the position where its equation is simplest. Its equation is

$$y = \frac{1}{\sqrt{2\pi}} e^{\left(\frac{-x^2}{2}\right)}.$$

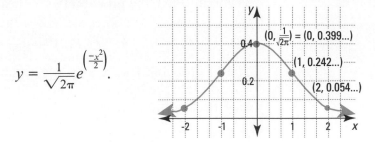

Shown is the Liberty Bell in Philadelphia.

The function which determines this graph is called a **normal distribution,** and the curve is called a **normal curve.** Notice that its equation involves the famous constants e ≈ 2.718 and π ≈ 3.14. Every normal curve is the image of the above graph under a composite of translations or scale changes. Thus the graph of $y = \frac{1}{\sqrt{2\pi}} e^{\left(\frac{-x^2}{2}\right)}$ is sometimes called the **standard normal curve.**

Normal curves are models for many natural phenomena. The graph of the function:

height to the nearest inch → number of men in the U.S. with that height

is very close to a normal curve. The curve has its highest point around 5′9″ or 5′10″.

Normal curves are often good mathematical models for the distribution of scores on an exam. The graph below shows an actual distribution of scores on a 40-question test given by one of the authors to 209 geometry students. (It was a hard test!) A possible corresponding normal curve is shown in red.

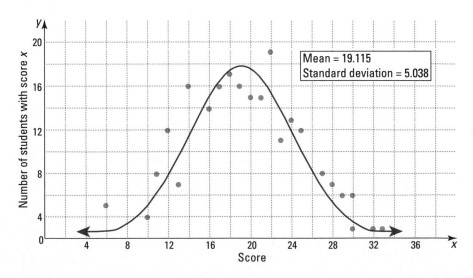

For other tests, the scores are **standardized** or **normalized**. This means that a person's score is not the number of correct answers, but some score chosen so that the distribution of scores is a normal curve. One advantage of normalizing scores is that you need to know no other scores to know how a person's score compares with the scores of others.

College Board SAT scores are standardized with a mean of 500 and a standard deviation of 100. Many IQ tests are normalized so that the mean IQ is 100 and the standard deviation is 15.

The next graph shows percents of scores in certain intervals of a normal distribution with mean m and standard deviation s. Each percent gives the probability of scoring in a particular interval. Actual values for endpoints of these intervals are given below the graph for particular applications.

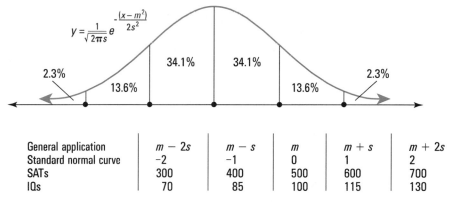

$$y = \frac{1}{\sqrt{2\pi}s} e^{-\frac{(x-m^2)}{2s^2}}$$

2.3% 13.6% 34.1% 34.1% 13.6% 2.3%

General application	$m - 2s$	$m - s$	m	$m + s$	$m + 2s$
Standard normal curve	-2	-1	0	1	2
SATs	300	400	500	600	700
IQs	70	85	100	115	130

For instance, the graph indicates that on an IQ test with mean 100 and standard deviation 15, about 34.1% of IQ's are between 100 and 115. Thus, about 68.2% of IQ scores are between 85 and 115. Other information may be similarly read from the graph.

QUESTIONS

Covering the Reading

1. Let $P(n) = \frac{\binom{5}{n}}{32}$.
   a. Calculate $P(3)$ and indicate what it could represent.
   b. What kind of function is P?

2. a. What is the domain of the function $P: x \to \frac{\binom{10}{x}}{1024}$ graphed in this lesson?
   b. What is the range of this function?

3. If a fair coin is tossed 10 times, what is the probability of 5 heads?

4. Let $P(n) =$ the probability of getting n heads in 6 tosses of a fair coin.
  a. Make a table of values for P.
  b. Graph P.

5. Give an equation for the standard normal curve.

6. Give one application of normal curves.

7. a. What does it mean for scores to be standardized?
  b. What is one advantage of doing this?

8. Approximately what percent of scores on a normal curve are within one standard deviation of the mean?

9. On an IQ test with mean of 100 and standard deviation of 15, approximately what percent of people have IQs below 85?

10. Approximately what percent of people score above 700 on an SAT test with mean 500 and standard deviation 100?

Applying the Mathematics

11. If you repeatedly toss 10 fair coins, about what percent of the time would you expect to get from 4 to 6 heads?

12. Some tests are standardized so that the mean is the grade level at which the test is taken and the standard deviation is 1 grade level. So, for students who take a test at the beginning of 10th grade, the mean is 10.0 and the standard deviation is 1.0.
  a. On such a test taken at the beginning of 10th grade, what percent of students would be expected to score below 8.0 grade level?
  b. If a test is taken in the middle of 8th grade (grade level 8.5), what percent of students score between 7.5 and 9.5?

13. In the fall of 1993, total ACT scores for seniors ranged from 1 to 36 with a mean near 21 and a standard deviation near 5. About what percent of students had an ACT score above 26?

14. Let $y = \dfrac{1}{\sqrt{2\pi}} e^{\left(\frac{-x^2}{2}\right)}$. Estimate y to the nearest thousandth when $x = 1.5$.

15. In a normal distribution, 0.13% of the scores lie more than 3 standard deviations away from the mean in each direction. This implies that about 1 out of __?__ people has an IQ over __?__.

Review

16. To play the Tri-West Lotto, a lottery sponsored by Idaho, Montana, and South Dakota, a player picks six numbers from 1 to 41. To win the jackpot, the player must match all six winning numbers.
  a. What is the probability of winning the jackpot?
  b. What are the odds against winning the jackpot? *(Lesson 13-9)*

17. In Washington's Quick Pick, the computer picks three numbers each from 0 to 9 for a player. The numbers must match in order. What is the probability of winning this game? *(Lesson 13-9)*

18. *True or false.* The probability of getting exactly 25 heads in 50 tosses of a fair coin is less than $\frac{1}{10}$. Justify your answer. *(Lesson 13-8)*

19. Evaluate $_nC_0$. *(Lesson 13-7)*

20. a. Expand $(1 - x^2)^3$.
　　b. Factor $(1 - x^2)^3$. *(Lessons 11-3, 13-6)*

21. Find the mean, median, mode, and standard deviation of these scores: 83, 85, 88, 92, 92, 94. *(Lesson 13-4)*

22. a. Graph $\{(x, y): x^2 + (y - 5)^2 = 1\}$.
　　b. Describe in words the graph of $\{(x, y): x^2 + (y - 5)^2 < 1\}$. *(Lessons 12-2, 12-3)*

In 23 and 24, solve **a.** exactly; **b.** to the nearest hundredth.
(Lessons 6-7, 9-10, 11-5)

23. $3^x = 10$ 　　　　　　　　　　　　**24.** $5x^2 + 3x = 10$

25. A hot air balloon is sighted in the same direction from two points P and Q on level ground at the same elevation. From point P the angle of elevation of the balloon is 21°. From point Q the angle of elevation is 15°. If points P and Q are 10.2 km apart, how high is the balloon? *(Lessons 10-1, 10-7, 10-9)*

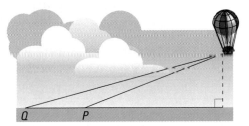

Exploration

26. Together with some other students, toss 12 coins at least 200 times and count the number of heads each time. Let $P(h)$ = the number of times h heads appear.
　　a. How close is P to a normal distribution?
　　b. What is the mean of the distribution (the mean number of heads)?
　　c. Estimate the standard deviation of the distribution.

Park 'n' study. *Many schools allow students in certain classes to drive to school and park in the school parking lot. In 1993, over 6 million teenagers under the age of 18 were licensed to drive.*

What is Sampling?

In a large high school with 510 seniors, the administration wanted to know the percent of seniors who owned cars. The principal walked into a senior homeroom and polled the 25 seniors there. Four of the students said they owned cars. Since $\frac{4}{25} = 16\%$, the principal estimated that about 16% of the seniors in the school owned cars. The principal made an inference based on *sampling*.

In sampling, the **population** is the set of all people or events or items that could be sampled. A **sample** is a subset of the population that is actually studied. Above, the population is the set of 510 seniors. The sample is the set of 25 seniors polled by the principal.

Example

A company is testing light bulbs to see how long they shine before burning out. What is the sample and what is the population?

Solution

The population is the set of all light bulbs that could be tested, perhaps all the light bulbs that have been or will be made by this company. The sample is the set of light bulbs actually tested.

Reasons for Sampling

The principal could have polled all seniors, but perhaps there was no time. Sampling is often used to save time. However, in the Example, sampling is absolutely necessary because the ability to sell a light bulb is destroyed when it is sampled, and so the manufacturer cannot sample all light bulbs.

Sampling is also necessary when the population is infinite. For instance, suppose a coin is tossed 100 times to determine whether or not it is fair. The population is the set of all tosses that could be made. The sample is the set of 100 tosses actually used.

A use of sampling familiar to you is getting ratings of television programs. Ratings are percents of households tuned to programs. The higher the rating for a program, the more a television station can charge for advertising. Because there are so many people who watch television, polling everyone would be too costly. So ratings companies use a sample of households, usually from 1000 to 3000 in number. The population for TV ratings is the set of all households with televisions. If 23.1% of all households sampled are tuned to a particular show, then the rating is 23.1

Shown is a scene from the TV series DeGrassi Junior High. *The series, which chronicled the life of junior high school students in an inner-city school, was popular among young teens.*

Random Samples

The reliability of a sample depends on its being representative of the population. The only sure way to make it representative is for each element of the population to have the same probability of being selected for the sample. We then call the sample a **random sample.** If seniors in the school described above are assigned to homerooms according to extracurricular interests, the principal's sample may not have been a random sample. Coin tossing is closer to random.

TV stations often want to split the ratings sample to determine whether teenagers or senior citizens or other groups are watching. Random sampling may not give them enough people in each of these smaller samples. So they *stratify* the sample, often by age. A **stratified sample** is a sample in which the population has first been split into subpopulations and then, from each subpopulation, a sample is selected. A **stratified random sample** occurs when the smaller samples are chosen randomly from the subpopulations.

How many ways can a sample be chosen? What is the probability that a sample will have particular characteristics? How large must a sample be in order to give accurate results? The answers involve combinations and the normal distribution. To see this, examine the table on the next page. This table is part of a larger **table of random numbers,** so-called because it was constructed so that each digit from 0 to 9 has the same probability of being selected, each pair of digits from 00 to 99 has the same probability of being there, each triple of digits from 000 to 999 has the same probability of being there, and so on.

You can use this table of random numbers to *simulate* what the principal might find if 20% of the seniors actually owned cars. Let each senior be represented by a digit. To simulate the 20%, a 0 or 1 will mean that the senior owns a car. A digit of 2 through 9 means the senior does not.

To use such a table, you must start randomly as well. With your eyes closed, point to a pair of digits on the page; use that pair as the row. Then point again to a pair of digits to use as the column. If you point to 32 and then to 07, start at the 32nd row, 7th column. If you point to a pair of digits whose number does not refer to a row or column, point again.

Now suppose you begin at the digit in the 32nd row, 7th column. It is a 7. Examining the next 25 numbers is like going into a homeroom and asking 25 seniors whether they own a car. Now choose a direction to go in—up, down, left or right—perhaps by rolling a die. We go right. The next 25 numbers are 7, 6, 2, 2, 2, 3, 6, 0, 8, 6, 8, 4, 6, 3, 7, 9, 3, 1, 6, 1, 7, 6, 0, 3, 8. Since 4 of the digits are either 0 or 1, in this sample 4 of the seniors own a car. If there are 510 digits (seniors) to choose from, there are $\binom{510}{25}$ potential samples, a very large number (over 10^{42}). These samples would have from 0 to 25 seniors who own cars, but more samples will have 5 who own cars than any other. Slightly fewer samples will have 4 or 6 who own cars. Again slightly fewer will have 3 or 7. A small percentage of the 10^{42} possible samples will have 0, and a very, very tiny percentage will have near 25 who own cars.

Random Number Table

col. row	1	2	3	4	5	6	7	8	9	10	11	12	13	14
1	10480	15011	01536	02011	81647	91646	69719	14194	62590	36207	20969	99570	91291	90700
2	22368	46573	25595	85393	30995	89198	27982	53402	93965	34095	52666	19174	39615	99505
3	24130	48360	22527	97265	76393	64809	15179	24830	49340	32081	30680	19655	63348	58629
4	42167	93093	06423	61680	17856	16376	39440	53537	71341	57004	00849	74917	97758	16379
5	37570	39975	81837	16656	06121	91782	60468	81305	49684	60672	14110	06927	01263	54613
6	77921	06907	11008	42751	27756	53498	18602	70659	90655	15053	21916	81825	44394	42880
7	99562	72905	56420	69994	98872	31016	71194	18738	44013	48840	63213	21069	10634	12952
8	96301	91977	05463	07972	18876	20922	94595	56869	69014	60045	18425	84903	42508	32307
9	89579	14342	63661	10281	17453	18103	57740	84378	25331	12566	58678	44947	05585	56941
10	85475	36857	43342	53988	53060	59533	38867	62300	08158	17983	16439	11458	18593	64952
11	28918	69578	88231	33276	70997	79936	56865	05859	90106	31595	01547	85590	91610	78188
12	63553	40961	48235	03427	49626	69445	18663	72695	52180	20847	12234	90511	33703	90322
13	09429	93969	52636	92737	88974	33488	36320	17617	30015	08272	84115	27156	30613	74952
14	10365	61129	87529	85689	48237	52267	67689	93394	01511	26358	85104	20285	29975	89868
15	07119	97336	71048	08178	77233	13916	47564	81056	97735	85977	29372	74461	28551	90707
16	51085	12765	51821	51259	77452	16308	60756	92144	49442	53900	70960	63990	75601	40719
17	02368	21382	52404	60268	89368	19885	55322	44819	01188	65255	64835	44919	05944	55157
18	01011	54092	33362	94904	31272	04146	18594	29852	71585	85030	51132	01915	92747	64951
19	52162	53916	46369	58586	23216	14513	83149	98736	23495	64350	94738	17752	35156	35749
20	07056	97628	33787	09998	42698	06691	76988	13602	51851	46104	88916	19509	25625	58104
21	48663	91245	85828	14346	09172	30168	90229	04734	59193	22178	30421	61666	99904	32812
22	54164	58492	22421	74103	47070	25306	76468	26384	58151	06646	21524	15227	96909	44592
23	32639	32363	05597	24200	13363	38005	94342	28728	35806	06912	17012	64161	18296	22851
24	29334	27001	87637	87308	58731	00256	45834	15398	46557	41135	10367	07684	36188	18510
25	02488	33062	28834	07351	19731	92420	60952	61280	50001	67658	32586	86679	50720	94953
26	81525	72295	04839	96423	24878	82651	66566	14778	76797	14780	13300	87074	79666	95725
27	29676	20591	68086	26432	46901	20849	89768	81536	86645	12659	92259	57102	80428	25280
28	00742	57392	39064	66432	84673	40027	32832	61362	98947	96067	64760	64584	96096	98253
29	05366	04213	25669	26422	44407	44048	37937	63904	45766	66134	75470	66520	34693	90449
30	91921	26418	64117	94305	26766	25940	39972	22209	71500	64568	91402	42416	07844	69618
31	00582	04711	87917	77341	42206	35126	74087	99547	81817	42607	43808	76655	62028	76630
32	00725	69884	62797	56170	86324	88072	76222	36086	84637	93161	76038	65855	77919	88006
33	69011	65797	95876	55293	18988	27354	26575	08625	40801	59920	29841	80150	12777	48501
34	25976	57948	29888	88604	67917	48708	18912	82271	65424	69774	33611	54262	85963	03547
35	09763	83473	73577	12908	30883	18317	28290	35797	05998	41688	34952	37888	38917	88050

The Distribution of Sample Means

Let $P(x)$ be the probability that a sample of 25 from 510 random digits contains x digits that are 0s or 1s. That is, $P(x)$ = the probability that a sample of size 25 contains x seniors with cars. An important theorem from statistics, the *Central Limit Theorem,* is that the function P is very closely approximated by a normal distribution whose mean is the mean of the population, in this case 20%, and whose standard deviation is $\sqrt{25 \cdot 20\% \cdot 80\%}$, which in this case is 2.

> **Central Limit Theorem**
> Suppose samples of size n are chosen from a population in which the probability of an element of the sample having a certain characteristic is p. Let $P(x)$ equal the probability that x elements have the characteristic. Then P is approximated by a normal distribution with mean np and standard deviation $\sqrt{np(1-p)}$.

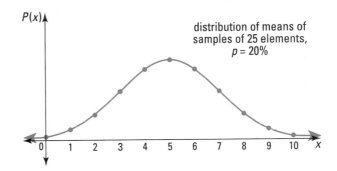

distribution of means of
samples of 25 elements,
$p = 20\%$

The normal distribution for the function P, when $n = 25$ and $p = 20\%$, is graphed here.

Recall the percents within given standard deviations for a normal distribution. If 20% of the seniors own cars and the principal polled 25 seniors at random, about 68% of the time the principal would find that from 3 to 7 seniors in a sample owned cars. That is what happened here. In these cases, the principal would infer that 12% to 28% of the seniors owned cars. That isn't too far off even with a sample of 25. About 95% of the time the principal would find that from 1 to 9 seniors in a sample (between 4% and 36% of the seniors) owned cars. That's a wider interval, but the principal could be 95% confident of the results.

Now let us turn to the TV polling example. Suppose that in reality 20% of households are tuned in to a particular show. Consider all the random samples of 1600 people. These samples will have a mean of 320 people (20%) tuned to the show. The standard deviation of these samples will be $\sqrt{1600 \cdot 20\% \cdot 80\%}$, or 16. That means that 68% of the time the samples will have between 304 and 336 people (between 19% and 21%) watching the show. The sample percentage will be within 1% of the actual. In fact, 95% of the time the sample will have between 288 and 352 people watching the show, that is, between 18% and 22%. So 95% of the time, the sample is within 2% of the actual amount. This accuracy is good enough for the networks.

distribution of means of
samples of 1600 elements,
$p = 20\%$ and s.d. $= 16$

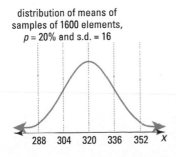

Final Thoughts

The Central Limit Theorem is an appropriate topic with which to end this book, because it involves so many of the ideas you have studied. It models samples of data, so it is fundamental in applications. The function in it, the normal distribution, has an equation with square roots, π, e, and negative exponents. Its graph is a translation and scale-change image of the parent curve $y = \frac{1}{\sqrt{2\pi}} e^{\left(-\frac{x^2}{2}\right)}$. The normal distribution is a probability distribution that is used on tests that for some people help to determine the college they will attend. It shows how interrelated are the ideas of mathematics, and how important are many of the ideas you learned this year.

QUESTIONS

Covering the Reading

1. Give two reasons for sampling.

In 2 and 3, identify the population and the sample in the sampling situation.

2. sampling to obtain TV ratings

3. polling potential voters to see which candidate is favored

4. What is the size of the samples often used in TV ratings?

5. What is the difference between a random sample and one that is not random?

6. Why are stratified samples often used to obtain TV ratings?

7. What does a TV rating of 18.6 mean?

8. *Copy and complete.* If 1600 people are polled randomly for TV ratings, 68% of the time the rating will be within __?__% of the actual percent of people watching the program.

9. *Multiple choice.* In this lesson, 20% of a senior class of 510 owned cars. The principal walked into a class and found that 4 of 25, or 16%, of the seniors he polled owned cars. What is the *best* reason that the percents are not equal?
   (a) The sample was not random.
   (b) Sample percents vary.
   (c) Students may not have been telling the truth.

10. Means of samples of size n from a distribution in which the probability of a characteristic is p approximate a normal distribution with what mean and what standard deviation?

In 11 and 12, a fair coin is tossed 1000 times.

11. The mean number of heads in such samples is __?__ and the standard deviation is __?__.

12. This implies that 68% of the time, from __?__ to __?__ heads are expected.

In 13 and 14, consider that in BASIC, a function named RND generates random numbers with decimal values between 0 and 1. RND always has the argument 1 so in programs you must use RND(1) to generate such a number.

13. a. Run the program below and describe its output.

```
10 FOR N = 1 TO 10
20 PRINT RND(1)
30 NEXT N
40 END
```

   b. Run the program again and write a sentence or two comparing its output to that in part a.

14. The following program simulates tossing a coin.

```
10 REM COIN TOSS SIMULATION
20 REM NMTOS = NUMBER OF TOSSES
30 REM X = A RANDOM NUMBER
40 REM H = NUMBER OF HEADS, T = NUMBER OF TAILS
50 INPUT "HOW MANY TOSSES?"; NMTOS
60 FOR I = 1 TO NMTOS
70 LET X = RND(1)
80 IF X < .5 THEN H = H + 1 ELSE T = T + 1
90 NEXT I
100 PRINT H; "HEADS AND"; T; "TAILS"
110 END
```

   a. Run the program for 50 tosses, and record the output.
   b. Run the program for 500 tosses, and record the output.
   c. Calculate the percent heads and percent tails for each run above. Which run more closely approximates the probability of getting a head on a toss of one coin?

15. You call a classmate to find out if he or she thinks there will be a test on Chapter 13 next Friday. Treated as a sampling situation, what is the population and what is the sample?

16. Construct a data set with mean 10, median 9, and mode 8.
    (Lesson 13-4)

17. Give the standard deviation of $\{2, 4, 6, 8, 10, 12, 14, 16, 18\}$.
    (Lesson 13-4)

18. Graph the binomial distribution for tossing a fair coin 3 times. *(Lesson 13-10)*

19. What is the probability of answering exactly 5 questions correctly on a 6-question test in which you have a 50% chance of getting each question correct? *(Lesson 13-8)*

20. *True or false.* Justify your answer. $3\left(\sum\limits_{n=1}^{4} n^2\right) = \sum\limits_{n=1}^{4}(3n^2)$ *(Lesson 13-3)*

21. If $2^x = 45$, what is x? *(Lesson 9-10)*

22. Solve: $t^{\frac{-1}{2}} = 81$. *(Lesson 7-8)*

23. Find equations for two parabolas congruent to $y = x^2$ and having vertex $(6, 5)$. *(Lesson 6-3)*

24. Simplify $\sqrt{4} \cdot \sqrt{9} + \sqrt{-4} \cdot \sqrt{9} + \sqrt{-4} \cdot \sqrt{-9} + \sqrt{4} \cdot \sqrt{-9}$. *(Lessons 6-2, 6-8)*

25. Give an equation for the right angle graphed here. *(Lesson 6-2)*

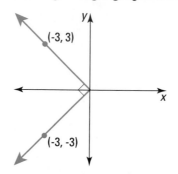

Exploration

26. Sample at least 25 people on a controversial issue of concern to you. From the size of your sample and the results you find, make reasonable inferences using the Central Limit Theorem.

A project presents an opportunity for you to extend your knowledge of a topic related to the material of this chapter. You should allow more time for a project than you do for typical homework questions.

1

Convergent and Divergent Geometric Series

In Lesson 13-2, you were given a formula for the sum of a geometric series.

a. Choose 5 different values for r such that $|r| > 1$. For each of these geometric series, let the first term g_1 equal 2. Determine the sum of each of these geometric series for 10 terms, 50 terms, and 100 terms. Organize your data into a chart.

b. Choose 5 different values for r such that $|r| < 1$. For each of these geometric series, let the first term g_1 equal 2. Determine the sum of each of these geometric series for 10 terms, 50 terms, and 100 terms. Organize your data into a chart.

c. The word *converge* means to approach or draw near to a particular value. For example, as x gets very large, $\frac{1}{x}$ converges to 0. The word *diverge* means does not converge. For example, as x gets large, x^2 diverges; as x gets large, x^2 does not approach any particular value. Make a conjecture about the conditions under which a geometric series will converge. Test your conjecture using $g_1 \neq 2$ and some of the values of r. Think about the formulas for geometric series. Write a mathematical argument to support your conjecture.

2

Genetics

In the 19th century, Gregor Mendel founded what has become the field of *genetics* by crossbreeding various kinds of peas and other plants. Mendel noticed that some peas were smooth and others wrinkled. In crossbreeding them he was able to predict the percent of each by assuming that there was a *gene* that determined whether a pea was smooth or not. The probabilities of peas being smooth in the 2nd, 3rd, and later generations can be calculated using ideas from this chapter. Look in a biology book or a reference book and write a short essay describing how these probabilities can be calculated.

3

Factorial Function

The values of $n!$ from $n = 1$ to $n = 10$ are: 1, 2, 6, 24, 120, 720, 5040, 40,320, 362,880, 3,628,800.

a. Notice that 5! is the first factorial to end in one zero, and 10! is the first factorial to end in two zeros. Find the relationship between n and z, the number of zeros in which $n!$ ends. (Note: You may think that $z = \left\lfloor \frac{n}{5} \right\rfloor$, but 100! ends in 24 zeros, not 20.)

b. Describe a pattern in the rightmost nonzero digit.

c. Summarize the methods you used to find the patterns in parts **a** and **b**.

4 Binomial Theorem for Other Exponents

The Binomial Theorem stated in Lesson 13-6 applies to integer exponents $n \geq 0$. In the 17th century, Isaac Newton extended this theorem to apply to negative and fractional exponents. Report on how the Binomial Theorem can be used to expand expressions such as

$(a + b)^{-3}$, $(a + b)^{\frac{1}{2}}$, and $(a + b)^{-\frac{4}{3}}$.

5 The Koch Curve

The snowflake curve called a Koch curve results from the following recursive process. Begin with an equilateral triangle. To create the $(n + 1)$th figure, take each segment on the nth figure, trisect it, and replace it with the following pattern:

a. Draw the first four stages of the Koch snowflake curve. (Hint: Drawing is easier if you use dot paper and a triangle with sides of 27 units to start, as shown below.)

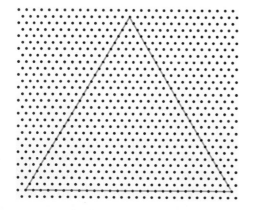

b. Count the number of segments in the Figure at each stage and record it in a table. Imagine repeating this process indefinitely. Visualize and describe how the figure changes at each stage.

c.

stage	1	2	3	4	...	n
number of segments	3	?	?	?		?

d. Let the length of a side of the first triangle be 1. Find the perimeter of the Koch curve at each stage. Generalize to find the perimeter of the figure at the nth stage.

stage	1	2	3	4	...	n
perimeter	3	?	?	?		?

e. How does the area of the snowflake grow? To be able to generalize the pattern, leave answers in radical form.

stage	1	2	3	4	...	n
area	$\frac{\sqrt{3}}{4}$	?	?	?		?

f. Suppose the process for creating the Koch curve is repeated infinitely many times. Is the perimeter of the snowflake finite or infinite? Why? Is the area finite or infinite? Why?

g. The Koch curve is an example of a *fractal*. Look in other books and copy an example of another fractal.

SUMMARY

Some sums and products are denoted by special symbols. For instance, the sum $x_1 + x_2 + \ldots + x_n$ is represented by

$$\sum_{i=1}^{n} x_i.$$

The product $n(n - 1)(n - 2) \cdot \ldots \cdot 2 \cdot 1$ is represented by $n!$.

A series is an indicated sum of terms of a sequence. Values of finite arithmetic or geometric series may be calculated from the following formulas:

For an arithmetic sequence $a_1, a_2, \ldots, a_n$ with common difference d,

$$\sum_{i=1}^{n} a_i = \tfrac{1}{2}n(a_1 + a_2) = \tfrac{n}{2}[2a_1 + (n - 1)d].$$

For a geometric sequence $g_1, g_2, \ldots, g_n$ with common ratio r,

$$\sum_{i=1}^{n} g_i = g_1 \frac{(1 - r^n)}{1 - r} = g_1 \frac{(r^n - 1)}{r - 1}.$$

A statistical measure is a number which is used to describe a data set. The mean and the median are measures of central tendency. The standard deviation is a measure of spread or dispersion. For the data set $x_1, \ldots, x_n$, the mean m is

$$\frac{\sum\limits_{i=1}^{n} x_i}{n}$$

and the standard deviation is

$$\sqrt{\frac{\sum\limits_{i=1}^{n} x_i (x_i - m)^2}{n}}.$$

The mode is the most common element in the data set, but it is not necessarily near the center of the set. A data set may have more than one mode.

Pascal's triangle is a 2-dimensional sequence. The $(r + 1)$st element in the nth row is denoted by $\binom{n}{r} = \frac{n!}{r!(n - r)!}$. The expression $\binom{n}{r}$, also denoted $_nC_r$, appears in several other important applications. It is the coefficient of $a^{n-r}b^r$ in the binomial expansion of $(a + b)^n$. It is the number of subsets, or combinations, with r elements taken from a set with n elements. And if a situation consists of n trials with two equally likely outcomes (say heads/tails on the toss of a coin), then the probability of getting exactly one of these outcomes r times is $\dfrac{\binom{n}{r}}{2^n}$.

The number of permutations of n letters is $n!$. By using permutations and combinations, the probabilities of winning many games of pure chance, such as lotteries, can be calculated.

Distributions of numbers such as test scores often resemble the graphs of probability values related to Pascal's triangle. As the number of the row of Pascal's triangle increases, the distribution takes on a shape more and more like a normal curve. Some tests are standardized so that their scores fit that shape. In a normal distribution, about 68% of the data are within one standard deviation of the mean, and about 95% are within two standard deviations.

Sampling is a procedure by which one tries to describe a larger set (the population) by looking at a smaller set. Statistics calculated from samples are used as estimates of population statistics. If the sample is random, its mean can be compared to other possible means because the distribution of means is close to a normal distribution. This information can be used to obtain the accuracy of a sample of a particular size.

VOCABULARY

Below are the most important terms and phrases for this chapter. You should be able to give a definition for those terms marked with a *. For all other terms you should be able to give a general description or a specific example.

Lesson 13-1
*series
*arithmetic series

Lesson 13-2
*geometric series

Lesson 13-3
Σ, sigma
Σ-notation, sigma notation, summation
    notation
index variable, index
!, factorial symbol
permutation

Lesson 13-4
*mean
*median
*mode
statistical measure
measure of center or of central tendency
standard deviation

Lesson 13-5
Pascal's triangle
Pascal's Triangle Explicit Formula Theorem

Lesson 13-6
binomial expansion
Binomial Theorem
binomial coefficients

Lesson 13-7
subset
*combination

Lesson 13-8
*independent events
*mutually exclusive events
binomial experiment
trial
Binomial Probability Theorem

Lesson 13-9
lottery

Lesson 13-10
probability function
probability distribution
binomial distribution, binomial probability
    distribution
normal distribution
normal curve
standardized scores, normalized scores

Lesson 13-11
*population
*sample
random sample
stratified sample
random numbers
Central Limit Theorem

PROGRESS SELF-TEST

Take this test as you would take a test in class. Then check your work with the solutions in the Selected Answers section in the back of the book.

1. Write using summation notation:
$$1^3 + 2^3 + 3^3 + \ldots + 20^3.$$

2. Evaluate $\sum\limits_{i=0}^{1000} (3i)$.

In 3–5, consider Sheila's scores on math quizzes this term: 80, 80, 88, 90, 93. Find the

3. mode.   **4.** mean.

5. score needed on the next quiz to bring her average up to 88.

6. a. Write rows zero through five of Pascal's triangle.

    b. If the top row is considered the 0th row, what is the sum of the numbers in the nth row?

In 7 and 8, expand.

7. $(x + y)^7$ **8.** $(x^2 - 3)^4$

9. In how many ways can 8 cheerleaders line up for a routine?

10. A pizza restaurant menu contains 15 possible ingredients for pizza. You order 3 of them. How many such combinations are possible?

In 11–13, evaluate.

11. $_8C_0$ **12.** $\binom{40}{38}$ **13.** $\dfrac{n!}{(n-1)!}$

14. A concert hall has 30 rows. The first row has 12 seats. Each row thereafter has two more seats than the preceding row. How many seats are in the concert hall?

15. Consider the sequence defined as follows.
$$\begin{cases} t_1 = 48 \\ t_n = \tfrac{1}{4}t_{n-1}, \text{ for integers } n \geq 2 \end{cases}$$

    a. Write the first four terms of the sequence.

    b. Find the sum of the first 15 terms of the sequence.

16. Let $P(n) = \dfrac{\binom{6}{n}}{2^6}$.

    a. Make a table of values for this function for integers n from 0 to 6.

    b. Graph the function.

    c. Describe in words what $P(n)$ represents in the context of tossing a coin.

17. In a certain lottery, you win if you pick 6 numbers that match 6 numbers drawn from the integers 1 to 55. If your chances of winning are $\dfrac{1}{n}$, what is n?

18. Consider that on a recent administration of the ACT, composite scores had a mean of 18.8 and a standard deviation of 5.9. Assume that these scores are normally distributed.

    a. About what percent of scores are within two standard deviations of the mean?

    b. About what percent of scores are at or above 24.7?

19. A poll of 1000 registered voters shows that 60% favor a school referendum. What is the population and what is the sample?

20. Why is sampling necessary to test the fairness of a coin?

CHAPTER REVIEW

Questions on SPUR Objectives

SPUR stands for **S**kills, **P**roperties, **U**ses, and **R**epresentations. The Chapter Review questions are grouped according to the SPUR Objectives for this chapter.

SKILLS DEAL WITH THE PROCEDURES USED TO GET ANSWERS.

Objective A: *Calculate values of a finite arithmetic series.* *(Lesson 13-1)*

In 1–4, evaluate.

1. $1 + 2 + 3 + \ldots + 60$

2. the sum of the first 71 even integers

3. $3 + 7 + 11 + \ldots + 87$

4. the sum of the first 10 terms of the sequence
$\begin{cases} a_1 = 100 \\ a_n = a_{n-1} - 5, \text{ for integers } n \geq 2. \end{cases}$

5. If the sum of integers $1 + 2 + 3 + \ldots + k = 630$, what is the value of k?

Objective B: *Calculate values of finite geometric series.* *(Lesson 13-2)*

In 6–9, evaluate.

6. $6 + 1.2 + .24 + \ldots + 6(.2)^8$

7. the sum of integer powers of 2 from 2^0 to 2^{19}

8. $4 - 12 + 36 - 108 + \ldots + 236196$

9. The sum of the first 8 terms of the sequence defined as follows:
$\begin{cases} g_1 = 9 \\ g_n = \frac{2}{3} g_{n-1}, \text{ for integers } n \geq 2. \end{cases}$

10. A geometric series has 12 terms; the constant ratio is 1.05; and the first term is 1000. What is the sum?

Objective C: *Use summation (Σ) or factorial (!) notation.* *(Lesson 13-3)*

In 11 and 12, a series is given. **a.** Write the terms of the series; and **b.** Evaluate.

11. $\sum\limits_{n=1}^{6} (2n - 5)$ 12. $\sum\limits_{i=-2}^{3} (7 \cdot 10^i)$

13. *Multiple choice.* Which equals the sum of squares $1 + 4 + 9 + 16 + \ldots + 100$?

(a) $\sum\limits_{n=1}^{10} n$ (b) $\sum\limits_{n=1}^{10} 2^n$ (c) $\sum\limits_{n=1}^{10} n^2$ (d) $\sum\limits_{n=1}^{100} n^2$

14. Suppose $a_1 = 15$, $a_2 = 16$, $a_3 = 16$, $a_4 = 17$, $a_5 = 18$. Evaluate $\dfrac{\sum\limits_{i=1}^{5} a_i}{5}$.

In 15 and 16, rewrite using Σ-notation.

15. $2 + 4 + 6 + \ldots + 144$

16. $M = \dfrac{x_1 + x_2 + \ldots + x_n}{n}$

17. If $f(n) = n!$, calculate $f(2) + f(6)$.

18. *Multiple choice.* $\dfrac{(n + 1)!}{n!} =$

(a) 1   (b) n (c) $n + 1$ (d) $n - 1$

Objective D: *Calculate permutations and combinations.* *(Lessons 13-3, 13-5, 13-6, 13-7)*

19. Copy and complete. The symbol $\binom{n}{r}$ represents the _?_ element in the _?_ row of Pascal's triangle, and is read _?_.

20. *Multiple choice.* Which of the following does not equal $\dfrac{12!}{9! \cdot 3!}$?

(a) $_{12}C_3$ (b) $\binom{12}{3}$

(c) $\binom{12}{9}$ (d) $12 \cdot 11 \cdot 10$

In 21 and 22, consider the set $\{A, B, C, \ldots, Y, Z\}$ of letters in the English alphabet.

21. How many permutations of the letters of the English alphabet are possible?

22. a. How many subsets have
    (i) 1 element?
    (ii) 3 elements?
    (iii) 20 elements?

b. What is the total number of subsets that can be formed?

In 23–26, evaluate.

23. $\binom{10}{5}$ **24.** $\binom{4}{4}$ **25.** $_7C_0$ **26.** $_{100}C_{99}$

Objective E: *Expand binomials.*   *(Lesson 13-6)*

In 27–30, expand.

27. $(x + y)^4$ **28.** $(p - 8)^7$

29. $(3n^2 - 4)^3$ **30.** $\left(\frac{a}{2} + 2b\right)^5$

In 31 and 32, *true or false.*   If false, rewrite the statement to make it true.

31. One term of the binomial expansion of $(8x + y)^{17}$ is $(8x)^{17}$.

32. One term of the binomial expansion of $(4n - p)^{10}$ is $\binom{10}{2}(4n)^8(-p)^2$.

33. *Multiple choice.*   Which equals $\sum_{r=0}^{n}\binom{n}{r}x^{n-r}6^r$?

  (a) $(x + n)^6$ (b) $(x + r)^n$
  (c) $(x + 6)^r$ (d) $(x + 6)^n$

PROPERTIES DEAL WITH THE PRINCIPLES BEHIND THE MATHEMATICS.

Objective F: *Recognize properties of Pascal's triangle.*   *(Lessons 13-5, 13-6)*

In 34–38, remember that the top row in Pascal's triangle is considered to be the 0th row.

					row
		1			0th
	1		1		1st
1		2		1	2nd
		⋮			

34. Write the 8th row of Pascal's triangle, and give one of its applications.

35. What is the sum of the numbers in the 5th row?

36. What is the sum of the numbers in the nth row?

37. Which entry in Pascal's triangle is the coefficient of $a^{n-r}b^r$ in the binomial expansion of $(a + b)^n$?

38. *True or false.*   For all positive integers n, $\binom{n}{1} = \binom{n}{n-1}$. Justify your answer.

USES DEAL WITH APPLICATIONS OF MATHEMATICS IN REAL SITUATIONS.

Objective G: *Solve real-world problems using arithmetic or geometric series.*   *(Lessons 13-1, 13-2)*

39. A pile of logs has one log in the top layer, 2 logs in the next layer, 3 logs in the 3rd layer, and so on.

a. If the pile contains 12 layers of logs, how many logs are there in the pile?

b. If you need to stack 210 logs as described above, how many logs will you need to put in the bottom layer?

40. A student saved 10¢ on January 1st, 20¢ on January 2nd, 30¢ on January 3rd. Each day the student saved 10¢ more than the previous day.

a. How much did the student save on January 31?

b. How much did the student save during the month of January?

c. About how many days would the student have to save to accumulate $100.00?

41. A ball on a pendulum moves 50 cm on its first swing. On each succeeding swing back or forth, it moves 90% of the distance of the previous swing. What is the total distance the ball travels in 12 swings of the pendulum?

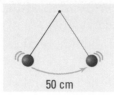

50 cm

42. A student is offered two summer jobs. Job A pays $100 the first week, with a raise of $5 per week beginning with the 2nd week. Job B pays $5 the first week with the salary doubled each week thereafter. If the student plans to work for 10 weeks in the summer, which job will result in the greatest summer salary? Justify your reasoning.

Objective H: *Solve problems involving permutations or combinations.* (*Lessons 13-3, 13-7*)

43. There are 12 notes in a musical octave: A, A#, B, C, C#, D, D#, E, F, F#, G, and G#. In some twelve-tone music, a theme uses each of these notes exactly once. Ignoring rhythm, how many themes are possible?

44. In how many ways can the letters of the word NICELY be arranged?

45. Ten people are in a room. Each decides to shake hands with everyone else exactly once. How many handshakes will take place?

In 46–48, consider that in the 104rd Congress there were 47 Democratic and 53 Republican senators.

46. How many choices are there for forming seven-person committees with members from either party?

47. How many four-member committees is it possible to form that are entirely Democratic?

48. What is the total number of possible committees with more than one member that are entirely Republican?

Objective I: *Use measures of central tendency or dispersion to describe data or distributions.* (*Lessons 13-4, 13-10*)

In 49–51, consider the test scores: 90, 68, 75, 80, 90, 68, 99, 87.

49. Find the mean, median, and mode.

50. Give possible ranges of values for the mean, median, and mode if a ninth score (ranging from 0 to 100) is added to the data set.

51. Calculate the standard deviation.

52. Consider the populations of the ten largest metropolitan areas in the world (1991 estimates). (Source: *1994 World Almanac;* data rounded to the nearest 100,000.)

Tokyo-Yokohama	27,200,000
Mexico City	20,900,000
São Paolo	18,700,000
Seoul	16,800,000
New York	14,600,000
Osaka-Kobe-Kyoto	13,900,000
Bombay	13,900,000
Calcutta	11,900,000
Rio de Janeiro	11,700,000
Buenos Aires	11,700,000

For this data set find the indicated statistic.

a. mean   **b.** median   **c.** mode

53. Consider the following heights of the starting five on a basketball team: 6'8", 6'10", 6'4", 6'8", 6'1". For this data set find the indicated statistic.

a. mode   **b.** mean   **c.** standard deviation

54. John played one round of golf each day during his vacation. If for the first six days his average was 90, what would he need to score on the 7th day to bring his average to 88?

In 55–57, use the following data reported by the College Entrance Examination Board on the SAT I math tests for the 1993–94 testing year.

	n	mean	standard deviation
juniors	848,264	500	121
seniors	794,407	466	120

55. Which group, juniors or seniors, shows a greater dispersion of scores?

56. The mean score of all students taking this test was 480, which is not the average of the means of the juniors and seniors. Why not?

57. Assume mathematics scores are normally distributed among juniors. Within what interval would you expect the middle 68% of scores of juniors to occur?

Objective J: *Solve problems using probability.*
(Lessons 13-8, 13-9)

In 58–60, consider that a fair coin is tossed 5 times. Calculate the probability of each event occuring.

58. exactly 1 head

59. exactly 3 heads

60. exactly 5 tails

61. Assume a student takes a true-false test with 10 questions and that the student guesses on each question. Find the probability of getting

  a. exactly 7 items correct.

  b. 7 or more items correct.

62. In Florida's Lotto game, a player chooses six numbers from 1 to 49. To win the jackpot, the player must match all six winning numbers. What is the probability of this occuring?

63. The Virginia Pick 4 game requires that a player choose four numbers, each a digit from 0 to 9. To win the grand prize, a player must get all four numbers correct and in order. What is the probability of winning this game?

In 64 and 65, in the POWERBALL™ game a player picks five numbers from 1 to 45 and a powerball number also from 1 to 45. In Lesson 13-9 you calculated the probability of winning the jackpot in this game. Smaller prizes are given for other situations. Find the probability of the following occuring.

64. getting the five winning numbers but not the powerball number

65. getting four of the five winning numbers and the powerball number

---

REPRESENTATIONS DEAL WITH PICTURES, GRAPHS, OR OBJECTS THAT ILLUSTRATE CONCEPTS.

Objective K: *Give reasons for sampling.*
(Lesson 13-10)

66. What is an advantage of using a random sample over one that is not random?

67. To find the ratings of a television show in a small town, a network uses a sample rather than the population.

  a. What is the population in this situation?

  b. Why might a sample be preferred over the population?

Objective L: *Graph and analyze binomial and normal distributions.* *(Lesson 13-10)*

68. Consider the function $P(n) = \dfrac{\binom{8}{n}}{2^8}$.

  a. Evaluate $P(n)$ for integers 0, 1, . . . , 8.

  b. Graph this function.

  c. What name is given to this function?

69. Below is pictured a normal distribution with mean m and standard deviation s.

  a. What percent of the data are greater than or equal to m?

  b. About what percent of the data are between $m - s$ and $m + s$?

  c. About what percent of the data are more than two standard deviations away from m?

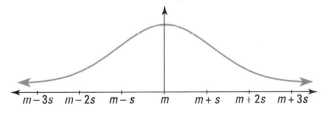

Selected Properties of Real Numbers

For any real numbers a, b, and c:

Postulates of Addition and Multiplication (Field Properties)

Algebra Properties from Earlier Courses

	Addition	*Multiplication*
Closure properties	$a + b$ is a real number.	ab is a real number.
Commutative properties	$a + b = b + a$	$ab = ba$
Associative properties	$(a + b) + c = a + (b + c)$	$(ab)c = a(bc)$
Identity properties	There is a real number 0 with $0 + a = a + 0 = a$.	There is a real number a with $1 \cdot a = a \cdot 1 = a$.
Inverse properties	There is a real number $-a$ with $a + -a = -a + a = 0$.	If $a \neq 0$, there is a real number $\frac{1}{a}$ with $a \cdot \frac{1}{a} = \frac{1}{a} \cdot a = 1$.
Distributive property	$a(b + c) = ab + ac$	

Postulates of Equality

Reflexive property	$a = a$
Symmetric property	If $a = b$, then $b = a$.
Transitive property	If $a = b$ and $b = c$, then $a = c$.
Substitution property	If $a = b$, then a may be substituted for b in any arithmetic or algebraic expression.
Addition property	If $a = b$, then $a + c = b + c$.
Multiplication property	If $a = b$, then $ac = bc$.

Postulates of Inequality

Trichotomy property	Either $a < b$, $a = b$, or $a > b$.
Transitive property	If $a < b$ and $b < c$, then $a < c$.
Addition property	If $a < b$, then $a + c < b + c$.
Multiplication property	If $a < b$ and $c > 0$, then $ac < bc$. If $a < b$ and $c < 0$, then $ac > bc$.

Postulates of Powers

For any nonzero bases (a ≠ 0, b ≠ 0) and integer exponents m and n:

Product of Powers property	$b^m \cdot b^n = b^{m+n}$
Power of a Power property	$(b^m)^n = b^{mn}$
Power of a Product property	$(ab)^m = a^m b^m$

▶

▸ Quotient of Powers property $\qquad \dfrac{b^m}{b^n} = b^{m-n}$, for $b \neq 0$

Power of a Quotient property $\qquad \left(\dfrac{a}{b}\right)^m = \dfrac{a^m}{b^m}$, for $b \neq 0$

Selected Theorems Of Graphing

The set of points (x, y) satisfying $Ax + By = C$, where A and B are not both 0, is a line.

The line with equation $y = mx + b$ has slope m and y-intercept b.

Two non-vertical lines are parallel if and only if they have the same slope.

Two non-vertical lines are perpendicular if and only if the product of their slopes is -1.

The set of points (x, y) satisfying $y = ax^2 + bx + c$ is a parabola.

Selected Theorems of Algebra

For any real numbers a, b, c, and d (with denominators of fractions not equal to 0):

Multiplication Property of 0 $\qquad\qquad 0 \cdot a = 0$

Multiplication Property of -1 $\qquad\qquad -1 \cdot a = -a$

Opposite of an Opposite Property $\qquad -(-a) = a$

Opposite of a Sum $\qquad\qquad\qquad -(b + c) = -b + -c$

Distributive Property of
Multiplication over Subtraction $\qquad a(b - c) = ab - ac$

Addition of Like Terms $\qquad\qquad ac + bc = (a + b)c$

Addition of Fractions $\qquad\qquad\qquad \dfrac{a}{c} + \dfrac{b}{c} = \dfrac{a + b}{c}$

Multiplication of Fractions $\qquad\qquad \dfrac{a}{b} \cdot \dfrac{c}{d} = \dfrac{ac}{bd}$

Equal Fractions $\qquad\qquad\qquad\qquad \dfrac{ac}{bc} = \dfrac{a}{b}$

Means-Extremes $\qquad\qquad$ If $\dfrac{a}{b} = \dfrac{c}{d}$, then $ad = bc$.

Binomial Square $\qquad\qquad (a + b)^2 = a^2 + 2ab + b^2$

Extended Distributive Property $\qquad$ To multiply two polynomials, multiply each term in the first polynomial by each term in the second.

Zero Exponent $\qquad\qquad$ If $b \neq 0$, $b^0 = 1$.

Negative Exponent $\qquad\qquad$ If $b \neq 0$, then $b^{-n} = \dfrac{1}{b^n}$.

Zero Product Theorem $\qquad ab = 0$ if and only if $a = 0$ or $b = 0$.

Absolute Value-Square Root $\qquad \sqrt{a^2} = |a|$

Product of Square Roots $\qquad$ If $a \geq 0$ and $b \geq 0$, then $\sqrt{ab} = \sqrt{a} \cdot \sqrt{b}$.

Quadratic Formula $\qquad$ If $ax^2 + bx + c = 0$ and $a \neq 0$, then $x = \dfrac{-b \pm \sqrt{b^2 - 4ac}}{2a}$.

Geometry Properties from Earlier Courses

In this book, we use many measurement formulas. The following symbols are used.

A = area
a = length of apothem
a, b, and c are lengths of sides (when they appear together)
b_1 and b_2 are lengths of bases
B = area of base
C = circumference
d = diameter
d_1 and d_2 are lengths of diagonals
h = height
L = lateral area

ℓ = length or slant height
n = number of sides
p = perimeter
P = perimeter of base
r = radius
S = total surface area
s = side
θ = measure of angle
T = sum of measures of angles
V = volume
w = width

Two-Dimensional Figures	Perimeter, Length, and Angle Measure	Area
n-gon	$T = 180(n - 2)$	
regular n-gon	$p = ns$ $\theta = \dfrac{180(n - 2)}{n}$	$A = \frac{1}{2}ap$
triangle	$p = a + b + c$	$A = \frac{1}{2}bh$ $A = \sqrt{\frac{p}{2}\left(\frac{p}{2} - a\right)\left(\frac{p}{2} - b\right)\left(\frac{p}{2} - c\right)}$ (Hero's formula)
right triangle	$c^2 = a^2 + b^2$ (Pythagorean theorem)	$A = \frac{1}{2}ab$
equilateral triangle	$p = 3s$	$A = \dfrac{\sqrt{3}}{4}s^2$
trapezoid		$A = \frac{1}{2}h(b_1 + b_2)$
parallelogram		$A = bh$
rhombus	$p = 4s$	$A = \frac{1}{2}d_1d_2$
rectangle	$p = 2\ell + 2w$	$A = \ell w$
square	$p = 4s$	$A = s^2$
circle	$C = \pi d = 2\pi r$	$A = \pi r^2$

Three-Dimensional Figures	Lateral Area and Total Surface Area	Volume
prism		$V = Bh$
right prism	$L = Ph$ $S = Ph + 2B$	$V = Bh$
box	$S = 2(\ell w + \ell h + hw)$	$V = \ell wh$
cube	$S = 6s^2$	$V = s^3$
pyramid		$V = \frac{1}{3}Bh$
regular pyramid	$L - \frac{P\ell}{2}$ $S = \frac{P\ell}{2} + B$	$V = \frac{1}{3}Bh$
cylinder		$V = Bh$
right circular cylinder	$L = 2\pi rh$ $S = 2\pi rh + 2\pi r^2$	$V = \pi r^2 h$
cone		$V = \frac{1}{3}Bh$
right circular cone	$L = \pi r\ell$ $S = \pi r\ell + \pi r^2$	$V = \frac{1}{3}\pi r^2 h$
sphere	$S = 4\pi r^2$	$V = \frac{4}{3}\pi r^3$

Selected Theorems of Geometry

Parallel Lines

Two lines are parallel if and only if:

> corresponding angles are congruent.
> alternate interior angles are congruent.
> alternate exterior angles are congruent.
> they are perpendicular to the same line.

Triangle Congruence

Two triangles are congruent if:

SSS   three sides of one are congruent to three sides of the other.

SAS   two sides and the included angle of one are congruent to two sides and the included angle of the other.

ASA   two angles and the included side of one are congruent to two angles and the included side of the other.

AAS   two angles and a non-included side of one are congruent to two angles and the corresponding non-included side of the other.

SsA   two sides and the angle opposite the longer of the two sides of one are congruent to two sides and the angle opposite the corresponding side of the other.

Angles and Sides of Triangles

Triangle Inequality	The sum of the lengths of two sides of a triangle is greater than the length of the third side.
Isosceles Triangle	If two sides of a triangle are congruent, then the angles opposite those sides are congruent.
Unequal Sides	If two sides of a triangle are unequal in length, then the angle opposite the larger side is larger than the angle opposite the smaller side.
Unequal Angles	If two angles of a triangle are unequal in measure, then the side opposite the larger angle is larger than the side opposite the smaller angle.
Pythagorean Theorem	In a right triangle with legs a and b and hypotenuse c, $c^2 = a^2 + b^2$.
30-60-90 Triangle	In a 30-60-90 triangle, the sides are in the extended ratio $x : x\sqrt{3} : 2x$.
45-45-90 Triangle	In a 45-45-90 triangle, the sides are in the extended ratio $x : x : x\sqrt{2}$.

Parallelograms

A quadrilateral is a parallelogram if and only if:

> one pair of opposite sides are congruent and parallel.
> both pairs of opposite sides are congruent.
> both pairs of opposite angles are congruent.
> its diagonals bisect each other.

Quadrilateral Hierarchy

If a figure is of any type in the hierarchy pictured here, it is also of all types above it to which it is connected.

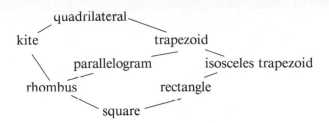

Properties of Transformations

A-B-C-D | Every isometry (composite of reflections) preserves angle measure, betweenness, collinearity, and distance.

Two-Reflection | The composite of two reflections over intersecting lines is a rotation whose center is the intersection of the lines and whose magnitude is twice the measure of the nonobtuse angle formed by the lines in the direction from the first line to the second. The composite of two reflections over parallel lines is a translation whose direction is perpendicular to the lines from the first line to the second and whose magnitude is twice the distance between the lines.

Isometry | Every isometry is a reflection, rotation, translation, or glide reflection.

Size Change | Every size change with magnitude k preserves angle measure, betweenness, and collinearity; a line is parallel to its image; distance is multiplied by k.

Fundamental Theorem of Similarity

If two figures G and G' are similar with ratio of similitude k, then:

angle measures in G' = corresponding angle measures in G;
lengths in G' = $k \cdot$ corresponding lengths in G;
perimeters in G' = $k \cdot$ corresponding perimeters in G;
areas in G' = $k^2 \cdot$ corresponding areas in G;
volumes in G' = $k^3 \cdot$ corresponding volumes in G.

Triangle Similarity

Two triangles are similar if:

three sides of one are proportional to three sides of the other (SSS).
two sides of one are proportional to two sides of the other and the included angles are congruent (SAS).
two angles of one are congruent to two angles of the other (AA).

Coordinate plane formulas

For all $A = (x_1, y_1)$ and $B = (x_2, y_2)$:

Distance formula $\qquad AB = \sqrt{(x_2 - x_1)^2 + (y_2 - y_1)^2}$

Midpoint formula $\qquad$ The midpoint of $\overline{AB}$ is $\left(\frac{x_1 + x_2}{2}, \frac{y_1 + y_2}{2}\right)$.

For all points (x, y):

reflection over the x-axis	$(x, y) \rightarrow (x, -y)$
reflection over the y-axis	$(x, y) \rightarrow (-x, y)$
reflection over $y = x$	$(x, y) \rightarrow (y, x)$
size change of magnitude k, center $(0,0)$	$(x, y) \rightarrow (kx, ky)$
translation h units horizontally, k units vertically	$(x, y) \rightarrow (x+h, y+k)$

An asterisk (*) preceding a theorem indicates that the theorem is also found in a previous UCSMP text.

Theorems of UCSMP Advanced Algebra

Chapter 1

Vertical-Line Test for Functions: No vertical line intersects the graph of a function in more than one point.

***Opposite of a Sum Theorem:** For all real numbers a and b, $-(a + b) = -a + -b$.

Chapter 2

The Fundamental Theorem of Variation:
a. If y varies directly as x^n (that is, $y = kx^n$), and x is multiplied by c, then y is multiplied by c^n.
b. If y varies inversely as x^n $\left(\text{that is, } y = \frac{k}{x^n}\right)$, and x is multiplied by a nonzero constant c, then y is divided by c^n.

Theorem: The graph of the direct-variation function $y = kx$ has constant slope k.

Converse of the Fundamental Theorem of Variation:
a. If multiplying every x-value of a function by c results in multiplying the corresponding y-value by c^n, then y varies directly as the nth power of x. That is, $y = kx^n$.
b. If multiplying every x-value of a function by c results in dividing the corresponding y-value by c^n, then y varies inversely as the nth power of x. That is, $y = \frac{k}{x^n}$.

Chapter 3

***Theorem:** If two lines have the same slope, then they are parallel.

***Theorem:** If two non-vertical lines are parallel, then they have the same slope.

***Theorem:** The graph of $Ax + By = C$, where A and B are not both 0, is a line.

Point-Slope Theorem: If a line contains (x_1, y_1) and has slope m, then it has equation $y - y_1 = m(x - x_1)$.

Theorem: The sequence defined by the recursive formula
$$\begin{cases} a_1 \\ a_n = a_{n-1} + d, \text{ for integers } n \geq 2, \text{ is the arithmetic sequence with} \end{cases}$$
first term a_1 and constant difference d.

Theorem (nth Term of an Arithmetic Sequence) The nth term a_n of an arithmetic sequence with first term a_1 and constant difference d is given by the explicit formula $a_n = a_1 + (n - 1)d$.

Chapter 4

Theorem: $\begin{bmatrix} k & 0 \\ 0 & k \end{bmatrix}$ is the matrix for S_k.

Theorem: $\begin{bmatrix} a & 0 \\ 0 & b \end{bmatrix}$ is the matrix for $S_{a,b}$.

Theorem: $\begin{bmatrix} -1 & 0 \\ 0 & 1 \end{bmatrix}$ is the matrix for r_y.

Theorem: $\begin{bmatrix} 1 & 0 \\ 0 & -1 \end{bmatrix}$ is the matrix for r_x.

Theorem: $\begin{bmatrix} 0 & 1 \\ 1 & 0 \end{bmatrix}$ is the matrix for $r_{y=x}$.

Matrix Basis Theorem: Suppose T is a transformation represented by a 2×2 matrix. If T: $(1, 0) \rightarrow (x_1, y_1)$ and

T: $(0, 1) \rightarrow (x_2, y_2)$ then T has the matrix $\begin{bmatrix} x_1 & x_2 \\ y_1 & y_2 \end{bmatrix}$.

Theorem: If M_1 is the matrix for transformation T_1 and M_2 is the matrix for transformation T_2, then $M_2 M_1$ is the matrix for $T_2 \circ T_1$.

Theorem: A rotation of $b°$ following a rotation of $a°$ with the same center results in a rotation of $(a + b)°$. In symbols: $R_b \circ R_a = R_{a+b}$.

Theorem: $\begin{bmatrix} 0 & -1 \\ 1 & 0 \end{bmatrix}$ is the matrix for R_{90}.

Theorem: $\begin{bmatrix} 1 & 0 \\ 0 & -1 \end{bmatrix}$ is the matrix for R_{180}.

Theorem: $\begin{bmatrix} 0 & 1 \\ -1 & 0 \end{bmatrix}$ is the matrix for R_{270}.

*__Theorem:__ If two lines with slopes m_1 and m_2 are perpendicular, then $m_1 m_2 = -1$.
*__Theorem:__ If two lines have slopes m_1 and m_2 and $m_1 m_2 = -1$, then the lines are perpendicular.
*__Theorem:__ Under a translation, a line is parallel to its image.

Chapter 5

Inverse Matrix Theorem: If $ad - bc \neq 0$, the inverse of

$\begin{bmatrix} a & b \\ c & d \end{bmatrix}$ is $\begin{bmatrix} \dfrac{d}{ad - bc} & \dfrac{-b}{ad-bc} \\ \dfrac{-c}{ad - bc} & \dfrac{a}{ad-bc} \end{bmatrix}$.

System-Determinant Theorem: A 2×2 system has exactly one solution if and only if the determinant of the coefficient matrix is *not* zero.

Linear-Programming Theorem: The feasible region of every linear-programming problem is convex, and the maximum or minimum quantity is determined at one of the vertices of this feasible region.

Chapter 6

Binomial Square Theorem: For all real numbers x and y:
$(x + y)^2 = x^2 + 2xy + y^2; (x - y)^2 = x^2 - 2xy + y^2$.

*__Absolute Value—Square Root Theorem:__ For all real numbers x, $\sqrt{x^2} = |x|$.

Graph-Translation Theorem: In a relation described by a sentence in x and y, the following two processes yield the same graph:
(1) replacing x by $x - h$ and y by $y - k$;
(2) applying the translation $T_{h,k}$ to the graph of the original relation.

Corollary: The image of the parabola $y = ax^2$ under the translation $T_{h,k}$ is the parabola with the equation $y - k = a(x - h)^2$.

Theorem: The graph of the equation $y = ax^2 + bx + c$ is a parabola congruent to the graph of $y = ax^2$.

Theorem: To complete the square on $x^2 + bx$, add $\left(\frac{1}{2}b\right)^2$. So $x^2 + bx + \left(\frac{1}{2}b\right)^2 = \left(x + \frac{1}{2}b\right)^2 = \left(x + \frac{b}{2}\right)^2$.

*__Quadratic Formula:__ If $ax^2 + bx + c = 0$ and $a \neq 0$, then $x = \dfrac{-b \pm \sqrt{b^2 - 4ac}}{2a}$.

Theorem: If $k > 0$, $\sqrt{-k} = i\sqrt{k}$.

Discriminant Theorem: Suppose a, b, and c are real numbers with $a \neq 0$. Then the equation $ax^2 + bx + c = 0$ has: (i) two real solutions if $b^2 - 4ac > 0$; (ii) one real solution if $b^2 - 4ac = 0$; (iii) two complex conjugate solutions if $b^2 - 4ac < 0$.

Chapter 7

*__Zero Exponent Theorem:__ If b is a nonzero real number, $b^0 = 1$.

*__Negative Exponent Theorem:__ For any positive base b and real exponent n, or any nonzero base b and integer exponent n, $b^{-n} = \frac{1}{b^n}$.

*__Annual Compound Interest Formula:__ Let P be the amount of money invested at an annual interest rate of r compounded annually. Let A be the total amount after t years. Then $A = P(1 + r)^t$.

General Compound Interest Formula: Let P be the amount invested at an annual interest rate r compounded n times per year. Let A be the amount after t years. Then $A = P\left(1 + \frac{r}{n}\right)^{nt}$.

Recursive Formula for a Geometric Sequence: The sequence defined by the recursive formula
$$\begin{cases} g_1 \\ g_n = rg_{n-1}, \text{ for integers } n \geq 2, \text{ where } r \text{ is a} \end{cases}$$
nonzero constant, is the geometric, or exponential, sequence with first term g_1 and constant multiplier r.

Explicit Formula for a Geometric Sequence: In the geometric sequence with first term g_1 and constant ratio r, $g_n = g_1(r)^{n-1}$, for integers $n \geq 1$.

$\frac{1}{n}$ **Exponent Theorem:** When $x \geq 0$ and n is an integer greater than 1, $x^{\frac{1}{n}}$ is an nth root of x.

Number of Real Roots Theorem: Every positive real number has 2 real nth roots when n is even, and 1 real nth root when n is odd. Every negative real number has 0 real nth roots when n is even, and 1 real nth root when n is odd.

Rational Exponent Theorem: For any nonnegative real number x and positive integers m and n, $x^{\frac{m}{n}} = \left(x^{\frac{1}{n}}\right)^m$, the mth power of the positive nth root of x, and $x^{\frac{m}{n}} = (x^m)^{\frac{1}{n}}$, the positive nth root of the mth power of x.

Chapter 8

Inverse Relation Theorem: Suppose f is a relation and g is the inverse of f. Then:
(1) A rule for g can be found by switching x and y.
(2) The graph of g is the reflection image of the graph of f over the line $y = x$.
(3) The domain of g is the range of f, and the range of g is the domain of f.

Horizontal-Line Test for Inverses Theorem: The inverse of a function f is itself a function if and only if no horizontal line intersects the graph of f in more than one point.

Inverse Functions Theorem: Two functions f and g are inverse functions if and only if: (1) For all x in the domain of f, $g \circ f(x) = x$, and (2) for all x in the domain of g, $f \circ g(x) = x$.

Power Function Inverse Theorem: If $f(x) = x^n$ and $g(x) = x^{\frac{1}{n}}$ and the domains of f and g are the set of nonnegative real numbers, then f and g are inverse functions.

Root of a Power Theorem: For all positive integers $m > 1$ and $n \geq 2$, and all nonnegative real numbers x, $\sqrt[n]{x^m} = \left(\sqrt[n]{x}\right)^m = x^{\frac{m}{n}}$.

Root of a Product Theorem: For any nonnegative real numbers x and y, and any integer $n \geq 2$: $(xy)^{\frac{1}{n}} = x^{\frac{1}{n}} \cdot y^{\frac{1}{n}}$ (power form); $\sqrt[n]{xy} = \sqrt[n]{x} \cdot \sqrt[n]{y}$ (radical form).

Theorem: When $\sqrt[n]{x}$ and $\sqrt[n]{y}$ are defined and are real numbers, then $\sqrt[n]{xy}$ is also defined and $\sqrt[n]{xy} = \sqrt[n]{x} \cdot \sqrt[n]{y}$.

Chapter 9

Exponential Growth Model: If a quantity a grows by a factor b ($b > 0$, $b \neq 1$) in each unit period, then after a period of length x, there will be ab^x of the quantity.

Continuously Compounded Interest Formula: If an amount P is invested in an account paying an annual interest rate r compounded continuously, the amount A in the account after t years will be $A = Pe^{rt}$.

Log of 1 Theorem: For every base b, $\log_b 1 = 0$.

$\log_b$ of b^n Theorem: For every base b and any real number n, $\log_b b^n = n$.

Product Property of Logarithms: For any base b, and positive real numbers x and y, $\log_b (xy) = \log_b x + \log_b y$.

Quotient Property of Logarithms: For any base b, and for any positive real numbers x and y, $\log_b \left(\frac{x}{y}\right) = \log_b x - \log_b y$.

Power Property of Logarithms: For any base b, and for any positive real number x, $\log_b (x^n) = n \log_b x$.

Change of Base Property: For all positive real numbers a, b, and t, $b \neq 1$ and $t \neq 1$, $\log_b a = \dfrac{\log_t a}{\log_t b}$.

Chapter 10

Complements Property: For all θ between $0°$ and $90°$, $\sin \theta = \cos(90° - \theta)$ and $\cos \theta = \sin(90° - \theta)$.

Pythagorean Identity: For all θ between $0°$ and $90°$, $(\cos \theta)^2 + (\sin \theta)^2 = 1$.

Tangent Theorem: For all θ between $0°$ and $90°$, $\tan \theta = \dfrac{\sin \theta}{\cos \theta}$.

Exact-Value Theorem: $\sin 30° = \cos 60° = \frac{1}{2}$; $\sin 45° = \cos 45° = \frac{\sqrt{2}}{2}$; $\sin 60° = \cos 30° = \frac{\sqrt{3}}{2}$.

Law of Cosines: In any $\triangle ABC$, $c^2 = a^2 + b^2 - 2ab \cos C$.

Law of Sines: In any $\triangle ABC$, $\frac{\sin A}{a} = \frac{\sin B}{b} = \frac{\sin C}{c}$.

Supplements Theorem: For all θ in degrees, $\sin \theta = \sin(180° - \theta)$.

Conversion Factors for Degrees and Radians: To convert radians to degrees, multiply by $\frac{180 \text{ degrees}}{\pi \text{ radians}}$. To convert degrees to radians, multiply by $\frac{\pi \text{ radians}}{180 \text{ degrees}}$.

Chapter 11

***Extended Distributive Property:** To multiply two polynomials, multiply each term in the first polynomial by each term in the second.

***Binomial-Square Factoring Theorem:** For all a and b, $a^2 + 2ab + b^2 = (a + b)^2$; $a^2 - 2ab + b^2 = (a - b)^2$.

***Difference-of-Squares Factoring Theorem:** For all a and b, $a^2 - b^2 = (a + b)(a - b)$.

Discriminant Theorem for Factoring Quadratics: Suppose a, b, and c are integers with $a \neq 0$, and let $D = b^2 - 4ac$. Then the polynomial $ax^2 + bx + c$ can be factored into first degree polynomials with integer coefficients if and only if D is a perfect square.

***Zero-Product Theorem:** For all a and b, $ab = 0$ if and only if $a = 0$ or $b = 0$.

Factor Theorem: $x - r$ is a factor of a polynomial $P(x)$ if and only if $P(r) = 0$.

Rational-Zero Theorem: Suppose that all the coefficients of the polynomial function defined by $f(x) = a_n x^n + a_{n-1} x^{n-1} + \ldots + a_2 x^2 + a_1 x + a_0$ are integers with $a_n \neq 0$ and $a_0 \neq 0$. Let $\frac{p}{q}$ be a rational number in lowest terms. If $\frac{p}{q}$ is a zero of f, then p is a factor of a_0 and q is a factor of a_n.

Fundamental Theorem of Algebra: Every polynomial equation $P(x) = 0$ of any degree with complex number coefficients has at least one complex number solution.

Number of Roots of a Polynomial Equation Theorem: Every polynomial equation of degree n has exactly n roots, provided that multiplicities of multiple roots are counted.

Polynomial-Difference Theorem: $y = f(x)$ is a polynomial function of degree n if and only if, for any set of x-values that form an arithmetic sequence, the nth differences of corresponding y-values are equal and the $(n - 1)$st differences are not equal.

Chapter 12

Theorem: The graph of $y = ax^2$ is the parabola with focus $\left(0, \frac{1}{4a}\right)$ and directrix $y = -\frac{1}{4a}$.

Center-Radius Equation for a Circle Theorem: The circle with center (h, k) and radius r is the set of points (x, y) that satisfy $(x - h)^2 + (y - k)^2 = r^2$.

Corollary: The circle with center at the origin and radius r is the set of points (x, y) that satisfy the equation $x^2 + y^2 = r^2$.

Interior and Exterior of a Circle Theorem: Let C be the circle with center (h, k) and radius r. Then the interior of C is described by $(x - h)^2 + (y - k)^2 < r^2$. The exterior of C is described by $(x - h)^2 + (y - k)^2 > r^2$.

Equation for an Ellipse: The ellipse with foci $(c, 0)$ and $(-c, 0)$ and focal constant $2a$ has equation $\frac{x^2}{a^2} + \frac{y^2}{b^2} = 1$, where $b^2 = a^2 - c^2$.

Theorem: In the ellipse with equation $\frac{x^2}{a^2} + \frac{y^2}{b^2} = 1$, $2a$ is the length of the horizontal axis, and $2b$ is the length of the vertical axis.

Theorem: The image of the unit circle with equation $x^2 + y^2 = 1$ under $S_{a,b}$ is the ellipse with equation $\left(\frac{x}{a}\right)^2 + \left(\frac{y}{b}\right)^2 = 1$.

Graph Scale-Change Theorem: In a relation described by a sentence in x and y, the following two processes yield the same graph:
(1) replacing x by $\frac{x}{a}$ and y by $\frac{y}{b}$;
(2) applying the scale change $S_{a,b}$ to the graph of the original relation.

Theorem: An ellipse with axes of lengths $2a$ and $2b$ has area $A = \pi ab$.

Equation for a Hyperbola: The hyperbola with foci $(c, 0)$ and $(-c, 0)$ and focal constant $2a$ has equation $\frac{x^2}{a^2} - \frac{y^2}{b^2} = 1$, where $b^2 = c^2 - a^2$.

Theorem: The asymptotes of the hyperbola with equation $\frac{x^2}{a^2} - \frac{y^2}{b^2} = 1$ are $\frac{y}{b} = \pm \frac{x}{a}$ $\left(\text{or } y = \pm \frac{b}{a} x\right)$.

Theorem: The graph of $y = \frac{k}{x}$ or $xy = k$ is a hyperbola. When $k > 0$, this is the hyperbola with foci $(\sqrt{2k}, \sqrt{2k})$ and $(-\sqrt{2k}, -\sqrt{2k})$ and focal constant $2\sqrt{2k}$.

Chapter 13

Theorem: The sum of the integers for 1 to n is $\frac{1}{2} n(n + 1)$.

Theorem: Let $S_n = a_1 + a_2 + \ldots + a_n$ be an arithmetic series. Then $S_n = \frac{n}{2} (a_1 + a_n)$.

Corollary: Let $S_n = a_1 + a_2 + \ldots + a_n$ be an arithmetic series with constant difference d. Then $S_n = \frac{n}{2}\{2a_1 + (n - 1)d\}$.

Theorem: Let S_n be the sum of the first n terms of the geometric sequence with first term g_1 and constant ratio $r \neq 1$. Then $S_n = \frac{g_1(1 - r^n)}{1 - r}$ or $S_n = \frac{g_1(r^n - 1)}{r - 1}$.

Theorem: There are $n!$ permutations of n distinct objects.

Pascal's Triangle Explicit-Formula Theorem: If n and r are integers with $0 \leq r \leq n$, then $\binom{n}{r} = \frac{n!}{r!(n - r)!}$.

Binomial Theorem: For all complex numbers a and b, and for all integers n and r with $0 \leq r \leq n$,
$$(a + b)^n = \sum_{r=0}^{n} \binom{n}{r} a^{n-r}b^r.$$

Theorem: The number of subsets of r elements which can be formed from a set of n elements is $\frac{n!}{r!(n - r)!}$, the binomial coefficient $\binom{n}{r}$.

Theorem: A set with n elements has 2^n subsets.

Binomial Probability Theorem: Suppose a binomial experiment has n trials. If the probability of success is p, and the probability of failure is $q = 1 - p$, then the probability that there are r successes in n trials is $\binom{n}{r} p^r q^{n-r}$.

Corollary: Suppose a trial in a binomial experiment has probability $\frac{1}{2}$. Then the probability of r successes in n trials of the experiment is $\frac{\binom{n}{r}}{2^n}$.

Central Limit Theorem: Suppose samples of size n are chosen from a population in which the probability of an element of the sample having a certain characteristic is p. Let $P(x)$ equal the probability that x elements have the characteristic. Then P is approximated by a normal distribution with mean np and standard deviation $\sqrt{np(1 - p)}$.

Programming Languages

COMMANDS

The BASIC commands used in this course, their translation into one calculator language, and examples of their use are given below.

LET . . .   A value is assigned to a given variable. Some versions of BASIC allow you to omit the word LET in the assignment statement.

LET A = 5	$5 \rightarrow A$

The number 5 is stored in a memory location called A.

LET N = N + 2	$N + 2 \rightarrow N$

The value in the memory location called N is increased by 2 and then restored in the location called N. (N is replaced by N + 2.)

PRINT . . .   The computer/calculator displays on the screen what follows the PRINT command. If what follows is a constant or variable, the value of that constant or variable is displayed. If what follows is in quotes, the quote is displayed exactly.

PRINT A	Disp A

The computer prints the number stored in memory location A.

PRINT "X = " A/B	Disp "X = ", A/B

Displayed is X = (value of A/B). Notice that the space after the equal sign in the quotes is transferred into a space after the equal sign in the displayed sentence. On some calculators, the display will place X = and the value on separate lines.

INPUT . . .   The computer asks the user to give a value to the variable named, and stores that value.

INPUT X	Input X

When the program is run, the computer/calculator will prompt you to give it a value by displaying a question mark, and then store the value you type in memory location X.

INPUT "HOW OLD"; AGE	Input "How Old", Age

The computer/calculator displays HOW OLD? and stores your response in memory location AGE.

REM . . .   This command allows remarks to be inserted in a program. These may describe what the variables represent, what the program does, or how it works. REM statements are often used in long complex programs or in programs others will use.

REM PYTHAGOREAN THEOREM

The statement appears when the LIST command is given, but it has no effect when the program is run. Some calculators have no corresponding command.

FOR . . .
NEXT . . .
STEP . . .   The FOR command assigns a beginning and ending value to a variable. The first time through the loop, the variable has the beginning value in the FOR command. When the program hits the line reading NEXT, the value of the variable is increased by the amount indicated by STEP. The commands between FOR and NEXT are then repeated.

10 FOR N = 3 TO 10 STEP 2	For (N, 3, 10, 2)
20 PRINT N	Disp N
30 NEXT N	End
40 END	

The program assigns 3 to N and then displays the value of N. On reaching NEXT, the program increases N by 2 (the STEP amount), and prints 5. The next N is 7, then 9, but ▶

► 11 is too large, so the program executes the command after NEXT, ending itself. The NEXT command is not needed in some calculator languages. The output from both programs is given here.

3
5
7
9

IF . . . THEN . . . The program performs the consequent (the THEN part) only if the antecedent (the IF part) is true. When the antecedent is false, the program *ignores* the consequent and goes directly to the next line of the program.

IF X > 100 THEN END   If X ≤ 100
PRINT X   Then Disp X
END   End

If the X value is greater than 100, the program goes to the end statement. If the X value is less than or equal to 100, the computer/calculator displays the value stored in X.

GOTO . . . The program goes to whatever line of the program is indicated. GOTO statements are generally avoided because they interrupt program flow and make programs hard to interpret.

5 (Command)   5 (Command)
10 GOTO 5   Goto 5

The program goes to line 5 and executes that command.

END . . . The computer stops running the program. A program should have only one end statement.

END   End

FUNCTIONS

A large number of functions are built in to most versions of BASIC and to all calculators. They are the same functions used outside of programming. Each function name must be followed by a variable or constant enclosed in parentheses. Here are some examples of the uses of functions in programs.

ABS The absolute value of the number that follows is calculated.

LET A = ABS(-10)   ABS(-10) → A

The computer calculates $|-10| = 10$ and assigns the value 10 to memory location A.

Like the absolute value function, the trigonometric functions SIN, COS, and TAN are identified in the same way in virtually all programming languages. Other functions are also similar.

INT The greatest integer less than or equal to the number that follows is calculated.

B = INT(N + .5)   INT(N + .5) → B

The program adds .5 to the value of N, calculates $\lfloor N + .5 \rfloor$, and stores the result in B.

Some functions are identified differently in different languages. Here are two examples.

LOG or LN The natural logarithm (logarithm to base e) of the number that follows, is calculated.

LET J = LOG(6)   LN(6) → J

The program calculates $ln\ 6$ and assigns the value 1.791759469228 to memory location J. It may display only some of these decimal places.

SQR The square root of the number or expression that follows is calculated.

C = SQR(A*A + B*B) $\sqrt{(A*A + B*B)}$ → C

The program calculates $\sqrt{A^2 + B^2}$ using the values stored in A and B and stores the result in C.

LESSON 1-1 (pp. 6–11)

19. (b) **21.** (e) **23.** (d) **25.** Sample: Mark is building a deck requiring x feet of lumber. He already has y feet of lumber. How much more lumber should he buy? **27.** Sample: A meteor falls x meters in y seconds. At what rate is it approaching the Earth? **29.** -0.12 **31.** 120 mg **33.** (a) **35. a.** $x = -9$ **b.** Does $3 \cdot (-9) = 5 \cdot (-9) + 18$? Does $-27 = -45 + 18$? $-27 = -27$. Yes, it checks.

LESSON 1-2 (pp. 12–17)

21. a. the set of positive real numbers **b.** 43.3 cm^2 **23. a.** Yes; each year corresponds to exactly one value. **b.** No; $M = 24.3$ corresponds to both $F = 21.3$ and $F = 21.5$. **25.** 9600 **27.** $a + b$ **29.** ab **31.** (b)

LESSON 1-3 (pp. 19–23)

13. $g(15)$; $g(15) = 32{,}768$; $h(15) = 225$; $32{,}768 > 225$ **15.** 50 **17. a.** 57.5 **b.** Sample: The increase in the average price of one gal of unleaded gasoline from 1978 to 1980 is about 58¢. **19.** $x = 1984$ **21.** Yes; each value of x corresponds to only one value of y.
23. a. \$428 **b.** \$$(278 + 25n)$ **25. a.** $n = \frac{1}{2}$ **b.** $\frac{1}{2} + 10 = 3\left(\frac{1}{2}\right) + 9$? $10\frac{1}{2} = 1\frac{1}{2} + 9$? Yes, it checks.

LESSON 1-4 (pp. 24–29)

13. $B(5) \approx 42$ **15.** $t = 1, 8, 14$ **17. a.** Sample: $(0, 0); (1, 1); (2, 4)$ **b. Sample: See below. c.** Yes **19. See below. 21.** ≈ 460 acres **23.** $\approx 2{,}105{,}000$ **25. a.** 25 **b.** -23 **c.** $1 + \frac{3n^3}{8}$ **27. a.** $f(3) = 0$; $f(6) = 9$ **b.** $\{3, 4, 5, 6, \ldots\}$ **29.** $n = 244$; $\frac{1}{2}(244) - 72 = 50$? $122 - 72 = 50$? Yes.

17. b.

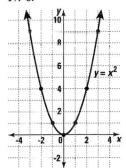

19.

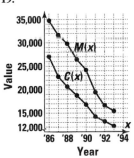

LESSON 1-5 (pp. 30–35)

13. 162.5 **15.** $y = 9$ **17.** $y = \frac{12}{5} = 2.4$ **19.** $x = 10$ **21. a.** $f(18) = \frac{1}{20}$ **b.** $n = 97$ **23.** $\approx$ \$100 billion **25.** approximately 1975 to 1992 **27.** ≈ 10.19 ft

LESSON 1-6 (pp. 36–40)

9. (a) **11. a.** $\pi = \frac{C}{D}$ **b.** Sample: π is the ratio of the circumference of a circle to its diameter. **13.** Sample: $P = \frac{1.1 \text{ cm}}{.5 \cdot 4.2 \text{ cm}} \approx 0.52$ **15. a.** $S = \frac{2R}{P}$ **b.** 60 ft **17.** $n = \frac{t - a}{d} + 1$ **19.** $x = 12$ **21.** 1.42 sec **23.** (c)

LESSON 1-7 (pp. 42–47)

17. 18,000; 18,900; 19,845 **19. a.** n **b.** s_n **c.** Sample: the set of whole numbers less than 40 **21.** (c) **23.** $n = \frac{S}{180} + 2$ or $\frac{S + 360}{180}$ **25. a.** $P(x) = 3x + 4$ **b.** $P(40) = 124$ **c.** $x = 12$ **27.** 70%

LESSON 1-8 (pp. 48–54)

13. a. $\begin{cases} s_1 = 15 \\ s_n = \text{previous term} + 2, \text{ for integers } n \geq 2 \end{cases}$ **b.** $s_{10} = 33$
15. a. 3, 12, 48, 192 **b.** 3, 12, 48, 192 **c.** True **17.** Sample: A recursive sequence is dependent on previous terms. Therefore, an initial value is needed in order to use the formula.
19. a. $c_3 = 225{,}000$ **b.** \$3,844,335.94 **21. a.** 30,782,000 people **b.** 2006 **23. a.** 16 square units **b.** $8\sqrt{5} \approx 17.9$ units **c. See below.**

23. c. Sample:

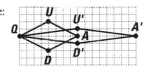

LESSON 1-9 (pp. 55–60)

13. a. The first term is 3. Each term after the first is found by adding 11 to the previous term.
b. $\begin{cases} t_1 = 3 \\ t_n = t_{n-1} + 11, \text{ for integers } n \geq 2 \end{cases}$
15. a. 100, 102, 105, 109, 114 **b.** Sample: If 99 is subtracted from each term, the result is the triangular number sequence.
17. a. 29, 47, 76, 123 **b.** $\begin{cases} L_1 = 1 \\ L_2 = 3 \\ L_n = L_{n-2} + L_{n-1}, \text{ for integers } n \geq 3 \end{cases}$
19. a. $a_4 = 40$; This means after 4 weeks Antonio has \$40.00 in his account. **b.** 50th term **21. a.** $V = \frac{nRT}{P}$ **b.** $T = \frac{PV}{nR}$
23. a. $\{x: 0 \leq x \leq 50\}$ **b.** $\{y: 10 \leq y \leq 80\}$ **25.** the set of all real numbers

CHAPTER 1 PROGRESS SELF-TEST (pp. 64–65)

1. $t_1 = 5 + 7(1) = 12$; $t_2 = 5 + 7(2) = 19$; $t_3 = 5 + 7(3) = 26$; $t_4 = 5 + 7(4) = 33$ **2. a.** $S_1 = 5$; $S_2 = 5 + 7 = 12$; $S_3 = 12 + 7 = 19$; $S_4 = 19 + 7 = 26$ **b.** The first term is 5. Each term after the first is found by adding 7 to the previous term. **3.** $t_8 = 5 + 7(8) = 61$ **4.** $S_5 = 26 + 7 = 33$; $S_6 = 33 + 7 = 40$; $S_7 = 40 + 7 = 47$; $S_8 = 47 + 7 = 54$ **5.** $f(3) = 9(3)^2 - 11(3) = 81 - 33 = 48$ **6.** $T(12) = \frac{(12)(12 + 1)}{2} = \frac{(12)(13)}{2} = 78$ **7.** (b) **8.** $s = \frac{12}{t}$ miles per hour **9.** $4(3x - 8) = 11$; $12x - 32 = 11$; $12x = 43$; $x = \frac{43}{12}$ **10.** $1.2y = 0.9$; $y = .75$ **11.** Multiply both sides by 100; $12p + 8(15{,}000 - p) = 148{,}000$; $12p + 120{,}000 - 8p = 148{,}000$; $4p + 120{,}000 = 148{,}000$; $4p = 28{,}000$; $p = 7000$ **12.** $\left(\frac{1}{3}\right)(\pi)(4)^2(6) = 100.53 \approx 101$ cm^3 **13.** A reasonable domain is the set of positive real numbers since length can be any positive number. **14.** $h = \frac{3V}{\pi r^2}$ **15.** domain: $\{1, 3, 5\}$; range: $\{2, 4, 6\}$ **16.** (b), (c), (d) **17.** Yes, it is a function because each x value corresponds to only one y value. **18.** No, since $x = 1$ corresponds to $y = 1$ and $y = -1$, it is not a function.

19. $\begin{cases} m_1 = 20 \\ m_n = \text{previous amount} + 10 \text{ or } \boxed{\text{ANS}} + 10, \text{ for integers } n \geq 2 \end{cases}$

or $\begin{cases} m_1 = 20 \\ m_n = m_{n-1} + 10, \text{ for integers } n \geq 2. \end{cases}$ **20. a.** Multiply through
by 10; $5z + 2z + 900 = 10z$; $900 = 3z$; $z = 300$ **b.** Sample:
Mary had a certain amount of money. She spent half of it on
clothes and $\frac{1}{5}$ of it on groceries. She has 90 dollars left. How much
did she originally have? **21.** $P(1) = 2\pi \cdot \sqrt{\frac{1}{9.8}} \approx 2$ seconds
22. $B = 45w$; $250 = 45w$; $w = 5.56$; after about 6 weeks, the class
will have earned a pizza party. **23.** $C(1975) \approx 10.5$ million. The
number of students enrolled in college in 1975 was about
10.5 million. **24.** $x = 1988$. In 1988, the number of students
enrolled in college was equal to the number of students enrolled in
high school. **25.** $\{x: 1970 \geq x \leq 2000, \text{ where } x \text{ is an integer}\}$

The chart below keys the **Progress Self-Test** questions to the objectives in the **Chapter Review** on pages 66–69 or to the **Vocabulary** (Voc.) on page 63. This will enable you to locate those **Chapter Review** questions that correspond to questions students missed on the **Progress Self-Test**. The lesson where the material is covered is also indicated on the chart.

Question	1	2	3	4	5	6	7	8	9	10
Objective	E	F	E	E	B	B	D	I	C	C
Lesson	1-7	1-8, 1-9	1-7	1-8, 1-9	1-3	1-3	1-6	1-1	1-5	1-5
Question	11	12	13	14	15	16	17	18	19	20
Objective	C	A	H	D	H	M	G	G	J	C, K
Lesson	1-5	1-1	1-2	1-6	1-2	1-4	1-2	1-2	1-8, 1-9	1-5
Question	21	22	23	24	25					
Objective	J	K	L	L	L					
Lesson	1-2, 1-3	1-3	1-4	1-4	1-4					

CHAPTER 1 REVIEW (pp. 66–69)

1. 119 **3.** 6973.57 **5.** 5 **7.** −243 **9.** $g(5) = 11$; $g(6) = 13$ **11.** 6;
$\frac{3}{2}(6) = 9$? Yes, it checks. **13.** $\frac{3}{4}$; $\frac{6}{\frac{3}{4}} = 8$? $6 \cdot \frac{4}{3} = 8$? Yes, it checks.

15. $\frac{1}{5}$; $3 - \left(\frac{1}{5} + 2\right) = 4\left(\frac{1}{5}\right)$? $3 - \frac{11}{5} = \frac{4}{5}$? Yes, it checks. **17.** 60;
$\frac{60}{2} + \frac{60}{2} + 10 = 60$? $30 + 20 + 10 = 60$? Yes, it checks.
19. $n = \frac{t - 4}{-5}$ **21.** $\frac{g}{2}$ **23.** (a), (d) **25.** 23, 26, 29, 32, 35 **27.** 10, 3,
−4, −11, −18 **29. a.** 16, 11, 6, 1, −4
b. $\begin{cases} t_1 = 16 \\ t_n = t_{n-1} - 5, \text{ for integers } n \geq 2 \end{cases}$
31. a. The first term is 6. All other terms are obtained by adding 6
to the previous term.
b. $\begin{cases} t_1 = 6 \\ t_n = t_{n-1} + 6, \text{ for integers } n \geq 2 \end{cases}$
33. No; $x = -1$ corresponds to $y = 1$ and $y = -1$.

35. a. Sample:

x	0	1	1	4
y	0	1	−1	2

b. No

37. a. Sample:

x	−5	−5	−5	−5
y	0	1	3	3

b. No

39. a. $\{-9, -\frac{1}{2}, 2, 3\}$ **b.** $\{-3, -2, \frac{1}{2}, 9\}$ **41. a.** the set of real numbers
b. No **c.** the set of nonnegative real numbers **43.** $100x$ cm
45. $\frac{B}{M}$ blocks per minute **47.** $20 - c$ sticks **49. a.** $26,000,
$27,560, $29,213.60, $30,966.42, $32,824.40 **b.** $78,665.59
51. 315 feet **53.** the population of Baltimore in 1950 **55. a.** −51,000
b. Sample: In Baltimore, the population dropped by 51,000 people
from 1980 to 1990. **57. a.** $14,000 = 4000 + 200d$ **b.** $d = 50$ days
59. 170 minutes or about 3 hours **61.** domain: $\{x: -6 \leq x \leq 6\}$;
range: $\{y: -1 \leq y \leq 1\}$ **63.** 0 **65.** No; a vertical line intersects the
graph twice. **67.** Yes; no vertical line intersects the graph in more
than one point.

LESSON 2-1 (pp. 72–77)

13. a. $P = kw^3$; P, dependent; w, independent **b.** 32.4 watts
15. a.

r	1	2	3	4	5	6	7	8	9	10
A	π	4π	9π	16π	25π	36π	49π	64π	81π	100π

b. 4 times as large **c.** 4 times as large **d.** 4 times as large
e. quadruples **f.** is multiplied by 9 **17.** 3.9 **19.** 3^6 **21.** 3^{15}
23. $150,000 **25.** 1988-1989; 1990-1991; 1991-1992 **27. See right.**

27.

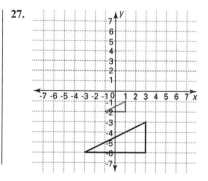

LESSON 2-2 (pp. 78–82)

9. $t = \frac{k}{w}$ **11. a.** $I = \frac{k}{D^2}$ **b.** $k = 1346.7$ lumens $\cdot$ m^2
c. ≈ 3.37 lumens **13.** directly

15. a.

s	1	2	3	4	5	6	7	8	9	10	11	12
t	36	18	12	9	$7\frac{1}{5}$	6	$5\frac{1}{7}$	$4\frac{1}{2}$	4	$3\frac{3}{5}$	$3\frac{3}{11}$	3

b. is halved **c.** is multiplied by $\frac{1}{3}$ **17.** 117.63 pounds **19.** $y = 80$
21. x^8 **23. a. See below. b.** $y = -5$ **c.** $x = 2$

23. a.

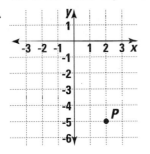

LESSON 2-3 (pp. 84–88)

7. y is multiplied by 81. By the Fundamental Theorem of
Variation, if x is multiplied by 3, then y is multiplied by 3^4 or 81.
9. y is multiplied by $\frac{1}{2}$. **11.** y is divided by $\frac{1}{2}$ or y is multiplied by 2.
13. 2 : 1 **15.** multiplied by 8; $V = \frac{4}{3}\pi r^3$ for a sphere, so if the
radius is doubled, the volume is multiplied by 2^3 or 8.
17. ≈ 45.4 lb **19.** (c) **21. a.** cp cents **b.** $\frac{rd}{2}$ cents

23. See right.

23.

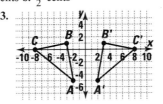

LESSON 2-4 (pp. 89–93)

9. $\frac{3}{50}$ **11.** 7 **13. a.** $c = 1.25$ g **b.** domain: nonnegative real
numbers; range: nonnegative real numbers **c.** Sample: (4, 5),
(7, 8.75), (14, 17.5) **d.** Sample: **See below. e.** 1.25 dollars/gal
f. For every gallon increase, the cost increases by $1.25.
15. $\triangle ABC \sim \triangle DEF$ by Angle Angle Similarity. Because they are
similar, their corresponding sides are proportional. So, the ratios
between the vertical and horizontal legs, or slopes, are equal.
17. 8 : 1 **19.** direct variation **21.** inverse variation **23. a.** $Z = \frac{V}{I}$
b. $V = ZI$

13. d.

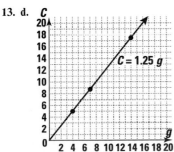

LESSON 2-5 (pp. 96–102)

9. When $x = 0$, $y^2 = k(0)^2 = 0$. Therefore, when $x = 0$, $y = 0$.

11. a.

d	0	10	20	30	40
N	0	50	200	450	800

b. Sample: 610 houses **c. See below. d.** ≈ 625 houses **e.** They are
close. **f.** 612 homes **13. a. See below. b.** 4; $\frac{1}{4}$, $-\frac{1}{4}$, -4 **c.** $y = 4x$
and $y = -\frac{1}{4}x$; or $y = -4x$ and $y = \frac{1}{4}x$ **15. a.** $I = \frac{k}{d^2}$ **b.** The intensity
of sound will be divided by 16. **17. a.** The set of natural numbers.
b. 5, 20, 43, 74, 113 **19. a.** $\triangle DEF$, $\triangle KLM$ **b.** $\triangle DEF \cong \triangle ABC$
by ASA Congruence Theorem; $\triangle KLM \cong \triangle ABC$ by SSS
Congruence Theorem

11. c.

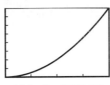

$0 \le x \le 40$, x scale = 10
$0 \le y \le 800$, y scale = 100

13. a.

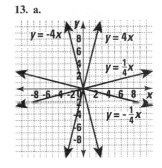

LESSON 2-6 (pp. 104–109)

11. a. $k = 200$ ft-lb; $d = \frac{200}{w}$

b.

w	10	20	30	40	50	60	70	80	90	100
d	20	10	$6\frac{2}{3}$	5	4	$3\frac{1}{3}$	$2\frac{6}{7}$	$2\frac{1}{2}$	$2\frac{2}{9}$	2

c. Sample: **See below. d.** Yes; w can be any positive real number.
13. No. If $f(x) = 0$, then $\frac{1}{x^2} = 0$ or $x^2 \cdot 0 = 1$ which is impossible.
Therefore, $f(x) \ne 0$ for all values of x. **15.** $P_2 = -6x^2$ **17.** No;
a direct variation graph must go through the origin. **19. a.** Mikki
cleared the fractions by multiplying both sides of the equation by 6,
the least common denominator. **b.** $x = 6$

11. c.

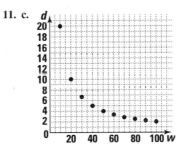

LESSON 2-7 (pp. 110–115)

9. a. (II) **b.** $k = \frac{1}{2}$ **c.** When $x = 4$, $y = \frac{1}{2}(4)^2 = \frac{1}{2}(16) - 8$, Yes, it
checks. **d.** 18 **11. a. See below. b.** $d = kt^2$; using the point (2, 19.6),
we find that $k = 4.9$. Check: for $t = 5$, $d = (4.9) \cdot 5^2 = 122.5$; it
checks. **c.** 99.225 meters **13.** (e) **15.** (d) **17.** y is multiplied by $\frac{1}{16}$.
19. about 1¢

11. a.

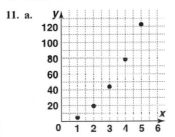

9. (c) **11.** $P = \frac{kw}{h^2}$ **13.** none **15.** $x = 0$; $y = 0$ **17. a.** the set of all real numbers **b.** $\{y: y \le 0\}$ **19.** line; $\frac{3}{4}$ **21. a.** The first term is -3. Each term after the first is equal to the square of the previous term plus two. **b.** -3; 11; 123; $15{,}131$; $228{,}947{,}163$

11. a. The volume of a rectangular solid varies jointly as its length, width, and height. **b.** 1 **13. a.** $H = kA(T_I - T_O)$ **b.** $k \approx 1.13 \frac{\text{Btu}}{\text{ft}^2 - {}^\circ\text{F}}$ **c.** $H \approx 1.13\, A(T_I - T_O)$ **d.** ≈ 1446.4 Btu **15.** $f(3)$ **17.** False

19. a. Sample

x	y
-3	90
-2	40
-1	10
0	0
1	10
2	40
3	90

b. See below. **c.** parabola **d.** 30 **e.** No; a parabola does not have a constant rate of change. **21.** $f(\pi) = 4\pi^2 + 2 \approx 41.48$ **23.** Adding Fractions Property

19. b.

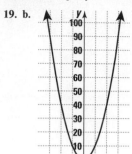

CHAPTER 2 PROGRESS SELF-TEST (pp. 132–133)

1. $n = \frac{k}{d^2}$ **2.** $w = \frac{kd^4}{L^2}$ **3.** y varies directly as the fifth power of x.
4. $S = kp^2$ is the variation equation. Find k. $10 = k(3)^2$; $k = \frac{10}{9}$.
Rewrite the variation equation using the constant. $S = \frac{10}{9}p^2$;
$S = \frac{10}{9}(8)^2$; $S = \frac{640}{9} \approx 71.1$ **5.** The Fundamental Theorem of Variation says that if $y = kx^2$, and x is multiplied by 2, y is multiplied by $2^2 = 4$. So, the y-value is quadrupled. Sample: $(1, 3)$ and $(2, 12)$. **6.** The Fundamental Theorem of Variation says that if $y = \frac{6}{x}$ and x is multiplied by c, then y is divided by $c^1 = c$.
7. At $x = 3$, $y = 9$, and at $x = 4$, $y = 16$. The rate of change is $\frac{16 - 9}{4 - 3} = \frac{7}{1} = 7$. **8.** False **9.** parabola, $k > 0$ **10.** x can be any real number except zero, so the domain $= \{x: x \ne 0\}$. **11. a.** neither inversely nor directly **b.** inversely **c.** S.A. $= 4\pi r^2$ which is the form S.A. $= kr^2$, with $k = 4\pi$. The surface area of a sphere varies directly with the square of its radius.

12. a. Sample:

x	y
-2	10
-1	5
0	0
1	-5
2	-10

b. See right. **c.** -5; slope is determined from the slope-intercept form or by finding the rate of change from two pairs in the table.
13. a. See right. **b.** the x-axis and y-axis **14.** (c) **15. a.** See right. **b.** If $F = \frac{k}{L}$, when $L = 3$ and $F = 120$, $k = 360$ in.-lb and $F = \frac{360}{L}$. This formula is valid for the rest of the data. So $F = (k, L)$ is a good model. **c.** Since $F = \frac{360}{L}$, when $L = 12$ in., $F = (360, 12) = 30$ lb of force to turn the bolt. **16.** Sample: The number of CD's a student could buy with a week's paycheck varies directly as the number of hours worked. **17.** The variation equation is $C = kd^3$. First, find k. $.79 = k(2)^3$; $k = .09875 \frac{\text{cents}}{\text{inches}^3}$

Now, rewrite the variation equation using k. $C = .09875(d)^3$; $C = .09875(3)^3$; $C = \$2.67$. **18.** (d) **19.** $V = khg^2$ **20.** The variation equation is $S = kPr^4$. First, find k. $.09604 = k(100)(.065)^4$; $k = 53.8$. The unit of k is $\frac{1}{\text{cm-sec-mmHg}}$. Now, rewrite the variation equation using k. $.09604 = 53.8(P)(.05)^4$; $P = 285.6$ mmHg (millimeters of mercury).

12. b.

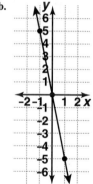

13. a.

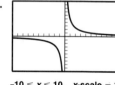

$-10 \le x \le 10$, x-scale $= 1$
$-10 \le y \le 10$, y-scale $= 1$

15. a.

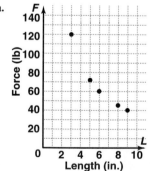

The chart below keys the **Progress Self-Test** questions to the objectives in the **Chapter Review** on pages 132–133 or to the **Vocabulary** (Voc.) on page 131. This will enable you to locate those **Chapter Review** questions that correspond to questions you missed on the **Progress Self-Test**. The lesson where the material is covered is also indicated on the chart.

Question	1	2	3	4	5	6	7	8	9	10
Objective	A	A	A	B	D	D	C	E	E	E
Lesson	2-2	2-9	2-1	2-1	2-3	2-3	2-5	2-6	2-5	2-6

Question	11	12	13	14	15	16	17	18	19	20
Objective	F	C, I	E, I	J	H	F	G	K	H	G
Lesson	2-1, 2-2	2-4	2-6	2-4, 2-6	2-7	2-1	2-1	2-5	2-8	2-9

CHAPTER 2 REVIEW (pp. 134–137)

1. $y = kx^2$ **3.** $z = kxt$ **5.** directly; the square of r **7.** y varies inversely as the square of x. **9.** 21 **11.** −32 **13.** $\frac{9}{5} = 1.8$ **15.** −15
17. $-\frac{3}{4} = -.75$ **19.** y is tripled. **21.** p is divided by $2^2 = 4$.
23. divided by c^n **25.** not affected **27.** (0, 0) **29.** (b) **31.** domain: the set of real numbers; range: the set of nonnegative real numbers
33. $x = 0$ **35.** $n = \frac{k}{r^3}$ **37.** $p = \frac{km}{d^2}$ **39.** inversely **41.** directly
43. $13.50 **45.** 3 min **47.** ≈ 1020.2 lb of force **49. a.** See below.
b. $L = kS^2$ **c.** Sample: $k ≈ .045 \frac{\text{ft}}{(\text{mph})^2}$; $L = .045S^2$ **d.** Sample:
220.5 ft **51. a. i.** See below. **ii.** P varies directly as R. Sample: The slope between any two points is constant. **b. i. See right.**
ii. P varies directly as I^2. **c.** $P = kRI^2$ **d.** $k = 1$; $P = RI^2$
e. $P = 1600$ watts
53. a. Sample: **b. See right. 55. a.** Sample: **b. See right.**

x	y
−2	−1
−1	−0.5
0	0
1	0.5
2	1

x	y
−2	−8
−1	−2
0	0
1	−2
2	−8

57. See right. 59. (c) **61.** (b) **63.** positive real **65. See right.**
67. a. Sample: $-10 ≤ x ≤ 10$; $-5 ≤ y ≤ 10$ **b.** Sample: $-20 ≤ x ≤ 0$; $-5 ≤ y ≤ 10$

49. a.

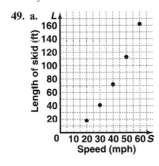

51. ai.

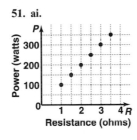

51. bi.

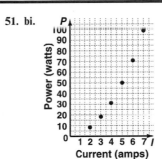

53. b.

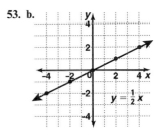

55. b.

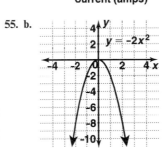

57.

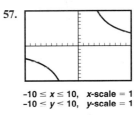

−10 ≤ x ≤ 10, x-scale = 1
−10 < y < 10, y-scale = 1

65.

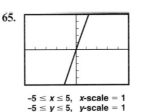

−5 ≤ x ≤ 5, x-scale = 1
−5 ≤ y ≤ 5, y-scale = 1

LESSON 3-1 (pp. 140–145)

19. a. Sample: (6, 92), (9, 88), (12, 84) **b.** $C = 100 - \frac{4}{3}h$
c. 30 hours **d.** No, it would take 75 hours to sell all the cars.
21. a. 1.2 miles **b.** 30 min or .5 hr **c.** −0.8 mi/hr **d.** Sample: How long does Carmen work after school? Answer: $1\frac{3}{4}$ hrs.

23. a. See right. **b.** $y = \frac{1}{4}x$ **25.** $x = 15$

23. a.

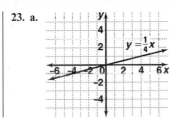

LESSON 3-2 (pp. 146–151)

13. a. See below. **b.** $y = \frac{2}{5}x - 6$ **c.** $x = 22\frac{1}{2}$; Does $3 = \frac{2}{5}(22\frac{1}{2}) - 6$?
Does $3 = 9 - 6$? Yes, it checks. **15. a.** See below. **b.** The rate of
change (slope) for both lines is the constant 2, so the lines are
parallel. **17. a.** See below. **b.** $y = -2.5x + 50$ **19.** negative
21. zero **23.** $S_4 = 54$

13. a.

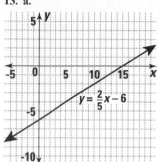

15. a.

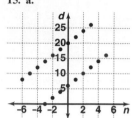

17. a.

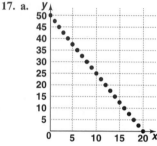

LESSON 3-3 (pp. 152–156)

7. a. $3.23 **b.** $.49Z + .59T$
c. $.49Z + .59T = 5$
9. $2.4S + 112B + 26.2R$ minutes
11. See right. **13.** $\frac{2}{3}$
15. a. $R = 100 - \frac{3}{4}d$ **b.** after
120 days **17.** False; you can't tell
whether Terry went uphill or not.
19. $\sqrt{(a - c)^2 + (b - d)^2}$

11.

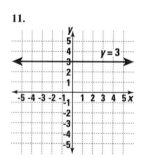

LESSON 3-4 (pp. 157–161)

15. a. oblique; neither A nor B is 0. **b.** x-intercept = 9;
y-intercept = -6 **c.** See below. **17.** $10x - 5y = 1$
19. a. $1.2 = 0.1N + 0.2Y$ **b.** See below. **c.** 2 oz **21. a.** domain: {2}
b. range: the set of real numbers **23.** $10.20x + 12.35y$
25. a. $2500P + 3000S = 310,000$ **b.** 112

15. c.

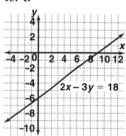

19. b

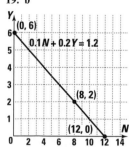

LESSON 3-5 (pp. 162–167)

13. a. Sample: $c - 1290 = 3(b - 30)$ **b.** $1200 **c.** $1500
15. a. $13.15; $13.15; $13.85 **b.** See below.
c. $\begin{cases} C = 13.15, n \le 80 \\ C = 13.15 + 0.035(n - 80), n > 80 \end{cases}$ **17. a.** See below.
b. $\overline{PQ}$: $x = 3$; $\overline{QR}$: $y = -5$; $\overline{RS}$: $x = -2$; $\overline{SP}$: $y = 4$ **c.** 45 units^2
19. ≈ 87 lb **21.** 80

15. b.

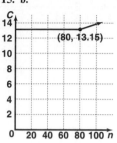

17. a.

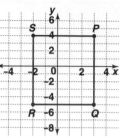

LESSON 3-6 (pp. 169–174)

15. a. ≈ 92 **b.** ≈ 86.7 **17.** (d)
19. (b) **21.** Sample:
$S - 180 = 180(n - 3)$
23. See right.
25. a. $.4x + .2y = 10$
b. 22.45 ounces **27.** $6.50

23.

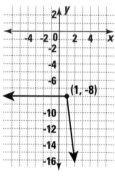

LESSON 3-7 (pp. 175–179)

9. $\begin{cases} a_1 = 13 \\ a_n = a_{n-1} + 6, \text{ for integers } n \ge 2 \end{cases}$ **11.** 85 in.
13. Sample:
```
10 LET A = 1
20 FOR N = 1 TO 1000
30 PRINT A
40 A = A + 2
50 NEXT N
60 END
```
15. a., c. See right. **b.** about
$y = .14x - 10.25$ **d.** ≈ 2.35 million
e. ≈ 4.03 million **f.** Sample: better
reporting of child-abuse cases; increased
awareness of the problem of child abuse
17. They are identical; both lines have
slope $-\frac{3}{4}$ and y-intercept 3. **19.** The
number of tiles needed is divided by 4.
21. a. (iv) **b.** (ii) **c.** (iii) **d.** (i)

15. a, c.

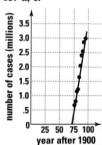

LESSON 3-8 (pp. 180–185)

9. a. explicit; you can find any specific term without relying on
previous terms. **b.** 626 **11. a.** 8.1, 9.8, 11.5 **b.** $a_n = 8.1 +$
$(n - 1)1.7$ **13.** 2.5 mi/week **15.** 49 **17. a.** $7x - 5y = 4$
b. x-intercept = $\frac{4}{7}$; y-intercept = $-\frac{4}{5}$ **19.** $H = \frac{kz}{w}$ **21.** (c)
23. a. $-8, -7, -6$ **b.** See below.

23. b.

$$-\frac{21}{4}$$

LESSON 3-9 (pp. 186–191)

13. (b) **15. a.** $500 **b.** $w = 275 + 75 \cdot \left\lfloor \dfrac{d}{300} \right\rfloor$ **17. a.** $2.79;

$9.27; $11.43 **b. See right.**

19. a. $\begin{cases} a_1 = 46 \\ a_n = a_{n-1} - 3.5 \text{ for integers } n \geq 2 \end{cases}$

b. $a_n = 46 - (n-1) \cdot 3.5$ **21.** Sample: $y + 1 = 2(x - 3)$

23. See right.

17. b.

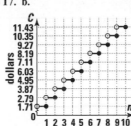

23.

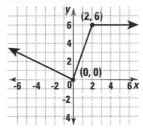

CHAPTER 3 PROGRESS SELF-TEST (pp. 195–196)

1. y-intercept $= -5$; $m = 3$ **See right. 2.** vertical
3. a. $4x - 5y = 12$; $-5y = -4x + 12$; $y = \frac{4}{5}x - \frac{12}{5}$; slope is $\frac{4}{5}$.
b. Let $y = 0$, then $4x = 12$; $x = 3$; 3 is the x-intercept.
Let $x = 0$, then $-5y = 12$; $y = -\frac{12}{5}$; $-\frac{12}{5}$ is the y-intercept.
4. $\{m: m < 0\}$ **5.** $m = \frac{3-2}{-5-4} = -\frac{1}{9}$; Sample: $y - 2 = -\frac{1}{9}(x - 4)$
6. Since the line is parallel to line $y = \frac{5}{3}x + 4$, $m = \frac{5}{3}$; $y + 1 = \frac{5}{3}(x - 5)$. **7. a.** vertical **b.** horizontal **8.** $36S + 48L$ in.
9. $d = -40 + .8t$; $-10 = -40 + .8t$; $30 = .8t$; $t = 37.5$ seconds
10. a. 1, 6, 11, 16, 21, 26, 31, 36, 41, 46, 51, 56 **b.** Yes. There is a constant difference of 5 between successive terms. **11. a. See right. b.** By reading the graph in part **a**, the cost for a call of more than 6 minutes but less than or equal to 7 minutes is $14.50.
12. No. An arithmetic sequence must have a constant difference between successive terms, but $8 - 5 \neq 2 - (-3)$.
13. $10 \cdot \left\lfloor \dfrac{(365 + 5)}{10} \right\rfloor = 10 \cdot \left\lfloor \dfrac{370}{10} \right\rfloor = 10 \cdot \lfloor 37 \rfloor = 10 \cdot 37 = 370$

14. $m = \dfrac{80 - 0}{100 - 0} = .8$; $R - 0 = .8(C - 0)$; $R = .8C$ **15.** $a_1 = -7$;
$d = -10 - (-7) = -3$; $a_n = -7 - 3(n - 1)$
16. $\begin{cases} a_1 = -7; \\ a_n = a_{n-1} - 3, \text{ for integers } n \geq 2 \end{cases}$ **17.** (a) **18. a.** Yes; the points all lie relatively close to a line. **b.** Sample: about
$y = -.83x + 73.07$ **c.** $y = -.83(42) + 73.07$; $y = -34.86 + 73.07$;
$y = 38.21$; a 42-year old has about 38 expected years of life left.
19. a. $m = \dfrac{800 - 400}{900 - 600} = \dfrac{400}{300} = \dfrac{4}{3}$ **b.** The left has the steepest slope for the first 200 feet of horizontal distance.

1.

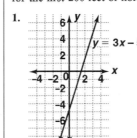

$y = 3x - 5$

11. a.

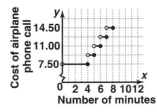

The chart below keys the **Progress Self-Test** questions to the objectives in the **Chapter Review** on pages 197–201 or to the **Vocabulary** (Voc.) on page 194. This will enable you to locate those **Chapter Review** questions that correspond to questions missed on the **Progress Self-Test.** The lesson where the material is covered is also indicated on the chart.

Question	1	2	3	4	5	6	7	8	9	10
Objective	L	E	A	E	B	B	E	H	G	D, F
Lesson	3-2	3-4	3-2, 3-4	3-1	3-5	3-5	3-4	3-3	3-1	3-7
Question	11	12	13	14	15	16	17	18	19	
Objective	M	F	C	I	D	D	M	J	K	
Lesson	3-9	3-7	3-9	3-5	3-8	3-7	3-1, 3-9	3-6	3-1	

CHAPTER 3 REVIEW (pp. 197–201)

1. a. 7 **b.** -2 **3. a.** 0 **b.** 4 **5. a.** -4.7 **b.** none **7.** $y - 75 = 8(x - 40)$ **9.** Sample: $y - 4 = -\frac{2}{3}(x - 2)$ **11.** Sample: $y - 2 = -\frac{3}{2}(x + 1)$ **13. a.** 16 **b.** -17 **15.** -17 **17. a.** $a_n = 7 + 5(n - 1)$
b. $\begin{cases} a_1 = 7 \\ a_n = a_{n-1} + 5, \text{ for integers } n \geq 2 \end{cases}$ **c.** 377
19. $\begin{cases} a_1 = -9 \\ a_n = a_{n-1} + 2, \text{ for integers } n \geq 2 \end{cases}$ **21.** Sample: **a.** 1, 3, 5, 7, 9
b. $\begin{cases} a_1 = 1 \\ a_n = a_{n-1} + 2, \text{ for integers } n \geq 2 \end{cases}$ **c.** $a_n = 1 + 2(n - 1)$
23. a. 10 **b.** 5; 7; 9; 11; 13 **25.** (b) **27.** domain: the set of all real numbers; range: the set of all real numbers **29.** slopes are equal **31.** horizontal **33. a.** horizontal **b.** oblique **c.** vertical **35.** False **37.** The graph is a set of collinear points. **39.** Yes **41.** No **43.** No

45. a. $w = 3 + .2n$ **b.** 7.4 kg **47.** 240 ft **49. a.** $1.7F + 2.5S = 250$ **b.** 32 **51.** $124,000 **53. a. See page 904. b.** Sample: No; the data do not seem to lie on a line. **c.** about $y = .74x - 9.39$
d. The correlation coefficient is .203. This indicates a weak fit of the data to this line. **55. a. See page 904. b.** Sample: $y - 334 = 14.07(x - 1970)$ **c.** Sample: 530,980 physicians **d.** about $y = 14.45x - 28144.96$ **e.** Sample: Increase in number of physicians per year. **f.** about 523.84 thousands or 523,840 physicians
g. Sample: The correlation coefficient is .998, which indicates a very strong linear relationship between the regression equation and the data. **57. a.** $19.62 **b.** $84.24 **59. a.** 30 mi **b.** 45 mph
c. 210 mi **61.** (c) **63. See page 904. 65. See page 904.**
67. positive **69.** negative **71. a. See page 904. b.** domain: the set of all real numbers; range: $\{y: y \leq 3\}$ **73. See page 904.**
75. a. See page 904. b. False

53. a.

55. a.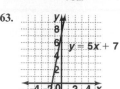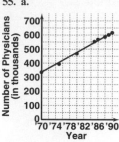

63.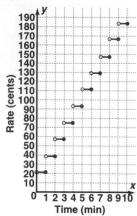

65.

71. a.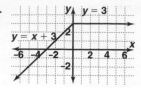

73.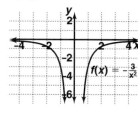

75. a.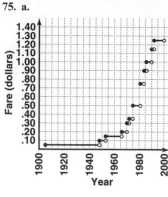

LESSON 4-1 (pp. 204–208)

15. 7; 8 **17. a.** Sample:

	from A	from B	from C
to A	0	2	5
to B	3	0	3
to C	1	4	0

b. Adam, 4; Barbara, 6; Clem, 8 **19.** Sample: When $k = -100$, the graph is a line slanting downward from left to right with a very steep slope. As k increases, but remains less than 0, the slant is still downward from left to right, but the slope is less steep. When $k = 0$, the slope is 0 and the line becomes horizontal. For positive values of k, the slant of the line is upward from left to right. The line gets steeper as the value of k increases. When $k = 100$, the graph is a line slanting upward from left to right with a very steep slope. **See below. 21.** $(5 + x)(3 + y) = 15 + 3x + 5y + xy$
23. For all real numbers a, b, and c, $a + (b + c) = (a + b) + c$.

19.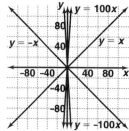

LESSON 4-2 (pp. 209–213)

9. a.
$$M = \begin{bmatrix} -9 & 3 & 6 & -12 \\ -13 & 15 & -2 & -28 \\ -7 & -1 & 8 & -6 \\ 5 & -4 & -1 & 9 \\ 1 & -4 & 3 & 5 \\ 4 & -9 & 5 & 13 \end{bmatrix}$$

b. how many more points each team had in 1993-94 than in 1992-93 **c.** how many more wins, losses, ties, and total points Boston had in 1993-94 than in 1992-93

11. $\begin{bmatrix} ka & kb \\ kc & kd \end{bmatrix}$ **13.** $a = 8$; $b = -\frac{3}{5}$; $c = 21$; $d = 24.5$ **15. a., c. See below. b.** rhombus **17. a. See below. b.** domain: set of all real numbers except 0; range: set of all negative real numbers **19.** reflection **21.** rotation **23.** $PQ = 17$

15. a., c.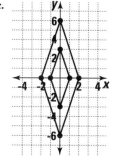

17. a.

LESSON 4-3 (pp. 214–219)

13. Sample:

	band uniforms	basketball uniforms	track uniforms	swimsuits
a. east	10	20	15	10
central	20	0	10	0
west	15	20	0	10

b. $\begin{bmatrix} 10 & 20 & 15 & 10 \\ 20 & 0 & 10 & 0 \\ 15 & 20 & 0 & 10 \end{bmatrix} \begin{bmatrix} 90 \\ 40 \\ 50 \\ 25 \end{bmatrix} = \begin{bmatrix} 2700 \\ 2300 \\ 2400 \end{bmatrix}$; East $2700,

Central $2300, West $2400 **15.** $x = \frac{1}{3}$ **17.** $2 \times n$ **19.** The distance between $(2, -1)$ and $(7, 6)$ is $\sqrt{74}$; the distance between $(2, -1)$ and $(-5, 4)$ is $\sqrt{74}$; the distance between $(7, 6)$ and $(-5, 4)$ is $\sqrt{148}$. The triangle is isosceles. **21.** $f(20) = \frac{1}{2}$
23. 3

LESSON 4-4 (pp. 221–226)

15. a. $P' = (7.5, 10)$; $OP' = \sqrt{(7.5 - 0)^2 + (10 - 0)^2} =$
$\sqrt{156.25} = 12.5$, and $OP = \sqrt{3^2 + 4^2} = \sqrt{25} = 5$; $\frac{OP'}{OP} = \frac{12.5}{5}$
$= 2.5$ **b.** $y = \frac{4}{3}x$ **17. a.** $\frac{7}{2}$ **b.** $\frac{7}{2}$ **c.** Yes, because the slopes are equal.
19. $a = -\frac{1}{2}$; $b = -\frac{7}{6}$ **21.** from the 35th to the 50th second
23. 56.25 cm^2

LESSON 4-5 (pp. 227–231)

9. The slope of $\overline{AB}$ is $\frac{7}{2}$ and the slope of $\overline{A'B'}$ is $\frac{35}{4}$. Since the slopes
are different, the lines cannot be parallel. **11. a.** rectangle
b. $\begin{bmatrix} 0 & 12 & 12 & 0 \\ 0 & 0 & 12 & 12 \end{bmatrix}$ **c.** square **11. d.**

d. See right. **13. a.** $S_3 = \begin{bmatrix} 3 & 0 \\ 0 & 3 \end{bmatrix}$

b. $\begin{bmatrix} 3 & 0 \\ 0 & 3 \end{bmatrix}\begin{bmatrix} 5 & 4 \\ 1 & 2 \end{bmatrix} = \begin{bmatrix} 15 & 12 \\ 3 & 6 \end{bmatrix}$

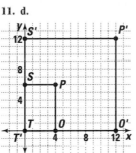

15. a. 5,263,904 **b.** Sample: Texas
has the biggest problem because
it has the greatest number of
uninsured residents.
17. a. $x = 225$ **b.** $y = 25$

LESSON 4-6 (pp. 232–238)

13. $\begin{bmatrix} 0 & 1 \\ 1 & 0 \end{bmatrix}\begin{bmatrix} x \\ y \end{bmatrix} = \begin{bmatrix} 0 \cdot x & + & 1 \cdot y \\ 1 \cdot x & + & 0 \cdot y \end{bmatrix} = \begin{bmatrix} y \\ x \end{bmatrix}$ **15.** The area is
multiplied by 9.
17. Factory X: $72 for a
trick ski, $48 for a slalom
ski, $55 for a cross-country
ski; Factory Y: $64.50 for a
trick ski, $43 for a slalom
ski, $49.25 for a cross-
country ski **19.** See right. **19.**

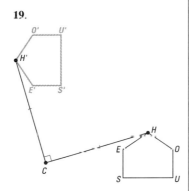

LESSON 4-7 (pp. 240–245)

13. a. $\begin{bmatrix} 0 & 1 \\ -1 & 0 \end{bmatrix}$ **b.** R_{270} **c.** not the same; $r_{y=x} \circ r_x = R_{90}$
15. a. (i) identity transformation (ii) identity transformation
(iii) identity transformation **b.** If a point is reflected over a line and
then its image is reflected over the same line, the final image is the
original point. **17.** $M_k \cdot C = \begin{bmatrix} k & 0 \\ 0 & k \end{bmatrix}\begin{bmatrix} m & n \\ p & q \end{bmatrix} = \begin{bmatrix} km & kn \\ kp & kq \end{bmatrix}$

$C \cdot M_k = \begin{bmatrix} m & n \\ p & q \end{bmatrix}\begin{bmatrix} k & 0 \\ 0 & k \end{bmatrix} = \begin{bmatrix} km & kn \\ kp & kq \end{bmatrix}$

19. a., d. See right.
b. isosceles triangle
c. $\begin{bmatrix} -28 & 28 & 0 \\ 0 & 0 & 7 \end{bmatrix}$

e. isosceles **f.** No. Sample:
the angles at the
corresponding vertices do
not have the same measure.
21. a. 1994. **b.** about
10,450,000 **23.** (a)

19. a., d.

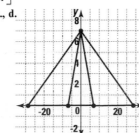

LESSON 4-8 (pp. 246–250)

11. $\begin{bmatrix} -1 & 0 \\ 0 & -1 \end{bmatrix}\begin{bmatrix} -1 & 0 \\ 0 & -1 \end{bmatrix} = \begin{bmatrix} 1 & 0 \\ 0 & 1 \end{bmatrix}$ $R_{180} \circ R_{180} = R_{360}$, which shows
that the matrix for R_{360} is the identity matrix. **13.** See below.
15. a., b. See below. **17.** $\begin{bmatrix} 1 & 0 \\ 0 & 1 \end{bmatrix}$ **19.** $\begin{bmatrix} 1 & 0 \\ 0 & 3 \end{bmatrix}$ **21. a., b.** See below.
c. reflection about x axis and size change of magnitude 2
23. $a^2 + b^2 = c^2$ **25. a.** similar and congruent **b.** rotation **27.** $-\frac{1}{2}$

13.

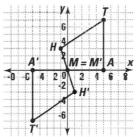

15. a., b.

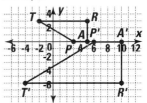

21. a., b.

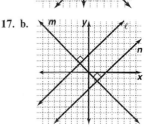

LESSON 4-9 (pp. 251–255)

7. a. 3 **b.** $y + 2 = 3(x - 7)$ **9.** The reciprocal of 0 is undefined.
11. $x = 6$ **13. a.** $A' = (3, -7)$; $B' = (1, 4)$ **b.** See below. **c.** $\overleftrightarrow{AB}: \frac{2}{11}$;
$\overleftrightarrow{A'B'}: -\frac{11}{2}$ **d.** Their product is -1; the lines are perpendicular. **e.** 90°

15. a. $\perp$ **b.** See below. **17. a.** $//$ **b.** See below. **19.** $M = \begin{bmatrix} \frac{4}{3} & 0 \\ 0 & \frac{3}{4} \end{bmatrix}$

21. a. $\begin{bmatrix} 89.82 \\ 99.43 \end{bmatrix}$ **b.** $99.43 **23. a.** $x = \frac{w - u}{v - y}$ **b.** when $v = y$

13. b.

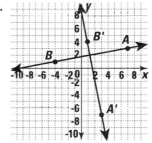

15. b.

17. b.

LESSON 4-10 (pp. 256–260)

9. No. The contrapositive of the theorem must hold true; that is,
if a line is not parallel to its image, then they are not translation
images of each other. **11.** $T_{10, 14}$ **13.** $y = -2x + 12$ **15.** $-\frac{3}{2}$
17. a. $H' = (-5, 1)$; $I' = (3, -1)$ **b.** Both equal $\sqrt{68}$. **c.** No; the
product of their slopes is not -1. **19. a.** $\begin{bmatrix} 1 & 0 \\ 0 & -1 \end{bmatrix}$ **b.** $(a, -b)$

21. a. $\begin{bmatrix} 1 & 0 \\ 0 & 6 \end{bmatrix}$ **b.** $(a, 6b)$ **23. a.** $\begin{bmatrix} 0 & 1 \\ 1 & 0 \end{bmatrix}$ **b.** (b, a) **25.** $\begin{bmatrix} 0 & -1 \\ -5 & 2 \end{bmatrix}$

1. See right. 2. 1st
$$\begin{array}{c} & P & B & G \\ \text{1st} & \begin{bmatrix} 14 & 3 & 8 \\ 120 & 190 & 250 \end{bmatrix} \\ \text{Econ} \end{array}$$

3. Two matrices with dimensions $m \times n$ and $n \times p$ can be multiplied—the number of columns in the left matrix matches the number of rows in the right matrix. AB would be a 3×2 times a 2×2 matrix; the product does exist. BA would be a 2×2 times a 3×2 matrix; the product does not exist. **4.** $BC = \begin{bmatrix} 2 & 0 \\ 1 & 5 \end{bmatrix}\begin{bmatrix} 8 & 6 \\ -2 & 2 \end{bmatrix} = \begin{bmatrix} 16+0 & 12+0 \\ 8-10 & 6+10 \end{bmatrix} =$

$\begin{bmatrix} 16 & 12 \\ -2 & 16 \end{bmatrix}$ **5.** $B - C = \begin{bmatrix} 2 & 0 \\ 1 & 5 \end{bmatrix} - \begin{bmatrix} 8 & 6 \\ -2 & 2 \end{bmatrix} = \begin{bmatrix} 2-8 & 0-6 \\ 1--2 & 5-2 \end{bmatrix} =$

$\begin{bmatrix} -6 & -6 \\ 3 & 3 \end{bmatrix}$ **6.** $7B = 7\begin{bmatrix} 2 & 0 \\ 1 & 5 \end{bmatrix} = \begin{bmatrix} 14 & 0 \\ 7 & 35 \end{bmatrix}$ **7.** $\begin{bmatrix} 1 & 0 \\ 0 & 1 \end{bmatrix}\begin{bmatrix} a & b \\ c & d \end{bmatrix} =$

$\begin{bmatrix} 1a+0c & 1b+0d \\ 0a+1c & 0b+1d \end{bmatrix} = \begin{bmatrix} a & b \\ c & d \end{bmatrix}$ and $\begin{bmatrix} a & b \\ c & d \end{bmatrix}\begin{bmatrix} 1 & 0 \\ 0 & 1 \end{bmatrix} =$

$\begin{bmatrix} a\cdot1+b\cdot0 & a\cdot0+b\cdot1 \\ c\cdot1+d\cdot0 & c\cdot0+d\cdot1 \end{bmatrix} = \begin{bmatrix} a & b \\ c & d \end{bmatrix}$. The result of multiplying

any 2×2 matrix by $\begin{bmatrix} 1 & 0 \\ 0 & 1 \end{bmatrix}$ is the original matrix. In other words, the image of any point under that transformation is the same as the preimage. **8.** The slope of the line $y = 5x - 3$ is 5, so the slope of the line perpendicular to this line is $-\frac{1}{5}$. Using the point $(3, -2.5)$ and this slope in $y - y_1 = m(x - x_1)$, the desired equation is $y - (-2.5) = -\frac{1}{5}(x - 3)$, or $y + 2.5 = \frac{1}{5}(x - 3)$. **9.** $r_x \circ R_{270} =$

$\begin{bmatrix} 1 & 0 \\ 0 & -1 \end{bmatrix}\cdot\begin{bmatrix} 0 & 1 \\ -1 & 0 \end{bmatrix} = \begin{bmatrix} 0+0 & 1+0 \\ 0+1 & 0+0 \end{bmatrix} = \begin{bmatrix} 0 & 1 \\ 1 & 0 \end{bmatrix}$ **10.** Each point is moved 8 units to the right and 8 units down, so the translation is $T_{8,-8}$. **11.** Under a translation, a line is parallel to its image, so $\overleftrightarrow{FI} \parallel \overleftrightarrow{F'I'}$. This can be shown directly. $F = (-6, 6)$ and $I = (-4, 7)$. So the slope of $\overleftrightarrow{FI}$ is $\frac{6-7}{-6--4} = \frac{1}{2}$. $F' = (2, -2)$ and $I' = (4, -1)$. So the slope of $\overleftrightarrow{F'I'}$ is $\frac{-2--1}{2-4} = \frac{1}{2}$.

12. $\begin{bmatrix} -1 & 0 \\ 0 & 1 \end{bmatrix}\begin{bmatrix} -6 & -4 & -2 & -4 & -6 & -4 \\ 6 & 7 & 4 & 1 & 2 & 4 \end{bmatrix} = \begin{bmatrix} 6 & 4 & 2 & 4 & 6 & 4 \\ 6 & 7 & 4 & 1 & 2 & 4 \end{bmatrix}$

13. Prices

	Deckshoes	Pumps	Sandals	Boot
Los Angeles	23	8	10	5
Tucson	11	5	10	15
Santa Fe	2	3	15	15

$\cdot$
Deckshoes	18
Pumps	58
Sandals	12
Boots	76

The product is: $\begin{array}{c} 3 \times 4 \\ 4 \times 1 \end{array}$

$\begin{bmatrix} 23\cdot18 + 8\cdot58 + 10\cdot12 + 5\cdot76 \\ 11\cdot18 + 5\cdot58 + 10\cdot12 + 15\cdot76 \\ 2\cdot18 + 3\cdot58 + 15\cdot12 + 15\cdot76 \end{bmatrix} =$

$\begin{bmatrix} 1378 \\ 1748 \\ 1530 \end{bmatrix}$. The revenues are: \$1,378,000 for Los Angeles; \$1,748,000 for Tucson; and \$1,530,000 for Santa Fe. **14.** The

sum of the two matrices is $\begin{bmatrix} 8 & 11 \\ 5 & 4 \\ 15 & 16 \\ 2 & 0 \end{bmatrix} + \begin{bmatrix} 10 & 14 \\ 11 & 13 \\ 7 & 9 \\ 0 & 3 \end{bmatrix} =$

$\begin{bmatrix} 8+10 & 11+14 \\ 5+11 & 4+13 \\ 15+7 & 16+9 \\ 2+0 & 0+3 \end{bmatrix} = \begin{bmatrix} 18 & 25 \\ 16 & 17 \\ 22 & 25 \\ 2 & 3 \end{bmatrix}$ **15.** This is a scale change

represented by $S_{a,b}$ with matrix $\begin{bmatrix} a & 0 \\ 0 & b \end{bmatrix}$. So the matrix for a

horizontal stretch of 2 and vertical shrink of $\frac{1}{4}$ is $\begin{bmatrix} 2 & 0 \\ 0 & \frac{1}{4} \end{bmatrix}$.

16. $(x, y) \rightarrow (y, x)$ is a reflection over $y = x$, so the matrix is

$\begin{bmatrix} 0 & 1 \\ 1 & 0 \end{bmatrix}$. **17.** $\begin{bmatrix} 1 & 0 \\ 0 & -1 \end{bmatrix}\begin{bmatrix} 7 & -1 & 3 \\ 6 & 2 & -4 \end{bmatrix} = \begin{bmatrix} 7 & -1 & 3 \\ -6 & -2 & 4 \end{bmatrix}$ means that the reflection of $\triangle ABC$ over the x-axis is the image $\triangle A'B'C'$ with vertices $A' = (7, -6)$, $B' = (-1, -2)$, and $C' = (3, 4)$. **18.** $A' = (-6, 7)$, $B' = (-2, -1)$, and $C' = (4, 3)$. **See below. 19.** Sample: Two points on the line $x + 2y = 5$ are $(5, 0)$ and $(1, 2)$. The matrix describes the translation $T_{3,2}$. Apply the translation to the two points to get image points $(8, 2)$ and $(4, 4)$. So an equation of the image is $y - 2 = \frac{4-2}{4-8}(x - 8)$ or $y = -\frac{1}{2}x + 6$.

20. a. $\begin{bmatrix} 3 & 0 \\ 0 & 3 \end{bmatrix}\begin{bmatrix} 5 & 2 & -3 \\ -3 & 2 & -5 \end{bmatrix} = \begin{bmatrix} 15 & 6 & -9 \\ -9 & 6 & -15 \end{bmatrix}$

b. $\begin{bmatrix} 4 & 0 \\ 0 & 4 \end{bmatrix}\left(\begin{bmatrix} 3 & 0 \\ 0 & 3 \end{bmatrix}\begin{bmatrix} 5 & 2 & -3 \\ -3 & 2 & -5 \end{bmatrix}\right) = \begin{bmatrix} 60 & 24 & -36 \\ -36 & 24 & -60 \end{bmatrix}$

1.
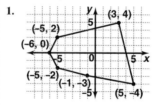

18.
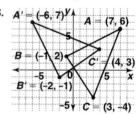

The chart below keys the **Progress Self-Test** questions to the objectives in the **Chapter Review** on pages 266-269 or to the **Vocabulary** (Voc.) on page 264. This will enable you to locate those **Chapter Review** questions that correspond to questions missed on the **Progress Self-Test**. The lesson where the material is covered is also indicated on the chart.

Question	1	2	3	4	5	6	7	8	9	10
Objective	I	G	D	B	A	A	D	C	F	F
Lesson	4-1	4-1	4-3	4-3	4-2	4-2	4-3	4-9	4-7, 4-8	4-10

Question	11	12	13	14	15	16	17	18	19	20
Objective	E	F	H	H	F	F	F	I	F	F
Lesson	4-10	4-6	4-3	4-2	4-5	4-6	4-6	4-8	4-10	4-4, 4-7

CHAPTER 4 REVIEW (pp. 266-269)

1. $\begin{bmatrix} 11 & 6 \\ 4 & -8 \\ 8 & 2 \end{bmatrix}$ **3.** $\begin{bmatrix} 5 & 0 & 8 \\ 16 & 13 & -1 \\ 6 & 13 & 2 \end{bmatrix}$ **5.** $a = 1.6; b = -10.1$ **7.** [59]

9. $\begin{bmatrix} -18 & -24 & -30 \\ 11 & 13 & 15 \\ 40 & 50 & 60 \end{bmatrix}$ **11.** $a = 5; b = -3$ **13.** $y + 1 = 2(x - 3)$

15. $y - 2 = 4(x - 2)$ **17. a.** True **b.** Sample: $\begin{bmatrix} 2 & 3 \\ 1 & -4 \end{bmatrix} + \begin{bmatrix} 1 & 7 \\ 5 & 6 \end{bmatrix} =$

$\begin{bmatrix} 1 & 7 \\ 5 & 6 \end{bmatrix} + \begin{bmatrix} 2 & 3 \\ 1 & -4 \end{bmatrix} = \begin{bmatrix} 3 & 10 \\ 6 & 2 \end{bmatrix}$ **19. a.** No **b.** Yes **21.** $\begin{bmatrix} 1 & 0 \\ 0 & 1 \end{bmatrix}$

23. (a) **25.** (d) **27.** -0.2 **29.** $\begin{bmatrix} 1 & 0 \\ 0 & 1 \end{bmatrix}$; it is the identity

transformation. **31. a.** $\begin{bmatrix} -1 & 0 \\ 0 & 1 \end{bmatrix}$ **b.** r_y **33. a.** $T_{2,-1}$ **b.** Because the

matrix for *PEAR* is $\begin{bmatrix} 0 & 6 & 7 & 2 \\ 6 & 7 & -2 & 2 \end{bmatrix}$, the translation matrix has

four columns represented by $\begin{bmatrix} 2 & 2 & 2 & 2 \\ -1 & -1 & -1 & -1 \end{bmatrix}$. Then

$\begin{bmatrix} 2 & 2 & 2 & 2 \\ -1 & -1 & -1 & -1 \end{bmatrix} + \begin{bmatrix} 0 & 6 & 7 & 2 \\ 6 & 7 & -2 & 2 \end{bmatrix} = \begin{bmatrix} 2 & 8 & 9 & 4 \\ 5 & 6 & -3 & 1 \end{bmatrix}$ which is

the matrix for the vertices of the image $P'E'A'R'$. **35.** $\begin{bmatrix} 4 & 0 \\ 0 & 6 \end{bmatrix}$

37. $\begin{bmatrix} 1 & 0 & -4 & 0 \\ 3 & .5 & -1 & 5 \end{bmatrix}$ **39.** $\begin{bmatrix} 3 & 4 & 1 \\ 0 & 2 & 0 \end{bmatrix}$

41.
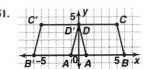

	Boys	Girls
freshmen	490	487
sophomores	402	416
juniors	358	344
seniors	293	300

43. the price for a dozen of eggs, a pound each of plums, peaches,

and bananas at Market II **45.** $[270 \quad 320 \quad 210]\begin{bmatrix} 15 & 6 \\ 10 & 8 \\ 2 & 1 \end{bmatrix} =$

$[7670 \quad 4390]$ Factory 1: \$7,670,000; Factory 2: \$4,390,000

47. a. 0.6 **b.** $\begin{bmatrix} 249.00 & 403.20 & 154.80 \\ 118.80 & 236.40 & 65.40 \end{bmatrix}$ **49.** $\begin{bmatrix} 1 & 3 & 9 & 6 \\ 1 & 7 & 1 & -6 \end{bmatrix}$

51. See below.

51.
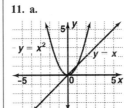

LESSON 5-1 (pp. 272-278)

17. a. See below. **b.** $\{x: 45 \le x \le 55\}$ **19.** He should use "or" instead of "and," since "$x = 2$ and $x = -2$" has no solutions.
21. no more than 500 miles **23. a.** $y = x + 250$ **b.** $y = -50$
25. a. y is divided by 64. **b.** y is multiplied by 8.

17. a.

LESSON 5-2 (pp. 279-284)

11. a. two solutions **b.** $(2, 1)$, $\left(-\frac{1}{2}, -4\right)$ **c.** Check $(2, 1)$: Is $2 \cdot 1 = 2$?
Yes. Is $2 \cdot 2 - 1 = 3$? Yes. Check $\left(-\frac{1}{2}, -4\right)$: Is $\left(-\frac{1}{2}\right) \cdot (-4) = 2$? Yes.
Is $2 \cdot \left(-\frac{1}{2}\right) - (-4) = 3$? Yes. **13. a.** See below. **b.** 2 solutions
c. $(-4.3, 9.3)$, $(2.3, 2.7)$ **15. a.** $y = 9x$ **b.** father: $9m$; daughter:
45 m **c.** 10 seconds after he started **d.** 90 m from the start
17. a. $(-1.28, 4.89)$ **b.** $(2.61, 20.44)$ **19. a.** See below. **b.** See
below. **21. a.** 45

b. $S_n = \frac{n(n + 1)}{2}$ or $\begin{cases} t_1 = 1 \\ t_n = t_{n-1} + n \end{cases}$ for integers $n \ge 2$

23. a. Yes; all squares are similar. Similar figures with equal areas must be congruent. **b.** No; a counterexample is a right triangle with length 3 in. and height 4 in. and a right triangle with length 2 in. and height 6 in. Both have area of 6 in², but they are not congruent.

13. a.
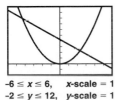
$-6 \le x \le 6, \quad x\text{-scale} = 1$
$-2 \le y \le 12, \quad y\text{-scale} = 1$

19. a.
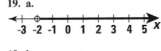

19. b.
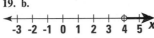

LESSON 5-3 (pp. 285-291)

11. a. See below. **b.** consistent **c.** $(0, 0)$, $(1, 1)$

13. a. $\begin{cases} y = .3x \\ y = 2 + .25x \end{cases}$ **b.** 40 prints **c.** \$12.00 **15. a.** See

below. **b.** 100 **c.** Under the incentive plan, anyone selling fewer than 100 gadgets receives a pay cut. **d.** \$500; \$550.
e. Sample: Salespeople have an incentive to sell more gadgets because they make more money by selling more gadgets.

17. $\begin{bmatrix} 58 & 14 & -4 \\ -10 & 7 & -1 \end{bmatrix}$ **19. a.** $40a + 50p \le 500$ **b.** 0 apples and

10 pears, 5 apples and 6 pears, or 10 apples and 2 pears
21. $\angle ROC = 70°$; $\angle CKR = 60°$; $\angle OCK = 150°$; $\angle KRO = 80°$

11. a.
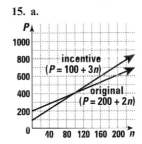

15. a.

LESSON 5-4 (pp. 292-297)

13. 90¢ **15.** $(u, v) = (-2, -2)$ **17.** design: \$50; editing: \$80;
production: \$60 **19.** (c)

21. a. $\begin{cases} u_1 - 112 \\ a_n = a_{n-1} + 8, \text{ for integers } n \ge 2 \end{cases}$

b. $a_n = 112 + (n - 1) \cdot 8$ or $a_n = 104 + 8n$

23. $\begin{bmatrix} 7 & 8 & 9 \\ \sqrt{5} & \sqrt{6} & \sqrt{7} \\ -1 & 0 & 1 \end{bmatrix}$

LESSON 5-5 (pp. 299–304)

15. Sample: $\begin{bmatrix} 2 & 6 \\ 1 & 3 \end{bmatrix}$ **17. a.** $\begin{bmatrix} 0 & 1 \\ -1 & 0 \end{bmatrix}$ **b.** Sample: The matrix in part **a** represents R_{270} or R_{-90}. Geometrically, R_{-90} and R_{90} are opposite transformations. **19.** $\frac{1}{3}$ kg for bread, $\frac{1}{6}$ kg for pizza crust **21.** (2, 8) **23.** in 20 hours

LESSON 5-6 (pp. 305–311)

13. a. $\begin{cases} u + b + 3t = 561 \\ 7b + t = 906 \\ 5u + 2b + 7t = 1758 \end{cases}$ **b.** unicycle: \$183; bicycle: \$117; tricycle: \$87

15. $n = -3$ **17. a.** Sample: $\begin{bmatrix} 2 & 3 \\ 3 & 5 \end{bmatrix}$ **b.** Sample: $\begin{bmatrix} 5 & -3 \\ -3 & 2 \end{bmatrix}$

19. (17, −7) **21.** 1.6 oz **23.** See below.

23.

```
      6   8  10  12  14   x
```

LESSON 5-7 (pp. 312–318)

15. $y < -3$ **17. a.** $10x + 15y < 75$ **b.** See below. **c.** 24 solutions

19. $y < -\frac{3}{4}x - 4$

17. b.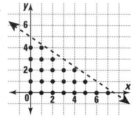

21. a. $\begin{cases} 3x + 4y = 0 \\ x + 2y = -2 \end{cases}$

b. (4, −3) **23.** −100

25. (a) and (c)

LESSON 5-8 (pp. 319–324)

13. a. $\begin{cases} x \geq 0 \\ y \leq x + 1 \\ y \leq \frac{1}{2}x + 3 \\ x \leq 6 \\ y \geq 0 \end{cases}$ **b.** True **15. a.** $\begin{cases} 0 \leq x \leq 1000 \\ 0 \leq y \leq 600 \\ 20x + 30y \leq 24{,}000 \end{cases}$

c. (4, 5)

b. See above right.

17. a. $R \leq \dfrac{D - 200}{9}$

b. \$144.44 **19.** (2, 4)

21. $5x + 8y$ dollars

23. a. $8n + 64$ **b.** $n^2 + 8n$

c. $n^2 + 16n + 64$

d. $n^2 + 16n + 64$

15. b.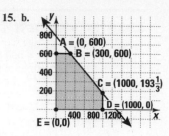
A = (0, 600)
B = (300, 600)
C = (1000, $193\frac{1}{3}$)
D = (1000, 0)
E = (0,0)

LESSON 5-9 (pp. 325–330)

11. T. Sample: The following table shows the values of P corresponding to each of the vertices of the feasible region. The profit value is the greatest at vertex T. By the Linear-Programming Theorem, this vertex maximizes profit over the feasible set. **13. a.** (ii) **b.** (iii) **c.** (iv) **d.** (i) **15. a.** (0, 0), (0, 50), (40, 10), (48, 0) **b.** $P = 75C + 85S$ **c.** (0, 50) **17.** (b)

Vertex	P
Q	250
R	460
S	580
T	680
U	660
V	610

19. $S_{-\frac{1}{3}} = \begin{bmatrix} -\frac{1}{3} & 0 \\ 0 & -\frac{1}{3} \end{bmatrix}$ **21.** $y \geq x^2$

LESSON 5-10 (pp. 331–335)

9. a. See below. **b.** $\left(-\frac{10}{3}, \frac{23}{3}\right)$ **11.** $x = -1$, $y = 4$ **13.** 32 kg of 25% aluminum; 128 kg of 75% aluminum **15. a.** 10; 9.5; 9

b. $\begin{cases} a_1 = 10 \\ a_n = a_{n-1} - .5, \text{ for integers } n \geq 2 \end{cases}$

c. $-2a_n + 21 = n$ **d.** 99th term

9. a.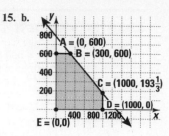
$\left(\frac{-10}{3}, \frac{23}{3}\right)$

CHAPTER 5 PROGRESS SELF-TEST (p. 339)

1. $n > -2$, See page 909. **2.** See page 909. **3.** (−1.7, −2.8), (1.2, −1.4)
4. a. inconsistent **b.** The graphs of these equations are both vertical lines, but they have different x-intercepts. Since the lines do not intersect, the system is inconsistent. **5.** Sample: In the third equation, substitute r with $t + 11$ and s with $4t$. This gives $3(t + 11) - 8(4t) = 4$. Solve this equation to get $t = 1$. Substitute $t = 1$ into the original system's first and second equation; $s = 4(1) = 4$, $r = 1 + 11 = 12$; $(r, s, t) = (12, 4, 1)$
6. Sample: $-3x + 3y = 2$ (multiply by 2) $\quad -6x + 6y = 4.$
$\qquad\quad -4x - 2y = 3$ (multiply by 3) $\quad \dfrac{-12x - 6y = 9}{}$
$\qquad\qquad\qquad\qquad\qquad$ (add) $\qquad\qquad -18x = 13$
$\qquad\qquad\qquad\qquad\qquad\qquad\qquad\qquad x = -\dfrac{13}{18}$

Substitute $x = -\dfrac{13}{18}$ into the first equation; $-3\left(-\dfrac{13}{18}\right) + 3y = 2$; $y = -\dfrac{1}{18}$; $\left(-\dfrac{13}{18}, -\dfrac{1}{18}\right)$. **7.** Let e = the price for 1 egg; let s = the price for one sausage link. The system $\begin{cases} 2e + 2s = 2.78 \\ 3e + 4s = 4.99 \end{cases}$ describes our situation.

$2e + 2s = 2.78$ (multiply by −2)
$-4e - 4s = -5.56$
$\dfrac{3e + 4s = 4.99}{}$
$-e \qquad\quad = -.57 \qquad$ (add)
$e = .57$
They might charge 57¢ for one egg.

8. a. The coefficient matrix is $\begin{bmatrix} 8 & 3 \\ 6 & 5 \end{bmatrix}$.

b. The inverse of $\begin{bmatrix} 8 & 3 \\ 6 & 5 \end{bmatrix}$ is $\dfrac{1}{8 \cdot 5 - 6 \cdot 3} \begin{bmatrix} 5 & -3 \\ -6 & 8 \end{bmatrix}$

$= \dfrac{1}{22}\begin{bmatrix} 5 & -3 \\ -6 & 8 \end{bmatrix} = \begin{bmatrix} \frac{5}{22} & \frac{-3}{22} \\ \frac{-6}{22} & \frac{8}{22} \end{bmatrix}$.

c. $\begin{bmatrix} \frac{5}{22} & \frac{-3}{22} \\ \frac{-6}{22} & \frac{8}{22} \end{bmatrix}\begin{bmatrix} 8 & 3 \\ 6 & 5 \end{bmatrix}\begin{bmatrix} x \\ y \end{bmatrix} = \begin{bmatrix} \frac{5}{22} & \frac{-3}{22} \\ \frac{-3}{11} & \frac{4}{11} \end{bmatrix}\begin{bmatrix} 41 \\ 39 \end{bmatrix}$;

$\begin{bmatrix} 1 & 0 \\ 0 & 1 \end{bmatrix}\begin{bmatrix} x \\ y \end{bmatrix} = \begin{bmatrix} \frac{88}{22} \\ \frac{33}{11} \end{bmatrix}$; $\begin{bmatrix} x \\ y \end{bmatrix} = \begin{bmatrix} 4 \\ 3 \end{bmatrix}$; The solution to the system is

$x = 4$ and $y = 3$. **9. a.** Sample: $\begin{bmatrix} 2 & 4 \\ 1 & 2 \end{bmatrix}$

b. The determinant of the matrix is $2 \cdot 2 - 4 \cdot 1 = 0$.
10. See right. 11. Pick a point in the shaded region; $(0, -1)$
satisfies only (c) and (d); since the boundary $y = x$ is included,
the answer is (c). **12. a.** Let $c =$ the number of chairs built per
day, and $s =$ the number of sofas built per day. The number of
hours the carpenters can work per day satisfies $7c + 4s \le 133$,
and the upholsterers' hours satisfy $2c + 6s \le 72$. Since the
manufacturer cannot make a negative number of chairs or sofas,
$c \ge 0$ and $s \ge 0$.
$$\begin{cases} 7c + 4s \le 133 \\ 2c + 6s \le 72 \\ c \ge 0 \\ s \ge 0 \end{cases}$$
b. See right. The vertices of the feasible region will be solutions to
the systems:
$$\begin{cases} s = 0 \\ 7c + 4s = 133 \end{cases} \qquad \begin{cases} 7c + 4s = 133 \\ 2c + 6s = 72 \end{cases}$$
$$\begin{cases} 2c + 6s = 72 \\ c = 0 \end{cases} \qquad \begin{cases} s = 0 \\ c = 0 \end{cases}$$
The vertices of the feasible set are the points $(0, 12)$, $(0, 0)$,
$(19, 0)$, and $(15, 7)$. **c.** The profit equation is $P = 80c + 70s$.

By the Linear-Programming Theorem, we know that P is
maximized at one of the vertices of the feasible region. So check
the vertices to find which one gives the maximum value.
$$80(0) + 70(12) = 840$$
$$80(0) + 70(0) = 0$$
$$80(19) + 70(0) = 1520$$
$$80(15) + 70(7) = 1690$$
The profit is maximized in the feasible region at the point
$(15, 7)$. The manufacturer can maximize profits, earning \$1690,
by producing 15 chairs and 7 sofas per day.

1.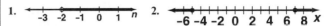

2.

10.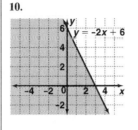

12. b.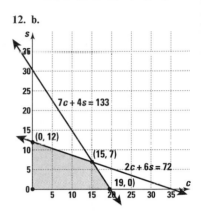

The chart below keys the **Progress Self-Test** questions to the objectives in the **Chapter Review** on pages 340–343 or to the **Vocabulary** (Voc.) on page 338. This will enable you to locate those **Chapter Review** questions that correspond to questions missed on the **Progress Self-Test**. The lesson where the material is covered is also indicated on the chart.

Question	1	2	3	4	5	6	7	8	9
Objective	H	H	I	D	A	A, C	F	B, C	B
Lesson	5-1	5-1	5-2	5-3	5-3	5-3, 5-4, 5-6	5-3, 5-4, 5-6	5-5, 5-6	5-5

Question	10	11	12
Objective	J	E, K	G
Lesson	5-7	5-8	5-9, 5-10

CHAPTER 5 REVIEW (pp. 340–343)

1. (a) **3.** inconsistent or no solutions **5.** $(-6.5, -10.5)$;
$-10.5 = -6.5 - 4$ and $2(-6.5) + 10.5 = -2.5$ **7.** $\left(\frac{3}{2}, \frac{1}{2}, \frac{1}{5}\right)$;
$2\left(\frac{3}{2}\right) + 15\left(\frac{1}{5}\right) = 6$, $\frac{3}{2} = 3\left(\frac{1}{2}\right)$, and $\frac{1}{5} = \frac{2}{5}\left(\frac{1}{2}\right)$

9. a. 2 **b.** $\begin{bmatrix} \frac{1}{2} & 0 \\ 0 & 1 \end{bmatrix}$ **11. a.** 0 **b.** The inverse does not exist.

13. $\begin{bmatrix} \frac{6}{69} & -\frac{9}{69} \\ \frac{7}{69} & \frac{1}{69} \end{bmatrix}$

15. Since the determinant, $3 \cdot 4 - 6 \cdot 2$, equals zero, the matrix
has no inverse. **17.** $(7, 0)$ **19.** $m = 39$, $n = 28$ **21.** Yes; the
solution $(7, 2)$ satisfies both systems. **23.** inconsistent
25. a. consistent **b.** infinitely many solutions **27. a.** consistent
b. two solutions **29.** $\frac{21}{4}$ **31.** False **33.** (a), (c) **35.** Yes; it is one

of the vertices of the feasible region. **37.** Sugar-O's: $8\frac{1}{3}$ g;
Health-Nut: $16\frac{2}{3}$ g **39. a.** about 53 **b.** more than 53
41. \$790 first class, \$700 for business class, and \$550 for tourist
class **43.** 88 Olde English shelves; 112 Cool Contemporary shelves
45. (c) **47.** $x < \frac{-5}{2}$, See below. **49.** See below. **51.** See below.
53. $x \le -1$ or $x > 3$ **55.** See page 910. **57.** See page 910. **59.** See
page 910. **61.** $y \ge -\frac{5}{8}x + 5$ **63.** See page 910. **65.** $y \le -1$ and
$x \le -3$

47.

49.

51.

55.

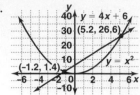

57.

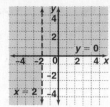

59.

63.

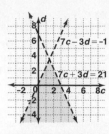

LESSON 6-1 (pp. 346–350)

17. $140w + 4w^2$ m^2 **19.** $\frac{1}{2}n^3 + n^2 + \frac{1}{2}n$ **21.** $-3x^2 + 12x - 12$
23. $h = -1$ **25. a.** 7.475 million visitors/year
b. 0.13 billion dollars/year **27. a.** See below. **b.** They have the same vertex (0, 0) and they are both symmetric to the y-axis.

27. a.

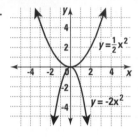

LESSON 6-2 (pp. 351–355)

23. a. See below. **b.** They coincide everywhere. **25.** $x - 3$ or $3 - x$
27. a. $\pm\sqrt{13}$ **b.** $\pm\sqrt{14}$ **29.** False; for instance, when $n = 1$,
$(1 + 11)^2 = 144$ and $1^2 + 11^2 = 122$; so $(n + 11)^2 = n^2 + 11^2$ is not true. **31.** under r_x: $y = -3x^2$; under r_y: $y = 3x^2$

23. a.

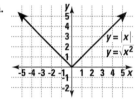

LESSON 6-3 (pp. 357–362)

15. a. $y - 3 = -1.4x^2$ **b.** $y = -1.4(x - 3)^2$ **17.** $x = 11$
19. $n = -0.25$ or -0.75 **21. a.** 4500 sq ft
b. $6000 - 320x + 4x^2$ sq ft **23. a.** $2x^2 + 16x + 32$
b. $2x^2 + 16x + 35$ **25.** $\left\lfloor\frac{s}{40}\right\rfloor + 1$

LESSON 6-4 (pp. 363–369)

13. x-intercepts (1, 0), (3, 0), y-intercept (0, 3), vertex (2, −1)
See above right. 15. a. See above right. **b.** $x = 2.5$
c. (2.5, −12.25) **17. a.** $h = -4.9t^2 + 6700$ **b.** See above right.
c. $\approx$ 37 seconds **19.** $y = -.5x^2 - 2x + 2$ **21. a.** See above right.
b. The graph of $y + 4 = |x - 5|$ is the image of $y = |x|$ under the
translation $T_{5, -4}$. **23. a.** $96 + 40w + 4w^2$ sq. in. **b.** $40 + 8w$ in.
25. $a = 12$ **27. a.** $y \approx -411x + 823{,}031$ **b.** $\approx$ \$2675
c. The truck does not continue to depreciate in value at the same
rate. In this case, one gets a negative value for the truck in the year
2006 if the line of best fit is used.

13.

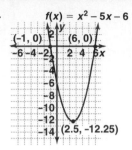

15. a.

$f(x) = x^2 - 5x - 6$

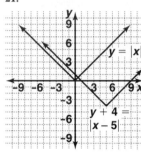

17. b.

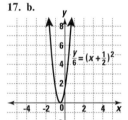

21.

LESSON 6-5 (pp. 370–375)

13. a. the set of all real numbers
b. (−2, 4) **c.** $\{y: y \geq 4\}$
15. a. $h = -4.9t^2 + 629$
b. about 11.3 seconds
17. a. $y = 6x^2 + 6x + \frac{3}{2}$
b. See right. c. $\left(0, 1\frac{1}{2}\right)$
19. $y = \pm\sqrt{20} \approx \pm 4.47$
21. −1

17. b.

LESSON 6-6 (pp. 376–381)

9. a. Sample: $h = -16t^2 + 55t + 325$ **b.** 325 ft **c.** $\approx$ 5.8 sec
d. $\approx$ 6.5 sec **11. a.** $(x + 3)(x + 7)$ **b.** $x^2 + 10x + 21$
13. a. (0, −20) **b.** $y + 92 = 2(x + 6)^2$ **c.** (−6, −92)

15.
$$\begin{bmatrix} 100 & 10 & 1 \\ 400 & 20 & 1 \\ 900 & 30 & 1 \end{bmatrix}\begin{bmatrix} a \\ b \\ c \end{bmatrix} = \begin{bmatrix} 19 \\ 42 \\ 73 \end{bmatrix}; \begin{bmatrix} 1 & 0 & 0 \\ 0 & 1 & 0 \\ 0 & 0 & 1 \end{bmatrix}\begin{bmatrix} a \\ b \\ c \end{bmatrix} =$$

$$\begin{bmatrix} \frac{1}{200} & -\frac{1}{100} & \frac{1}{200} \\ -\frac{1}{4} & \frac{2}{5} & -\frac{3}{20} \\ 3 & -3 & 1 \end{bmatrix}\begin{bmatrix} 19 \\ 42 \\ 73 \end{bmatrix}; \begin{bmatrix} a \\ b \\ c \end{bmatrix} = \begin{bmatrix} \frac{1}{25} \\ \frac{11}{10} \\ 4 \end{bmatrix} = \begin{bmatrix} .04 \\ 1.1 \\ 4 \end{bmatrix}$$

17. 6 **19.** 17 **21.** True; $\dfrac{4 \pm \sqrt{12}}{6} = \dfrac{4 \pm 2\sqrt{3}}{6} = \dfrac{2 \pm \sqrt{3}}{3}$.

LESSON 6-7 (pp. 382–387)

17. $n = 3$ **19. a.** $x = -3 + \sqrt{10} \approx .16$ or $x = -3 - \sqrt{10} \approx -6.16$; x-intercepts **b.** $(-3, -10)$ **c. See below. d.** $x = -3$ **21. a. See below. b.** Sample: $r = -.0025i^2 + .825i$ **c.** Sample: $r = .6i + 3.75$ **d.** Sample: The quadratic model is more appropriate because the points are closer to the quadratic model than the linear model.
23. a. See below. b. preimage: $y = x$, $y = -x$; image: $y = x - 2$, $y = -x - 2$ **25. a.** $2808 + 192w + 2w^2$ in^2 **b.** 3608 in^2

19. c.

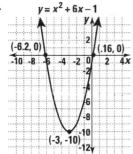

$y = x^2 + 6x - 1$
$(-6.2, 0)$ $(.16, 0)$
$(-3, -10)$

21. a.

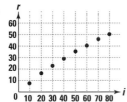

23. a.

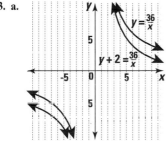

$y = \frac{36}{x}$

$y + 2 = \frac{36}{x}$

LESSON 6-8 (pp. 388–392)

23. $659i$ **25.** False; Sample: $2i + (-2i) = 0$.
27. $x = \dfrac{-0 \pm \sqrt{0^2 - 4 \cdot 1 \cdot 1}}{2 \cdot 1} = \dfrac{\pm \sqrt{-4}}{2} = \dfrac{2i}{2} = \pm i$
29. a. $x = \pm 3i$ **b.** $x = 3 \pm 3i$ **31.** $x = -6 \pm \sqrt{38}$
33. $2\pi rh + \pi h^2$

LESSON 6-9 (pp. 393–398)

21. $1 + 0i$ **23.** $2a$ **25. a.** $1 \pm 2i$ **b.** $(1 + 2i)^2 - 2(1 + 2i) + 5 = 0$?; $1 + 4i - 4 - 2 - 4i + 5 = 0$?; $0 = 0$? Yes.
$(1 - 2i)^2 - 2(1 - 2i) + 5 = 0$?; $1 - 4i - 4 - 2 + 4i + 5 = 0$?; $0 = 0$? Yes. **27.** Ms^4p^2 **29.** $\dfrac{1 \pm \sqrt{3}}{2}$

LESSON 6-10 (pp. 400–407)

13. See below. 15. False; if the discriminant $= 0$, the graph has exactly one x-intercept. **17. a.** error message in line 300 **b.** add lines:
```
110  IF A = 0, THEN 610
610  PRINT "NOT A QUADRATIC
     EQUATION"
620  GOTO 999
```
19. 9 **21.** -48 **23.** $5 - 3i$ **25.** $56 - 20i$ **27. a.** $y = -\dfrac{7}{6}x^2 - \dfrac{3}{2}x + \dfrac{47}{3}$ **b. See below.**

13.

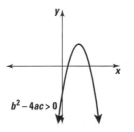

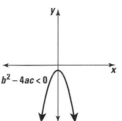

$b^2 - 4ac > 0$ $b^2 - 4ac = 0$

$b^2 - 4ac < 0$

27. b.

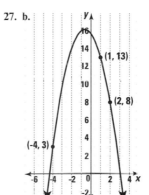

$(1, 13)$
$(2, 8)$
$(-4, 3)$

CHAPTER 6 PROGRESS SELF-TEST (p. 412)

1. $y - 12 = x^2 - 8x$; $y - 12 + 16 = x^2 - 8x + 16$; $y + 4 = (x - 4)^2$ **2.** $(4, -4)$ **3.** $0 + 4 = (x - 4)^2$; $\pm\sqrt{4} = x - 4$; $x = 4 \pm 2$; $x = 6$ or $x = 2$. Or let $y = 0$, that is, $x^2 - 8x + 12 = 0$;
$x = \dfrac{8 \pm \sqrt{64 - 48}}{2} = \dfrac{8 \pm 4}{2}$; $x = 2$ or $x = 6$. **4.** $2i \cdot i = 2i^2 = -2$
5. $\sqrt{-8} \cdot \sqrt{-2} = 2i\sqrt{2} \cdot i\sqrt{2} = i^2 \cdot 2(\sqrt{2})^2 = -4$ **6.** $\dfrac{4 + \sqrt{-8}}{2} = \dfrac{4 + 2i\sqrt{2}}{2} = 2 + i\sqrt{2}$ **7.** $18i^2 + 12i - 12i - 8 = -18 - 8 = -26$

8. $2 - 4i - (1 + 5i) = 1 - 9i$ **9.** The vertex of $y - 2 = -(x + 1)^2$ is $(-1, 2)$. This is a translation $T_{-1, 2}$ of the preimage $y = -x^2$. (c)
10. See page 912. 11. $a = 3$, $b = 14$, $c = -5$; $x = \dfrac{-14 \pm \sqrt{14^2 - 4 \cdot 3 \cdot (-5)}}{2 \cdot 3}$; $x = \dfrac{-14 \pm \sqrt{196 + 60}}{6}$; $x = \dfrac{-14 \pm \sqrt{256}}{6}$; $x = \dfrac{-14 \pm 16}{6}$; $x = \dfrac{1}{3}$ or $x = -5$ **12.** $m + 40 = \pm\sqrt{2}$; $m = -40 \pm \sqrt{2}$
13. $3x^2 - 18x + 75 = 0$; $x^2 - 6x + 25 = 0$; $x = \dfrac{6 + \sqrt{6^2 - 4 \cdot 1 \cdot 25}}{2} = \dfrac{6 \pm \sqrt{-64}}{2} = \dfrac{6 \pm 8i}{2} = 3 \pm 4i$; $x = 3 + 4i$ or $x = 3 - 4i$. **14.** True

15. **a.** one **b.** two **16. a.** There is one real root. **b.** There are two complex conjugate roots. **17.** $(a^2 - 6a + 9) + (a^2 + 6a + 9) = 2a^2 + 18$ **18.** $64v^2 + 16v + 1$ **19.** $h = -4.9t^2 + 10t + 20$
20. $h = -16(.5)^2 + 12(.5) + 4 = 6$; after .5 second, the ball is 6 feet high. **21.** $0 = -16t^2 + 12t + 4$; $t = \frac{-12 \pm \sqrt{144 - 4(4)(-16)}}{2(-16)} = \frac{-12 \pm 20}{-32}$; $t = -\frac{1}{4}$ or $t = 1$. The ball hits the ground after 1 second.
22. $A = (30 - 2s)(40 - 2s) = 1200 - 140s + 4s^2 = 4s^2 - 140s + 1200$ **23.** (a) **24. a. See right. b.** The quadratic model is $H = an^2 + bn + c$; choose three points, (1, 0), (3, 3), and (5, 10), and substitute them into the model.
$$\begin{cases} 0 = a + b + c \\ 3 = 9a + 3b + c \\ 10 = 25a + 5b + c; \end{cases}$$
subtract each equation from the one below it;
$$\begin{cases} 3 = 8a + 2b \\ 7 = 16a + 2b; \end{cases}$$
subtract again to obtain $4 = 8a$, or $a = \frac{1}{2}$.

Substituting $a = \frac{1}{2}$ into $3 = 8a + 2b$ gives $b = -\frac{1}{2}$; substituting into $0 = a + b + c$ gives $c = 0$, $H = \frac{1}{2}n^2 - \frac{n}{2}$. **c.** 4950 **25.** $y - 4 = 2x^2 - 6x$; $\frac{y - 4}{2} = x^2 - 3x$; $\frac{y - 4}{2} + \frac{9}{4} = x^2 - 3x + \frac{9}{4}$; $y - 4 + \frac{9}{2} = 2\left(x - \frac{3}{2}\right)^2$; $y + \frac{1}{2} = 2\left(x - \frac{3}{2}\right)^2$; the parabola is the image of $y = 2x^2$ under the translation $T_{\frac{3}{2}, -\frac{1}{2}}$. So it opens upward, has the vertex $\left(\frac{3}{2}, -\frac{1}{2}\right)$, and has an axis of symmetry $x = \frac{3}{2}$.

10.

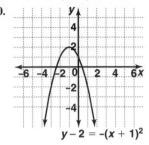

$y - 2 = -(x + 1)^2$

24. a.

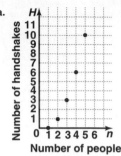

The chart below keys the **Progress Self-Test** questions to the objectives in the **Chapter Review** on pages 413–415 or to the **Vocabulary** (Voc.) on page 411. This will enable you to locate those **Chapter Review** questions that correspond to questions missed on the **Progress Self-Test**. The lesson where the material is covered is also indicated on the chart.

Question	1	2	3	4	5	6	7	8	9	10
Objective	B	B	C	D	D	D	D	D	I	J
Lesson	6-5	6-5	6-7	6-8	6-8	6-8	6-9	6-9	6-3	6-3
Question	11	12	13	14	15	16	17	18	19	20
Objective	C	C	C	E	K	F	A	A	G	G
Lesson	6-7	6-7	6-7	6-2	6-10	6-10	6-1	6-1	6-7	6-7
Question	21	22	23	24	25					
Objective	G	G	I	H	B					
Lesson	6-7	6-1	6-3	6-6	6-5					

CHAPTER 6 REVIEW (pp. 413–415)

1. $a^2 + 2ax + x^2$ **3.** $9x^2 + 24x + 16$ **5.** $9t^2 - 90t + 225$
7. $y = 3x^2 + 12x + 12$ **9.** $y + 31 = (x + 5)^2$ **11.** $y + 63 = 3(x - 5)^2$ **13.** $x = 3$ **15.** $z = \pm 2i\sqrt{2}$ **17.** $x = -\frac{1}{2} \pm \frac{\sqrt{5}}{2}$
19. $z = 4$ **21.** $k = 2 \pm \sqrt{6}$ **23.** $x = \pm 5i$ **25.** $p = \frac{1}{2}$ or 3 **27.** 1
29. -28 **31.** $50i\sqrt{2}$ **33.** $2 \pm 2i\sqrt{5}$ **35.** $1 + 12i$ **37.** $-6 + 10i$
39. $\frac{3}{2} + 3i$ **41.** $23 + 19i$ **43.** $4 - 11i$ **45.** $\frac{7}{89} - \frac{29}{89}i$ **47.** all nonpositive numbers **49.** 9 **51.** -4 **53. a.** 0 **b.** 1 **c.** rational
55. a. 45 **b.** 2 **c.** irrational **57.** Two; the discriminant is 9 which is greater than 0. **59.** 5.5 m × 11 m **61.** 4.52 sec **63. a.** 51, 70
b. $p(n) = \frac{3}{2}n^2 - \frac{1}{2}n$ **c.** 3725 **65.** $y + 5 = (x + 7)^2$ **67.** (d)
69. The solutions to $(k - 1)^2 = 36$ are 1 more than the solutions to $k^2 = 36$. **71.** $a = 9$ **73. See right.** vertex (-2, 4), x-intercepts $(-2 \pm 2\sqrt{3}, 0)$ **75. See right.** vertex (1, 1), x-intercepts $(1 \pm \frac{\sqrt{2}}{2}, 0)$
77. (c) **79. See right.** **81.** 2 **83.** Yes; the discriminant of the equation $6x^2 - 12x = -5$ is $(-12)^2 - 4 \cdot 6 \cdot 5 = 24 > 0$. Therefore, this equation has two real solutions which means the parabola $y = 6x^2 - 12x$ intersects the line $y = -5$ at two points.

73.

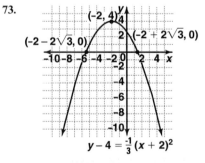

(-2, 4); $(-2 - 2\sqrt{3}, 0)$; $(-2 + 2\sqrt{3}, 0)$
$y - 4 = \frac{-1}{3}(x + 2)^2$

75.

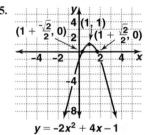

$(1 + \frac{-\sqrt{2}}{2}, 0)$; (1, 1); $(1 + \frac{\sqrt{2}}{2}, 0)$
$y = -2x^2 + 4x - 1$

79.

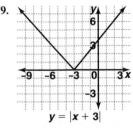

$y = |x + 3|$

LESSON 7-1 (pp. 418–425)

11. True **13. a.** $q = \left(\frac{p}{100}\right)^4$; **b.** the set of real numbers from 0 to 1
15. $y = x^6$ **17.** Yes. An even nth power function is symmetric to the y-axis. Thus if (x, y) is on the graph, $(-x, y)$ is also on the graph.
19. a. Let $x =$ the number of yards of silk ribbon produced in an hour and $y =$ the number of yards of cotton ribbon produced in an hour.
$$\begin{cases} .70x + .40y \leq 2000 \\ x + y \leq 4000 \\ 0 \leq x \leq 2500 \\ 0 \leq y \end{cases}$$
b. See below. c. Yes; by the Linear Programming Theorem, the maximum of $P = .15x + .08y$ will be determined at one of the vertices of the feasible region. These vertices are at (0, 0), (2500, 0), (2500, 625), $\left(1333\frac{1}{3}, 2666\frac{2}{3}\right)$ and (0, 4000). The point that maximizes $.15x + .08y$ is (2500, 625). **21.** R_{270} **23.** $x = 6$
25. a. r_y **b.** 10:25

19. b.

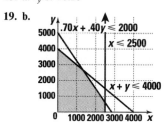

LESSON 7-2 (pp. 426–432)

23. x and x^7, x^3 and x^5, x^4 and x^4, x^0 and x^8 **25.** $x = 3$ **27.** z^6
29. y^2 **31. a.** $\approx 2.01 \cdot 10^8$ **b.** the surface area of the earth in square miles **33.** ≈ 95.14 lb per person **35. a.** $f(x) = x^3$ **b.** -125
37. $1.05b$ dollars

LESSON 7-3 (pp. 433–437)

19. $\frac{1}{5}$ **21.** False, $-3x^{-2} = -3 \cdot \frac{1}{x^2} = \frac{-3}{x^2} \neq \frac{1}{3x^2}$. **23.** larger; Since
$0 < x < 1$, $x^{-2} = \frac{1}{x^2} > \frac{1}{x} >$ **25. a.** $I = \frac{k}{d^2}$ **b.** $I = kd^{-2}$ **27.** $\frac{1}{n^{2001}}$
29. If $c = 0$, then $\frac{c^m}{c^n} = \frac{0^m}{0^n} = \frac{0}{0}$, which is undefined. **31. a.** $14 + 0i$
b. $0 + 14i$ **c.** $5 + 7i$

LESSON 7-4 (pp. 438–443)

15. \$12 **17. a.** He is charging simple interest because the amount to be paid back increases by the same amount each year. **b.** 6%
19. $\frac{16}{125}$ **21.** $1024z^{10}$ **23.** about 0.027 **25. a.** 4, 12, 36, 108, 324
b. No, the difference between consecutive terms is not constant.

LESSON 7-5 (pp. 444–449)

11. a. 486 **b.** $g_n = 2 \cdot 3^{n-1}$ for integers $n \geq 1$
13. a. $\begin{cases} g_1 = 40 \\ g_n = (-1) \cdot g_{n-1}, \text{ for integers } n \geq 2 \end{cases}$
b. $g_n = 40(-1)^{n-1}$ for integers $n \geq 1$ **15.** \$4477.12 **17.** (a)
19. $216a^{21}b^{16}c^4$ **21. a.** $x = 3 \pm \sqrt{5}$ **b.** The graph intersects the x-axis at $x = 3 - \sqrt{5}$ and $3 + \sqrt{5}$. **23.** 6

LESSON 7-6 (pp. 450–456)

21. the 22nd root of 2, or $2^{\frac{1}{22}}$ **23.** $<$ **25.** $>$ **27. a.** $\frac{2}{3}$ **b.** Check;
Does $\left(\frac{2}{3}\right)^4 = \frac{16}{81}$? Yes, it checks. **29. a.** This could be a geometric
sequence. **b.** $g_n = 100 \cdot \left(\frac{9}{10}\right)^{n-1}$ for all integers $n \geq 1$
c. $\begin{cases} g_1 = 100 \\ g_n = \left(\frac{9}{10}\right)g_{n-1}, \text{ for all integers } n \geq 2 \end{cases}$
31. a. This could be an arithmetic sequence. **b.** $a_n = 100 + (n-1) \cdot (-10)$ for all integers $n \geq 1$
c. $\begin{cases} a_1 = 100 \\ a_n = a_{n-1} - 10, \text{ for all integers } n \geq 2 \end{cases}$
33. $\frac{-5x^3}{9}$ **35.** $0 + -1i$

LESSON 7-7 (pp. 458–463)

21. a. -2 **b.** 2 **c.** No **23.** $\frac{10,000}{2401}$ **25.** $\approx 228,670,000$ km **27. a.** 3;
2; ≈ 3.61 **b.** No **29. a.** -2 **b.** 80 **c.** $a_n = 5 \cdot (-2)^{n-1}$ for integers
$n \geq 1$ **d.** 81,920 **31.** -1 **33.** Product of Powers Property
35. ≈ 4123 cm^2

LESSON 7-8 (pp. 464–468)

13. positive **15.** negative **17. a.** $64^{-1} = .015625 = \frac{1}{64}$;
$64^{-\frac{5}{6}} = .03125 = \frac{1}{32}$; $64^{-\frac{4}{6}} = .0625 = \frac{1}{16}$; $64^{-\frac{3}{6}} = .125 = \frac{1}{8}$;
$64^{-\frac{2}{6}} = .25 = \frac{1}{4}$; $64^{-\frac{1}{6}} = .50 = \frac{1}{2}$; $64^0 = 1$; $64^{\frac{1}{6}} = 2$; $64^{\frac{2}{6}} = 4$;
$64^{\frac{3}{6}} = 8$; $64^{\frac{4}{6}} = 16$; $64^{\frac{5}{6}} = 32$; $64^1 = 64$; **b.** Sample: Every power of 64 can be rewritten as a power of 2, because $64 = 2^6$.
Specifically, $64^x = (2^6)^x = 2^{6x}$. As x increases by $\frac{1}{6}$, the power of 2
increases by $6\left(\frac{1}{6}\right) = 1$. **19.** x^{-1}; Sample check: Let $x = 2$.
$(2^3)^{-\frac{1}{3}} = 8^{-\frac{1}{3}} = \frac{1}{8^{\frac{1}{3}}} = \frac{1}{2}$; $x^{-1} = 2^{-1} = \frac{1}{2}$. It checks. **21.** $2x^{\frac{13}{6}}$; Sample
check: Let $x = 64$. $\frac{x}{3x^{-\frac{2}{3}}} \cdot 6x^{\frac{1}{2}} = \frac{64}{3\left(64^{-\frac{2}{3}}\right)} \cdot 6 \cdot 64^{\frac{1}{2}} = \frac{64}{3} \cdot \frac{1}{16} \cdot$
$48 = 64 \cdot 16^2 = 16,384$; $2x^{\frac{13}{6}} = 2\left(64^{\frac{13}{6}}\right) = 2 \cdot 2^{13} = 2^{14} = 16,384$.
It checks. **23. a.** $(a + b)^{-1}$ **b.** $\frac{1}{10}$ **25.** ≈ 65 **27.** ≈ 1.46. If $f(x) = x^7$ and $f(x) = 14$, then $x^7 = 14$ since every positive real number has only 1 real nth root when n is odd, $x = 14^{\frac{1}{7}} \approx 1.46$.
29. a. 14 ft **b.** $14(.7)^{n-1}$ **31.** Sample: $(x^1)(x^4)$, $(x^6)(x^{-1})$, $(x^7)(x^{-2})$, $(x^{12})(x^{-7})$, $(x^8)(x^{-3})$, $(x^{10})(x^{-5})$

CHAPTER 7 PROGRESS SELF-TEST (p. 472)

1. $3^{-4} = \frac{1}{81}$, $-3^4 = -81$, $3^{\frac{1}{4}} \approx 1.32$; therefore, $3^{\frac{1}{4}} > 3^{-4} > -3^4$.
2. $9^{-2} = \frac{1}{9^2} = \frac{1}{81}$ **3.** $\left(\frac{1}{32}\right)^{-\frac{6}{5}} = 32^{\frac{6}{5}} = \left(32^{\frac{1}{5}}\right)^6 = 2^6 = 64$
4. $(11,390,625)^{\frac{1}{6}} = 15$ **5.** $\left(625x^4y^8\right)^{\frac{1}{4}} = 625^{\frac{1}{4}}x^{\frac{4}{4}}y^{\frac{8}{4}} = 5xy^2$
6. $\frac{-96x^{15}y^3}{4x^3y^{-5}} = -24 \cdot x^{15-3}y^{3-(-5)} = -24x^{12}y^8$ **7.** $\frac{(2x)^5}{16x} = \frac{2^5(x)^5}{16x} =$
$\frac{32x^{20}}{16x} = 2x^{20-1} = 2x^{19}$ **8.** $9x^4 = 144$; $x^4 = 16$; By the definition of nth root, the real solutions to $x^4 = 16$ are the real 4th roots of 16.

So one solution is $16^{\frac{1}{4}} = 2$. By the Number of Real Roots Theorem, there are two real roots when n is even. The second real root is -2. **9.** $\left(c^{\frac{3}{2}}\right)^{\frac{2}{3}} = 64^{\frac{2}{3}} = 16$; $c = 16$ **10.** $5^n \cdot 5^{21} = 5^{29}$;
$5^{n+21} = 5^{29}$; $n + 21 = 29$; $n = 8$ **11.** $200 \cdot \left(1 + \frac{0.0375}{365}\right)^{365 \times 5}$
$\approx \$241.24$ **12.** (a); $a^{-\frac{4}{5}} = \frac{1}{a^{\frac{4}{5}}} = \frac{1}{\left(a^4\right)^{\frac{1}{5}}} = \left(\frac{1}{a^4}\right)^{\frac{1}{5}}$ **13.** (c); The graph is symmetric to the y-axis, so the exponent must be even. Because the graph goes through the point (0, 0), the exponent must be positive.

913

14. $r = \frac{8}{2} = \frac{32}{8} = \frac{128}{32} = \ldots = 4$; $g_n = 2 \cdot (4)^{n-1}$ for intergers ≥ 1
15. $g_1 = 12$; $g_2 = \frac{1}{2} g_1 = 6$; $g_3 = \frac{1}{2} g_2 = 3$; $g_4 = \frac{1}{2} g_3 = \frac{3}{2}$
16. $\frac{2.1 \cdot 10^2}{10^{-3}} = 2.1 \cdot 10^{2-(-3)} = 2.1 \cdot 10^5 = 210{,}000$
17. $B = A \cdot r^2 = 440 \cdot \left(2^{\frac{1}{12}}\right)^2 \approx 494$ hertz **18.** $\ell^3 \cdot s^3 \cdot c^4$

19. $h(15) = 349 \cdot 10^{-.02(15)} = 349 \cdot 10^{-.3} \approx 175$ hours
20. $x^4 = 81$; $x^2 = \pm 9$; $x = \pm 3$ **21.** False; $64^{\frac{1}{6}}$ means the positive 6th root of 64. So $64^{\frac{1}{6}} = 2$. **22.** 8 years ago means $t = -8$. $5000 \cdot \left(1 + \frac{5.4\%}{12}\right)^{-8 \cdot 12} \approx \$3249.19$

The chart below keys the **Progress Self-Test** questions to the objectives in the **Chapter Review** on pages 473–474 or to the **Vocabulary** (Voc.) on page 471. This will enable you to locate those **Chapter Review** questions that correspond to questions missed on the **Progress Self-Test**. The lesson where the material is covered is also indicated on the chart.

Question	1	2	3	4	5	6	7	8	9	10
Objective	A	A	A	A	B	B	B	D	D	B
Lesson	7-2,7-3,7-6	7-3	7-6	7-8	7-2,7-7	7-2	7-2	7-6	7-7	7-2
Question	11	12	13	14	15	16	17	18	19	20
Objective	G	A	I	C	C	B	H	F	F	E
Lesson	7-4	7-8	7-1	7-5	7-5	7-2,7-3	7-6	7-2	7-8	7-6
Question	21	22								
Objective	E	G								
Lesson	7-6	7-4								

CHAPTER 7 REVIEW (pp. 473–474)

1. 1 **3.** $\frac{34}{10000}$ **5.** $\frac{3}{2}$ **7.** 10 **9.** $\frac{1}{6}$ **11.** 18.57 **13.** 512 **15.** False; $(117{,}649)^{\frac{1}{6}}$ is the positive 6th root of 117,649 only. **17.** $x = 8$
19. $n = \frac{3}{2}$ **21.** $-64x^6$ **23.** $\frac{16b}{81a}$ **25.** $\frac{3p^5}{4q^3}$ **27. a.** $g_n = \left(-\frac{3}{8}\right) \cdot (-2)^{n-1}$ for all integers $n \geq 1$
b. $\begin{cases} g_1 = -\frac{3}{8} \\ g_n = -2 \cdot g_{n-1}, \text{ for all integers } n \geq 2 \end{cases}$
c. 768 **29.** ≈ 65.53 **31.** 5, 20, 80, 320 **33.** 10, -20, 40, -80
35. $x = \pm 8$ **37.** $x \approx 2.29$ **39.** $y = 125$ **41.** $m = \frac{1}{9}$ **43. a.** ± 15
b. 15 **45.** True **47.** (a) **49.** (b), (c) **51.** when n is odd; 2 solutions
53. $\frac{1}{2}$ **55.** $c^{10} \cdot d^5$ **57. a.** $P = \frac{k}{d^2}$ **b.** $P = kd^{-2}$ **59.** ≈ 1.25 mm

61. $\left(\frac{A_1}{A_2}\right)^{\frac{3}{2}}$ **63.** \$209.78 **65.** \$4918.48 **67.** $\approx 4.59\%$ **69. a.** 61.44 cm
b. ≈ 16.1 cm **c.** $\approx 120(.8)^{n-1}$ **71. a.** $P_n = .9(.9)^{n-1}$ **b.** ≈ 14 strokes
73. ≈ 587 hertz **75. a. See below. b.** domain: the set of all real numbers; range: the set of all real numbers **c.** rotation symmetry
77. See below. There is no intersection of the graph of $y = x^8$ and that of $y = -10$.

75. a.

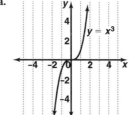

77.

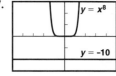

$-5 \leq x \leq 5$, x-scale = 1
$-15 \leq y \leq 15$, y-scale = 5

LESSON 8-1 (pp. 478–483)

15. a. 8 **b.** does not exist **17. a.** x **b.** when $x = 0$
19. a. $r \circ s(x) = 1000 + \sqrt{2(2500 + \sqrt{x})}$ **b.** ≈ 1095 barracuda
21. $\begin{bmatrix} \frac{a}{6} & \frac{1}{3} \\ -\frac{1}{2} & 0 \end{bmatrix}$ **23.** False **25. a.** $-\frac{4}{5}$ **b.** $\frac{5}{4}$

LESSON 8-2 (pp. 484–489)

13. a., c. See right. b. $y = \frac{x - 9}{4}$
d. Yes **e.** The slopes are reciprocals.
15. a. $U \approx 0.294M$ **b.** $M = 3.4U$
17. a. 15 **b.** x **c.** identity function
19. False; because when x is negative, $|x| = -x$. **21.** $m\angle 1 = 30°$, $m\angle 2 = 50°$, $m\angle 3 = 80°$

13. a, c

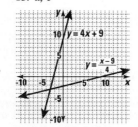

LESSON 8-3 (pp. 490–494)

11. a. See below. b. The graphs are reflection images of each other across the line with equation $y = x$. **c.** $h(g(x)) = \left(x^{\frac{2}{3}}\right)^{\frac{3}{2}} = x$; $g(h(x)) = \left(x^{\frac{3}{2}}\right)^{\frac{2}{3}} = x$ **d.** Yes; because $h(g(x)) = g(h(x)) = x$
13. f; by the Inverse Functions Theorem, the inverse of the inverse of a function would yield the original function, given the inverse exists. **15. a.** $y = \frac{x - 25000}{.4}$ **b.** the number of tickets y she has to sell to get a specific income x **17.** $3x^2 - 4$ **19.** $9x^2 - 24x + 16$
21. -1

11. a.

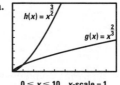

$0 \leq x \leq 10$, x-scale = 1
$0 \leq y \leq 10$, y-scale = 1

LESSON 8-4 (pp. 495–499)

19. (b) **21. a.** $\sqrt[3]{V}$ cm **b.** $V^{\frac{1}{3}}$ cm **23.** 2 **25.** $\sqrt[4]{x}$
27. a. 1000.012 cm **b.** about 2.01 seconds **c.** $T = 2\pi\sqrt{\dfrac{100 + .0003C}{980}}$
d. $f(g(C))$ **29.** 343 **31.** $-3, 2, 4$

LESSON 8-5 (pp. 500–505)

15. 30 **17.** $\sqrt[7]{2} + \sqrt[5]{3}$ **19.** False. Sample: $\sqrt[7]{x} \cdot \sqrt[5]{y} = x^{\frac{1}{7}} \cdot y^{\frac{1}{5}} =$
$x^{\frac{5}{35}} \cdot y^{\frac{7}{35}} = (x^5)^{\frac{1}{35}} \cdot (y^7)^{\frac{1}{35}} = \sqrt[35]{x^5 y^7} \neq \sqrt[35]{xy}$ **21.** $3abc\sqrt[4]{5bc^2}$
23. $n = \frac{3}{5}$ **25.** False. The range of an even power function is the
set of nonnegative numbers, which must be the domain of its
inverse. Here the domain includes $x < 0$. **27.** $\approx$ \$5.999 trillion
29. $\frac{1}{29} + \frac{17}{29}i$

LESSON 8-6 (pp. 506–509)

11. a. $7 + 4\sqrt{3}$ **b.** $\dfrac{2 + \sqrt{3}}{2 - \sqrt{3}} \approx 13.93$; $7 + 4\sqrt{3} \approx 13.93$
13. $\dfrac{BC}{AB} = \dfrac{\sqrt{3}}{3}$ **15.** $\dfrac{2\pi\sqrt{Lg}}{g}$ **17.** $3w^2\sqrt[3]{12v^2w^2}\sqrt{2v}$, or $6vw^2\sqrt[6]{18vw^4}$
19. a. $\sqrt{30}$ **b.** $\sqrt[6]{30}$ **21.** Two x values have the same y value.
Sample: When $x = \pm 1$, then $y = 5$. **23.** \$6,515.99 **25.** $\dfrac{1}{p^{\frac{1}{2}}q}$
27. PD

LESSON 8-7 (pp. 511–516)

15. 3 **17.** -4 **19. a.** $(2 + 2i)^4 = [2(1 + i)]^4 = 2^4[(1 + i)^2]^2 =$
$2^4 \cdot (2i)^2 = 16 \cdot (-4) = -64$ **b.** $(-2 - 2i)^4 = [-(2 + 2i)]^4 =$
$(-1)^4(2 + 2i)^4 = 1(-64) = -64$ **c.** $(2 - 2i)^4 = [2(1 - i)]^4 =$
$2^4[(1 - i)^2]^2 = 16 \cdot (-2i)^2 = 16 \cdot (-4) = -64$ **d.** The other
one is $-2 + 2i$. $(-2 + 2i)^4 = [-(2 - 2i)]^4 = (2 - 2i)^4 = -64$
e. $\sqrt[4]{-64}$ is not defined since it would not represent a unique value.
21. $f \circ g(x) = \sqrt[15]{x}$ **23.** $\dfrac{20 - 4\sqrt{3}}{11}$ **25. a.** 2
b. $\approx (4.123 - 3.873)(4.123 + 3.873) - 0.25 \cdot 7.996 = 1.999 \approx 2$
27. about 1.6 times **29.** 20

LESSON 8-8 (pp. 517–521)

11. $(-4, 14), (-4, 6)$ **13. a.** $\approx$ 796 mph **b.** 45 feet **15.** $z = 1$
17. $20ab^2\sqrt{a}$ **19.** True; $\dfrac{1}{\sqrt{5}} = \dfrac{1 \cdot \sqrt{5}}{\sqrt{5} \cdot \sqrt{5}} = \dfrac{\sqrt{5}}{5}$ **21.** (b)
23. Sample: The points $(2, -1)$ and $(-2, -1)$ are on the graph of the
function, so it does not pass the Horizontal-Line Test.
25. $\begin{cases} x \geq 0 \\ y \leq 4x + 30 \\ y \geq 6x \end{cases}$

CHAPTER 8 PROGRESS SELF-TEST (p. 526)

1. $g(-1) = 2(-1) - 3 = -5$; $f(-5) = (-5)^2 + 6 = 31$ **2.** $f(g(x)) =$
$f(2x - 3) = (2x - 3)^2 + 6$; $f(g(x)) = 4x^2 - 12x + 15$
3. a. $x = 5y - 6$; $x + 6 = 5y$; $\frac{1}{5}x + \frac{6}{5} = y$ **b.** This is a function
because for every value of x, there is exactly one value of y.
4. See right. **5.** Sample: If the domain is restricted to be the set
of positive real numbers, then the inverse of the function is also a
function. **6.** False. The radical expression $\sqrt[6]{64}$ stands for the
positive 6th root of 64; it cannot equal a negative number.
7. $1 = 2\pi\sqrt{\dfrac{L}{980}}$; $\dfrac{1}{2\pi} = \sqrt{\dfrac{L}{980}}$; $\dfrac{1}{4\pi^2} = \dfrac{L}{980}$, $L = \dfrac{980}{4\pi^2} \approx 24.82$ cm
8. $\sqrt[4]{5^4 \cdot x^4 \cdot y^8} = \sqrt[4]{5^4} \cdot \sqrt[4]{x^4} \cdot \sqrt[4]{y^8} = 5xy^2$
9. $\sqrt[5]{(-2)^5 \cdot 3 \cdot x^{15} \cdot y^3} = \sqrt[5]{(-2)^5} \cdot \sqrt[5]{x^{15}} \cdot \sqrt[5]{3y^3} = -2x^3\sqrt[5]{3y^3}$
10. $\dfrac{2x^4}{\sqrt{16x}} = \dfrac{2x^4}{4\sqrt{x}} = \dfrac{x^4}{2\sqrt{x}} \cdot \dfrac{\sqrt{x}}{\sqrt{x}} = \dfrac{x^4\sqrt{x}}{2x} = \dfrac{x^3}{2}\sqrt{x}$, or $\dfrac{1}{2}x^{\frac{7}{2}}$
11. All three answers are correct. Sample justification: $\dfrac{9\sqrt{2}}{2} =$
$\dfrac{\sqrt{81}\sqrt{2}}{\sqrt{4}} = \dfrac{\sqrt{162}}{\sqrt{4}} = \sqrt{\dfrac{162}{4}} = \sqrt{\dfrac{81}{2}} = \dfrac{\sqrt{81}}{\sqrt{2}} = \dfrac{9}{\sqrt{2}}$, so the
answers are equal. Then check any one of them using the
Pythagorean Theorem.

Amanda: $\left(\sqrt{\dfrac{81}{2}}\right)^2 + \left(\sqrt{\dfrac{81}{2}}\right)^2 = \dfrac{81}{2} + \dfrac{81}{2} = 81$.
Bruce: $\left(\dfrac{9}{\sqrt{2}}\right)^2 + \left(\dfrac{9}{\sqrt{2}}\right)^2 = \dfrac{81}{2} + \dfrac{81}{2} = 81$.
Carlos: $\left(\dfrac{9\sqrt{2}}{2}\right)^2 + \left(\dfrac{9\sqrt{2}}{2}\right)^2 = \dfrac{162}{4} + \dfrac{162}{4} = 81$. **12.** $\left(7^{\frac{1}{4}}\right)^{\frac{1}{3}} =$
$7^{\frac{1}{4} \cdot \frac{1}{3}} = 7^{\frac{1}{12}}$ **13.** $\dfrac{6}{\sqrt{10} - \sqrt{4}} = \dfrac{6}{\sqrt{10} - 2} \cdot \dfrac{\sqrt{10} + 2}{\sqrt{10} + 2} =$
$\dfrac{6(\sqrt{10} + 2)}{10 - 4} = \dfrac{6(\sqrt{10} + 2)}{6} = \sqrt{10} + 2$ **14.** $p = \left(\dfrac{s}{6.5}\right)^7 = \left(\dfrac{20}{6.5}\right)^7 \approx$
2611 horsepower **15.** $\dfrac{400}{9} = \sqrt{3t}$; $\left(\dfrac{400}{9}\right)^2 = (\sqrt{3t})^2$; $\dfrac{160000}{81} = 3t$;
$t = \dfrac{160000}{243}$, or ≈ 658.44 **16.** $\sqrt[3]{x + 5} = 9 - 12$; $(\sqrt[3]{x + 5})^3 =$
$(-3)^3$; $x + 5 = -27$; $x = -32$ **17.** for n an odd integer ≥ 1
18. domain = $\{x : x \geq 0\}$; range = $\{y : y \geq 0\}$

4.

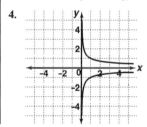

The chart below keys the **Progress Self-Test** questions to the objectives in the **Chapter Review** on pages 527–529 or to the **Vocabulary**
(Voc.) on page 525. This will enable you to locate those **Chapter Review** questions that correspond to questions missed on the **Progress Self-Test**. The lesson where the material is covered is also indicated on the chart.

Question	1	2	3	4	5	6	7	8	9	10
Objective	A	A	B, F	I	I	C, G	H	D	D	D
Lesson	8-1	8-1	8-2	8-2	8-2	8-4	8-8	8-5	8-5	8-6

Question	11	12	13	14	15	16	17	18
Objective	D	D	D	H	E	E	F	G
Lesson	8-6	8-4	8-6	8-8	8-8	8-7, 8-8	8-3	8-4, 8-7

CHAPTER 8 REVIEW (pp. 527–529)

1. g **3. a.** 20 **b.** $x^2 - 5$ **5.** Yes, $f \circ g(x) = -\frac{2}{7} \cdot \left(-\frac{7}{2}x\right) = x$;
$g \circ f(x) = -\frac{7}{2} \cdot \left(-\frac{2}{7}x\right) = x$ **7.** $f^{-1}(x) = \frac{1}{4}x + \frac{1}{2}$ **9.** f^{-1}:
$x \to \frac{1}{2}x - \frac{7}{2}$ **11.** $g^{-1}(x) = -\sqrt{-x}$ for $x \le 0$ **13.** 5 **15.** $\frac{4}{25}$
17. 1.41 **19.** −4.31 **21.** a^3 **23.** $2xy\sqrt[6]{2x^2y}$ **25.** $-b^2c^6\sqrt[5]{b^4}$
27. $\sqrt[8]{h}$ **29.** $\sqrt[7]{7}$ **31.** $\frac{3(\sqrt{5}+1)}{4}$ **33.** $a = 3.4$ **35.** $y = -3127$
37. no real solutions **39.** True **41.** True **43.** (k, a) **45.** when n
is an odd integer >2 **47.** When a is negative, the right side of
the equation is not defined but the left side is. **49.** $\sqrt[6]{(-1)^6} =$
$\sqrt[6]{1} = 1 \ne -1$ **51.** True **53.** The larger balloon's diameter is
about 2.15 times as long as the smaller balloon's diameter.
55. $\approx$ 180 feet **57. See right.** **59. a.** (a) **b.** Sample: Let the
domain be the set of all nonnegative numbers. **61. a. See right.**
b. the set of all real numbers. **63. a. See right. b.** domain =
the set of all real numbers; range = the set of all real numbers

57.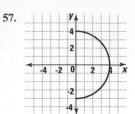

61. a.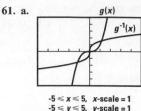

$-5 \le x \le 5$, x-scale = 1
$-5 \le y \le 5$, y-scale = 1

63. a.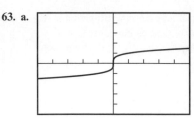

$-5 \le x \le 5$, x-scale = 1
$-5 \le y \le 5$, y-scale = 1

LESSON 9-1 (pp. 532–538)

17. a. 600 bacteria/hour **b.** 1200 bacteria/hour
c. 2400 bacteria/hour **d.** Sample: The rate of change between
$x = n$ and $x = n + 1$ is $600 \cdot 2^{n-1}$. **19.** $\approx$ 2.5%

21. a. $f^{-1}(x) = \frac{x+9}{4}$ **b. See below. c.** True **23. a.** $\begin{bmatrix} 1 & 1 & 5 \\ -3 & -2 & -7 \end{bmatrix}$

b. See below. 25. a. $12,750 **b.** $12,000 **c.** $15,000 \cdot \left(1 - \frac{r}{100}\right)$

21. b.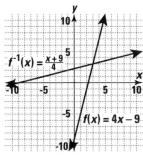

23. b.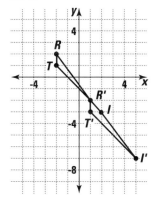

LESSON 9-2 (pp. 539–545)

13. a.

t	1	2	3	4
N	8000	6400	5120	4096

b.

t	1	2	3	4
N	9000	8000	7000	6000

c. After 4 years under the linear model, the car would have a value
of $6000, while under the exponential model it would only have a
value of $4096. **d.** Sample: An extension of the above table shows
that after 9 years, the car would have greater value under the
exponential model than the linear model.

t	8	9	10
N (linear model)	2,000	1,000	0
N (exponential model)	1,678	1,342	1,074

15. 2 **17.** domain: the set of all real numbers **19.** $\approx$ 27,253,000
21. a. 6.17% **b.** 5.54% $\le r \le$ 6.80% **23. a.** $10^{2.4}$ **b.** $\approx$ 251.19

LESSON 9-3 (pp. 547–551)

11. $<$ **13. a.** 25% **b.** $25,404.00
15. a.

t(half-life)	1	2	3	4	5
A(%)	50	25	12.5	6.25	3.125

b. $A = 100(0.5)^x$ **c.** $A = 100(0.5)^{\frac{t}{1600}}$ **17.** $2 + \sqrt{3}$
19. a. 56, 28, 14, 7, $\frac{7}{2}$, $\frac{7}{4}$ **b.** geometric sequence **21.** (c)

LESSON 9-4 (pp. 552–556)

7. a. A year in the middle of the range, 1890, was chosen as a
starting point, which led to an equation model of the form
$y = ab^{x-1890}$. Substituting (1890, 63) into the equation gives a value
of 63 for a. The value for b, the annual growth factor, was
calculated by taking the tenth root of the decade growth factor,
which is $\frac{63}{50} = 1.26$. So $b = \sqrt[10]{1.26} = 1.02$, and thus
$y = 63(1.02)^{x-1890}$. **b.** After 1910, the population increased more
slowly than before 1910, so the annual growth factor of about 2%
would no longer apply. **9. a. See below. b.** domain: set of all real
numbers; range: set of all positive numbers **11. a.** $947.43
b. $948.83 **c.** $948.84 **13.** 81 **15.** $-3a^2 + 3b^2$

9. a.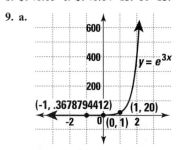

LESSON 9-5 (pp. 557–562)

21. .0012 is between .01 and .001
which can be rewritten as 10^{-2}
and 10^{-3}; so log .0012 is between
−3 and −2. **23.** x; because f and g
are inverses of each other.
25. a. Both the domain and the
range are the set of all real numbers.
b. See right. c. Yes, because it
passes the Vertical-Line Test.
d. $y = x^{\frac{1}{3}}$ or the cube root function
27. 10^4 **29.** $\approx$ 2.697 watts

25. b.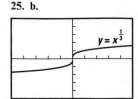

$-5 \le x \le 5$, x-scale = 1
$-5 \le y \le 5$, y-scale = 5

LESSON 9-6 (pp. 563–569)

15. 10 **17.** about 2.5 **19.** $\log_{10}100000$ means 10 to what power equals 100,000. Since $100,000 = 10^5$, $\log_{10}100000 = \log_{10}10^5 = 5$. **21. a.** $x = 100,000$ **b.** $x \approx .69897$ **23.** Sample: $y = .5^x$ **See below. 25.** domain: the set of positive real numbers; range: the set of all real numbers **27.** qr^4

23.

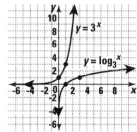

LESSON 9-7 (pp. 570–575)

19. a. See below. b. True **c.** Yes; $x = 0$ **21. a.** $\log_5 125 = 3$; $\log_{125} 5 = \frac{1}{3}$ **b.** $\log_4 16 = 2$; $\log_{16} 4 = \frac{1}{2}$ **c.** $\log_a b = \frac{1}{\log_b a}$ **23.** $m_1 = -6.83$ **25.** 5 **27. a.** $y = (0.5)^{\frac{x}{3.82}}$, where y = the amount of ^{222}Rn, x is the number of days, and we assume the initial amount of ^{222}Rn is 1 unit. **b. See below. c.** ≈ 12.69 days **29. a.** x^{12} **b.** x^{-8} **c.** x^{20} **d.** $\left| x^5 \right|$

19. a.

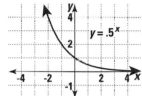

27. b.

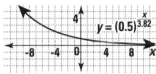

LESSON 9-8 (pp. 576–582)

19. a. $pH = 6.1 + \log B - \log C$ **b.** $\log C = 6.1 + \log B - pH$; $C \approx 1.906$ **21. a.** $\log \left(\frac{40,000,000,000,000}{149,600,000} \right) = \log 267379.6791 \approx 5$ **b.** $\log (40,000,000,000,000) - \log (149,600,000) \approx 13.602 - 8.175 \approx 5$ **23. a.** 1 **b.** $\frac{1}{2}$ **c.** $\frac{1}{6}$ **d.** 0 **e.** -1 **f.** $\frac{-1}{2}$ **25. a.** $y = 0.1$ **b.** does not exist **27. a. See above right. b.** Sample: because the graph looks a little like an exponential curve. **c. See below.**

d. Sample: $1980 \le x \le 1990$ **e.** Sample: $y = 6.12(1.096)^x$, because after 1990, the rate grows faster. If using $y = 5.48(1.096)^x$, the rate produced in 2000 will be near 31.47, which is only slightly higher than the rate in 1992 and not consistent with an exponential graph. **29.** (b)

27. a.

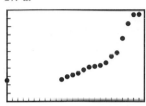

$1970 \le x \le 1995$, x-scale = 1
$0 \le y \le 33$, y-scale = 3

27. c.

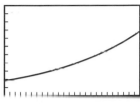

$1970 \le x \le 1995$, x-scale = 1
$0 \le y \le 33$, y-scale = 3

LESSON 9-9 (pp. 583–587)

15. Sample: $\approx (0.5, -0.69)$; by tracing along the graph of $y = \ln x$ until the x-coordinate equals .5. **17.** 1 **19.** $p \approx 63.6\%$ **21. a.** Yes **b.** (1, 0) **23.** False; $\log(1.7 \times 10^3) = \log(1.7) + \log(10^3)$ **25.** $pH = 5$ **27.** $50e^{.02(10)} \approx 61.07$ million people. This projection fell short by about 6 million people. **29.** 50,000 cm^3

LESSON 9-10 (pp. 588–592)

11. about 42 minutes **13. a.** 2006 **b.** about 130 million **c.** Mexico **15.** $y \approx 2.36$ **17.** $V \approx 2311.8$ m/sec, or ≈ 2.312 km/sec **19.** $x = 6$; Check: Does $1.80618 = 6(.30103)$? Yes. **21.** $\log p + 2\log q$ **23.** -1 **25. See below.** The two graphs are reflection images of each other across the y-axis.

25.

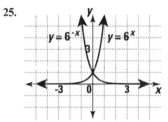

CHAPTER 9 PROGRESS SELF-TEST (pp. 597–598)

1. $43500 \times 1.17^3 \approx \$69,700$ **2.** $5 \times 2^{\frac{24}{0.5}} = 5 \times 2^{48} \approx 1.41 \times 10^{15}$ bacteria **3. a. See page 918. b.** $4^\pi \approx 77.9$ **c.** Yes; $y = 0$ **4. a.** $\ln(42.7) \approx 3.75$ **b.** Does $e^{3.75} \approx 42.7$? $42.52 \approx 42.7$. Yes, it checks. **5.** $\log(1,000,000) = \log(10)^6 = 6 \log 10 = 6$ by the Power Property of Logarithms **6.** $\log_4\left(\frac{1}{16}\right) = \log_4\left(\frac{1}{4^2}\right) = \log_4(4)^{-2} = -2$ by the $\log_b$ of b^n Theorem **7.** $\ln(e)^{-6} = -6$ by the $\log_b$ of b^n Theorem **8.** By the Log of 1 Theorem, $\log_2 1 = 0$. **9. a.** $40 \cdot \left(\frac{1}{2}\right)^3 = 5$ mg **b.** Let A = the amount of ^{14}C left. $A = 40 \cdot \left(\frac{1}{2}\right)^t$ **10.** $8 - x^{\frac{3}{4}}$; $(8)^{\frac{4}{3}} = \left(x^{\frac{3}{4}}\right)^{\frac{4}{3}}$; $x = 16$ **11.** $y \log 6 = \log 32$; $y = \frac{\log(32)}{\log(16)} \approx 1.93$ **12.** True; $\log\left(\frac{M}{N^2}\right) = \log M - \log(N^2)$ by the Quotient Property of Logarithms. $\log M - \log(N^2) = \log M - 2 \log N$ by the Power Property of Logarithms. **13.** False; By the Product Property of Logarithms, $\log_3 (x \cdot y) = \log_3 x + \log_3 y$. **14.** $3000 = 1500e^{0.04t}$; $2 = e^{0.04t}$; $\ln 2 = 0.04t$; $t = \frac{\ln 2}{0.04} \approx \frac{.69315}{0.04} \approx 17.329$. After about 17 years, the investment will be valued at $3000. **15.** $-26 + 13 = -2.5 \cdot \log\left(\frac{I_1}{I_2}\right)$; $-13 = -2.5 \cdot \log\left(\frac{I_1}{I_2}\right)$; $5.2 = \log\left(\frac{I_1}{I_2}\right)$; $\frac{I_1}{I_2} = 10^{5.2} \approx 158500$; so the sun is about 160,000 times as bright as the moon. **16.** $125 - 95 = 30$dB. So the 125 dB sound is $10^{\frac{30}{10}} = 10^3 = 1000$ times more intense. **17.** $12000 \cdot (0.92)^8 \approx 6159$ **18.** (b); Since the growth factor is between 0 and 1, the equation represents exponential decay. **19. a.** Use $p = p_0 a^t$; use the data for $t = 7$ and $t = 17$ to find the decade growth factor: $\frac{30.5}{26.4} \approx 1.14015$. So $a^{10} = 1.14015$, and $a \approx 1.0132$, the annual growth factor. The starting point is 1993, $p_0 = 22.9$ million. Therefore, $p = 22.9 \cdot (1.01)^t$ million. **b.** Using the equation from part **a**, $p = 22.9 \cdot 1.01^{32} \approx 31.5$ million, or $\approx 31,500,000$. **c.** Sample: The projected birthrate may decline, or mortality may increase. Also, migration may increase or immigration may decrease. **20. a.** Sample: (1, 0), (3, 1), (9, 2) **b.** domain: set of positive real numbers; range: set of all real numbers **c., e. See page 918. d.** $x = \log_3 y$; $y = 3^x$

3. a.

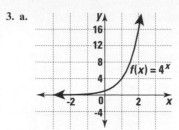

20. c, e.

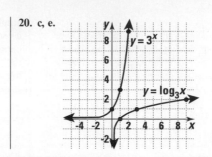

The chart below keys the **Progress Self-Test** questions to the objectives in the **Chapter Review** on pages 599–601 or to the **Vocabulary** (Voc.) on page 596. This will enable you to locate those **Chapter Review** questions that correspond to questions missed on the **Progress Self-Test**. The lesson where the material is covered is also indicated on the chart.

Question	1	2	3	4	5	6	7	8	9	10
Objective	F	F	I, D	A, E	A	A	A	A	F	C
Lesson	9-1	9-1	9-1	9-9	9-5, 9-8	9-7, 9-8	9-9, 9-8	9-7, 9-8	9-2	9-7
Question	11	12	13	14	15	16	17	18	19	20
Objective	B	E	E	F	H	H	F	I	G	D, J, E, I
Lesson	9-10	9-8	9-8	9-3, 9-10	9-6	9-6	9-2	9-2	9-4	9-7

CHAPTER 9 REVIEW (pp. 599–601)

1. 3 **3.** 9 **5.** 15 **7.** −3 **9.** 4.99 **11.** 4.47 **13.** undefined
15. $x = 3$ **17.** $n \approx 14.21$ **19.** $z \approx 3.09$ **21.** $a \approx 1.78$ **23.** $x = 11$
25. $z = 10{,}000$ **27.** $x = 225$ **29.** $x = 4$ **31.** domain = the set of all real numbers; range = $\{y: y > 0\}$ **33.** $a > 1$ **35.** x-axis
37. $f^{-1}(x) = \ln x$ **39.** True **41.** $6^{-3} = \frac{1}{216}$ **43.** $a = 10^b$
45. $\log 0.0631 \approx -1.2$ **47.** $\log_x z = y$ **49.** Product Property of Logarithms **51.** Power Property of Logarithms **53.** $\log_b b^n = n$
55. $\log \frac{x}{y}$ **57.** ≈ 29.681 million **59.** about 2001 **61.** ≈ 3.10 hours
63. a. 0.3125 g **b.** $5 \cdot (0.5)^{\frac{t}{29}}$g **65.** $y = 9.99 \cdot (1.45)^x$ **67. a. See right. b.** about 3 days **c.** $y = 1001(.80)^x$ **d.** about 11.54 g
69. $I = 10^{-3}$ w/m^2 **71.** $\approx 71{,}621$ feet **73. See right.**
75. a. $y = 3^x$ **b.** $y = 2^x$ **c.** Because $2 < e < 3$, the graph of $y = e^x$ will lie between the graph $y = 2^x$ and $y = 3^x$. **77.** Sample: $\left(\frac{1}{5}, -1\right)$, (1, 0), (5, 1), (11.2, 1.5), (25, 2) **See right. 79. a.** Sample: (1, 0), $(e, 1)$, $(e^2, 2)$, $\left(\frac{1}{e}, -1\right)$, $\left(\frac{1}{e^2}, -2\right)$ **See right. b.** $y = e^x$
81. x-intercept is 1.

67. a.

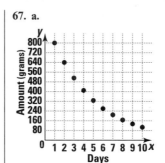

73.

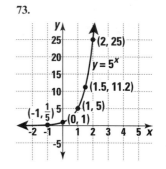

77.

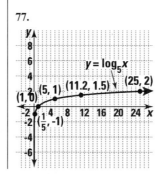

79. a.

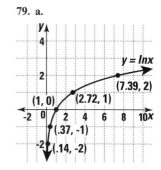

LESSON 10-1 (pp. 604–610)

15. a. See right. b. ≈ 6.2 feet **17.** ≈ 7.4 cm
19. $p = 31$ **21. a.** 5 **b.** x **23.** True
25. a. 75° **b.** 15°

15. a.

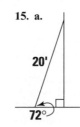

20'

72°

LESSON 10-2 (pp. pp. 611–615)

13. $\approx .5°$ **15.** $\approx 15°$ **17.** ≈ 401 ft **19.** $R_{270}(S) = (.2, 0)$; $R_{270}(K) = (1, -.6)$; $R_{270}(Y) = (.2, -.6)$ **21.** $\approx 37°$ **23.** $x\sqrt{2}$

LESSON 10-3 (pp. 616–621)

11. $AC = \sqrt{h^2 - \frac{1}{4}h^2} = \frac{\sqrt{3}}{2}h$ $\sin 60° = \frac{AC}{h} = \frac{\frac{\sqrt{3}}{2}h}{h} = \frac{\sqrt{3}}{2}$

$\cos 30° = \frac{AC}{h} = \frac{\frac{\sqrt{3}}{2}h}{h} = \frac{\sqrt{3}}{2}$ **13. a.** $IT = \frac{10}{\sqrt{3}}$ or $\frac{10\sqrt{3}}{3}$

b. $PT = \frac{20}{\sqrt{3}}$ or $\frac{20\sqrt{3}}{3}$ **15.** $\frac{1}{3}$ **17.** $\approx 9.5°$ **19.** See below.

a. $\begin{cases} x \approx -1.6 \\ y = 5 \end{cases}$ or $\begin{cases} x \approx 3.6 \\ y = 5 \end{cases}$ **b.** Sample: because $y = 5$,

$5 = x^2 - 2x - 1$; $x^2 - 2x - 6 = 0$; $x = \frac{2 \pm \sqrt{4 + 24}}{2} = 1 \pm \sqrt{7}$

$x \approx -1.646$ or $x \approx 3.646$ It checks. **21. a.** 65,000
b. $59,000 + 6000n$ **c.** 6 times

19.

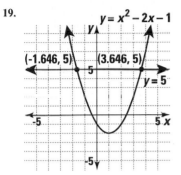

LESSON 10-4 (pp. 623–627)

17. c **19.** c **21.** a **23. a.** 1 **b.** -1 **25.** $\sin(-90°) = -1$ and $\cos(90° - -90°) = \cos(180°) = -1$ **27.** $\theta = 76°$ **29. a.** $\frac{\sqrt{3}}{2}$ **b.** $\frac{1}{2}$
c. $\frac{\sqrt{3}}{3}$ **31.** $\approx 4.8°$ **33.** ≈ 14.56

LESSON 10-5 (pp. 628–632)

13. negative; positive **15.** $\approx 104.5°$ **17.** negative; When $270° < \theta < 360°$, $\sin \theta$ is negative and $\cos \theta$ is positive, so $\tan \theta = \frac{\sin \theta}{\cos \theta}$ is negative. **19.** $-\sqrt{3}$ **21.** First find the magnitude between $0°$ and $360°$ corresponding to the given magnitude. $-630° + 2 \cdot 360° = 90°$, so $R_{-630°} = R_{90°}$; $R_{630}(1, 0) = R_{90}(1, 0) = (0, 1)$, so $\sin(-630°) = 1$. **23. a.** about 10.36% **b.** about 9.85% **25.** $x = 7$ or 1 **27.** $\sqrt{(a + 6)^2 + (b - 5)^2}$

LESSON 10-6 (pp. 633–638)

9. about $1.59p$ **11. a.** ≈ 6.4 light-years **b.** No, Sirius would be brighter in the sky as seen from Alpha Centauri because the distance between them is about 6.4 light-years, less than the distance between Sirius and the Earth, which is about 8.8 light-years. **13. a.** ≈ 49 mm **b.** $\approx 28°$ **15. a.** $c = 25$ **b.** $\triangle ABC$ is a right triangle, so by the Pythagorean Theorem, $7^2 + 24^2 = c^2$ and so $c = 25$ **17.** $\frac{\sqrt{3}}{2}$ **19.** $\frac{1}{2}$ **21.** (a) **23.** ≈ 6.9 years
25. a. $p = \frac{qx}{y}$ **b.** $x = \frac{py}{q}$

LESSON 10-7 (pp. 639–644)

9. a. ≈ 42 mm **b.** $\approx 55°$ **c.** $\approx 55°$ **d.** True; in an SAS situation, once you find the third side, you have enough information to use either the Law of Cosines or the Law of Sines. **11.** $\approx 2.25 \times 10^8$ m/sec
13. a. m$\angle ABD = 142°$; m$\angle ADB = 13°$ **b.** ≈ 282 m **c.** ≈ 174 m
15. a. $\cos \theta = \frac{3}{5}$ or $\cos \theta = -\frac{3}{5}$ **b.** See below. **17.** $-\frac{\sqrt{2}}{2}$
19. a. $h = -16t^2 + 30t + 12$ **b.** See below. **c.** ≈ 26 feet **d.** ≈ 2.2 seconds after being thrown **21. a.** $y = 0.8x$ **b.** direct variation

15. b.   **19. b.**

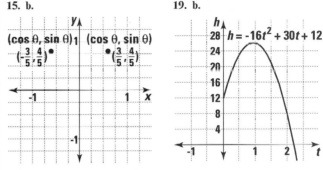

LESSON 10-8 (pp. 647–651)

13. (a) **15.** No **17. a.** Yes **b.** 4 **19.** m$\angle J \approx 70.2°$ **21.** $\cos \theta$
23. a. ≈ 1.806 **b.** ≈ 4.159 **c.** 6.000

LESSON 10-9 (pp. 653–657)

7. m$\angle L \approx 37.8°$, m$\angle K \approx 12.2°$, $k \approx 5.5$ **9.** Law of Sines
11. Law of Sines **13.** By the Law of Sines, $\sin F \approx 3.23$, which is impossible. **15.** $\frac{\sin B}{AC} = \frac{\sin C}{AB} = \frac{\sin F}{DE} = \frac{\sin E}{DF}$; Since $AC = DF$, then $\sin B = \sin E$. Thus m$\angle B =$ m$\angle E$ or m$\angle B = 180 -$ m$\angle E$. Since m$\angle E$ is smaller than m$\angle F$ and $\angle B$ is smaller than $\angle C$, both $\angle B$ and $\angle E$ are acute. Thus m$\angle B =$ m$\angle E$. Thus, $\triangle ABC = \triangle DEF$ by AAS. **17.** Yes; sample: $x = 0$. **19.** $\approx 11,450$ ft. **21.** $\frac{1}{2}$
23. 2π units **25.** 105°

LESSON 10-10 (pp. 658–663)

15. a. 60° **b.** $\frac{\pi}{3}$ **17.** $-\frac{1}{2}$ **19. a.** 5π units **b.** 5π units **21.** 4π feet
23. $\sin C = \frac{24 \cdot \sin 16°}{10} \approx 0.662$; when $\sin \theta$ is positive, there are two possible values of θ in the domain $0° < \theta < 180°$. So m$\angle C = \approx 41.4°$ or 138.6°. **25.** 1 **27.** The right triangle is constructed with legs of length 2 units and 1 unit. If the angle needed is θ, then $\tan \theta = \frac{1}{2}$, so $\theta \approx 26.5°$.

CHAPTER 10 PROGRESS SELF-TEST (p. 668)

1. $\frac{24}{25} = 0.960$ **2.** $\frac{7}{25} = 0.280$ **3.** $\tan \theta = \frac{1}{2}$, so $\theta \approx 26.6°$.
4. a; $(\cos 70°, \sin 70°) = (a, b)$ **5.** d; $(\cos 160°, \sin 160°) = (c, d)$
6. 123°; $\sin \theta° = \sin(180 - \theta°)$ for all θ. **7.** $(\cos 210°, \sin 210°) = R_{180}(\cos 30°, \sin 30°) = (-\cos 30°, -\sin 30°)$. So $\sin 210° = -\sin 30° = -\frac{1}{2}$ **8.** Use the Law of Cosines. $c^2 = 85^2 + 110^2 - 2(85)(110)\cos 40°$; $c^2 \approx 7225 + 12100 - 18700(.7660)$; $c^2 \approx 5000.8$; $c \approx 70.7$. The runners are about 71 m apart.
9. Use the Law of Cosines. $8^2 = 5^2 + 11^2 - 2 \cdot 5 \cdot 11 \cos x$; $64 = 25 + 121 - 110 \cos x$; $-82 = -110 \cos x$; $.7455 \approx \cos x$;

$x \approx 42°$. **10.** Use the Law of Sines. The angle opposite x is $(180 - 83 - 40)°$, or 57°. $\frac{\sin 83°}{2.7} = \frac{\sin 57°}{x}$; $\frac{.9925}{2.7} = \frac{.8387}{x}$; $x = \frac{2.7(.8387)}{.9925}$; $x \approx 2.3$ **11.** Use the Law of Sines: $\frac{\sin 31°}{421} = \frac{\sin x}{525}$; $\sin x = \frac{525(.5150)}{421}$; $\sin x \approx .6422$; $x \approx 40°$ or 140°. m$\angle S \approx 40°$ or 140°. **12.** Height of the cliff is 8 feet (for the sign) + 2 feet (for the eagle) + x. $\sin 70° = \frac{x}{130}$; $x \approx 122$ feet. So the height of the nest is ≈ 132 feet above the ground. **13.** (d); Sample: Let $\theta = 60°$. Is $\cos(60° + 180°) = \cos 60°$? $-0.5 \neq 0.5$
14. $24° = 24° \cdot \frac{\pi \text{ radians}}{180°} = \frac{2}{15}\pi$ radians, or $\approx .42$ radians

15. $\frac{5\pi}{6}$ radians $\left(\frac{180°}{\pi \text{ radians}}\right) = 150°.$ (cos 150°, sin 150°) $=$
$r_y(\cos 30°, \sin 30°) = r_y\left(\frac{\sqrt{3}}{2}, \frac{1}{2}\right) = \left(-\frac{\sqrt{3}}{2}, \frac{1}{2}\right).$ So cos 150° $=$
$-\frac{\sqrt{3}}{2}.$ **16.** 360° **17.** 1; 0 **18. See right. 19.** $-1 \le y \le 1$
20. Using a calculator, one solution is found to be $\theta \approx 58°$; using
$\sin \theta = \sin (180° - \theta)$, the second solution is about 122°.

18.

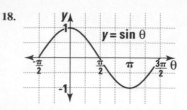

The chart below keys the **Progress Self-Test** questions to the objectives in the **Chapter Review** on pages 669–671 or to the **Vocabulary** (Voc.) on page 667. This will enable you to locate those **Chapter Review** questions that correspond to questions missed on the **Progress Self-Test.** The lesson where the material is covered is also indicated on the chart.

Question	1	2	3	4	5	6	7	8	9	10
Objective	A	A	F	I	I	E	B	G	H	H
Lesson	10-1	10-1	10-2	10-4	10-5	10-9	10-5	10-6	10-6	10-7

Question	11	12	13	14	15	16	17	18	19	20
Objective	H	F	E	D	B, D	J	J	J	J	C
Lesson	10-9	10-2	10-3	10-10	10-3,10-10	10-8	10-8	10-8	10-8	10-9

CHAPTER 10 REVIEW (pp. 669–671)

1. 0.292 **3.** -0.766 **5.** -0.500 **7.** 0.976 **9.** 4.444 **11.** $\frac{\sqrt{2}}{2}$
13. $\frac{\sqrt{3}}{2}$ **15.** $\frac{\sqrt{2}}{2}$ **17.** 45° or 135° **19.** $\approx 120°$ **21.** $\approx .730$ rad
23. -.6 **25.** $\frac{7\pi}{12}$ **27.** 3π **29.** 405° **31.** -22.5° **33.** True **35.** False;
$\tan \theta = \frac{\sin \theta}{\cos \theta}$ **37.** 41° **39.** ≈ 671 km **41.** $\approx 9.5°$ **43.** *B;* about
10.8 miles **45.** about 12.4′ long **47.** ≈ 8.9 **49.** ≈ 19.9
51. $m\angle E \approx 46.1°$, $m\angle T = 21.9°$, $t = 3.6$ **53.** .191 **55.** *b* **57.** *c*
59. a. See right. b. 2π **c.** $\left(\frac{-3\pi}{2}, 0\right), \left(\frac{-\pi}{2}, 0\right), \left(\frac{\pi}{2}, 0\right)$ and $\left(\frac{3\pi}{2}, 0\right)$
61. Sample: $T_{-90°,0}$

59. a.

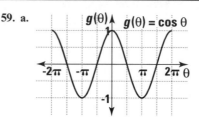

LESSON 11-1 (pp. 674–679)

15. 2 **17. a.** *g* **b.** *f* **c.** Sample: A polynomial function does not have a variable in the exponent. An exponential function does.
19. $x = 3, y = -1, z = 1$ **21.** 44,550 people per year **23.** $2x^2 + 2x$

LESSON 11-2 (pp. 680–685)

15. $11,250 - 300b - 75a - 75c + 2ab + 2bc$ **17.** $a^2 - b^2 - c^2 + 2bc$ **19. a.** 5 **b.** 4000 **c.** 4000 **d.** 1905 to 1925 **e.** 1925
21. \$19,800

LESSON 11-3 (pp. 686–691)

21. (d); $4q^2 + r^2 - 4qr = 4q^2 - 4qr + r^2 = (2q - r)^2$
23. a. $4x(x^2 - 22x + 120) = 4x(x - 12)(x - 10)$ **b.** Sample:
$4x(x - 12)(x - 10) = 4x(x^2 - 22x + 120) = 4x^3 - 88x^2 + 480x$
25. (d) **27.** $V(h) = h^3 + 7h^2 + 10h$
29. a.

x	$f(x)$
-5	-90
-4	-32
-3	0
-2	12
-1	10
0	0
1	-12
2	-20
3	-18
4	0
5	40

b., c. See right. 31. -1

29. b., c.

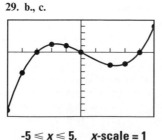

-5 ≤ x ≤ 5, x-scale = 1
-100 ≤ y ≤ 50, y-scale = 10

LESSON 11-4 (pp. 692–697)

9. To solve $P(x) = k$ by graphing, graph the system $y = P(x)$ and $y = k$ on the same set of axes, and identify the x-coordinate of the point(s) of intersection, if any. **11. a.** difference of squares
b. $(7x + 5y)(7x - 5y)$ **13.** $x^2(x + 5)(x - 4)$; Sample check: The graphs of $f(x) = x^4 + x^3 - 20x^2$ and $g(x) = x^2(x + 5)(x - 4)$ are identical. **15.** $a^2 - b^2 + ac - bc$ **17.** Sample: $27d^3 - 8$ **19.** slope of $\overline{MT} = \frac{0 - a}{a - 0} = -1$, slope of $\overline{AH} = \frac{a - 0}{a - 0} = 1$; $(-1)(1) = -1$

LESSON 11-5 (pp. 699–705)

17. $a, b, -c$, and $-d$ **19.** $(\approx.61, 2), (\approx-.61, 2)$ **21.** $(a + 7b)^2$
23. $20x$ **25.** $\frac{25}{4}$ **27. a.** $V(x) = 4x^3 - 42x^2 + 108x$ in³
b. $S(x) = -4x^2 + 108$ in²

LESSON 11-6 (pp. 706–710)

15. a. See page 921. The zeros of f are $x = -3$ or 3.
b. $(x - 3)(x - 3)(x + 3)(x + 3) = (x - 3)^2(x + 3)^2$ **17.** $-3, \frac{9}{2}, 7$
19. a. Sample: $P(x) = (x + 3)(x - 5)(x - 8)$; $Q(x) = x(x + 3)(x - 5)(x - 8)$ **b. See page 921. c.** $R(x) = k(x + 3)(x - 5)(x - 8)$, where k is a constant or polynomial equation in x. **21.** $x = 9.3$ m **23. a.** geometric sequence **b.** $a_n = 2^n$
$\begin{cases} a_1 = 2 \\ a_n = 2 \cdot a_{n-1}, \text{ for integers } n \ge 2 \end{cases}$

15. a.

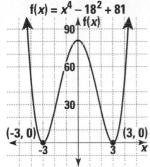

$f(x) = x^4 - 18^2 + 81$

19. b.

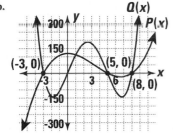

LESSON 11-7 (pp. 711–716)

9. a. $a_0 = 0$ so there would be only 1 rational root: 0
b. Sample: $x^3 - 4x = x(x^2 - 4) = x(x + 2)(x - 2)$; 0, 2, or −2 are rational zeros. **11.** 4 **13.** $(2x - 10)(2x + 10)$ or $4(x + 5)(x - 5)$ **15.** Sample: $V(x) = (x + 3)(x - 4)(x + 2)$
17. a. 2.5% **b.** $0 **c.** $768.80 **19.** 29

LESSON 11-8 (pp. 717–722)

15. $x = \frac{7}{3}i$ **17.** $(z^2 + 1)(z + 1)(z - 1) = 0$; $z = i, -i, 1, -1$
19. $t = 0$, $t = 7$, and $t = -1$ **21.** $\approx .75$ **23.** $x = 100,000$
25. a. See below. **b.** F varies inversely with r. **c.** $F = \frac{k \cdot w \cdot S^2}{r}$

25. a.

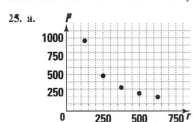

LESSON 11-9 (pp. 725–729)

13. a. 6, 6, 6, 6 **b.** 1 **c.** See below. **d.** $y = 6x + 5$ **e.** The first differences are equal to the slope of the line that models the data; Yes. **15. a.** $f(5) = 55$; $f(6) = 91$ **b.** 3 **17.** 3, since the degree is 3 **19.** −2 and 0 **21.** No, $b^2 - 4ac > 0$. **23.** (c) **25.** $(x, y, z) = \left(28, \frac{1}{2}, 8\right)$

13. c.

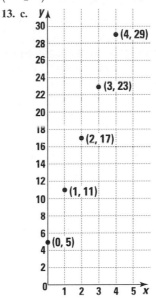

LESSON 11-10 (pp. 730–736)

13. a. $t_n = f(n) = \frac{1}{2}n^2 + \frac{1}{2}n$ **b.** Yes. $\frac{n(n + 1)}{2} = \frac{n^2}{2} + \frac{n}{2}$ **c.** Samples: The 8th triangular number is 36. It takes 36 dots to make an equilateral triangle pattern with side 8. **15. a.** Yes, the second differences are equal. **b.** 2 **17.** $x^2(5x - 9)(2x + 5)$
19. $3x^2 + 15x + 19$ **21.** $329.59 **23.** $33\frac{1}{3}$%

CHAPTER 11 PROGRESS SELF-TEST (p. 740)

1. The first money saved earns interest for 5 years. The total is $750x^5 + 600x^4 + 925x^3 + 1075x^2 + 800x$. **2.** If $x = 1.04$, then $750x^5 + 600x^4 + 925x^3 + 1075x^2 + 800x = 750(1.04)^5 + 600(1.04)^4 + 925(1.04)^3 + 1075(1.04)^2 + 800(1.04) \approx 912.49 + 701.92 + 1040.50 + 1162.72 + 832 = 4649.63$.
3. $V(x) = x(60 - 2x)(40 - 2x) = x(2400 - 120x - 80x + 4x^2) = x(2400 - 200x + 4x^2) = 4x^3 - 200x^2 + 2400x$
4. $(a^2 + 3a - 7)(5a + 2) = a^2(5a + 2) + (3a)(5a + 2) - 7(5a + 2) = 5a^3 + 2a^2 + 15a^2 + 6a - 35a - 14 = 5a^3 + 17a^2 - 29a - 14$ **5.** $10s^7t^2 + 15s^3t^4 = 5s^3t^2(2s^4 + 3t^2)$ **6.** $25y^2 + 60y + 36 = (5y + 6)(5y + 6) = (5y + 6)^2$ **7.** $z^3 - 216z = z(z^2 - 216) = z(z - \sqrt{216})(z + \sqrt{216}) = 0$. So the zeros are $z = 0$ and $z = \pm\sqrt{216} = \pm6\sqrt{6}$. **8. a.** The degree is 5. **b.** none of these

9. $P(x)$ can be rewritten in the form $a_nx^n + a_{n-1}x^{n-1} + \ldots + a_2x^2 + a_1x + a_0$, where $n = 5$, $a_n = -8$ and $a_0 = -3$. The factors of −3 are ±3 and ±1. By the Rational Zero Theorem, if $\frac{p}{q}$ is a rational zero of P, then p is a factor of a_0 and q is a factor of a_n. So the possible zeros of P are ±1, $\pm\frac{1}{2}$, $\pm\frac{1}{4}$, $\pm\frac{1}{8}$, ±3, $\pm\frac{3}{2}$, $\pm\frac{3}{4}$, and $\pm\frac{3}{8}$.
10. See page 922. **11.** If $h(x) = 4x^3(5x - 11)(x + \sqrt{7}) = 0$, then $x^3 = 0$, $5x - 11 = 0$, or $x + \sqrt{7} = 0$. So $x = 0$, $x = \frac{11}{5}$, or $x = -\sqrt{7}$. **12. a.** If $f(x) = 3x^4 - 12x^3 + 9x^2 = 0$, then $3x^2(x^2 - 4x + 3) = 0$ and $3x^2(x - 3)(x - 1) = 0$.
b. There are 3 zeros: 0, 3, or 1 **13.** Since $y = f(x)$ has degree 3, it has at most 3 real zeros. **14. a.** The table shows one zero at $x = 3$. The other zeros are between $x = -2$ and $x = -1$, and between $x = 1$ and $x = 2$, because those are the intervals in which the values of the polynomial change signs. **b.** By using an automatic grapher, the

921

noninteger roots can be estimated: $x \approx 1.7$, $x \approx -1.7$. Therefore, $x \approx -1.7$ is the smallest. **15.** (c). Never; a polynomial of degree 11 has 11 complex roots. **16.** If r is a root or zero of a function, then $x - r$ is a factor. Since we know that $f(2) = 0$, then 2 is a root and $(x - 2)$ is a factor of the polynomial. That is option (d). **17.** Since the zeros are -2, 1, 3, and 5, the factors for the function are $x - (-2)$, $x - 1$, $x - 3$, and $x - 5$. Thus the function is $P(x) = k(x + 2)(x - 1)(x - 3)(x - 5) = k(x^2 + x - 2)(x^2 - 8x + 15) = k(x^2(x^2 - 8x + 15) + x(x^2 - 8x + 15) - 2(x^2 - 8x + 15)) = k(x^4 - 8x^3 + 15x^2 + x^3 - x^2 + 15x - 2x^2 + 16x - 30) = k(x^4 - 7x^3 + 5x^2 + 31x - 30)$, where k is any nonzero constant or polynomial. **18. a.** Since the second differences are equal, the data points can be modeled with a polynomial function. **b.** The degree of that function is 2.

n	1	2	3	4	5	6	7	8
t	2	5	9	14	20	27	35	44
1st diff		3	4	5	6	7	8	9
2nd diff			1	1	1	1	1	1

19. Since the second differences are equal, the general equation is $z = f(x) = ax^2 + bx + c$.

x	-2	-1	0	1	2	3	4
z	12	4	0	0	4	12	24
1st diff		-8	-4	0	4	8	12
2nd diff			4	4	4	4	4

Using these data, $f(1) = a(1)^2 + b(1) + c = 0$, $f(2) = a(2)^2 + b(2) + c = 4$, $f(3) = a(3)^2 + b(3) + c = 12$. Then
$$\left. \begin{array}{r} 9a + 3b + c = 12 \\ 4a + 2b + c = 4 \\ a + b + c = 0 \end{array} \right\} \left. \begin{array}{r} 5a + b = 8 \\ 3a + b = 4 \end{array} \right\} 2a = 4;$$

Thus $a = 2$. From $3a + b = 4$, $3(2) + b = 4$, $6 + b = 4$, $b = -2$; and from $a + b + c = 0$, $2 + (-2) + c = 0$, $c = 0$. The polynomial function is $z = f(x) = 2x^2 - 2x$. **20.** Every polynomial equation $P(x) = 0$ of any degree with complex number coefficients has at least one complex number solution; Karl Gauss first proved it in 1797.

10.

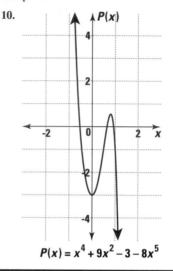

$$P(x) = x^4 + 9x^2 - 3 - 8x^5$$

The chart below keys the **Progress Self-Test** questions to the objectives in the **Chapter Review** on pages 741–745 or to the **Vocabulary** (Voc.) on page 739. This will enable you to locate those **Chapter Review** questions that correspond to questions missed on the **Progress Self-Test**. The lesson where the material is covered is also indicated on the chart.

Question	1	2	3	4	5	6	7	8	9	10
Objective	H	H	I	A	B	B	C	E	G	J
Lesson	11-1	11-1	11-2	11-2	11-3	11-3	11-6	11-2	11-7	11-7
Question	11	12	13	14	15	16	17	18	19	20
Objective	C	B, C	K	K	F	F	K	D	D	L
Lesson	11-5	11-3, 11-5	11-4	11-4	11-8	11-5	11-5	11-9	11-10	11-8

CHAPTER 11 REVIEW (pp. 741–745)

1. $x^3 + 2x - 3$ **3.** $8y^3 + 60y^2 + 150y + 125$ **5.** $6x^3 + 2x^2y - 3xy - y^2$ **7.** $a^3; -9b^2$ **9.** $(x - 7)^2$ **11.** $(r^2s^2 + 9)(rs + 3)(rs - 3)$ **13.** $(x - 7)(x - 2)$ **15.** $(2p + 5)(2p - 3)$ **17.** $(x + 4)^2(x - 4)^2$ **19.** $(z + i\sqrt{27})(z - i\sqrt{27})$ **21.** $x = .5, -\frac{1}{3}, 0$ **23.** $x = 0, -4, -\frac{7}{9}$; no multiple roots **25.** $n = -8$ (double root), 0 **27.** Yes; $y = .5x^2 + 3.5x + 3$ **29.** Yes; $a_n = f(n) = -6n + 11$ **31.** $p(x) = 24x^4 + 278x^3 - 119x^2 + 12x$; $q(x) = 48x^4 + 556x^3 - 238x^2 + 24x$ **33. a.** 5 **b.** 7 **35.** (b) **37.** (d) **39.** Sample: $3x^6 - 4x^4 + 5x^2$ **41.** The product is not equal to zero. **43.** n; multiple **45.** True **47.** (b) **49.** (d) **51. a.** $\pm 1, \pm\frac{1}{3}, \pm 2, \pm\frac{2}{3}$ **b.** $\frac{-1}{3}, 2$ **53.** $(z - 1)(z + 1)(z - 2)(z + 2)(z - 3)(z + 3)$ **55. a.** $150x^7 + 150x^6 + 150x^5 + 150x^4 + 150x^3 + 150x^2 + 150x + 150$ **b.** $\$1484.62$ **c.** $\approx 14\%$ **57. See right.** A reasonable domain would be $0 \leq n \leq 29$. **59. a.** $G = n^2 - n$ **b.** $G(20) = 380$ **61.** $(11 - 2x)(17 - 2x) = 4x^2 - 56x + 187$ **63.** $S(x) = 187 - 4x^2$ **65.** $S(x) = 1.5 - 4x^2$ **67. a. See page 923. b.** $f(x) = (x - 3)(x + 3)(2x - 1)$ **69. a.** Sample: $f(x) = x^3 + 4x^2 - 11x - 30$

b. See page 923. c. $f(x) = k(x^3 + 4x^2 - 11x - 30)$, where k is any nonzero constant **d.** Sample: They all intersect the x-axis at the same points and no others. **71.** $x \approx -.8$ **73.** False **75. a.** 1 **b.** 3 **c.** 3.1 **d.** None of the x-intercepts is rational since if one were rational, it would be a factor of 18. **77.** $-10 < x < -9$, $-4 < x < -3$ **79.** (c)

57.

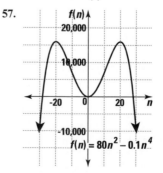

$$f(n) = 80n^2 - 0.1n^4$$

67. a.

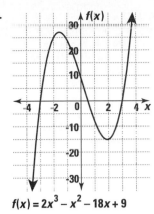

$f(x) = 2x^3 - x^2 - 18x + 9$

69. b.

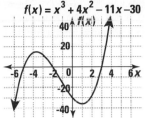

$f(x) = x^3 + 4x^2 - 11x - 30$

LESSON 12-1 (pp. 748–753)

15. Focus is $\left(0, -\frac{1}{4}\right)$; directrix is $y = \frac{1}{4}$. **17. a.** down **b.** $\left(0, -\frac{1}{16}\right)$ **19.** (b) **21. a.** $y = \pm\sqrt{7}$ **b.** $y = -2 \pm \sqrt{7}$ **c.** $y = 49$ **d.** $y = 47$ **23. a.** $\frac{3}{5}$ **b.** 100

LESSON 12-2 (pp. 754–758)

11. No, there are two possible locations where the shopping mall could be located according to this information. **13. a.** $x^2 + y^2 - 300x - 200y + 15,600 = 0$ **b.** $A = 1$; $B = 0$; $C = 1$; $D = -300$; $E = -200$; $F = 15,600$ **15. a.** Yes. All the points on the parabola are ordered pairs. **b.** No. A vertical line drawn through the parabola will intersect the parabola in more than one point. **17.** -4, -1, 1, 4 **19.** ≈ 6% **21.** (c)

LESSON 12-3 (pp. 759–763)

9. See below. **11. a.** See below. **b.** 2 **c.** $x^2 + y^2 = 4$ **d.** 30 and 40 **13. a.** $x^2 + y^2 \le r^2$ **b.** $x^2 + y^2 > r^2$ **15.** A parabola is the set of all points in a plane equidistant from a fixed point (its focus) and a fixed line (its directrix). **17. a.** $r = 5$ **b.** $x^2 + y^2 = 25$

19. $2x^2 + y^2 - 2x\sqrt{x^2 + y^2}$ **21.** See below.

9.

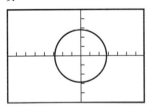

$-8 \le x \le 7$, x-scale = 1
$-5 \le y \le 5$, y-scale = 1

11. a.

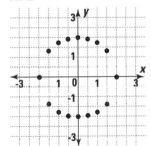

21.

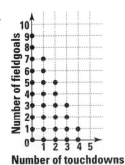

LESSON 12-4 (pp. 765–770)

17. a. $\frac{x^2}{49} + \frac{y^2}{36} = 1$ **b.** $\frac{x^2}{49} + \frac{y^2}{36} < 1$ **19. a.** Answers should look like the drawing. **b.** The sum of the distance from the tacks to the pencil's tip is equal to the length of the string. Since the length is constant, by the definition of ellipse, the curve is an ellipse. **c.** the length of the string. **21. a.** $\frac{x^2}{2256.25} + \frac{y^2}{1521} = 1$ **b.** ≈ 54 ft **c.** ≈ 20 ft **23.** $x^2 + (y - 4)^2 = 16$ **25.** 8π units **27.** ≈ 19.5

LESSON 12-5 (pp. 771–775)

9. Proof: Find an equation for the image of the circle $x^2 + y^2 = 1$ under $S_{a, b}$. Let (x', y') be the image of (x, y). Since $S_{a, b}$: $(x, y) \rightarrow (ax, by)$, $ax = x'$ and $by = y'$. So $x = \frac{x'}{a}$ and $y = \frac{y'}{b}$. We know $x^2 + y^2 = 1$. Substituting $\frac{x'}{a}$ for x and $\frac{y'}{b}$ for y in that equation, an equation for the image is $\left(\frac{x'}{a}\right)^2 + \left(\frac{y'}{b}\right)^2 = 1$. Since (x', y') represents a point on the image, the equation can be rewritten as $\frac{x^2}{a^2} + \frac{y^2}{b^2} = 1$ **11.** $171\pi \approx 537$ square meters **13.** Sample: For $e = \frac{1}{2}$, you could have $c = 4$, $a = 8$, $b = 4\sqrt{3} \approx 6.9$ See right. **15. a.** See below. **b.** $\frac{x^2}{64} + \frac{y^2}{28} = 1$ **17.** (b) **19.** (c) **21.** $235,225 < x^2 + y^2 < 250,000$ **23.** $y + 1 = 2(x - 3)$

13.

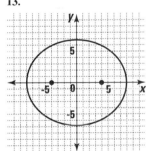

15. a.

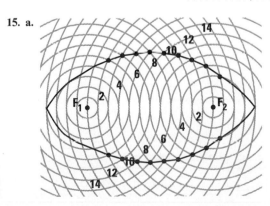

11. See below. 13. a. See below. b. Yes. It is the reflection image of the hyperbola $x^2 - y^2 = 1$ over the line $y = x$, so it is a hyperbola with the same focal constant and with foci that are the reflection images of the foci of the hyperbola $x^2 - y^2 = 1$. **15. a.** They both have foci at $(\sqrt{34}, 0)$ and $(-\sqrt{34}, 0)$. **b.** The vertices of $\frac{x^2}{25} - \frac{y^2}{9} = 1$ are at (5, 0) and (−5, 0); the vertices of $\frac{x^2}{9} - \frac{y^2}{25}$ are at (3, 0) and (−3, 0). The asymptotes of $\frac{x^2}{25} - \frac{y^2}{9} = 1$ are $y = \pm\frac{3}{5}x$; the asymptotes of $\frac{x^2}{9} - \frac{y^2}{25}$ are $y = \pm\frac{5}{3}x$. **17.** ≈ 2.2 ft **19.** See below. **21.** $\frac{169}{4} = 42.25$ **23. a.** $L = 100 - \frac{1}{2}N$ **b.** 200

11.

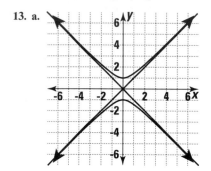

$-6 \le x \le 6$, *x*-scale = 1
$-6 \le y \le 6$, *y*-scale = 1

13. a.

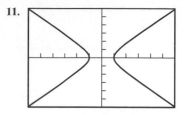

19.

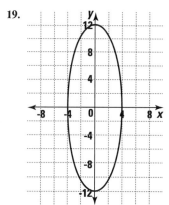

15. $9x^2 - 4y^2 - 36 = 0$; $A = 9$, $C = -4$, $F = -36$, $B = D = E = 0$
17. (60, 60), (−60, −60) **19. a.** See below. **b.** See below. **21.** See below. **23.** (c) **25.** focus **27.** $x^2 + (y - 240)^2 = 57600$

19. a.

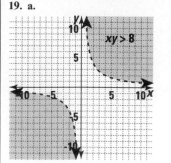

19. b.

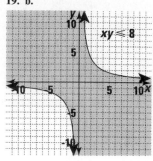

21.

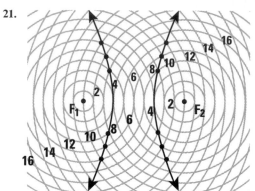

11. a. $2x + 2y = 150$; $xy = 1300$ **b.** See below. Sample: Estimated solutions: (25, 50) or (50, 25) **c.** $x \approx 27.2$, $y \approx 47.8$ or $x \approx 47.8$, $y \approx 27.2$ **13. a.** $\left(\frac{4 + \sqrt{41}}{5}, \frac{2 - 2\sqrt{41}}{5}\right)$, $\left(\frac{4 - \sqrt{41}}{5}, \frac{2 + 2\sqrt{41}}{5}\right)$
b. Does $\left(\frac{4 + \sqrt{41}}{5}\right)^2 + \left(\frac{2 - 2\sqrt{41}}{5}\right)^2 = 9$ and $2\left(\frac{4 + \sqrt{41}}{5}\right) + \left(\frac{2 - 2\sqrt{41}}{5}\right) = 2$? Yes. Does $\left(\frac{4 - \sqrt{41}}{5}\right)^2 + \left(\frac{2 + 2\sqrt{41}}{5}\right)^2 = 9$ and $2\left(\frac{4 - \sqrt{41}}{5}\right) + \left(\frac{2 + 2\sqrt{41}}{5}\right) = 2$? Yes. **15.** one branch of a hyperbola **17. a.** $(-\sqrt{61}, 0)$, $(\sqrt{61}, 0)$ **b.** (−6, 0), (6, 0) **c.** $\frac{y}{5} = \pm\frac{x}{6}$
19. a. ≈ $5.43 \cdot 10^9$ km **b.** ≈ $1.387 \cdot 10^9$ km **21. a.** 12 **b.** $20\sqrt{6}$
c. $\frac{4 \pm 10\sqrt{6}}{3}$

11. b.

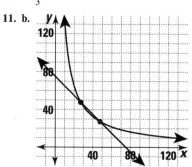

LESSON 12-9 (pp. 794–798)

9. (3, 0) **11.** The first equation represents a circle with center (3, 0) and radius 2, and the second equation represents a circle with center (−3, 0) and radius 2. Since the circles do not intersect, there are no real solutions. **13.** 2 **15. See below.** Estimated solutions: (1.1, 9.3), (−2.3, −4.3) **17.** circle **19.** one branch of a hyperbola **21.** 96π square units **23. a.** 5.25 m **b.** Yes

15.

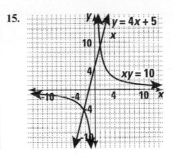

CHAPTER 12 PROGRESS SELF-TEST (p. 803)

1. $(x + 3)^2 + (y − 13)^2 = 100$ **2. See below. 3.** Since the image of $x^2 + y^2 = 1$ under $S_{a,b}$ is $\frac{x^2}{a^2} + \frac{y^2}{b^2} = 1$, the image under $S_{3,4}$ is $\frac{x^2}{9} + \frac{y^2}{16} = 1$. **4.** The vertices of $\frac{x^2}{a^2} + \frac{y^2}{b^2} = 1$ are $(−a, 0), (a, 0), (0, b),$ and $(0, −b)$, so the vertices of $\frac{x^2}{9} + \frac{y^2}{16} = 1$ are $(−3, 0), (3, 0), (0, 4),$ and $(0, −4)$. **See right. 5.** Since $a = 13$ and $b = 5$, an equation is $\frac{x^2}{169} + \frac{y^2}{25} = 1$. **6.** The area of the ellipse $= \pi ab = (13)(5)\pi = 65\pi \approx 204$. **7.** From the graph, the intersections are about (3.5, 1.5) and (−.5, −2.5). **See right. 8.** Since $x − 2 = 4x − x^2$, then $x^2 − 3x − 2 = 0$ and $x = \frac{3 \pm \sqrt{3^2 − 4(−2)(1)}}{2} = \frac{3 \pm \sqrt{17}}{2}$. The two points of intersection are $\left(\frac{3 + \sqrt{17}}{2}, \frac{−1 + \sqrt{17}}{2}\right)$ and $\left(\frac{3 − \sqrt{17}}{2}, \frac{−1 − \sqrt{17}}{2}\right)$. **9.** The length of the major axis is 2.8 + 4.6 = 7.4 billion miles. **10. See right.**
11. a. $y = 2(x + 4)^2 − 9 = 2(x^2 + 8x + 16) − 9 = 2x^2 + 16x + 32 − 9 = 2x^2 + 16x + 23$ **b.** It represents a parabola.
12. a. $\begin{cases} x^2 + y^2 = 400^2 \\ (x − 400)^2 + y^2 = 500^2 \end{cases}$ **b.** Expand $(x − 400)^2 + y^2 = 500^2$ to get $x^2 − 800x + 400^2 + y^2 = 500^2$. Rearrange terms: $x^2 + y^2 − 800x = 500^2 − 400^2 = 90000$. Substitute 400^2 for $x^2 + y^2$ (from the first equation) to get $400^2 − 800x = 90000$. Solve for x: $x = 87.5$. Substitute for x in first equation and solve for y: $y = \pm\sqrt{400^2 − 87.5^2} \approx \pm390$. The possible locations for the fire are either 87.5 m east and 390 m north or 87.5 m east and 390 m south of the first station. **13.** $xy = 2$ is equivalent to $y = \frac{2}{x}$. All equations of the form $y = \frac{k}{x}$ have the x- and y-axes as asymptotes, so the equations for the asymptotes are $y = 0$ and $x = 0$. **14.** This is a hyperbola in standard form. $a^2 = 9$, so $a = 3$, and $b^2 = 4$, so $b = 2$. Thus the vertices are $(−3, 0)$ and $(3, 0)$. Asymptotes are $\frac{y}{b} = \pm\frac{x}{a}$ or $y = \pm\frac{2}{3}x$. **See right. 15.** The distance from the parabola's vertex to the directrix is 1 unit. So the distance from the vertex to the focus must also be 1 unit along the parabola's axis of symmetry ($x = 3$). Since the parabola opens downward the focus must be below the vertex. Thus the focus has coordinates (3, 4).

2.

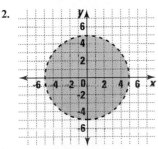

4.

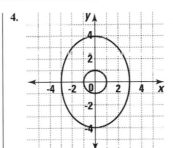

7.

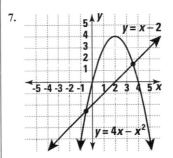

10.

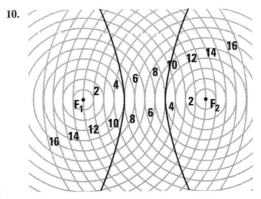

14.

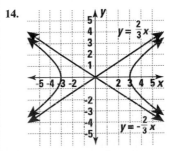

The chart below keys the **Progress Self-Test** questions to the objectives in the **Chapter Review** on pages 804–807 or to the **Vocabulary** (Voc.) on page 802. This will enable you to locate those **Chapter Review** questions that correspond to questions missed on the **Progress Self-Test**. The lesson where the material is covered is also indicated on the chart.

Question	1	2	3	4	5	6	7	8	9	10
Objective	B	J	G	J	J	C	K	D	H	E
Lesson	12-2	12-3	12-5	12-4,12-5	12-4	12-5	12-8	12-8	12-4	12-6
Question	11	12	13	14	15					
Objective	A,G	I	F	J	F					
Lesson	12-4,12-1	12-9	12-7	12-6	12-1					

CHAPTER 12 REVIEW (pp. 804–807)

1. $x^2 + 0xy + y^2 - 6x + 14y - 42 = 0$ **3.** $5x^2 + 0xy + 6y^2 + 0x + 0y - 30 = 0$ **5.** $4x^2 + 0xy + 0y^2 - 16x - y + 22 = 0$
7. $x^2 + y^2 = 36$ **9.** $y_2 = \sqrt{20 - x^2}$ **11.** $\frac{x^2}{75} + \frac{y^2}{25} > 1$
13. $\frac{x^2}{9} + \frac{y^2}{36} = 1$ **15.** $\frac{x^2}{16} - \frac{y^2}{33} = 1$ **17.** $50\pi \approx 157$ **19.** $4\pi \approx 12.6$
21. (4.5, 14), (−2, 27) **23.** no solution **25.** (5, 4), (5, −4), (−5, 4),
(−5, −4) **27.** (1, 2), $\left(-\frac{5}{4}, \frac{25}{8}\right)$ **29.** See below. **31.** See right.

33. center (0, 0), radius $\sqrt{5}$ **35. a.** $\left(0, \frac{1}{2}\right)$ **b.** (0, 0) **c.** $y = -\frac{1}{2}$
37. a. (−4, 0), (4, 0) **b.** $\frac{y}{2} = \pm\frac{x}{4}$ **39.** ellipse **41. A:** hyperbola;
B: parabola; **C:** ellipse; **D:** circle **43.** True **45.** True
47. At 5 feet from the center line, the tunnel has a height of
$\sqrt{12^2 - 5^2} \approx 10.9$ ft. so the truck will fit. **49. a.** 3.1 million miles
b. 185.9 million miles **51.** $100 < (x - 200)^2 + (y - 100)^2 < 900$
53. 10′ by 20′ **55.** The epicenter is 10.8 miles east and 48.8 miles
south of station 1. **57. a.** $14 **b.** 420 **59.** See right. **61.** See
right. **63.** $\frac{x^2}{49} + \frac{y^2}{16} = 1$ **65.** (b) **67. a.** See right. Estimated
solution from graph: $(x, y) = (0, -9)$ **b.** $0^2 + (-9)^2 = 81$
$0^2 + (-9 + 18)^2 = 81$ **69.** Sample: **See right.**

29.

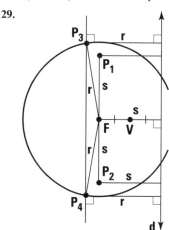

31.

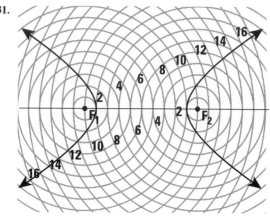

59.

$$\frac{x^2}{16} - \frac{y^2}{81} = 1$$

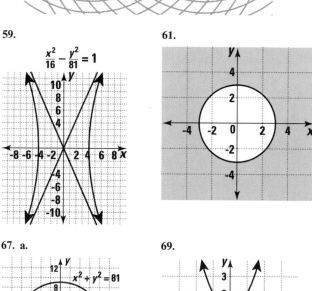

61.

67. a.

$x^2 + y^2 = 81$

$x^2 + (y + 18)^2 = 81$

69.

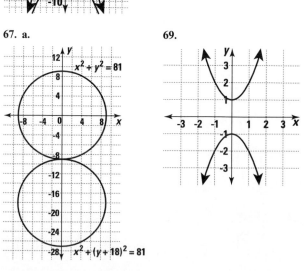

LESSON 13-1 (pp. 810–816)

13. a. 15, 52, 465 **b.** $T = 10 + 3(n - 1)$, $S = \frac{3}{2}n^2 + \frac{17}{2}n$

c. Change the following lines:
20 LET TERM = 8
45 FOR N = 2 TO 30
50 TERM = TERM + 11

15. 4230 **17. a.** 27 **b.** 57 **c.** 3500 **19. a.** $-\frac{3}{2}$ **b.** $\frac{81}{2}$

c. $a_n = 8 \cdot \left(-\frac{3}{2}\right)^{n-1}$ for all integers $n \geq 1$ **21.** $y = -2x^2 + 8x + 10$

23. a. $\begin{bmatrix} \frac{n}{3n-2} & \frac{-2}{3n-2} \\ \frac{-1}{3n-2} & \frac{3}{3n-2} \end{bmatrix}$ **b.** $\frac{2}{3}$

LESSON 13-2 (pp. 817–822)

13. a. ≈ 2.85 inches **b.** ≈ 55.46 inches **15. a. See below.**
b. ≈ 20.78 m **17. a.** 250,000 **b.** 250,500 **c.** $\frac{1000 \cdot (1000 + 1)}{2} =$
$500,500 = 250,000 + 250,500$ **19.** $f(-2) = \frac{1}{64}$; $f(0) = 1$; $f\left(\frac{2}{3}\right) = 4$
21. $y = \log_8 x$ **23. a.** trapezoid **b.** True. Sample proof (algebraic):
The length of one diagonal is
$\sqrt{[a - (-b)]^2 + (0 - c)^2} = \sqrt{(a + b)^2 + c^2}$ and
the length of the other is
$\sqrt{(-a - b)^2 + (0 - c)^2} = \sqrt{(a + b)^2 + c^2}$.
So the diagonals of this quadrilateral have the same length. Sample proof (geometric): The vertices of the quadrilateral are reflection images of each other over the y-axis. Since reflections preserve distance, the diagonals are congruent.

15. a.

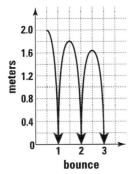

LESSON 13-3 (pp. 823–829)

19. $\sum_{i=1}^{8} 9 \cdot 2^{i-1}$ **21. a.** $\sum_{i=1}^{n} i^3 = \left(\sum_{i=1}^{n} i\right)^2$ **b.** $\sum_{i=1}^{4} i^3 = 1^3 + 2^3 + 3^3 +$
$4^3 = 1 + 8 + 27 + 64 = 100$; $\left(\sum_{i=1}^{4} i\right)^2 = (1 + 2 + 3 + 4)^2 = 10^2 =$
100 **23. a.** 1, 2, 6, 24, 120, 720, 5040 **b.** the factorial sequence
25. a. $1.52 **b.** $355.29 **27. See below. 29.** 160 kg

27.

LESSON 13-4 (pp. 830–834)

11. mean $\approx$ $21,363.64; median = $20,000; mode = $20,000
13. $\frac{2x + 3y}{5}$ **15. a.** Sample: The two teams have the same
mean (73.5″) and median (73″) height. **b.** Sample: The Rocketeers'
heights are spread out more than the Sunbursts' heights. For the
Rocketeers s.d. ≈ 3.4 inches and for the Sunbursts s.d. ≈ 1.2 inches.
17. a. 144 **b.** 30 **19.** 0.6561 m **21. a.** 10! = 3,628,800
b. 8! = 40320 **23.** y is multiplied by 3^4 or 81

LESSON 13-5 (pp. 835–840)

17. True; $\frac{n!}{(n-2)!} = n(n - 1)$ if $n \geq 2$. **19.** $a = 9$, $b = 1$ or
$a = 9$, $b = 8$ **21.** the 3rd elements in each row **23.** $\binom{n}{4} + \binom{n}{5} =$
$\frac{n!}{4!(n-4)!} + \frac{n!}{5!(n-5)!} = \frac{n!}{4!(n-5)!(n-4)} + \frac{n!}{4! \cdot 5(n-5)!} =$
$\frac{5n!}{4! \cdot 5(n-5)!(n-4)} + \frac{n!(n-4)}{4! \cdot 5(n-5)!(n-4)} = \frac{5n! + n!(n-4)}{5!(n-4)!} =$
$\frac{n!(5 + n - 4)}{5!(n-4)!} = \frac{n!(n+1)}{5!(n-4)!} = \frac{(n+1)!}{5!(n+1-5)!} = \binom{n+1}{5}$ **25.** ≈ 13.15
27. a. 10 **b.** $n - r$ **29. a.** $a^2 + 2ab + b^2$ **b.** $a^3 + 3a^2b + 3ab^2 + b^3$
c. $a^4 + 4a^3b + 6a^2b^2 + 4ab^3 + b^4$

LESSON 13-6 (pp. 841–844)

11. $(y + 2a)^n$ **13. a.** $a^4 + 4a^3b + 6a^2b^2 + 4ab^3 + b^4$
b. $(a + b)^4$ because $(a^2 + 2ab + b^2)^2 = [(a + b)^2]^2$
15. 1.010045120210252 **17.** (c) **19.** (d) **21. a.** 10^5 **b.** $10^{\frac{1}{2}} \approx 3.16$
c. 10.5 **23.** 3825 lb

LESSON 13-7 (pp. 845–850)

17. 512 **19.** 1 **21. a.** 52 **b.** 1326 **c.** $\frac{1}{270,725}$
23. $8x^3 - 36x^2y + 54xy^2 - 27y^3$ **25. a.** 165 **b.** (i) **27.** (c)

LESSON 13-8 (pp. 851–856)

13. $\frac{1}{256}, \frac{8}{256}, \frac{28}{256}, \frac{56}{256}, \frac{70}{256}, \frac{56}{256}, \frac{28}{256}, \frac{8}{256}, \frac{1}{256}$ **15. a.** $\frac{1}{16}$ **b.** $\frac{4}{16}$ **c.** $\frac{6}{16}$
d. $\frac{4}{16}$ **e.** $\frac{1}{16}$ **f.** $\frac{11}{16}$ **17.** $\frac{22}{64} \approx .34$ **19.** $p^7 + 7p^6q + 21p^5q^2 + 35p^4q^3$
$+ 35p^3q^4 + 21p^2q^5 + 7pq^6 + q^7$ **21.** $n = 8$

LESSON 13-9 (pp. 857–861)

7. a. $\frac{1}{575,757}$ **b.** The player must pick a different set of five
numbers on each of the 3 boards. **9.** $\frac{1}{2,598,960}$ **11.** Sample:
Advantages of getting a smaller amount each year are: (1) Unless
the winner already earns a lot of money, this will lower the amount
the winner pays in tax; (2) the winner has more time to plan how
to invest or spend the prize money. Some disadvantages include:
(1) The winner cannot receive the compounded interest that would
accrue each year on the entire sum; (2) if the economy is
experiencing inflation, the money received in later years is actually
worth less than if it had all been received at once. It's probably not
fair to the winner, but lotteries are meant to raise money for the
states that run them. So overall, it's probably fair. **13.** $\approx .53$
15. $\frac{1}{6}n^3 - \frac{1}{2}n^2 + \frac{1}{3}n$ **17.** 666.666 **19. a.** $\pm 1, \pm 2, \pm\frac{1}{3}, \pm\frac{2}{3}$ **b.** 1, $\frac{2}{3}$

LESSON 13-10 (pp. 862–867)

11. $\approx 66\%$ **13.** $\approx 15.9\%$ **15.** 769; 145 **17.** $\frac{1}{1000}$ **19.** 1
21. mean = 89; median = 90; mode = 92; s.d. = 4 **23. a.** $\frac{\log 10}{\log 3}$
b. 2.10 **25.** ≈ 9.1 km

11. 500; ≈ 15.8 13. a. Answers will vary; lists of 10 random numbers between 0 and 1. b. Each output should be different.

15. the class; the classmate you called 17. ≈ 5.16 19. $\frac{6}{64} = \frac{3}{32}$
21. $\frac{\log 45}{\log 2} \approx 5.49$ 23. $y - 5 = (x - 6)^2$; $y - 5 = -(x - 6)^2$
25. $x = -|y|$

PROGRESS SELF-TEST (p. 879)

1. $\sum\limits_{i=1}^{20} i^3$ 2. $a_1 = 0$, $n = 1001$, $a_n = 3000$, $S_{1001} = \frac{n}{2}(a_1 + a_n) = 1{,}501{,}500$ 3. The mode is the most frequent score, which is 80.
4. The mean is the sum of the scores divided by the number of scores, or $\frac{431}{5} = 86.2$. 5. To bring her average for 6 scores up to 88, she needs a total of $(6)(88) = 528$ on the 6 scores. Since she already has a total of 431 for the first five scores, she needs $528 - 431 = 97$ on the next quiz.

6. a.
```
              1
            1   1
          1   2   1
        1   3   3   1
      1   4   6   4   1
    1   5   10   10   5   1
```
b. 2^n 7. First, fill in powers of x and y, following from the Binomial Theorem. $(x + y)^7 = \_\_x^7 + \_\_x^6y + \_\_x^5y^2 + \_\_x^4y^3 + \_\_x^3y^4 + \_\_x^2y^5 + \_\_xy^6 + \_\_y^7$ Second, put in coefficients. $(x + y)^7 = \binom{7}{0}x^7 + \binom{7}{1}x^6y + \binom{7}{2}x^5y^2 + \binom{7}{3}x^4y^3 + \binom{7}{4}x^3y^4 + \binom{7}{5}x^2y^5 + \binom{7}{6}xy^6 + \binom{7}{7}y^7$ Lastly, evaluate the coefficients referring to 7th row of Pascal's triangle or by using the formula $\binom{n}{r} = \frac{n!}{r!(n - r)!}$. $(x + y)^7 = x^7 + 7x^6y + 21x^5y^2 + 35x^4y^3 + 35x^3y^4 + 21x^2y^5 + 7xy^6 + y^7$ 8. Use $(a + b)^4 = a^4 + 4a^3b + 6a^2b^2 + 4ab^3 + b^4$ with $a = x^2$ and $b = -3$: $(x^2 - 3)^4 = (x^2)^4 + 4(x^2)^3(-3) + 6(x^2)^2(-3)^2 + 4(x^2)(-3)^3 + (-3)^4 = x^8 - 12x^6 + 54x^4 - 108x^2 + 81$. 9. $8! = 40{,}320$ 10. $\binom{15}{3} = \frac{15!}{3! \, 12!} = \frac{15 \cdot 14 \cdot 13 \cdot 12!}{3 \cdot 2 \cdot 1 \cdot 12!} = 5 \cdot 7 \cdot 13 = 455$ 11. $_8C_0 = \frac{8!}{8! \, 0!} = 1$ 12. $\binom{40}{38} = \frac{40!}{38! \, 2!} = \frac{40 \cdot 39 \cdot 38!}{2 \cdot 1 \cdot 38!} = 20 \cdot 39 = 780$ 13. $\frac{n!}{(n - 1)!} = \frac{n \cdot (n - 1)!}{(n - 1)!} = n$

14. This is an arithmetic series with $a_1 = 12$, $n = 30$, $d = 2$. So $S_n = \frac{n}{2}[2a_1 + (n - 1)d] = 1230$. 15. a. 48, 12, 3, $\frac{3}{4}$ b. This is a geometric series with $g_1 = 48$, $r = \frac{1}{4}$, and $n = 15$. So $S_n = \frac{g_1(1 - r^n)}{1 - r} \approx 64$. 16. $P(n) = \frac{\binom{6}{n}}{26}$; use elements of row 6 of Pascal's triangle to evaluate $\binom{6}{n}$

a.
n	0	1	2	3	4	5	6
$P(n)$	$\frac{1}{64}$	$\frac{3}{32}$	$\frac{15}{64}$	$\frac{5}{16}$	$\frac{15}{64}$	$\frac{3}{32}$	$\frac{1}{64}$

b. See below. c. $P(n)$ represents the probability of obtaining exactly n heads when a fair coin is tossed 6 times.
17. $\binom{55}{6} = \frac{55!}{49! \, 6!} = \frac{55 \cdot 54 \cdot 53 \cdot 52 \cdot 51 \cdot 50}{6 \cdot 5 \cdot 4 \cdot 3 \cdot 2 \cdot 1} = 28{,}989{,}675$
18. a. Since 34.1% of the scores are within one standard deviation of the mean, in each direction, and another 13.6% of the scores are within a second standard deviation, in each direction, the percent within two standard deviations is about $2(34.1 + 13.6) = 95\%$.
b. A score of 24.7 is 5.9 above the mean of 18.8, which is one standard deviation above the mean. The percent of scores at or above one standard deviation above the mean is $13.6 + 2.3$ or about 16%. 19. the set of all registered voters; the set of the 1000 voters polled 20. The population of all possible tosses is infinite.

16. b.

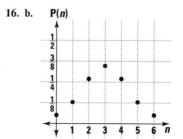

The chart below keys the **Progress Self-Test** questions to the objectives in the **Chapter Review** on pages 880-883 or to the **Vocabulary** (Voc.) on page 878. This will enable you to locate those **Chapter Review** questions that correspond to questions missed on the **Progress Self-Test**. The lesson where the material is covered is also indicated on the chart.

Question	1	2	3	4	5	6	7	8	9	10
Objective	C	A, C	I	I	I	F	E	E	H	H
Lesson	13-3	13-1, 13-3	13-4	13-4	13-4	13-5	13-6	13-6	13-3	13-7

Question	11	12	13	14	15	16	17	18	19	20
Objective	D	D	C	G	B	L	J	I	K	K
Lesson	13-7	13-5	13-3	13-1	13-2	13-10	13-9	13-10	13-11	13-11

CHAPTER 13 REVIEW (pp. 880–883)

1. 1830 3. 990 5. 35 7. 1,048,575 9. ≈ 25.95 11. a. $(-3) + (-1) + 1 + 3 + 5 + 7$ b. 12 13. (c) 15. $\sum\limits_{n=1}^{72} 2n$ 17. 722
19. $(r + 1)$st; nth; n choose r 21. 26! 23. 252 25. 1
27. $x^4 + 4x^3y + 6x^2y^2 + 4xy^3 + y^4$ 29. $27n^6 - 108n^4 + 144n^2 - 64$
31. True 33. (d) 35. $2^5 = 32$ 37. nth row and $(r + 1)$st element

39. a. 78 b. 20 41. ≈ 358.8 cm 43. $12! = 479{,}001{,}600$ 45. 45
47. 178,365 49. mean = 82.125; median = 83.5; mode = 90
51. ≈ 10.53 53. a. $6'8''$ b. $6'6.2''$ c. $\approx 3.25''$ 55. juniors
57. 379 to 621 59. $\frac{5}{16}$ 61. a. $\frac{15}{128}$ b. $\frac{11}{64}$ 63. $\frac{1}{10,000}$ 65. $\frac{1}{274,896}$
67. a. All households in town with at least one TV. b. It would be difficult or too expensive or take too much time to poll the entire population, even in a small town. 69. a. 50% b. 68.2% c. 4.6%

absolute value   The operation or function defined by
$$|x| = \begin{cases} x \text{ when } x \geq 0 \\ -x \text{ when } x < 0 \end{cases}.$$ The distance of x from 0 on a number line. (351)

absolute value function   The function with equation $f(x) = |x|$. (352)

acceleration   The rate of change of the velocity of an object. (366)

acidic   A substance whose pH is between 0 and 7. (565)

algebraic expression   A combining of numbers, variables and operations in a way that stands for a number. Sometimes called simply an *expression*. (6)

algebraic sentence   A sentence in which expressions are related by equality or inequality. (6)

alkaline   A substance whose pH is between 7 and 14. (565)

angle of depression   The angle between the line of sight and the horizontal when the line of sight points down. (612)

angle of elevation   The angle between the line of sight and the horizontal when the line of sight points up. (612)

annual yield   The rate of interest earned after all the compoundings have taken place in one year. Also called *effective annual yield*. (440)

annual growth factor   For a given decade, it is the positive number b such that b^{10} gives the decade growth factor. Also called *yearly growth factor*. (554)

argument of a function   A value of the domain variable in a function. (20)

arithmetic mean   The result of adding the n numbers in a data set and dividing the sum by n. Also called the *average* or *mean*. (502)

arithmetic sequence   A sequence with a constant difference between consecutive terms. Also called *linear sequence*. (175)

arithmetic series   An indicated sum of successive terms of an arithmetic sequence. (813)

arrow notation   The notation $f: x \rightarrow y$ used to describe y as a function of x. Also called *mapping notation*. (21)

asymptote   A line approached by the graph of a function. (778)

asymptotes of a hyperbola   Two lines which are approached by the points on the branches of a hyperbola as the points get farther from the foci. The asymptotes of the hyperbola with equation $\frac{y^2}{a^2} - \frac{y^2}{b^2} = 1$ are $\frac{y}{b} = \pm \frac{x}{a}$. (777)

automatic grapher   A calculator or computer program that can automatically display graphs when an equation is entered. (94)

average   The result of adding the n numbers in a data set and dividing the sum by n. Also called *arithmetic mean* or *mean*. (502)

axis of symmetry of a parabola   The line with equation $x = h$ of the parabola with equation $y - k = a(x - h)^2$; the line containing the focus of the parabola perpendicular to the directrix. (359, 748)

base b in the expression b^n. (418)

bearing   The direction of an object, as measured clockwise from due north. (607)

bel   A unit of sound intensity; 10 bels is a decibel. (563)

binomial   A polynomial with two terms. (680)

binomial coefficients   The coefficients of terms in the expansion of $(a + b)^n$. (843)

binomial distribution   A probability function in which the values of the function are proportional to binomial coefficients. Also called *binomial probability distribution*. (862)

binomial expansion   The result of writing the power of a binomial as a sum. (841)

binomial experiment   A situation in which n independent trials occur, and each trial has exactly two mutually exclusive outcomes. (852)

binomial probability distribution   A probability function in which the values of the function are proportional to binomial coefficients. Also called *binomial distribution*. (862)

boundary   A line or curve separating a plane or part of a plane into two regions. (312, 760)

branches of a hyperbola   The two separate parts of the graph of a hyperbola. (105)

calculator key sequence   A list of keystrokes to be performed on a calculator. (14)

center of a circle   The fixed point from which the set of points of the circle are at a given distance. (754)

center of an ellipse   The intersection of the axes of the ellipse. (767)

circle   The set of all points in a plane at a given distance from a fixed point. (754)

clearing fractions.   The process of multiplying both sides of an equation by a common multiple of the denominators to eliminate the fractions. (32)

closure   One of the field properties; if you multiply two 2×2 matrices, the result is a 2×2 matrix. (240)

coefficient matrix   A matrix which represents the coefficients of the variables of a system. (305)

coefficients of a polynomial   The numbers $a_n, a_{n-1}, a_{n-2}, \ldots, a_0$ in the polynomial $a_n x^n + a_{n-1} x^{n-1} + a_{n-2} x^{n-2} + \ldots + a_0$. (674)

column   A vertical list in a table, rectangular array, or spreadsheet. (204)

combination   Any choice of r objects from n objects. (847)

combined variation   A situation in which direct and inverse variations occur together. (122)

common logarithm   A logarithm to the base 10. (558)

common logarithm function The function with equation $y = \log_{10} x$ or $y = \log x$. Also called *logarithm function with base 10*. (560)

completing the square A technique used to transform a quadratic from $ax^2 + bx + c$ form to $a(x - h)^2 + k$ form. (371)

complex conjugate The complex conjugate of $a + bi$ is $a - bi$. (395)

complex number A number that can be written in the form $a + bi$, where a and b are real numbers and $i = \sqrt{-1}$; a is called the real part and b the imaginary part. (393)

composite of f and g The composite $g \circ f$ of two functions f and g is the function that maps x onto $g(f(x))$, and whose domain is the set of all values in the domain of f for which $f(x)$ is in the domain of g. (479)

composition of functions The function that results from first applying one function, then another; denoted by the symbol $\circ$. (480)

composite of transformations The transformation $T_2 \circ T_1$ that maps a figure F onto F'' if transformation T_1 maps figure F onto figure F', and transformation T_2 maps figure F' onto figure F''. (242)

compound sentence A sentence in which two clauses are connected by the word "and" or by the word "or." (273)

compounding The process of earning interest on the interest of an investment. (438)

concentric circles Two or more circles with the same center. (754)

congruent figures Two figures such that one is the image of the other under a composite of isometries (reflections, rotations, translations, glide reflections). (213)

conic graph paper Graph paper consisting of two intersecting sets of concentric circles. (764)

conic section A cross-section of a double cone; the intersection of a double cone with a plane. Also called *conic*. (747)

conjecture An educated guess. (62)

conjugate For any expression of the form $a + \sqrt{b}$, the conjugate is the expression $a - \sqrt{b}$. (507)

consistent system A system that has one or more solutions. (287)

constant difference In an arithmetic sequence, the difference of two consecutive terms. (175)

constant matrix A matrix which represents the constants in a system of equations. (305)

constant multiplier In a geometric sequence, the ratio of two consecutive terms. Also called *constant ratio*. (444)

constant of variation The non-zero real constant k in the equation $y = kx^n$ or $y = \frac{k}{x^n}$ for a variation function. (72)

constant ratio In a geometric sequence, the ratio of two consecutive terms. Also called *constant multiplier*. (444)

constant-decrease situation A situation in which a quantity y decreases by a constant amount for every fixed increase in x. (141)

constant-increase situation A situation in which a quantity y increases by a constant amount for every fixed increase in x. (140)

constraint A restriction on a variable or variables in a situation. (271)

Continuous Change Model The equation $N(t) = N_0 e^{rt}$, where N_0 is the initial amount and r is the growth factor over a time t. (548)

continuous compounding The limit of the process of earning interest with periods of compounding approaching zero. Also called *instantaneous compounding*. (547)

convex region A region of the plane in which any two points of the region can be connected by a line segment which lies entirely in the region. (324)

coordinate plane A plane in which there is a one-to-one correspondence between the points in the plane and the set of ordered pairs of real numbers. (24)

corollary A theorem that follows immediately from another theorem. (358)

correlation coefficient A number between –1 and 1 that indicates how well a linear or other equation fits data. (171)

$\cos^{-1} x$ The number between 0 and 180°, or between 0 and π, whose cosine is x. (611, 653)

cosine function The correspondence $\theta \to \cos \theta$ that associates θ with the x-coordinate of the image of (1, 0) under R_θ. (647)

cosine of θ (cos θ) In a right triangle with acute angle θ, $\cos \theta = \frac{\text{length of leg adjacent to } \theta}{\text{length of hypotenuse}}$. The first coordinate of R_θ (1, 0). (605, 624)

counting number A member of the set {1, 2, 3, 4, 5, . . .}. Also called the *natural numbers*. (14)

cube of x The third power of x, denoted by x^3. (418)

cube root A cube root t of x, denoted by $\sqrt[3]{x}$, is a solution to the equation $t^3 = x$. (450)

cubic polynomial A polynomial of a single variable with degree 3, such as $ax^3 + bx^2 + cx + d$. (675)

cubing function The powering function f defined by $f(x) = x^3$. (420)

data set A collection of elements in which an element may appear more than once. (830)

decade growth factor The ratio of an amount in a specific year to the amount ten years earlier. (554)

decibel (dB) A unit of sound intensity. $\frac{1}{10}$ of a bel. (563)

default window The window that is set on an automatic grapher by the manufacturer. (94)

degree of a polynomial in several variables The largest sum of the exponents of the variables in any term in a polynomial expression. (680)

degree of a polynomial in a single variable The largest exponent of the variable in the polynomial. (674)

dependent variable A variable whose values always depend on the value(s) of other variable(s). (12)

depreciation The decrease in value over time of manufactured goods. (539)

determinant of a 2×2 matrix For the matrix $M = \begin{vmatrix} a & b \\ c & d \end{vmatrix}$, the number $ad - bc$. (301)

difference of matrices The result of subtracting two matrices. (210)

dimensions $m \times n$ A descriptor of a matrix with m rows and n columns has dimensions $m \times n$; a rectangle with adjacent sides of lengths m and n. (204)

direct variation A function mapping x onto y with an equation of the form $y = kx^n$, $k \neq 0$, and $n > 0$. (72)

directly proportional Two variables x and y so related that when one of the variables is multiplied by k, so is the other. (72)

directrix A line associated with a parabola such that the distance from it to any point on the parabola is equal to the distance from that point to the focus. (748)

discrete graph A graph that is made up of unconnected points. (105)

discriminant of a quadratic equation For the equation $ax^2 + bx + c = 0$, the value of $b^2 - 4ac$. (403)

domain of a function The set of values which are allowable substitutions for the independent variable. (13)

double cone The surface generated by a line rotating about an axis that contains a point on the line. (747)

double root The root of a quadratic equation when the discriminant is 0; a root with multiplicity of 2. (718)

e The constant 2.718281828459. . . that the sequence of numbers of the form $\left(1 + \frac{1}{n}\right)^n$ approaches as n increases without bound. The base of natural logarithms. (547, 583)

eccentricity The ratio of the distance between the foci to the focal constant in an ellipse or hyperbola. (774)

element of a matrix The object in a particular row and column of a matrix. (203)

ellipse The set of points P in a plane which satisfy $PF_1 + PF_2 = d$, where F_1 and F_2 (its *foci*) are any two points and d (its *focal constant*) is a constant with $d > F_1F_2$. (765)

equal complex numbers Two complex numbers with equal real parts and equal imaginary parts. $a + bi = c + di$ if and only if $a = c$ and $b = d$. (393)

equal matrices Two matrices which have the same dimensions and in which corresponding elements are equal. (205)

equation A sentence stating that two expressions are equal. (8)

Euler's $f(x)$ notation Notation that represents the value of a function f with argument x as $f(x)$. (20)

evaluating an expression Substituting for the variables in an expression and calculating a result. (8)

expanding a polynomial Writing a power of a polynomial or the product of polynomials as a polynomial. (674)

explicit formula for nth term A formula which describes any term in a sequence in terms of its position in the sequence. (43)

exponent n in the expression b^n. (418)

exponential curve A graph of an exponential equation. (535)

exponential decay A situation described by an exponential function where the growth factor is between 0 and 1. (539)

exponential function A function with the independent variable in the exponent. A function with an equation of the form $y = ab^x$. (535)

exponential growth A situation described by an exponential function where the growth factor is greater than one. (535)

Exponential Growth Model If a quantity a has growth factor b for each unit period, then after a period of length x, there will be ab^x of the quantity. (541)

exponential sequence A sequence with a constant ratio between consecutive terms. Also called *geometric sequence*. (444)

exponentiation An operation by which a variable is raised to a power. Also called *powering*. (418)

expression A combining of numbers, variables and operations in a way that stands for a number. (6)

exterior of a circle The region outside a circle. (760)

extraneous solution A solution that is gained but does not check in the original equation. (518)

f^{-1} The symbol for the inverse of function f. (491)

$f(x)$ notation The notation used to describe functions, read "f of x." (20)

factor: A number or expression which evenly divides a given expression. (686)

factored form The product of two or more factors which equals the given expression. (686)

factorial function The function defined by the equation $f(n) = n!$ where $n!$ is the product of the integers from n to 1. (826)

factoring The rewriting of a polynomial as a product of factors. (686)

fair coin A coin that has an equal probability of landing on either side. (853)

feasible region The set of solutions to a system of linear inequalities. Also called *feasible set*. (319)

Fibonacci sequence The sequence 1, 1, 2, 3, 5, 8, 13, ... A recursive definition is
$$\begin{cases} F_1 = 1, \\ F_2 = 1 \\ F_n = t_{n-1} + t_{n-2} \end{cases} \text{for } n \geq 3. \text{ (57)}$$

field properties The assumed properties of addition and multiplication for real numbers. (Appendix A)

floor function The function that maps x onto $\lfloor x \rfloor$, the greatest integer less than or equal to x. Also called *greatest integer function* or *rounding down function*. (187)

focal constant The constant sum of the distances from a point on an ellipse to the two foci of the ellipse. The absolute value of the difference of the distances from a point on a hyperbola to the two foci of the hyperbola. (765, 777)

focus (plural foci) In a parabola, the point along with the directrix from which a point is equidistant. The two points from which the sum (ellipse) or difference (hyperbola) of distances to a point on the conic section is constant. (748, 761, 777)

formula A sentence stating that a single variable is equal to an expression with one or more different variables. (8)

function A relation in which for each ordered pair the first coordinate has exactly one second coordinate. (12)

function composition The function that results from first applying one function, then another; denoted by the symbol ∘. (480)

general form of a quadratic relation An equation of the form $Ax^2 + Bxy + Cy^2 + Dx + Ey + F = 0$, where A, B, C, D, E and F are real numbers and at least one of A, B, or C is not zero. (786)

geometric mean The nth root of the product of n numbers. (502)

geometric sequence A sequence with a constant ratio between successive terms. Also called *exponential sequence*. (444)

geometric series An indicated sum of successive terms of a geometric sequence. (817)

gravitational constant The acceleration of a moving object due to gravity, often denoted by g. Near the Earth's surface, $g \approx 32$ ft/sec^2 $\approx$ 9.8 m/sec^2. (366)

greatest integer function The function that maps x onto $\lfloor x \rfloor$, the greatest integer less than or equal to x. Also called *floor function* or *rounding down function*. (187)

growth factor In the exponential function $y = ab^x$, the amount b by which y is multipled for every unit increase in x. (535)

half-life The amount of time required for a quantity in an exponential decay situation to decay to half its original value. (540)

half-plane Either of the two sides of a line in a plane. (312)

hierarchy A diagram that shows how various ideas are related, with a direction that moves from more specific to more general. (889)

horizontal asymptote A horizontal line that is approached by the graph as the values of x get very large (or very small). (107)

horizontal line A line with an equation of the form $y = b$. (148)

horizontal magnitude The number a in the scale change that maps (x, y) onto (ax, by). (227)

horizontal scale change The stretching or shrinking of a figure in only the horizontal direction. A transformation which maps (x, y) onto (kx, y). (227)

hyperbola The graph of every function with an equation of the form $y = \frac{k}{x}$, where $k \neq 0$; the set of points P in a plane which satisfy $|PF_1 - PF_2| = d$, where F_1 and F_2 are any two points and d is a constant with $0 < d < F_1F_2$. (104, 777)

i One of the two square roots of -1, denoted by $\sqrt{-1}$. (389)

identity function The function defined by $f(x) = x$. (420)

2 × 2 identity matrix The matrix $\begin{bmatrix} 1 & 0 \\ 0 & 1 \end{bmatrix}$. (219)

3 × 3 identity matrix: The matrix $\begin{bmatrix} 1 & 0 & 0 \\ 0 & 1 & 0 \\ 0 & 0 & 1 \end{bmatrix}$. (219, 309)

identity transformation The transformation in which each point coincides with its image. (223)

image The result of applying a transformation to a preimage. (221)

imaginary number A number which is the square root of a negative real number. (388)

imaginary part In the complex number $a + bi$, the real number b. (393)

imaginary unit The number i. (389)

in terms of A sentence which is written with one variable in terms of another has the form of the first variable set equal to an expression with one or more terms involving the second variable. (36)

inconsistent system A system with no solutions. (287)

independent events Two or more events whose outcomes do not affect each other. (851)

independent variable In a formula, a variable upon whose value other variables depend. (12)

index The subscript used for a term in a sequence indicating the position of the term in the sequence. The variable under the Σ sign in summation notation. (44, 824)

index variable The variable under the Σ sign in summation notation; also called *index*. (824)

inequality An open sentence containing one of the symbols $<$, $>$, $\leq$, $\geq$, $\neq$ or $\approx$. (272)

initial condition The starting point in a situation. (141)

input A value of an independent variable. (13)

integer An element of the set $\{0, 1, -1, 2, -2, 3, -3, \ldots\}$. (14)

interior of a circle The region inside a circle. (760)

intersection of two sets The set consisting of those values common to both sets. (273)

interval A solution to an inequality of the form $x \leq a$ or $a \leq x \leq b$, where the $\leq$ can be replaced by $<$, $>$, or $\geq$. (272)

inverse of a matrix Matrices M and N are inverse matrices if and only if their product is the identity matrix. (299)

inverse of a relation The relation obtained by reversing the order of the coordinates of each ordered pair in the relation. (485)

inverse trigonometric functions One of the functions $\cos^{-1}$, $\sin^{-1}$, or $\tan^{-1}$. (611)

inverse-square curve The graph of $y = \frac{k}{x^2}$. (106)

inverse-square variation A variation that can be described by the equation $y = \frac{k}{x^2}$, with $k \neq 0$. (80)

inverse-variation function A function with a formula of the form $y = \frac{k}{x^n}$, with $k \neq 0$, *and* $n > 0$. (78)

inversely proportional to A relationship between two variables whose product is a constant. Also called *varies inversely as*. (78)

irrational number A real number which cannot be written as a ratio of integers. (354)

irreducible polynomial A polynomial that cannot be factored into polynomials of lower degree with coefficients in the same domain as the coefficients of the given polynomial. Also called *prime polynomial*. (689)

joint variation A situation in which one quantity varies directly as the product of two or more independent variables, but not inversely as any variable. (124)

lattice point A point with integer coordinates. (758)

leading coefficient The coefficient of the variable of highest power in a polynomial in a single variable. (674)

least squares line A line that best fits the data. Also called *regression line* or *line of best fit*. (170)

limit A number or figure which the terms of a sequence approach as n gets larger. (54)

line of best fit A line that best fits the data. Also called *regression line* or *least squares line*. (170)

line of reflection The line over which a figure is reflected. (232)

line of sight An imaginary line from one position to another, or in a particular direction. (612)

line of symmetry For a figure F, a line m such that the reflection image of F over m equals F itself. (99)

linear combination An expression of the form $Ax + By$ is called a linear combination of x and y. (152)

linear function A function f with the equation $f(x) = mx + b$, where m and b are real numbers. (141)

linear inequality An inequality in which both sides are linear expressions. (272)

linear polynomial A polynomial of the first degree, such as $y = mx + b$. (675)

linear scale A scale with units spaced so that the difference between successive units is the same. (565)

linear sequence A sequence with a constant difference. Also called *arithmetic sequence*. (176)

linear-combination method A method of solving systems which involves adding multiples of the given equations. (294)

linear-combination situation A situation in which all variables are to the first power and are not multiplied or divided by each other. (154)

linear-programming problem A problem which leads to a system of linear inequalities in which the goal is to maximize or minimize a linear combination of the solutions to the system. (326)

log x The logarithm of x to the base 10. The exponent to which 10 must be raised to equal x. (558)

logarithm function to the base 10 The function with equation $y = \log_{10} x$ or $y = \log x$. See also *common logarithm function*. (560)

logarithm function to the base b The function with equation $y = \log_b x$. (570)

logarithm of m to the base b Let $b > 0$ and $b \neq 1$. Then n is the logarithm of m to the base b, written $n = \log_b m$, if and only if $b^n = m$. (570)

logarithm of x to the base 10 y is the logarithm of x to the base 10, written $y = \log x$, if and only if $10^y = x$. (558)

logarithmic curve The graph of a function of the form $y = \log_b x$. (558, 571)

logarithmic equation An equation of the form $y = \log_b x$. (560)

logarithmic scale A scale in which the units are spaced so that the ratio between successive units is the same. (565)

lottery A game or procedure in which prizes are distributed among people by pure chance. (857)

magnitude of a size change In the size change that maps (x, y) onto (kx, ky), the number k. Also called *size change factor*. (221)

major axis of an ellipse The segment which contains the foci and has two vertices of an ellipse as its endpoints. (767)

mapping notation The notation $f: x \rightarrow y$ for a function f. Also called *arrow notation*. (21)

mathematical model A graph, sentence, or other mathematical idea that describes an aspect of a real-world situation. (111)

matrix A rectangular arrangement of objects, its *elements*. (203)

matrix addition If two matrices A and B have the same dimensions, their sum $A + B$ is the matrix in whose element in each position is the sum of the corresponding elements in A and B. (209)

matrix form of a system A representation of a system using matrices. The matrix form for
$$\begin{cases} ax + by = e \\ cx + dy = f \end{cases} \text{ is}$$
$$\begin{bmatrix} a & b \\ c & d \end{bmatrix} \begin{bmatrix} x \\ y \end{bmatrix} = \begin{bmatrix} e \\ f \end{bmatrix}. \ (305)$$

matrix multiplication Suppose A is an $m \times n$ matrix and B is an $n \times p$ matrix. The product $A \cdot B$ or AB is the $m \times p$ matrix whose element in row i and column j is the product of row i of A and column j of B. (215)

matrix subtraction If two matrices A and B have the same dimensions, their difference $A - B$ is the matrix whose element in each position is the difference of the corresponding elements in A and B. (210)

maximum The largest value in a set. (99)

mean The result of adding the n numbers in a data set and dividing the sum by n. Also called *arithmetic mean* or *average*. (830)

measure of center A number which in some sense is at the "center" of a data set; the mean or median of a data set. Also called *measure of central tendency*. (831)

measure of spread A number, like standard deviation, which describes the extent to which elements of a data set are dispersed or spread out. (831)

median When the terms of a data set are placed in increasing order, if the set has an odd number of terms, the middle term; if the set has an even number of terms, the average of the two terms in the middle. (830)

method of finite differences A technique used to determine whether a data set can be modeled by a polynomial function. If taking differences of consecutive y-values eventually produces differences which are constant, then the data set can be modeled by a polynomial function. (726)

midpoint formula the midpoint of the segment with endpoints (x_1, y_1) and (x_2, y_2) is $\left(\dfrac{x_1 + x_2}{2}, \dfrac{y_1 + y_2}{2}\right)$. (60)

minimum The smallest value in a set. (99)

minor axis of an ellipse The segment which has two vertices of an ellipse as its endpoints and does not contain the foci. (767)

mode The number or numbers which occur most often in a data set. (830)

model for an operation A pattern that describes many uses of that operation. (7)

monomial A polynomial with one term. (680)

multiplicity of a root For a root r of a polynomial equation $P(x) = 0$, the highest power of $x - r$ that appears as a factor of $P(x)$. (718)

mutually exclusive events Two or more events which cannot happen at the same time. (851)

Napierian logarithm Another name for natural logarithm. (583)

natural logarithm A logarithm to the base e, written ln. Also called *Napierian logarithm*. (583)

natural number An element of the set $\{1, 2, 3, 4, 5, \ldots\}$. Also called *counting number*. (14)

neutral A substance whose pH is 7; a substance which is neither acidic or alkaline. (565)

normal curve The curve of a normal distribution. (864)

normal distribution A function whose graph is the image of the graph of $y = \dfrac{1}{\sqrt{2}} e^{\frac{-x^2}{2}}$ under a composite of translations or scale transformations. (864)

normalized scores Scores whose distribution is a normal curve. Also called *standardized scores*. (865)

***n*th power function** The function defined by $f(x) = x^n$, where n is a positive integer. (420)

***n*th root** Let n be an integer greater than one. Then b is an nth root of x if and only if $b^n = x$. (450)

***n*th term** The term occupying the nth position in the listing of a sequence. The general term of a sequence. (43)

oblique line A line that is neither horizontal or vertical. (159)

one-to-one correspondence A mapping in which each member of one set is mapped to a distinct member of another set, and vice versa. (221)

open sentence A sentence that may be true or false depending on what values are substituted for the variables in it. (272)

opens down A description of the shape of a parabola whose vertex is a maximum; a parabola whose equation is of the form $y = ax^2 + bx + c$, where $a < 0$. (99)

opens up A description of the shape of a parabola whose vertex is a minimum; a parabola whose equation is of the form $y = ax^2 + bx + c$, where $a > 0$. (99)

order of operations Rules used to evaluate expressions worldwide. 1. Perform operations within grouping symbols from inner to outer. 2. Take powers. 3. Do multiplications or divisions from left to right. 4. Do additions or subtractions from left to right. (8)

output A value of the dependent variable in a function. (13)

parabola The set consisting of every point in the plane of a line ℓ (its *directrix*) and a point F not on ℓ (its *focus*) whose distance from F equals its distance from ℓ. (99, 748)

paraboloid A three-dimensional figure created by rotating a parabola in space around its axis of symmetry. The set of points equidistant from a point F (its focus) and a plane P. (751)

Pascal's triangle The sequence satisfying $\binom{n}{0} = \binom{n}{n} = 1$ for all integers $n \geq 0$, and $\binom{n+1}{r+1} = \binom{n}{r} + \binom{n}{r+1}$, where n and r are any integers with $0 \leq r \leq n$. The triangular array

where if x and y are located next to each other on a row, the element just below and directly between them is $x + y$. (47, 837)

perfect-square trinomial A trinomial of the form $a^2 + 2ab + b^2$ or $a^2 - 2ab + b^2$. (370)

period The horizontal translation of smallest positive magnitude that maps the graph of a function onto itself. (649)

periodic function A function whose graph can be mapped to itself under a horizontal translation. (649)

permutation An arrangement of n different objects in a specific order. (826)

pH scale A logarithmic scale used to measure the acidity of a substance. (565)

piecewise-linear graph A graph made of parts, each of which is a piece of a line. (142)

pitch The measure of the steepness of the slant of a roof. (39)

point matrix A 2×1 matrix. (205)

point-slope form of a linear equation An equation of the form $y - y_1 = m(x - x_1)$, where (x_1, y_1) is a point on the line with slope m. (163)

polynomial equation An equation of the form $y = a_nx^n + a_{n-1}x^{n-1} + \ldots + a_1x^1 + a_0$, where n is a positive integer and $a_n \neq 0$. (673)

polynomial function A function f of the form $f(x) = a_nx^n + a_{n-1}x^{n-1} + \ldots + a_1x^1 + a_0$, where n is a positive integer and $a_n \neq 0$. (675)

polynomial in x An expression of the form $a_nx^n + a_{n-1}x^{n-1} + a_{n-2}x^{n-2} + \ldots + a_1x^1 + a_0$, where n is a positive integer and $a_n \neq 0$. (674)

polynomial model A polynomial equation which fits a data set. (730)

population In a sampling situation, the set of all people, events, or items that could be sampled. (868)

Power of a Power Postulate For any nonnegative bases and nonzero real exponents or any nonzero base and integer exponents, $(b^m)^n = b^{mn}$. (427)

Power of a Product Postulate For any positive bases and real exponents or any nonzero bases and integer exponents, $(ab)^m = a^mb^m$. (428)

Power of a Quotient Postulate For any positive bases and real exponents, or any nonzero bases and integer exponents, $\left(\frac{a}{b}\right)^m = \frac{a^m}{b^m}$. (429)

power The expression x^n; the result of the operation of exponentiation or powering. (418)

powering An operation by which a variable is raised to a power. Also called *exponentiation*. (418)

preimage An object to which a transformation is applied. (221)

prime polynomial A polynomial that cannot be factored into polynomials of lower degree with coefficients in the same domain as the coefficients of the given polynomial. Also called *irreducible polynomial*. (689)

principal The amount of money invested in an investment. (438)

probability of an event If a situation has a total of t equally likely possibilities and e of these possibilities satisfy conditions for a particular event, then the probability of the event is $\frac{e}{t}$. (851)

probability distribution A function which maps a set of events onto their probabilities. Also called *probability function*. (862)

Product of Powers Postulate For any nonnegative bases and nonzero real exponents, or any nonzero bases and integer exponents, $b^m \cdot b^n = b^{m+n}$. (426)

quadratic An expression, equation, or function that involves sums of constants and first and second powers of variables, but no higher power. (346)

quadratic equation An equation which involves quadratic expressions. (346)

quadratic equation in two variables An equation of the form $Ax^2 + Bxy + Cy^2 + Dx + Ey + F = 0$, where A, B, C, D, E, and F are real numbers and at least one of A, B, or C is not zero. (747)

quadratic expression An expression which contains one or more terms in x^2, y^2, or xy, but no higher powers of x or y. (346)

quadratic form An expression of the form $Ax^2 + Bxy + Cy^2 + Dx + Ey + F$. (346)

quadratic function The function with equation $f(x) = ax^2 + bx + c$. (346)

quadratic model A quadratic equation which fits a set of data. (376)

quadratic polynomial A polynomial of a single variable with degree 2, such as $ax^2 + bx + c$. (675)

quadratic relation in two variables The sentence $Ax^2 + Bxy + Cy^2 + Dx + Ey + F = 0$ (or the inequality using one of the symbols $>$, $<$, $\geq$, $\leq$) where A, B, C, D, E, and F are real numbers and at least one of A, B, or C is not zero. (747)

quadratic system A system that involves at least one quadratic sentence. (789)

quadratic-linear system A system that involves linear and quadratic sentences. (789)

quadratic-quadratic system A system that involves two quadratic sentences. (794)

quartic equation A fourth degree polynomial equation. (717)

quartic polynomial A polynomial of a single variable with degree 4, such as $ax^4 + bx^3 + cx^2 + dx + e$. (675)

quintic equation A fifth degree polynomial equation. (717)

Quotient of Powers Property For any positive bases and real exponents, or any nonzero bases and integer exponents: $\frac{b^m}{b^n} = b^{m-n}$. (429)

r_m The reflection over line m. (232)

r_x The reflection over the x-axis. (234)

r_y The reflection over the y-axis. (233)

$r_{y=x}$ The reflection over line $y = x$. (234)

R_{90} The rotation of magnitude 90° counterclockwise with center at the origin. (247)

R_{180} The rotation of magnitude 180° counterclockwise with center at the origin. (243)

R_{270} A rotation of magnitude 270° counterclockwise with center at the origin. (248)

R_x A rotation of magnitude x counterclockwise with center at the origin. (246)

radian (rad) A measure of an angle, arc, or rotation such that π radians = 180 degrees. (658)

radical notation $\sqrt[n]{x}$ The notation for the nth root of an expression. (495)

radical sign $\sqrt{x}$ The symbol for the square root of x. (495)

radius The distance between any point on a circle and the center of the circle. (754)

random numbers Numbers which have the same probability of being selected. (869)

random sample A sample in which each element has the same probability as every other element in the population of being selected for the sample. (869)

range of a function   The set of values of the function. (13)

rate of change   Between two points, the quantity $\frac{y_2 - y_1}{x_2 - x_1}$. For a line, its slope. (89)

ratio of similitude   In two similar figures, the ratio between a length in one figure and the corresponding length in the other. (222)

rational number   A number which can be written as a simple fraction. A finite or infinitely repeating decimal. (14)

rationalizing the denominator   When a fraction has irrational or complex numbers in its denominator, the process of rewriting a fraction without irrational or complex numbers in its denominator. (507)

real numbers   Those numbers that can be represented by finite or infinite decimals. (14)

real part   In a complex number of the form $a + bi$, the real number a. (393)

rectangular hyperbola   A hyperbola with perpendicular asymptotes. (786)

recursive formula   A set of statements that indicates the first term of a sequence and gives a rule for how the nth term is related to one or more of the previous terms. Also called *recursive definition*. (49)

reflecting line   In a reflection, the perpendicular bisector of the line segment connecting a preimage point and its image. (232)

reflection   A transformation under which the image of a point P over a reflecting line m is (1) P itself, if P is on m; (2) the point P' such that m is the perpendicular bisector of the segment connecting P with P', if P is not on m. (232)

reflection-symmetric figure   A figure which coincides with a reflection image of itself. (99)

refraction   When a beam of light in air strikes the surface of water it is refracted or bent. (643)

regression line   A line that best fits a set of data. Also called *line of best fit* or *line of least squares*. (170)

relation   A set of ordered pairs. (26)

repeated multiplication model for powering   If b is a real number and n is a positive integer, then $b^n = \underbrace{b \cdot b \cdot b \cdot b \cdot \ldots \cdot b}_{n \text{ factors}}$. (418)

Richter scale   A logarithmic scale used to measure the magnitude of intensity of an earthquake. (568)

root of an equation   A solution to an equation. (403)

rotation   A transformation with a center O under which the image of O is O itself and the image of any other point P is the point P' such that m$\angle POP'$ is a fixed number (its *magnitude*). (246)

rounding down function   The function, denoted by $\lfloor x \rfloor$, whose values are the greatest integer less than or equal to x. Also called *greatest integer function* or *floor function*. (187)

row   A horizontal list in a table, rectangular array, or spreadsheet. (204)

sample   In a sampling situation, the subset of the population actually studied. (868)

sampling   Using a subset of a population to estimate a result for an entire population. (868)

scalar   A real number by which a matrix is multiplied. (210)

scalar multiplication   An operation leading to the product of a scalar k and a matrix A, namely the matrix kA in which each element is k times the corresponding element in A. (210)

scale change   The stretching or shrinking of a figure in either a horizontal direction only, in a vertical direction only, or in both directions. The horizontal scale change of magnitude a and a vertical scale change of magnitude b maps (x, y) onto (ax, by), and is denoted by $S_{a,b}$. (227)

scatterplot   A plot with discrete points used to display a data set. (169)

scientific calculator   A calculator which performs arithmetic using algebraic order of operations, and with keys such as those for exponents, powering, logarithms, inverses, and trigonometric functions. (15)

sequence   An ordered list. (41)

series   An indicated sum of terms in a sequence. (813)

shrink   A scale change in which a magnitude in some direction has absolute value less than one. (227)

sigma notation (Σ-notation)   A shorthand notation used to restate a series. Also called *summation notation*. (824)

similar figures   Two figures such that one is the image of the other under a composite of isometries (reflections, rotations, translations, glide reflections) and size changes. (222)

simple fraction   A fraction of the form $\frac{a}{b}$, where a and b are integers and $b \neq 0$. (354)

simple interest   The amount of interest I earned when calculated using the formula $I = Prt$, where P is the principal, r is the annual rate, and t is the time in years. (442)

simplified form   An expression rewritten so that like terms are combined, fractions are reduced, and only rational numbers are in the denominator. (501)

simplify an nth root   The process of factoring the expression under the radical sign into perfect nth powers and then applying the Root of a Product Theorem. (501)

simulation   A procedure used to answer questions about real-world situations by performing experiments that closely model them. (869)

$\sin^{-1} x$ The number between $-90°$ and $90°$, or between $-\frac{\pi}{2}$ and $\frac{\pi}{2}$, whose sine is x. If $\sin u = v$, then on a restricted domain, $\sin^{-1} v = u$. (611, 653)

sine function   The correspondence $\theta \rightarrow \sin \theta$ that associates θ with the y-coordinate of the image of $(1, 0)$ under R_θ. (647)

sine of θ (sin θ) In general, the second coordinate of $R_\theta(1, 0)$. In a right triangle with acute angle θ, $\sin\theta = \frac{\text{length of leg opposite }\theta}{\text{length of hypotenuse}}$. (605, 624)

sine wave A graph which can be mapped onto the graph of $g(\theta) = \sin\theta$ by any composite of reflections, translations, and scale changes. (649)

sinusoidal situations Situations that lead to sine waves. (649)

size change For any $k \neq 0$, the transformation that maps the point (x, y) onto (kx, ky); a transformation with center O such that the image of O is O itself and the image of any other point P is the point P' such that $OP' = k \cdot OP$ and P' is on ray OP if k is positive, and on the ray opposite ray OP if k is negative. (221)

size change factor In the size change $(x, y) \to (kx, ky)$, the number k. (221)

slope The slope determined by two points (x_1, y_1) and (x_2, y_2) is $\frac{y_2 - y_1}{x_2 - x_1}$. Also called *rate of change*. (90)

slope-intercept form of a linear equation A linear equation of the form $y = mx + b$, where m is the slope and b is the y-intercept. (141)

solution set for a system The intersection of the solution sets of the individual sentences of a system. (279)

solving a sentence Finding all solutions to a sentence. (30)

solving a triangle The use of trigonometry to find all the missing measures of sides and angles of a triangle. (655)

square matrix A matrix with the same number of rows and columns. (299)

square of x The second power of x, denoted by x^2. (418)

square root A square root of t is a solution to $y^2 = t$. The positive square root of a positive number x is denoted $\sqrt{x}$. (352, 450)

square root function The function f with equation $f(x) = \sqrt{x}$, where x is a nonnegative real number. (14)

squaring function The powering function f defined by $f(x) = x^2$. (420)

standard deviation Let S be a data set of n numbers $\{x_1, x_2, \ldots, x_n\}$. Let m be the mean of S. Then the standard deviation (s.d.) of S is $\sqrt{\frac{\sum_{i=1}^{n}(x_i - m)^2}{n}}$. (832)

standard form of a linear equation An equation for a line in the form $Ax + By = C$, where A and B are not both zero. (157)

standard form of a polynomial: A polynomial written in the form $a_n x^n + a_{n-1} x^{n-1} + \ldots + a_1 x^1 + a_0$, where n is a positive integer and $a_n \neq 0$. (674)

standard form of a quadratic equation An equation of the form $ax^2 + bx + c = 0$, with $a \neq 0$. (385)

standard form of a quadratic relation An equation in the form $Ax^2 + Bxy + Cy^2 + Dx + Ey + F = 0$ where A, B, C, D, E, and F are real numbers and at least one of A, B, or C is nonzero. (786)

standard form of an equation for a hyperbola An equation for a hyperbola in the form $\frac{x^2}{a^2} - \frac{y^2}{b^2} = 1$, where $b^2 = c^2 - a^2$, the foci are $(c, 0)$ and $(-c, 0)$ and the focal constant is $2a$. (779)

standard form of an equation for a parabola An equation for a parabola in the form $y = ax^2 + bx + c$, where $a \neq 0$. (363)

standard form of an equation for an ellipse An equation for an ellipse in the form $\frac{x^2}{a^2} + \frac{y^2}{b^2} = 1$, where $b^2 = a^2 - c^2$, with foci $(c, 0)$ and $(-c, 0)$ and focal constant $2a$. (767)

standard position for an ellipse (or hyperbola) A location in which the origin of a coordinate system is midway between the foci with the foci on an axis. (765, 779)

standardized scores Scores whose distribution is a normal curve. Also called *normalized scores*. (865)

statistical measure A single number which is used to describe an entire set of numbers. (830)

step function A graph that looks like a series of steps, such as the graph of the function with equation $y = \lfloor x \rfloor$. (186)

stratified random sample A sample that is the union of samples chosen randomly from subpopulations of the entire population. (869)

stratified sample A sample in which the population has first been split into subpopulations and then, from each subpopulation, a sample is selected. (869)

stretch A scale change $(x, y) \to (ax, by)$ in which a or b is greater than one. (227)

subscript A number or variable written below and to the right of a variable. (44)

subscripted variable A variable with a subscript. (44)

subset A set whose elements are all chosen from a given set. (845)

subtraction of matrices Given two matrices A and B having the same dimensions, their difference $A - B$ is the matrix whose element in each position is the difference of the corresponding elements in A and B. (210)

sum of cubes pattern For all a and b, $a^3 + b^3 = (a + b)(a^2 - ab + b^2)$. (691)

summation notation A shorthand notation used to restate a series. Also called Σ-*notation* or *sigma notation*. (824)

symbol manipulator Computer software of a calculator preprogrammed to perform operations on variables. (675)

system A set of conditions joined by the word "and"; a special kind of compound sentence. (279)

tan^{-1} The number between $0°$ and $180°$, or between 0 and π whose tangent is x. If $\tan u = v$, then on a restricted domain, $\tan^{-1} v = u$. (611)

tangent of θ (tan θ) In general, $\tan\theta = \frac{\sin\theta}{\cos\theta}$, provided $\cos\theta \neq 0$. In a right triangle with acute angle θ, $\tan\theta = \frac{\text{length of leg opposite }\theta}{\text{length of leg adjacent to }\theta}$. (605, 618)

tangent line A line that intersects a circle or ellipse in exactly one point. (791)

term of a sequence An element of a sequence. (42)

theorem   In a mathematical system, a statement that has been proved. (27)

transformation   A one-to-one correspondence between sets of points. (221)

translation   A transformation that maps (x, y) onto $(x + h, y + k)$, denoted by $T_{h,k}$. (256)

trial   One occurrence of an experiment. (852)

triangular number   An element of the sequence 1, 3, 6, 10, . . ., whose nth term is $\frac{n(n+1)}{2}$. (42)

triangulation   The process of determining the location of points using triangles and trigonometry. (641)

trigonometric ratios   The ratios of the lengths of the sides in a right triangle. (605)

trinomial   A polynomial with three terms. (680)

union of two sets   The set consisting of those elements in either one or both sets. (274)

unit circle   The circle with center at the origin and radius 1. (623)

value of a function   If $y = f(x)$, the value of y. (20)

variable   A symbol that can be replaced by any one of a set of numbers or other objects. (6)

varies directly as   The situation that occurs when two variables x and y are so related that when one of the variables is multiplied by k, so is the other. Also called *directly proportional to*. (72)

varies inversely as   The situation that occurs when two variables x and y are so related that when one of the variables is multiplied by k, the other is divided by k. Also called *inversely proportional to*. (78)

velocity   The rate of change of distance with respect to time. (366)

vertex form of an equation of a parabola   An equation of the form $y - k = a(x - h)^2$ where (h, k) is the vertex of the parabola. (359)

vertex of a parabola   The intersection of a parabola and its axis of symmetry. (359, 748)

vertical asymptote   A vertical line that is approached by the graph of a relation. (107)

vertical line   A line with an equation of the form $x = h$. (158)

vertical magnitude   In the scale change $(x, y) \rightarrow (ax, by)$, the number b. (227)

vertical scale change   A transformation that maps (x, y) onto (x, by). (227)

vertices of a hyperbola   The points of intersection of the hyperbola and the line containing its foci. (777)

vertices of an ellipse   The endpoints of the major and minor axes of the ellipse. (767).

whole number   An element of the set {0, 1, 2, 3, 4, 5, . . . }. (14)

window   The part of the coordinate grid shown on the screen of an automatic grapher. (94)

x-axis   The line in the coordinate plane in which the second coordinates of points are 0. (24)

x-intercept   The value of x at a point where a graph crosses the x-axis. (159)

y-axis   The line in the coordinate plane in which the first coordinates of points are 0. (24)

y-intercept   The value of y at a point where a graph crosses the y-axis. (141)

yield   The rate of interest earned after all the compoundings have taken place in one year. Also called *effective annual yield or annual yield*. (440)

zero of a function   For a function f, a value of x for which $f(x) = 0$. (700)

zoom   A feature on an automatic grapher which enables the window of a graph to be changed without keying in interval endpoints for x and y. Also called *rescaling*. (282)

$A \cap B$	intersection of sets A and B	f^{-1}	inverse of a function f
$A \cup B$	union of sets A and B	$\log_b m$	logarithm of m to the base b
$f(x)$	function notation read "f of x"	e	2.71828 . . .
$f{:}x \rightarrow y$	function notation read "f maps x onto y"	$x!$	x factorial
A'	image of A	$\ln x$	natural logarithm of x
S_k	size change of magnitude k	$\sin \theta$	sine of θ
$S_{a,b}$	scale change with horizontal magnitude a and vertical magnitude b	$\cos \theta$	cosine of θ
		$\tan \theta$	tangent of θ
		rad	radian
r_x	reflection over the x-axis	a_n	"a sub n"; the nth term of a sequence
r_y	reflection over the y-axis		
$r_{y=x}$	reflection over the line $y = x$	$\sum\limits_{i=1}^{n} i$	the sum of the integers from 1 to n
$T_2 \circ T_1$	composite of transformations T_1 and T_2	S_n	the partial sum of the first n terms of a sequence
R_θ	rotation of magnitude θ counterclockwise	$\binom{n}{r}, {}_nC_r$	the number of ways of choosing r objects from n objects
$T_{h,k}$	translation of h units horizontally and k units vertically	INT (X)	the BASIC equivalent for $\lfloor x \rfloor$
$\begin{bmatrix} a & b \\ c & d \end{bmatrix}$	2×2 matrix	$\boxed{\sqrt[x]{}}, \boxed{\sqrt[x]{y}}$	calculator nth root key
M^{-1}	inverse of matrix M	$\boxed{x!}$	calculator factorial key
det M	determinant of matrix M	$\boxed{y^x}$	calculator powering key
$\sqrt{}$	radical sign; square root	$\boxed{x^{-1}}$	calculator reciprocal key
$\sqrt[n]{x}$	the real nth root of x	$\boxed{\log}$	calculator common logarithm key
i	$\sqrt{-1}$	$\boxed{e^x}$	calculator e^x key
$\sqrt{-k}$	a solution of $x^2 = -k$, $k > 0$	$\boxed{\ln}$	calculator natural logarithm key
$a + bi$	a complex number, where a and b are real numbers	$\boxed{\sin}$	calculator sine key
$g \circ f$	composite of functions f and g	$\boxed{\cos}$	calculator cosine key
$|x|$	absolute value of x	$\boxed{\tan}$	calculator tangent key
$\lfloor x \rfloor$	greatest integer less than or equal to x		

Acknowledgments

Unless otherwise acknowledged, all photographs are the property of Scott, Foresman & Company. Page abbreviations are as follows: (T)top, (C)center, (B)bottom, (L)left, (R)right, (INS)inset.

COVER & TITLE PAGE: Steven Hunt (c) 1994 vi(l) William J. Warren/West Light vi(r) Tim Laman/Adventure Photo vii(l) Peticdas/Megna/Fundamental Photographs vii(r) Jack Krawczyk/Panoramic Images, Chicago viii Index Stock International ix Telegraph Colour Library/FPG x Steve Chenn/West Light 3 Brooks Kraft/Sygma 4C Steve Vance/Stockworks 4BL David Phillips/Photo Researchers 5C William J. Warren/West Light 5BR Pfetschinger/Peter Arnold, Inc. 6 Sidney Harris 8 David Joel/Tony Stone Images 10 Courtesy Tsakurshori, Second Mesa, AZ/Jerry Jacka Photography 12 Tony Freeman/Photo Edit 19 Milt & Joan Mann/Cameramann International, Ltd. 22 Myrleen Cate/Photo Edit 23 David Ahrenberg/Tony Stone Images 30 Milt & Joan Mann/Cameramann International, Ltd. 32 Christopher Brown/Stock Boston 34 Michael Newman/Photo Edit 36 Eric Neurath/Stock Boston 39 Mark Segal/Tony Stone Images 40 Tony Freeman/Photo Edit 42 David Carriere/Tony Stone Images 45 CNRI/SPL/Photo Researchers 48 John D. Cunningham/Visuals Unlimited 54 Ed Simpson/Tony Stone Images 57 Oxford Scientific Films/ANIMALS ANIMALS 58 Sidney Harris 61ALL Ron Kimball 62 Claude Nuridsany & Marie Perennou/Photo Researchers 70TL NASA 70-71T R.Kord/H. Armstrong Roberts 70-71C H.D.Thoreau/West Light 70B Tim Laman/Adventure Photo 71T Tom Tracy/The Stock Shop 72 David Joel/Tony Stone Images 73 Guido A. Rossi/The Image Bank 75 Milt & Joan Mann/Cameramann International, Ltd. 77 Phyllis Picardi/Stock Boston 78 James Shaffer/Photo Edit 80 NASA 82 Al Tielemans/Duomo Photography Inc. 84 John Chellman/ANIMALS ANIMALS 86 Tom McHugh/Natural History Museum of Los Angeles County/Photo Researchers 89 Tom Ives 92 Courtesy General Dynamics, Electric Boat Division 101 VU/Carlyn Galati/Visuals Unlimited 102 Milt & Joan Mann/Cameramann International, Ltd. 104 National Optical Astronomy Observatories & Lowell Observatory 110 Darryl Torckler/Tony Stone Images 113 Bob Newman/Visuals Unlimited 116 Edward Lee/Tony Stone Images 122 David Young-Wolff/Photo Edit 124 Milt & Joan Mann/Cameramann International, Ltd. 125 Brent Jones 126 Courtesy Andersen Windows 138TL The Stock Market 138-139T Tim Brown/Profiles West 138C Peticolas/Megna/Fundamental Photographs 138-139B Art Wolfe/Tony Stone Images 139C Lance Nelson/The Stock Market 140 Jean Francois Causse/Tony Stone Images 141 John Cancalosi/Stock Boston 143 Linc Correll/Stock Boston 145 David Ball/The Stock Market 151 Patti Murray/ANIMALS ANIMALS 152 Focus On Sports 155 John Curtis/The Stock Market 157 Stephen Frisch/Stock Boston 159 V.Jane Windsor/St.Petersburg Times 161 Bob Strong/The Image Works 164 Francis Lepine/Valan Photos 166 Andrew Sacks/Tony Stone Images 167 Don Mason/Susan Havel/The Stock Market 169 James Marshall/The Stock Market 170 Milt & Joan Mann/Cameramann International, Ltd. 171 Milt & Joan Mann/Cameramann International, Ltd. 175 Laima Druskis/Stock Boston 176 Milt & Joan Mann/Cameramann International, Ltd. 178 Michael Keller/The Stock Market 182 Don Dubroff/Tony Stone Images 184 Charles Gupton/Stock Boston 185 Ariel Skelley/The Stock Market 186 Fujifotos/The Image Works 191T DOLLEY MADISON by Bass, Otis (c)1817, The New-York Historical Society, New York City 191B ANNA ELEANOR ROOSEVELT, detail, Copyright by the White House Historical Association 192 The Stock Shop 193T David Madison 202-203T Jack Krawczyk/Panoramic Images, Chicago 202CL P.George/H. Armstrong Roberts 202CR Index Stock International 202-203B C.Ursillo/H. Armstrong Roberts 203C Douglas Pulsipher/The Stock Solution 204 Courtesy of United Musical Instruments U.S.A.Inc. 207 U. S. Army Photo Center of Military History 210 John Colwell/Grant Heilman Photography 212 Alan Carey/The Image Works 213 Courtesy The Kohler Company 214 Michael Newman/Photo Edit 216 M.Granitsas/The Image Works 218T Jerry Jacka Photography 218B Chip & Rosa Maria de la Cueva Peterson 223ALL Courtesy of Jim Jennings, Jennings Chevrolet & Geo Inc., Glenview, IL 225ALL Everett Collection 226 Nubar Alexanian/Stock Boston 229 Photo Edit 230 Terry Donnelly/Tony Stone Images 232 Leo Keeler/Earth Scenes 238 Dirk Gallian 240 Milt & Joan Mann/Cameramann International, Ltd. 246 Brent Jones 249 Jeffrey Muir Hamilton/Stock Boston 254 Alex MacLean/Landslides 255 Rosemary Finn 261T Wiley/Wales/Profiles West 261BL Library of Congress 261BR Library of Congress 268B Courtesy R.R.Donnelley & Sons Co. 269 David R. Frazier Photolibrary 270-271T Gary Mirando/New England Stock Photo 270C Color Box/FPG 270B Index Stock International 271C SuperStock, Inc. 271BL Color Box/FPG 272 David Madison 274 Michael Newman/Photo Edit 278 Bettmann 283 Suzanne Murphy/Tony Stone Images 284 Robert Frerck/Odyssey Productions, Chicago 286 Giuliano Colliva/The Image Bank 296 Mark Antman/The Image Works 302 Bettmann 304 Doug Miner/Sygma 305 Ralph Mercer/Tony Stone Images 311 John Eastcott/YVA Momatiuk/The Image Works 312 Randy G. Taylor/Leo de Wys 313 B. & J. Heaton/Stock Boston 315 David Falconer/David R. Frazier Photolibrary 318 Brooks Kraft/Sygma 320 Henkel-Harris Furnit/Stock Boston 326 John Eastcott/YVA Momatiuk/Stock Boston 329 Thomas Hovland/Grant Heilman Photography 331 Eastcott/Momatiuk/The Image Works 333T Bettmann 335 Courtesy McDonnell-Douglas 336-337 John Kelly/Tony Stone Images 336INS Stockworks 337INS Headhunters 344T Travelpix/FPG 344C Backgrounds/West Light 344BL Gary A. Bartholomew/West Light 344-345BR&B Ron Watts/West Light 345T Sipa/Fritz/Leo de Wys 346 Greig Cranna/Stock Boston 350 Jeff Gnass 355 Vandystadt/Photo Researchers 357 Bob Amft 362 Nancy Pierce/Photo Researchers 363 Jeff Gnass 364 Bettmann Archive 368 Zalman Usiskin 372 Focus On Sports 376 Bob Daemmrich/Stock Boston 378 Jeff Gnass 381 Bettmann Archive 386 Focus On Sports 389 Sidney Harris 392 Neg.A91033/Field Museum of Natural History, Chicago 393 From CHAOS by J.Glieck,

(c)1987 Viking Press 395 Philippe Plailly/SPL/Photo Researchers 405 Hulton Deutsch Collection Ltd. 408T Jay Silverman/The Image Bank 408B David Madison 409T Arthur Tilley/FPG 409C H.M.Gousha, a division of Simon & Schuster, Inc. All rights reserved. Used by permission. 409B West Light 416T R.Price/West Light 416-417B Michael Schimpf/Mon Tresor/Panoramic Images, Chicago 417T Ralph A.Clevenge/West Light 417C Roberto Villa/Leo de Wys 418 Everett Collection 423 Breck P. Kent/ANIMALS ANIMALS 424 Porterfield/Chickering/Photo Researchers 426 NASA 429 National Optical Astronomy Observatories 433 Biophoto Associates/Photo Researchers 437 Copyright by the White House Historical Association, Photo: National Geographic Society 441 Jose L.Pelaez/The Stock Market 443 Focus On Sports 444 Joel Gordon Photography 448 Lee Boltin 450 Teri Bloom 452 David Spangler 455 Miro Vintoniv/Stock Boston 461 Instituto E. Museo di Storia Della Scienza 462 John Gerlach/Earth Scenes 464 Grant Heilman/Grant Heilman Photography 467 Catherine Koehler 468 James W.Kay 469T Bill Losh/FPG 469C Sussane Kaspar/Leo de Wys 469B Jon Feingersh/The Stock Market 470L Rob Bolster/Stockworks 470R Charles Waller/Stockworks 476–477 Pete Turner, Inc./The Image Bank 476C Bill Ross/West Light 476B (c)1991 Cindy Lewis 477C Mark Harwood/Tony Stone Images 477B Steven M. Rollman/Natural Selection 478 Mugshots/The Stock Market 482 Zig Leszczynski/ANIMALS ANIMALS 483 Tony Freeman/Photo Edit 489 Campbell 494 Photo Courtesy Ringling Brothers and Barnum & Bailey Combined Shows, Inc. 495 Sidney Harris 499 Mike Mazzaschi/Stock Boston 500 PhotoFest 502, 503 NASA 506 Lawrence Migdale 509 David Wells/The Image Works 511 Bob Amft 516, 517 Milt & Joan Mann/Cameramann International, Ltd. 521T Drake Well Museum 521B Milt & Joan Mann/Cameramann International, Ltd. 523B Mary Evans Picture Library 530TL Telegraph Colour Library/FPG 530TR Medichrome/The Stock Shop 530C Ed Honowitz/Tony Stone Images 530B Chuck Davis/Tony Stone Images 531B Derek Trask/Leo de Wys 532 Dr.Kari Lounatmaa/SPL/Photo Researchers 536 David R. Frazier Photolibrary 538 Luis Villotai/The Stock Market 539 Bernard Boutrit/Woodfin Camp & Associates 540 Jean Clottes/Sygma 545 Robert Frerck/Woodfin Camp & Associates 548 R.Bossu/Sygma 550 D.Gontier/The Image Works 551 Courtesy Maxine Waters 552 Doug Wechsler/Earth Scenes 555 St.Joseph Museum, St.Joseph, Missouri 556T E.J.Camp/Outline Press Syndicate Inc. 556C Everett Collection 556B Everett Collection 562 USDA/SS/Photo Researchers 563 Brooks Kraft/Sygma 568 AP/Wide World 575 Dion Ogust/The Image Works 576 Stuart Franklin/Magnum Photos 580 Dennis Cox/ChinaStock 583, 585 NASA 587 Viviane Holbrooke/The Stock Market 589 Larry House/Tony Stone Images 593 (c) 1991 Cindy Lewis 594 M.Barrett/H. Armstrong Roberts 602-603T Randy Faris/West Light 602C Dennis O'Clair/Tony Stone Images 602BL Per Eriksson/Leo de Wys 602-603B Telegraph Colour Library/FPG 603CL Henryk Kaiser/Leo de Wys 603R Tom Van Sant/The Stock Market 604 Jan Kanter 607 U.S.Defense Department 609 Daniel Forster/Stock Newport 611 David Pollack/The Stock Market 615 Alex Quesada/Woodfin Camp & Associates 620 Joe Sohm/The Image Works 623 Villota/The Stock Market 625 Everett Collection 627 Joe Bator/The Stock Market 631 David Spangler 632 Courtesy Todd-Page Construction 637 Everett Collection 639 Chuck Nacke/Woodfin Camp & Associates 643, 647 Milt & Joan Mann/Cameramann International, Ltd. 651 Julie Houck/Stock Boston 653 Bob Daemmrich/Stock Boston 657 Cary Wolinsky/Stock Boston 658 Peter Beck/The Stock Market 664T Steve Kahn/FPG 664B Ron Watts/West Light 665 Kevin Alexander/Profiles West 672-673T Marvy!/The Stock Market 672C V.Cody/West Light 672B Charles Bowman/Leo de Wys 673C Will & Deni McIntyre/Tony Stone Images 673B David Bishop/Phototake 674 PhotoFest 676 Tony Freeman/Photo Edit 679 Terry Murphy/ANIMALS ANIMALS 680 Rhoda Sidney/Photo Edit 683 Richard Lord 684 Ken Krueger/Tony Stone Images 692 David R. Frazier Photolibrary 706 Biophoto Associates/SS/Photo Researchers 717 Bettmann Archive 722 Alan Oddie/Photo Edit 734 Library of Congress 737–738 Mark Segal/Panoramic Images, Chicago 738INS Gary A. Bartholomew/West Light 746-747T Tecmap/West Light 746C Bill Ross/West Light 746B Dennis Welsh/Adventure Photo 747C Craig Aurness/West Light 747B Craig Aurness/West Light 751L Runk/Schoenberger/Grant Heilman Photography 751R Milt & Joan Mann/Cameramann International, Ltd. 753 Bettmann Archive 754T A.T.Willett/The Image Bank 754B January 31, 1994/Newsweek Magazine 757 (c)Woodfield Associates. Reprinted with permission. 759 David Delossy/The Image Bank 765 Robert Llewellyn 768 Matthew Neal McVay/Stock Boston 771 Benn Mitchell/The Image Bank 775 Dave Bartruff/Stock Boston 787 Cynthia Clampitt 789 Courtesy British Airways 793 David Young-Wolff/Photo Edit 795 John Conger 797 Library of Congress 798 M.Harker/G&J Images/The Image Bank 799T J.Blank/H. Armstrong Roberts 799B R.Kord/H. Armstrong Roberts 800T ChromoSohm/Sohm 800B Richard J.Wainscoat/Peter Arnold, Inc. 808T Peter Steiner/The Stock Market 808C SuperStock, Inc. 808B Steve Chenn/West Light 809T Al Francekevich/The Stock Market 809B Rick Gayle/The Stock Market 814T David Ball/Tony Stone Images 814B David Spangler 816 Ron Spomer/Visuals Unlimited 818 Larry Lefever/Grant Heilman Photography 821 William J. Weber/Visuals Unlimited 822 Tom McCarthy/Photo Edit 826 Bob Daemmrich 829 Everett Collection 833 Chris Arend/AlaskaStock Images 847 Tony Freeman/Photo Edit 849 Rosemary Finn 850 Martin Rogers/Tony Stone Images 857 AP/Wide World 862 Anup & Manuj Shah/ANIMALS ANIMALS 864 Independence National Historical Park, Philadelphia, PA. 868 Michael Newman/Photo Edit 869 Everett Collection 875T Japack/Leo de Wys 875B Bettmann Archive 876 Weinberg/Clark/The Image Bank